AF333835

POPULATIONS OF HIGH ENERGY SOURCES IN GALAXIES

IAU SYMPOSIUM No. 230

COVER ILLUSTRATION: M33 AT X-RAYS

Image of the Local Group spiral galaxy M33, obtained with the XMM-Newton satellite. The main part of the body of this galaxy is shown, featuring many individual X-ray sources that can be examined in detail with present-day satellite X-ray observatories. The centre is in the middle of the picture, marked by the brightest X-ray source in this image, which is the only 'Ultra Luminous X-ray source' in the Local Group.

Image courtesy of W. Pietsch (MPE, Garching, Germany).

INTERNATIONAL ASTRONOMICAL UNION

UNION ASTRONOMIQUE INTERNATIONALE

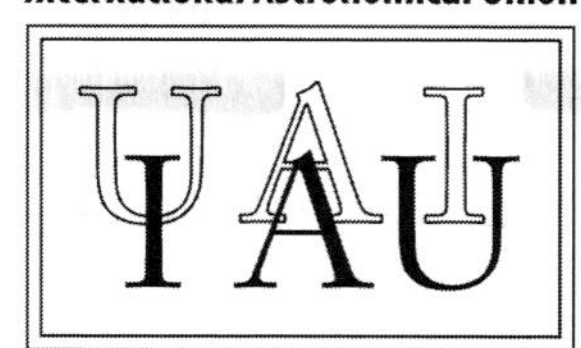

POPULATIONS OF HIGH ENERGY SOURCES IN GALAXIES

PROCEEDINGS OF THE 230th SYMPOSIUM OF THE
INTERNATIONAL ASTRONOMICAL UNION
HELD IN DUBLIN, IRELAND
15–19 AUGUST 2005

Edited by

E.J.A. MEURS
DUNSINK OBSERVATORY, DIAS, DUBLIN, IRELAND

and

G. FABBIANO
*HARVARD-SMITHSONIAN CENTER FOR ASTROPHYSICS,
CAMBRIDGE, MA, USA*

CAMBRIDGE
UNIVERSITY PRESS

CAMBRIDGE UNIVERSITY PRESS
The Edinburgh Building, Cambridge CB2 2RU, UK
40 West 20th Street, New York, NY 10011–4211, USA
477 Williamstown Road. Port Melbourne, VIC 3207, Australia
Ruiz de Alarcón 13, 28014 Madrid, Spain
Dock House, The Waterfront, Cape Town 8001, South Africa

© International Astronomical Union 2006

First published 2006

Printed in the United Kingdom at the University Press, Cambridge

Typeset in System LaTeX 2_ε

A catalogue record for this book is available from the British Library

Library of Congress Cataloguing in Publication data

ISBN – 13 978 0521 85201 2
ISBN – 10 0521 85201 3
ISSN 1743-9213

Table of Contents

INTRODUCTION

Session 1. KEY SOURCE CATEGORIES IN OUR GALAXY

Session 2. HIGH ENERGY PROCESSES IN THE ISM

Session 3. DETAILED POPULATION STUDIES IN THE NEARER GALAXIES

Session 4. SOURCE CLASSES THAT EMERGE FROM SAMPLING OVER GALAXIES

Session 5. OVERALL POPULATION CHARACTERISTICS

Session 6. HIGH ENERGY POPULATION SYNTHESIS

Session 7. THE HIGH-REDSHIFT CONTEXT

CLOSING

INDICES

Preface

This Symposium emphasizes the emerging field of studies of individual X-ray sources in other galaxies, as well as of the overall populations of X-ray sources in those galaxies, based on the information obtained for the individual sources in their populations. At the same time, our knowledge of high-energy sources in our own Galaxy is reaching new levels of sophistication at γ-rays (thanks to the INTEGRAL satellite) and at TeV energies (since the HESS array became operational). Going deeper into the Universe, we have to work with integrated galaxy emission properties, which still may be understood with the help of the detailed knowledge on individual sources and global population descriptions obtained from studies of galaxies near to us. But also, when redshifts become substantial, we will be looking at younger galaxies and high-energy source populations will be affected accordingly. An individual source that nevertheless may still stand out in objects of the deep Universe is the possibly very prominent one in the nucleus of galaxies. With these considerations in mind IAU Symposium No 230 was organized, entitled "Populations of high-energy sources in galaxies".

The contributions to the Symposium are published in these Proceedings, with only a few contributions missing. Both oral and poster contributions are included, as well as a record of the discussion after oral contributions (as far as Q/A sheets were returned alright!). All contributions were subjected to a refereeing process. Abstracts are included when no contribution was received, with discussion for talks if available, so as to give an impression of the overall programme.

An important point to note is that some contributions contain illustrations that, to a greater or lesser degree, use colour to clarify the information that is presented. These Proceedings are published in black and white, but, when desirable, colour versions of many illustrations in this volume can be seen on the electronic Web site (for the first 18 months on the Web pages of the Publisher, Cambridge University Press, after which copies of the Proceedings files will be transferred to the ADS.

The Symposium itself was run smoothly, thanks to the efforts of many people. The Secretaries of the Department of Experimental Physics in University College Dublin (Catherine Handley) and Dunsink Observatory (Carol Woods) took care of many tasks, with substantial additions by other LOC members (Brian McBreen, Laura Norci, Cóilín Ó Maoiléidigh). Competent assistance was received from the staff of the Dublin Castle Conference Centre. The Symposium was opened officially, on Monday 15 August 2005, by the Irish Minister for Education and Science, Mrs Mary Hanafin, TD. The Sessions of the scientific programme were supported by several students (Ronan McSwiney, Gavin Behan, Diarmuid Byrne, Sinead McGlynn, Susan Foley, Elaine Winston, Gary Melady, John French, Paul Ward) who handled microphones and distributed Q/A sheets. Several SOC members, and renowned colleagues not carrying the burden of presenting an oral contribution themselves, cordially chaired the Sessions (Jean Clavel, Pepi Fabbiano, Katsuji Koyama, Vladimir Lipunov, Brian McBreen, Evert Meurs, George Miley, Zhen-Ru Wang, Mike Watson). Technicians of the Department of Experimental Physics of University College Dublin constructed and transported the poster boards.

The Conference Dinner was hosted by the Minister for Education and Science, in the plush State Appartments in Dublin Castle, in the splendid St. Patrick's Hall. At the end of the Dinner a short demonstration was given of modern Irish Dancing (not by the otherwise highly committed LOC, but by four professional dancers).

The overall Symposium programme featured also an event for the general public. On Tuesday evening, 16 August 2005, a public talk was delivered by renowned astrophysicist Prof. Geoffrey Burbidge.

Great help with the substantial editing task of these Proceedings was received from Dunsink Observatory PhD student, Cóilín Ó Maoiléidigh, who got the best out of the contributions as of latex performance, and Secretary Carol Woods, who transcribed the Q/A sheets. It should be noted that Cóilín Ó Maoiléidigh also had taken on the care for the Symposium Web pages†.

The Editors (November 2005)

† http://www.dunsink.dias.ie/IAUS230/

THE ORGANIZING COMMITTEE

Scientific

E.J.A. Meurs (Ireland, Chair)
G. Fabbiano (USA, Co-Chair)
L. Bassani (Italy)
B. McBreen (Ireland)
H.-Y. Chu (USA)
C. Done (UK)
G. Hasinger (Germany)

G. Koenigsberger (Mexico)
K. Koyama (Japan)
V. Lipunov (Russia)
M. Mas-Hesse (Spain)
Th. Montmerle (France)
G. Romero (Argentina)
Z. Wang (China)

Local

B. McBreen (Chair)
C. Handley
C. Ó Maoiléidigh

E.J.A. Meurs
L. Norci
C. Woods

Acknowledgements

The symposium is supported by the IAU Divisions XI (Space and High Energy Astrophysics), VI (Interstellar Matter), VII (Galactic System), and VIII (Galaxies and the Universe); and by the IAU Commissions No. 44 (Space and High Energy Astrophysics), No. 28 (Galaxies), No. 33 (Structure and Dynamics of the Galactic System), and No. 34 (Interstellar Matter).

Sponsorship was obtained from:
International Astronomical Union,
Department of Education and Science,
Dublin Institute for Advanced Studies,
University College Dublin,
European Space Agency,
Faílte Ireland,
Royal Irish Academy,
and is gratefully acknowledged.

The participants gathered on the courtyard of Dublin Castle for the Conference Photograph. Count the number of ties.

Participants

Pedro Augusto, Universidade da Madeira, Funchal, Portugal — augusto@uma.pt
Aya Bamba, Cosmic Radiation Laboratory, RIKEN, Wako, Saitama, Japan — bamba@crab.riken.jp
Amy Barger, Astronomy, University of Wisconsin-Madison, Madison, WI, USA — barger@astro.wisc.edu
Robin Barnard, Physics and Astronomy, Open University, Milton Keynes, UK — R.Barnard@open.ac.uk
Michael Bauer, MPE, Garching, Germany — mbauer@mpe.mpg.de
Angela Bazzano, IASF–INAF, Rome, Italy — Angela@rm.iasf.cnr.it
Gavin Behan, School of Physical Sciences, DCU, Dublin, Ireland — gavin.behan3@mail.dcu.ie
Marina Kaufman Bernadó, Instituto Argentino de Radioastronomy, Buenos Aires, Argentina — mar@vianw.net.ar
Arash Bodaghee, I.S.D.C., Versoix, Switzerland — arash.bodaghee@obs.unige.ch
Valenti Bosch-Ramon, Astronomia i Meteorologia, Universitat de Barcelona, Barcelona, Spain — vbosch@am.ub.es
Nicola Brassington, School of Physics and Astronomy, University of Birmingham, Edgbaston, Birmingham, UK — njb@star.sr.bham.ac.uk
Ines Brott, ISDC, Versoix, Switzerland — Ines.Brott@obs.unige.ch
Geoffrey Burbige, Physics, University of California, San Diego, USA — gburbige@ucsd.edu
Carlos del Burgo, Dunsink Observatory, Dublin 15, Ireland — cburgo@dunsink.dias.ie
Andrey Bykov, Theoretical Astrophysics, A.F. Ioffe Institute of Physics and Technology, St.Petersburg, Russia — byk@astro.ioffe.ru
Diarmuid Byrne, Physics Department, Trinity College, Dublin, Ireland — byrnedc@gmail.com
Marion Cadolle Bel, DSM/DAPNIA/SAp, CEA-Saclay, Gif-Sur-Yvette, France — mcadolle@cea.fr
Paul Callanan, Department of Physics, University College Cork, Cork, Ireland — paulc@ucc.ie
Stefania Carpano, IAAT, Tuebingen, Germany — stefania.carpano@uni-tuebingen.de
Maria Chernyakova, ISDC, Versoix, Switzerland — Masha.Chernyakova@obs.unige.ch
Jean Clavel, ESTEC/SCI-SA, Noordwijk, The Netherlands — jean.clavel@esa int
Peter Cogan, Experimental Physics, UCD, Dublin, Ireland — peter@ferdia.ucd.ie
Christopher Copperwheat, Space and Climate Physics (MSSL), UCL, Dorking, UK — cmc@mssl.ucl.ac.uk
Robin Corbet, X-ray Astrophysics Laboratory, NASA GSFC/USRA, Greenbelt, MD, USA — corbet@gsfc.nasa.gov
Michael Corcoran, Exploration of the Universe Division, USRA, GSFC, Greenbelt, MD, USA — corcoran@barnegat.gsfc.nasa.gov
Thierry Courvoisier, ISDC, Versoix, Switzerland — thierry.courvoisier@obs.unige.ch
John Cunniffe, Information Technology, NUIG, Galway, Ireland — jc@it.nuigalway.ie
Eamonn Cunningham, Physics, DCU, Dublin, Ireland — eamonn.cunningham@dcu.ie
Michael Daniel, Experimental Physics, UCD, Dublin, Ireland — michael.daniel@ucd.ie
Anthony Dean, Physics and Astronomy, University of Southampton, Southampton, UK — ajd@astro.soton.ac.uk
Patrick Elebert, Physics, University College Cork, Cork, Ireland — p.elebert@ucc.ie
Martin Elvis, HEAD, Harvard-Smithsonian Center for Astrophysics, Cambridge, MA, USA — elvis@cfa.harvard.edu
Michael Eracleous, Astronomy and Astrophysics, Pennsylvania State University, University Park, PA, USA — mce@astro.psu.edu
Pepi Fabbiano, HEAD, Harvard-Smithsonian Center for Astrophysics, Cambridge, MA, USA — pepi@cfa.harvard.edu
Sergei Fabrika, Special Astrophysical Observatory, Karachai-Cherkess Republic, Russia — fabrika@sao.ru
Suzanne Foley, Experimental Physics, UCD, Dublin, Ireland — sfoley@bermuda.ucd.ie
John French, Experimental Physics, UCD, Dublin, Ireland — jfrench@bermuda.ucd.ie
Michael Garcia, HEAD, Harvard-Smithsonian Center for Astrophysics, Cambridge, MA, USA — garcia@cfa.harvard.edu
Pranab Ghosh, Department of Astronomy & Astrophysics, Tata Institute of Fundamental Research, Mumbai, India — pranab@mailhost.tifr.res.in
Omaira Gonzalez-Martin, Instituto de Astrofísica de Andalucía, Granada, Spain — omaira@iaa.es
Lindsey Shaw Greening, Open University, Milton Keynes, UK — L.Shaw-Greening@open.ac.uk
Richard Griffiths, Physics, Carnegie Mellon University, Pittsburgh, PA, USA — griffith@astro.phys.cmu.edu
Hans-Jakob Grimm, SAO/CFA, Cambridge, MA, USA — hgrimm@head.cfa.harvard.edu
Fabien Grise, Observatoire Astronomique de Strasbourg, Strasbourg, France — grise@astro.u-strasbg.fr
Carlos M. Gutierrez, Instituto de Astrofísica de Canarias, La Laguna, Tenerife, Spain — cgc@ll.iac.es
Juan Antonio Zurita Heras, ISDC, Versoix, Geneva, Switzerland — Juan.Zurita@obs.unige.ch
Edward van den Heuvel, Sterrenk. Inst. 'Anton Pannekoek', University of Amsterdam, Amsterdam, The Netherlands — edvdh@science.uva.nl
Ann Hornschemeier, Laboratory for X-ray Astrophysics, NASA GSFC, Greenbelt, MD, USA — annh@milkyway.gsfc.nasa.gov
Rene Hudec, HEA Group, Astronomical Institute, Observatory Ondrejov, Ondrejov, Czech Republic — rhudec@asu.cas.cz
Natalia Ivanova, Physics and Astronomy, North-western University, Evanston, IL, USA — nata@northwestern.edu
Roy Kilgard, SAO, Cambridge, MA, USA — rkilgard@cfa.harvard.edu
Dong-Woo Kim, SAO, Cambridge, MA, USA — kim@cfa.harvard.edu
Albert Kong, Kavli Institute for Astrophysics and Space Research, MIT, Cambridge, MA, USA — akong@space.mit.edu
Katsuji Koyama, Department of Physics, Kyoto University, Kyoto, Japan — koyama@cr.scphys.kyoto-u.ac.jp
Peter Kretschmar, RSSD, ESAC, Madrid, Spain — Peter.Kretschmar@esa.int
Roman Krivonos, Space Research Institute, Russian Academy of Science, Moscow, Russia — kris@hea.iki.rssi.ru
Volodymyr Kryvdyk, Astronomy, Taras Shevchenko Kyiv National University, Kyiv, Ukraine — kryvdyk@univ.kiev.ua
Tatiana Larchenkova, Astro Space Center of P.N.Lebedev Physical Institute, Moscow, Russia — tanya@lukash.asc.rssi.ru
Jean-Christophe Leyder, GAPHE, Universite de Liège, Liège, Belgium — leyder@astro.ulg.ac.be
Xiangdong Li, Department of Astronomy, Nanjing University, Nanjing, P.R. China — lixd@nju.edu.cn
Vladimir Lipunov, Relativistic Astrophysics Dep., Sternberg Astronomical Inst. and Moscow State Univ., Moscow, Russia — lipunov@sai.msu.ru
Qingzhong Liu, Purple Mountain Observatory, Nanjing, P.R. China — qzliu@pmo.ac.cn
Alexander Lutovinov, High Energy Astrophysics, Space Research Insitute, Moscow, Russia — aal@hea.iki.rssi.ru
Tom Maccarone, Sterrenkundig Instituut 'Anton Pannekoek', University of Amsterdam, Amsterdam, The Netherlands — tjm@science.uva.nl
Josefa Masegosa, Instituto de Astrofísica de Andalucía, CSIC, Granada, Spain — pepa@iaa.es
Andrew McCann, Experimental Physics, UCD, Dublin, Ireland — andrew@ferdia.ucd.ie
Brian McBreen, Experimental Physics, UCD, Dublin, Ireland — brian.mcbreen@ucd.ie
Sheila McBreen, RSSD, ESTEC, Noordwijk, The Netherlands — smcbreen@rssd.esa.int
Sinead McGlynn, Experimental Physics, UCD, Dublin, Ireland — smcglynn@bermuda.ucd.ie
Ronan McSwiney, School of Physical Sciences, DCU, Dublin 9, Ireland — ronanmcswiney@hotmail.com
Gary Melady, Experimental Physics, UCD, Dublin, Ireland — gmelady@bermuda.ucd.ie
Evert Meurs, Dunsink Observatory, Dublin, Ireland — ejam@dunsink.dias.ie
George Miley, Sterrewacht Leiden, Leieden, The Netherlands — miley@strw.Leidenuniv.nl
Jon Miller, HEAD, Harvard-Smithsonian CfA, Cambridge, MA, 02138, USA — jmmiller@cfa.harvard.edu
Felix Mirabel, European Southern Observatory, Santiago, Chile — fmirabel@eso.org
Zdenka Misanovic, MPE, Garching, Germany — zdenka@mpe.mpg.de
Alberto Moretti, Brera Observatory, Milan-Merate, Italy — moretti@merate.mi.astro.it
Andreas Müller, MPE, Garching, Germany — amueller@mpe.mpg.de
Kevin Nolan, Tallaght Institute of Technology, Dublin, Ireland
Louisa Nolan, School of Physics and Astronomy, University of Birmingham, Edgbaston, UK — lan@star.sr.bham.ac.uk
Laura Norci, School of Physical Sciences, DCU, Dublin 9, Ireland — lno@physics.dcu.ie
Cóilín Ó Maoiléidigh, Dunsink Observatory, Dublin, Ireland — melodies@dunsink.dias.ie
Manfred Pakull, Observatoire de Strasbourg, Strasbourg, France — pakull@astro.u-strasbg.fr
Thomas Pannuti, Spitzer Science Center, JPL/California Institute of Technology, Pasadena, CA, USA — tpannuti@ipac.caltech.edu
Wolfgang Pietsch, MPE, Garching, Germany — wnp@mpe.mpg.de
V. Francesco Polcaro, IASF–INAF, Rome, Italy — polcaro@rm.iasf.cnr.it
David Pooley, Astronomy, University of California, Berkeley, CA, USA — dave@astron.berkeley.edu
Andrea Prestwich, CfA, Cambridge, MA, USA — andreap@head.cfa.harvard.edu

John Quinn, Experimental Physics, UCD, Dublin, Ireland   john.quinn@ucd.ie
Andrew Read, Physics and Astronomy, University of Leicester, Leicester, UK   amr30@star.le.ac.uk
Mark Reynolds, Physics, University College Cork, Cork, Ireland   m.reynolds@ucc.ie
Frank M. Rieger, Mathematical Physics, UCD, Dublin, Ireland   frank.rieger@ucd.ie
Timothy Roberts, Physics and Astronomy, University of Leicester, Leicester, UK   tro@star.le.ac.uk
Jerome Rodriguez, Service d'Astrophysique, CEA, Saclay DSM/DAPNIA/SAP, Gif sur Yvette, France   jrodriguez@cea.fr
Craig Sarazin, Department of Astronomy, University of Virginia, Charlottesville, VA, USA   sarazin@virginia.edu
Norbert Schartel, XMM-Newton SOC, ESA, Madrid, Spain   Norbert.Schartel@sciops.esa.int
Eric M. Schlegel, HEAD, SAO, Cambridge, MA, USA   eschlegel@cfa.harvard.edu
Atsushi Senda, Cosmic Radiation Laboratory, RIKEN, Wako City, Saitama, Japan   senda@crab.riken.jp
Rosanne di Stefano, Harvard-Smithsonian Center for Astrophysics, Cambridge, MA, USA   rd@cfa.harvard.edu
Douglas Swartz, MSFC, Huntsville, AL, USA   doug.swartz@msfc.nasa.gov
Heinz Völk, MPI for Nuclear Physics, Heidelberg, Germany   Heinrich.Voelk@mpi-hd.mpg.de
Vojtech Simon, HEA Group, Astron. Inst. and Acad. of Sciences of the Czech Republic, Ondrejov, Czech Republic   simon@asu.cas.cz
Gregory Sivakoff, Astronomy, University of Virginia, Charlottesville, VA, USA   grs8g@virginia.edu
Roberto Soria, Harvard-Smithsonian Center for Astrophysics, Cambridge, MA, USA   rsoria@cfa.harvard.edu
Alina Streblyanska, MPE, Garching, Germany   alina@mpe.mpg.de
Martin Stuhlinger, ESAC, Madrid, Spain   Martin.Stuhlinger@sciops.esa.int
Tomonori Totani, Department of Astronomy, Kyoto University, Kyoto, Japan   totani@kusastro.kyoto-u.ac.jp
Ginevra Trinchieri, INAF-OABrera, Milano, Italy   ginevra@brera.mi.astro.it
Pietro Ubertini, IASF–INAF, Rome, Italy   ubertini@rm.iasf.cnr.it
Masaru Ueno, Department of Physics, Tokyo Institute of Technology, Tokyo, Japan   masaru@hp.phys.titech.ac.jp
Susanna Vergani, Paris, France   vergani@merate.mi.astro.it
Jacco Vink, SRON, Utrecht, The Netherlands   j.vink@sron.nl
Rasmus Voss, MPA, Garching, Germany   voss@mpa-garching.mpg.de
Roland Walter, ISDC, Versoix, Switzerland   Roland.Walter@obs.unige.ch
Zhen-Ru Wang, Department of Astronomy, Nanjing University, Nanjing, China   zrwang@nju.edu.cn
Martin Ward, Physics, University of Durham, Durham, UK   m.ward@durham.ac.uk
Paul Ward, Dunsink Observatory, Dublin, Ireland   pward@dunsink.dias.ie
Michael Watson, Physics and Astronomy, University of Leicester, Leicester, UK   mgw@star.le.ac.uk
Georg Weidenspointner, CESR, Toulouse, France   Georg.Weidenspointner@cesr.fr
Rudy Wijnands, Sterrenk. Inst. 'Anton Pannekoek ', Univ. of Amsterdam, Amsterdam, The Netherlands   rudy@science.uva.nl
Dave Willis, ISDC, Versoix, Switzerland   David.Willis@obs.unige.ch
Elaine Winston, Experimental Physics. UCD, Dublin, Ireland   ewinston@bermuda.ucd.ie
Diane Wong, Astrophysics, UC Berkeley, Berkeley, CA, USA   dianew@astro.berkeley.edu
Yueheng Xu, Department of Physics and Astronomy, University of Leicester, Leicester, UK   yx12@star.le.ac.uk
Tahir Yaqoob, Physics and Astronomy, Johns Hopkins University, Baltimore, MD, USA   yaqoob@pha.jhu.edu
Andreas Zezas, Harvard-Smithsonian Center for Astrophysics, Cambridge, MA, USA   azezas@cfa.harvard.edu

Introduction

Opening by the Irish Minister for Education and Science, Mrs. Mary Hanafin, TD. Behind the table, from left to right: Pepi Fabbiano (SOC Co-Chair), Evert Meurs (SOC Chair) and Brian McBreen (LOC Chair).

Populations of High Energy Sources in Galaxies
Proceedings IAU Symposium No. 230, 2005
E. J. A. Meurs & G. Fabbiano, eds.

© 2006 International Astronomical Union
doi:10.1017/S174392130600768X

Scope of this Symposium

E. J. A. Meurs

Dunsink Observatory, Dublin Institute for Advanced Studies, Dublin, Ireland
ejam@dunsink.dias.ie

The original idea for this IAU Symposium arose from realizing that present-day X-ray satellites, XMM-Newton and Chandra, are now allowing us to conduct studies of individual X-ray sources in other galaxies, much like this was until recently mostly confined to sources in our own Galaxy and in the Magellanic Clouds. In addition, γ-ray astronomy is catching up as it were, now being able to study well-defined sources in our own Galaxy with the INTEGRAL satellite, and also the highest-energy sources accessible, at TeV, with the newly constructed Cherenkov receiver array(s).

Observations employing high-energy radiation (X-ray, γ-ray) offer fascinating views of galaxies. As for a more everyday example, a human subject, different wavelengths of observation put the person in a different light, and highlight different aspects of the subject. Images of human faces in normal (visible) light are familiar enough, but when employing X-ray radiation we can focus on notably the underlying bone structure (the skull), and faces look different again at, for instance, UltraViolet or InfraRed wavelengths. The InfraRed mapping of a face, interestingly, can be used for deriving specific diagnostics of health conditions elsewhere in the body.

Similarly, when different wavelengths are used, galaxies show up in different ways. (For an example, compare optical and Xray images of the nearby galaxy M51, Figure 1 in contribution of Kilgard in these Proceedings.) The galaxian components highlighted at X-rays are stars in special, usually evolutionary advanced, stages of their development, e.g. X-ray binaries, SuperNova Remnants, colliding wind binaries, but also possibly a core source indicating some level of nuclear activity. Some galaxies feature noticeable diffuse emission components, such as the combined effects of stellar winds and supernova ejections. The individual sources (apart from a core source) represent comparatively short lasting stages in stellar evolution and the X-ray observations therefore highlight a particularly interesting, limited selection out of the 10^{11} or so members of the stellar population, thus being of specific diagnostic value.

Our ability nowadays to examine these X-ray sources in other galaxies in considerable detail, is a result of the improvements in the instrumentation of X-ray satellite observatories since X-ray astronomy began, 43 years ago with the discovery of the first X-ray source outside the solar system (Sco X-1).

As an example, it is instructive to follow how the X-ray observations became more informative for the nearest normal galaxy, M31 (the Andromeda Nebula). (For convenience of illustration we focus here on imaging data, but obviously spectral information has developed enormously as well.) In 1972 M31 appeared as an entry in the First Uhuru catalogue, a source with a 90% confidence area of 17 square degrees. The Uhuru satellite employed a collimator, that provided directional information but had no imaging capabilities. Around 1980 X-ray images were obtained, with the Einstein satellite of the central region of M31 in a HRI detector exposure (Van Speybroeck *et al.* 1979). ROSAT's HRI detector gave a still better image in the nineties (Figure 1). In 2004 Chandra obtained the sharpest picture to date; the correspondence with, and the improvement upon, the

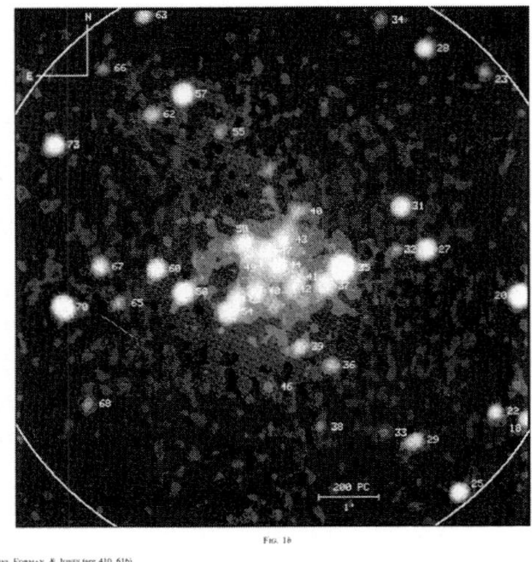

Figure 1. Central region of M31 with the ROSAT HRI (Primini *et al.* 1993).

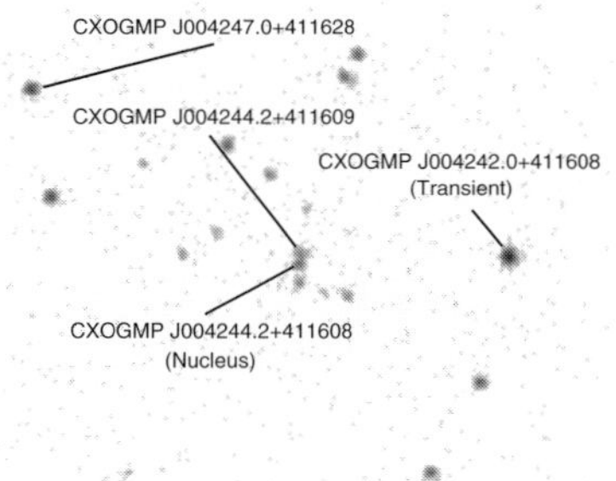

Figure 2. Central region of M31 with Chandra (Garcia *et al.* 2000).

comparatively much more diffuse individual source brightness distributions in the ROSAT HRI exposure is evident in Garcia *et al.* (2000; their Figure 1).

It is exciting that, at this moment, similarly important improvements in imaging capability are taking place at γ-rays (INTEGRAL) and at TeV energies (HESS array). These advances are also described in these Proceedings.

References

Garcia, M.R., Murray, S.S., Primini, F.A., *et al.* 2000, ApJ 537, L23
Primini, F.A., Forman, W. & Jones, C. 1993, ApJ 410, 615
Van Speybroeck, L., *et al.* 1979, ApJ 234, L45

Session 1

Key source categories in our Galaxy

LOC members making a preparatory visit to Dublin Castle. From right: Cóilín Ó Maoiléidigh, Laura Norci, Evert Meurs. Photo courtesy of C. Woods.

Populations of High Energy Sources in Galaxies
Proceedings IAU Symposium No. 230, 2005
E. J. A. Meurs & G. Fabbiano, eds.

© 2006 International Astronomical Union
doi:10.1017/S1743921306007691

Key Source Categories in the Galaxy

Paul J. Callanan[1]

[1]Department of Physics, University College, Cork, Ireland.
email: paulc@ucc.ie

Abstract. The proliferation of X-ray astronomy missions over the last 10 years has had a major impact on our knowledge of Galactic X-ray binaries. More recently, Chandra and XMM-Newton have provided a dramatically improved census of extra-galactic X-ray binaries. In this talk I will provide an overview of the various populations and observational properties of X-ray binaries in our Galaxy, in an effort to "set the scene" for a comparison with their extra-galactic counterparts.

Keywords. X-rays: binaries, binaries: close, stars: neutron, black hole physics.

1. Introduction

"A round tower of fire was seen in the air over Ros Ela on Sunday the feast of St. George for five hours of the day" – Irish monks record the Crab supernova in 1054 AD.

The appearance of the Crab Supernova in 1054 was a propitious moment in astronomy – a mysterious harbinger of discoveries to come (the quotation above comes from observations made by monks in Tullamore in Ireland, one of the very few written European accounts of this event: see Polcaro in these proceedings for a detailed discussion of similar historical records). However, it took more than 900 years before X-ray astronomy could finally provide the tool required to comprehensively study the formation and evolution of such compact objects in our Galaxy and beyond. Currently, astronomers inhabit a "golden age" of multiwavelength astronomy – an astronomy that is problem driven, rather than its modus operandi of the past, dominated by observations at a fixed (e.g. optical) wavelength range.

X-ray astronomers currently find themselves with a multitude of observatories at their disposal, including Chandra, XMM-Newton, INTEGRAL, Swift, RXTE, etc. Of all the objects in the menagerie of sources accessible to X-ray astronomy, Galactic X-ray binaries have led to particular insights into the formation and evolution of compact objects in accreting binaries. Here I give a brief overview of some of the salient properties of these binaries in our Galaxy: whilst a review of this type will be, by necessity, no-where near complete, hopefully it can still serve to "set the scene" for the extra-galactic (X-ray binary) discussions to follow in the rest of these proceedings.

2. The View from Afar: a very brief overview

Currently, there are almost 300 X-ray binaries known in our Galaxy, including both the persistently bright and the quiescent systems (e.g. see Fig. 1). These systems have traditionally been divided into the so-called High Mass X-ray Binaries (HMXBs), with a secondary mass $M_2 > 10$ M$_\odot$, and Low Mass X-ray Binaries (LMXBs), with $M_2 \leqslant 1$ M$_\odot$ – which, for the purpose of this review, also includes the so-called Intermediate Mass X-ray Binaries (e.g. Pfahl *et al.* 2003). Accretion generally occurs via either a stellar

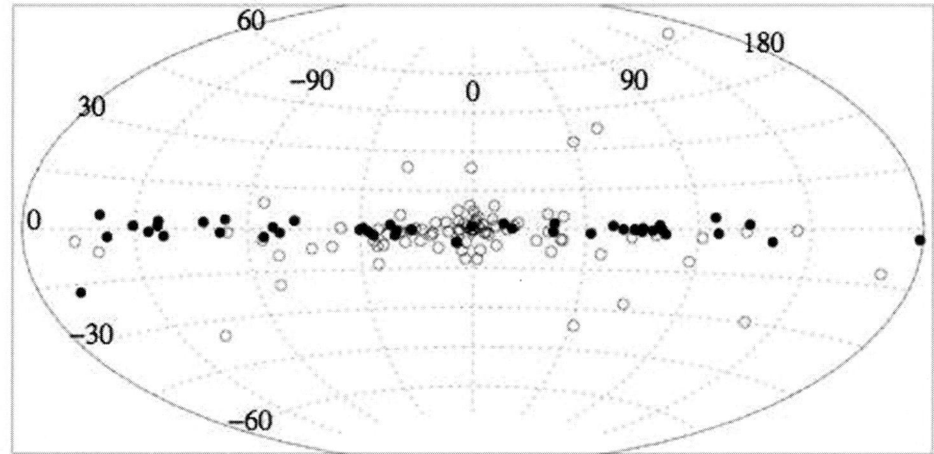

Figure 1. The Galactic distribution of X-ray sources: from Grimm *et al.* (2002). The open circles denote LMXBs, and the filled ones HMXBs.

wind or circumstellar disk for the HMXBs, and via Roche Lobe overflow for the LMXBs. A large fraction of both types of X-ray binary consists of transient sources: whereas the recurrence timescale of the LMXB transients is highly irregular, many of the HMXB transients recur periodically.

The X-ray emission of our Galaxy is dominated by the LMXBs in the bulge: Grimm *et al.* (2002) find that the (average) total LMXB X-ray luminosity (L_x) of our Galaxy is $\sim$2–3 $\times$ 10^{39} ergs s^{-1}, compared to $\sim$2–3 $\times$ 10^{38} ergs s^{-1} for the HMXBs.

For our Galaxy and others, it has also been shown that a universal (X-ray) luminosity function exists for both LMXBs (scaling with galaxy mass) and HMXBs (scaling with star formation rate), see Grimm *et al.* (2002, 2003 and these proceedings). This seems remarkable, considering the transient nature of many LMXBs and HMXBs as observed in our Galaxy and others, and the fact that the appearance of our Galaxy is dominated by only the 5–10 most luminous sources.

Nearer the Galactic Centre, a large fraction of the observed "diffuse" X-ray emission is likely due to quiescent X-ray transients, absorbed HMXBs and magnetic cataclysmic variables (CVs), in addition to genuinely diffuse emission (e.g. Muno *et al.* 2003, Hands *et al.* 2004).

Finally, some 13 LMXBs, both persistently bright and transient, are known to exist in Galactic globular clusters, in addition to fainter populations of sources including CVs, millisecond pulsars, magnetically active binaries, etc. (e.g. Verbunt and Lewin, 2006). Uniquely amongst all the binaries we discuss here, many of these globular cluster systems are thought to be created by 2 and 3-body tidal interactions between stars in the dense cluster cores (e.g. Pooley *et al.* 2003).

We will now discuss each of these key types of compact binary in more detail.

3. The High Mass Systems

The majority (some $\sim$96) of the 130 known HMXBs pulsate, with spin periods ranging from 69 msec to 1400 s. Their X-ray spectra are characterised by a flat power-law with an energy index 0–1, and a high energy cut-off of 10–20 keV. A soft excess is often seen below 1 keV (see White, Nagase and Parmar 1995 for more details).

Grimm *et al.* (2002) suggested that HMXBs tend to avoid the inner 3–4 kpc of the Galactic disk, although Kuulkers (2005) suggests that many of the recently discovered, highly absorbed Integral sources sources may be obscured HMXBs and hence closer to the Galactic Centre.

The SMC contains a disproportionately large population of HMXBs: this is probably due to tidal interactions between the SMC and the LMC in the past (see e.g. Corbet, these proceedings).

There are two main categories:

3.1. *Systems with Supergiant Secondaries (Of, or earlier than B2)*

These comprise $\sim25\%$ of all known Galactic HMXBs (Charles and Coe, 2006). With mass loss dominated by stellar wind from the secondary (although Roche lobe overflow can also play a role), L_x, although persistent, is low, $\sim10^{34}$–10^{35} ergs s^{-1}, and hence the optical flux is completely dominated by the secondary star. Many of these systems exhibit an ellipsoidal modulation, and five undergo X-ray eclipses, and as such are ideal for accurately measuring neutron star masses – if sufficiently high resolution optical spectra can be obtained (e.g. van der Meer *et al.* 2005).

3.2. *Systems with Be Star Secondaries*

These constitute some 60% of all known Galactic HMXBs. Accretion occurs via an equatorial region of material generated by anisotropic wind loss from the (rapidly rotating) secondary star. Combined with the eccentric orbit of the compact object, this results in a highly variable X-ray luminosity, often showing periodic X-ray outbursts (recurring with the orbital period), with L_x 10^{37}–10^{38} ergs s^{-1} (at maximum). Typical quiescent X-ray luminosities range between 10^{33}–10^{34} ergs s^{-1}. An IR excess due to emission from circumstellar disk is often observed (e.g. Charles and Coe, 2006). Finally, we note that a strong correlation exists between orbital and spin periods for the Be systems (Corbet, 1984).

With their relatively bright secondary stars, the optical counterparts to HMXBs should be identifiable in nearby galaxies: whilst several such systems have been found, relatively few HMXB transients appear to have been identified so far (e.g. Williams *et al.* 2006).

4. The Low Mass Systems

More than half of all known Galactic X-ray binaries are LMXBs, with orbital periods ranging from 11 mins to 16.5 days. For those systems that are persistently bright, $L_x \sim 10^{36}$–10^{38} ergs s^{-1}. LMXB X-ray spectra can be crudely characterised by power-law/bremsstrahlung emission (e.g. due to Comptonization), with a black body component for the more luminous ($L_x > 10^{37}$ ergs s^{-1}) sources. Lower luminosity sources often exhibit thermonuclear (Type-I) X-ray bursts, and harder, power-law spectra with an energy index ~1 and an exponential high energy cut-off (again, see White, Nagase and Parmar, 1995 for more details).

Many of the techniques used to study CVs have been successfully applied to LMXBs (e.g. Doppler tomography), particularly those LMXBs in a quiescent state – e.g. radial velocity and ellipsoidal studies of the secondary star. It has been found, for example, that the shorter period LMXBs have accretion geometries comparable to those found in CVs – e.g. Roche-lobe overflow mediated accretion, relatively high mass ratios ($= M_1/M_2$, where M_1 is the compact object mass), outer disk "bulges" where the accretion stream

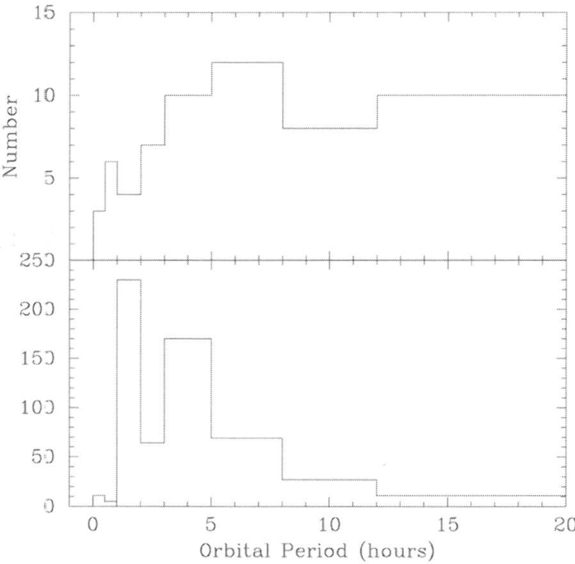

Figure 2. The CV (lower) and LMXB (upper) orbital period distribution.

meets the disk, precessing accretion disks, etc. In addition, the accretion disk instability that drives the outbursts of so many CVs is also thought to be responsible for LMXB transient behaviour (see next section, and Charles and Coe, 2006 for more details).

In Fig. 2, we plot the orbital period distribution for 596 CVs and 81 LMXBs, from the most recent catalogue of Ritter and Kolb (2005) over the range 0–20 hrs. The orbital period bins are rather wide, in an attempt to improve on the statistics for the LMXB distribution. Although the LMXB numbers are low, the distributions are clearly quite different. The data suggest that the CV period gap (at 2–3 hrs) is not present in the LMXB orbital distribution, perhaps because of the effect of X-ray irradiation on the orbital evolution of these systems. The relative fraction of longer period LMXBs (in comparison to CVs) should also be noted: this may be due again to the effect of X-ray irradiation on the secondary star, enabling it to fill its Roche lobe and maintain mass-transfer for the longer period systems.

A quite spectacular Type-I bursting phenomenon has recently been discovered – the "superbursts" (e.g. Kuulkers, 2005). Lasting several hours and with total fluences of $\sim 10^{42}$ ergs, such bursts (some of which may be due to C burning) should be observable from extragalactic (i.e. beyond SMC/LMC) distances.

4.1. *X-ray Novae (aka Soft X-ray Transients)*

Approximately half of all known LMXBs are transient, with typical recurrence times 10–50 years. In outburst, these are the brightest sources in the X-ray sky, with $L_x \geqslant 10^{39}$ ergs s^{-1} for the more extreme examples. Their quiescent luminosities, on the other hand, are equally dramatic, reaching values as low as 10^{30} ergs s^{-1} in some cases. Quiescent neutron star systems appear to be systematically brighter than this, strongly suggesting that the fainter black hole systems possess an event horizon across which material is accreted before it can radiate (e.g. Garcia *et al.* 2001).

With orbital periods $\sim$4 hrs – 33.5 days, and secondary stars that can be spectroscopically studied in quiescence, these systems have yielded 15 binaries for which strongest evidence exists for the presence of stellar mass black holes in our Galaxy (see Charles and Coe, 2006 for more details).

4.2. *Accreting Binary Millisecond Pulsars*

Of all the X-ray binaries discussed in this review, this is the class of object that is expanding at the fastest rate. Since the discovery of the prototypical system in 1996, SAX J1808.4-3658 – the all-singing, all-dancing, bursting and pulsating X-ray binary – 6 more systems have been discovered at the time of writing (see Wijnands, 2005). With orbital periods ranging from 40 mins to 4.3 hours, and spin periods from 1.67-5.4 msec, these systems finally provide firm confirmation of the link between accreting LMXBs and millisecond pulsars. Many of these transients would have escaped detection in the past, as they have relatively meagre maximum outburst luminosities, $\sim 10^{36}$ ergs s^{-1}. Why X-ray pulsations can be so easily observed for this class of system, and so much harder to see directly for the vast majority of LMXBs, is still unclear (although it may be due to the lower accretion rates present in these short period transients: e.g. Wijnands, 2005).

The best studied of these systems in quiescence remains SAX J1808.4-3658: interestingly, Campana *et al.* (2004) suggest that strong heating of the secondary star is maintained even in this state, caused by irradiation due to the spin-down luminosity of the neutron star, similar to the case of the "black-widow" binary millisecond pulsar PSR B1957+20 (e.g. Callanan *et al.* 1995).

Between these and the burst oscillations observed in some LMXBs (and, of course, the persistently bright pulsators), some 20 LMXBs now exist for which the spin period of the neutron star is known.

4.3. *X-ray Reprocessing*

In contrast to the HMXBs, the optical flux of most LMXBs (except those with the longest orbital periods) is dominated by X-ray processing. As discussed by van Paradijs and McClintock (1994), this allows us constrain the orbital period of a given system, if we know the X-ray luminosity and absolute magnitude (M_V): this is particularly useful in the case of extragalactic LMXBs (see e.g. Williams *et al.* 2005). It is interesting to investigate the effect of the nature of the compact object on this correlation: for example, systems with a higher mass ratio ($q = M_1/M_2$) might be expected to have larger, and hence brighter, accretion disks (e.g. Garcia 2005). Indeed, taking two values of $q = 10$ and 1 – corresponding to a 1 $M_\odot$ secondary with a black hole and neutron star primary respectively – we expect a difference in Roche-lobe radius of the primary star of ~ 2 (for the same orbital period and X-ray luminosity).

In Fig. 3, we plot the relationship derived by van Paradijs and McClintock (1994) between M_V and logΣ, where $\Sigma = P^{2/3}(L_x/L_{Edd})^{1/2}$ (with P the orbital period and L_{Edd} the Eddington luminosity), for some 12 LMXBs for which these parameters can be measured with reasonable confidence. In addition to their best fit, we superimpose a line ~ 1 mag higher, corresponding to the offset we expect between neutron star and black hole systems (of comparable X-ray luminosity and orbital period). It can be seen that this offset is perilously close to the scatter in the data, indicating that M_V, L_X and P measurements alone do not appear to provide sufficient discrimination between black hole and neutron star systems in the case of our Galaxy. However, as some of the scatter in this plot is undoubtedly due to uncertainty in the distance to the systems, longer term monitoring of extragalactic LMXBs (i.e. systems at known distance) may reveal a systematic difference in M_V between the black hole and neutron star systems.

4.4. *Globular Cluster Sources*

Of the 13 luminous LMXBs in Galactic globular clusters, 5 are transient. Furthermore, 4–5 are likely to have ultrashort ($\leqslant 1$ hr) orbital periods, as inferred either from direct measurement, from the L_x M_V relationship discussed above, or other properties (e.g.

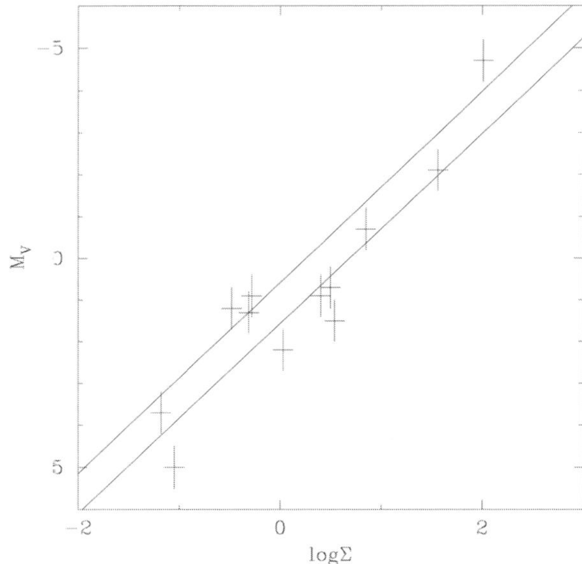

Figure 3. The absolute magnitude vs Σ, from van Paradijs and McClintock (1994); see Section 4.3 for the meaning of the two lines.

Deutsch *et al.* 2000, Verbunt & Lewin, 2005). These compact binaries are important for the overall evolution of the host cluster (e.g. providing it with energy against collapse), and suggest that clusters not only make binaries via 2 and 3-body interactions (see Fig. 4), but also create ultra-compact systems. Many other types of binaries are also present in globular clusters (CVs, millisecond pulsars, magnetically active binaries, etc). Bildsten and Deloye (2004) suggest that most cluster LMXBs are ultracompact, with lifetimes of a few $\times 10^6$ yrs: this has the effect of increasing the overall birthrate of binaries in globular clusters, bringing it into agreement with the observed population of the (descendent) millisecond pulsars.

Have globular clusters made most, if not all, of the Galactic LMXBs we see today? This has been suggested by, for example, Grindlay (1984), and more recently by White, Sarazin and Kulkarni (2002). As pointed out by Verbunt and Lewin (2005), in our Galaxy and M31 there are 10 disk LMXBs for each cluster LMXB, whereas for elliptical galaxies this ratio of cluster to non-cluster LMXBs is 1:1: this suggests that most Galactic LMXBs formed in the disk. However, the argument is not clearcut: Gnedin and Ostriker (1996) raise the possibility that the current population of Galactic globular clusters are but a very small fraction of the initial population, with many clusters in the past completely disrupted as they orbit through the plane of the Galaxy. On the other hand, elliptical galaxies, or course, do not have the same perturbing influence on their clusters. This issue appears to have been resolved by Irwin (2005), who provides strong arguments for the creation of the majority of "field" LMXBs in the field itself, rather than through cluster disruption or destruction.

5. X-ray Population of the Galactic Centre

Finally, we briefly discuss the population of X-ray sources recently uncovered near the Galactic Centre. Muno *et al.* (2003) discovered some $\sim$2000 point-like sources within $\sim$20 pc of Galactic Centre, with $L_x \sim 10^{31} - 10^{33}$ ergs s^{-1}. As many of these have relatively hard X-ray spectra (kT>8 keV), they are similar to magnetic CVs – as postulated,

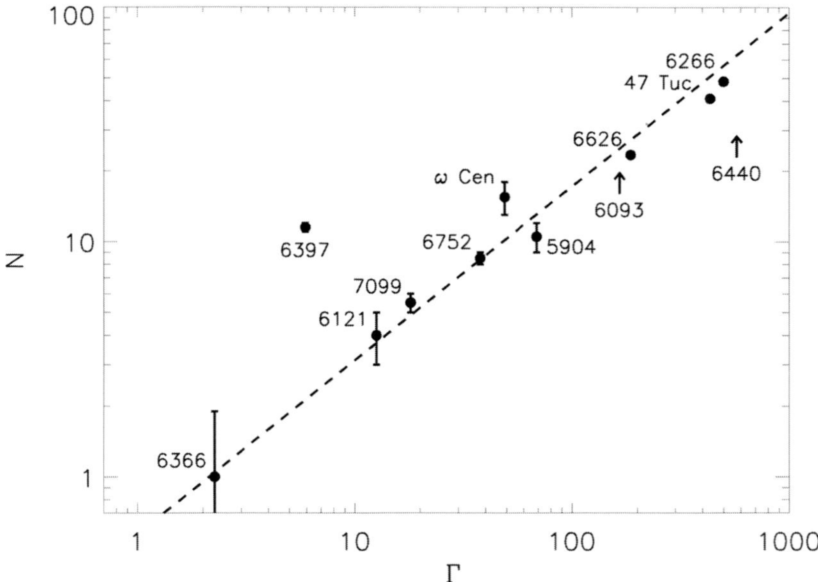

Figure 4. The number of faint globular cluster sources vs encounter rate Γ ($\propto$ (central density)$^{1.5}$(core radius)^2) for a sample of globular clusters. See Pooley *et al.* (2003) for more details.

for example, by Watson (1999). These sources may contribute to a large fraction of the diffuse emission near the Galactic Centre. In addition, Muno *et al.* (2005) report 7 transients, some of which vary by a factor of $\sim$10, with L_x (maximum) $> 5 \times 10^{33}$ ergs s^{-1}: 4 of these lie within 1 pc of Sgr A*. These could be HMXBs, or, intriguingly, LMXBs produced by 3 body interactions (in the environment near the Galactic Center where the stellar density is $\sim$240–900 M$_\odot$ pc^{-3}: see e.g. Muno *et al.* 2003, 2004).

6. Conclusions

In summary, we note that:

(1) The combination of RXTE Galactic monitoring, and high resolution imaging of Chandra and XMM, has finally allowed us (at least attempt) to unite the Galactic and extragalactic populations of X-ray binaries.

(2) Optical observations are an essential tool in the study of X-ray binary populations in our Galaxy and others. We note, however, that once the demise of the HST occurs, and its successor launched, the kind of imaging previously possible with the HST will only be available in the near and far IR, where its successor will be optimised. Hence more effort spent now on characterising the IR properties of Galactic X-ray binaries, for comparison with future IR observations of their extra-galactic cousins, will be time well spent.

(3) Whatever about the need for a future space based optical facility, there is an obvious and pressing need for a future imaging X-ray astronomy mission with larger throughput/effective area than either Chandra or XMM (e.g. White, 2005). However, there is also a very strong case for (cheaper !) lower resolution, but wider field monitoring missions. The expanding number of Galactic black hole X-ray binaries, and the even more rapidly increasing number of accreting binary millisecond pulsars are but a foretaste of what could be delivered by such a mission.

Acknowledgements

I would like to acknowledge support from Science Foundation Ireland, and comments from many contributors to the Symposium.

References

Bildsten, K. and Deloye, C.J. 2004 ApJ, 607 L119

Campana, S. *et al.* 2004, ApJ, 614, L49

Callanan, P.J., van Paradijs, J. and Rengelink, R. 1995, ApJ, 439, 928

Charles, P.A. and Coe, M.J. 2006, in Compact Stellar X-ray Sources, W.H.G. Lewin and M. van der Klis, eds, CUP.

Corbet, R.H.D. 1986, MNRAS 220 1947

Deutsch, E.W., Margon, B. and Anderson, S.F. 2000, ApJ 530 21.

Garcia, M. R., McClintock, J. E., Narayan, R., Callanan, P.J., Barret, D. and Murray, S. S. 2001, ApJ, 553, L47

Garcia, M.R. 2005, priv. comm.

Grindlay, J. E. 1984, Adv. Space Res., 3, 19

Grimm, H.-J., Gilfanov, M. and Sunyaev, R., 2002, A&A 391, 923

Grimm, H.-J., Gilfanov, M. and Sunyaev, R., 2003, MNRAS 339, 793

Hands A.D.P.,Warwick R.S., Watson M.G. and Helfand D.J. 2004, MNRAS

Irwin, J. 2005, ApJ, 631, 511

Kuulkers, E., 2004, Nuclear Physics B Proceedings Supplements, Volume 132, p. 466-475

Kuulkers, E., 2005, astro-ph/0504625

Muno, M.P., *et al.* ApJ 2003 ApJ 589 225

Muno, M.P., *et al.* 2004 ApJ 613 1179

Muno, M.P., Pfahl, E., Baganoff F.K., Brandt, W.N., Ghez,A., Lu, J. and Morris M.R., 2005 ApJ 622 L113

Pfahl, E., Rappaport, S. and Podsiadlowski, P. 2003, ApJ, 597 1036

Pooley, D., *et al.* 2003, ApJ, 591 L131

Ritter, H. and Kolb, U., 2005 (updated version of 2003 A&A 404 301)

van der Meer, A., Kaper, L., van Kerkwijk, M.H. and van den Heuvel, E.P.J., 2005, astro-ph/0502313

van Paradijs, J. and McClintock, J.E. 1994, A&A 290 133

Verbunt, F. and Lewin, W.H.G. 2006, in Compact Stellar X-ray Sources, W.H.G. Lewin and M. van der Klis, eds, CUP.

Watson, M.G., 1999 in "Annapolis workshop on Magnetic Cataclysmic Variables", ASP Conf Series Vol 157, Coel Hellier and Koji Mukai, eds.

White, N.E., Nagase, F. and Parmar, A.N. 1995 in "X-ray Binaries", W.H.G. Lewin, J. van Paradijs and E.P.J. van den Heuvel, CUP, p27.

White, R. E., Sarazin, C. L. and Kulkarni, S. R. 2002, ApJ, 571, L23

White, N.E. 2005, Adv. Space Res., 35, 96

Wijnands, R. 2005, astro-ph/0501264

Williams, B.F., Garcia, M.R., McClintock, J.E., Primini, F.A. and Murray, S.S. 2005, ApJ., 632 1086.

Williams, B., Garcia, M.R., Naik, S. and Callanan, P.J., 2006, submitted to ApJ.

Discussion

ERACLEOUS: Do we know any example in our Galaxy of systems that are HMXBs accreting via genuine RLOF (i.e. $L_x \sim 10^{38}$ erg/s and a supergiant companion)?

CALLANAN: We know of a number of RLOF systems of this type but their L_x are several orders of magnitude lower.

MIRABEL: GRS 1915 is one such example.

CALLANAN: Yes, that is the exception to many rules.

The Symposium audience listening. How many are sleeping? Notice the bouncer at the door.

Populations of High Energy Sources in Galaxies
Proceedings IAU Symposium No. 230, 2005
E. J. A. Meurs & G. Fabbiano, eds.

© 2006 International Astronomical Union
doi:10.1017/S1743921306007708

High Energy Emission from Supernova Remnants

Jacco Vink[1]

[1]Astronomical Institute, University of Utrecht, PO Box 80000, NL-3508TA,Utrecht,
The Netherlands
j.vink@astro.uu.nl

Abstract. This paper discusses several aspects of current research on high energy emission from supernova remnants, covering the following main topics: 1) The recent evidence for magnetic field amplification near supernova remnant shocks, which makes that cosmic rays are more efficiently accelerated than previously thought. 2) The evidence that ions and electrons in some remnants have very different temperatures, and only equilibrate through Coulomb interactions. 3) The evidence that the explosion that created Cas A was asymmetric, and seems to have involved a jet/counter jet structure. And finally, 4), I will argue that the unremarkable properties of supernova remnants associated with magnetar candidates, suggest that magnetars are not formed from rapidly ($P \approx 1$ ms) rotating proto-neutron stars. It is therefore more likely that they are formed from massive progenitor stars with high magnetic fields.

Keywords. (ISM:) supernova remnants, (ISM:) cosmic rays, X-rays: ISM.

---

1. Introduction

Supernova remnants (SNRs) are important high energy sources. Not only are they thought to be the major source of Galactic cosmic rays of energies up to at least 10^{15} eV, they are, until a supernova occurs in the Galaxy, also the only Galactic sources with which can get a direct view on supernova explosions and nucleosynthesis.

Moreover, SNRs are sources in which interesting physical processes take place, some of which can now be spatially or spectroscopically resolved with the current generation of high spatial and spectral resolution X-ray telescopes, *Chandra*, and *XMM-Newton*.

2. Collisionless shocks

That the hot plasma of SNRs that we observe with X-ray telescopes exists is somewhat of a surprise. The reason is that SNR shocks take place in the tenuous interstellar medium, with typical densities of $n \sim 1$ cm^{-3}. Two body atomic collisions in such a medium are so rare that, were it not for the generation of plasma waves, plasmas could move through each other without hardly any noticeable interaction. Because of this lack of two body collisions (Coulomb interactions) SNR shocks are called collisionless.

The small collision rates in SNRs has two consequences. One of them, non-equilibrium ionization (NEI), has been well established since the eighties (see e.g. Liedahl 1999, for an introduction). SNR plasmas are usually in NEI, meaning that the plasma has not yet had time to reach ionization equilibrium. In other words the plasma is still ionizing. NEI plasmas are usually characterized by the parameter $n_{\mathrm{e}}t$, i.e. the product of electron density and time. Plasmas with temperatures of $kT_{\mathrm{e}} \sim 1$ keV are out of equilibrium when $n_{\mathrm{e}}t \lesssim 10^{12}$ cm^{-3} s.

Less well known is that there is considerable uncertainty how collisionless plasmas are heated by shocks, and in particular whether this leads immediately downstream of the

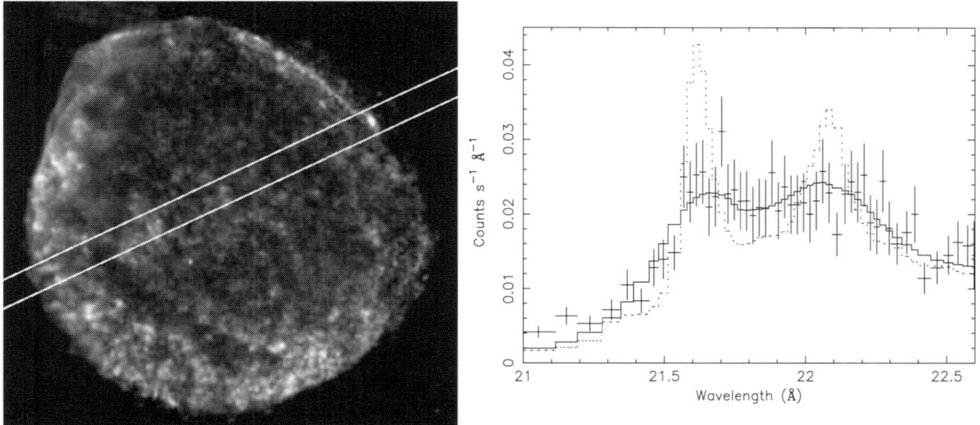

Figure 1. On the left: Map of O VII emission of SN1006 made from several *Chandra*-ACIS observations. The lines indicate the region observed by the *XMM-Newton* RGS instrument. The target was the bright knot in the northwest. Right: Detail of the RGS1 spectrum of the northwestern knot, showing O VII Heα line emission. The dashed line is the best fit model without line broadening, whereas the solid line shows the model including thermal line broadening (Vink *et al.* 2003).

shock to equal temperatures for all particles (electrons, ions) involved. Basic considerations, such as conservation of mass, momentum and energy, lead to the Rankine-Hugoniot relations, but it is not clear whether collisionless shock heating leads to equal temperatures for all particles, or whether it heats different plasma elements to different temperatures. In the latter case one expects that the temperature of each species is proportional to $kT_i \propto m_i v_s^2$, with m_i the particle mass and v_s the shock speed. Far downstream of the shock the particles will eventually equilibrate their energies. Also the Coulomb equilibration of electrons and ions is governed by the parameter $n_e t$. Equilibration is reached when $n_e t \lesssim 10^{12}\,\mathrm{cm}^{-3}\,\mathrm{s}$.

Over the last decade several observations in the optical (Ghavamian *et al.* 2001, 2003), UV (Raymond *et al.* 1995; Laming *et al.* 1996; Korreck *et al.* 2004), and X-rays (Vink *et al.* 2003) have been used to address this issue. They all rely on the fact that ion temperatures can be measured by observing thermal Doppler broadenings. For SNRs this can only be done for the limb brightened edges of the shells, because far inside the shell, bulk motions in the line of sight dominate the line broadening.

In Fig. 1 X-ray evidence is shown that for SN 1006 the ion temperature is much hotter, $kT_{\mathrm{OVII}} \sim 500$ keV, than the electron temperature $kT_e \approx 1.5$ keV. Of course what is normally measured is the electron temperature, as this determines the X-ray continuum shape and emission line ratios.

SN1006 is a very suitable target for such a study, as its plasma is very far out of equilibrium, $\log n_e t = 9.5$. However, optical, Hα, measurements of line broadening do not rely on low $n_e t$ values, as Hα emission only occurs very close to the shock front. Using Hα measurements of several SNRs, Rakowski *et al.* (2003) have shown that slow equilibration of temperatures is probably a function of shock velocity or Mach number, as only the fastest shocks appear to have substantial differences between electron and proton temperatures.

3. Cosmic ray acceleration and magnetic field amplification

One of the important findings of *Chandra* concerning SNRs was that with its CCD detectors it was able to pick out thin, non-thermal X-ray emitting filaments (Hwang *et al.*

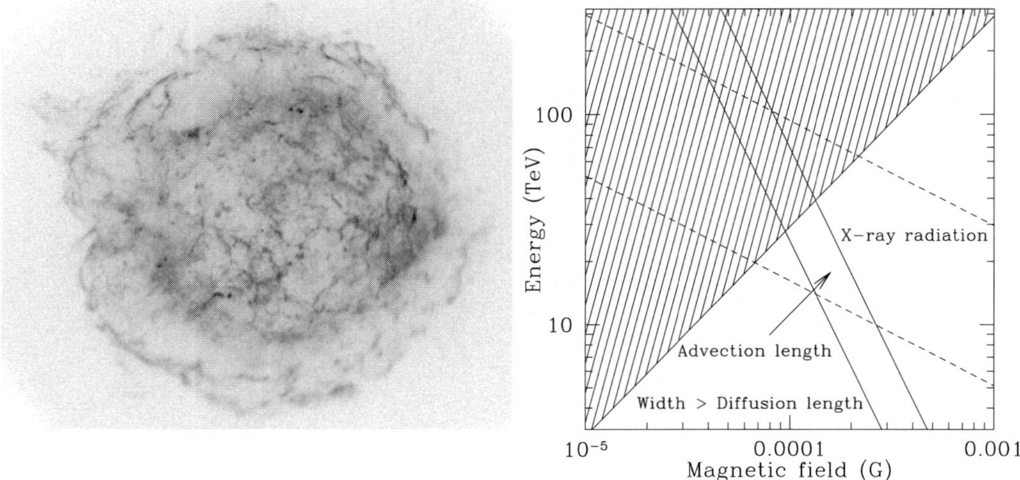

Figure 2. A deep *Chandra* image of Cas A (Hwang *et al.* 2004) in the 4-6 keV continuum band (left). Note the thin filaments, marking the border of the remnant (NB the point spread function is not uniform). The remnant has a radius of about $2.5'$. Right: Determination of the maximum electron energy versus magnetic field strength for the region just downstream of Cas A's shock front, as determined from the thickness of the filaments. The shaded area is excluded, because the filament width cannot be smaller than the minumum possible diffusion length (c.f. Vink & Laming 2003).

2002; Gotthelf *et al.* 2001). Vink & Laming (2003) showed that these filaments probably emit synchrotron emission, with electron energies $\gtrsim 10$ TeV. The narrow widths of these filaments are then best interpreted as the result of synchrotron losses.

The reason is that the plasma downstream of the shock sweeps the relativistic particles away from the shock. At the same time the electrons rapidly lose energy, so that at some point away from the shock front the electrons only emit synchrotron radiation at energies below the X-ray band. For standard high Mach number shocks the plasma velocity with respect to the shock front is given by $\Delta v = \frac{1}{4}v_s$. So for the width of the filaments one can write $\Delta r = \frac{1}{4}v_s \tau_{loss}$, with the synchrotron loss time given by $\tau_{loss} = 635/(B^2 E)$. In order to disentangle the electron energy E, and the average downstream magnetic field strength B one has to use the fact that the peak photon energy as a result of synchrotron radiation is $\epsilon = 7.4 E^2 B$ keV. Fig. 2 shows graphically what for Cas A the possible values for B and E are. It turns out that the magnetic field is high $B = 200$–500 μG for Cas A, but also for other young SNRs (e.g. Bamba *et al.* 2005; Ballet 2005), which is much stronger than might be expected if the magnetic field is just the shock compressed mean Galactic field.

This may be surprising, but it is a nice confirmation of recent theoretical work that indicates that strong cosmic ray streaming close to fast SNR shocks may lead to non-linear amplification of magnetic fields (Bell & Lucek 2001; Bell 2004).

In fact this solves a piece of the puzzle concerning cosmic ray acceleration. SNRs were for a long time thought to be the most plausible sources of Cosmic Rays up to or beyond 3×10^{15} eV, at which energy the cosmic ray spectrum has a break. However, with only mean Galactic magnetic field values SNRs are not able to efficiently accelerate particles up to even 10^{14} eV (Lagage & Cesarsky 1983). It looks now that *Chandra* has solved this problem.

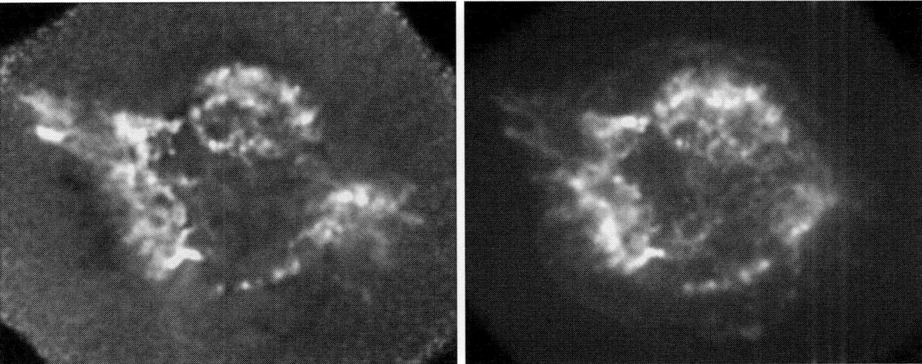

Figure 3. Left: Image based on the deep *Chandra* observation of Cas A, which has been processed to bring out the jet/counter jet structure (Vink 2004a; Hwang *et al.* 2004). Right: For comparison a *Chandra* X-ray image taken in the band of Si XIII line emission.

Although X-ray observations show that one necessary ingredient, large, turbulent magnetic fields are present (see also Vink 2004b), there is still no direct evidence that SNRs accelerate also ions up to high energies. Note that ions are the main ingredient of the cosmic rays that bombard the earth atmosphere. In that respect the many SNRs that have been observed with Cherenkov telescopes, in particular H.E.S.S. (Aharonian *et al.* 2004), are very promising. The reason is that collisions of relativistic ions result in the production of pions, and π^0 particles decay into two photons, giving rise to γ-ray emission. However, also inverse Compton scattering of background photons by relativistic electrons produces γ-ray emission. In that case X-ray and Cherenkov γ-ray telescopes may observe the same electron cosmic ray population. It has not yet been resolved which mechanism is responsible for the γ-ray emission from SNRs.

4. Asymmetric supernova explosions: a link with GRBs?

The origin of cosmic rays may be an almost century old problem, but an equally fascinating, but more recent problem is the nature of gamma-ray bursts (GRBs). It is becoming more and more clear that long duration GRBs are probably associated with core collapse supernovae of subclass Type Ibc (Stanek *et al.* 2003). However, the mechanism that generates the powerful relativistic jets that we observe as GRBs is not well known. The collapsar model (MacFadyen *et al.* 2001) is one of the most popular models. In this model the stellar core collapses into a black hole that accretes matter, and generates jets. An alternative model is magneto-rotational jet formation (Akiyama *et al.* 2003).

In this light it is interesting that it was recently discovered that the bright Galactic SNR Cas A seems to have exploded with a jet/counter jet. A normal X-ray image of Cas A does not immediately reveal this, but dividing a narrow band image dominated by Si XIII line emission by a narrow band image with Mg XI line emission brings out a clear jet/counter jet structure (Fig. 3, Vink 2004a; Hwang *et al.* 2004). The spectra of the jet reveal an apparent absence of Ne and Mg. The dominant elements seem to be Si, S, and Ar, but some Fe seems also present. The emission measure of the jet combined with the average velocity of the plasma suggest quite a high kinetic energy in the jet, $\sim 5 \times 10^{50}$ erg, about 25% of the total explosion energy.

So in terms of the total explosion energy the jets seem to contain a substantial, but not a dominant fraction of the energy. The jets seem not to have been relativistic jets, like those

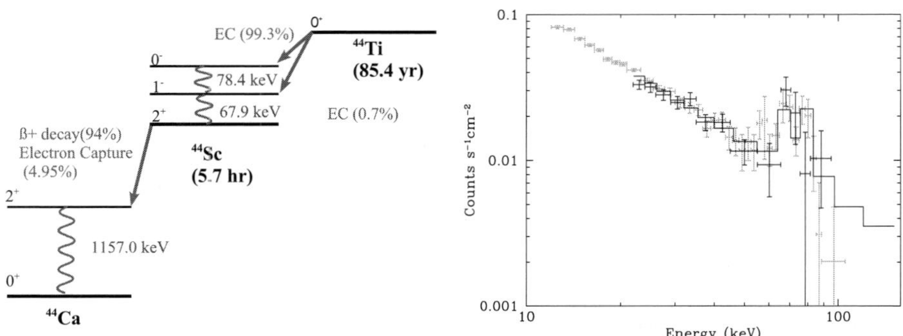

Figure 4. Left: The radio active decay scheme of ^{44}Ti. Right: hard X-ray spectrum of Cas A, as observed by *BeppoSAX* (in gray) and *INTEGRAL*, both showing clear signs of emission around 68 keV and 78 keV due to ^{44}Ti decay lines (Vink 2005).

of GRBs. Nevertheless, perhaps the same underlying mechanism produces both types of jets. In that respect it is interesting that Cas A is likely the result of a Type Ib explosion, i.e. it belonged to the same supernova subclass with which GRBs are associated. However, it is unlikely that the collapsar model in its present form is responsible for the jets in Cas A, because there is the simple fact that the explosion appears to have resulted in the formation of a recently detected neutron star (Tananbaum 1999) rather than a black hole.

Interestingly, not only the presence of a jet/counter jet makes Cas A an interesting SNR from the point of view of the explosion mechanism. Equally interesting is that Fe-rich knots seem to have been ejected with greater speed than the Si-rich material synthesized further away from the core. This is clear from the presence of Fe-rich knots in the southeast of the remnant, outside of the main Si-rich shell (Hughes *et al.* 2000; Hwang & Laming 2003).

In the north the Fe is projected to the inside of the Si-rich shell. However, this appears to be a projection effect, because the measured Doppler velocities of Fe in the north are higher than for Si (Willingale *et al.* 2002). It is not clear how much of the Fe in Cas A is still unshocked, but some of the shocked Fe must have been ejected with velocities of up to 7800 km s^{-1}. There is no obvious symmetry to the Fe-rich ejecta, so their emergence is probably related to hydrodynamical instabilities close to the core of the explosion (Kifonidis *et al.* 2003).

The 3D morphology of Cas A as reconstructed from the Doppler imaging obtained from *XMM-Newton* gives further evidence that the explosion was deviating from spherical (Willingale *et al.* 2002). Apart from the jets and Fe-rich knots, the ejecta can best be described by a donut shape. This morphology also appears to describe best the high resolution spectroscopy data of the Small Magellanic Cloud remnant 1E0102.2-7219, obtained by the grating spectrometers of *Chandra* (Flanagan *et al.* 2004).

There is one other observational fact concerning Cas A that may point to an intrinsically asymmetric explosion: the presence of ^{44}Ti, which suggest a high initial ^{44}Ti yield of $\sim 10^{-4}$ $M_\odot$. ^{44}Ti is exclusively an explosive nucleosynthesis product, and is synthesized close to the core of the explosion. It is an alpha-rich freeze out product (Arnett 1996), and as such very sensitive to explosion energy and explosion asymmetries.

The decay of ^{44}Ti ($\tau = 86$ yr) is accompanied by three strong γ-ray lines (Fig. 4), which have been detected by *CGRO-COMPTEL* (the 1157 keV line, Iyudin *et al.* 1994), and *BeppoSAX* (the 68 and 78 keV lines, Vink *et al.* 2001), with lines fluxes of $\sim 2 \times 10^{-5}$ ph cm^{-2}s^{-1}. The fact that the line emission is caused by radio-active decay makes

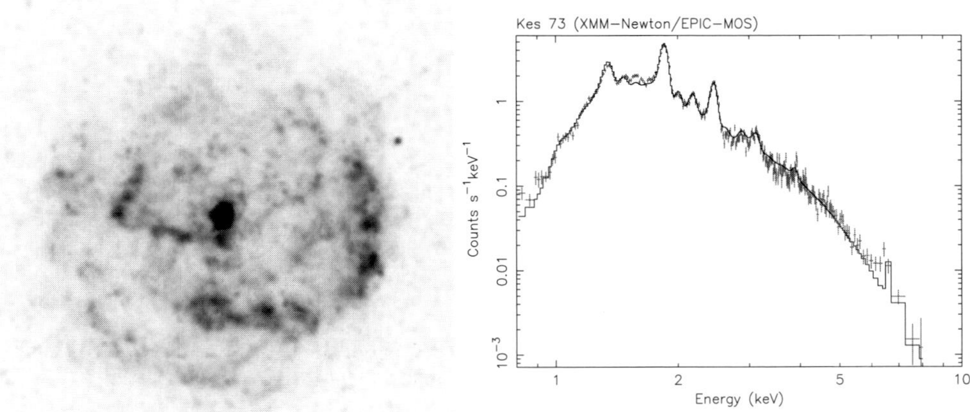

Figure 5. Left: *Chandra* image of Kes 73. The AXP itself is highly saturated and shows up as a dark spot in the center of the remnant. Right: *XMM-Newton* EPIC-MOS spectrum of Kes 73, with the best fit Sedov model shown as a solid line. From this Sedov model (Borkowski *et al.* 2001) one can estimate the total kinetic energy of the supernova (Vink *et al.* in preparation).

that with ^{44}Ti we can also probe the unshocked ^{44}Ti ejecta. Cas A is therefore a target of ESA's γ-ray observatory INTEGRAL. Preliminary results of the observations have been published in Vink (2005) (Fig. 4).

5. The remarkable magnetars and the unremarkable supernova remnants in which they reside

Core collapse supernovae obtain their energy from the gravitational potential energy released when the core of a massive star collapses. The classical view is that core collapse would either result in the creation of a radio pulsar with typical magnetic fields of $B_{dip} \sim 10^{12}$ G, or otherwise in the formation of a black hole. However, over the last decade it has become clear that some neutron stars appear to have very high magnetic fields, up to $B_{dip} \sim 10^{15}$ G. Such neutron stars are called "magnetars". It is thought that two classes of X-ray pulsars are manifestations of magnetars. Depending on the presence or absence of (soft) γ-ray flashes, these pulsars are either labeled Soft-gamma-ray-repeaters (SGRs), or Anomalous X-ray Pulsars (AXPs) (see Kaspi & Gavriil 2004; Woods & Thompson 2004, for reviews).

It is not clear what causes the creation of a magnetar, but two mechanisms have been proposed. Perhaps the most popular explanation is that magnetars are created during the core collapse of massive stars with a high angular momentum. This results in the formation of a proto-neutron star which rotates non-uniformly with an average rotation period close to the break up limit $P \sim 1$ ms. This allows for the efficient operation of an $\alpha - \Omega$ dynamo, which rapidly amplifies the magnetic field (Duncan & Thompson 1992). Once the high magnetic field is in place, a magnetar will lose most of its angular momentum due to magnetic breaking in less than a few hunderd seconds, i.e. during the supernova explosion itself (e.g. Thompson *et al.* 2004). This means that most of the rotational energy will be rapidly pumped into the supernova ejecta. A neutron star with $P = 1$ ms has a rotational energy of $\sim 3 \times 10^{52}$ erg, which should be compared to an average core collapse supernova energy of $\sim 10^{51}$ erg.

What interests us here is the fact that of the dozen or so magnetar candidates, three are associated with bright SNRs: Kes 73, N49, and CTB 109. So one would expect that the kinetic energy of these SNRs is exceptionally high, as the ejecta have been energized by the rapid spin down of the rapidly rotating magnetar. However, the kinetic energies of these SNRs are unremarkable: A study of the energetics with *XMM-Newton*-EPIC spectroscopy reveals that the kinetic energies of Kes 73, N49, and CTB 109 are resp. 0.8×10^{51} erg, 2.0×10^{51} erg, and 0.7×10^{51} erg (Vink *et al.* 2005, in preparation, and Sasaki *et al.* 2004). This is far short of the $\sim 3 \times 10^{52}$ erg expected if magnetars owed their existence to the operation of an $\alpha - \Omega$ dynamo (Duncan & Thompson 1992). In fact one can put a lower limit on the initial spin period, by equating the observed SNR energies to the initial rotational energy of the pulsar. This gives $P_i > 5.6\sqrt{(E_{SNR}/10^{51} \text{ erg})}$ ms, which is closer to the classical initial spin period of radio pulsars, 10 ms, than to the break up limit of a neutron star.

So the unremarkable energies of SNRs associated with magnetars imply that magnetars are the result of collapses of the cores of massive stars with high magnetic fields (see also Ferrario & Wickramasinghe 2005), rather than from stars with a high angular momentum.

6. Summary

I have shown that X-ray studies of SNRs provides us with important information on collisionless shock physics, cosmic ray acceleration and magnetic field amplification by SNRs.

Moreover, by studying SNRs we can learn about the details of the supernova explosions that caused them. For example, the kinematics and spatial distribution of metals in Cas A reveal that the explosion was intrinsically asymmetric, and was accompanied by the emergence of a jet/counter jet system. And I have discussed that the most remarkable property of SNRs associated with magnetars is that they are unremarkable: Their energies are similar to those of other SNRs, which suggests that magnetars were *not* formed from proto-neutron stars with period close to 1 ms.

References

Aharonian, F. A. *et al.* 2004, Nat, 432, 75
Akiyama, S., Wheeler, J. C., Meier, D. L. & Lichtenstadt, I. 2003, ApJ, 584, 954
Arnett, D. 1996, Supernovae and Nucleosynthesis (NJ: Princeton Univ. Press)
Ballet, J. 2005, astro-ph/0503309
Bamba, A., Yamazaki, R., Yoshida, T., Terasawa, T. & Koyama, K. 2005, ApJ, 621, 793
Bell, A. R. 2004, MNRAS, 353, 550
Bell, A. R. & Lucek, S. G. 2001, MNRAS, 321, 433
Borkowski, K. J., Lyerly, W. J. & Reynolds, S. P. 2001, ApJ, 548, 820
Duncan, R. C. & Thompson, C. 1992, ApJ, 392, L9
Ferrario, L. & Wickramasinghe, D. T. 2005, MNRAS, 356, 615
Flanagan, K. A., Canizares, C. R., Dewey, D., *et al.* 2004, ApJ, 605, 230
Ghavamian, P., Rakowski, C. E., Hughes, J. P. & Williams, T. B. 2003, ApJ, 590, 833
Ghavamian, P., Raymond, J., Smith, R. C. & Hartigan, P. 2001, ApJ, 547, 995
Gotthelf, E. V. *et al.* 2001, ApJ, 552, L39
Hughes, J. P., Rakowski, C. E., Burrows, D. N. & Slane, P. O. 2000, ApJ, 528, L109
Hwang, U., Decourchelle, A., Holt, S. S. & Petre, R. 2002, ApJ, 581, 1101
Hwang, U. & Laming, J. M. 2003, ApJ, 597, 362
Hwang, U. *et al.* 2004, ApJ, 000, 000
Iyudin, A. F. *et al.* 1994, A&A, 284, L1
Kaspi, V. M. & Gavriil, F. P. 2004, Nuclear Physics B Proceedings Supplements, 132, 456
Kifonidis, K., Plewa, T., Janka, H.-T. & Müller, E. 2003, A&A, 408, 621
Korreck, K. E., Raymond, J. C., Zurbuchen, T. H. & Ghavamian, P. 2004, ApJ, 615, 280

Lagage, P. O. & Cesarsky, C. J. 1983, A&A, 125, 249

Laming, J. M., Raymond, J. C., McLaughlin, B. M. & Blair, W. P. 1996, ApJ, 472, 267

Liedahl, D. A. 1999, Lecture Notes in Physics, Berlin Springer Verlag, 520, 189

MacFadyen, A. I., Woosley, S. E. & Heger, A. 2001, ApJ, 550, 410

Rakowski, C. E., Ghavamian, P. & Hughes, J. P. 2003, ApJ, 590, 846

Raymond, J. C., Blair, W. P. & Long, K. S. 1995, ApJ, 454, L31

Sasaki, M., Plucinsky, P. P., Gaetz, T. J., *et al.* 2004, ApJ, 617, 322

Stanek, K. Z. *et al.* 2003, ApJ, 591, L17

Tananbaum, H. 1999, IAU circ., 7246, 1

Thompson, T. A., Chang, P. & Quataert, E. 2004, ApJ, 611, 380

Vink, J. 2004a, New Astronomy Review, 48, 61

Vink, J. 2004b, astro-ph/0409517

Vink, J. 2005, Adv. Space Res., 35, 976

Vink, J. & Laming, J. M. 2003, ApJ, 584, 758

Vink, J., Laming, J. M., Gu, M. F., Rasmussen, A. & Kaastra, J. 2003, ApJ, 587, 31

Vink, J. *et al.* 2001, ApJ, 560, L79

Willingale, R., Bleeker, J. A. M., van der Heyden, K. J., Kaastra, J. S. & Vink, J. 2002, A&A, 381, 1039

Woods, P. & Thompson, C. 2004, astro-ph/0406133

Discussion

UBERTINI: Do you think the hard x-ray continuum in the Cas A spectrum below the Ti line could be due by electron synchrotron emission at the shock front of the expanding SNR?

VINK: The continuum could be either non-thermal emission from ~100 keV electrons, accelerated by forward or internal shocks, or X-ray synchrotron from TeV electrons. In order to get synchrotron emission at ~50 keV you need very high (>2000 Km/s) shock velocities, which are likely to exist only at the forward shock.

Populations of High Energy Sources in Galaxies
Proceedings IAU Symposium No. 230, 2005
E. J. A. Meurs & G. Fabbiano, eds.

© 2006 International Astronomical Union
doi:10.1017/S174392130600771X

Detection of Nonthermal X-ray structures near the Galactic Center with *Chandra*

Atsushi Senda[1], Katsuji Koyama[2], Ken Ebisawa[3], Jun Kataoka[4], and Yoshiaki Sofue[5]

[1]Cosmic Radiation Laboratory, RIKEN (The Institute of Physical and Chemical Research),
2-1 Hirosawa, Wako, Saitama 351-0198, Japan
email: senda@crab.riken.jp

[2]Cosmic Ray Group, Department of Physics, Graduate School of Science, Kyoto University

[3]Center for Planning and Information syntems, ISAS/JAXA

[4]Department of physics, Faculty of Science, Tokyo Institute of Technology

[5]Institute of Astronomy, University of Tokyo

Abstract. We have discovered a number of nonthermal X-ray features within central 40 pc region of the Galactic center by analysing 600-ksec observations of *Chandra* archival data. Most of the detected X-ray structures exhibit small-scale knot-like morphologies and their spectra are well reproduced by single hard power-law with photon indices of 1-2. Among them, the most outstanding features are the three X-ray knots which are aligned on a straight line from the potition of Sgr A* to north-northwest direction. The X-ray properties of these knots lead us to suspect that they are X-ray jets ejected from Sgr A* in the recent past. In addition, we have obtained an indication that the summed flux of nonthermal diffuse X-rays within 30 pc of the GC seems to be smoothly connected to the 20-100 keV flux detected with *INTEGRAL* IBIS/ISGRI. These results suggest that the origin of GC hard X-rays (or High energy Gamma-rays) is not (or partly) from the Galactic nucleus.

Keywords. Galaxy: center, X-rays: ISM, ISM: jets and outflows.

1. Introduction

A number of nonthermal radio emissions from linear filaments has been detected around the Galactic center (GC) (e.g. Yusef-Zadeh *et al.* (1984), Morris (1994), LaRosa *et al.* (2000), LaRosa *et al.* (2004)). Their spatial distribution seems concentrated within only the central hundred parsecs and their spectral properties suggest that their emission mechanism should be synchrotron radiation from high energy electrons. These characteristics lead us to suspect that the GC region holds some particle accelerators. In addition, it also suggests that the magnetic field strength around the GC is extremely strong ($B \sim 1$ mG) inferred from the extreme linearity of these radio filaments.

Recently, discovery of nonthermal X-ray filaments has been reported with *Chandra* and *XMM-Newton* observations (Wang *et al.* (2002), Sakano *et al.* (2003), Lu *et al.* (2003)). Although their emission mechanism and connection to the radio filaments are still unclear. It is partly because of poor samples of X-ray filaments.

Here we present the first systematic analysis of nonthermal X-ray structures within the $17' \times 17'$ of the Galactic center. We also analyzed the INTEGRAL IBIS/ISGRI data and found possible relation with Chandra nonthermal diffuse X-rays. A possible interpretation of emission mechanism of diffuse hard nonthermal X-rays detected with Chandra/INTEGRAL is briefly discussed.

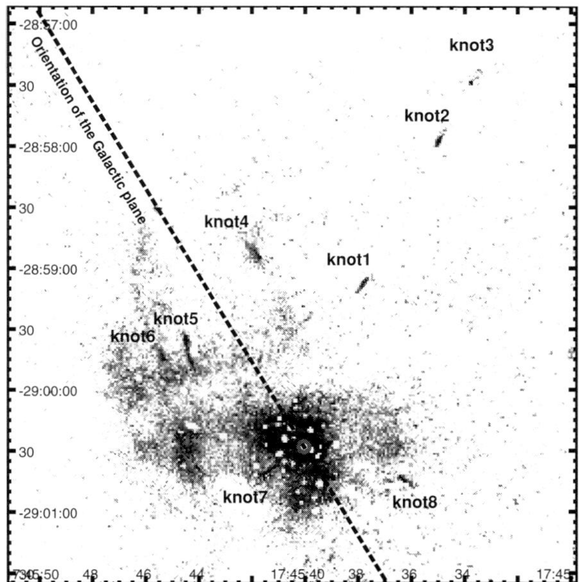

Figure 1. Continuum X-ray image of the Galactic center region with *Chandra*. The selected energy band is 3.42 keV–6.30 keV and the gray scale is logarithmically. Source ID of X-ray knots is also shown in the image.

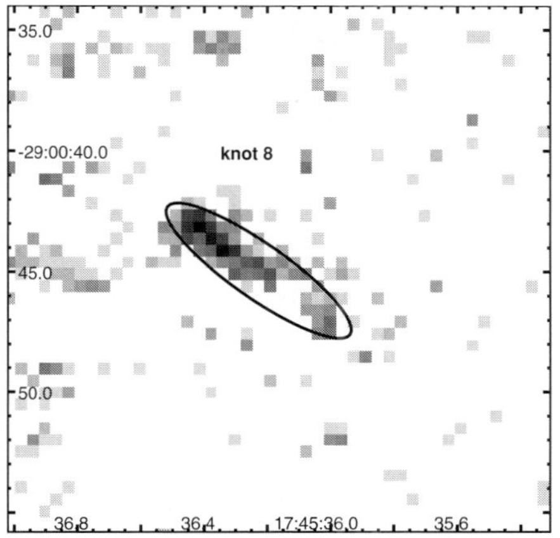

Figure 2. Close up X-ray image of knot 8 with *Chandra* ACIS-I. X-ray image is constructed from 3 to 300 counts per pixel with a logarithmic scale. All of the detected point sources are excluded from the image.

2. Analyses and Results

2.1. *Imaging analyses*

To extract results presented in the following sections, we analyzed eleven archival *Chandra* ACIS-I datasets. All of these observations were performed in Timed-Exposure Faint mode except for Obs ID=242, which was done in Timed-Exposure Very Faint mode. The total effective exposure of the combined event file amounts to 598 ksec.

To resolve diffuse nonthermal X-ray structures, we constructed continuum band image, in which there is no prominent emission line contained. From the continuum band image as is shown in Figure 1 we resolved eight prominent knot-like features, then we named them from 1 to 8. The position of the each sources and their identifications are shown in Figure 1. A sample of Close-up X-ray image of one of the knots (knot8) is shown in Figure 2.

2.2. *Spectral Analyses*

As a source region of each X-ray knot, we selected an elliptical region along their shape while the background spectrum is taken from the surrounding annulus with its center coordinate and width $1 < r < 3$ times of the radius of each source region, respectively. All of their spectra exhibit no emission lines from ions, hence we modelled each spectrum with a powerlaw model with absoption of interstellar medium taken into account. The fitting results of these knots are shown in Table 1. The best-fit values for column density are not smaller than $N_{\mathrm{H}} \sim 1.0 \times 10^{22}$ cm^{-2}, which is consistent with the value when they are located at the distance of the GC, and the possibility that these knots are foreground object is robustly rejected. These results lead us to conclude that knots 1–8 are nonthermal X-ray sources located near the Galactic center.

We also tried to fit each spectrum with a thin-thermal plasma model (MEKAL). The model could make an acceptable fit, however, the best-fit value of an electron temperature was determined to be >70 keV, which is too high a temperature to be explained by any energetic events. Therefore we thought that the application of a thermal model is unrealistic and the emission mechanism should be of nonthermal origin.

Table 1. Best-fit parameters for X-ray knot1–8 with a powerlaw model

Source ID	Source counts[a] (counts)	Column Density ($N_{\rm H}$) (10^{22} H cm^{-2})	Photon index (Γ) ($N(E) \propto E^{-\Gamma}$)	Flux[b] (ergs cm^{-2} s^{-1})
knot 1	501	8.5 (4.3–14)	1.3 (0.3–2.2)	4.1×10^{-14}
knot 2	485	16 (9.1–21)	2.3 (1.7–3.1)	3.4×10^{-14}
knot 3	295	10 (5.3–23)	1.8 (0.3–2.7)	3.8×10^{-14}
knot 4	800	12.9 (8.8–16.6)	1.8 (1.3–2.1)	5.3×10^{-14}
knot 5	562	21.4 (15.6–30.2)	3.2 (2.2–4.6)	3.7×10^{-14}
knot 6	782	8.7 (6.0–12.5)	3.1 (2.8–4.3)	3.4×10^{-14}
knot 7	183	1.1 (< 37.0)	-2.4 (< 1.5)	5.2×10^{-14}
knot 8	419	10.2 (7.0–18.6)	0.9 (-0.4–1.2)	3.7×10^{-14}

All the errors given in parentheses indicate the 90% confidence limits.
[a] Total counts from the source region after subtracting background counts. The subtracted background counts are normalized with the angular size of the region.
[b] An absorbed X-ray flux of 2.0–10.0 keV.

2.3. *A Possible Connection to Hard X-rays detected with INTEGRAL*

Bélanger *et al.* (2004) reported on the discovery of significant hard X-ray excess near the Galactic nulceus with *INTEGRAL* IBIS (Imager on Board the *INTEGRAL* Satellite) observations, which was named IGR J1745.6−2901. However, whether the excess of IGR J1745.6−2901 can be attributed to point-like source(s) or extended emission is still debatable.

For the purpose of investigating possible connection between nonthermal X-rays detected with *Chandra* and IGR J1745.6−2901, we constructed the 15–500 keV spectrum of IGR J1745.6−2901 from *INTEGRAL* IBIS/ISGRI data and then performed combined spectral fitting with 1.0–10 keV *Chandra* data. For the *Chandra* data, we use integrated spectrum of all nonthermal components detected within $17' \times 17'$ region with reference to the position of Sgr A*. The fitting results are shown in Table 2. At first, we fit both data sets with a powerlaw convolved with interstellar absorption. Because $N_{\rm H}$ should not be influential in the INTEGRAL data above 10 keV, the value of column density ($N_{\rm H}$) is fixed to the best-fit value obtained from *Chandra* data. We cannot obtain acceptable fit with single power-law model, since the observed photon index is very different between *Chandra* and *INTEGRAL*, hence we attempt two alternative models; cutoff-powerlaw and broken-powerlaw. Both phenomenological models can successfully reproduce the 1.0–500 keV X-ray spectrum statistically acceptable. Both of the fits exhibit characteristic energy cutoff/break around $E_0 \sim 25$ keV.

Table 2. Best-fit parameters for nonthermal hard X-rays of the Galactic center

Model	Photon index 1 ($N(E) \propto E^{-\Gamma_1}$)	Cutoff/Break energy (E_0 keV)	Photon index 2 ($N(E) \propto E^{-\Gamma_2}$)	Reduced χ^2 ($\chi^2/d.o.f.$)
powerlaw	2.61 (2.51–2.75)	—	—	154.3/84
cutoff-powerlaw	1.42 (1.36–1.57)	24.8 (21.2–28.2)	—	59.0.3/83
broken-powerlaw	1.53 (1.44–1.65)	25.3 (22.3–28.7)	3.35 (3.10–3.60)	53.4 /82

All the errors given in parentheses indicate the 90% confidence limits.

If the suprathermal electrons exist near the Galactic center, they should emit X-rays via bremsstrahlung. Energy loss rate by Coulomb interaction and bremsstrahlung are given as follows (Hayakawa(1969));

$$-\left(\frac{dE}{dt}\right)_{\mathrm{cou}} = 7.62 \times 10^{-9} n_{\mathrm{H}}(3\ln\gamma + 18.8) \ \mathrm{eV \ s}^{-1}$$

$$-\left(\frac{dE}{dt}\right)_{\mathrm{bremss}} = 1.37 \times 10^{-16} n_{\mathrm{H}}(\ln\gamma + 0.36)E \ \mathrm{eV \ s}^{-1}$$

Since the X-ray emitting electrons have energies near MeV, energy loss of Coulomb interaction must dominate. But the cooling time of the nonthermal electrons can be expressed as follows (Rephaeli (1979));

$$\tau_{\mathrm{cou}} \sim 4.3 \times 10^4 \beta \left(\frac{n}{1 \ \mathrm{cm}^{-3}}\right)^{-1} \left(\frac{E}{1 \ \mathrm{MeV}}\right) \ \mathrm{yr}$$

$$\tau_{\mathrm{bremss}} \sim 4.3 \times 10^8 \left(\frac{n}{1 \ \mathrm{cm}^{-3}}\right)^{-1} \left(\frac{E}{1 \ \mathrm{MeV}}\right) \ \mathrm{yr}$$

Thus, Coulomb interaction dominates the cooling time of suprathermal electrons in a typical condition of the Interstellar medium. By equating $\tau_{\mathrm{cou}} = \tau_{\mathrm{age}}$, we can obtain the break energy of the electron;

$$E_{\mathrm{cou}} \sim 23.2\beta \frac{n}{1 \ \mathrm{cm}^{-3}} \frac{\tau_{\mathrm{age}}}{10^3 \ \mathrm{yr}} \ \mathrm{keV}$$

Masai *et al.* (2002) proposed that the suprathermal electrons can be generated from cooler ($kT_e \sim 0.2 keV$) thermal plasma via stochastic accerelation. The hard component of the Galactic diffuse X-ray emission can be explained by the emission by the electrons of "suprathermal" populations.

Acknowledgements

We express particular thanks to K. Makishima and A. Bamba for their fruitful comments. This work is financially supported by Special Postdoctoral fellowship program of RIKEN.

References

Bélanger, G., *et al.*, 2004, *A&A*, 601, L163
Ekers, R. D., van Gorkom, J. H., Schwarz, U. J., & Goss, W. M., 1983, *A&A*, 122, 143
Hayakawa, S., 1969, Cosmic Ray Physics, Nuclear and Astrophysical Aspects (New York: Wiley-Interscience)
Lang, C. C., Morris, M., & Echevarria, L., 1999, *ApJ*, 526, 727
LaRosa, T. N., Kassim, N. E., Lazio, T. J. W., & Hyman, S. D. 2000, *AJ*, 119, 207
LaRosa, T. N., Nord, M. E., Lazio, T. J. W., & Kassim, N. E., 2004, *ApJ*, 607, 302
LaRosa, T. N., Brogan, C. L., Shore, S. N., Lazio, T. J., Kassim, N. E., & Nord, M. E., 2005, *ApJ*, in press, astro-ph/0505244
Lu, F. J., Wang, Q. D., & Lang, C., 2003, *AJ*, 126, 319
Masai, K., Dogiel, V. A., Inoue, H., Schönfelder, V., and Strong, A. W., 2002, *ApJ*, 581, 1071
Morris, M., 1994, in the Nuclei of Normal Galaxies, ed. R. Genzel & A. I. Harris (NATO ASI Ser. C, 445; Dordrecht: Kluwer), 185
Rephaeli, Y., 1979, *ApJ*, 227, 364
Sakano, M., Warwick, R. S., Decourchelle, C., & Predehl, P., 2003, *MNRAS*, 340, 747
Wang, Q. D., Lu, F., & Lang, C., 2002, *ApJ*, 581, 1148
Yusef-Zadeh, F., Morris, M., & Chance, D. 1984, *Nature*, 310, 557

Discussion

BOSCH-RAMON: Did you infer the kinetic luminosity of the ejection (when "knots" were ejected)?

SENDA: It is rather difficult because the X-ray spectra of "knots" exhibit completely non-thermal features, hence we cannot estimate the emission measure (and then temperature) of these knots from our analyses.

PIETSCH: Can you explain the difference between the histogram (intensity) in the observations 2 years apart?

SENDA: I think the difference is due to statistical fluctuation because the intensity of each "knot" is very faint (Photon counts of each "knot" is typically only 300–500 cnts/600 ksec).

Populations of High Energy Sources in Galaxies
Proceedings IAU Symposium No. 230, 2005
E. J. A. Meurs & G. Fabbiano, eds.

A new age and distance indicator of SNRs with nonthermal X-ray filaments

Aya Bamba[1], Ryo Yamazaki[2] and Junko S. Hiraga[3]

[1]RIKEN (The Institute of Physical and Chemical Research) 2-1, Hirosawa, Wako, Saitama
351-0198, Japan
email: bamba@crab.riken.jp

[2]Department of Earth and Space Science, Graduate School of Science, Osaka University,
Toyonaka, Osaka 560-0043, Japan
email: ryo@vega.ess.sci.osaka-u.ac.jp

[3]Department of High Energy Astrophysics Institute of Space and Astronautical Science (ISAS)
Japan Aerospace Exploration Agency (JAXA) 3-1-1 Yoshinodai, Sagamihara, Kanagawa
229-8510, Japan
email: jhiraga@astro.isas.jaxa.jp

Abstract. It is discovered that young SNRs have very thin filaments emitting nonthermal X-rays from accelerated electrons. In this paper, a new age and distance indicator is proposed using the spectral and spatial features of nonthermal filaments. We applied this method to the Vela Jr. SNR, for which age and distance are still unknown, and estimated that this SNR is one of the nearest and youngest SNR in our Galaxy: the estimated distance and age are 0.33 (0.26–0.50) kpc and 660 (420–1400) years, respectively.

Keywords. methods: data analysis, supernova remnants, X-rays: individual (Vela Jr.).

---

1. Introduction

Recently, it was discovered that young SNRs have very thin filaments emitting non-thermal X-rays from accelerated electrons (Bamba *et al.* 2003), indicating that cosmic ray acceleration occurs efficiently in local regions, and as a result, the SNR age characterizes the spatial and spectral feature of filaments most effectively. Bamba *et al.* (2005) found the relation,

$$\mathcal{B} \equiv \nu_{roll}/\theta_d{}^2 = 2.6^{+1.2}_{-1.4} \times 10^{27} t_{age}{}^{-2.96^{+0.11}_{-0.06}} \text{ Hz pc}^{-2}, \tag{1.1}$$

where ν_{roll}, θ_d, t_{age} are the roll-off frequency of synchrotron X-rays (Reynolds 1998), scale length in the downstream region of the shock, and the SNR age, respectively. In this paper, we make a method to estimate the distance and age of SNRs with $\mathcal{B}$, and apply it to a SNR, Vela Jr., with *Chandra* observations. This SNR is very curious because of a possible detection of 1.157 MeV γ-ray line emitted by ^{44}Ti (Iyudin *et al.* 1998). The other fact which makes this SNR famous is the existence of nonthermal X-rays from the rims (Slane *et al.* 2001), which implies that the rims of Vela Jr. are cosmic ray accelerators. However, nobody knows its distance and age. The details of the discussion are in Bamba *et al.* (2005b).

2. Results

Figure 1 are *Chandra* images of north-western (NW) shell of Vela Jr. We can see clear thin filaments in the hard band image, which is very similar to those in SN 1006. We chose three filaments, No. 1–3 in Figure 1(b) for the following analysis. The spatial and

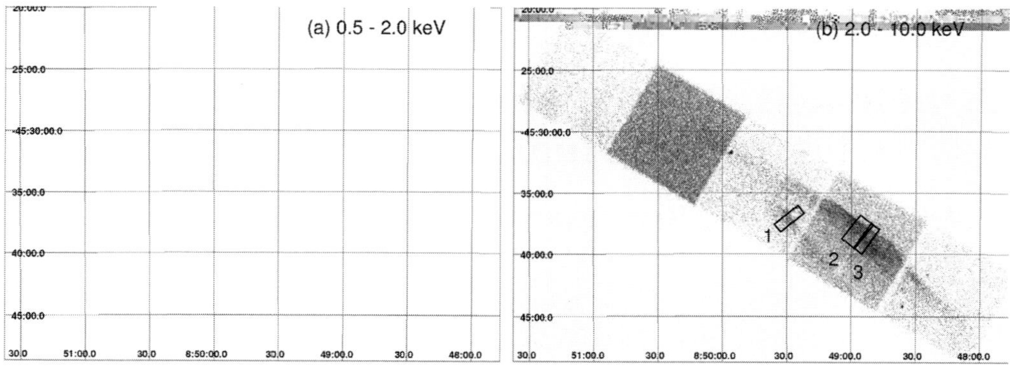

Figure 1. *Chandra* images of the NW rim of Vela Jr. in the 0.5–2.0 keV band (a) and 2.0–10.0 keV band (b).

Table 1. Spatial and spectral parameters of the filaments

	1	2	3
θ_u [arcsec]	19.0 (12.1–31.4)	3.68 (2.69–5.75)	—
θ_d [arcsec]	31.8 (23.7–47.1)	65.0 (38.2–145)	37.1 (26.5–60.2)
ν_{roll} [10^{16}Hz]	3.7 (23.7–47.1)	5.3 (3.6–7.3)	4.3 (3.4–5.3)
Flux$_{2-10 \ \mathrm{keV}}$ [ergs cm^{-2}s^{-1}]	7.4×10^{-13}	1.8×10^{-12}	7.6×10^{-13}

spectral analysis have been done in the same way to the SN 1006 case (Bamba *et al.* 2003), and we got the best-fit parameters as seen in Table 1. The spectral index at 1 GHz is assumed to be 0.3 (Combi *et al.* 1999). The spectral feature indicates that the filaments emit synchrotron X-rays.

For the distance and age estimation, it needs another relation between the radius and age. We chose the SNR evolution model by Truelove & McKee (1999) (see eq. (1) and (2), and Table 6 and 7), with the condition of constant interstellar medium of 0.1 cm^{-3}, the ejecta mass of $1.4M_\odot$, and the kinetic explosion energy of 10^{51} ergs. Together with the evolution model and the function $\mathcal{B}$, the distance and age are estimated to be 0.33 (0.26–0.50) kpc and 660 (420–1400) years, respectively. These results imply that Vela Jr. is one of the nearest and youngest SNRs in our Galaxy.

References

Bamba, A., Ueno, M., Koyama, K., & Yamauchi, S. 2003, *ApJ*, 589, 253

Bamba, A., Yamazaki, R., Yoshida, T., Terasawa, T., & Koyama, K. 2005, *ApJ*, 621, 793

Bamba, A., Yamazaki, R., & Hiraga, J. S. 2005, *ApJ*, in press (astro-ph/0506331)

Combi, J. A., Romero, G. E., & Benaglia, P. 1999, *ApJL*, 519, L177

Iyudin, A. F., *et al.* 1998, *Nature*, 396, 142

Reynolds, S. P. 1998, *ApJ*, 493, 375

Slane, P., Hughes, J. P., Edgar, R. J., Plucinsky, P. P., Miyata, E., Tsunemi, H., & Aschenbach, B. 2001, *ApJ*, 548, 814

Truelove, J. K., & McKee, C. F. 1999, *ApJS*, 120, 299

Populations of High Energy Sources in Galaxies
Proceedings IAU Symposium No. 230, 2005
E. J. A. Meurs & G. Fabbiano, eds.

© 2006 International Astronomical Union
doi:10.1017/S1743921306007733

A Detailed Observation of a LMC SNR, DEM L241, with *XMM-Newton*

Aya Bamba[1], Masaru Ueno[2], Hiroshi Nakajima[3], Koji Mori[4] and Katsuji Koyama[3]

[1]RIKEN (The Institute of Physical and Chemical Research) 2-1, Hirosawa, Wako, Saitama
351-0198, Japan
email: bamba@crab.riken.jp

[2]Department of physics, Faculty of Science, Tokyo Institute of Technology 2-12-1,
Oo-okayama, Meguro-ku, Tokyo 152-8551, Japan
email: masaru@hp.phys.titech.ac.jp

[3]Department of Physics, Graduate School of Science, Kyoto University, Sakyo-ku, Kyoto
606-8502, Japan
email: nakajima@cr.scphys.kyoto-u.ac.jp, koyama@cr.scphys.kyoto-u.ac.jp

[4]Department of Applied Physics, Faculty of Engineering University of Miyazaki, 1-1 Gakuen
Kibana-dai Nishi Miyazaki, 889-2192, Japan
email: mori@astro.miyazaki-u.ac.jp

Abstract. We report on an *XMM-Newton* observation of the supernova remnant (SNR) DEM L241 in the Large Magellanic Cloud. In the soft band image, the emission shows an elongated structure, like a killifish (Head and Tail), with a central point source, named as XMMU J053559.3–673509 (Eye). The Eye's spectrum is well reproduced with a power-law model. The source has neither significant coherent pulsations nor time variabilities. Its luminosity and spectrum remind us that the source might be a pulsar and/or pulsar wind nebula in DEM L241. The spectra of Head and Tail are well reproduced by a non-equilibrium ionization plasma model with over-abundant Ne and under-abundant Fe, suggesting that the progenitor of DEM L241 is a very massive star.

Keywords. supernova remnants, X-rays: individual (DEM L241), X-rays: individual (XMMU J053559.3−673509).

---

1. Introduction

Supernovae and supernova remnants (SNRs) shape and enrich the chemical and dynamical structure of the interstellar medium and clouds. The Large Magellanic Cloud (LMC) is the best galaxy for the systematic study of SNRs, thanks to the known distance (Feast 1999). In the LMC, we pointed out DEM L241 (0536−67.6), which was identified by Mathewson *et al.* (1985), and observed with *XMM-Newton*. The total exposure is 45 ks for MOS and 43 ks for pn, respectively. The details are in Bamba *et al.* (2005).

2. Results

Figure 1 shows the *XMM-Newton* MOS $1+2$ images of DEM L241 in the (a) 0.5–2.0 keV and (b) 2.0–9.0 keV bands. In the soft band image, we can see a diffuse structure elongated from southeast to northwest with the size of $\sim 1.5' \times 3'$, corresponding to 22 pc×44 pc at 50 kpc. The shape is like a killifish, with double peaked feature on its "Head" and "Tail". In addition to the body of the fish, there is a point source like an "Eye" of the fish. On the other hand, only Eye can be seen in the hard band image. We found no candidate of the counterpart of Eye in the SIMBAD database, and named it as XMMU J053559.3−673509.

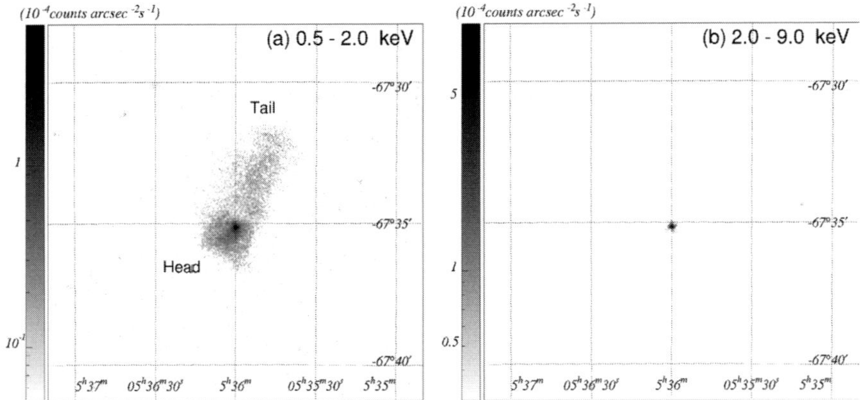

Figure 1. MOS $1 + 2$ images in the (a) 0.5–2.0 keV and (b) 2.0–9.0 keV bands.

Table 1. Best-fit parameters for the Head, Tail, and Eye

Parameters	Head	Tail	Eye
kT_e/Γ [keV/—]...............	0.54 (0.46–0.57)	0.43 (0.38–0.50)	1.57 (1.51–1.62)
$n_e t_p$ [10^{11}cm^{-3}s]...............	2.4 (1.9–4.3)	3.4 (2.1–5.2)	—
[Ne/H]........................	0.64 (0.60–0.69)	0.58 (0.54–0.63)	—
[Si/H]........................	(<0.05)	(<0.07)	—
[Fe/H]........................	0.09 (0.08–0.10)	0.08 (0.07–0.09)	—
$E.M.$ [10^{58}cm^{-3}]...............	2.5 (2.2–3.0)	2.1 (1.4–2.8)	—
$N_H{}^{LMC}$ [10^{21}cm^{-2}]............	4.2 (3.3–5.2)	1.6 (0.6–2.3)	3.3 (2.7–4.0)
Flux$_{0.5-10keV}$ [ergs cm^{-2}s^{-1}]...	3.2×10^{-13}	3.6×10^{-13}	6.4×10^{-13}

The spectrum of Eye is hard and has no line-like structure, and is well fitted with an absorbed power-law model (Table 1). The intrinsic luminosity is 2.2×10^{35} ergs s^{-1} in 0.5–10.0 keV. The central position and hard spectrum may indicate that Eye is a pulsar and/or pulsar wind nebula (PWN) of DEM L241. We searched for but could not find any coherent pulsations and time variabilities. If Eye is a pulsar and/or PWN, it belongs to the bright and hard class (Gotthelf & Olbert 2002).

The Head and Tail emission have, on the other hand, soft and line-rich spectra. Both are well fitted with a non-equilibrium ionization collisional plasma emission model (Borkowski *et al.* 2001) ver. 2.0 as seen in Table 1. The over-abundant Ne and less-abundant Fe relative to the average LMC value (0.3; Russel & Dopita 1992) indicate that the progenitor of DEM L241 is a very massive star. The estimated total plasma mass ($\sim 200 M_\odot$) and the thermal energy (5×10^{50} ergs) also suggest the massive star origin, together with the existence of the central point source, Eye, and the OB star association, LH 88 (Chu & Kennicutt 1988).

References

Bamba, A., Ueno, M., Nakajima, H., Mori, H., & Koyama, K. 2005, submitted to *A&A*

Borkowski, K. J., Lyerly, W. J., & Reynolds, S. P. 2001, *ApJ*, 548, 820

Chu, Y. & Kennicutt, R. C. 1988, *AJ*, 96, 1874

Gotthelf, E. V. & Olbert, C. M. 2002, ASPC 271, 171

Mathewson, D. S., Ford, V. L., Tuohy, I. R., Mills, B. Y., Turtle, A. J., & Helfand, D. J. 1985, *ApJS*, 58, 197

Russell, S. C. & Dopita, M. A. 1992, *ApJ*, 384, 508

Populations of High Energy Sources in Galaxies
Proceedings IAU Symposium No. 230, 2005
E. J. A. Meurs & G. Fabbiano, eds.

© 2006 International Astronomical Union
doi:10.1017/S1743921306007745

Discovery and study of the accreting pulsar 2RXP J130159.6-635806

M. Chernyakova[1,2]†, **A. Lutovinov**[3], **J. Rodriguez**[4,1] **and M. Revnivtsev**[5,3]

[1]INTEGRAL Science Data Centre, Chemin d'Écogia 16, 1290 Versoix, Switzerland
email:Masha.Chernyakova@obs.unige.ch

[2]Geneva Observatory, 51 ch. des Maillettes, CH-1290 Sauverny, Switzerland

[3]Space Research Institute, 84/32 Profsoyuznaya Street, Moscow 117997, Russia

[4]CEA Saclay, DSM/DAPNIA/Service d'Astrophysique (CNRS UMR 7158 AIM), 91191 Gif sur Yvette, France

[5]Max-Planck-Institut für Astrophysik, Karl-Schwarzschild-Str. 1, D-85740 Garching bei München, Germany

Abstract. We report on analysis of the poorly studied source 2RXP J130159.6-635806 at different epochs with *ASCA*, *BeppoSAX*, *XMM-Newton* and *INTEGRAL*. The source shows coherent X-ray pulsations at a period $\sim$700 s with $\dot{\nu} \sim 2 \times 10^{-13}$ Hz s^{-1}. A broad band (1–60 keV) spectral analysis of 2RXP J130159.6-635806 based on almost simultaneous *XMM-Newton* and *INTEGRAL* data demonstrates that the source spectrum is an absorbed power law with a photon index $\Gamma \sim 0.5 - 1.0$ and a cut-off energy of $\sim$25 keV. We also report on the identification of the likely infrared counterpart to 2RXP J130159.6-635806. The interstellar reddening does not allow us to strongly constrain the spectral type of the counterpart. The latter is, however, consistent with a Be star, the kind of which is often observed in accretion powered X-ray pulsars.

Keywords. pulsars: individual: 2RXP J130159.6-635806 – gamma rays: observations – X-rays: binaries – X-rays: individual: 2RXP J130159.6-635806.

On February 7, 2004 the *INTEGRAL* observatory detected a source which was not in the *INTEGRAL* reference catalog. This source was also clearly detected by *XMM-Newton* during its PSR B1259–63 monitoring programme (Chernyakova *et al.*, 2004). The best coordinates we derive for 2RXP J130159.6-635806 are RA$_{J2000} = 13^h 01^m 58^s.8$, DEC$_{J2000} = -63°58'10''$. This position is about 6$''$ from the best *ROSAT* position of 2RXP J130159.6-635806. Taken into account the uncertainties of the localisation we conclude that most likely *XMM-Newton* source and the *ROSAT* one are the same.

The 1993–2004 time history of the 2–10 keV flux from 2RXP J130159.6-635806 as observed by *ASCA* and *XMM-Newton* is shown in the upper panel of Figure 1. While during the *ASCA* (1994–1995) and the first half of the *XMM-Newton* observations (2001–2003) the flux of the source was practically constant at a value $\sim$2.5 $\times 10^{-11}$ ergs cm^2 s^{-1}, an outburst occurred between the end of January and the beginning of February 2004. During this period the source flux increased by a factor of more than 5. This outburst was also detected by *INTEGRAL* in the 20–60 keV energy range. While the *ASCA* and *XMM-Newton* data are well fitted with a simple power law modified by photoabsorption, *INTEGRAL* data show a presence of a high-energy cut-off at about $\sim$25 keV, typical for accreting X-ray pulsars. We fitted the *XMM-Newton* and *INTEGRAL* joint

† on leave from Astro Space Center of the P.N. Lebedev Physical Institute, Moscow, Russia

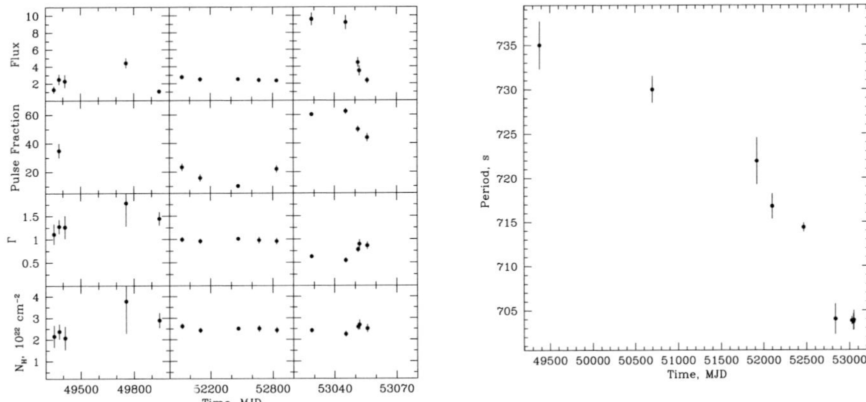

Figure 1. (*left*) Time evolution of the spectral parameters of 2RXP J130159.6-635806 and the $2-10$ keV pulse fraction (in %). Flux is given in units of 10^{-11} erg/s/cm^2. (*right*) Time evolution of 2RXP J130159.6-635806 pulse period.

spectrum with an absorbed cut-off power law. The best fit parameters obtained are: $N_{\mathrm{H}} = (2.55 \pm 0.13) \times 10^{22}$ cm^{-2}, $\Gamma = 0.69 \pm 0.05$, $E_{cut} = 24.3 \pm 3.4$ keV, $E_f = 8.5 \pm 3.3$ keV.

Analyzing the light curve of 2RXP J130159.6-635806 we found that it demonstrates near coherent strong variations with a characteristic time about 700 s. The evolution of the pulse period is shown in right panel of Figure 1. An average spin up rate changes from $\dot{P} \simeq -6 \times 10^{-8}$s s^{-1} in 1994 – 2001, to $\dot{P} \simeq -2 \times 10^{-7}$s s^{-1} in 2001 – 2004.

The discussed above long term behaviour of the source, its spectral and timing properties, tend to indicate a high mass X-ray binary with a Be companion. It seems that this source belongs to a rather small group of persistent Be/X-rays binaries with rather low X-ray luminosity ($< 10^{35}$erg/s) and relatively long (> 200s) pulse periods (Reig & Roche, 1999). In order to check this we used the results of DSS and 2MASS surveys. In the 2MASS catalog we found a source with coordinates (equinox 2000) RA $= 13^h 01^m 58^s.7$, DEC $= -63°58'09''$ (at $\sim 1.1''$ from the best *XMM-Newton* position) and magnitudes $J = 12.96 \pm 1.33$, $H = 12.05 \pm 0.03$, $K_s = 11.35 \pm 0.09$. The good agreement between both positions would tend to suggest that this source is the likely counterpart to 2RXP J130159.6-635806. Using the value of Galactic absorption $N_{\mathrm{H}} = 1.7 \times 10^{22}$ cm^{-2} we estimate the de-reddened magnitudes $J_{der} = 10.73 \pm 1.33$, $H_{der} = 10.72 \pm 0.03$, $K_{s\,der} = 10.51 \pm 0.09$. If the companion star is a Be main sequence star with surface temperature around 10000 K and the radius around 6-10 $R_\odot$ we can expect to see its infrared brightness $J, H, K \sim 10 - 11$ if the binary system is at the distance ~ 4–7 kpc. An additional tentative argument in favour of such source distance is the source location in the direction to the Crux spiral arm tangent. At such a distance the unabsorbed intrinsic luminosity of 2RXP J130159.6-635806 is about $\sim 5 \times 10^{34}$–10^{35} erg/s, *i.e.* compatible with the typical luminosities of the persistent Be/X-ray binaries.

More detailed information on this source is given in Chernyakova *et al.* (2005).

References

Chernyakova, M., A. Lutovinov, J. Rodriguez, & M. Revnivtsev, 2005,
    accepted by *MNRAS* (astro-ph/0508515)
Chernyakova, M., Shtykovskiy, P., Lutovinov, A., *et al.*, 2004, *The Astronomer's Telegram*,
    251, 1
Reig, P. & Roche, P., 1999, *MNRAS*, 306, 100

Populations of High Energy Sources in Galaxies
Proceedings IAU Symposium No. 230, 2005
E. J. A. Meurs & G. Fabbiano, eds.

© 2006 International Astronomical Union
doi:10.1017/S1743921306007757

Colliding Wind Binary X-ray Sources

**Michael F. Corcoran[1], Kenji Hamaguchi[1],
A. M. T. Pollock[2], J. M. Pittard[3], I. R. Stevens[4], D. B. Henley[4],
A. F. J. Moffat[5], and S. Marchenko[6]**

[1]Universities Space Research Association & XAL-NASA/GSFC,
Code 662 GSFC, Greenbelt, MD, 20771
email: corcoran@milkyway.gsfc.nasa.gov, kenji@milkyway.gsfc.nasa.gov

[2]European Space Astronomy Centre, Apartado 50727,
Villafranca del Castillo, 28080 Madrid, Spain
email: andy.pollock@esa.int

[3]School of Physics and Astronomy, University of Leeds,
Woodhouse Lane, Leeds LS2
email: jmp@ast.leeds.ac.uk

[4]School of Physics and Astronomy, University of Birmingham,
Edgbaston, Birmingham B15 2TT
email: irs@star.sr.bham.ac.uk

[5]Département de physique, Université de Montréal,
Montreal, QC H3C 3J7, Canada
email: moffat@astro.umontreal.ca

[6]Department of Physics and Astronomy, Western Kentucky University,
Bowling Green, KY 42101-3576
email: sergey@astro.wku.edu

Abstract. Very massive stars ($\gtrsim 20$ M$_\odot$) are rare but important components of galaxies. Products of core nucleosynthesis from these stars are distributed into the circumstellar environment via wind-driven mass loss. Explosive nucleosynthesis after core collapse further enriches the galactic medium. Clusters of such stars can produce galactic chimneys which can pierce the galactic disk and chemically enrich intergalactic space. Such processes are vitally important to the chemical evolution of the early Universe, when the stellar mass function was much more weighted to massive stars.

Very massive stars are difficult to study, since they are formed in distant clusters which yield problems of sensitivity and source crowding. A relatively new tool for studying these systems is via high spatial, spectral and temporal resolution observations in the X-ray band. In this note we describe some recent progress in studying mechanisms by which very massive stars produce X-ray emission.

Keywords. X-ray emission, binary stars, winds.

1. Introduction

The advent of modern X-ray optics and detectors has revolutionized our astrophysical understanding of high energy processes in galaxies. While X-ray luminosity functions in normal galaxies are dominated by the end-products of stellar evolution (black holes, neutron stars and supernova remnants) the study of the fainter stellar X-ray source populations has increased our knowledge of the evolution of these sources, and how they contribute to the evolution of the galactic and intergalactic medium. A particularly rare but important population of such relatively faint sources are the "colliding wind" X-ray binaries. In these objects, which are composed of a pair of massive stars (O stars, Wolf-Rayet stars, or some combination) undergoing strong mass loss, X-ray emission primarily arises in the "bow-shock" where the winds from the stars collide. The emission from these systems provides an extremely useful probe of the nature of the shock, and by inference the nature of the mass-loss process, itself important since mass loss drives the subsequent evolution of the star and determines the final outcome for it (a neutron star or black hole, a supernova or a hypernova).

Problems in observing X-ray emission from such "colliding-wind" binaries (CWBs) arise due to limitations in spatial resolution (many such systems occur in crowded cluster cores), in spectral resolution (the emission tends to be thermal and thus X-ray line properties provide important information about the flow) and in monitoring (the X-ray emission is variable due to changes in emission measure &/or intervening absorption). The availability of the high resolution and high sensitivity detectors on the *Chandra* and *XMM-Newton* X-ray observatories, along with the scheduling flexibility of the *Rossi X-ray Timing Explorer* (*RXTE*) has allowed detailed examination of some of the bright CWBs.

2. A Brief Review of Current Observations

Observations of massive clusters with *Chandra* and *XMM-Newton* have shown that many of the bright stars in the core either show X-ray excesses &/or X-ray variability. Examples include studies of NGC 3603 by Moffat *et al.* (2002), the Galactic Center clusters (Law & Yusaf-Zadeh 2004) and NGC 6231 (Sana *et al.* 2004). These observations reveal emission at a level of $\gtrsim 10^{34}$ ergs s^{-1} in NGC 3603, far exceeding the canonical relation $L_x/L_{bol} \approx 10^{-7}$ by at least an order of magnitude. The identification of long-period binaries via their X-ray excesses can have important implications on understanding the formation of extremely massive stars (e.g., Bonnell & Bate 2002).

High resolution X-ray spectroscopy has been obtained by *Chandra* of the "canonical" CWB, WR 140, at two phases around its X-ray minimum (Pollock *et al.* 2005). High-resolution spectroscopy by *Chandra* of η Car was obtained at 5 phases around its X-ray minimum (Corcoran *et al.* 2006, in prep.). These X-ray observations show broad, blueshifted X-ray emission line profiles at least at some orbital phases, while the grating spectra of η Car reveals clear changes in line width and line centroid vs. orbital phase. These effects are probably dominated by the changing viewing geometry of the shock cone as it follows the weaker-wind star in its orbit. There are also indications that at least some of the hot plasma is out of collisional ionization equilibrium. If so this has important implications for models of the shock emission.

Detailed monitoring of the X-ray emission from η Car (Corcoran 2005) and (recently) WR 140 (Pollock *et al.* 2006, in prep.) has been obtained, showing in detail the shape of the X-ray "eclipse" in both systems. The durations of the minima are different, in that the minimum in η Car lasts about 3 months, whereas the minimum in WR 140 lasts only a few days. This indicates differences in viewing geometry between the two systems, differences in the stability of the colliding wind shock, or some combination.

References

Bonnell, I. A. & Bate, M. R., MNRAS, 336, 659
Corcoran, M. F. 2005, AJ, 129, 2018
Law, C., & Yusef-Zadeh, F. 2004, ApJ, 611, 858
Moffat, A. F. J., *et al.* 2002, ApJ, 573, 191
Pollock, A. M. T., *et al.* 2005, ApJ, 629, 482
Sana, H., *et al.* 2004, MNRAS, 350, 809

Populations of High Energy Sources in Galaxies
Proceedings IAU Symposium No. 230, 2005
E. J. A. Meurs & G. Fabbiano, eds.

© 2006 International Astronomical Union
doi:10.1017/S1743921306007769

Spin changes in X-ray pulsars

H.-L. Dai and X.-D. Li

Department of Astronomy, Nanjing University, Nanjing 210093, P. R. China
email: hldai@nju.edu.cn, lixd@nju.edu.cn

Abstract. The conventional picture of disk accretion onto magnetized neutron stars has been challenged by the spin changes observed in a few X-ray pulsars, and by theoretical results from numerical simulations of disk-magnetized star interactions. Here we present a model for the torque exerted by accretion disks on magnetized neutron stars, assuming accretion continues even for fast rotators.

Keywords. accretion disk, pulsars, magnetic field.

1. Introduction

Magnetic fields play an important role in transferring angular momentum between neutron stars and accretion disks. In the model developed by Ghosh & Lamb (1979) the stellar magnetic field lines are assumed to penetrate the accretion disk, and become twisted because of the differential rotation between the star and the disk. The resulting torque can spin up or down the central star, depending on the star's rotation, magnetic field strength, and mass accretion rate. However, this standard model has been challenged by recent observational and theoretical investigations on accreting X-ray pulsars (Bildsten *et al.* 1997; Galloway *et al.* 2002; Romanova *et al.* 2003, 2004). The puzzling issues include (1) abrupt spin reversals in X-ray pulsars, (2) spin-down in millisecond X-ray pulsars throughout the outburst, (3) spin-down rates increasing with accretion rate, and (4) accretion in the "propeller" regime.

2. The torque

We adopt the following assumptions during both accretion and propeller phases.

(1) The azimuthal component B_ϕ of the field on the surface of the disk is generated by rotation shear (Wang 1995)

$$\frac{B_\phi}{B_z} = \begin{cases} \gamma(\Omega_* - \Omega_{\rm K})/\Omega_{\rm K}, & \Omega_* \leqslant \Omega_{\rm K} \\ \gamma(\Omega_* - \Omega_{\rm K})/\Omega_*, & \Omega_* > \Omega_{\rm K}, \end{cases} \tag{2.1}$$

where B_z is the vertical component of the field, the parameter γ is of order unity, Ω_* and $\Omega_{\rm K}$ are the stellar spin rate and the Keplerian angular velocity in the disk.

(2) The inner radius R_0 of the disk is at the magnetospheric radius $R_{\rm m}$ (Davidson & Ostriker 1973),

$$R_0 \simeq R_{\rm m} = \left(\frac{\mu^4}{2GM\dot{M}^2} \right)^{1/7}, \tag{2.2}$$

where $\mu = B_* R_*^3$ is the magnetic moment of the neutron star.

(3) In the magnetosphere the angular momentum transfer rate is given by,

$$N_0 = \xi\dot{M}(GMR_0)^{1/2}(1 - \omega), \tag{2.3}$$

where $\omega = \Omega_*/\Omega_{\rm K}(R_0)$ and $0 < \xi < 1$.

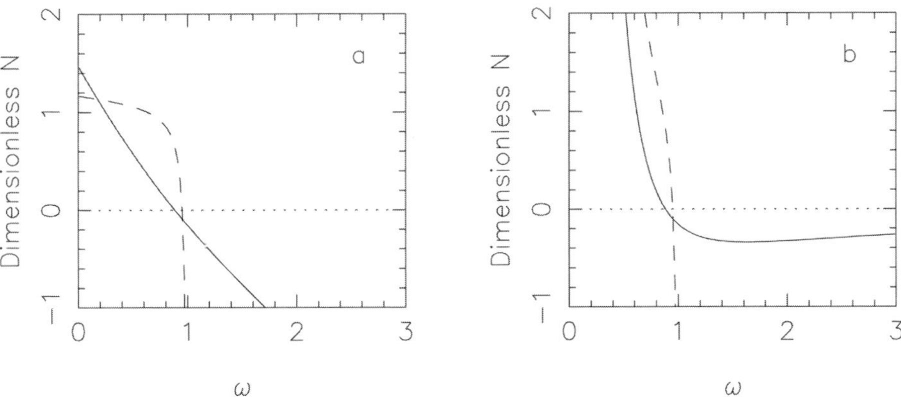

Figure 1. The solid lines show the dimensionless torque (a) $N/\dot{M}(GMR_0)^{1/2}$ and (b) $N/(\mu^2/3R_c^3)$ as a function of the fastness parameter ω. The dashed lines represent the standard model.

The total torque exerted on the star by the disk is then

$$
N = \begin{cases} \dot{M}(GMR_0)^{1/2}[\xi(1-\omega) + \frac{\sqrt{2}\gamma}{3}(1 - 2\omega + \frac{2\omega^2}{3})], & \omega \leqslant 1 \\ \dot{M}(GMR_0)^{1/2}[\xi(1-\omega) + \frac{\sqrt{2}\gamma}{3}(\frac{2}{3\omega} - 1)], & \omega > 1. \end{cases}
\tag{2.4}
$$

Figures 1a and 1b show the dependence of the dimensionless torque $N/\dot{M}(GMR_0)^{1/2}$ and $N/(\mu^2/\sqrt{2}R_c^3)$ on ω in solid curves with $\xi = \gamma = 1$, respectively. The dashed curves in the figures represent the results in the disk model of Wang (1995) for comparison. In Fig. 1a there is no singularity problem in this work when $\omega \to 1$, the parameter space range for ω for spin-down with accretion is also much wider. The critical fastness parameter ω_c, at which $N = 0$, is $\omega_c \simeq 0.884$. Furthermore, it indicates that, for constant $\dot{M}$, the spin-down torque increases with Ω_* ($\propto \omega$), in line with the numerical simulation results by Romanova *et al.* (2004). Figure 1b shows how the torque varies with $\omega \propto \dot{M}^{-3/7}$ when μ and Ω_* are invariant. One can see that the spin-down torque is not a monotonous function of ω (or $\dot{M}$) when $\omega > \omega_c$. The spin-down torque takes the maximum value when $\omega \simeq 1.634$. Thus a given spin-down torque can correspond to two values of ω. *When $\omega > 1.634$, the spin-down torque increases with $\dot{M}$,* which is opposite to the dependence in the standard model, but consistent both with the observational fact in the X-ray pulsar GX $1+4$ that the X-ray flux appears to be increasing with the spin-down torque (Chakrabarty 1995), and with the numerical simulation results by Romanova *et al.* (2004).

Acknowledgements

This work was supported by NSFC under grant number 10025314 and MSTC under grant number NKBRSF G19990754.

References

Bildsten, L. *et al.* 1997, ApJS, 113, 367
Chakrabarty, D. 1995, Ph.D. Thesis, California Inst. Tech.
Davidson, K. & Ostriker, J. P. 1973, ApJ, 179, 585
Galloway, D. K. *et al.* 2002, ApJ, 576, L137
Ghosh, P. & Lamb, F. K. 1979a, ApJ, 232, 259
Romanova, M. M. *et al.* 2003, ApJ, 595, 1009
Romanova, M. M. *et al.* 2004, ApJ, 616, L151
Wang, Y. M. 1995, ApJ, 449, L153

Populations of High Energy Sources in Galaxies
Proceedings IAU Symposium No. 230, 2005
E. J. A. Meurs & G. Fabbiano, eds.

© 2006 International Astronomical Union
doi:10.1017/S1743921306007770

High–Energy Astrophysics with Lobster Eye X-ray ASM

R. Hudec[1], V. Simon[1], L. Sveda[2], L. Pina[2] and A. Inneman[3]

[1]Astronomical Institute, Academy of Sciences of the Czech Republic, CZ–251 65 Ondrejov, Czech Republic
email: rhudec@asu.cas.cz

[2]Czech Technical University, Faculty of Nuclear Physics, Prague, Czech Republic,

[3] Center for Advanced X-ray Technologies, Reflex sro, Prague, Czech Republic

Abstract. We report on astrophysical aspects of fully innovative very wide–field X-ray telescopes with high sensitivity. They are expected to contribute essentially to study of various astrophysical objects such as AGN, SNe, Gamma-ray bursts (GRBs), X-ray flashes (XRFs), galactic binary sources, stars, CVs, X-ray novae, various transient sources, etc.

Keywords. X-ray telescopes, All-Sky Monitor

Introduction

The X-ray sky is highly variable, rich in variable and transient sources of both galactic as well as extragalactic origin. However, since many of these transient events cannot be predicted, and are relatively rare, very wide-field instruments are required. They must achieve high sensitivities and provide precise localizations in order to effectively study the objects. Wide field X-ray telescopes with imaging optics are expected to represent an important tool in future space astronomy projects in general, especially those for deep monitoring and surveys in X-rays over a wide energy range. The Lobster–Eye (LE) wide field X-ray optics has been suggested in 70ies by Schmidt (Schmidt, 1975, orthogonal stacks of reflectors) and by Angel (Angel, 1979, array of square cells). Up to 180 deg FOV may be achieved. This novel X-ray optics offers an excellent opportunity to achieve very wide fields of view (FOV, 1000 square degrees and more) while the widely used classical Wolter grazing incidence mirrors are limited to roughly 1 deg FOV (Priedhorsky *et al.*, 1996, Inneman *et al.*, 2000).

Science

Deep (limiting flux of 10^{-12} erg cm^{-2} s^{-1} can be easily achieved for daily scanning observation) X-ray sky monitoring with large FOVs (e.g. FOV of 6 × 180 deg can be easily assembled on the space station ISS) is expected to contribute significantly to various fields of modern astrophysics. A few most important examples are listed below.

(1) Gamma Ray Bursts (GRBs). Detection rates of nearly 20 GRBs/year can be obtained for the prompt X-ray emission of GRBs, taking into account the expected GRB rate 300/year. (2) X-ray flashes. Detection rates of nearly 8 X-ray flashes/year are expected, assuming XRF rate of 100/year. (3) X-ray binaries. Because of their high variability in X-rays they will be one of major targets in LE observations. LE will be able to observe their short-time outbursts by long-term extended monitoring. Almost all galactic XRB are expected to be within the detection limits. (4) Stars. Because of the low X-ray luminosity of ordinary stars, only nearby stars are expected to be observable. We estimate the lower limit of ordinary stars observable by the LE telescope as 600. The sampling

rate of LE observations will be sufficient enough to observe sudden X-ray flux increases during flares while still having the capability of monitoring the variability on time scales of years.(5) Supernovae. The LE telescope should be able to detect the theoretically predicted thermal flash lasting for $\sim$1000 sec for the first time. Together with the optical SNe detection rate and estimates of the LE FOV we estimate the total number of SNe thermal flashes observed by the LE experiment to $\sim$10/year. (6) AGNs. Active Galactic Nuclei will surely be one of the key targets of the LE experiment. LE will be able to monitor the behavior of the large ($\sim$1000) sample of AGNs providing long-term observational data with good time sampling (hours). (7) X-ray transients. The LE experiment will be ideal to observe X-ray transients of various nature due to its ability to observe the whole sky several times a day for a long time with a limiting flux of about 10^{-12} erg cm^{-2} s^{-1}. More and fainter X-ray transients are expected to be detected by the LE sky monitor enabling the detailed study of these phenomena. (8) Cataclysmic Variables. Cataclysmic Variables (CVs) are very active galactic objects, often showing violent long-term activity in both the optical and X-ray passband (outbursts, high/low state transitions, nova explosions) as well as rapid transitions between the states of activity. Search for the relation of the optical and X-ray activity is very important – monitoring of a large number of CVs is necessary to catch them in various states of activity. Most up to now X-ray observations of CVs: (i) Snapshots catching selected CVs in a particular state of activity, (ii) In most cases the transitions between the states are not covered, and (iii) Poor statistics of phenomena and objects (deeper studies available for only a few CVs). Important classes of CVs for LOBSTER are Non-magnetic dwarf novae (DNe), Supersoft X-ray sources (SSXSs), Classical novae (CNe), and Polars with soft X-ray excess.

Conclusions

Analysis and simulations of Lobster–Eye X-ray telescopes have been carried out. They have indicated that these innovative devices will be able to monitor the X-ray sky at an unprecedented level of sensitivity, an order of magnitude better than any previous X-ray all–sky monitor. Limits as faint as 10^{-12} erg cm^{-2} s^{-1} for daily scanning observation as well as the angular resolution < 4 arcmin in soft X-ray range are expected to be achieved allowing monitoring of all classes of X-ray sources, not only X-ray binaries, but also fainter classes such as AGNs, coronal sources, cataclysmic variables, as well as fast X-ray transients including gamma–ray bursts and the nearby type II supernovae. The various prototypes of both Schmidt as well as Angel arrangements have been produced and tested successfully, demonstrating the possibility to construct these lenses by innovative but feasible technologies. Both very small Schmidt lenses (3×3 mm) as well as large lenses (300×300 mm) have been developed, constructed, and tested. This makes the proposals for space projects with very wide field lobster eye optics possible for the first time.

Acknowledgements

We acknowledge the support provided by the Ministry of Industry and Trade of the Czech Republic, FD-K3/052, and partly by the Grant Agency of the AS CR (A3003206) and Grant Agency of the Czech Rep. (205/05/2167).

References

Angel J. R. P., 1979, ApJ, 364, 233
Inneman A. et al., 2000, Proc. SPIE, 4138, 94
Priedhorsky, W. C. et al., 1996, MNRAS 279, 733
Schmidt, W. K. H., 1975, NucIM, 127, 285

Populations of High Energy Sources in Galaxies
Proceedings IAU Symposium No. 230, 2005
E. J. A. Meurs & G. Fabbiano, eds.

© 2006 International Astronomical Union
doi:10.1017/S1743921306007782

HMXBs in the Galaxy and the MCs

Q. Z. Liu

Purple Mountain Observatory, Academia Sinica, Nanjing 210008, China
email: qzliu@pmo.ac.cn

Abstract. The population of HMXBs in the MCs and the Galaxy is investigated. The SMC has an over-abundant population of HMXBs. This provides the clues that the SMC should be more active in massive star formation. The pulse periods in the SMC are slightly shorter than those in the Galaxy, implying that most of the pulsars in Be/X-ray binaries are not spin-up.

Keywords. Massive X-ray binaries, the Galaxy, the Magellanic Clouds

1. Introduction

High-mass X-ray binaries (HMXBs) consist of a neutron star or a black hole, orbiting a massive star. The X-ray emission in these sources is due to accretion of matter from the early-type star by the compact companion. Conventionally HMXBs can be further divided into two subgroups: those in which the primary is a Be star and those in which the primary is a supergiant.

2. HMXB populations in the MCs and the Galaxy

X-ray observations have revealed that the MCs contain an unexpectedly large number of HMXBs, especially in the SMC. Up to now, 92 known or probable HMXB sources in the SMC and 36 HMXBs in the LMC have been identified (Liu *et al.* 2005) and they continue to be discovered, although only a small fraction of these are active at any one time because of their transient nature. Of the 128 HMXBs, 4 of the objects are supergiant systems, and all the rest are likely Be/X-ray binaries. To compare the HMXB populations in the MCs and in our Galaxy, we collected the HMXBs in the three galaxies from our previous X-ray binary catalogues (Liu *et al.* 2000, 2005) and recent literature.

Table 1 summarizes the HMXB populations (and the LMXB populations as well (Liu *et al.* 2001 and the updated version)) in the three galaxies. Apparently the number of HMXBs normalized by the mass of the galaxies is much higher in the SMC than in our Galaxy and the LMC. Unusually, all but one of the X-ray binaries so far discovered in the MCs are HMXBs. The number ratio of HMXBs to LMXBs exhibits a striking difference between the MCs and our Galaxy, as is already pointed out by Schmidtke *et al.* (1999). Such a high density of HMXB population must provide clues about the star formation rates in the SMC, since HMXBs could be an indicator of star formation rate (Grimm *et al.* 2003) while LMXBs are generally considered to comprise an older population than HMXBs. The MCs should be very active in massive star formation, which is consistent with the results that MCs have experienced encounters with the Galaxy in past and the encounters likely triggered bursts of star formation (see e.g. Yoshizawa & Noguchi 2003).

In contrast to the large number of HMXBs in the MCs, it is very unusual that there are only two black hole candidates in the LMC and no black hole X-ray binaries in the SMC, while more than 40 black hole candidates were found in the Galaxy. However, Zhang, Li & Wang (2004) found that it is very difficult to detect black hole/Be systems, due to the efficient truncation of the Be circumstellar disk by black holes.

Table 1. Source populations in the MCs and the Galaxy

Galaxy	HMXBs			LMXBs		Ref.
	Total	Pulsar	BHC	Total	BHC	
SMC	92	47	0	0	0	Liu *et al.* 2005
LMC	36	7	2	1	0	Liu *et al.* 2005
Galaxy	$\sim$110	57	2(?)	$\sim$200	34(?)	Liu *et al.* 2000, 2001

3. Pulse period distribution of HMXBs

A simple scaling based on the relative masses of the MCs and the Galaxy predicts only 2 pulsars in the SMC and 20 in the LMC. However, currently 49 X-ray pulsars are known in the SMC (Coe *et al.* 2005), and 7 in the LMC. All pulsars except 2 in the SMC are most likely in HMXB systems as suggested by their X-ray properties. Fig. 1 shows the pulse period distributions of the X-ray pulsars in the MCs and our Galaxy. On average, the pulse periods in the SMC are slightly shorter than those in the Galaxy. Since the pulsars in the SMC are likely much younger than those in the Galaxy, it is likely that most of the pulsars in Be/X-ray binaries are not spin-up.

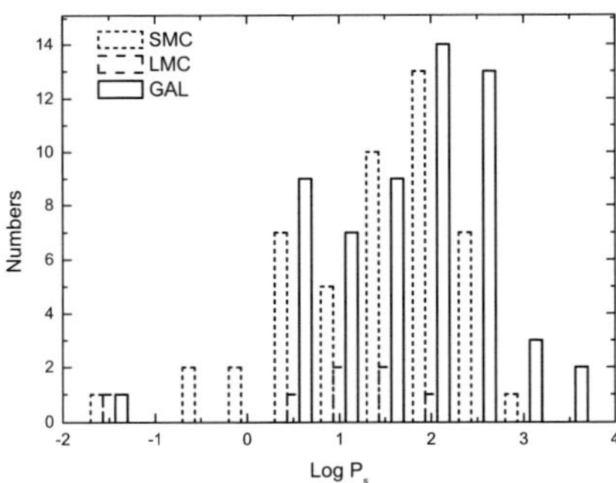

Figure 1. A comparison of the pulse period distributions of the X-ray pulsars in the 3 galaxies.

Acknowledgements

I thank the support from NSFC under Grants No. 10173026 and No. 10433030.

References

Coe, M. J. *et al.* 2005, MNRAS, 356, 502
Grimm, H.-J., Gilfanov, M., & Sunyaev, R. 2003, MNRAS, 339, 793
Liu, Q. Z., van Paradijs, J. & van den Heuvel, E. P. J. 2000, A&AS, 147, 25
Liu, Q. Z., van Paradijs, J. & van den Heuvel, E. P. J. 2001, A&A, 368, 1021
Liu, Q. Z., van Paradijs, J. & van den Heuvel, E. P. J. 2005, A&A, in press
Schmidtke, P. C., Cowley, A. P., Crane, J. D. *et al.* 1999, AJ, 117, 927
Yoshizawa, A. M. & Noguchi, M. 2003, MNRAS, 339, 1135
Zhang, F., Li, X.-D. & Wang, Z.-R. 2004, ApJ, 603, 663

Populations of High Energy Sources in Galaxies
Proceedings IAU Symposium No. 230, 2005
E. J. A. Meurs & G. Fabbiano, eds.

© 2006 International Astronomical Union
doi:10.1017/S1743921306007794

Astrometric detection of Neutron Star Companions to High Mass X-ray Binaries

C. Ó Maoiléidigh[1], E. J. A. Meurs[1] and L. Norci[2]

[1]Dunsink Observatory, Castleknock, Dublin 15, Ireland
email: melodies@dunsink.dias.ie; ejam@dunsink.dias.ie

[2]Dublin City University, Department of Physics, Glassnevin, Dublin 15, Ireland
email: lno@physics.dcu.ie

Abstract. We consider the feasibility of detecting neutron star companions to High Mass X-ray Binaries (HMXBs) using astrometric techniques, specifically for accuracies expected for the upcoming Gaia satellite. The direct determination of orbital parameters of HMXBs will increase the census of measured neutron star masses.

Keywords. X-rays: binaries, Stars: neutron, Astrometry

The brightest X-ray sources in the Galaxy are X-ray binaries, where a compact object such as a black hole, neutron star or white dwarf accretes matter from a normal stellar companion either by Roche Lobe overflow or a stellar wind. The mass of the companion and the parameters of the system can be derived or estimated if an eclipse occurs or through radial velocity variations. However, a more accurate determination is provided by a direct astrometric measurement of the non-linear motion of the stellar companion due to the hidden compact object. What we seek to do is to use the very high positional accuracies which the Gaia satellite will achieve for a large number of HMXRBs to investigate the masses of the compact companions. We take a number of representative HMXRBs and estimate their masses and distances. We then estimate their initial masses and a corresponding expected range of possible orbital separations. Finally we determine the range of orbital separation within which Gaia's positional accuracy would be sufficient to infer the presence of a hidden companion for these specific cases. Hence we can determine a probability of neutron star detection.

Table 1. HMXBs with known orbital periods

Name	Name	Name	Name	Name
QV Nor	V801 Cen	V635 Cas	0834-430	J0535.0-6700(LMC)
Vela X-1	Cen X-3	BQ Cam	1657-415	SMC 34
LS 5039	BP Cru	V572 Pup	J2103.5+4545	SMC 37
V725 Tau	X Per	V830 Cen	0535-668	J0053.8-7226(SMC)
V441 Pup	V662 Cas	V850 Cen	LMC X-4	SMC X-1
V884 Sco	γ Cas	J2239.3+6116	LMC 218	SMC 58
V615 Cas				

[a]Orbital period from (Harmanec *et al.* 2000). [b]Values taken from (McSwain *et al.* 2004). We adopt distances of 49kpc to the LMC and 58kpc to the SMC (Mateo 1998).

As an illustrative example and comparative base we use the the HMXB catalogue of Liu *et al.* (2000) and restrict ourselves to those systems for which an orbital period is known. These include two systems for which we found the orbital period listed in the literature: γ Cas (Harmanec *et al.* 2000) and LS 5039 (McSwain *et al.* 2004). Three of the

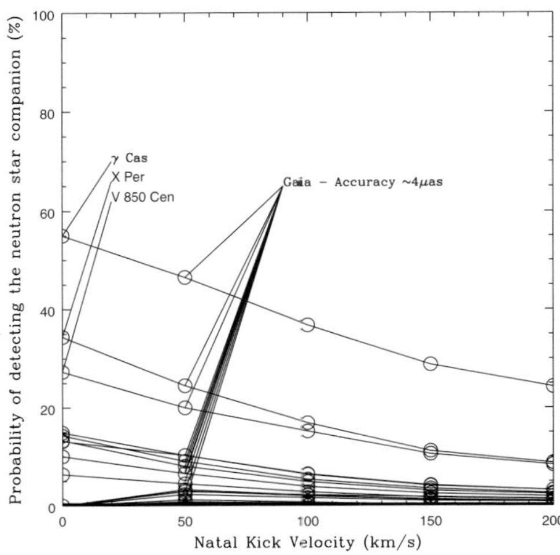

Figure 1. Percentage probability of Gaia detecting a neutron star companion to each of the HMXB sample stars, considering a theoretical set of possible system configurations that survive the supernova. The probabilities are shown as a function of the natal kick velocity to the neutron star, due to asymmetries in the supernova explosion

objects are left out even though they meet our criteria: Cyg X-1, LMC X-1, LMC X-3, which are black-hole binaries, for the reason that the ideas developed for determining the theoretical range of separations apply specifically to neutron star companions. This gives a total of 31 objects, 22 galactic and 9 extragalactic, see Table 1. Using their observed quantities such as magnitude, colour index and spectral type we calculated their approximate distances and estimated their masses.

From the estimated masses we can calculate what the likely initial masses were. Then, using an initial distribution for the semi-major axis we determine what the likely final separation distribution might be due to the effects of mass transfer and the supernova explosion of the primary (initially most massive star). Next we can determine what range of detectable separations Gaia will have. The presence of a compact companion can be inferred due to the non-linear motion of the bright stellar component of the binary. This means that the semi-major axis must be greater than some threshold value in order for the perturbation to normal straight line motion to be detectable. There is also a maximum semi-major axis due to the finite length of the mission resulting in possibly only a partial part of the orbit being sampled within this time. Thus giving us the limits for our detection range specific to each of our sample cases. For more details see Ó Maoiléidigh *et al.* (2005). Figure 1 illustrates the percentage probability of detection for our sample stars.

As is evident from Figure 1 the probability of detection is significant for a number of systems. Since we have known orbital periods for our sample we can see which system would actually be detected. γ Cas and X-Per have a sufficiently long period and are close enough to be detected astrometrically. This is about a 6% detection rate from our sample. However if we only consider galactic objects then this improves to about 9%.

References

Harmanec, P., Habuda, P., Štefl, S., Hadrava, P., Korčáková, D., Koubský, P., Kritička, J., Kubát, J., Škoda, P., Šlechta, M. & Wolf, M. 2000, A&A, 364, 85

Liu, Q. Z., van Paradijs, J. & van den Heuvel, E.P.J. 2000, A&AS, 147, 25

Mateo, M., L. 1998, ARA&A, 36, 435

McSwain, M. V., Gies, D. R., Huang, W., Wiita, P. J. & Wingert, D. W. 2004, ApJ, 600, 927

Ó Maoiléidigh, C., Meurs, E. J. A. & Norci, L., 2005, NewA, 10, 591

Populations of High Energy Sources in Galaxies
Proceedings IAU Symposium No. 230, 2005
E. J. A. Meurs & G. Fabbiano, eds.

© 2006 International Astronomical Union
doi:10.1017/S1743921306007800

X-ray hardness ratios for stars of different spectral types

E. J. A. Meurs[1], P. Casey[1] and L. Norci[2]

[1] Dunsink Observatory, Dublin Institute for Advanced Studies, Castleknock, Dublin 15, Ireland
[2] School of Physics, Dublin City University, Dublin, Ireland

Abstract. A substantial number of stars was detected in the ROSAT All Sky Survey. Analysis of these data indicates a systematic shift in the distribution of their X-ray hardness ratios, from late type dwarfs to late type giants to early type stars, that can be attributed to systematic differences in line-of-sight absorption.

Keywords. Stars: X-rays

Extensive surveys with the Einstein and ROSAT satellites have established that X-ray emission is a common feature of normal stars. The stellar X-ray characteristics separate basically into OB stars, as a result of shocks in their strong stellar winds, and FGKM stars, as a result of coronal processes. It is uncertain whether the A stars are X-ray emitters of any importance (Haisch *et al.* 1992; Berghöfer *et al.* 1994).

The two different processes of X-ray emission in normal stars are, in principle, likely to be reflected in X-ray spectral differences. Using the ROSAT All Sky Survey (RASS) data, Motch *et al.* (1998) showed that normal stars of all types exhibit a broad distribution in a Hardness Ratio – Hardness Ratio diagram. However, Meurs *et al.* (2005) found that OB stars tend to be found at one extreme end of this overall distribution. We decided therefore to conduct a more comprehensive investigation of stellar HRs on the basis of the RASS.

The RASS produced the biggest catalogue of X-ray sources, detected at soft X-rays (0.1–2.4 keV). We use the Bright Source Catalogue, whith count rates >0.05 ct s^{-1}. Compared to the deeper RASS Faint Source Catalogue, this choice restricts (i) contamination by other source categories and (ii) large errors on the HRs.

Lists of stars were taken from the Yale Bright Star Catalogue (Hoffleit & Warren 1991) as available in SIMBAD. The spectral types O–M were divided in luminosity classes I, III and V. ROSAT detections were collected, based on positional coincidence within 1 arcmin; a comparison with the diagram in Motch *et al.* (1998) shows that conceivable misidentifications are insignificant.

The detected stars have ROSAT HRs defined as:
$$HR1 = ((0.5\text{-}2.0)\text{-}(0.1\text{-}0.4))/((0.5\text{-}2.0)+(0.1\text{-}0.4))$$
$$HR2 = ((0.9\text{-}2.0)\text{-}(0.5\text{-}0.9))/((0.9\text{-}2.0)+(0.5\text{-}0.9))$$
where the intervals between parentheses are in keV. If at least one of the two HRs has errors >0.4 the measurement is discarded, which removes effectively occasional outliers in HR1–HR2 plots.

The results of this excercise confirm a segregation for the HRs of OB and F–M stars. The OB group includes, for the B types, only B0-B2 since there is evidence that a change in behaviour occurs after B2 (Meurs *et al.* 1992; Berghöfer *et al.* 1997).

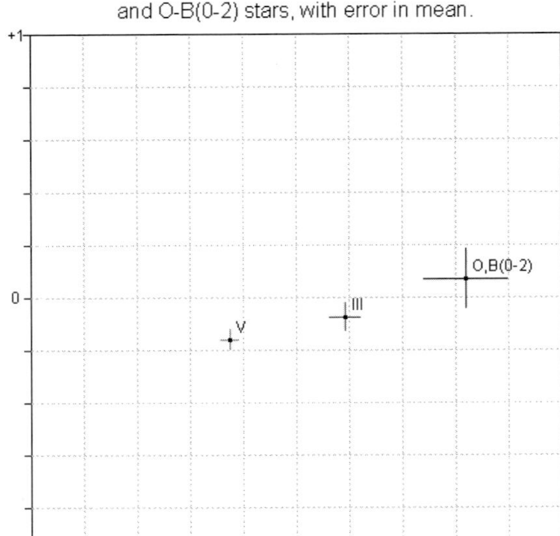

Figure 1. Median Hardness Ratios for late type dwarfs, late type giants and early type stars (from left to right).

In addition, the F–M dwarfs and giants occupy different ranges, mostly in HR1 and to a lesser extent in HR2. The centroids for the OB stars and the late-type giants and dwarfs are plotted in Figure 1, with the standard deviations for each of these cases.

Comparing with a N_H-T grid in HR1-HR2 space for thermal spectral models (Carr 2003), the shift of centroids implies a greater difference in absorbing column N_H than in temperature T. For the case of Einstein data the effect of absorption on the HR was doubted (see e.g. Vaiana 1990), but the ROSAT data are in a softer window and are more strongly sensitive to line-of-sight absorption. The progressive shift of the centroids (from late-type dwarfs via late-type giants to OB stars) corresponds to an increase in luminosity, thus also of increasing average distance and of absorbing column N_H (which is always most noticeable in HR1). The actual increases in N_H, as deduced from the grid for thermal models, imply a typical absorption of 0.5 magn/kpc, which is a realistic value. Thus, differences in N_H may be mainly responsible for the OB versus F–M stars separation, rather than different X-ray emission processes as such.

References

Berghöfer, T. W., *et al.* 1994, A&A 292, L5
Berghöfer, T. W., *et al.* 1997, A&A 322, 167
Carr, M. 2003, PhD Thesis, Trinity College Dublin, Dublin, Ireland
Haisch, B., *et al.* 1992, Nature 360, 239
Hoffleit, D. & Warren, W. H. Jr 1991, 5th "Bright Star Catalogue", Yale University Press
Meurs, E. J. A., *et al.* 1992, A&A 265, L41
Meurs, E. J. A., *et al.* 2005, ApJ 624, 307
Motch, C., *et al.* 1998, AAS 132, 341
Vaiana, G. S. 1990, in "Imaging X-ray Astronomy", M. Elvis (Ed.), CUP, p.61

Populations of High Energy Sources in Galaxies
Proceedings IAU Symposium No. 230, 2005
E. J. A. Meurs & G. Fabbiano, eds.

© 2006 International Astronomical Union
doi:10.1017/S1743921306007812

Long-term activity of the Rapid Burster

Vojtěch Šimon[1]

[1]Astronomical Institute, Academy of Sciences of the Czech Republic, 251 65 Ondřejov,
Czech Republic

Abstract. The recurrence time T_C of outbursts of the remarkable and unique system, the Rapid
Burster (MXB 1730–335), observed by *ASM/RXTE*, is analyzed by the method of the O–C
residuals. The variations of T_C are large and occur all the time, but generally they are not
chaotic; the mean value of T_C is 160 days between the years 1996–2005 but a large shortening of
T_C, accompanied by a large decrease of the maximum intensity $I_{\max}$ and the relative energy RE
(energy output) of most outbursts, occurred in this interval. The outbursts are found to display
a correlation between RE and $I_{\max}$, but no correlation with the outburst duration. The observed
behaviour is discussed in terms of the thermal instability of the accretion disk. A comparison
of this prototype with other neutron star soft X-ray transients, like Aql X-1 and 4U 1608–52,
helps us find the common links in the disk behaviour in such systems.

Keywords. accretion, accretion disks, instabilities, binaries (including multiple): close, circumstellar matter, stars: dwarf novae, stars: neutron, X-rays: binaries

1. Introduction

MXB 1730–335 (Rapid Burster, RB) is a remarkable system, discovered in 1976 (Lewin
et al. 1976). Two types of X-ray bursts are observed during the outbursts of the RB:
Type I – thermonuclear runaway of the accreted matter on the neutron star (NS), and
Type II – spasmodic accretion (e.g. Lewin *et al.* 1995). The RB is a prototype of a group
of such systems and is the only one which displays both types of burst. The outbursts
of the RB (e.g. Masetti 2002) are similar to those of soft X-ray transients (SXTs) (see
e.g. Chen et al. 1997). Outbursts of SXTs are attributed to the thermal instability of the
accretion disk (similar to that in dwarf novae), modified by the irradiation of the disk
(e.g. Dubus *et al.* 2001).

2. Data source, analysis, results

The 1.5–12 keV sum band daily means of observations from the All Sky Monitor (ASM)
onboard *RXTE* (Levine et al. 1996) (`http://xte.mit.edu`) were used. The method of
the O–C residuals (e.g. Vogt 1980, Šimon 2000) was applied. It enables us to determine
T_C and study its variations. Typical error with which the maximum of outburst can be
determined is about 1–2 days (much smaller than the recurrence time T_C of outbursts).

It emerges that the variations of T_C of outbusts of the RB are large, but generally not
chaotic. The method of the O–C residuals reveals that T_C varies all the time; the O–C
curve displays a more complicated profile of the variations of T_C than found by Masetti
(2002). We find that the variations of T_C in the RB are quite similar to those of the
outbursts in the NS SXTs Aql X-1 and 4U 1608–52 (Šimon 2002a, Šimon 2004) and in
dwarf novae (e.g. Vogt 1980, Šimon 2000, Šimon 2002b). The prevailing long-term trend
in the O–C diagram for the outbursts of the RB shows that the individual outbursts
are really dependent on each other. This suggests that there remains a relatively large
amount of matter in the disk after every outburst. The "clock" governing the length

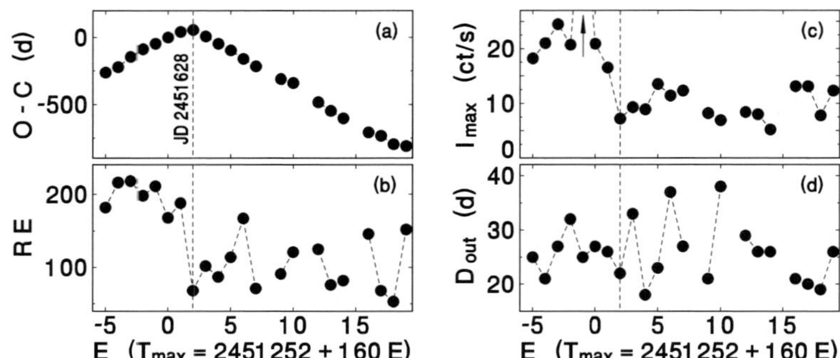

Figure 1. (a) O–C diagram for the moments of the outburst maxima. Time is expressed in epochs E. The long vertical line denotes the moment of the shortening of $T_{\rm C}$. (b) Variations of the relative energy RE (integration of the outburst light curve (in dimensionless units)). (c) Maximum intensity $I_{\rm max}$ of outburst. (d) Outburst duration $D_{\rm out}$.

of $T_{\rm C}$ appears to reside in the outer regions of the disk; this situation appears to be analogous to that of Aql X-1 and 4U 1608–52 and enables us to put the RB to the context of NS SXTs. A remarkable change of the O–C curve (shortening of $T_{\rm C}$) around JD 2451628 was accompanied by an abrupt and large drop of both $I_{\rm max}$ and RE (Fig. 1). This suggests that the amount of matter accreted during the outburst around JD 2451628 was considerably smaller than in the previous events, and also smaller than in most subsequent events.

$I_{\rm max}$ of outbursts of the RB fluctuates less than RE, which suggests that the density of matter arriving to the NS fluctuates less than the total amount of matter accreted during a given outburst. The mean outburst duration $D_{\rm out}$ remains almost unchanged prior to and after JD 2451628, only considerably larger fluctuations appeared – this suggests that the time interval given by the propagation of the heating front and the cooling front (plus a possible state when the whole disk is in the hot state near the maximum of outburst) does not change considerably. The time averaged mass inflow from the donor star appears to remain unchanged because the relation $RE/T_{\rm C}$ is almost the same prior to and after JD 2451628. An increase of the disk viscosity may be a more plausible explanation than the variations of the mass outflow from the donor.

Acknowledgements

This research has made use of the observations provided by the *ASM/RXTE* team. This study was supported by the grant 205/05/2167 of the Grant Agency of the Czech Republic and the project ESA PRODEX INTEGRAL 14527.

References

Chen, W. *et al.* 1997, *ApJ* 491, 312
Dubus, G. *et al.* 2001, *A&A* 373, 251
Levine, A.M. *et al.* 1996, *ApJ* 469, L33
Lewin, W.H.G. *et al.* 1976, *ApJ* 207, L95
Lewin, W.H.G., *et al.* 1995, *in X-ray Binaries*, Cambridge Univ. Press, p. 175
Masetti, N. 2002, *A&A* 381, L45
Šimon, V. 2000, *A&A* 354, 103
Šimon, V. 2002a, *A&A* 381, 151
Šimon, V. 2002b, *A&A* 382, 910
Šimon, V. 2004, *A&A* 418, 617
Vogt, N. 1980, *A&A* 88, 66

Populations of High Energy Sources in Galaxies
Proceedings IAU Symposium No. 230, 2005
E. J. A. Meurs & G. Fabbiano, eds.

© 2006 International Astronomical Union
doi:10.1017/S1743921306007824

The Be/X-ray transient HD34921

V. F. Polcaro[1], L. Norci[2], E. J. A. Meurs[3], S. Bernabei[4] and A. S. Miroshnichenko[5]

[1] IASF-INAF, Rome, Italy
email: polcaro@rm.iasf.cnr.it

[2] School of Physics, Dublin City University, Dublin, Ireland

[3] Dunsink Observatory, Dublin Institute for Advanced Studies, Castleknock, Dublin 15, Ireland

[4] Osservatorio Astronomico, Bologna, Italy

[5] University of Toledo, USA

Abstract. HD34921 has been identified as the counterpart of the X-ray source 4U0515+38 (=1H0521+373). The InfraRed properties are reminiscent of the B[e] system CI Cam. Optical short-term variability suggests a compact companion. We discuss how this system fits in the overall framework of Be stars and X-ray binaries.

Keywords. Stars: B[e], Stars: X-rays, Stars: Neutron

HD34921 (BD+37°1160) was identified as counterpart of the X-ray source 4U0515+38 by Polcaro *et al.* (1990; P90). After the original Uhuru observations, this X-ray transient was detected again by HEAO1 as 1H0521+373, at a much lower flux level. Subsequent balloon-borne hard X-ray telescopes also detected this source (at still higher energies), in 1980 and 1981 (Ubertini *et al.* 1982; Polcaro *et al.* 1984). Following upper limits with EXOSAT's ME detectors (P90), 4U0515+38 has not been seen anymore at high energies. Data for the 2-7 keV band are shown in Figure 1.

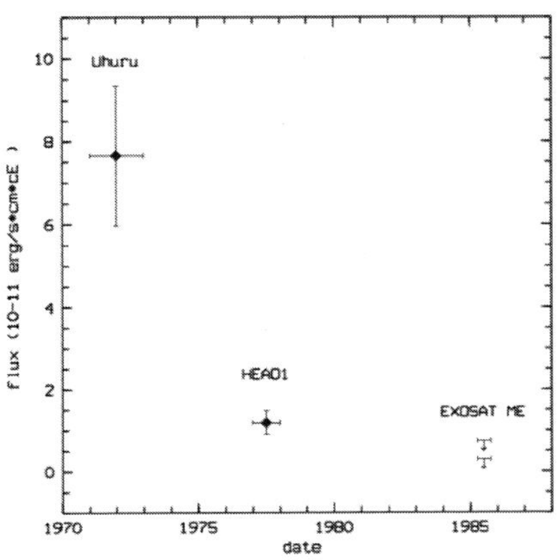

Figure 1. X-ray lightcurve of HD34921 at 2-7 keV.

The optical counterpart, HD 34921, is a known variable star. P90 classified it as B0III-IVpe and estimated a distance of 1.7 kpc. A number of spectroscopic studies has been carried out since its association with the X-ray source 4U0515+38.

Spectra taken in 1990 show a variable HeII λ4686 emission line, which is a recognized indicator of an accretion disk and led to the identification of HD 34921 as the counterpart of 4U0515+38 (P90). Classical Be stars do not have HeII in emission, while it is present in other X-ray transients like X Per and γ Cas. At the same time, the emission lines of HeII and H$_\alpha$ as well as H$_\beta$ were found to vary on short timescales, of order of 300 s for V/R (violet/red wing) variations (Rossi *et al.* 1991). Both H$_\alpha$ and H$_\beta$ were split, as is typical for a disk structure, and since the same variability timescale is observed for HeII (which points at an accretion disk), the split H$_\alpha$ and H$_\beta$ profiles are thought to refer similarly to the surroundings of a compact object. We note that variability due to e.g. hotspots in a Be star (excretion) disk would have timescales of months. H$_\alpha$ and (permitted) FeII emission lines, varying in phase with H$_\beta$, have been seen split on occasions as well.

Spectroscopic data taken in 2003 (Polcaro *et al.*, in preparation), suggest a third, non-variable component in the H_β profiles. Polcaro *et al.* (2005) infer that this component also filled in the top of the H_α profile in 2003 due to a second emission region besides the accretion disk, which presumably features lower temperature and higher density than the disk. Periods of high Equivalent Width and non-split H_α may signify shell ejection phases.

In the spectral energy distribution (SED) of HD 34921 a Near-InfraRed excess has been observed in the J, H, K bands (P90) that appears as a thermal free-free component, which has been interpreted as an extended atmosphere. The Far-InfraRed source IRAS 051921+3737 has also been associated with this star (P90), suggesting a very cold dust envelope, which is uncommon for HMXBs. Clark *et al.* (1999) noticed strong analogies between the Near-InfraRed spectra of HD 34921 and CI Cam, confirming a very complex circumstellar environment. As for CI Cam, these authors proposed a B[e] classification also for HD 34921, based on the observed analogies in the InfraRed. The two stars would be the only X-ray binary counterparts of the B[e] spectral type.

Considering the B[e] classification criteria adopted by Lamers *et al.* (1998), in HD34921 the required forbidden emission lines of [FeII] and [OI] in the optical spectrum are absent, while rather than a strong Near-/Mid-IR excess due to hot circumstellar dust it features a cold dust envelope. Therefore, the B[e] classification of HD34921 is subject to some doubt (Norci *et al.* 2006).

The absence of forbidden lines in the optical spectrum of HD34921 means that the circumstellar environment has (regions of) high density. This could be e.g. ring-shaped shell ejections. In CI Cam only forbidden emission lines are seen produced in the envelope, while the photospheric lines are not visible (unlike HD34921). The observation that in HD34921 also non-Balmer, permitted emission lines follow the 300 seconds variation suggests that these too are linked with the probable accretion disk, rather than with a stellar excretion disk, for which there is little direct evidence.

We note that HD34921 is an OB star, outside an association. The proper motions correspond to a velocity of 42 km/s, which qualifies HD34921 as a runaway star. The high velocities of OB runaway stars are thought to have been caused by either dynamical ejection from a dense cluster (thus in the early stages of development of OB associations) or a supernova event in a binary system. In the case of HD 34921, where there is evidence for a collapsed companion, the supernova in a binary mechanism provides the likely explanation for the high velocity (and the system survived the supernova event, as a good fraction of systems are likely to do; see also O'Maoileidigh *et al.* 2005). Thus, binary evolution has likely played a role in the evolution of HD 34921. Due to the presence of a companion, outer disk parts (with lesser density) may become truncated due to Roche lobe cut-off, which interestingly provides an other way to suppress forbidden line emission.

References

Clark, J.S., Steele, I.A., Fender, R.P. & Coe, M.J. 1999, A&A 348, 888

Lamers, H.J.G.L.M., Zickgraf, F.-J., de Winter, D., *et al.* 1998, A&A 340, 117

Norci, L., *et al.* 2006, ASP Conf. Series, in press

O'Maoileidigh, C., Meurs, E.J.A. & Norci, L. 2005, NewA 10, 591

Polcaro, V.F., *et al.* 1984, A&A 131, 229

Polcaro, V.F., *et al.* 1990, A&A 231, 354 (P90)

Polcaro, V.F., Norci, L., Meurs, E.J.A. & Bernabei, S. 2005, in press

Rossi, C., Norci, L. & Polcaro, V.F. 1991, A&A 249, L19

Ubertini, P., *et al.* 1982, A&A 106, 174

Populations of High Energy Sources in Galaxies
Proceedings IAU Symposium No. 230, 2005
E. J. A. Meurs & G. Fabbiano, eds.

© 2006 International Astronomical Union
doi:10.1017/S1743921306007836

The possible explanation of low-frequency noise of pulsars in globular clusters

Tatiana I. Larchenkova[1] and Sergei M. Kopeikin[2]

[1]Astro Space Center of P.N.Lebedev Physical Institute, Leninskii Prospect 53, Moscow
117924, Russia
email: tanya@lukash.asc.rssi.ru

[2]Department of Physics and Astronomy, University of Missouri-Columbia,
Columbia, MO 65211, USA
email: kopeikins@missouri.edu

Abstract. The low-frequency (LF) timing noise contaminates residuals of time-of-arrivals (TOAs) of pulsar signals on long time intervals and in some cases can be explained by external astronomical factors not intrinsically related to the pulsar itself. A number of millisecond pulsars located in globular clusters show the LF noise presence in their rotational phase. We discuss a possible origin of this noise as caused by random time variations in the Shapiro time delay caused by flybys of stars of the globular cluster passing near the pulsar line of sight. The Shapiro time delay is integrated over space of parameters characterizing statistical ensemble of stars in the globular cluster and its long-term indeterministic time variation is obtained. We use this result for numerical simulations of the autocovariance function of the LF timing noise and show it can be used for the measuring of the density profile distribution in the globular cluster.

Keywords. Pulsars, globular clusters, relativity

The long-term monitoring of millisecond pulsars has revealed the presence of non-white component of the LF noise of an astrophysical origin being not intrinsically related to the pulsar itself. One of the problems of current pulsar timing analysis is the detection of such a correlated ("red") noise in the spectrum of its TOA residuals and the measuring of its amplitude and spectral index. Now a high accuracy of pulsar timing observations and the exceptionally precision of the data processing algorithms allow to solve this problem. The globular cluster pulsars show the LF noise presence in their rotational phase. We propose the Shapiro effect, i.e. the relativistic time delay in the propagation of pulsar signals passing through the gravitational field of the astrophysical body (the stars of the globular cluster), as a possible cause of the LF timing noise of pulsars in globular clusters. For example, knowledge of the theoretical power spectrum of this effect and the comparison with the observational spectrum allow to obtain the additional information about the globular cluster structure and extend the mass function aside the low mass.

The exact expressions for the relativistic time delay of propagation of electromagnetic signals passing through the non-stationary gravitational field have been derived by Kopeikin & Schafer (1999) in the form of instantaneous function of the retarded time. As appears from this paper the position of gravitating bodies (stars) should not be taken at the moment of the impulse emission but at its corresponding retarded moment of time.

Let us assume that the origin of the coordinate system is at the barycenter of the globular cluster. And t_0 - moment of emission of photon, $\vec{x}$ - barycentric coordinates of the observer, $\vec{x}_a$ - barycentric coordinates of a-th star, $\vec{x}_0$ - barycentric coordinates of the pulsar. Then we can obtain the analytic expression of the stochastic noise process as the result of random time variations in the Shapiro time delay caused by flybys of

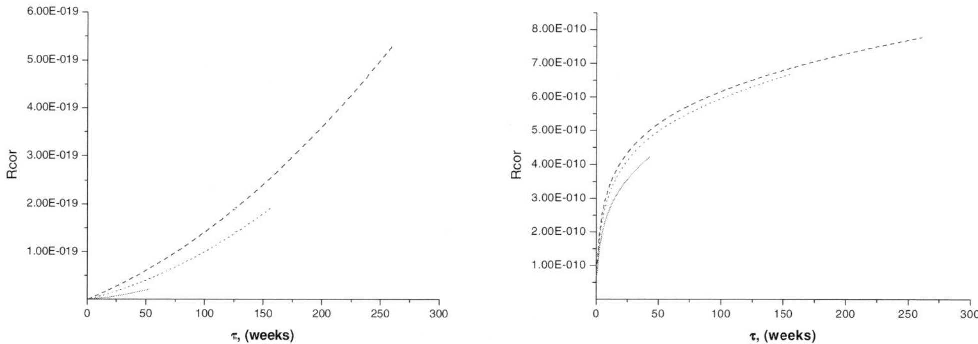

Figure 1. *(a)* The autocovariance function of stochastic Shapiro effect as a function of the time for the isothermal-sphere model. *(b)* The same as in the panel *(a)*, but in the case of small impact parameters ($|\vec{d}_a| < 1AU$). The pulsar is located in the cluster center for both cases

stars of the globular cluster passing near the pulsar line of sight (see Larchenkova & Kopeikin (2005) for details). Neglecting all small aberration terms (v/c) we have derived the formula for the stochastic relativistic time delay of the emission of pulsars located in the globular cluster

$$\varepsilon(t) = \sum_{a=1}^{N} \frac{2GM_a}{c^3} \cdot \left[\ln(1 + 2(t-T)\frac{\vec{d}_a(T_0)\vec{v}_a(T_0)}{\vec{d}_a(T_0)\vec{x}_a(t_0)}) - \ln(1 + \frac{(-\vec{q}-\vec{K}_0)\vec{V}_{0a}(T_0)}{Q_{0a} + \vec{K}_0\vec{Q}_{0a}}(t-T)) \right]$$

where G is the gravitational constant, c is the light velocity, T is the beginning of observations, T_0 defines the moment of the impulse emission corresponding to the time T, M_a and $\vec{v}_a$ are the mass and the velocity of a-th star, $\vec{K}_0$ is the unit vector, $\vec{V}_{0a} = \vec{v}_0(T_0) - \vec{v}_a(T_0)$, $\vec{Q}_{0a} = \vec{x}_a(T_0) - \vec{x}_0(T_0)$, $\vec{q} = \frac{\vec{Q}_{0a}}{|\vec{Q}_{0a}|}$, $\vec{v}_0$- the pulsar velocity, $\vec{d}_a$ - the impact parameter.

Assuming that the globular cluster stars are uncorrelated, it was shown that the stochastic Shapiro effect manifests itself by the LF noise of the pulsar rotational frequency for large values of the impact parameter. For the pulsar located in the globular cluster 47 Tuc the numerical analysis was done in the case of small values of impact parameters (the short-term part of this stochastic process). The isothermal sphere model and King model for the cluster density distribution, the Salpeter function for the mass distribution of the cluster stars and the Maxwell function for the velocity distribution in a star cluster were used in our analysis. The numerical simulations show that the autocovariance function of the stochastic Shapiro effect can be good approximated by a second-power polynomial in the case of large impact parameters (see Figure 1a). The power spectral index is equal to ~ -1.8 for the King model as well as for the isothermal-sphere model. The autocovariance function of the stochastic Shapiro effect for small impact parameters has logarithmic behavior and can be interpreted by flicker noise (see Figure 1b). The power spectral index for this case is ~ -1.5.

We have shown the rotational phase modulation of the pulsar located in globular cluster can be explained by relativistic time delay fluctuations of the pulsar signal propagation in the gravitational field of arbitrarily moving globular cluster stars.

References

Kopeikin, S.M. & Schafer, G. 1999, *Physical Review D* 60, 124002
Larchenkova, T.I. & Kopeikin, S.M. 2005, *Astronomy Letters* 31, N12

Populations of High Energy Sources in Galaxies
Proceedings IAU Symposium No. 230, 2005
E. J. A. Meurs & G. Fabbiano, eds.

© 2006 International Astronomical Union
doi:10.1017/S1743921306007848

An Optical Outburst of the Cataclysmic Variable in M22

A. Hourihane[1], P. J. Callanan[1] and A. M. Cool[2]

[1]Department of Physics, University College Cork, Ireland
email: a.hourihane@ucc.ie,paulc@ucc.ie

[2]Department of Physics and Astronomy, San Francisco State University, 1600 Holloway
Avenue, San Francisco, CA 94132, USA
email: cool@sfsu.edu

Abstract. Cataclysmic variables (CVs) formed through close encounters may be the most
abundant class of compact binaries in globular clusters. As part of a systematic search for CVs
undergoing dwarf nova eruptions (DN) in globular clusters, our 2004 monitoring program of
M22 detected an outburst of a CV candidate during May. We present a light curve for the May
2004 period, obtained using the *ISIS* image subtraction routine. Our ground based results are in
good agreement with previous *HST* measurements, and confirm the DN nature of the outburst
and the object's CV status. Further application of the *ISIS* software will enable us to identify
other DN candidates in the core of M22 with the aim of characterizing the properties of these
systems, as opposed to similar ones in the field.

1. Introduction

Compact binary systems in globular clusters are an important target for observations.
The process of their formation, unlike primordial field binaries, is thought to be three-
body collisions and tidal capture (e.g. Hut & Verbunt, 1983), leading in general to very
close binaries, with perhaps different properties relative to their counterparts in the field.
So far relatively few DN outbursts of globular cluster CVs have been observed (e.g. Shara,
Bergeron, Gilliland, *et al.*, 1996; Ó Tuairisg, Butler, Shearer, *et al.*, 2003) compared to
the predicted populations of such systems and it is of significant interest to determine if
this is due to an underlying difference in the nature of the cluster CVs compared with
field CVs. Furthermore, the presence of HeII emission lines in the *HST* spectra of these
systems suggests a possible magnetic nature (Grindlay, Cool, Callanan, *et al.*, 1995).

To characterise these systems we present results from a long-term monitoring program
of the globular cluster M22 during which a DN outburst of a cluster CV was observed.

A $\sim$3 magnitude brightening of the same source was originally interpreted by Sahu,
Casertano, Livio, *et al.* (2001) as a gravitational microlensing of a background bulge star
by a low mass foreground cluster object. Subsequent analysis of *HST* data by Anderson,
Cool & King (2003) prompted reclassification of the object as a cluster CV, *CV1*.

2. Observations and Data Analysis

Observations obtained from the 1.3m telescope at the *Cerro Tololo Inter-American*
Observatory (*CTIO*) in Chile cover the period of March - November 2004. The dataset
consists of 6x6 arcminute optical V-band images centred on the core of M22.

Photometry was performed on the crowded-core CV using the *ISIS* image subtraction software (Alard & Lupton, 1998; Alard, 2000). The light curve shows the $\sim$15 day

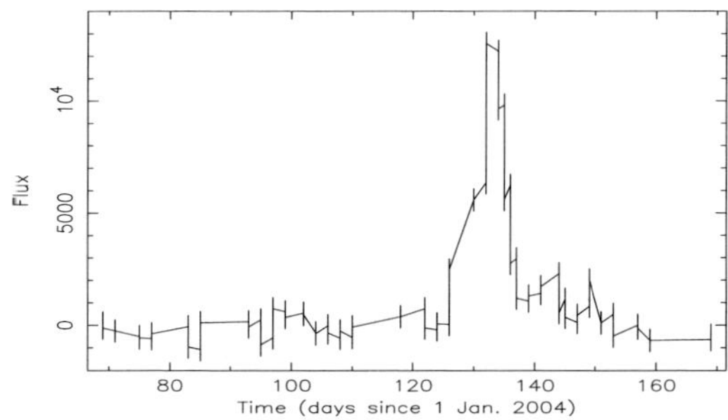

Figure 1. The *ISIS* differential flux light curve for the DN eruption of the M22 CV.

May outburst (Figure 1), the only large outburst seen in nine months of monitoring. Subsequent data showed several spikes above noise-level which may be shorter-timescale outbursts, but better sampling is necessary to confirm this.

3. Discussion

The outburst characteristics are consistent with the approximate 3.5 magnitude, 18 day 1999 outburst reported by Sahu *et al.* (2001). It also fits easily into the DN category, and the repeated outbursts together with the proper motion studies of Anderson *et al.* (2003) effectively confirm the CV nature of the source.

The next step is to search for lower-amplitude, shorter-timescale outbursts from *CV1* and other globular cluster CVs. Such outbursts would suggest an Intermediate Polar (IP) type. IPs are weak-field magnetic CVs harbouring truncated accretion disks which have been shown to lead to shorter outburst durations, especially in the V-band, compared to non-truncated disks. Even though outbursts starting at the inner disk edge are predicted to be rare in IPs, those starting at the outer edge are as frequent as in non-magnetic systems, hence small DN eruptions may not be uncommon in this class. Magnetic systems are also predicted to show a shorter outburst decline time (Angelini & Verbunt, 1989), so careful monitoring of the outburst lightcurve morphology during the decay from maximum brightness could help discriminate between a magnetic and non-magnetic system.

Although our data do not provide conclusive evidence as to the IP nature of *CV1*, we believe that ground-based studies of DN outburst morphologies have a role to play in constraining the nature of globular cluster CVs. Further monitoring is necessary to confirm a possible magnetic nature for the CV and for CVs in other globular clusters.

References

Alard, C. 2000, *A&AS* 144, 363

Alard, C. & Lupton, R.H. 1998, *ApJ* 503, 325

Anderson, J., Cool, A.M. & King, I.R. 2003, *ApJ* 597, 137

Angelini, L. & Verbunt, F. 1989, *MNRAS* 238, 697

Grindlay, J.E., Cool, A.M., Callanan, P.J., Bailyn, C.D., Cohn, H.N. & Lugger, P.M. 1995, *ApJ* 455, 47

Hut, P. & Verbunt, F. 1983, *Nature* 301, 587

Ó Tuairisg, S., Butler, R.F., Shearer, A., Redfern, R.M., Butler, D. & Penny, A. 2003, *MNRAS* 345, 960

Sahu, K.C., Casertano, S., Livio, M., Gilliland, R.L., Panagia, N., Albrow, M.D. & Potter, M. 2001, *Nature* 411, 1022

Shara, M., Bergeron, L., Gilliland, R., Saha, A. & Petro, L. 1996, *ApJ* 471, 804

Populations of High Energy Sources in Galaxies
Proceedings IAU Symposium No. 230, 2005
E. J. A. Meurs & G. Fabbiano, eds.

© 2006 International Astronomical Union
doi:10.1017/S174392130600785X

The optical & IR lightcurve of PSR B1957+20

M. T. Reynolds[a], P. J. Callanan[a], A. S. Fruchter[b], M. A. P. Torres[c], M. E. Beer[d] and R. A. Gibbons[e]

[a]Department of Physics, University College Cork, Ireland
email : m.reynolds@ucc.ie
[b]Space Telescope Science Institute, 3700 San Martin Drive, Baltimore, MD 21218, USA
[c]Center for Astrophysics, 60 Garden Street, Cambridge, MA 02138, USA
[d]Department of Physics and Astronomy, University of Leicester, Leicester LE1 7RH, England
[e]University of Maryland, College Park, MD 20742, USA

Abstract. We present a new analysis of the light curve of the secondary star in the PSR B1957+20 system. Combining previous data and new data points at minimum from the *Hubble Space Telescope*, we have 100% coverage in the R-band. We also have a number of new K_s band data points, which we use to constrain the IR magnitude of the system. We model this with the Eclipsing Light Curve code. From the modeling we obtain colour information about the secondary at minimum in BVRI & K. For our best fit model we are able to constrain the system inclination to 66.6 ± 2.1^o for pulsar masses ranging from $1.35 - 1.9$ $M_\odot$. The pulsar mass is unconstrained. We also find that the secondary is *not* filling its Roche lobe, which has important consequences for evolutionary models of this system. The temperature of the un-irradiated side of the companion is in agreement with previous estimates.

Keywords. X-rays: binaries, pulsars: individual (PSR B1957+20).

1. Introduction

The binary millisecond pulsar *PSR B1957+20* (Fruchter *et al.* 1988) is the original and one of the best studied members of its class. It consists of a 1.6 ms radio pulsar, which is eclipsed for approximately 10% (in the radio) of its 9.17 hr orbit, by a companion of mass no less than $0.022 M_\odot$. The eclipsing region is considerably larger than the Roche lobe of the companion star, suggesting a wind of material from the secondary, due to ablation by the impinging pulsar radiation (Fruchter *et al.* (1988)).

2. Modeling

The data, consisting of optical photometry from the *WHT* and the *HST* and infra-red data from the *Magellan* telescope, were modeled with the ELC code (Orosz & Hauschildt 2000).

2.1 Inclination

We modeled the system for a number of prospective pulsar masses ranging from $1.35 - 1.9$ $M_\odot$. For a given pulsar mass the inclination was constrained to within $\sim 2^\circ$, i.e. for a pulsar of mass 1.40 $M_\odot$, $i = 67.4^\circ \pm 1.2^\circ$ (see Fig. 1). Overall for the above range of pulsar masses we find the inclination of the system to be $i = 66.6^\circ \pm 2.1^\circ$, at the 2σ level.

2.2 Roche lobe filling factor

At no point in our attempts to model this system were we able to obtain an acceptable fit for a secondary filling its Roche lobe. For our models using the NEXTGEN model

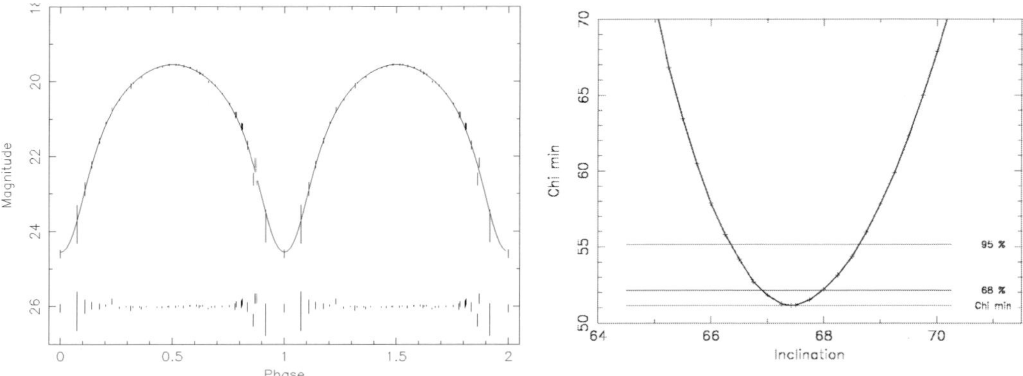

Figure 1. LEFT: The best fit to the combined R-band data with residuals. The residuals are to scale. Two orbital phases are displayed for added clarity. RIGHT: The best fit inclination is $i = (67.4 \pm 1.2)°$ with a $\chi^2_\nu = 1.16$. The pulsar mass is 1.40 M$_\odot$.

atmospheres the Roche lobe filling factor, f, was approximately constant, $0.80 \leqslant f \leqslant 0.84$ (2σ level), Hence the secondary is tightly constrained as *not* currently filling its Roche lobe. This appears to rule out Roche lobe overflow as the primary means of mass loss in this system.

2.3 Temperature of the Secondary @ max/min

We obtained a value of $T = 2980 \pm 150$ K (2σ), for the effective temperature of the unilluminated side of the companion star for pulsar masses in the range $1.35 - 1.90$ M$_\odot$. For individual pulsar masses the 2σ error was only ± 70 K. The corresponding temperature at maximum is $T = 8500 \pm 200$ K (2σ).

2.4 Modeling the Temperature Gradient

The temperature gradient across the surface of the low mass secondary was simulated using a modified version of the code described in Beer *et al.* (2002). It was found that the heated material extended beyond the directly irradiated region but that not all of the unilluminated portion of the star was heated. Consequently a large temperature gradient between the illuminated and unilluminated sides was found to exist.

We now plan to obtain phase resolved spectroscopy of the companion in an effort to measure the mass of the Neutron star. As the error on the inclination is so low, this will ensure that any mass determination will be limited primarily by the accuracy of the radial velocity measurements.

Acknowledgements

We thank Jerome Orosz for kindly providing us with the ELC code.
MTR & PJC wish to acknowledge financial support from Science Foundation Ireland.

References

Beer, M.E., Podsiadlowski, Ph. 2002, MNRAS, 335, 358
Bessell, M.S., AJ, 101, 662
Covey, K.R., Hawley, S.L., Knapp, G.R., & Anderson, S.F. 2003, csss, 12, 658
Fruchter, A.S., Stinebring, D.R., & Taylor, J.H. 1988, Nature, 333, 237
Hansen, B.M.S. 1998, ApJ, 520, 680
Leggett, S.K. 1992, ApJS, 82, 351
Oroszl, J.A. & Hauschildt, P.H. 2000, A&A 364, 265.

Populations of High Energy Sources in Galaxies
Proceedings IAU Symposium No. 230, 2005
E. J. A. Meurs & G. Fabbiano, eds.

© 2006 International Astronomical Union
doi:10.1017/S1743921306007861

Optical spectroscopy of the 2005 outburst of the X-ray transient XTE J1118+480

P. Elebert[1], P. J. Callanan[1] and M. A. P. Torres[2]

[1]Department of Physics, University College Cork, Ireland
email: p.elebert@ucc.ie

[2]Harvard–Smithsonian Center for Astrophysics,
60, Garden St., Cambridge, MA 02138, USA.

Abstract. We present time-resolved spectroscopy from the January 2005 outburst of the X-ray nova XTE J1118+480. X-ray observations show this outburst to be about half as bright as the previous outburst in 2000. This suggests that the accretion rate was lower than that in the 2000 outburst. Our spectroscopic analysis shows that the emission from the Hα line occurs at a higher velocity in the recent outburst compared with the 2000 outburst, more consistent with quiescence/near-quiescence values. This is surprising, considering the $\sim$5 magnitude brightening of the accretion disk relative to its quiescent state.

Keywords. X-rays: binaries, accretion disks, techniques: spectroscopic

1. Introduction

The X-ray nova (XRN) XTE J1118+480, a black hole candidate, was discovered in March 2000 by the All-Sky Monitor on the Rossi X-ray Timing Explorer (Remillard *et al.* 2000). The optical counterpart was found to have brightened by $\sim$6 magnitudes compared with the pre-outburst luminosity. Further analysis showed that the system underwent a previous mini-outburst in January 2000.

The primary has been determined to have a minimum mass of 6 M$_\odot$ (McClintock *et al.* 2001) which is greater than the theoretical maximum mass for a neutron star (Rhoades and Ruffini 1974). An upper limit of 0.36 M$_\odot$ has been estimated for the mass of the secondary, a K5 dwarf (McClintock *et al.* 2003).

The system remained in quiescence until January 2005, when it again increased in brightness both optically and in X-rays. This recent outburst closely resembles the mini-outburst of January 2000 in terms of the length of the outburst. However, the maximum X-ray flux from the 2005 outburst is approximately half as strong as that from the 2000 outburst. Due to the rarity of their outbursts, there are relatively few detailed studies of XRN accretion disks during outburst.

2. Data

Our primary data consist of optical spectra acquired with the FAST spectrograph (Fabricant *et al.* 1998), attached to the 1.5m Tillinghast Telescope at the Fred L. Whipple Observatory, Arizona. These data were acquired as part of the CfA ToO program on X-ray Novae/Transients. Use of a 3 arcsec slit and the 300 line mm^{-1} grating provided a wavelength coverage of $\lambda\lambda$3480–7400 Å with a dispersion of $\sim$1.46 Å pix^{-1}. The spectra were extracted and wavelength calibrated using standard IRAF tasks as part of the CfA spectroscopic data pipeline. These data were obtained over 3 nights in January and February 2005.

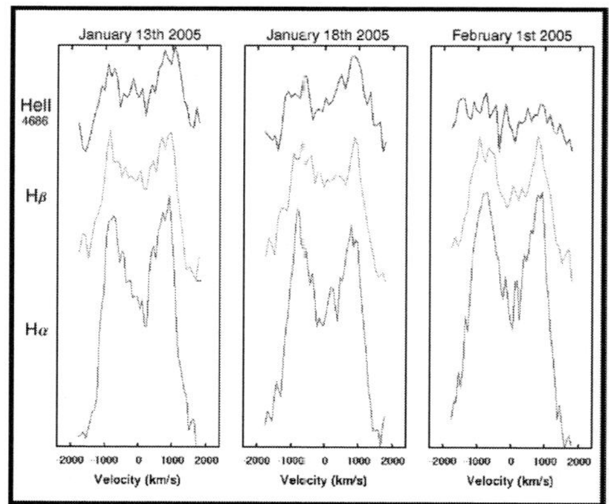

Figure 1. Averaged spectra for Hα, Hβ and HeII λ4686

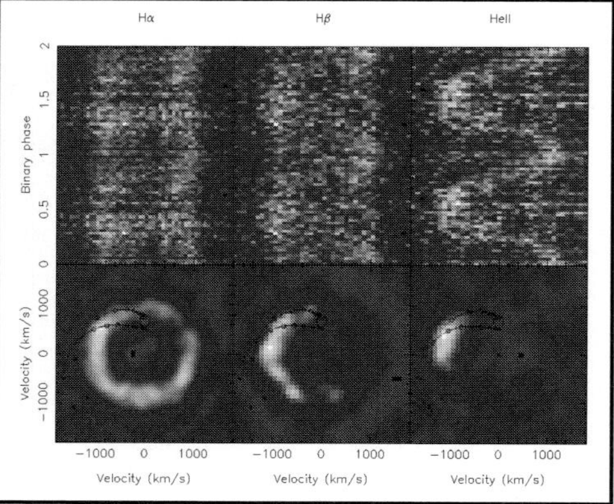

Figure 2. Trailed spectra and Doppler maps for Jan 13 2005

3. Results and Discussion

In Figure 1, the equivalent width of the HeII line is lower on the third night than on the first two nights. This may be consistent with the reduced X-ray flux (that powers this line) on the third night. 2-Gaussian fits to the averaged Hα lines gives a peak-to-peak separation of ∼1600 km s^{-1}, also evident in the emission rings in the Doppler maps in Figure 2 (Marsh and Horne 1988).

In the 2000 outburst, the peak-to-peak separation of the Hα line was ∼1200 km s^{-1}, and remained close to this value during that outburst (Torres *et al.* 2002). In quiescence however, the peak-to-peak velocity increased to ∼1660–1830 km s^{-1} (Torres *et al.* 2004). The higher peak-to-peak separation suggests that the outer part of the disk is less luminous during the recent outburst. reducing the amount of lower velocity material contributing to the line profiles, consistent with our observations. This suggests a considerably lower accretion rate in the January 2005 outburst, which is surprising, considering the ∼5 magnitude optical brightening of the accretion disk relative to its quiescent state.

Acknowledgements

The ToO program at CfA also involved the following CoIs: M. R. Garcia, D. Steeghs, J. E. McClintock, J. M. Miller and P. Zhao. We acknowledge the use of *molly, doppler* and *trailer* software packages developed by T. R. Marsh, University of Warwick.

P. E. and P. J. C. acknowledge support from Science Foundation Ireland.

References

Fabricant, D., Cheimets, P., Caldwell, N. & Geary, J. 1998 *PASP* 110, 79

Marsh, T. R. & Horne, K. 1988 *MNRAS* 236, 269

McClintock, J. E., Garcia, M. R., Caldwell, N., Falco, E. E., Garnavich, P. M. & Zhao, P. 2001 *ApJ* 551, 147

McClintock, J. E., Narayan, R., Garcia, M. R., Orosz, J. A., Remillard, R. A. & Murray, S. S. 2003 *ApJ* 593, 435

Remillard, R., Morgan, E., Smith, D. & Smith, E. 2000 *IAU Circ.* 7389

Rhoades, C. E. & Ruffini, R. 1974, *Phys. Rev. Lett.* 32, 324

Torres, M. A. P., Callanan, P. J., Garcia, M. R., McClintock, J. E., Garnavich, P., Balog, Z., Berlind, P., Brown, W. R., Calkins, M. & Mahdavi, A. 2002 *ApJ* 569, 423

Torres, M. A. P., Callanan, P. J., Garcia, M. R., Zhao, P., Laycock, S. & Hong, A. K. H. 2004 *ApJ* 612, 1026

Populations of High Energy Sources in Galaxies
Proceedings IAU Symposium No. 230, 2005
E. J. A. Meurs & G. Fabbiano, eds.

© 2006 International Astronomical Union
doi:10.1017/S1743921306007873

The INTEGRAL mission – an overview

P. Kretschmar[1], **C. Winkler**[2], **T. J.-L. Courvoisier**[3,15], **G. Di Cocco**[4], **R. Diehl**[5], **N. Gehrels**[6], **S. Grebenev**[7], **W. Hermsen**[7], **J. M. Mas-Hesse**[8], **F. Lebrun**[10], **N. Lund**[11], **G. G. C. Palumbo**[12], **J. Paul**[10], **J.-P. Roques**[13], **R. Sunyaev**[7,16], **B. Teegarden**[6] and **P. Ubertini**[14]

[1]ESA-ESAC, P.O. Box 50727, 28080 Madrid, Spain

[2]ESA-ESTEC, RSSD, Keplerlaan 1, 2201 AZ Noordwijk, The Netherlands

[3]INTEGRAL Science Data Centre, Chemin d'Écogia 16, 1290 Versoix, Switzerland

[4]IASF/CNR, Sezione di Bologna, via Gobetti 101, 40129 Bologna, Italy

[5]Max-Planck Institut fr extraterr. Physik, Postfach 1603, 85740 Garching, Germany

[6]NASA-GSFC, Code 661, Greenbelt, MD 20771, USA

[7]Space Research Institute, Profsoyuznaya 84/32, 117810 Moscow, Russia

[8]SRON-Utrecht, Sorbonnelaan 2, 3584 CA Utrecht, The Netherlands

[9]LAEFF-INTA, PO Box 50727, 28080 Madrid, Spain

[10]Service d'Astrophysique, CEA-Saclay, 91191 Gif-sur-Yvette Cedex, France

[11]DSRI, Juliane Maries Vej 30, 2100 Copenhagen OE, Denmark

[12]Dip. di Astronomia, Universita degli Studi di Bologna, via Ranzani 1, 40127 Bologna, Italy

[13]CESR, Boite Postale 4346, 31029 Toulouse, France

[14]IASF/CNR, via del Fosso del Cavaliere, 00133 Rome, Italy

[15]Geneva Observatory, Ch. des Maillettes, 1290 Sauverny, Switzerland

[16]Max-Planck Institut fr Astrophysik, Postfach 1317, 85741 Garching, Germany

Abstract. The ESA observatory INTEGRAL (International Gamma-Ray Astrophysics Laboratory) is dedicated to fine imaging and spectroscopy in the energy range 15 keV to 10 Mev with concurrent X-ray (3-35 keV) and optical monitoring. It was launched on October 17, 2002 and has been succesfully operating ever since. Its two main instruments the spectrometer SPI – optimized for high resolution spectroscopy – and the imager IBIS – optimized for for high resolution imaging – are complemented by the X-ray monitor JEM-X and the optical monitor OMC. All the high energy instruments use coded mask techniques, allowing imaging in the gamma-ray range and combining wide fields of view with high spatial resolution. The presentation gives an overview of the unique properties of INTEGRAL.

Keywords. gamma rays: observations, X-rays: stars.

1. Introduction

The gamma-ray observatory INTEGRAL (INTErnational Gamma-Ray Astrophysics Laboratory) is ESA's second gamma-ray mission after COS-B in 1975. It was selected in June 1993 as the next medium-size scientific mission within ESA's "Horizon 2000" programme, with important contributions from Russia (PROTON launcher) and NASA (Deep Space Network ground station).

INTEGRAL has been operating very succesfully since its launch on October 17, 2002. The ESA Science Programme Committee has approved a rolling mission extension, beyond the first two years, until at least 2008. Efforts are ongoing to extend the mission life further; at the current rate of consumption consumables would hold for >10 more years.

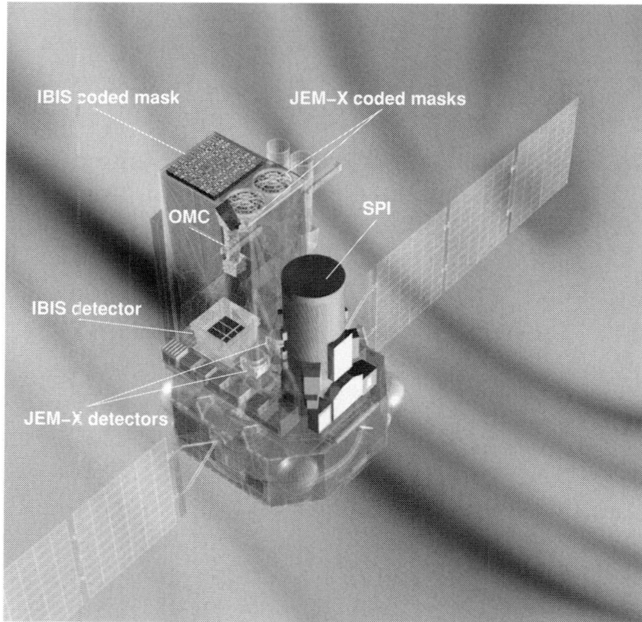

Figure 1. Artists impression of the INTEGRAL spacecraft. The dimensions are 5×2.8×3.2 m; the deployed solar panels are 16 m across.

2. Mission Overview

The spacecraft (Jensen *et al.* 2003) consists of a service module (bus) containing all spacecraft subsystems and a payload module containing the scientific instruments. The service module has been developed in parallel for two ESA scientific missions, INTEGRAL and XMM-Newton. The spacecraft has been built under ESA contract by a large industrial consortium, led by Alenia Spazio as prime contractor. The total launch mass was about 4 t.

Launched by a four-stage PROTON from Baikonur/Kazakhstan on 17 October 2002, INTEGRAL was inserted into a geosynchronous highly eccentric orbit with high perigee in order to provide long periods of uninterrupted observation with nearly constant background and away from trapped radiation (electron and proton radiation belts). The initial orbital parameters are: 72-hour orbit with an inclination of 52.2°, a height of perigee of 9000 km and a height of apogee of 154 000 km. Owing to background radiation effects in the high-energy detectors, scientific observations are carried out while the satellite is above a nominal altitude of 60 000 km (approaching radiation belts) and above 40 000 km (leaving radiation belts). This means that >80% of the time – or about 60 hours per orbit – can be used for scientific observations (real-time, 108 kbps science telemetry).

3. Payload

INTEGRAL carries two main gamma-ray instruments, the spectrometer SPI (Vedrenne *et al.* 2003) – optimized for the high-resolution gamma-ray line spectroscopy (20 keV-8 MeV), and the imager IBIS (Ubertini *et al.* 2003) – optimized for high-angular resolution imaging (15 keV-10 MeV). Two monitors, JEM-X (Lund *et al.* 2003) in the (3-35) keV X-ray band, and OMC (Mas-Hesse *et al.* 2003) in optical Johnson V-band complement the payload. An on-board particle radiation monitor allows an assessment of the radiation environment local to the spacecraft.

Table 1. Key parameters of the INTEGRAL payload.

Parameter	SPI	IBIS
Detector	19^a Ge detectors (6×7 cm) cooled to 85 K	16384 CdTe pixels ($4\times4\times2$ mm) 4096 CsI pixels ($8.4\times8.4\times30$ mm)
Detector area (cm^2)	500	2600 (CdTe), 2890 (CsI)
Spectral resolution (FWHM)	3 keV @ 1.7 MeV	8 keV @ 100 keV
Field of view (fully coded)	$16°$ (corner to corner)	$9° \times 9°$
Angular resolution (FWHM)	$2.5°$ (point source)	12'
Source location radius	$\leqslant 1.3°$	$\leqslant 1'$ (10σ sources)
Absolute timing accuracy	$\leqslant 200\ \mu$s	$\leqslant 200\ \mu$s
Mass (kg)	1309	746

Parameter	JEM-X	OMC
Energy range	4 keV – 35 keV	500 nm – 600 nm (*V*-filter)
Detector	Microstrip Xe/CH$_4$-gas detector (1.5 bar)	CCD (2061×1056 pixels) 1024×1024 pixels iaging area
Detector area (cm^2)	500 per detector^b	
Spectral resolution (FWHM)	2 keV @ 22 keV	–
Limiting magnitude	–	17.8 (3σ, 5000 s)
Field of view (fully coded)	$4.8°$ (corner to corner)	$5° \times 5°$
Angular resolution (FWHM)	3'	25"
Source location radius	$\leqslant 30$"	6"
Absolute timing accuracy	$\leqslant 200\ \mu$s	$\geqslant 1$ s
Mass (kg)	65	17

^{a} At the time of writing two detectors have failed
^{b} Currently only one of the two JEM-X detectors is being operated

The spectrometer, imager and X-ray monitor share a common principle of operation: they are all coded aperture mask telescopes. The coded mask technique is the key for imaging, which is all-important in separating and locating sources. It also provides good background subtraction because for any particular source direction the detector pixels can be considered to be split into two intermingled subsets, those capable of viewing the source and those for which the flux is blocked by opaque mask elements. Effectively the latter subset provide an exactly contemporaneous background measurement for the former, made under identical conditions.

The payload was extensively calibrated pre-launch on instrument and system level. During the early mission phase, the in-flight calibration was initially performed on Cyg X-1, and empty fields, and completed later using the Crab nebula and pulsar. Further in-flight calibration refinement is being done via regular observations of the Crab twice per year.

All instruments (Fig. 1 and Table 1) are co-aligned with overlapping fully coded field-of-views ranging from 4.8 diameter (JEM-X), 5 (OMC), to 9 (IBIS) and 16 corner-to-corner (SPI) – see also Fig. 3 – and they are operated simultaneously, so, an observer receives all data from all four instruments, allowing to combine data across instruments. Alternatively, many users with different scientific goals, e.g., point sources vs. diffuse emission, can make use of the same data.

4. Ground Segment

The ground segment consists of two major elements, the Operations Ground Segment (OGS) and the Science Ground Segment (SGS). The OGS, consisting of the ESA and NASA ground stations and the Mission Operations Centre at ESOC, implements the observation plan within the spacecraft system constraints into an operational command sequence. In addition, the OGS performs all standard spacecraft and payload operations and maintenance tasks. The SGS itself consists of two components, the INTEGRAL Science Operations Centre (ISOC, Much *et al.* 2003) and the INTEGRAL Science Data Centre (ISDC, Courvoisier *et al.* 2003). The ISOC processes the accepted observation proposals into an optimised observation plan which consists of a time line of target pointings plus the corresponding instrument configuration. ISOC is also responsible for the implementation of Target of Opportunity observations within the pre-planned observing programme. The ISDC receives the science telemetry plus the relevant ancillary spacecraft data from the OGS, usually within seconds of the actual data taking onboard. Final data products are distributed to the observer with a usual time delay of 10–12 weeks and archived for later use by the science community.

5. Observing programme

INTEGRAL was conceived from its initial study phase in 1989 as an observatory-type mission (nominal lifetime 2 years,extensions up to 5 years possible). Most of the total observing time (65% during year 1, 70% year 2, 75% year 2+) is awarded as the General Programme to the scientific community at large. Typical observations last from 100 ksec up to about two weeks. Proposals for observations are selected on their scientific merit only by a single Time Allocation Committee (TAC). These selected observations make up the general programme. The first call (AO-1) for observation proposals was issued on 1 November 2000; at the time of writing INTEGRAL is executing the AO-3 programme which will cover observations until mid August 2006.

As a return to those scientific collaborations and individual scientists who contributed to the development, design and procurement of INTEGRAL and who are represented in the INTEGRAL Science Working Team (ISWT), a portion of the total scientific observing time, the guaranteed time, is being used for their Core Programme (Winkler 2001) observations. Since the third year of operations, the share of guaranteed time is 25%. This split is agreed up to 2007, for 2008+ the share between guaranteed and open time is still to be discussed.

The Core Programm for the current AO-3 consists of several elements: scans of the Galactic Plane (2.1 Ms); surveys of the Galactic Center region (1.3 Ms), the Norma and the Scutum arm (0.8 and 1.4 Ms); a survey of an extragalactic field around 3C 273 (1 Ms); specific scans in the GC region from galactic latidue $b = -30$ to $b = +30$ in order to map the diffuse emission latidue profile (1.3 Ms) and 1.1 Ms for specific Target of Opportunity (ToO) follow-up observations.

Due to the highly variable nature of X-ray and gamma-ray sources, Target of Opportunity observations generally play an important role in INTEGRAL's observation programme. In many cases corresponding proposals have been submitted during the AO process and accepted by the TAC; these include cases of unknown and new sources. Some cases are also covered by the Core Programme as mentioned above. Finally it is always possible to request directly a ToO observation for an interesting astronomical object via the ISOC WWW pages (`http://integral.esac.esa.int/`). In all cases, the final decision to interrupt the current programme for the ToO lies with the Project Scientist. Until August 2005 there have been 23 ToO targets with ∼5 Ms total observing time.

Gamma-ray bursts falling into the FOV of INTEGRAL are considered a special class of ToO objects. While no dedicated follow-up observations are done, succesful proposers can obtain data rights to the relevant INTEGRAL data around the burst for their scientific purposes.

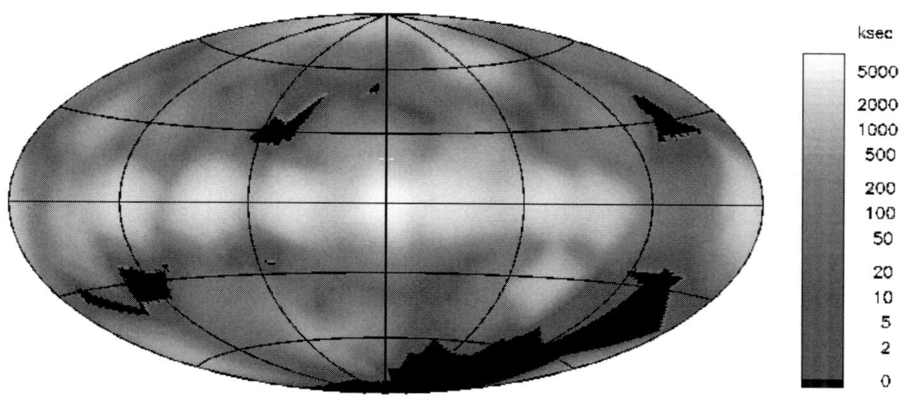

Figure 2. INTEGRAL exposure map in galactic coordinates after 1000 days in orbit (July 13, 2005) based on the IBIS partially coded FOV. Courtesy of E. Kuulkers (ESA-ESAC).

6. Gamma-Ray Burst detection

INTEGRAL has no on-board GRB detection capabilities, but still manages to detect bursts and disseminate positions with arcminute precision within less then a minute of the start of the burst. This is achieved via the INTEGRAL Burst Alert System (IBAS) running at the ISDC and analyzing the real-time data stream from the satellite, which usually arrives within seconds (Mereghetti & Gotz 2005).

Since its activation in December 2002 up to July 2005 IBAS has detected 30 GRBs within the FOV, corresponding to $\sim$1 burst per month, consistent to pre-launch estimates. 21 of these alerts have been disseminated rapidly (tens of seconds), the remaining are mostly from the early mission. The standard position uncertainty is 3' (90% error radius). The bursts detected by INTEGRAL are among the faintest for which good localization exists (see Fig. 3). In addition to the bursts in the FOV, 1–2 bursts per day are registered by the SPI anticoincidence shield. Lightcurves for ACS are also available at the ISDC and a catalogue has been published by Rau *et al.* 2005.

Since end 2004, IBAS also delivers alerts for bursts from known X-ray bursters, Soft Gamma-ray Repeaters and Anomalous X-ray Pulsars.

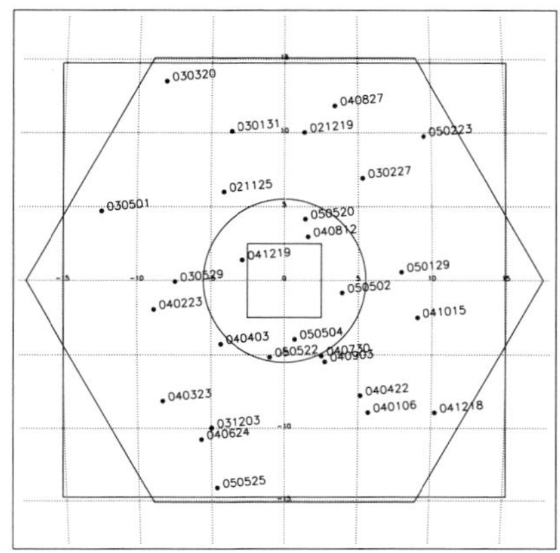
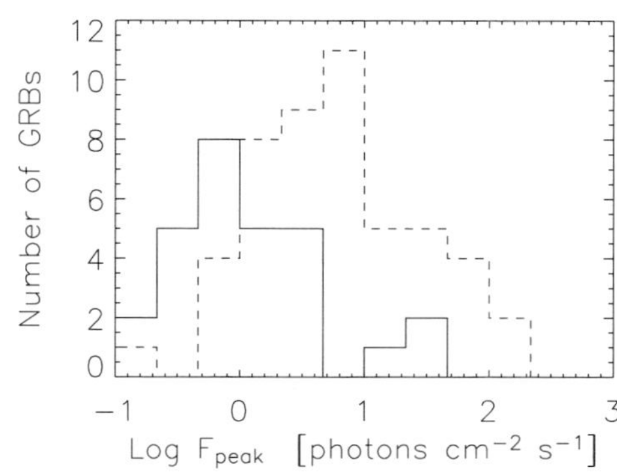

Figure 3. Left: positions of the GRBs localized with IBAS within the FOV of the INTEGRAL instruments: IBIS (large square), SPI (hexagon), JEM-X (circle) and OMC (small square). Right: Distribution of the peak fluxes for the GRBs detected with IBAS (solid line) and BeppoSAX (dashed). Both figures taken from Mereghetti & Gotz (2005).

7. Public data archives

The main archive for INTEGRAL data is located at the ISDC. A copy of the public data is available via the ISOC Science Data Archive using a different user interface similar to that used for other ESA missions.

The ISDC also provides access to summary scientific results with maps, lightcurves and spectra for ~180 sources based on the public data. In addition, the INTEGRAL Bright Source Catalog is maintained as collaboration between HEASARC and the ISDC. Optical lightcurves from the OMC can also be obtained via a specialized data archive at LAEFF/INTA.

PI led programs include, e.g., the Galactic Bulge Monitoring Program (Kuulkers *et al.* 2005) with quick-look results in near real time or the monitoring program on accreting X-ray pulsars (Sidoli *et al.* 2004).

For links to these services and more see: `http://integral.esac.esa.int/services.html`

8. Conclusions

Due to its sensitivity in the hard X-ray and gamma-ray bands, INTEGRAL's view of the universe is not affected by the interstellar absorption that can shroud sources from X-ray satellites. This ability has already led to, e.g., the detection of a new class of heavily absorbed sources and to new insights on the high energy emission for source classes like AXP's and accreting ms pulsars.

In comparison with previous gamma-ray missions, the high imaging resolution allows to disentangle close sources and avoid source confusion. At the same time, a wide field of view gives good coverage of areas not directly targeted leading to a large number of serendipitous detections.

These abilities are brought to good use in several survey programs both in the Core Programme and in the Open Time, besides the various studies of individual sources.

INTEGRAL's unique capaibilities concerning observing diffuse gamma-ray line emission have resulted in many important observations including the first 511 keV all-sky map (Knoedlseder *et al.* 2005).

In summary, the mission performs well and produces the scientific results it was built to do, in the important window of the hard X-ray to gamma-ray range. The prospects for more interesting results in the future are promising.

References

Jensen, P. L., *et al.* 2003, *A&A* 411, L7

Knoedlseder, J., *et al.* 2005, *A&A* 441, 513, 2005

Kuulkers, E., *et al.* 2005, *ATel*, 438

Lund, N., *et al.* 2003, *A&A* 411, L231

Mas-Hesse, J. M., *et al.* 2003, *A&A* 411, L261

Mereghetti, S. & Gotz, D. 2005, *astro-ph/0506198*

Rau, A., *et al.* 2005, *A&A*, 438, 1175

Sidoli, L., *et al.* 2004, *NuPhS*, 132, 648

Ubertini, P., *et al.* 2003, *A&A* 411, L131

Vedrenne, G., *et al.* 2003, *A&A* 411, L63

Winkler, C. 2001, *ESA SP-459: Exploring the Gamma-Ray Universe*, 471

Winkler, C., *et al.* 2003, *A&A* 411, L1

Discussion

GARCIA: 1. Why is only 1 X-ray telescope run, when there are 2?
2. How fast are ToOs done?

KRETSCHMAR: 1. To save the 2nd XRT for future use. There were instrumental problems at activation.
2. In $\sim 1 - 2$ days.

LIPUNOV: 1. Did you include in statistic of the detection GRB alerts from Soft Gamma Repeaters?
2. How much GRB alerts come from low galactic latitude

KRETSCHMAR: 1. No.
2. You can find this information on the INTEGRAL website.

Populations of High Energy Sources in Galaxies
Proceedings IAU Symposium No. 230, 2005
E. J. A. Meurs & G. Fabbiano, eds.

© 2006 International Astronomical Union
doi:10.1017/S1743921306007885

Cataclysmic variables and related objects with INTEGRAL†

Vojtěch Šimon[1], René Hudec[1], Filip Munz[1] and Jan Štrobl[1]

[1]Astronomical Institute, Academy of Sciences of the Czech Republic, 251 65 Ondřejov, Czech Republic

Abstract. We present selected results of observations and monitoring of cataclysmic variables (CVs) and related objects by the instruments IBIS and OMC onboard *INTEGRAL*. We analyse the behaviour of the intermediate polar (IP) V1223 Sgr in a state which we call a shallow low state. We present far X-ray spectra (E up to 60 keV), and the optical and far X-ray orbital modulation. The relation between far X-ray and optical flux appears to be stable over about 400 days. We also present far X-ray observations of some other CVs (or at least upper limits of their far X-ray fluxes) and show that the systems with magnetized white dwarf (like polars and IPs) are observable with *INTEGRAL*. The OMC light curves show that an outburst and a shallow low state of IX Vel are caused by the mass transfer variations, and not by the thermal instability of the disk, and that the amplitude of the flickering decreases with the decreasing intensity of the "stable" component.

Keywords. accretion, accretion disks, magnetic fields, binaries (including multiple): close, circumstellar matter, stars: dwarf novae, novae, cataclysmic variables, X-rays: binaries.

---

1. Introduction

Cataclysmic variables (CVs) are close binary systems in which mass accretion onto a white dwarf (WD) occurs. The observed X-ray and optical activity, luminosity and spectrum of CVs crucially depend on the mass accretion rate $\dot{m}$ and the strength of the magnetic field (MF) of the WD. This strength generally increases from non-magnetic CVs through intermediate polars (IPs) to polars. Bremsstrahlung is the dominant process for the X-ray radiation of CVs (see Warner (1995) for a review). The temperature (kT) of the X-ray spectrum largely separates the populations of magnetic and non-magnetic CVs (Fig. 1). IPs appear to have systematically harder spectra and their kT largely differs from system to system in comparison with non-magnetic CVs.

Symbiotic systems are a heterogeneous group and are the long-period cousins of CVs and X-ray binaries (e.g. Mikolajewska & Kenyon (1992)).

IBIS, SPI, JEM-X and OMC onboard *ESA INTEGRAL* (Winkler *et al.* (2003)) enable us to obtain simultaneous information in the optical, medium X-ray, far X-ray, and gamma-ray spectral region (or at least a suitable upper limit) for each CV in each scan or field. The CV analyses represent a part of the *INTEGRAL* Core Programme (CP, topic 5.5, responsible scientist R. Hudec).

† Based on observations with *INTEGRAL*, an ESA project with instruments and science data centre funded by ESA member states (especially the PI countries: Denmark, France, Germany, Italy, Switzerland, Spain), Czech Republic and Poland, and with the participation of Russia and the USA.

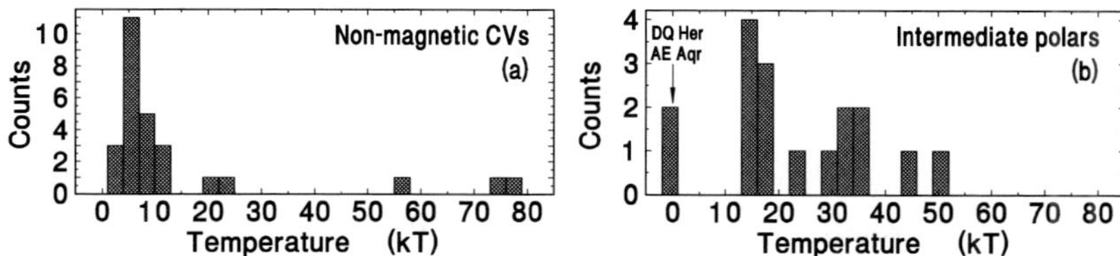

Figure 1. The statistical distributions of the temperatures (kT) in non-magnetic CVs and IPs. The data come from various published sources – mostly ASCA measurements in the 0.8–10 keV passband (e.g. Baskill *et al.* (2005), Hellier *et al.* (1996)).

2. Data source, analysis, results

The observations by *INTEGRAL* are carried out in the form of so-called science windows (ScW) which can be co-added to form a so-called mosaic. We used OSA software ver. 4.2. The fluxes were extracted by 2-D Gaussian fit using `mosaic_spec` (part of the new OSA release). For the spectral analysis, XSPEC ver. 11.3.0 and the recent response matrices (`rmf_0014`, `rmf_0007`) were employed. The results are summarized in Table 1 (the types of CVs were mostly taken from Downes *et al.* (2001)). The OMC image consists of $\sim$100 subwindows, each of them containing an object from a catalogue. It enables a precise photometry and provides us with the information on the optical brightness simultaneous to the far X-ray and gamma-ray band.

1223 Sgr: *INTEGRAL* caught this IP in a state of brightness lower than the average, which we call a shallow low state ($V \approx 13.4$–13.5). This is just the magnitude from which the episodes of deep low states occur (Garnavich & Szkody (1988)). We have a unique opportunity to investigate the relation between the activity in the optical and far X-ray region (Fig. 2a). In this state, when $\dot{m}$ is $\sim$3 times lower in comparison with the high state, the profile of the optical orbital modulation is still in good agreement with that in the high state (Jablonski & Steiner (1987)) (Fig. 2b). A period search in the residuals using the method of Scargle (1982) (code AVE) showed that the beat period (Steiner *et al.* (1981)) is still dominant over the rotational period of the WD. The orbital modulation in the 15–40 keV passband is flat, without statistically significant dips – this suggests that the blobs of overflowing matter have $N_{\rm H} < 10^{23}$ atoms cm^{-2}, taking the sizes of the error bars into account (Fig. 2c). It also implies that the observable emission region does not vary through the orbital cycle. The relation between the optical and X-ray flux was stable in all three observations over an interval of $\sim$400 days.

GK Per: IBIS observed this IP in quiescence between dwarf nova-type outbursts. A comparison of the measured flux with the synthetic quiescent bremsstrahlung spectrum

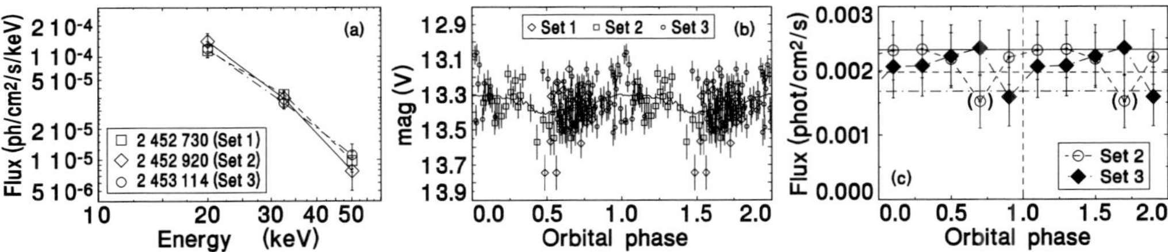

Figure 2. **(a)** Far X-ray spectra of V1223 Sgr from IBIS. The Julian Date of each spectrum is listed. **(b)** The OMC data folded with the orbital period $P_{\rm orb}$ (only images with 100 sec exp. time). The solid line represents the fit by the moving averages with the semi-interval of 0.15 phase. **(c)** IBIS 15–40 keV data binned with $P_{\rm orb}$. Several values of $N_{\rm H}$ are shown as the horizontal lines for illustration (from top to bottom: 0 atoms cm^{-2}, 5×10^{23} atoms cm^{-2}, 10^{24} atoms cm^{-2}).

Table 1. The intensities of CVs from the IBIS *INTEGRAL* observations. T gives the type (IP – intermediate polar, P – polar, DN – dwarf nova). S refers to the state of activity (SLS – shallow low state, HS – high state, Q – quiescence). Magnitudes from OMC and from AFOEV are abbreviated as o and a, respectively. The luminosities L assume the distances given by Beuermann *et al.* (2004), Watson *et al.* (1995) and Duerbeck (1981).

CV	T	S	JD (average)	Integr. t [sec]	Flux [10^{-4} photon/cm^2/s]	L [erg/s]	mean mag
V1223 Sgr	IP	SLS	2452730	66700	14.0 ± 2.8 (15–25 keV)	1.39×10^{33}	13.44 o
					5.9 ± 0.9 (25–40 keV)	9.5×10^{32}	
		SLS	2452920	105000	11.5 ± 1.7 (15–25 keV)	1.14×10^{33}	13.36 o
					6.5 ± 0.6 (25–40 keV)	1.05×10^{33}	
		SLS	2453114	96000	11.9 ± 2.1 (15–25 keV)	1.18×10^{33}	13.32 o
					5.3 ± 0.7 (25–40 keV)	8.5×10^{32}	
V2400 Oph	IP	HS	2452920	53760	9.4 ± 1.1 (15–40 keV)		14.6–14.9 o
V1432 Aql	P	HS	2452756	37160	8.8 ± 0.9 (15–40 keV)	1.4×10^{32}	14.6 o
GK Per	IP	Q	2452848	78980	2.7 ± 1.2 (25–40 keV)	4.6×10^{32}	13.1 a

Upper limits of fluxes of non-detected CVs [10^{-4} photon/cm^2/s] (13–40 keV): V603 Aql (Type: NL) Flux=1.7 (integr.time 425000 sec); V1500 Cyg (P) Flux=1.5 (390000 sec); EM Cyg (DN) Flux=3.6 (81400 sec); EY Cyg (DN) Flux=2.8 (262000 sec); V426 Oph (DN/IP?) Flux=3.6 (24600 sec); QQ Vul (P) Flux=4.3 (24300 sec); IX Vel (NL) Flux=1.6 (511000 sec).

with the parameters from Ishida *et al.* (1992) shows a relatively plausible agreement. Our observations were obtained during the time interval between the neighbouring outbursts $\Delta t = 973$ days ($\sim$42 percent of this interval, measured since the previous outburst). The measurements by Ishida *et al.* (1992) were obtained during $\Delta t = 983$ days ($\sim$29 percent of this interval). This can suggest that the amount of matter arriving to the WD and the parameters of the X-ray emitting region on the WD remained almost the same during these phases of quiescence.

IX Vel: OMC covered two important events – an outburst and a shallow low state (Fig. 3a). The latter set is fitted by the code HEC13, written by Dr. Harmanec and based on the method of Vondrák (1969) and Vondrák (1977). The time scales of the decaying and rising branches of both events can be taken as comparable.

We can determine the mechanism governing this activity. The nature of variations in nova-like CVs is controversial and the thermal instability of the outer disk region is supposed also to play a role (Honeycutt (2001)). Decay rate of the decaying branch is an important parameter in this regard. The disks of dwarf novae (DNe) evolve on the thermal time scale during the final decaying branches of outbursts, which is given by the propagation of the cooling front – the magnitude of a DN decays with τ_D here (Warner (1995); Fig. 3b). We also include τ_P of the viscous plateau of outbursts of DNe, i.e. near the top of the outburst (AFOEV data). IX Vel lies far above the $\tau_D - P_{orb}$ relation and its disk appears to evolve on the viscous time scale, comparable to that of DNe of a similar P_{orb}. The disk in IX Vel is thus fully ionized. Given the similar slope of the decaying and rising branch of the low state in IX Vel, we interpret this event in terms of the fluctuations of $\dot{m}$ from the donor star and to the response of the disk on the viscous time scale. The "quiescent" level before and after the outburst is the same as the bottom of the low state, which suggests that mass transfer variations can explain both events. We also find that the amplitude of the flickering decreases with the decreasing intensity of the "stable" component during the low state.

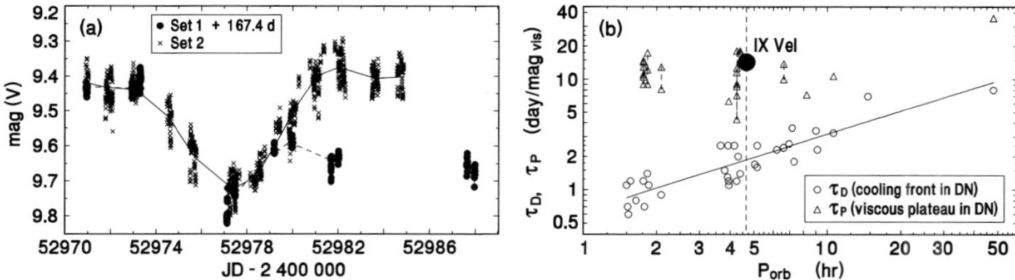

Figure 3. (a) Superposition of an outburst (duration < 14 days) and a short episode of a low state in IX Vel. The first packet of data preceding the outburst comes from JD 2452796 (out of the scale). **(b)** Decay rates of the decaying branches of outbursts of DNe as a function of P_{orb}. The position of IX Vel is marked.

V2116 Oph, CI Cyg and RS Oph: IBIS *INTEGRAL* proved to be suitable for the detection of the symbiotics with the hardest X-ray spectra, according to the classification by Mürset *et al.* (1996); V2116 Oph can serve as an example. OMC observations caught the symbiotic CI Cyg in quiescence ($V \approx 11$) and out of eclipse. We find the brightness to be stable at $V \approx 11$ on the time scale of days, with no flickering, in variance with RS Oph (Šimon *et al.* (2004)).

Acknowledgements

Our study was supported by the project ESA PRODEX INTEGRAL 14527 and partly by the grant 205/05/2167 of the Grant Agency of the Czech Republic. We have made use of data from the AFOEV database, Strasbourg, France. We thank F. Hroch for the preparation of a mosaic of V2400 Oph and P. Sobotka for a search in the OMC database. The code AVE developed by the GEA society is available at URL: `www.astrogea.org/soft/ave/aveint.htm`. We acknowledge using the code HEC13 written by Dr. P. Harmanec.

References

Baskill, D.S., Wheatley, P.J., & Osborne, J.P. 2005, *MNRAS* 357, 626
Beuermann, K., Harrison, Th.E., McArthur, B.E., *et al.* 2004, *A&A* 419, 291
Downes, R.A., Webbink, R.F., Shara, M.M., *et al.* 2001, *PASP* 113, 764
Duerbeck, H.W. 1981, *PASP* 93, 165
Garnavich, P. & Szkody, P. 1988, *PASP* 100, 1522
Hellier, C., Mukai, K., Ishida, M., & Fujimoto, R. 1996, *MNRAS* 280, 877
Honeycutt, R.K. 2001, *PASP* 113, 473
Ishida, M., Sakao, T., Makishima, K., *et al.* 1992, *MNRAS* 254, 647
Jablonski, F. & Steiner, J.E. 1987, *ApJ* 323, 672
Mikolajewska, J. & Kenyon, S. 1992, *MNRAS* 256, 177
Mürset, U., Jordan, S., & Wolff, B. 1996, *Supersoft X-Ray Sources, LNP* Vol. 472, p. 251
Scargle, J.D. 1982, *ApJ* 263, 835
Steiner, J.E., Schwartz, D.A., Jablonski, F.J., *et al.* 1981, *ApJ* 249, L21
Šimon, V., Hudec, R., & Hroch, F. 2004, *IBVS* 5562, 1
Vondrák, J. 1969, *BAIC* 20, 349
Vondrák, J. 1977, *BAIC* 28, 84
Warner, B. 1995, *Cataclysmic Variable Stars* Cambridge Univ. Press
Watson, M.G., Rosen, S.R., O'Donoghue, D., *et al.* 1995, *MNRAS* 273, 681
Winkler, C., Courvoisier, T.J.-L., Di Cocco, G., *et al.* 2003, *A&A* 411, L1

Populations of High Energy Sources in Galaxies
Proceedings IAU Symposium No. 230, 2005
E. J. A. Meurs & G. Fabbiano, eds.

© 2006 International Astronomical Union
doi:10.1017/S1743921306007897

Scientific results with INTEGRAL

Th. Courvoisier

ISDC Science Data Centre, Versoix, Switzerland
thierry.courvoisier@obs.unige.ch

Abstract.
Instruments will be briefly introduced and the main mission results to date will be described.

Discussion

MIRABEL: Has any spatial asymmetry or time variability been detected in the 511 keV emmission from the Galactic Centre region?

COURVOISIER: No, the emission is symmetrical (in the limits of the statistics) and has not been seen to be variable. The latter is expected from an extended source.

ELVIS: Is there any clue as to where the heavy obscuration toward the "Type 2" X-ray binaries lies?

COURVOISIER: Yes, the material hiding the X-ray source is local to the emitting system and such that the companion is not obscured.

Populations of High Energy Sources in Galaxies
Proceedings IAU Symposium No. 230, 2005
E. J. A. Meurs & G. Fabbiano, eds.

© 2006 International Astronomical Union
doi:10.1017/S1743921306007903

The Swift Mission

Alberto Moretti[1,2]

[1] INAF–Osservatorio Astronomico di Brera, via E. Bianchi, 23807 Merate(LC), Italy
[2] On behalf of the *Swift* Team

Abstract. The *Swift* Gamma-ray Burst Explorer mission, launched on 2004 November 20, is a multiwavelength observatory for gamma-ray burst (GRB) astronomy. The satellite carries three instruments: a new-generation wide-field gamma-ray (15–150 keV) detector that detects bursts, calculates 1–4 arcmin positions, and triggers autonomous spacecraft slews; a narrow-field X-ray telescope that gives 5 arcsec positions and performs spectroscopy in the 0.2 to 10 keV band; and a narrow-field UV/optical telescope that operates in the 170–600 nm band and provides 0.3 arcsec positions and optical finding charts. In the first 8 months of the mission (until the end of July 2005), *Swift* detected 54 GRBs performing detailed X-ray and UV/optical afterglow observations spanning timescales from 1 minute to several days after the burst. *Swift* has already collected a rich trove of early X-ray afterglow data and some interesting features are emerging. In particular early afterglow signatures reveal valuable and unprecedented information about GRBs, including the prompt emission – afterglow transition, GRB emission site, central engine activity, forward-reverse shock physics, and the GRB immediate environment.

Keywords. Swift satellite, Gamma Ray Burst.

1. Introduction

Gamma-ray bursts (GRBs) were discovered in the late 1960s in data from the Vela satellites. For over 25 years following their discovery GRBs remained one of the greatest mysteries in astrophysics, largely due to a frustrating inability to pinpoint their locations on the sky and to detect any counterparts at longer wavelengths. This situation changed dramatically in February 1997, when the *Beppo-SAX* satellite discovered the first X-ray afterglow of a GRB, confirming the teoretical predictions that X-ray, optical, and radio afterglows are produced by the expanding relativistic fireball that produces the burst itself. For long bursts, the discovery by BeppoSAX and ground-based observers of X-ray through radio afterglow allowed redshifts to be measured and host galaxies to be found, proving a cosmological origin. GRB correspond to a huge instantaneous energy release of 10^{51}–10^{52} ergs and are probably related to black hole formation, possibly related to endpoints of stellar evolution; they also represent bright beacons from the high redshift universe. By the end of 2004 *Beppo-SAX*, *RXTE*, *ASCA*, *Chandra*, and *XMM-Newton* had found 55 X-ray afterglows. The mission planning requirements of these satellites have typically led to delays of 6–24 hours between the burst and the afterglow detection, during which time the afterglow emission fades by many orders of magnitude. *Swift* (Gehrels *et al.* 2004) was designed to fill in the gap by making very early multiple wavebands observations of afterglows, beginning approximately a minute after the discovery of a GRB. Since afterglows fade quickly, typically as t^{-1} or t^{-2}, Swift's rapid $\sim$1 minute response allows observations when the emissions are orders of magnitude brighter than the previous few-hour response capabilities.

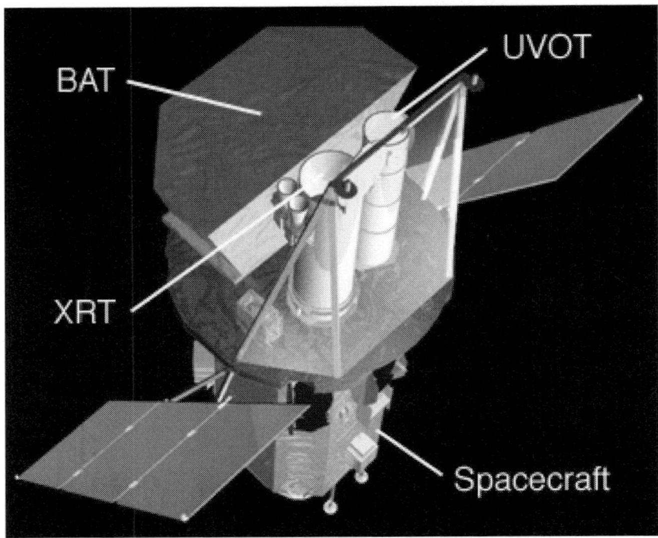

Figure 1. The *Swift* satellite.

2. Swift Mission

Swift's science payload consists of three instruments mounted onto an optical bench. These instruments, the Burst Alert Telescope (BAT), X-Ray Telescope (XRT) and Ultra-Violet/Optical Telescope (UVOT), are shown on the spacecraft in Figure 1. The launch was on a Delta 7320-10 on 2004 November 20 to a 22 degrees inclination, 600-km altitude orbit. Swift has a nominal lifetime of 2 years with a goal of 5 years and an orbital lifetime of ∼8 years. Normal data are downlinked in several passes each day over the Italian Space Agency (ASI) ground station at Malindi, Kenya. Tracking and Data Relay Satellite System (TDRSS) are used to send burst alert messages to the ground. Similarly, information about bursts observed by other spacecraft are uplinked through TDRSS for evaluation by Swift's on-board figure-of-merit (FoM) software. There is no on-board propulsion system. Pointing is provided by momentum wheels with momentum unloaded by magnetic torquers. Pointing knowledge is through the gyroscopes, star trackers, sun sensors, and magnetometers. The observation strategy is for the XRT and UVOT to be almost constantly observing positions of bursts previously detected by the BAT. The most recently detected burst have priority (although this can be adjusted as science dictates). Typically 5 sources are observed each orbit. When a burst is detected, an automated series of XRT and UVOT observations are performed lasting 20,000 s in exposure. Following that time, additional observations are scheduled via ground planning.

2.1. *Burst Alert Telescope*

The Burst Alert Telescope (BAT) is a highly sensitive, large FOV instrument designed to provide critical GRB triggers and 4-arcmin positions. It is a coded-mask instrument with a 1.4 steradian field-of-view (half coded). The energy range is 15–150 keV for imaging with a non-coded response up to 500 keV. Within the first ∼10 seconds of detecting a burst, the BAT calculates an initial position, decides whether the burst merits a spacecraft slew and, if worthy, sends the position to the spacecraft. Further information on the XRT is given by Barthelmy *et al.* (2004). Since the BAT coded FOV always includes the XRT and UVOT fields-of-view, long duration gamma-ray emission from the burst can be studied simultaneously with the X-ray and UV/optical emission. When a burst is detected, the

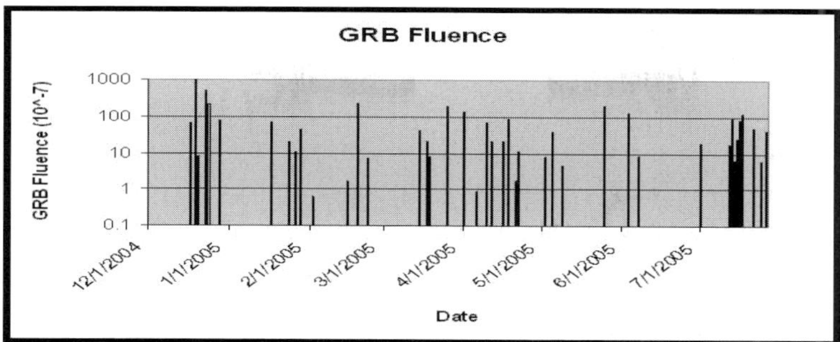

Figure 2. The temporal distribution of the GRB detected by *Swift*, with an indication of the flux measurements.

sky location and intensity are immediately sent to the ground and distributed to the community through the Gamma-Ray Burst Coordinates Network (GCN). The data from the BAT are also producing a sensitive hard X-ray all-sky survey over the course of Swift's two year mission.

2.2. *X-Ray Telescope*

Swift's X-Ray Telescope (XRT) is a focusing X-ray telescope with a 110 cm^2 effective area (at 1.5 keV), 23 arcmin FOV, 18 arcsec resolution (half-power diameter at 1.5 keV), and 0.2–10 keV energy range. XRT is designed to measure the fluxes, spectra, and lightcurves of GRBs and afterglows over a wide dynamic range covering more than 7 orders of magnitude in flux. The XRT pinpoints GRBs to 5-arcsec accuracy within 10 seconds of target acquisition for a typical GRB and studies the X-ray counterparts of GRBs beginning 50–200 seconds from burst discovery and continuing for days to weeks. Further information on the XRT is given by Burrows *et al.* (2004).

2.3. *Ultra-Violet/Optical Telescope*

The Ultra-Violet/Optical Telescope (UVOT) design is based on the Optical Monitor (OM) on-board ESA's XMM-Newton mission. UVOT is co-aligned with the XRT and carries an 11-position filter wheel, which allows low-resolution grism spectra of bright GRBs, and broadband UV/visible photometry. There is also a 4× field expander (magnifier) that delivers diffraction limited sampling of the central portion of the telescope FOV. Photons register on the microchannel plate intensified CCD (MIC). Further information on the UVOT is given by Roming *et al.* (2004).

3. Statistics in the first 8 months of the mission

Swift was successfully launched on 2004 November 20. The BAT first light was on the 3rd of December 2004, XRT first light was on the 11th of December 2004 and UVOT first light was on the 12th of January for UVOT. The first GRB was detected by BAT on the 17th of December 2004, while the first X-ray afterglow was detected by XRT on the 23rd of the same month. The first optical afterglow was detected by UVOT on the 15th of March 2005. The verification and calibration phase has ended on the 5th of April 2005 and since then all the data from the 3 instruments are publicly available together with the reduction and analysis software (http://swift.gsfc.nasa.gov/sdc/). In the first 8 months of the mission (until the end of July 2005), BAT detected 54 bursts (Fig. 2). Among them 3 were classified as short bursts, 7 as X-ray flash (XRF). The extrapolated detection number per year is 94±10. XRT detected 40 X-ray afterglows

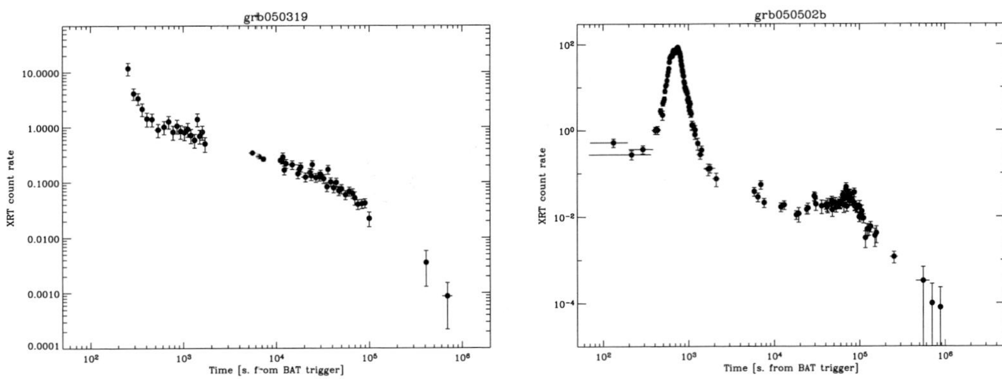

Figure 3. The lightcurve of the afterglows of the GRB 050319 (left, Cusumano *et al.* 2005) and 050502b (right, Falcone *et al.* 2005).

out of 41 pointings: 25 of them were observed within 300 seconds from the BAT trigger. In the same period UVOT detected 8 afterglows out of 37 pointings. For 21 bursts an optical counterpart was found by UVOT and/or optical ground telescopes. The average distance between the X-ray positions calculated from XRT data and the 21 afterglows with optical counterparts is 3.5 arcseconds. The typical distance between the BAT and XRT position is 100", much better than the expectations. For 12 bursts of this sample was possible to measure the redshift: the average redshift of this subsample is $\sim$2 (the average redshift of Beppo-SAX 55 bursts sample was $\sim$1) with 4.3 as maximum measured redshift.

4. Preliminary striking results: the X-ray afterglow light curves

Before *Swift*, X-ray afterglow emission was detected, in most cases, only several hours after the burst, by which time the flux typically showed a smooth single power law decay $\sim t^{-1}$. However, the *early* afterglow evolution – the first few hours, which can probe important questions such as the density profile of the external medium and the early radiative energy losses from the external shock, remained largely unexplored. From the observation of the first months of the *Swift* mission, a general trend starts to emerge that may become the standard to describe each GRB X-ray afterglow light curve (Chincarini *et al.* 2005, Nousek *et al.* 2005). Starting at the earliest XRT observations (approximately 10^2 s after the prompt gamma-rays), the X-ray flux F_ν follows a canonical behavior comprising three power law segments where $F_\nu \propto \nu^{-\beta} t^{-\alpha}$ (see Fig. 3): an initial steep decay slope (α_1) (Tagliaferri *et al.* 2005, Perri *et al.* 2005), which (at $t_{\mathrm{break},1}$) changes into a very flat decay (α_2), that in turn (at $t_{\mathrm{break},2}$) transitions to a slightly steeper slope (α_3). The spectrum remained constant throughout the breaks (within our available statistics) in all cases except two (GRBs 050315 and 050319) where the spectrum hardened (i.e. β decreased) across the first break (at $t_{\mathrm{break},1}$). It should be noted that the values of α_3 are consistent with those seen in previous missions since they typically started observations hours after the burst (Campana *et al.* 2005, Moretti *et al.* 2005).

The most promising explanation for the initial fast flux decay (α_1) is that it is the tail of the prompt gamma-ray emission which is emitted from large angles ($\theta > \Gamma^{-1}$) relative to our line of sight. This model produces a sharp flux decay with $\alpha_1 = 2 + \beta_1$, in rough agreement with observations, while $\alpha_1 > 2 + \beta_1$ might also be expected for $t/T_{\mathrm{GRB}} \lesssim$ a few. The shallow intermediate flux decay (α_2) is most likely caused by

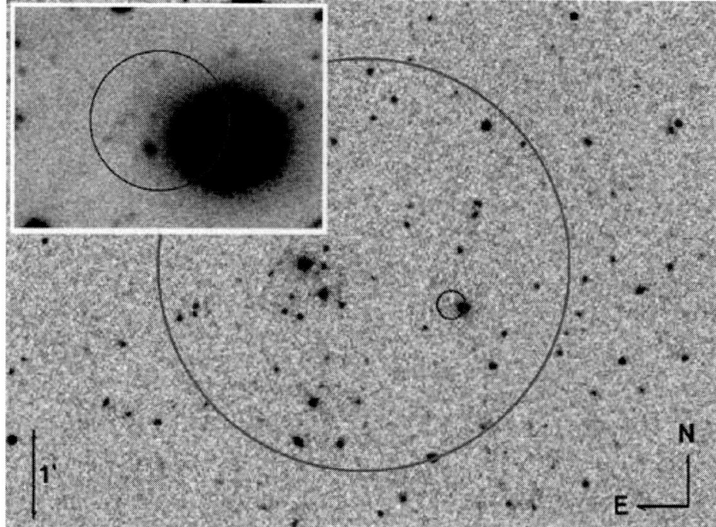

Figure 4. The XRT (small circle) and BAT (large circle) error circles of the short GRB 050509b overplotted on the optical image.

continuous energy injection into the forward shock. This energy injection is probably due to a decrease in the Lorentz factor, Γ, of the outflow toward the end of the prompt GRB, resulting in a monotonic increase of Γ with radius. This outflow gradually catches up with the afterglow shock, resulting in a smooth energy injection. The third power law segment of the light curve (α_3) is most likely the well known afterglow emission from a spherical adiabatic external shock. The observed values of the temporal index (α_3) and the spectral index (β_3) are consistent with this interpretation.

The early X-ray light curves obtained with *Swift* XRT often show flares (Burrows *et al.* 2005, Romano *et al.* 2005). The most prominent flare so far was in GRB 050502b (Fig. 3), where the flux increased by a factor of $\sim$500. Some of these flares have very sharp temporal features where the flux changes significantly on time scales $\Delta t \ll t$. Most flares have a very steep rise and decay (with very large temporal rise/decay indices when fitted to a power law). When the flare is bright enough to follow its spectral evolution, its hardness ratio evolves during the flare, and its spectral index is somewhat different than the one associated with the underlying power law decay of the X-ray light curve before and after the flare. Furthermore, the fluxes before and after the flare lie approximately on the same power law decay, suggesting that the flare originates from a distinct physical component than that responsible for the underlying power law decay. These flares are most likely caused by internal shocks within the outflow that is ejected from the central source at late times (very close to the time when these flares are seen). This implies the the central source quite often remains active for hours after the GRB.

5. The origin of the short GRBs

The duration of GRBs has long been known to show a bimodal distribution; moreover short bursts are predominantly harder than the longer variety. Most of the progress to date on understanding the origin of GRBs has been for long bursts. Before *Swift*, no short GRB had been accurately and rapidly located, although arcminute-sized error boxes had been searched unsuccessfully hours to days after the event. Theoretical predictions for the cause of short GRBs have included neutron star – neutron star (NS-NS) or NS – black hole

(BH) mergers, extragalactic soft gamma repeater giant flares, or collapsars/hypernovae with different explosion or beaming properties than those responsible for the long bursts.

Swift has provided the first accurate localization of a short GRB, the GRB 050509b (Gehrels *et al.* 2005). No optical afterglow was detected and Chandra did not detect the X-ray afterglow 2 days later. The center of the XRT error circle lies only 9.8" away from the center of the large E1 elliptical galaxy 2MASX J12361286+2858580 at a redshift of $z = 0.225$, which is located in the galaxy cluster NSC J123610+285901 (Fig. 4). The association with this galaxy seems unlikely to be coincidental. Infact, galaxies as bright as this are relatively rare; the comoving number density of galaxies at least this luminous is 5×10^{-5} Mpc^{-3}; the probability of lying within 10" of a random location at $z = 0.225$ is 10^{-4} . The likely association between GRB050509b and 2MASX J12361286+2858580 is difficult to understand if the GRB resulted from any mechanism involving massive stars and recent star formation, such as a massive-star collapsar or flares by a Soft Gamma Repeater (SGR) magnetar. Infact the galaxy type for the suggested host galaxy is very different than those found for long GRBs; their hosts are typically sub-luminous and blue and show strong emission lines associated with star formation. On the other hand, 2MASX J12361286+2858580 is a very propitious site for a NS-NS or NS-BH merger. In fact, as Chandra observations have shown, giant ellipticals, especially those dominant in their cluster, have very large populations of low mass X-ray binaries (LMXBs) containing accreting neutron stars and black holes. The X-ray afterglow of GRB050509b is the weakest in flux of any of the 15 GRBs that XRT has promptly (<5 min) observed. If the redshift is 0.225, then the afterglow is more than 100 times less luminous than that of typical long-burst afterglows. This might be a result of the lower burst energy; the isotropic energy is about 10^{-4} of that of typical long GRBs On the other hand, the weak afterglow is consistent with the lower density medium one would expect around an evolved compact binary progenitor. While it is difficult to draw conclusions based on a single object, the *Swift* localization of GRB050509b near a large elliptical galaxy in a cluster strongly suggests a binary merger origin of at least this one short GRB.

References

Barthelmy, S. D., 2004, SPIE, 5165

Burrows, D. N., *et al.*, 2004, SPIE, 5165

Burrows, D. N., *et al.*, 2005, Science, 309,1833

Campana, S., *et al.*, 2005, ApJ, 625, L23

Chincarini, G., *et al.*, 2005, ApJ, submitted, astro-ph/0506453

Cusumano, G., *et al.*, 2005, ApJ, submitted

Falcone, A., *et al.*, 2005, ApJ, submitted

Gehrels, N., *et al.*, 2004, ApJ, 611, 1005, astro-ph/0505630

Gehrels, N., *et al.*, 2005, Nature, in press

Moretti, A., *et al.*, 2005, A&A, submitted

Perri, M., *et al.*, 2005, A&A, in press, astro-ph/0509164

Nousek, J. A., *et al.*, submitted to ApJ (2005), astro-ph/0508332

Romano, P., *et al.*, 2005, A&A, submitted

Roming, P. W. A., *et al.*, 2004, SPIE, 5165

Roming, P. W. A., *et al.*, 2005, astro-ph/0509273

Tagliaferri, G., *et al.*, 2005, Nature, 436, 985

Discussion

VAN DEN HEUVEL: Could you tell us a little bit more about the associations between these 3 short bursts with elliptical galaxies, which you mentioned? How good are these associations?

MORETTI: Until the end of July we observed 3 short GRB afterglows, 2 detected by Swift (050509b,050724) and 1 detected by HETEII (050709); For 050709 and 050724 the detection of the optical transient and the confirmation of the X-ray afterglow by Chandra observation within a neraby galaxy make the association completely firm (note that the galaxy associated with 050709 is not an elliptical). For 050509b only Swift XRT detected a very faint X-ray afterglow; its position on the sky overlaps a nearby elliptical galaxy. In the error circle (radius 7") other objects are present; however the association of the short GRB with this galaxy seem to be highly probable.

MCBREEN: Do you have redshift measurements for the 3 short GRBs?

MORETTI: The host galaxies of 050509b, 050709, and 050724 are at $z = 0.225$, $z = 0.16$ and $z = 0.258$, respectively.

MEURS: Just to be certain about these short afterglows: These are all X-ray afterglows but no UVOT detections?

MORETTI: All the afterglow lightcurves I presented are X-ray afterglow light curves as measured by the Swift XRT.

Populations of High Energy Sources in Galaxies
Proceedings IAU Symposium No. 230, 2005
E. J. A. Meurs & G. Fabbiano, eds.

© 2006 International Astronomical Union
doi:10.1017/S1743921306007915

XMM-Newton observations of PSR B1259–63 near the 2004 periastron passage

M. Chernyakova[1,2] †, A. Lutovinov[3], J. Rodriguez[4,1] and S. Johnston[5]

[1]INTEGRAL Science Data Centre, Chemin d'Écogia 16, 1290 Versoix, Switzerland
email: Masha.Chernyakova@obs.unige.ch

[2]Geneva Observatory, 51 ch. des Maillettes, CH-1290 Sauverny, Switzerland

[3]Space Research Institute, 84/32 Profsoyuznaya Street, Moscow 117997, Russia

[4]CEA Saclay, DSM/DAPNIA/Service d'Astrophysique (CNRS UMR 7158 AIM), 91191 Gif sur Yvette, France

[5]Australia Telescope National Facility, CSIRO, P.O. Box 76, Epping, NSW 1710, Australia.

Abstract. We present here the results of new *XMM-Newton* observations of the PSR B1259–63 system during the beginning of 2004, as the pulsar approaches the disc of the Be star. We combine these results with earlier X-ray data from *Beppo*SAX, *XMM-Newton*, and *ASCA*. The X-ray light curve looks similar to the radio light curve with a rapid increase in the flux around the time of the disc crossing. This supports a model in which the X-ray emission from the system is due to inverse Compton scattering of the pulsar wind relativistic particles with moderate Lorentz factor $\gamma \sim 10$–100 on the Be star soft photons.

Keywords. pulsars : individual: PSR B1259–63 – X-rays: binaries – X-rays: individual: PSR B1259–63.

---

1. Introduction

PSR B1259–63 is a ~ 48 ms radio pulsar in a highly eccentric (e ~ 0.87), 3.4 year orbit with a Be star (SS 2883), located at a distance of about 2 kpc. Timing analysis of the PSR B1259–63 system shows that the disc of Be star is highly tilted with respect to the orbital plane and thus the pulsar crosses the Be star disc twice per orbit, just prior to and just after periastron (Wang *et al.* 2004). The properties of the radio emission from this system are very different before and after the periastron passage. Unpulsed radio and X-ray emission is observed from this system with a similar double peak lightcurve around the periastron, see details in Johnston *et al.* (2005), Hirayama *et al.* (1999).

2. *XMM-Newton* observations

X-ray data available up to 2004 are much more sparse than radio ones, and it was not clear whether similarly to radio data X-ray emission rapidly grows around the moment of the first disk crossing, or its behaviour is much smoother. To answer this question we have organized a set of *XMM-Newton* observations of the system, as the pulsar approaches the disk. Here we present the results of these observations along with the analysis of the previously unpublished *Beppo*SAX (1997) and *XMM-Newton* (2001–2003) observations. A simple power law with a photoelectrical absorption describes the data well. In Figure 1a the time variation of the X-ray luminosity (top panel), spectral slope (middle panel)

† on leave from Astro Space Center of the P.N. Lebedev Physical Institute, Moscow, Russia.

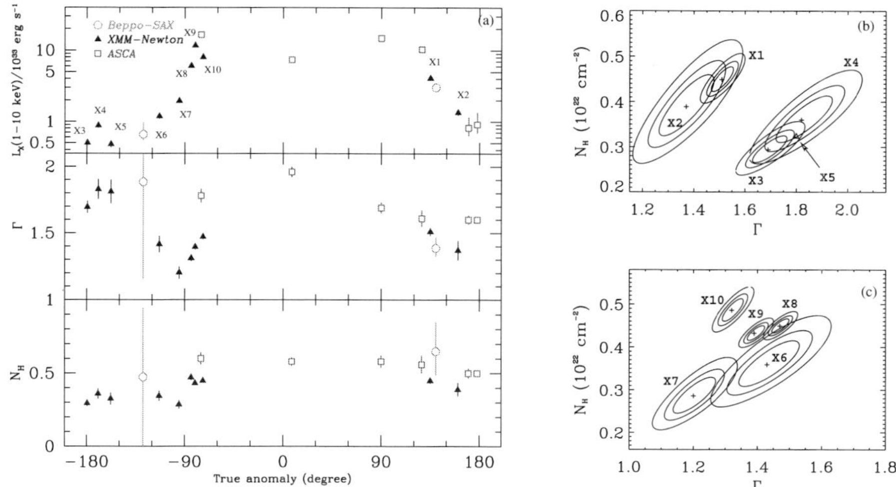

Figure 1. (*a*)Evolution of the PSR B1259–63 spectral parameters along the orbit; (*b,c*)Contour plots of N_H versus photon spectral index Γ.

and the column density (bottom panel) as a function of orbital phase are shown. In Figures 1b, 1c we show contour plots of N_H versus photon spectral index Γ, holding the 1–10 keV flux fixed at the values given in Figure 1a. The contours plotted are the 68%, 90%, and 99% confidence levels.

It is clearly seen that the behaviour of both L_X and N_H as a function of orbital phase is highly asymmetric, rising by a factor of several in five days at the time of the disc pre-periastron crossing. Following the empirical model of the Be-star equatorial disk (Waters *et al.* 1987) we found that the disk is dense enough to explain the observed extra absorption column gained within the system.

A sharp rise of the X-ray flux at the time of the disk crossing along with the increase of the column density strongly support the model of Chernyakova & Illarionov (1999, 2000), in which the observed X-ray emission is explained as an inverse Compton emission of the relativistic electrons of the pulsar wind (Lorentz factor $\gamma \sim 10$–100) on the Be star soft photons. Unpulsed radio emission in this model is due to the synchrotron emission of the relativistic particles. More details on the *XMM-Newton* 2004 observations can be found in Chernyakova *et al.* (2005).

References

Chernyakova, M. & Illarionov, A. 1999, *MNRAS* 304, 359

Chernyakova, M. & Illarionov, A. 2000, *ApSS* 274, 177

Chernyakova, M., Lutoviniv, A., Rodriguez, J., & Johnston, S. 2005, submitted to *MNRAS*.

Hirayama, M., Cominsky, L. R., Kaspi, V. M., Nagase, F., Tavani, M., Kawai, N., & Grove, J. E., 1999, *ApJ*, 521, 718

Johnston, S., Ball, L., Wang, N., & Manchester, R. N., 2005, *MNRAS*, 358, 1069

Wang, N., Johnston, S., & Manchester, R.N., 2004 *MNRAS*, 351, 599

Waters, L.B.F.M., Coté J., & Lamers, H.J.L.M., 1987, *A&A*, 185, 2006

Populations of High Energy Sources in Galaxies
Proceedings IAU Symposium No. 230, 2005
E. J. A. Meurs & G. Fabbiano, eds.

© 2006 International Astronomical Union
doi:10.1017/S1743921306007927

Optical observations of *IGR J00291+5934* in the post outburst phase

M. T. Reynolds[a], **P. Elebert**[a], **P. J. Callanan**[a], **B. Field**[a], **P. Tuite**[a], **M. A. P Torres**[b], **D. Steeghs**[b], **P. M. Garnavich**[c], **D. M. Terndrup**[d], **A. V. Filippenko**[e], **R. J. Foley**[e] & **E. T. Harlaftis**[f]

[a] Department of Physics, University College Cork, Ireland
email: m.reynolds@ucc.ie
[b] Center for Astrophysics, 60 Garden Street, Cambridge, MA 02138, USA
[c] Department of Physics, University of Notre Dame, Notre Dame, IN 46556-5670, USA
[d] Department of Astronomy, The Ohio State University, Columbus, OH 43210, USA
[e] Astronomy Department, University of California, Berkeley, California 94720, USA
[f] National Observatory of Athens, PO Box 20048, Athens 11810, Greece

Abstract. We present optical observations of the newly discovered accretion powered millisecond pulsar *IGR J00291+5934*, undertaken in the weeks following its outburst on 2nd December 2004. The decay to quiescence is seen to be highly variable with no indication of a modulation on the $\sim$2.46 hr orbital period apparent in the data, consistent with a system at low inclination. We also have a single *Keck LRIS* spectrum of the companion to *IGR J00291+5934* taken 10 days after outburst. Strong hydrogen and helium emission lines are observed confirming the identity of the counterpart.

Keywords. X-rays: binaries, pulsars: individual (IGR J00291+5934).

1. Introduction

IGR J00291+5934 was discovered by *INTEGRAL* on 2004 December 2*nd* (Shaw *et al.* 2005). The optical counterpart was identified two days later and was found to have a magnitude R $\approx$ 17.4 (Fox *et al.* 2004). It was quickly identified as the 6*th* member of the class of accretion powered millisecond X-ray pulsars to be discovered. It has the fastest spin period yet observed in this class $p_{spin} \sim$ 1.7ms and the second longest orbital period $\sim$2.46hr. We present here the first optical study of its outburst and decay.

2. Photometry

Two nights of R-band data were obtained on 2004 December 9*th* and 18*th* at the 1.3m MDM telescope at KittPeak and the 4m WIYN telescope respectively (see Fig. 1). We also obtained 2 nights of white light observations from the 1.2m telescope at the Kryoneri observatory in Greece. The resulting lightcurves were phased to the ephemeris of Markwardt *et al.* (2004). The lightcurves are observed to be highly variable with no significant modulation at the orbital period evident (Fig. 1). The observed fading is consistent with that observed by Bikmaev *et al.* (2005). The IR counterpart was first observed on December 8*th* (see Steeghs *et al.*, 2004), 6 days after the onset of the outburst.

3. Spectroscopy

In an effort to confirm the identification by Roelofs *et al.* (2004), a single 300s *LRIS* spectrum of the proposed optical counterpart was obtained 10 days post outburst (Filippenko *et al.*, 2004), Fig. 1. We observe broad (FWHM = 1200 km/s) emission lines of Hα 656 nm, Hβ 486 nm, and HeI 667.8 nm, as well as narrow (FWHM = 300 km/s),

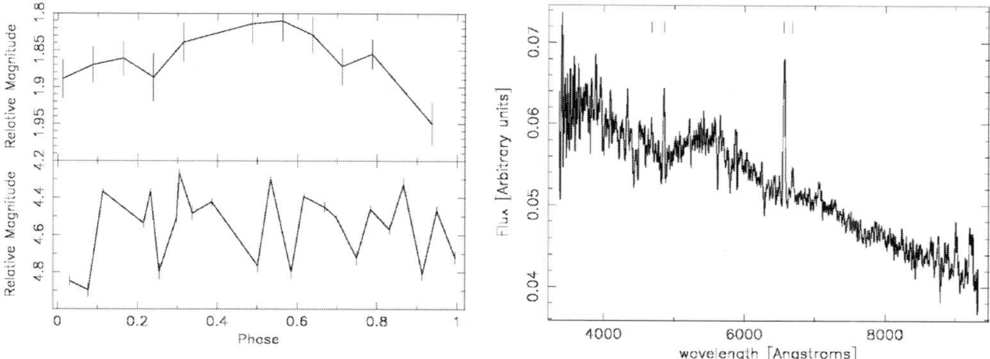

Figure 1. LEFT: The R-band light curve. The top panel is from 9.12.04 with the bottom panel obtained 9 days later. RIGHT: *Keck LRIS* spectrum of the optical counterpart taken on 12.12.04. The markers indicate, from left to right, the positions of the HeII 468.6 nm (EW = 0.06 nm), $H\beta$ 486 nm (EW = 0.54 nm), $H\alpha$ 656 nm (EW = 0.96 nm) & HeI lines 667.8 nm (EW = 0.1 nm).

very weak HeII 468.6 nm emission. The lines are superposed on a blue continuum. These are classic hallmarks of an X-ray transient in outburst and hence they firmly establish the counterpart of the accreting X-ray pulsar.

4. Discussion

IGR J00291+5934 is the 6*th* member of the class of accretion powered millisecond X-ray pulsars to be discovered and the third 'long' orbital period system. The decay of *IGR J00291+5934* is the fastest yet observed in this class of binary, with an e-folding time $\tau_e = 5.5$ days. In comparison, the e-folding time for *SAX J1808.4-3658* was $\sim$14 days. This may be due to the relatively small accretion disk present in these systems, because of (i) the short orbital period (and small system size), and (ii) the truncated inner disk due to the magnetosphere of the neutron star: this is similar to what happens during outbursts of intermediate polars (Angelini *et al.*, 1989). Recently Burderi *et al.* (2005) suggested that *IGR J00291+5934* could be a high inclination system; however the absence of any clear modulation in our data is inconsistent with this.

Acknowledgements

The data presented in this paper was obtained at the following observatories; *WIYN, MDM, Kryoneri, PAIRITEL & Keck I.*
MTR, PE & PJC wish to acknowledge financial support from Science Foundation Ireland. The authors wish to dedicate this paper to the memory of Emilios T. Harlaftis, who passed away in tragic circumstances earlier this year.

References

Angelini, L. & Verbunt, F., 1989, MNRAS, 238, 697

Bikmaev, I., Suleimanov, V., Galeev, A. & 7 co-authors, 2005, ATel, 395

Burderi, L., Di Salvo, T., Riggio, A. & 6 co-authors, astro-ph/0509224

Filippenko, A.V., Foley, R.J. & Callanan, P.J., 2004, ATel, 366

Fox, D.B. & Kulkarni, S.R., 2004, ATel, 354

Markwardt, C.B., Galloway, D.K., Chakrabarty, D. & 2 co-authors, 2004, ATel, 360

Roelofs, G., Jonker, J.G., Steeghs, D., Torres, M.A.P. & Nelemans, G., 2004, ATel, 356

Shaw, S.E., Mowlawi, N., Rodriguez, J., Ubertini, P. & 7 co-authors, astro-ph/0501507

Steeghs, D., Blake, C., Bloom, J.S., Torres, M.A.P., Jonker, P.G., & Starr, D., 2004, ATel, 363

Vanderspek, R., Morgan, E., Crew, G., Graziani, C., & Suzuki, M., 2005, ATel, 516

Populations of High Energy Sources in Galaxies
Proceedings IAU Symposium No. 230, 2005
E. J. A. Meurs & G. Fabbiano, eds.

© 2006 International Astronomical Union
doi:10.1017/S1743921306007939

Supporting the GLAST User Community

R. Corbet

X-ray Astrophysics Laboratory, NASA GSFC/USRA, Greenbelt, MD, USA
corbet@gsfc.nasa.gov

Abstract. The Gamma-ray Large Area Space Telescope (GLAST) is an international and multi-agency space mission that will study the cosmos in the energy range 10 keV – 300 GeV. The GLAST Science Support Center (GSSC) is the scientific community's interface with GLAST. The GSSC will provide data, analysis software and documentation. In addition, the GSSC will administer the guest investigator program for NASA HQ. Consequently, the GSSC will provide proposal preparation tools to assist proposers in assessing the feasibility of observing sources of interest.

Populations of High Energy Sources in Galaxies
Proceedings IAU Symposium No. 230, 2005
E. J. A. Meurs & G. Fabbiano, eds.

© 2006 International Astronomical Union
doi:10.1017/S1743921306007940

Polarimetry with SPI

D. R. Willis[1,2]†, D. J. Clark[3], R. Diehl[1], L. Hanlon[4], G. Kanbach[1], B. McBreen[4], S. McGlynn[4] and A. Strong[1]

[1]MPE, MPI, Garching, Munich, Germany

[2]ISDC, Chemin D'Ecogia 16, Versoix, CH-1290, Switzerland

[3]School of Physics and Astronomy, University of Southampton, SO17 1BJ, UK

[4]Space Science and Advanced Materials Lab, National University of Ireland, UCD, Dublin 4, Ireland

Abstract. Polarimetry in gamma-rays has the capability to enhance our understanding of compact object emission in our galaxy. In particular this diagnostic method could provide useful insight into the geometrical arrangement of these emitting objects and the roles that magnetic fields play in their emisson mechanisms. Gamma Ray Bursts have been studied in this way but the results, perhaps indicating a high degree of polarisation, remain unverified [Coburn & Boggs (2003), Wigger *et al.* (2004), Willis *et al.* (2005)]. The nature of GRBs solve many instrumental problems in polarimetry, however their true nature is less well defined and so a study of a better understood object such as the Crab Pulsar, for now, may reveal more as to the physics of the system.

Keywords. polarization, instrumentation: polarimeters, techniques: polarimetric, gamma rays: observations.

1. Introduction

Polar cap models for the Crab Pulsar give gamma-ray emission from pair-production cascades with either curvature radiation or inverse Compton scattered photons from primary particles accelerated in the polar caps of neutron stars, with magnetic photon splitting acting to attenuate in some models. All mechanisms apart form the pair-production exhibit charactericics of polarisation. **The outer gap** models involve particles accelerating in charge depletion layers further away from the magnetic poles. And give gamma-ray emission by synchrotron radiation or inverse Compton scattering. Again, clear polarisation signatures can be expected from both mechanisms. Phase resolved and multi-wavelength polarimetric measurements will distinguish between these models.

SPI has 19 hexagonal Germanium detectors, operating in the 20keV–8MeV band. By observing the angular distribution of the multiple events Compton scattered between the detectors and comparing this to the systematic angular distribution of these scattered events, a value for polarisation can be obtained.

2. Characterising the Scattering Response for Analysis

Simply histogramming up bins every 60 degrees for a staring observation has been unsuccessful in that any polarisation induced distribution is hidden by the systematic background. To search for this modulation both the background and the systematic structure needs to be removed. The chacterised multiple event response needs to be investigated. To do this a simple dual-detector model is constructed within the GEANT4

† Present address: ISDC, Chemin D'Ecogia 16, Versoix, CH-1290, Switzerland.

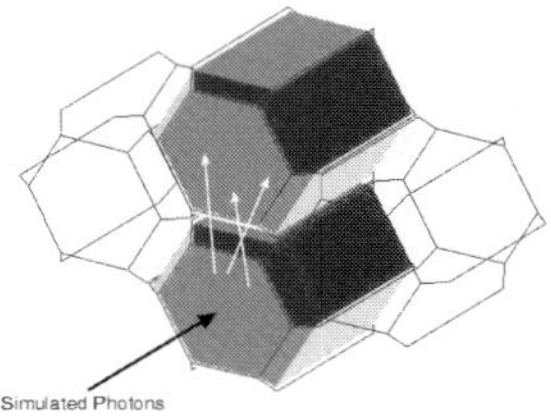

Figure 1. A simple GEANT model of two SPI germanium detector units with two passive units adjacent. Photons are fired into the lower units and the events selected on their interaction with the upper unit.

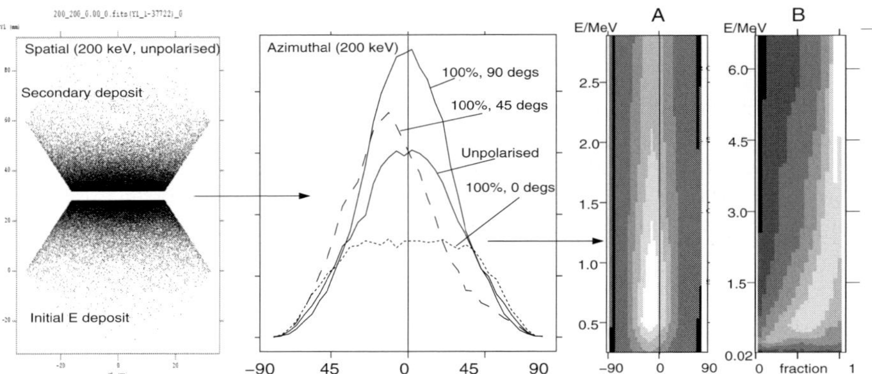

Figure 2. From the simple GEANT model each aspect to the scattering can be quantified. From the spatial distribution on the far left, to the polarisation dependent angular response, to a full energy matrix of the angular response (**A**). The energy fraction of the energy deposited in the inital detector is also characterised with energy into a matrix (**B**).

particle/photon simulation software (see Figure 1). Two adjacent detectors are also included as passive attenuators. Mono-energetic flux is fired into one of the detectors and events recorded when a multiple event occurs. Included in the model are simply the Germanium with their respective Aluminium casings. The spatial and so angular distributions can be studied, with respect to energy (see Figure 2).

Using this simple model, a 4-dimensional matrix can be built up dependent on energy, off-axis angle and the percentage and angle of polarisation. This matrix can be employed in one of two ways. Firstly, it can be used to filter all phase-resolved multiple events or used in pattern searching. Secondly, a full set of IRFs (SPI Instrument Response Functions) can be taken and altered for a hypothetical fully polarised flux at 0 and 90 degrees on the detector plane. These two IRFs can then be applied to SPI fitting software, such as SPIMODFIT. By fitting to two separate sources in identical positions, but each using a different polarised IRF, the ratio between the two will give both the degree and angle of polarisation.

References

Coburn, W. & Boggs, S E. 2003, *Nature* 423, 415
Wigger, C., *et al.* 2004, *ApJ* 613, 1088–1100
Willis, D. R., *et al.* 2005, *A&A* 439, 245–253
Smith, F. G., *et al.* 1988, *MNRAS* 233, 305–319

Populations of High Energy Sources in Galaxies
Proceedings IAU Symposium No. 230, 2005
E. J. A. Meurs & G. Fabbiano, eds.

© 2006 International Astronomical Union
doi:10.1017/S1743921306007952

Microquasars

F. Mirabel

European Southern Observatory, Santiago, Chile
fmirabel@eso.org

Abstract. I will review the discovery of microquasars and their importance in our understanding of High Energy Sources in Galaxies.

Discussion

BOSCH-RAMON: Comment: There is a recently uploaded paper to astro-ph by Casares, Ribó *et al.* pointing to LS5039 as harbouring a black-hole.

CLAVEL: Could Micro-QSOs be the sources of Ultra-high energy Cosmic Rays ($\geqslant 10^{19.5}$ eV), which we know cannot come from AGNs, since they interact with CMB photons on a length scale <20 Mpc?

MIRABEL: No model has been developed to produce 10^{20} eV particles by microquasars.

WARD: In terms of the prospects of testing the idea that some ULXs might be related to micro-quasars, what range in values of L_x/L_{radio} do the models allow?

MIRABEL: From the L_x/L_{radio} we expect radio fluxes below present detection limits. Microquasars could be detected but a lot of observing time would be required.

Populations of High Energy Sources in Galaxies
Proceedings IAU Symposium No. 230, 2005
E. J. A. Meurs & G. Fabbiano, eds.

© 2006 International Astronomical Union
doi:10.1017/S1743921306007964

Gamma-Ray Emission from Microquasars

M. M. Kaufman Bernadó[1,†] and G. E. Romero[1,⋆]

[1]Instituto Argentino de Radioastronomía, Villa Elisa, Argentina
†email: mmkb@fibertel.com.ar
⋆email: romero@iar.unlp.edu.ar

Abstract. We present models for gamma-ray production in microquasars and we propose them as possible parent populations for different groups of EGRET unidentified sources. These models are developed for a variety of scenarios taking into account several possible combinations, i.e. black holes or neutron stars as the compact object, low mass or high mass stellar companions, as well as leptonic or hadronic gamma-ray production processes.

Keywords. X-rays: binaries, radiation mechanisms: nonthermal, gamma rays: theory.

1. Introduction

It has been shown in Romero *et al.* (1999) that the distribution of the unidentified EGRET sources with galactic coordinates has a clear concentration of sources on the galactic plane plus a concentration in the general direction of the galactic center. This indicates a significant contribution from galactic sources. On the other hand, studying the galactic sources of the ASM catalog Grimm *et al.* (2002) found significant differences in the 3D spatial distribution of High-Mass X-Ray Binaries (HMXBs) and Low-Mass X-Ray Binaries (LMXBs) in our Galaxy. Whereas HXMBs are more concentrated towards the galactic plane with a vertical scale height of 150 pc, and clear indications of a distribution following the spiral structure, LMXBs have a strong tendency to concentrate towards the galactic bulge and their vertical distribution has a scale height of 410 pc.

2. High-Mass Microquasars – The Galactic Plane Population

2.1. *Leptonic Model: EC of the stellar photon field*

We propose that gamma-rays can be produced by external Compton (EC) scattering of UV stellar photons of the massive companions by relativistic leptons far from the base of the jet. We also take into account the interaction of these leptons with the photon fields of the disk and the corona. All the photon fields that interact with the jet are shown in Fig. 1. Let us consider a binary system where accretion onto the compact object results in the production of twin relativistic e^+e^--pair jets propagating in opposite directions. The relativistic leptons are considered to have an isotropic power-law density distribution. In the lab frame: $n(\gamma) = \frac{k}{4\pi}D^{2+p}\gamma^{-p}$, with γ, the Lorentz factor of the leptons, k, a constant, and D, the Doppler factor, $D = [\Gamma (1 \mp \beta \cos\phi)]^{-1}$ with Γ and β the bulk Lorentz factor and velocity in units of c respectively, and ϕ the viewing angle. The specific luminosity is then obtained by integrating the scattering rate over the particle energy distribution, and multiplying by the observed photon energy and the photon number density (See Kaufman Bernadó *et al.* 2002 for details).

Fig. 2 shows the SED for simple models that we have calculated (see specific values in the caption of the figure). It can be seen that luminosities of $\sim 10^{36-37}$ erg s^{-1} can be obtained in the observer frame at EGRET's energy range, i.e. 100 MeV – 20 GeV

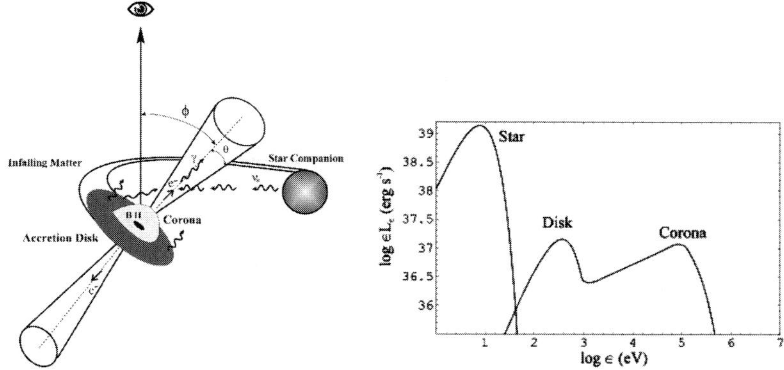

Figure 1. *Left*: The jet traverses photon fields created by the cold accretion disk, the hot corona, and the stellar companion. Inverse Compton up-scattering of some of these photons is unavoidable. Here θ is the half-opening angle of the cone defined by a possibly precessing jet. *Right*: Spectral enegy distribution of the external photon fields to which the jet is exposed.

(left panel of the figure) and therefore some variable unidentified EGRET sources in the galactic plane could be produced by microquasars with precessing jets. Another similar example can be seen in the right panel of Fig. 2 where the important difference with the former case is that the leptons of the jet are assumed to have a high energy cut-off in the TeV range. At TeV energies, where the spectral index is $\Gamma > 3$, the luminosity is $L(E > 1 \text{ TeV}) \sim 10^{32}$ erg s^{-1}. Hence, one of these microquasars located at a few kpc might be a detectable TeV gamma-ray source and thus they are potential targets for high-energy Cherenkov imaging telescopes like HESS or MAGIC.

We also studied the inverse Compton interactions with the other external photon fields due to the accretion disk and the corona (see Romero *et al.* 2002). We concluded that for high-mass microquasars (HMMQs) the predominant contribution comes in general from the scattering of the stellar photon field. This is explained by the higher luminosities obtained in this way for MeV gamma-ray energies. For higher energies, where the luminosities due to the upscattered disk and coronal photons might be important, the absorption by pair creation makes it difficult for these gamma-rays to escape in many cases.

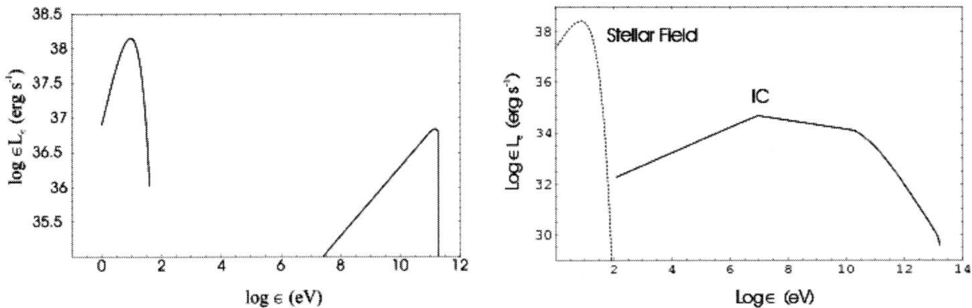

Figure 2. *Left*: spectral energy distribution (SED) of the scattered photons for a microquasar injecting a power-law spectrum of electrons in the photon field of the high-mass stellar companion. A cut-off at Lorentz factors $\gamma \sim 10^3$ has been assumed. The SED of the star is also shown (left top corner). *Right*: Inverse-Compton SED for a microquasar with a massive stellar companion. The jet electron index is $p = 2$ and electrons have a high-energy cut-off at multi-TeV energies. Notice the softening of the spectrum at high-energies due to the Klein-Nishina effect.

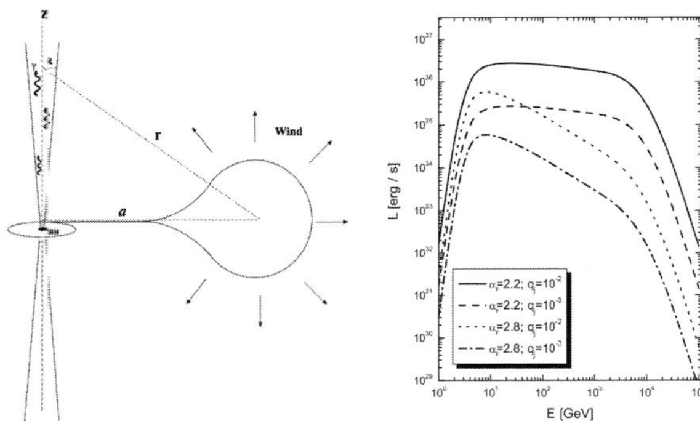

Figure 3. *Left*: In a HMMQ the stellar wind penetrates on a relativistic $e-p$ jet from the sides. The resulting interaction produces gamma-ray emission. *Right*: Spectral high-energy distribution for windy microquasars with proton index $\alpha_{\rm p} = 2.2$ and $\alpha_{\rm p} = 2.8$, for different jet/disk coupling constants (q_j). The jet inclination with respect to the line of sight is assumed to be 10 degrees. An angle of 30 degrees reduces the luminosity by about two orders of magnitude.

2.2. *Hadronic Model*

In this model the gamma-ray emission arises from the decay of neutral pions created in the inelastic collisions between relativistic protons in the jet and the ions of the stellar wind (see Fig. 3). The requisites for the model are a windy high-mass stellar companion and the presence of multi-TeV protons in the jet. The particle spectrum of the relativistic e-p flow is assumed to be a power law: $N'_{e,\,p}(E'_{e,\,p}) = K_{e,\,p}\,E'^{-\alpha_{\rm p}}_{e,\,p}$. Early-type stars, like OB stars, lose a significant fraction of their masses through very strong supersonic winds with typical mass loss rates $\sim 10^{-5} M_\odot \mathrm{yr}^{-1}$ and terminal wind velocities ~ 2500 kms^{-1}. Some protons pertaining to the wind will diffuse into the jet (see Romero *et al.* 2003 and 2005 for details).

The right panel of Fig. 3 shows the spectral high-energy distribution for models with different values of the proton index and of the jet/disk coupling parameter $q_{\rm j}$, with the results obtained using a numerical integration routine. The luminosities obtained at TeV gamma-ray energies imply that hadronic microblazars which are optically thin to pair production can be detected as unidentified, point-like sources with relatively hard spectra. This kind of sources could display variability. Ground-based Cherenkov telescopes like HESS and MAGIC might detect the signatures of such sources on the galactic plane. Hadronic microblazars might be part of this population, as well as of the parent population of low latitude unidentified EGRET sources. The proposed mechanism predicts as well possible neutrino detections with km-size detectors since these particles are produced in the successive decays that follow the $p-p$ interaction.

3. Low-Mass Microquasars – The Galactic Center Halo Population

In the case of the subset of unidentified EGRET sources whose spatial distribution forms a halo around the galactic center, we suggest that their possible counterparts could be low-mass microquasars (LMMQs) that have migrated away from the galactic plane or escaped from globular clusters. Only leptonic models can be considered for the halo sources since the hadronic ones require the presence of an early type of companion star in order to ensure the existence of strong stellar winds. We performed detailed calculations of the jet inverse Compton emission in the seed photon fields from the star, the accretion disk, and the hot coronal region, in different configurations of parameters such as jet Lorentz factors, powers, and angles with the line of sight (see Fig. 4, left panel).

These results show that even though the spectral index in the EGRET band matches that of the unidentified sources, the maximum predicted luminosity, $L_{max} \sim 4 \times 10^{29}$ $(E/100\ \mathrm{MeV})^{-0.5}\ \mathrm{erg\ s^{-1}\ sr^{-1}}$, is 5 orders of magnitude too faint to account for the typical halo source fluxes at distances of 5 to 10 kpc.

In contrast with high-mass systems where the external radiation energy density largely surpasses the magnetic one, Synchrotron-Self-Compton (SSC) emission in low-mass systems is likely to dominate even for a modest field strength of about 10G in the jet (see Fig. 4, right panel). The conclusion is that unlike the HMMQs case, the external Compton emission largely fails to produce the required luminosities observed from these unidentified EGRET sources and SSC emission appears as a promising alternative (see Grenier *et al.* 2005).

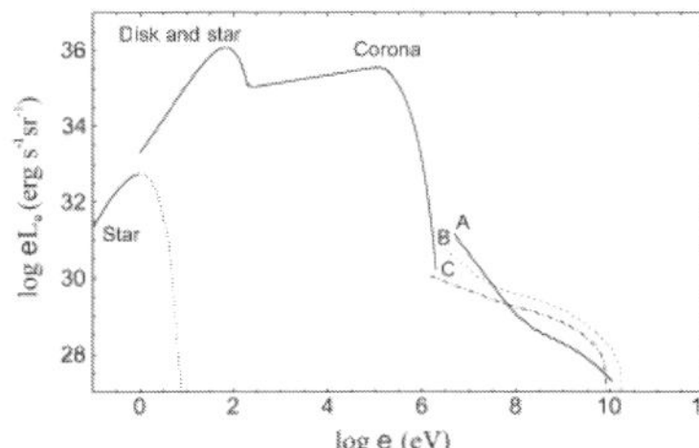

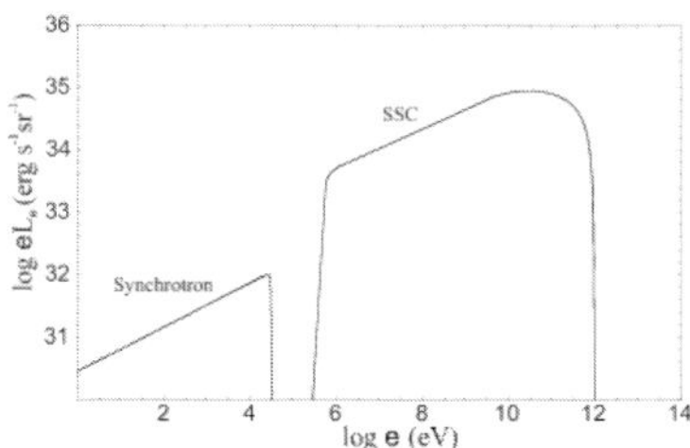

Figure 4. *Left*: SED of the EC emission from the jet of a microquasar with a K-M star companion, seen at angles of 5° (A), 15° (B), and 30° (C) from its axis, for $\Gamma = 3$, $q_{jet} = 10^{-2}$, and a jet electron index p=2.3 with $\gamma_e^{min} = 2$ and $\gamma_e^{max} = 10^4$. *Right*: SED of the synchrotron and SSC emission from the jet of an extreme microblazar, with a 10 G magnetic field, seen at 1° from its axis, for $\Gamma = 10$, $q_{jet} = 10^{-2}$, and a jet electron index p=2.3 with $\gamma_e^{min} = 10$ and $\gamma_e^{max} = 10^5$.

4. Prospects

In order to test these models in a statistically significant number of sources, a new generation of gamma-ray detectors is needed. These satellites are already planned for the quite near future, which is the case of the AGILE and GLAST missions. Their sensitivity is expected to be about 10 to 100 times better than the EGRET one. The existence of TeV emission in MQs might be soon confirmed by new and powerful instruments like MAGIC, HESS and VERITAS. In fact, the detection of microquasar LS 5039 at very high energy gamma rays has just been reported by the HESS collaboration (Aharonian *et al.* 2005). The microquasar LS5039 had already been associated with the EGRET source 3EG 1824-1514 by Paredes *et al.* (2000). On the other hand neutrino observations with IceCube and ANTARES will be crucial with respect to the hadronic model predictions.

References

Aharonian, F.A. *et al.* 2005, *Science* 309, 746
Grenier, I.A., Kaufman Bernadó, M.M., & Romero, G.E. 2005, *Ap&SS* 297, 109
Grimm, H.J., Gilfanov, M., & Sunyaev, R. 2002, *A&A* 391, 923
Kaufman Bernadó, M.M., Romero, G.E., & Mirabel, I.F. 2002, *A&A* 385, L10
Paredes, J.M., Martí, J., Ribó, M., & Massi, M. 2000, *Science* 288, 2340
Romero, G.E., Benaglia, P., & Torres, D.F. 1999, *A&A* 348, 868
Romero, G.E., Kaufman Bernadó, M.M., & Mirabel, I.F. 2002, *A&A* 393, L61
Romero, G.E., Torres, D.F., Kaufman Bernadó, M.M., & Mirabel, I.F. 2003, *A&A* 410, L1
Romero, G.E., Christiansen, H.R., & Orellana, M. 2005, *ApJ* 632, in press

Discussion

ERACLEOUS: In the cases of precessing jets, the disk has to be misaligned with the orbital plane of the binary. What could cause this?

KAUFMAN-BERNADO: We do not consider why the disk is misaligned. We just assume that they are.

Populations of High Energy Sources in Galaxies
Proceedings IAU Symposium No. 230, 2005
E. J. A. Meurs & G. Fabbiano, eds.

© 2006 International Astronomical Union
doi:10.1017/S1743921306007976

Leptonic emission from microquasar jets: from radio to very high-energy gamma-rays

Valentí Bosch-Ramon[1], Josep M. Paredes[1] and Gustavo E. Romero[2,3]

[1]Departament d'Astronomia i Meteorologia, Universitat de Barcelona, Av. Diagonal 647,
E-08028 Barcelona, Catalonia (Spain)
email: vbosch@am.ub.es

[2]Instituto Argentino de Radioastronomía, C.C.5, (1894) Villa Elisa, Buenos Aires (Argentina)

[3]Facultad de Ciencias Astronómicas y Geofísicas, UNLP, Paseo del Bosque, 1900 La Plata
(Argentina)

Abstract. Microquasars are sources of very high-energy gamma-rays and, very probably, high-energy gamma-ray emitters. We propose a model for a jet that can allow to give accurate observational predictions for jet emission at different energies and provide with physical information of the object using multiwavelength data.

Keywords. X-rays: binaries, gamma rays: theory, gamma rays: observations

---

Microquasars are X-ray binaries with relativistic jets whose emission extends from radio to gamma-rays. In particular, LS 5039, discovered by Paredes *et al.* (2000), turned out to be the first likely high-energy gamma-ray microquasar due to its possible association with the EGRET source 3EG J1824−1514 (Paredes *et al.* 2000). Further theoretical studies furnished with more reliability this association (Bosch-Ramon & Paredes 2004), which could be extended to other EGRET sources (Kaufman Bernadó *et al.* 2002, Romero *et al.* 2003, Bosch-Ramon *et al.* 2005a). Very recently, Aharonian *et al.* (2005) have published the detection of the microquasar LS 5039 at TeV energies, leaving no doubt about the gamma-ray emitting nature of these objects. The aim of the present work is to show that from the theoretical point of view microquasar jets can be sources of emission with energies covering the whole spectral band.

In our model, the jet is modelled as dynamically dominated by cold protons and radiatively dominated by relativistic leptons. The characteristics of the orbit and the companion star constrain the way in which the accretion takes place. This has been taken into accout for a consistent orbital variability treatment. The magnetic field energy density and the non-thermal particle maximum energy values along the jet depend on the cold matter energy density and the particle acceleration/energy loss balance, respectively, and the amount of relativistic particles within the jet is restricted by the efficiency of the shock to transfer energy to them. The model takes into account the external and internal photon and matter fields, which interact with relativistic particles in the magnetized jet, producing emission from radio to very high energies. Concerning the the physical parameters of the model, we have adopted the typical values for a massive X-ray binary system and the standard ones in jet and acceleration theory (OB star, semi-major axis of about 0.1 AU, strong stellar wind, jet matter/accretion matter rate ratio of about 0.1, magnetic field of about a 10% of the equipartition one, acceleration efficiencies 10^{-3}-10^{-1}qBc). For further details concerning the model, see Bosch-Ramon *et al.* (2005b).

This model has been applied to LS 5039. In Fig. 1, we show the computed SED, and the flux and photon index evolution along the orbit for this source. The model underpredicts the fluxes in the radio and TeV band. Although the difference is less than one order of magnitude, it could be interpreted as a hint that more intense high energy processes (particle acceleration and emission) are taking place already outside the binary system, since radiation is emitted mainly within the 100 GeV-photon absorption region close to the companion star. This could also lead to higher fluxes in radio, due to this radiation is optically thin when emitted outside the binary system. Concerning the variability of the source along the orbit, we have used a particular accretion model for this source consisting in a slow equatorial wind from a fast rotating stellar companion (Casares *et al.* 2005), altogether with a stream of matter ejected during the periastron passage and reaching the compact object at phase 0.8–0.9, when the X-ray (Bosch-Ramon *et al.* 2005c) and the TeV peaks (Aharonian *et al.* 2005, Casares *et al.* 2005) are observed. For more details concerning the application of the model to LS 5039, see the work of Paredes *et al.* (2005). It is worth noting that the total jet kinetic power required for producing a SED like the one plotted in Fig. 1 is few times 10^{36} erg/s. Therefore, the very likely nature of microquasars and, in particular, LS 5039 as GeV-TeV emitters appears well founded from theoretical grounds.

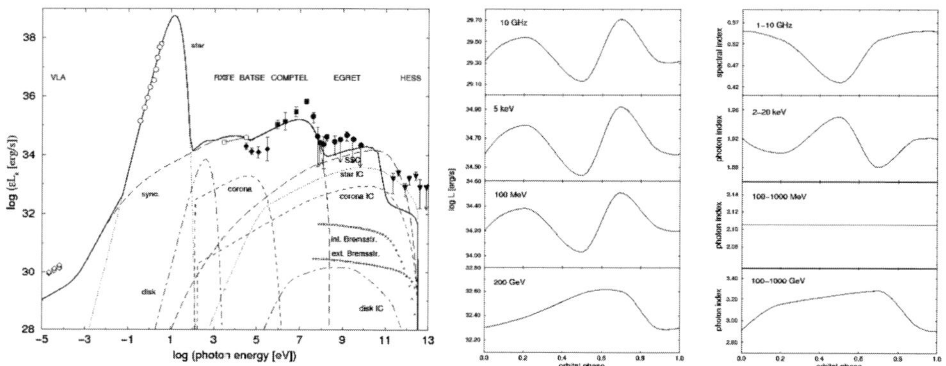

Figure 1. a) Computed SED for LS 5039 and b) its flux and photon index evolution at different energy bands.

Acknowledgements

V.B-R. and J.M.P. acknowledge partial support by DGI of the spanish Ministerio de Educación y Ciencia under grant AYA2004-07171-C02-01, as well as additional support from the European Regional Development Fund (ERDF/FEDER). During this work, V.B-R. has been supported by the DGI of the spanish Ministerio de Educación y Ciencia under the fellowship BES-2002-2699. G.E.R is supported by the Argentine Agencies CONICET and ANPCyT (PICT 03-13291).

References

Aharonian, F. *et al.* 2005, *Science* 309, 746
Bosch-Ramon, V. & Paredes, J. M. 2004, *A&A* 417, 1075
Bosch-Ramon, V. Romero, G. E. & Paredes, J. M. 2005a, *A&A* 429, 267
Bosch-Ramon, V. Romero, G. E. & Paredes, J. M. 2005b, *A&A*, submitted
Bosch-Ramon, V., Paredes, J. M., Ribó, M., Miller, J. M., Reig, P., Martí, J. 2005c, *ApJ* 628, 1
Casares, J., Ribó, M., Ribas, I., Paredes, J. M., Martí, J., Herrero, A. 2005, *MNRAS*, submitted
Kaufman Bernad, M., Romero, G. E. & Mirabel, I. F. 2002, *A&A* 385, 10
Paredes, J. M., Mart, J., Ribó, M. & Massi, M. 2000, *Science* 288, 2340
Paredes, J. M., Bosch-Ramon, V. & Romero, G. E. 2005, *A&A*, submitted
Romero, G. E., Torres, D. F., Kaufman Bernadó, M., Mirabel, I. F. 2003, *A&A* 410, 1

Populations of High Energy Sources in Galaxies
Proceedings IAU Symposium No. 230, 2005
E. J. A. Meurs & G. Fabbiano, eds.

© 2006 International Astronomical Union
doi:10.1017/S1743921306007988

INTEGRAL Observations of GRS 1758−258

Marion Cadolle Bel[1,2], Andrea Goldwurm[1,2] and Patrick Sizun[2]

[1]APC, UMR 7164, 11 place M. Berthelot, 75231 Paris, France
[2]Service d'Astrophysique, CEA-Saclay, Orme des Merisiers, 91191, Gif-Sur-Yvette, France
emails: mcadolle@cea.fr, goldwurm@cea.fr, sizun@cea.fr

Abstract. Since its launch in October 2002, the *INTEGRAL* satellite has observed the X-ray binary and black hole (BH) candidate GRS 1758−258, for more than 2 Ms. Between 2003 and 2004 *INTEGRAL* could follow its spectral and temporal behaviour: while it was weak and soft in spring 2003 it was detected up to 150 keV at the end of 2003 August, in a hard state similar to the one observed between 1990 and 1997 by previous high energy missions.

Keywords. Black hole physics − Stars: individual: GRS 1758−258 − Gamma rays: observations − X-rays: binaries − X-rays: general.

1. Introduction

GRS 1758−258 was discovered in 1990 with the *SIGMA*/GRANAT telescope (Mandrou 1990). Together with 1E 1740.7−2942, it is the brightest persistent source above 50 keV in the Galactic Bulge (Sunyaev *et al.* 1991). The strong and variable hard X-ray emission of GRS 1758−258 and its broad band spectral characteristics (Kuznetsov *et al.* 1999) suggest the presence of a BH in a binary system. Smith *et al.* (2002) measured an orbital period of 18.45 ± 0.10 days and suggested a red giant as companion star. GRS 1758−258 also shows two double-sided radio jets of ~ 1 arcminute size (Rodriguez *et al.* 1992) and it is often observed in the typical hard state (HS) of BH X-ray binaries (Kuznetsov *et al.* 1999, Sidoli *et al.* 2002). We present evidences of spectral evolution of this BH as seen by *INTEGRAL*.

2. Observations and Results

We have analyzed *INTEGRAL* public data of 3 observation periods of GRS 1758−258: epoch 1 (2003 Mar 22-Apr 21), 2 (2003 Sep 22–Oct 18) and 3 (2004 Feb 16–Apr 21). Figure 1(Left) shows the IBIS/ISGRI light curve of GRS 1758−258: the source was almost undetectable in epoch 1 but the average 20–140 keV count rate increased to the level of 16 cts s^{-1} (~ 80 mCrab) in epoch 2 and decreased slightly in epoch 3. Fitting the reconstructed spectra with a power law, we obtain a photon index of 2.55 ± 0.36 for epoch 1 and 1.90 ± 0.08 for epochs 2 and 3, indicating that the source transited to a harder state. As spectra for epochs 2 and 3 are generally compatible, we have modelled the data of begining of epoch 2 (2003 Sep 22–25) (for which simultaneous JEM-X, IBIS/ISGRI and SPI data were available) with an absorbed (fixed column density of 1.5×10^{22} atoms cm^{-2}, Mereghetti *et al.* 1997) multicolour disc black body (Mitsuda *et al.* 1984) and a Comptonization model (Titarchuk 1994, spherical geometry). Fig. 1 (Right) shows the derived count spectrum and best-fit model for these GRS 1758−258 data. The derived spectral parameters are consistent with the source being in a HS: $kT_{\mathrm{e}} = 40\ ^{+33}_{-3}$ keV, $\tau = 1.30\ ^{+0.13}_{-0.62}$ ($\chi^2_{red} = 1.10$, 187 degrees of freedom). The 5–200 keV flux is 2.74×10^{-9} ergs cm^2 s^{-1}

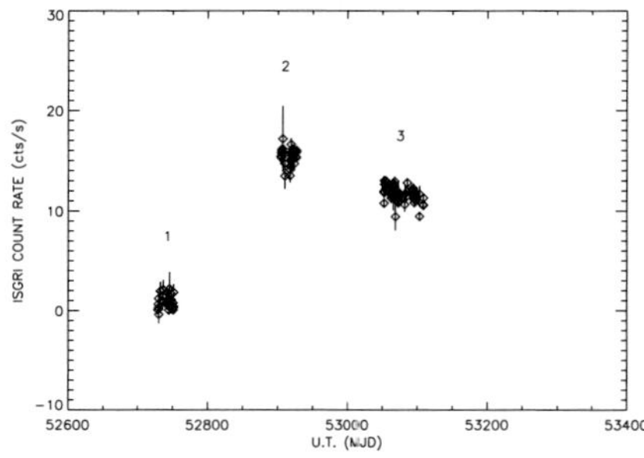 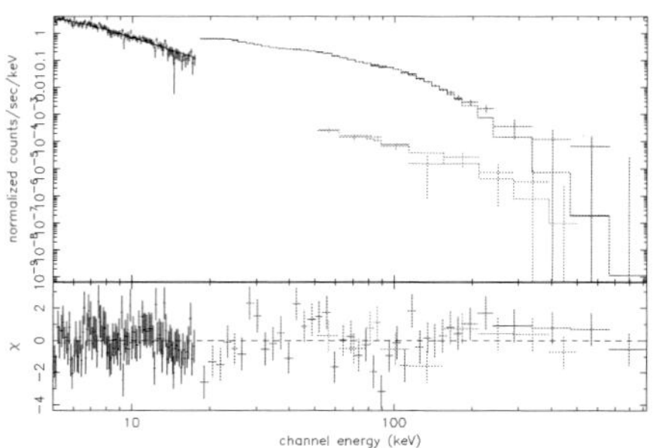

Figure 1. *Left:* The 20–140 keV IBIS/ISGRI light curve of GRS 1758−258 from 2003 Mar 22 up to 2004 Apr 21 (epochs indicated). *Right:* Count spectra of GRS 1758−258 in 2003 September 22–25 with the JEM-X (black), IBIS/ISGRI (red) and SPI (green) data along with the best-fit model. Residuals in σ units are also shown.

and the bolometric luminosity, for a distance of 8 kpc, is 2.70×10^{37} ergs s^{-1}. Our results are compatible with the values recently reported by Pottschmidt *et al.* (submitted).

3. Conclusions

We conclude that from epoch 1, when the source was much weaker and softer in hard X-rays but at a similar level in the 2–12 keV ASM/RXTE light curve, the source transited towards its typical hard state in epochs 2 and 3. It then showed spectral parameters and luminosities compatible with those observed by previous missions. Given the low ASM flux, the epoch 1 period was probably not dominated by bright soft disc emission, unlike the standard high soft state of BH binaries. GRS 1758−258 is indeed known to undergo dim soft states, without showing an anticorrelation between hardness and soft X-ray flux (Smith *et al.* 2002).

Acknowledgements

Based on observations with *INTEGRAL*, an ESA project with instruments and science data center funded by ESA member states (especially the PI countries: Denmark, France, Germany, Italy. Switzerland, Spain, Czech Republic and Poland, and with the participation of Russia and the USA).

References

Kuznetsov, S. I., *et al.* 1999, *Astronomy* (Letters) 25, 351
Mandrou, P. 1990, *IAUC* 5032
Mereghetti, S., *et al.* 1997 *ApJ* 476, 829
Mitsuda, K., *et al.* 1984, *PASJ* 36, 741
Pottschmidt, K., *et al.* 2005, submitted to *A&A*
Rodriguez, L. F., *et al.* 1992, *ApJ* (Letters) 401, 15
Sidoli, L., *et al.* 2002, *A&A* 388, 293
Smith, D. M., *et al.* 2002, *ApJ* (Letters) 578, 129
Sunyaev, R., *et al.* 1991, *A&A* (Letters) 247, 29
Titarchuk, L. G. 1994, *ApJ* 434, 570

Populations of High Energy Sources in Galaxies
Proceedings IAU Symposium No. 230, 2005
E. J. A. Meurs & G. Fabbiano, eds.

© 2006 International Astronomical Union
doi:10.1017/S174392130600799X

New Results from High Energy Gamma-Ray Astronomy

H. J. Völk

Max-Planck-Institut für Kernphysik, P. O. Box, 69029 Heidelberg, Germany
email: Heinrich.Voelk@mpi-hd.mpg.de

Abstract. High energy gamma-ray astronomy has recently made significant progresss through ground-based instruments like the *H.E.S.S.* array of imaging atmospheric Cherenkov telescopes. The unprecedented angular resolution and the large field of view has allowed to spatially resolve for the first time the morphology of gamma-ray sources in the TeV energy range. The experimental technique is described and the types of sources detected and still expected are discussed. Selected results include objects as different as a Galactic binary Pulsar, the Galactic Center and Supernova Remnants but they also concern the diffuse extragalactic optical/infrared radiation field. Finally, a scan of the Galactic plane in TeV gamma rays is described which has led to a significant number of new TeV sources, many of which are still unidentified in other wavelengths. The field has a close connection with X-ray astronomy which allows the study of the synchrotron emission from these very high energy sources.

Keywords. radiation mechanisms: nonthermal, pulsars, Galaxy: center, acceleration of particles, supernova remnants, gamma rays: observations, cosmology: observations.

---

1. Imaging Air Cherenkov Telescopes and the H.E.S.S. experiment

The principles of the Air Cherenkov Technique were invented in the 1950ies, with a first astronomical success in the late 1980ies, when the *Whipple* telescope in Arizona detected the Crab Nebula (Weekes *et al.* (1989)).

The recent progress in VHE γ-ray astronomy with the *H.E.S.S.* experiment in Namibia was possible through several new developments in the 1990ies (i) the stereoscopic technique, pioneered by the German/Spanish/Armenian *HEGRA* experiment on La Palma (ii) the development of fine pixel "cameras" as focal plane detectors (*CAT* telescope in the French Pyrenees) (iii) the realization of the spatial extension of most of the expected γ-ray sources and thus the necessity for wide field-of-view (FoV) detectors allowing also sky surveys, and (iv) the need for the detection of "many", spatially extended nearby sources to solve the problem of Cosmic Ray origin in our Galaxy. It is therefore advantageous to observe the Galactic disk in the southern sky which led to the choice of the Gamsberg area in Namibia in Southern Africa.

Fig. 1a shows two of the four telescopes of the *High Energy Stereoscopic System (H.E.S.S.)* at 1800 m.a.s.l. It is operated by a large collaboration led by the Max Planck Institute for Nuclear Physics in Heidelberg. The four telescopes operate in coincidence (for more technical details, see the Web page http://www.mpi-hd.mpg.de/hfm/ HESS/HESS/html).

Overall the resulting *H.E.S.S.* sensitivity (4 telescopes) is as follows: 1 hour of observation time for a detection of an energy flux density of 10^{-11} (10^{-12}) erg cm^{-2} s^{-1} at 100 GeV (1 TeV). With this performance the Crab Nebula can be detected at Zenith in $\sim$30 s. For comparison, the 1989 detection required $\sim$50 hr.

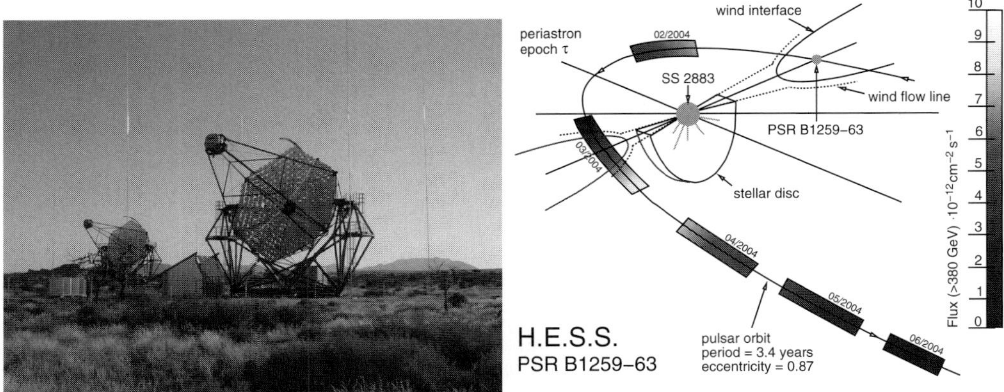

Figure 1. (Left:) The *H.E.S.S.* experiment in Namibia, near the famous Gamsberg. Two of the four 13m-telescopes are shown with their "cameras" in the focal point. (Right:) Orbital scheme of PSR B1259-63 with respect to the line of sight. The Pulsar approaches the stellar wind equatorial plane prior to periastron "behind" it.

2. Science with high energy gamma-ray astronomy

The two main fields of very high energy (VHE) γ-ray astronomy are High Energy Astrophysics and Observational Cosmology.

2.1. *High Energy Astrophysics*

High Energy Astrophysics concerns the most energetic and violent processes in the Universe, and in particular their nonthermal aspects. We have to expect that the nonthermal energy content U_{nonth} of relativistic baryonic particles is in most regions of the Universe comparable to the energy densities in the thermal gas U_{th} and the magnetic fields U_{mag}, i.e. $U_{nonth} \sim U_{th} \sim U_{mag}$. This should at least be true "everywhere" in galaxies and clusters of galaxies. But it holds probably also beyond, wherever cosmic structure formation with its violent, supersonic flows of baryonic matter has taken place or is still occurring. As a result of interactions of individual particles with collective excitations of the system, the particle sources and the associated nonthermal radiation should be characterized by power-law energy spectra rather than by thermal Maxwellian distributions

Because of its expected ubiquity I call this component the "Nonthermal Universe". Its study is intimately connected with that of stellar explosions, rapid outflows from galaxies, energy losses of extreme compact objects, and high-energy accretion processes up to the very largest spatial scales.

The types of VHE γ-ray sources found/*expected* in the Galaxy are Pulsar Nebulae, Supernova Remnants, X-ray Binaries ("Micro-Quasars"), Diffuse Galactic emission, *Molecular Clouds*, or possibly *new source types*. Extragalactic sources are Active Galactic Nuclei (e.g. Blazars), Radio Galaxies, *Gamma-ray Pair Halos, Starburst Galaxies, Galaxy Mergers, Clusters of Galaxies, and Gamma-Ray Bursts*.

2.2. *Observational Cosmology*

A major aspect of cosmology is cosmic structure formation. One of its consequences is the Extragalactic Background Light (EBL), the thermal, diffuse optical/infrared extragalactic background radiation from stars and Black Holes in galaxies and its re-radiation in the infrared. The EBL informs us about the epochs of galaxy formation and the history of their evolution.

TeV γ-quanta are absorbed by pair production on these low energy photons and therefore the spectra of distant extragalactic VHE γ-ray sources are expected to exhibit characteristic absorption features. They are the result of the magnitude and spectral variation of this background. Whereas a direct astronomical measurement of the EBL is very difficult, the $\gamma\gamma$-absorption is free of such complications. Its measurement for objects at different redshifts z should in principle allow even the resolution of the EBL in z. Very recently, measurements of two different Blazars with known redshift that were newly discovered by the *H.E.S.S.* Collaboration have indeed been used (Aharonian *et al.* (2005a); see also the talk by J. Quinn in these Proceedings) to give the most stringent upper limit on the EBL in the optical/near-infrared band to date. It appears significantly lower than expected from the current "direct" estimates and very close to the absolute lower limit represented by the integrated light of resolved galaxies (Madau & Pozetti (2000)). Apart from resolving the EBL largely into individual galaxies, this upper limit is especially in conflict with the claims of a high EBL flux at near-infrared wavelengths (Matsumoto *et al.* (2005); Cambresy *et al.* (2001)) which was often envisaged to be the result of radiation massively produced in the early Universe ($z \sim 10$) by the first stars (Pop III). Given the strong likelihood that Pop III stars rapidly enrich their environment with heavy elements and dust grains, such an assumption was in any case highly problematic.

Another major cosmological aspect of VHE γ-ray astronomy is an indirect Dark Matter search through the detection of annihilation radiation from the lowest-mass supersymmetric particles, called Neutralinos. Such weakly interacting massive particles are widely believed to constitute the Dark Matter in the Universe. From gravitational simulations they should be concentrated with a significant density increase in the central regions of Dark Matter halos, like the Galactic center (see below).

3. Selected results

3.1. *H.E.S.S. discovery of Pulsar B1259-630*

This 48 ms Pulsar is in a 3.4 yr-period highly excentric orbit around a Be star with blackbody temperature $T_* = 23.000$ K. The particles from the Pulsar Wind are expected to be ultrarelativistic electrons and positrons, emitting Inverse Compton (IC) radiation in the stellar radiation field. Indeed the TeV-spectrum has been successfully predicted in the synchrotron/IC context (Kirk *et al.* (1999)). *H.E.S.S.* has observed the source for $\sim$50 hrs in 2004 around perihelion; the flux was time-variable on the scales of days (Aharonian *et al.* (2005b)).

The Be star emits a strong disk-shaped wind that confines the Pulsar Wind Nebula into a "cometary" shape (where the analogy is meant morphologically rather than physically). The periastron lies behind the star (Fig. 1b). Unfortunately, moonlight did not allow observations during periastron itself. The data appear nevertheless to indicate a minimum of the VHE light curve near periastron (Fig. 2).

The IC losses, as measured by *H.E.S.S.* are about a factor of 10 weaker than the corresponding synchrotron losses, as inferred by *INTEGRAL* (Shaw *et al.* (2004)). This determines an effective magnetic field strength $B_{\mathrm{eff}} \approx 1$ G. In the expected Klein-Nishina regime the observed IC photon spectrum corresponds to a very hard differential electron spectrum with a power-law index $\alpha \geqslant 1.7 \pm 0.2$. It therefore cannot in turn come from a radiatively cooled electron population, and adiabatic losses must be dominant. This points to a rapidly and directionally expanding Pulsar Wind Nebula flow, resulting in a very interesting physical picture: The shocked Pulsar Wind is accelerated into a "cometary tail" flow, induced by the stellar wind. This flow is more strongly expanding outside the wind "disk" near periastron, where it is directed away from us. As a consequence its

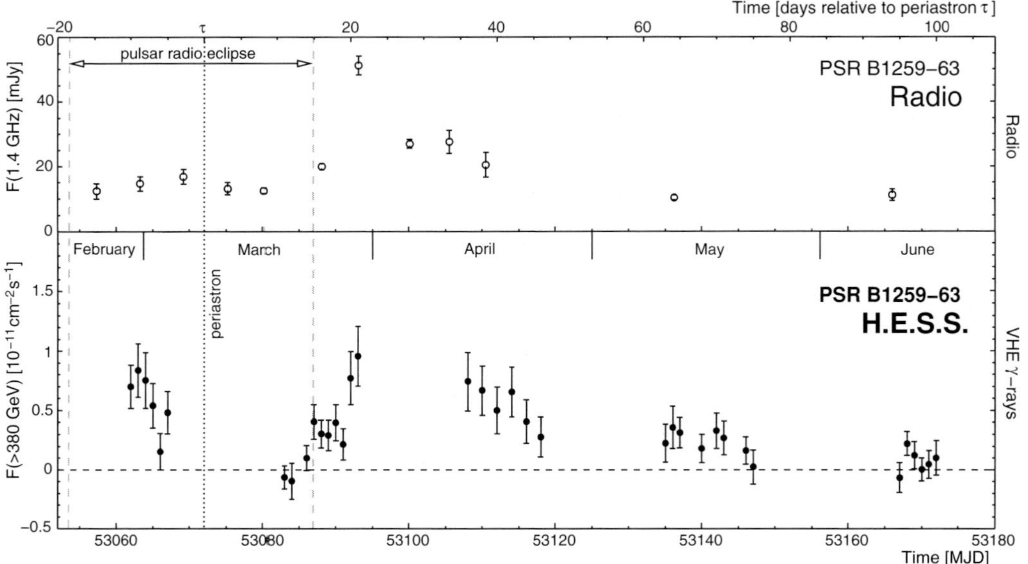

Figure 2. VHE γ-ray and unpulsed radio (Johnston *et al.* (2005)) light curves from PSR B1259-63 around its periastron passage (*dotted vertical line*).

emission is in addition diminished by Doppler dis-favoritism. Both effects together lead to a local minimum of the IC flux around periastron, despite the fact that the stellar radiation field peaks at this point.

3.2. *Galactic Center region*

The innermost region of our Galaxy has an abundance of nonthermal sources, many of them Supernova Remnants. *H.E.S.S.* has observed this region and has detected the neighborhood of Sgr A* with high significance (Aharonian *et al.* (2004a)). Two discrete sources were found, one around Sgr A* itself and another one about 1° away in the Galactic Disk (see Fig. 3a). This latter source is the Supernova Remnant (SNR) G0.9+0.1, dominated by a Pulsar Wind Nebula (Aharonian *et al.* (2005d)). Although the statistical angular resolution of the *H.E.S.S.* observation of the Sgr A* region is more than an order of magnitude better than the one achieved in other detections at TeV or GeV energies , the *H.E.S.S.* image center cannot yet be separated from Sgr A* itself within the present pointing accuracy of the system ($\sim$20 arcsec in both coordinates). The γ-ray source is apparently point-like. Therefore other properties like the energy spectrum or a possible time variability become important criteria to determine whether the γ-ray emission comes from the central Black Hole and/or its immediate environment (the accretion flow or jets), or from a more extended source in the neighborhood. No time variability has been found until now. The form of the differential γ-ray spectrum $\propto E^{-2.3}$ (Fig. 3b) corresponds to a relatively hard power-law, and is therefore also consistent with freshly produced Cosmic Rays interacting with dense gas (with a hydrogen density $N_H \sim 10^3 \mathrm{cm}^{-3}$), or diffusive shock acceleration of particles still confined in the adjacent SNR Sgr A East. Theoretically there is even the possibility of a continuous transition between these two last radiation mechanisms.

The most exotic scenario would be dominant steady-state Neutralino/WIMP annihilation in the expected Dark Matter Halo centered on Sgr A*. Apart from the present difficulties to model the observed TeV spectrum within supersymmetric theories, the rest energy of the Neutralino must then be larger than the maximum energy up to which the

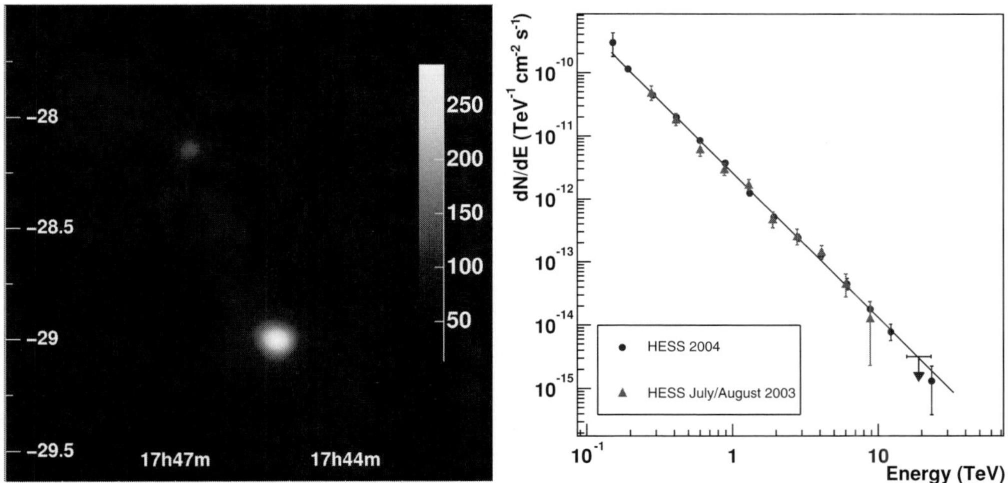

Figure 3. (Left:) *H.E.S.S.* VHE γ-ray image of the inner region of the Galaxy. The source on the lower right (*Galactic Center region*) is coincident within $1'$ of Sgr A*. The source on the upper left is the Supernova Remnant G0.9+0.1 in the Galactic plane. (Right:) VHE γ-ray spectrum of the *Galactic Center region* as obtained by the *H.E.S.S.* array. The energy spectrum is a power law and extends beyond 20 TeV.

supposed annihilation spectrum extends. This corresponds to at least 20 TeV. Since the largest particle accelerator to exist in the foreseeable future, the *LHC* in CERN, only reaches center-of-momentum energies of the order of a TeV, this particle could not possibly be discovered in accelerator experiments. Thus, if the detected γ-ray flux from the Galactic center is to be ascribed to Dark Matter annihilations, then only astrophysics can tackle this question – a very interesting perspective.

3.3. *Supernova Remnants and Cosmic Ray origin*

While the theoretical aspects are essentially understood (e.g. Völk (2004) for a recent overview), there is a shortage of γ-ray detections. The *EGRET* experiment could not unequivocally establish the detection of a shell-type SNR in the GeV-range. Also the widely publicised TeV γ-ray detection of SN 1006 by the *CANGAROO* experiment could not be confirmed by *H.E.S.S.* (Aharonian *et al.* (2005e)). The reason is in all probability the very low gas density $N_H < 0.1 \mathrm{cm}^{-3}$ (Ksenofontov *et al.* (2005)) in this object that is so bright in nonthermal X-rays (Koyama *et al.* (1995)). The second reason for the low TeV flux is the amplification of the magnetic field (Bell & Lucek (2001), Berezhko *et al.* (2003a)) which depresses the IC emission – for given synchrotron emission – compared to that in a more typical μG interstellar field strength. Meanwhile also the *CANGAROO* collaboration has announced that it could not any more detect SN 1006 in stereoscopic re-observations (Mori (2005)).

There are only three VHE detections until now: RX J1713.7-3946 (Muraishi *et al.* (2000); Enomoto *et al.* (2002); Aharonian *et al.* (2004b)), Cas A (Aharonian *et al.* (2001)), and RX J0852.0-4622 (Katagiri *et al.* (2005); Aharonian *et al.* (2005f)).

Cas A is a thoroughly studied object in all wavelength ranges below γ-ray energies and thus amenable to a detailed application of nonlinear acceleration theory (Berezhko *et al.* (2003b)). In addition, the observations of narrow filamentary structures in nonthermal X-rays (assumed to be synchrotron radiation) by Vink & Laming (2003) can be interpreted to imply an interior magnetic field strength of 500μG (Berezhko & Völk (2004)), consistent with the theoretical analysis of the spatially integrated synchrotron spectrum.

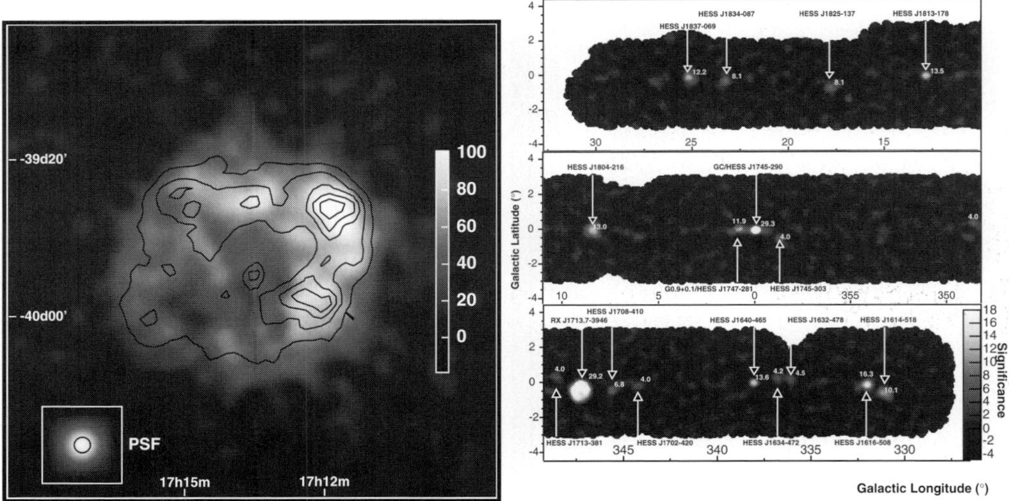

Figure 4. (Left:) Spatially resolved VHE γ-ray image of the SNR RX J1713.7-3936 as obtained by the *H.E.S.S.* array. The *ASCA* hard X-ray data (*contour lines*) are shown in addition. (Right:) *H.E.S.S.* scan of the Galactic Plane between $\pm 30°$ in longitude and $\pm 3°$ in latitude. The sources indicated are detected at a significance level greater than 4σ.

This strongly suggests Cas A as a source of nuclear Cosmic Rays, independent of all remaining uncertainties in the astronomical parameters.

The *H.E.S.S.* observations of RX J1713.7-3946 have for the first time resolved the morphology of an extended object in VHE gamma rays (Fig. 4a). The overall shell structure coincides closely in hard X-rays and gamma rays. Despite the complex structure this is unambiguous proof of the acceleration of charged particles to energies beyond 100 TeV. The most recent data show in addition that below about 10 TeV the photon spectrum can be approximated by a power law in energy with an index close to 2.0. Such a spectrum is clearly consistent with diffusive shock acceleration. More detailed studies are in progress.

RX J0852.0-4622, also called "Vela Jr", was, like RX J1713.7-3946, first detected with the *ROSAT* telesope in X-rays (Aschenbach (1998)). The object has a radius of $\sim 1°$, twice that of RX J1713.7-3946 and thus four times the radius of the full Moon. The TeV γ-ray morphology correlates again very well with the X-ray image, and the energy spectrum – not yet very precisely determined – can be fit by a power law with photon index 2.1. In this sense Vela Jr is similar to RX J1713.7-3946. Its apparently regular spherical shape and high γ-ray flux, comparable with that of the Crab Nebula, make it an ideal object for further studies.

4. The *H.E.S.S.* survey of the Galactic plane

In 2004 *H.E.S.S.* has completed the first 230 hour part of a VHE survey of the Galactic plane, covering $-30° < \ell < 30°$ in longitude and $-3° < b < 3°$ in latitude (Aharonian *et al.* (2005g)). The average flux sensitivity of the survey corresponds to ~ 3 percent of the Crab Nebula (Fig. 4a). Including recent re-observations of candidate sources from the initial survey fourteen previously unknown VHE sources have been found up to now (Aharonian *et al.* (2005h)). Most of the sources are still un-identified in other wavelength ranges.

In this manner for the first time a TeV instrument has investigated not only source candidates that were well known from other wavelength ranges, but did a successful blind search. The search for counterparts in other wavelength ranges is ongoing. Most known counterpart candidates are young Pulsar Wind Nebulae or SNRs. The new sources have triggered a worldwide activity with satellite X-ray instruments like *INTEGRAL, ASCA, XMM* and *Astro-E2*, and ground-based radio telesopes.

Acknowledgements

I would like to thank the members of the *H.E.S.S.* collaboration for many discussions on the observational results. I have learnt much from Felix Aharonian and Okkie de Jager about the physics of PSR B1259-63, even though I have of course the sole responsibility for the arguments in the text.

References

Aharonian, F.A. *et al.* (HEGRA Collaboration) 2001, *A&A* 370, 112

Aharonian, F.A. *et al.* (H.E.S.S. Collaboration) 2004a, *A&A* 425, L13

Aharonian, F.A. *et al.* (H.E.S.S. Collaboration) 2004b, *Nature* 432, 75

Aharonian, F.A. *et al.* (H.E.S.S. Collaboration) 2005a, *submitted*; arXiv:astro-ph/0508073

Aharonian, F.A. *et al.* (H.E.S.S. Collaboration) 2005b, *A&A*, in press; arXiv:astro-ph/0506280

Aharonian, F.A. *et al.* (H.E.S.S. Collaboration) 2005c, *A&A* 439, 1013

Aharonian, F.A. *et al.* (H.E.S.S. Collaboration) 2005d, *A&A* 432, L25

Aharonian, F.A. *et al.* (H.E.S.S. Collaboration) 2005e, *A&A* 437, 135

Aharonian, F.A. *et al.* (H.E.S.S. Collaboration) 2005f, *A&A*, in press; arXiv:astro-ph/05/0538

Aharonian, F.A. *et al.* (H.E.S.S. Collaboration) 2005g, *Science* 307, 1938

Aharonian, F.A. *et al.* (H.E.S.S. Collaboration) 2005h, *ApJ*, to appear

Aschenbach, B. 1998, *Nature* 396, 141

Bell, A.R. & Lucek, S.G. 2001, *MNRAS* 321, 433

Berezhko, E.G., Ksenofontov, L.T. & Völk, H.J. 2003a, *A&A* 412, L11

Berezhko, E.G., Pühlhofer, G. & Völk, H.J. 2003b, *A&A* 400, 971

Berezhko, E.G. & Völk, H.J. 2004, *A&A* 419, L27

Cambresy, L., Reach, W.T., Beichman, C.A., *et al.* 2001, *ApJ* 555, 563

Enomoto, R., Tanimori, T., Naito, T. *et al.* 2002, *Nature* 416, 823

Johnston, S., Ball, L., Wang, N., *et al.* 2005, *MNRAS* 302, 277

Katagiri, H., Enomoto, R., Ksenofontov, L.T. *et al.* 2005, *ApJ* 619, L163

Kirk, J.G., Ball, L. & Skjaeraasen, O. 1999, *Astropart. Phys.* 10, 31

Koyama, K., Petre, R., Gotthelf, E.V., *et al.* 1995, *Nature* 378, 255

Ksenofontov, L.T., Berezhko, E.G. & Völk, H.J. 2005, *A&A*, to appear; arXiv:astro-ph/0508318

Madau, P. & Pozetti, L. 2000, *Mon. Not. R. Astron. Soc.* 312L, 9

Matsumoto, T., Matsuura, S., Murakami, H., *et al.* 2005, *ApJ* 626, 31

Mori, M. 2005, in: B. Degrange (ed.), *Cherenkov2005*, Proceedings Workshop "Towards a Network of Atmospheric Cherenkov Detectors VII" (Palaiseau, April 2005), to be published

Muraishi, H., Tanimori, T., Yanagita, S. *et al.* 2000, *A&A* 354, L57

Shaw, S.E., Chernyakova, M., Rodriguez, J. *et al.* 2004, *A&A* 426, L33

Vink, J. & Laming, J. 2003, *ApJ* 548, 758

Völk, H.J. 2004, in: T. Kajita, Y. Asaoka, A. Kawachi, Y. Matsubara, & M. Sasaki (eds.), *Frontiers of Cosmic Ray Science*, Proceedings 28th Int. Cosmic Ray Conf., Universal Academy Press, Inc. (Tokyo), vol. 8, p. 29

Weekes, T. C., Cawley, M. F., Fegan, D. J., *et al.* 1989, *ApJ* 342, 379

Discussion

CHERNYAKOVA: Isn't the model of Kawacki *et al.* for PSRB 1259-63 better to explain the observed TeV emission, taking into account XMM observations of 2004 periastron passage, that are in favour of low Lorentz factor and IC mechanism for X-ray/Gamma-ray range?

VÖLK: Speaker believes in IC mechanism for both X-ray and TeV-ray domain. [Answer recorded by questioner.]

MEURS: You showed a number of Galactic Plane new TeV sources that probably are SNRs. One the other hand you showed only a few TeV detected unknown SNRs. Is there any dichotomy between TeV and non TeV SNRs?

VÖLK: I do not think that there is a dichotomy. Theoretically the non-detection of SN 1006 is the result of a very low external gas density ($N_H < 0.1 \mathrm{cm}^{-3}$). Tycho's SNR is also a TeV upper limit only. Theoretically this UL is still a factor of a few above expectations. We need indeed some more TeV SNRs, but I see no problem for them being the source population of the Galactic Cosmic Rays. On the contrary, we are close to resolve this 100 yr old question.

UBERTINI: How do you assess the evidence of few SNRs seen in hard X-rays (in particular with the INTEGRAL survey) & emitting TeV flux VS synchrotron/IC models.

VÖLK: The maximum electron energy p_{max} achieved by acceleration in the SNR blast wave is limited by synchrotron losses. Assuming the scattering mean free path $\sim$ gyro radius (Bohm limit), the synchrotron cut off frequency $\nu_{max} \propto V_s^2$, where the shock velocity V_s decreases in time, especially after the initial sweep-up phase. Therefore ν_{max} is in the hard X-ray range only for very young – and thus rare – SNRs (see Berezhko & Völk, A&A 427, 525 (2004), where this is discussed in detail). Thus we should not be surprised to see so few SNRs in hard X-rays.

Populations of High Energy Sources in Galaxies
Proceedings IAU Symposium No. 230, 2005
E. J. A. Meurs & G. Fabbiano, eds.

© 2006 International Astronomical Union
doi:10.1017/S1743921306008003

Ground Based Gamma-Ray Astronomy

Peter Cogan[1]†**, Michael K. Daniel**[1]**, David J. Fegan**[1]**, Stephen Gammell, Andrew McCann**[1]**, John Quinn**[1]

[1]School of Physics, University College Dublin
email: peter@ferdia.ucd.ie

Abstract. The field of ground based gamma-ray astronomy has seen rapid growth over the past thirty years with the development of the Imaging Atmospheric Cherenkov Technique to search for Very High Energy (VHE; E > 100 GeV) gamma radiation. This growth continues with the construction of four third generation telescope systems in Namibia, Australia, La Palma and the USA. These systems will search for VHE gamma radiation from such objects as AGN, SNRs, microquasars, dark matter and the galactic centre.

Keywords. TeV, gamma ray, IACT, AGN, SNR.

---

1. Imaging Atmospheric Cherenkov Technique

This paper provides a brief overview of the field of ground-based gamma-ray astronomy. In the first section, the technique will be described, and in the second section a selection of recent results will be summarised.

Upon striking the earth's atmosphere, a high-energy gamma ray initiates a relativistic cascade of electromagnetic particles known as an Extensive Air Shower (EAS). These particles produce Cherenkov radiation which may be detected on the ground using arrays of PMTs mounted in the focal plane of a large reflector. As the Cherenkov radiation produced is extremely faint, this can only be done on clear moonless nights, resulting in a duty cycle of $\sim$10 %. Due to the almost overwhelming hadronic background, a highly efficient rejection technique is required. This can be achieved as gamma-ray induced EAS produce small compact images whereas hadronic images are large and irregular. A parameterisation of the size and shape of the image (Hillas (1985)) allows discrimination between these images, with a background rejection of $\sim$98 % and a signal acceptance of $\sim$30 %.

Local muons produce images very similar to low energy ($<$200 GeV) gamma rays. These can be discriminated against using multiple telescopes with a large ($\sim$100 m) baseline operating with a coincidence trigger, as local muons only affect single (or adjacent) telescopes. Coupled with a large mirror area, this allows a reduction of the system's energy threshold to $\sim$100 GeV and results in arc minute resolution and an energy resolution close to 15 %. This technique, known as stereoscopy, is being employed by the four large ground based gamma-ray astronomy groups. These are CANGAROO-III (the Japanese-Australian collaboration sited in Woomera, Australia: Kubo *et al.* (2004)), HESS (the largely German-French collaboration sited in Namibia: (Benbow & Hess Collaboration (2005)), MAGIC (the German-Spanish-Italian collaboration sited in La Palma: Lorenz & MAGIC Collaboration (2005)) and VERITAS (the North American-European collaboration located in Arizona: Weekes (2003)). This new generation of telescopes is already producing spectacular new science results and has pushed the number of detected VHE sources beyond thirty (see Ong (2005) for details).

† Present address: School of Physics, University College Dublin.

2. Recent Results

The number of VHE gamma-ray source categories has risen to six in the past year. To date eleven AGN, twelve SNRs (six shell type and six plerionic) a binary pulsar, a microquasar, and possible detection of diffuse emission have been reported. There are also reports of three unidentified sources raising the possibility of a new family of dark emitters, although previously unidentified sources were subsequently associated with SNRs using deep Integral exposures and archival ASCA data.

Recently the SNR RX J1713.7-3946 was mapped in VHE gamma rays at one arc-minute resolution; this degree of resolution is unprecedented at any gamma-ray energy. VHE emission was found to correlate strongly with x-ray emission, which coupled with the hard spectrum, lent the first convincing evidence of SNRs as a source of cosmic rays (Berge *et al.* (2005)). The binary pulsar PSR B1259-63 which was observed near periastron, was the first VHE galactic variable to be detected (Beilicke *et al.* (2004)). VHE emission was detected from the vicinity of the galactic center - although the spectrum of this source has been measured (Hinton & Aharonian (2005)), it is not yet clear whether it is associated with the radio source SgrA*, or with at least one supernova remnant also in the field of view. As many as twelve sources, many extended, have been discovered using sky survey observations along the galactic plane (Funk & Lemiere (2005)). Many of these sources have firm SNR or EGRET unidentified associations, although there are some with no known association whatsoever. Future multiwavelength campaigns should result in the identification of these objects. The microquasar LS 5039 was detected at TeV energies (Aharonian *et al.* (2005)), leading to the exciting possibility of studying relativistic jets, similar to those produced by AGN, in our own galaxy. A selection of AGN, all but one blazars, has been detected. Multiwavelength campaigns have shed new light on AGN emission mechanisms, with both x-ray correlated and orphan TeV flares reported (Krawczynski *et al.* (2004)). These observations will help distinguish between hadronic and leptonic emission models. The detection of AGN at higher and higher redshifts has put new constraints on the optical/infra-red extra galactic (EBL) background, with recent observations suggesting this background may be less dense than previously thought. See Quinn (2005) for a more complete description of recent AGN observations.

References

Aharonian, F., *et al.* 2005, Science, 309, 746

Beilicke, M., Ouchrif, M., Rowell, G. & Schlenker, S. 2004, The Astronomer's Telegram, 249, 1

Benbow, W. & HESS Collaboration 2005, AIP Conf. Proc. 745: High Energy Gamma-Ray Astronomy, 745, 611

Berge, D., *et al.*, AIP Conf. Proc. 745: High Energy Gamma-Ray Astronomy, 745, 263

Funk & Lemiere 2005, 29th ICRC, Pune, *(in press)*

Hillas, A. M. 1985, 19th ICRC, La Jolla, 3, 445

Hinton, J. & Aharonian, F. 2005, Galactic Center Newsletter, 21, 5

Krawczynski, H., *et al.* 2004, *APJ*, 601, 151

Kubo, H., *et al.* 2004, New Astronomy Review, 48, 323

Lorenz, E., 2005, AIP Conf. Proc. 745: High Energy Gamma-Ray Astronomy, 745, 622

Ong, R. 2005, 29th ICRC, Pune, *(in press)*

Quinn, J. 2005, These proceedings

Weekes, T. C. 2003, Cherenkov 2005 Palaiseua, *(in press)*

Populations of High Energy Sources in Galaxies
Proceedings IAU Symposium No. 230, 2005
E. J. A. Meurs & G. Fabbiano, eds.

© 2006 International Astronomical Union
doi:10.1017/S1743921306008015

Status of VERITAS

M. K. Daniel[1] on behalf of the VERITAS collaboration[2]

[1] School of Physics, University College Dublin, Belfield, Dublin 4. Ireland.
email: michael.daniel@ucd.ie

[2] for collaboration members see http://veritas.sao.arizona.edu/VERITAS_members.html

Abstract. The Very Energy Radiation Imaging Telescope Array System (VERITAS) in its first phase of operation will consist of an array of 4 Imaging Atmospheric Cherenkov Telescopes (IACTs) arranged in a 'Mercedes' star configuration. To be located at a high, dark site in Southern Arizona the full array is expected to see first light in October 2006. In February of 2005 the first VERITAS telescope achieved first light at a temporary location near to the final site. This poster summarises the status of the VERITAS instruments as of summer 2005.

Keywords. telescopes, instrumentation: detectors, gamma rays: observations

1. Hardware

The VERITAS telescopes are of an f/1 Davies-Cotton design with a 12 m aperture. The tracking of the mount is measured to be accurate to $< 0.01°$ and has a slewing speed of $\sim 0.3°/s$, but will soon be upgraded to give a slew speed of $1°/s$. The reflector is segmented: consisting of 350 individual hexagonal mirror facets of $0.332\,\mathrm{m}^2$ each, providing a total mirror area of $\sim 100\,\mathrm{m}^2$. The mirrors are measured to have a reflectivity $>85\%$ from 280 to 450 nm. The point spread function (PSF) of the telescope for the observations outlined in this paper was $0.09°$ at $80°$ elevation, which is well within the pixel size of the camera. Even so, the telescope alignment has subsequently been further refined and the current PSF is $0.06°$.

The camera consists of 499 Photonis XP2970/02 photomultiplier tubes (PMTs), with a $0.15°$ spacing giving a total field of view of $3.5°$. Contained in the camera are pre-amplifier and anode current monitoring circuits. The signals from each PMT then pass through 50m of cable, where they are digitised by a custom designed 500 Mega-Sample-Per-Second Flash-ADC system.

The trigger is a 3 level system: the first level (L1) being a programmable constant fraction discriminator (CFD); the output of which is then passed to the second level (L2) pattern recognition system which is pre-programmed to recognise triggers resembling true Cherenkov light flashes; the final stage (L3) will be an array level trigger to discriminate coincident multiple telescope triggers. The observations detailed here have been performed with a conservative CFD threshold of 6.7 photo-electrons and a 3-fold adjacent pixel pattern trigger, which corresponds to a telescope raw trigger rate of $\sim 150\,\mathrm{Hz}$.

2. Observations & Data Analysis

A full simulation chain for Telescope 1 (T1) has been developed and shows good agreement with actual data: both in terms of image parameter distributions and in terms of raw trigger rates. For a Crab like spectrum (Hillas *et al.* (1998)) this results in T1 having a threshold energy of 150 GeV, giving a gamma-ray trigger rate of $\sim 22\gamma/\mathrm{minute}$. Since the muon background for a single telescope (Vacanti *et al.* (1994)) is very large some hard imaging cuts are made on single telescope data in order to maximise the significance of a signal, these in turn push up the threshold energy of T1 to $\sim$370 GeV.

Due to its steady signal the Crab nebula is the standard candle for TeV astronomy, therefore observations of this object are useful for gauging the performance of T1 and comparing its performance to current and previous instruments. A total of 3.9 hours on source observations taken at high elevation resulted in a $19.4\,\sigma$ detection at a rate of $\sim$2.5γ/min, which gives an idea of how harsh the cuts are that maximise significance. This shows that T1 has a sensitivity of $10\,\sigma/\sqrt{\text{hour}}$ for a Crab-like source, demonstrating the improvement in performance over its predecessor, the Whipple 10 m IACT (Finley et al. (2001)).

Mrk 421 is a Blazar at a redshift of $z \sim 0.031$. This is a highly variable source most easily seen when it is in a high flaring state, but close enough that it can still be observed in its quiescent phase. For 13.1 hours on source observations a signal was detected significant at the $12.5\,\sigma$ level.

Mrk 501 is another Blazar, at a similar distance to Mrk 421, with a redshift of $z \sim 0.034$. Again this a highly variable source that was not in a high flux state at the time of observations, but still yielded a $7.1\,\sigma$ significance for 6.7 hours of on source observation time.

3. Summary

The first of the VERITAS telescopes has been operating since February 2005 at a temporary site. Despite this site being at a lower elevation and having a brighter background light than is ideal, the telescope has met all technical specifications and is demonstrably an improvement over its predecessor.

The major mechanical components of all four telescopes have now been delivered and testing the array trigger will commence in late 2005 also at this temporary site. Further improvements foreseen for the system will be the addition of light cones, which will decrease the amount of dead space between the pixels in the camera and increase the photon collection area.

Most of the major infrastructure at the final Kitt Peak site is in place and, following a stop-gap site access issue, it is foreseen that the finished array should see first light in October 2006.

Acknowledgements

The technical assistance of E. Little and E. Roache are gratefully acknowledged in the running of the telescopes. VERITAS is funded in the United States by the *Department of Energy, National Science Foundation* and *Smithsonian Institution*; by *Science Foundation Ireland*; the *Natural Sciences and Engineering Research Council* of Canada; and the *Particle Physics and Astronomy Research Council* in the U.K.

References

Finley, J. P., *et al.* 2001, Proceedings of the 27th ICRC, 2827
Hillas, A. M., *et al.* 1998, ApJ 503, 744
Reynolds, P. T., *et al.* 1993 ApJ 404, 206
Vacanti. G., *et al.* 1994, Astropart. Phys. 2, 1

Populations of High Energy Sources in Galaxies
Proceedings IAU Symposium No. 230, 2005
E. J. A. Meurs & G. Fabbiano, eds.

© 2006 International Astronomical Union
doi:10.1017/S1743921306008027

The Search for pulsed TeV gamma-ray sources

Andrew McCann[1]†, Michael Daniel[1], David J. Fegan[1], Stephen Gammell, Peter Cogan[1] and John Quinn[1]

[1]School of Physics, University College Dublin
email: andrew@ferdia.ucd.ie

Abstract. The EGRET experiment detected pulsed emission from several pulsars in the HE energy range 20 MeV < E < 10 GeV. However, no pulsed emission has ever been detected above 300 GeV, which points to a cut-off in pulsed emission somewhere between $\sim$10 GeV and $\sim$300 GeV. Software has been developed at UCD which is capable of examining gamma-ray events in the energy range $E > 130$ GeV and detecting the signature periodicity of pulsars if it exists. Data recorded on the Crab pulsar with the Whipple Imaging Atmospheric Cerenkov Telescope has been analysed with this software and upper limits on pulsed emission have been obtained.

1. Introduction

The low flux of gamma rays above 10 GeV and the small collection area of gamma-ray satellites means that there have been no detections of pulsars above 10 GeV. The Imaging Atmospheric Cerenkov Technique (IACT) has limited sensitivity at energies below $\sim$300 GeV due to effects of the cosmic-ray background. However, it has a proved successful at detecting TeV gamma rays. To this date there has been no detection of TeV gamma-ray emission from pulsars by the IACT, which suggests that the emission mechanisms responsible for pulsed gamma rays are halted between $\sim$10 GeV and $\sim$300 GeV. This cut-off is a prediction of all current pulsar emission models, however the energy at which the cut-off occurs is unique to each model. Experimental determination of the cut-off energy, and its correlation with a particular theoretical model, provides the motivation for this work. For a review of pulsar emission models see Baring (2004) and Hirotani (2005) .

In this work data on the Crab pulsar, taken with the Whipple Telescope (Finley *et al.*, 2000), has been analysed with a gamma-ray selection procedure called *Kernel* analysis (Dunlea *et al.*, 2001). This procedure has proved successful at detecting gamma rays close to the telescope's energy threshold.

2. Results

No pulse profile is evident in the lightcurves determined from the gamma-ray data (see Figure 1) and χ^2 tests yielded no deviation from a uniform distribution. Using the method of Helene (1983), upper limits on peak area were determined, with the assumption that phases of emission at TeV energies are the same as those detected by EGRET. These upper limits were subsequently converted to flux upper limits, yielding an integral flux upper limit of $5.8 \times 10^{-7} m^{-2} s^{-1}$ for $E > 130 \pm 39$ GeV. This upper limit is compared with EGRET data and previous IACT investigations in Figure 2.

† Present address: School of Physics, University College Dublin.

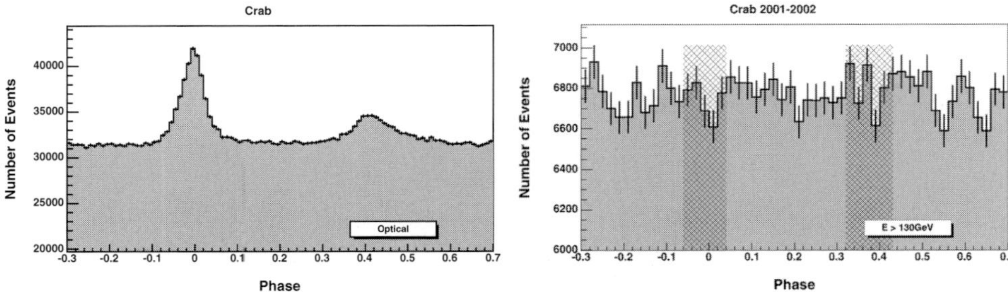

Figure 1. *Left*: The optical lightcurve of the Crab pulsar determined from data taken with the Multiple Mirror Telescope *Right*: Crab pulsar data taken with the Whipple telescope between 2000-2001 (E > 300 GeV). Both datasets were analysed by the same method. No pulse is evident in the gamma-ray data.

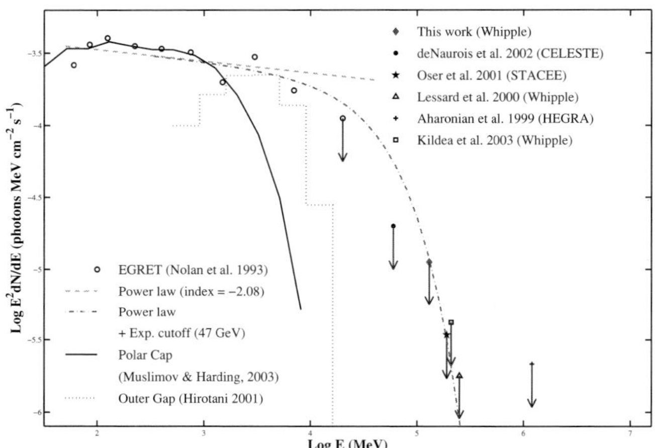

Figure 2. Pulsed photon spectrum of the Crab pulsar. Originally from Kildea *et al.* (2003).

3. Conclusion

This work has successfully probed pulsar emission at energies closer to the Whipple telescope energy threshold than any other investigation. This was made possible by applying a low-energy *Kernel* selection developed at UCD. In spite of these advances the results presented do not constrain any mainstream models of pulsed emission and it is clear that a further reduction in energy threshold will be required to do so. Other candidates for pulsed TeV emission are currently under study.

References

Baring, M.G. 2004, *Advances in Space Research*, 33, 552

Dunlea, S. *et al.*, 2000, in *ICRC*, Vol. 7, p. 2939

Finley, *et al.*, 2000, in *American Institute of Physics Conference Series*, p. 301

Kildea, J., Gammell, S., and VERITAS Collaboration 2003, in *ICRC*, p. 2377

Helene, O. 1983, *Nuclear Instruments and Methods* 212, 319

Hirotani, K. 2005, *Astrophysics and Space Science*, 297, 81

Session 2

High energy processes in the ISM

Questioning time.

Populations of High Energy Sources in Galaxies
Proceedings IAU Symposium No. 230, 2005
E. J. A. Meurs & G. Fabbiano, eds.

© 2006 International Astronomical Union
doi:10.1017/S1743921306008039

High-energy radiation generated by winds and shocks: SNRs and superbubbles

Andrei M. Bykov[1]

[1] A.F.Ioffe Institute for Physics and Technology, 194021, St.Petersburg, Russia
email: byk@astro.ioffe.ru

Abstract. We discuss populations of X-ray and γ-ray sources in star-forming regions (SFR). Interacting winds of massive stars and high supernova activity in SFRs can be powerful sources of high energy emission. Models of nonthermal particle acceleration in the vicinity of active SFRs are reviewed. A class of hard emission sources where a fast wind from a massive star collides with a supernova shell is described. Stellar winds of massive stars and core collapsed supernova explosions with great energy release in the form of multiple interacting shock waves inside the superbubbles are argued as favorable sites of nonthermal particle acceleration. Young stellar objects and supernova activity in the dense environment of starforming regions produce an another potentially abundant class of hard faint X-ray sources due to interaction of fast moving knots with the dense ambient medium. The knots could have very different physical nature, e.g. supernova ejecta fragments or Herbig-Haro-like objects. We argue that the sources may have rather steep logN—logS distribution and can contribute substantially to the galactic diffuse emission including the both low-ionized 6.4 keV and He-like Fe lines.

Keywords. X-rays: ISM, acceleration of particles, ISM: bubbles, supernova remnants

1. Introduction

Energetic outflow events with fast shocks in a dense ambient medium should result in a rapid conversion of kinetic power into emission. The star-forming regions in galaxies are generically associated with molecular clouds and contain a variety of energetic outflows at different stages of massive star evolution from proto-stellar accreting objects through fast winds of massive OB or WR stars to the most energetic supernova events.

A spectacular example is an interaction of a supernova remnant (SNR) with a molecular cloud that manifests itself by a number of appearances in a wide range of wavelengths from radio to gamma-rays (e.g. Shull 1980, Wheeler *et al.* 1980). Propagation of a radiative shock wave driven by a supernova shell through a molecular cloud leads to a substantial non-thermal emission both in hard X-rays and in γ-rays. The complex structure of a molecular cloud consisting of dense massive clumps embedded into the inter-clump medium could result in localized sources of hard X-ray/γ-ray emission correlated with bright molecular emission. It has been shown that a hard X-ray and γ-ray emission region should consist of an extended shell-like structure (appearing also in radio as a shell of a relatively flat spectrum) related to the radiative shock and localized sources corresponding to shocked molecular clumps (Chevalier 1999, Bykov *et al.* 2000). The shocked clumps would have (sub)parsec scales and emit very hard X-ray continuum spectra with no strong radio counterparts.

Here we shall discuss different phenomena of SFR activity in the dense environment – a potentially abundant class of hard faint X-ray sources due to interactions of fast moving supernova ejecta fragments with dense medium and non-thermal sources powered by SNR shell interactions with a fast wind of a massive star.

2. X-ray emission of fast moving knots in SFRs

2.1. *Supernova ejecta fragments in SFRs and their X-ray emission*

Multiwavelength studies of SNRs have revealed a complex structure of metal-rich ejecta with the presence of fast moving isolated fragments of SN ejecta, interacting with the surrounding media. In the optical a few hundred fast moving knots (FMKs) were observed with *HST* outside the main shell of Cas A (e.g. Fesen *et al.* 2002; 2005) and in some other SNRs. Optical FMKs in Cas A are very abundant in O-burning and Si-group elements. They have a broad velocity distribution around 6,000 km s^{-1} and apparent sizes below 0.01 pc.

An appearance and detection probability of individual fast moving knots in the X-rays depend strongly on the ambient density. The optical knots in Cas A should have $\mathcal{L}_x < 10^{29}$ erg s^{-1} because of the low ambient density. Similar FMKs in the dense environment of the Galactic Centre (GC) region would have $\mathcal{L}_x \gg 10^{29}$ erg s^{-1} with bright IR counterparts.

In a low density medium only relatively large FMKs can be observed in X-rays individually, but since they are long-lived, their collective emission could be substantial. *Chandra* and *XMM-Newton* observations revealed a head-tail structure with a prominent Si line in the Vela shrapnel A, indicating that the object is a fast ejecta fragment of the size $\sim$0.3 pc, velocity about 1,000 km s^{-1} and $\mathcal{L}_x \sim 10^{31}$ erg s^{-1} (see e.g. Aschenbach 2002).

In a dense cloud, fast moving knots will be bright, but short-lived. IC 443 – a SNR interacting with a molecular cloud is an obvious candidate to search for X-ray knots in a dense cloud. Some of the hard X-ray sources detected with *XMM-Newton* and *Chandra* in the "molecular" SNR IC 443 are likely to be fragments of the SN ejecta. In Fig. 1 (left panel) we present a *Chandra* image of XMMU J061804.3+222732 - an isolated source interacting with a molecular cloud border in IC 443 (Bykov, Bocchino & Pavlov 2005). The source has an extended morphology with two compact cores. The morphology is consistent with that expected for a crushed supernova ejecta fragment (see e.g. 2D simulations by Klein *et al.* 1994). The source size is below 0.1 pc and $\mathcal{L}_x \approx 10^{32}$ erg s^{-1} in the (0.5-7) KeV band (at 1.5 kpc distance). It has a very hard spectrum of a photon index $\Gamma \lesssim 1.3$ with a prominent thermal component. Further multi-wavelength high resolution observations will finally establish the nature of the source in IC 443.

2.2. *X-ray knot ensemble and the Galactic ridge emission*

The observed X-ray emission of the Galactic ridge is known to contain an extended component which can not be accounted for by the resolved X-ray point sources (e.g. Tanaka 2002, Ebisawa *et al.* (2000)). Two basic possibilities exist — unresolved faint sources or truly diffuse emission. Thus the problem of the origin of the observed large scale X-ray emission from the Galactic ridge requires a careful study of possible classes of abundant hard X-ray sources with $\mathcal{L}_x \sim 10^{29}$ erg s^{-1}.

Recently Ebisawa *et al.* (2000) analyzed deep *Chandra* observation of the Galactic plane at $(l,b) = (28.5°, 0.0°)$ with flux limit down to $\sim$3 $\times$ 10^{-15} erg cm^{-2} s^{-1} in 2-10 keV band and $\sim$2 $\times$ 10^{-16} erg cm^{-2} s^{-1} in 0.5-2 keV band. The number of sources detected in the hard band is consistent with the extrapolation of the extragalactic $logN - logS$ distribution of index about 1.3. They concluded that the sum of all detected point sources accounts only for 10% of the total observed flux and that the X-ray emission from the galactic plane has a truly diffuse origin. The later conclusion was drawn on the basis of the assumption that if there exists a large population of faint yet unresolved sources it should

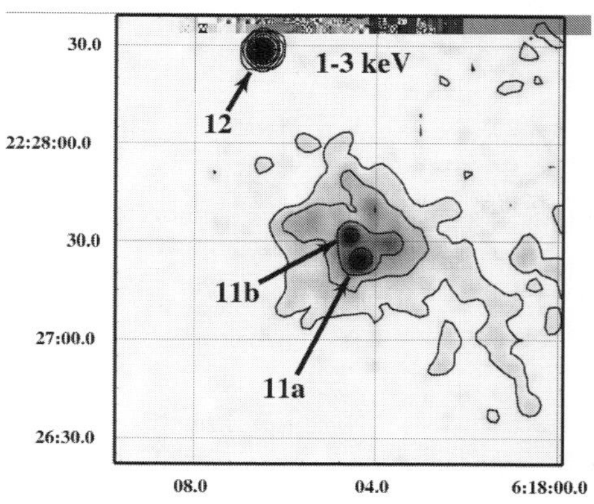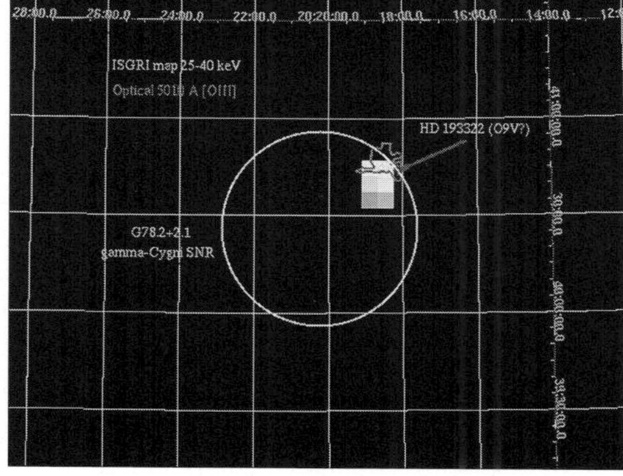

Figure 1. Left panel *Chandra* ACIS 1-3 keV image of XMMU J061804.3+222732 in the region of interaction of IC 443 SNR with a molecular cloud - a possible FMK.
Right panel *INTEGRAL ISGRI* 25-40 keV image of γ-Cygni SNR (Bykov *et al.* 2004). Optical [OIII] 5010 Å line contour (solid line) from Mavromatakis (2003). The SNR border (radio) is roughly indicated by the large solid circle. The arrow points to the position of the O9V star HD 193322 (20:18:07,+40:43:55). All maps are made for J2000. The source may originate from the SNR-O star wind interaction.

have rather flat $logN - logS$ distribution of index $\sim$1.3. We will show below, however, that some faint source populations may have much steeper distributions. Analyzing *XMM-Newton* observations covering three square degrees between $l = 19° - 22°$ and $b = \pm 0.6°$ Hands *et al.* (2004) concluded that the integrated contribution of the detected galactic sources plus the extragalactic signal accounts for up to 20% of the observed surface brightness.

The Galactic center region was deeply studied with *Chandra* observatory (Muno *et al.* 2003, 2004; Park *et al.* 2004). With an about 600 ks *Chandra* exposure of a $17' \times 17'$ field in the GC region Muno *et al.* (2003) catalogued 2357 X-ray sources limited by the luminosity $\mathcal{L}_x \gtrsim 10^{31}$ erg s^{-1} (2.0–8.0 keV). A substantial part of the positions of the *Chandra* sample sources is projected onto the Circum-Nuclear Disc (see e.g. Mezger *et al.* 1996) between 1.7 and 7 pc around the GC that contains $\sim 10^4 M_\odot$ of highly clumped matter. The sources have $logN - logS$ distribution of index $\alpha = -1.7 \pm 0.2$ which is substantially steeper than the indexes of the distribution at other galactic locations (c.f. Ebisawa *et al.* (2000)). A subclass of cataclismic variables - intermediate polars, was suggested by Muno *et al.* (2004) as a possible contributor of the sources in the sample, given their spectral properties. The properties of $logN - logS$ distribution of intermediate polars are not yet studied to draw the conclusion if the observed steep $logN - logS$ index can be extended down to the low luminosities ($\mathcal{L}_x \sim 10^{29}$ erg s^{-1}) to account for the substantial diffuse component.

In this respect we discuss here a potentially numerous ensemble of X-ray point sources associated with fast moving knots. A very important distinctive feature of the X-ray emission of fast moving knots is a strong dependence of the knot luminosity $\mathcal{L}_x$ on the ambient density. The knots can be metal-rich SN ejecta fragments (Bykov 2003) or alternatively they can result from some processes in accreting systems producing jets and ballistically moving isolated knots. An example is the Herbig-Haro flows in star formation regions. The flows may appear as parsec scale structures with the flow collimation depending on the young stellar object (YSO) mass. Most outflows from high-mass protostars are poorly collimated (e.g. Reipurth and Bally 2001).

An ensemble of unresolved knots can contribute substantially to the diffuse X-ray emission including the iron line emission observed from the Galactic Center region and the Galactic ridge.

An ensemble of hard X-ray point-like sources associated with fast moving supernova ejecta fragments is abundant and can account for the observed properties of the detected GC sources such as the hardness ratios and the logN–logS distribution.

We have simulated X-ray spectra of the FMKs of different velocities in the GC environment. In our model the hard X-ray emission of FMKs is due to both hot thermal postshock plasma and nonthermal particles accelerated at the bow shock. The particles propagate through a metal-rich clump, producing K-shell ionization line photons and hard nonthermal bremsstrahlung. It should be noted that most of continuum emission simulations made for the metal-rich supernova ejecta fragments are relevant to any kind of FMKs that do not necessarily originate from SN activity and thus have the standard composition. The knots could be ejected by compact objects like Sgr A*, YSOs or other objects. The X-ray line emission from metal-rich supernova ejecta fragments has specific features due to the substantial internal line resonant absorption effect. The efficiency of the bremsstrahlung non-thermal emission of the knot body is higher than that from the shocked ambient gas, because of high mean charge <Z> in the metal-rich knot.

The lifetime of an FMK in a dense media is an important factor in the study of the knot statistics. A fast moving knot decelerates due to the interaction with the ambient gas. The drag deceleration time of a knot of velocity v, mass $\mathcal{M}$ and radius $\mathcal{R}$ can be estimated as $\tau_\mathrm{d} \sim \mathcal{M}/(\rho_\mathrm{a} v \pi \mathcal{R}^2) \approx 10^3 \cdot \mathcal{M}_{-3}/(n_\mathrm{a2}\, v_8\, \mathcal{R}^2_{-2})$ years. Here $\mathcal{M}_{-3} = \mathcal{M}/10^{-3}\, M_\odot$ and $\mathcal{R}_{-2} = \mathcal{R}/(0.01\mathrm{pc})$. The number density n_a2 of the ambient matter is measured in 100 cm^{-3} and the FMK velocity v_8 is measured in 1,000 km s^{-1}. In the inner 20$'$ of the GC region the average number density $n_\mathrm{a} \gtrsim 100$ cm^{-3} (e.g. Mezger $et\ al.$ 1996) and the fragment deceleration time $\tau_\mathrm{d} \lesssim 10^3$years.

The high pressure gas behind the strong bow shock of a fast moving ejecta fragment could drive an internal shock resulting in the knot crush and fragmentation. In the case of an adiabatic bow shock the internal shock velocity $v_\mathrm{is} \approx v/\chi^{1/2}$, where the density contrast $\chi = \rho_\mathrm{k}/\rho_\mathrm{a}$, and ρ_a and ρ_k are the ambient gas and the dense fragment densities, respectively. The knot crushing time scale $\tau_\mathrm{c} = \mathcal{R}/v_\mathrm{is}$. There are 2D hydrodynamical simulations showing that the fast knot fragmentation occurs on the timescale of $\sim(3-4)\tau_\mathrm{c}$ (e.g. Klein $et\ al.$ 1994). For the FMKs of mass $\mathcal{M}_{-3} \sim 1$ and $\mathcal{R}_{-2} \sim 1$ the fragmentation time scale is a factor of $\sim$2.5 less than the drag deceleration time. However, the FMK will be a source of hard X-ray emission even at the fragmented stage if the strong forward shock is still driven by the FMK.

The hydrodynamical estimation of the inner shock velocity v_is given above assumes an efficient conversion of the bow shock ram pressure to the knot internal shock energy. The effects of nonthermal particle acceleration reduce the postshock gas pressure and the internal shock velocity, thus increasing τ_c. In that case the life time would be close to τ_d, unless the ablation processes. The effect of magnetic fields and nonthermal particles on the knot ablation is not yet studied in any details.

Massive star winds would change the circum-stellar environment, creating caverns of low density matter. The lifetime of fast fragments of a SN in the low density cavern should be longer than $R_\mathrm{b}/v_\mathrm{k}$, where R_b is a wind bubble size. Following the approach of Chevalier (1999) and accounting for the high pressure of the GC ambient matter one may estimate the typical scale $R_\mathrm{b} \sim$3 pc, for WR, while R_b is $\sim$1 pc for O9-B0 progenitor stars. The lifetimes of FMKs are $\sim$10^3 years. The collective effect of powerful stellar winds in compact associations could blow out a more extended cavern - superbubble, providing longer lifetimes for the fragments of the SNe exploded inside the cavern. Metal rich FMKs moving through the hot superbubble cavern can provide an efficient cosmic ray injection mechanism. We will discuss superbubbles in Section 3.

The X-ray spectra of the fragments are hard. They contain thermal and power law components with possible Fe lines. Both 6.7 and 6.4 keV lines are expected, depending on the FMK ionization structure and the relative strength of the nonthermal component. The total equivalent width is about 500 - 600 eV for a Fe mass of $\sim 10^{-4}\ M_\odot$.

Two or three SNRs during the last 1,000 years could produce an ensemble of more than 1,500 FMKs in the GC region providing a normalization of the logN–logS distribution consistent with the *Chandra* data in the Galactic Centre region. The filamentary features apparent in the *Chandra* images of the region could be also relevant to the SN activity.

Consider FMK ejecting events at a rate $\nu_{SN}(\vec{r}, t)$ with the total number $\mathcal{N}_\star$ of the ejected fragments per event. A simplified order-of-magnitude estimation of the summed contribution to the galactic X-ray emission from an ensemble of fast moving supernova ejecta fragments can be made using the relation $L_x \propto \mathcal{N}_\star \cdot \nu_{SN} \cdot \tau_d \times \mathcal{L}_x$. The thermal component of X-ray luminosity for a knot of velocity above 1,000 km/s scales with $\mathcal{L}_x \propto n_a^2 \mathcal{R}^3 T^{1/2}$. Then the total thermal X-ray luminosity of the FMK ensemble is

$$L_{xT} \sim 10^{37} \cdot \frac{\Delta M_{ej}}{M_\odot} \cdot \frac{\nu_{SN}}{0.03\ yr^{-1}} \cdot \mathcal{R}_{-2} \cdot n_{a2}\ \text{erg s}^{-1}.$$

The nonthermal emission component is more sensitive to some poorly constrained parameters, as the magnetic field in the cold ejecta fragment and the magnetic fluctuation spectrum. The nonthermal luminosity can be approximated with $\mathcal{L}_x \propto\ <Z> n_a \mathcal{R}^2 v^\zeta$. In a highly idealized case of a "thick target" where all the nonthermal electrons are stopped inside the knot we have $\zeta = 3$. That requires a relatively low particle diffusion coefficient in a dense knot of the mean atomic charge $< Z >$. Then

$$L_{xNT} \sim 2 \times 10^{37} \cdot \frac{\Delta M_{ej}}{M_\odot} \cdot \frac{\nu_{SN}}{0.03\ yr^{-1}} \cdot \frac{<Z>}{8} \cdot \left(\frac{v}{6,000\ \text{km s}^{-1}}\right)^2\ \text{erg s}^{-1}.$$

For typical parameters of the fast oxygen knots of velocity $\sim 6,000$ km s^{-1}, the mean charge $< Z >= 8$ and the total mass of the ejected knots $\Delta M_{ej} \sim 0.5 M_\odot$ we obtain $L_{xNT} \sim 10^{37}$ erg s^{-1}. If a SN ejects $\mathcal{N}_\star \sim 500$ FMKs of mass $10^{-3} M_\odot$ and velocity 6,000 km s^{-1}, the FMKs would carry about 15% of the SN kinetic energy. The "thick target" approximation is probably oversimplified, and $\mathcal{L}_x \propto v^\zeta$ with $\zeta < 3$ seems to be more realistic. On the other hand it is possible to develop a general approach to simulate the logN–logS distributions for an ensemble of ballistically moving sources providing a more rigorous description of their collective X-ray emission. The approach allows to study the logN–logS distribution of FMKs of supernova origin as well as ballistically moving objects ejected by accreting processes (e.g. jets and Herbig-Haro outflows).

2.3. *The* logN–logS *distribution of FMKs*

The X-ray luminosity $\mathcal{L}_x(v, M, \mathcal{R}, n_a)$ of an FMK depends on the knot velocity, mass, radius, and the ambient matter density. To simulate the logN–logS distribution of FMKs we should know the FMK velocity and size distributions. We consider here a simplified analytic model for logN–logS distributions. The differential distribution function of the FMKs velocities and radii $\mathcal{N}(\vec{r}, \vec{v}, \mathcal{R}, t)$ satisfies

$$\frac{\partial \mathcal{N}}{\partial t} + \dot{\vec{r}} \frac{\partial \mathcal{N}}{\partial \vec{r}} + \dot{\vec{v}} \frac{\partial \mathcal{N}}{\partial \vec{v}} + \dot{\mathcal{R}} \frac{\partial \mathcal{N}}{\partial \mathcal{R}} = q(\vec{r}, \vec{v}, \mathcal{R}, t), \tag{2.1}$$

where the source function $q(\vec{r}, \vec{v}, \mathcal{R}, t) = \nu \mathcal{N}_\star \psi(v, \mathcal{R})$. Here $\psi(v, \mathcal{R})$ is the probability distribution of the initial sizes and velocities of the FMKs. For a ballistically moving

FMK the drag deceleration in the ambient medium is $\dot{\vec{v}} \propto -\rho_a v \vec{v} \mathcal{R}^2$ (for a stretched knot $\mathcal{R}$ is the transverse radius).

We apply the Eq. (2.1) to some active region assuming a constant local ejection rate ν and a locally homogeneous ambient medium (on the FMK deceleration scale). Then the asymptotic averaged distribution is

$$\mathcal{N}(v, \mathcal{R}) = \int_v^\infty \frac{\bar{q}(v, \mathcal{R})}{\dot{v}} \, dv, \qquad (2.2)$$

where $\bar{q}(v, \mathcal{R}) \propto a^{-1} q(v, \mathcal{R}/a)$ and $a(\chi)$ is the knot expansion factor simulated by Klein et al. (1994) within an associated model of shocked cloud evolution.

It is instructive to estimate a power-law index of the flux distribution function $N(S)$. The index, defined as $\alpha = \log N / \log S$, is constant for a relatively narrow flux band. We shall consider both thermal and non-thermal emission models for FMKs described by Bykov (2003).

We assume that $\psi(v, \mathcal{R}) \propto \mathcal{R}^{-\beta} v^{-\eta}$ in a relatively narrow range of $\mathcal{R}$. Then, integrating the distribution $\mathcal{N}(v, \mathcal{R})$ over velocity (above the threshold $v_0 \propto \mathcal{R}^{-1/2}$) and using $\mathcal{L}_x \propto n_a^2 \mathcal{R}^3 T^{1/2}$, we obtain the local index estimation for *the thermal emission* $\alpha = -4/3 + \eta/6 - \beta/3$. The index is consistent with $\alpha = -1.7 \pm 0.2$ obtained by Muno et al. (2003) for the *Chandra* GC sample, if the initial distributions $\psi(v, \mathcal{R})$ satisfy $-3.4 < \eta - 2\beta < -1.0$. The value of $\eta \lesssim 1$ agrees with the velocity distribution of FMKs simulated by Kifonidis et al. (2003). The nonthermal flux distributions depend on the diffusion regime of the accelerated particles inside FMKs. The magnetic fluctuation spectrum inside an FMK and the dependence of the diffusion coefficients on the FMK size and velocity are yet poorly known. However, FMK thermal emission models and some simplified non-thermal models allow steep logN–logS distribution of $\alpha \lesssim -1.7$ with a large population of low luminosity objects. Their emission spectra contain hard power-law component of typical photon index Γ below 1.5. It remains to be checked if a numerous faint sources population can contribute seriously to the ridge diffuse emission component discussed above as it is proved to be the case in the soft γ-ray domain (see Lebrun et al. 2004).

3. Hard emission from SNR-wind interactions and superbubbles

Young massive star formation occurs in massive molecular clouds. A gravitational collapse of a giant molecular cloud of 10 to 30 pc size usually results in a birth of a group of several tens massive O and B stars, called an OB-association. As the O and B stars have short lifetimes (e.g. Maeder and Meynet 2000), the dispersion velocities within such a group are small, not exceeding 4-6 km s^{-1}, and the association remains compact, though it is not gravitationally bound. It is estimated that more than 60 percent of all the OB stars in the Galaxy belong to one of the OB-associations (Garmany, 1994). A typical size of such an association is about 35 pc. Thus, in an evenly distributed association containing 100 stars the distance between two neighbour stars is about 12 pc. Most of the OB-associations consist of substructures named OB-subgroups where the density is several times higher. For example, the R136 subgroup of the 30 Doradus association consists of 9 O stars within 3.4 pc (Walborn et al., 1999). One of the most massive young open stellar cluster in the Galaxy is Westerlund 1 that contains more than 200 massive stars with a very rich WR star population (e.g., Clark et al. 2005). Another rich cluster, the Arches cluster, is situated at a projected distance of 30 pc from the Galactic centre. The age of the Arches cluster is estimated as 2-2.5 Myr. The compact groups are the favorable sites of some new type of high energy sources produced by shocks and winds.

Colliding winds in early type binaries were suggested by Eichler and Usov (1993) as a possible strong sources of nonthermal particles and emission.

A region where an expanding shell of a SNR interacts with a fast powerful wind of a young massive star (or a star cluster) contains a converging MHD flow. Axisymmetric hydrodynamic simulations of SNR-wind collisions carried out by Velázques, Koenigsberger, Raga (2003) illustrate the basic properties of the multi-shock flow during the collision. Even before a direct collision the converging flow will exist between the fast star wind and the expanding supernova ejecta. The converging flow is argued to be a plausible site for GeV–TeV regime lepton acceleration. The kinetic models presented predict efficient particle acceleration with a hard spectrum of electrons at the TeV energy regime. Synchrotron emission of the TeV electrons may have a very hard spectrum with a photon index ~ 1 and a spectral break in between keV and MeV photon energies for extended SNRs. Because of that hard spectra of accelerated electrons the mechanism provides a high efficiency of conversion of SNR shell-fast wind kinetic luminosity into hard X-rays and γ-rays. The efficiency of synchrotron emission in hard X-rays is high. The X-ray luminosity may exceed 0.001 of the kinetic power of the shocks. In Fig.1 (right panel) we show a hard X-ray emission clump detected with *INTEGRAL-ISGRI* in a region containing γ-Cyg SNR and the stellar wind of the O9V star HD 193322. The hard X-ray source can be explained in this scenario. The new type of hard X-ray sources may be rather common in star forming regions. The Galactic center region could contain such kind of objects. The X-ray ridge seen by *Chandra* in the $15''$ vicinity of Sgr A* was suggested to be produced by interaction of Sgr A East SNR with collective wind from massive stars of the central cluster (Rockefeller *et al.* 2005). Recent hard X-ray observations with *INTEGRAL-ISGRI* of the Galactic Center region by Bélanger *et al.* (2005) revealed a number of hard X-ray sources in the close vicinity of Sgr A* some of which may have the same nature as that due to SNR-stellar wind interaction discovered in γ-Cyg SNR. The hard X-ray emission of synchrotron origin was suggested by Neronov *et al.* (2005) to be a source of hard diffuse emission in the Galactic Center region. The sources are promising candidates for observations with H.E.S.S. and other TeV telescopes.

At the later stages of the OB association evolution the kinetic energy release within the bubble created by a stellar association may reach a few times 10^{38} erg s^{-1} due to intense stellar winds and multiple SN explosions. The process is accompanied by formation of shocks, large scale flows and broad spectra of MHD fluctuations in a tenuous plasma with frozen-in magnetic fields. Vortex electric fields generated by the large scale motions of highly conductive plasma with shocks result in a non-equilibrial distribution of the charged nuclei. Non-linear models of temporal evolution of particle distribution function accounting for the feedback effect of the accelerated particles on the shock turbulence inside the superbubble demonstrated a high efficiency of the conversion of the large-scale turbulence energy to energetic particles on ~ 0.1 Myr time scale and soft-hard-soft evolution of the particle spectra in 10 Myrs with asymptotic power-law index below 3. The superbubbles should be very plausible sites of cosmic ray particle acceleration (e.g. Bykov 2001, Parizot *et al.* 2004).Thermal and non-thermal X-ray emission observed recently with *Chandra* in 30 Dor superbubble by Bamba *et al.* (2004) revealed the presence of particle acceleration processes there.

Acknowledgements

I would like to acknowledge my collaborators for fruitful cooperation. IAU travel support is gratefully acknowledged. The work was partially supported by RBRF grants 03-02-17433 and 04-02-16595.

References

Aschenbach, B. 2002, In: Neutron Stars, Pulsars and Supernova Remnants. Proc 270 WE-Heraeus seminar, eds. W.Becker *et al.*, MPE, p. 13

Bamba, A., Ueno, M., Nakajima, H. & Koyama, K. 2004, *ApJ* 602, 257

Bélanger, G., Goldwurm, A., Renaud, M. *et al.* 2005, *ApJ (in press), astro-ph/0508128*

Bykov, A.M. 2001, *Space Science Reviews* 99, 317

Bykov, A.M. 2003, *Astron. Astrophys.* 410, L5

Bykov, A.M., Chevalier, R.A., Ellison, D.C., Uvarov, Yu.A. 2000, *ApJ* 538, 203

Bykov, A.M., A. Krassilchtchikov, M. Yu. Uvarov, A. *et al.* 2004, *Astron. Astrophys.* 427, L31

Bykov, A. M., Bocchino, F. & Pavlov, G.G. 2005, *ApJ* (Letters) 624, L41

Chevalier, R.A. 1999, *ApJ* 511, 798

Clark, J. S., Negueruela, I., Crowther, P. A., Goodwin, S. P. 2005, *Astron. Astrophys.* 434, 949

Ebisawa, K., Tsujimotc, M., Paizis, A., *et al.* 2005, *ApJ (in press), astro-ph/0507185*

Eichler, D. & Usov, V. 1993, *ApJ* 402, 271

Fesen R.A., Morse, J.A., Chevalier, R.A. *et al.* 2002, *AJ*, 122, 2644

Fesen R.A., Hammell, M.C., Morse, J.A., *et al.* 2005, *ApJ (in press) astro-ph/0509067*

Garmany, D. 1994 *PASP, 106, 25*

Hands, A.D.P., Warwick, R.S., Watson, M.G. & Helfand, D.J. 2004, *MNRAS*, 351, 31

Kifonidis, K., Plewa, T., Janka, H.-Th. & Müller, E. 2003 *Astron. Astrophys.* 408, 621

Klein, R.I., McKee, C.F., & Colella, P. 1994, *ApJ*, 420, 213

Lebrun, F., Terrier, R., Bazzano, A., *et al.*, 2004, *Nature* 428, 293

Maeder, A. & Meynet, G. 2000, *Ann. Rev. Astron. Astrophys.* 38, 143

Mavromatakis, F. 2003, *Astron. Astroph.*, 408, 237

Mezger, P.G., Duschl, W.J. & Zylka, R. 1996, *Astron. Astroph. Rev.*, 7, 289

Muno, M.P., Baganoff, F.K., Bautz, M.W., Brandt, W.N., *et al.* 2003, *ApJ*, 589, 225

Muno, M.P., Arabadjis. J.S., Baganoff, F.K., Bautz, M.W., *et al.* 2004, *ApJ*, 613, 326

Neronov, A., Chernyakova,M., Courvoisier,T.J.-L. & Walter.R. 2005, *astro-ph/0506437*

Parizot, E., Marcowith, A., van der Swaluw, E. *et al.* 2004, *Astron. Astrophys.* 424, 747

Park, S.P., Muno, M.P., Baganoff, F.K., Maeda, Y., *et al.* 2004, *ApJ*, 603, 548

Reipurth, B. & Bally, J. 2001, *Ann. Rev. Astron. Astrophys.* 39, 403

Rockefeller, G., *et al.* 2005, *ApJ (in press) astro-ph/05006244*

Shull, J.M. 1980, *ApJ* 237, 769

Tanaka, Y. 2002, *Astron. Astrophys.* 382, 1052

Velázques, P.,Koenigsberger & G. Raga, A.C. 2003, *ApJ* 584, 284

Walborn, N.R., Barba, R.H., Brandner, W., Rubio, M., *et al.* 1999, *AJ* 117, 225

Wheeler, J.C., Mazurek,T.J. & Sivaramakrishnan, A. 1980, *ApJ* 237, 781

Discussion

COURVOISIER: What is the efficiency of the energetic particle acceleration in your super-bubble modeling?

BYKOV: The particle acceleration mechanism by multiple weak shock in the non-linear regime considered here provides efficient creation of a nonthermal nuclei population with a hard low-energy spectrum, containing a substantial part (about 30%) of the kinetic energy released by the winds of young massive stars and supernovae.

COURVOISIER: You just said that "a substantial fraction" of the energy dissipated in superbubbles comes out as non-thermal particles. This would mean that shocks deliver more non-thermal than thermal energy. Can you confirm and let us know how likely this is to be?

BYKOV: Yes, indeed the efficiency of non-thermal particle generation in a superbubble could be about 20-30% in the nonlinear model I mentioned. The relative efficiency of particle acceleration as compared with thermal heating is high enough in the case of weak shocks in magnetized plasma. In a weak MHD shock the entropy production (heating) is proportional to the cube of the density contrast, while the acceleration rate is proportional to the square of the amplitude.

Populations of High Energy Sources in Galaxies
Proceedings IAU Symposium No. 230, 2005
E. J. A. Meurs & G. Fabbiano, eds.

© 2006 International Astronomical Union
doi:10.1017/S1743921306008040

Recent nucleosynthesis results from INTEGRAL

G. Weidenspointner

Centre d'Etude Spatiale des Rayonnements, 9, avenue du Colonel-Roche, BP 4346, 31028
Toulouse Cedex 4, France
email: Georg.Weidenspointner@cesr.fr

Abstract. Since its launch in October 2002, ESA's INTEGRAL observatory has enabled significant advances to be made in the study of Galactic nucleosynthesis. In particular, the imaging Ge spectrometer SPI combines for the first time the diagnostic powers of high resolution gamma-ray line spectroscopy and moderate spatial resolution. This review summarizes the major nucleosynthesis results obtained with INTEGRAL so far. Positron annihilation in our Galaxy is being studied in unprecented detail. SPI observations yield the first sky maps in both the 511 keV annihilation line and the positronium continuum emission, and the most accurate spectrum at 511 keV to date, thereby imposing new constraints on the source(s) of Galactic positrons which still remain(s) unidentified. For the first time, the imprint of Galactic rotation on the centroid and shape of the 1809 keV gamma-ray line due to the decay of ^{26}Al has been seen, confirming the Galactic origin of this emission. SPI also provided the most accurate determination of the gamma-ray line flux due to the decay of ^{60}Fe. The combined results for ^{26}Al and ^{60}Fe have important implications for nucleosynthesis in massive stars, in particular Wolf-Rayet stars. Both IBIS and SPI are searching the Galactic plane for young supernova remnants emitting the gamma-ray lines associated with radioactive ^{44}Ti. None have been found so far, which raises important questions concerning the production of ^{44}Ti in supernovae, the Galactic supernova rate, and the Galaxy's chemical evolution.

Keywords. gamma rays: observations; Galaxy: general; nuclear reactions, nucleosynthesis, abundances

1. Introduction

Gamma-ray line astronomy has opened a new and unique window for studying nucleosynthesis in our Galaxy. The singular advantage of gamma-ray spectroscopy over other observations is that it offers the opportunity to detect directly and identify uniquely individual isotopes at their birthplaces. The rate at which radioactive decays proceed is in general unaffected by the physical conditions in their environment, such as temperature or density. The interstellar medium is not dense enough to attenuate gamma rays, so that radioactive decays can be observed throughout our Galaxy. Recent reviews on implications of gamma-ray observations for nucleosynthesis in our Galaxy can be found in Diehl, Prantzos, & von Ballmoos (2005) and Prantzos (2005).

2. Results

The nucleosynthesis results presented in the following have all been obtained from observations with the two main instruments on board the INTEGRAL observatory: the spectrometer SPI and the imager IBIS (for details regarding the instruments, see Kretschmar (2005) and references therein). These two instruments are complementary in their characteristics, providing an unprecedented view of the Universe at hard X-ray

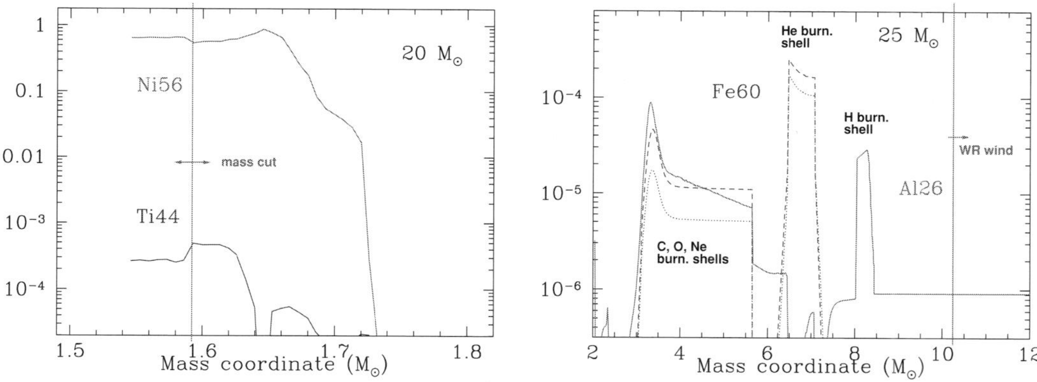

Figure 1. Left panel: radial abundance profiles (mass fractions) of ^{56}Ni and ^{44}Ti inside a 20 M$_\odot$ star after the passage of the shock front. Right panel: radial abundance profiles (mass fractions) of ^{26}Al and ^{60}Fe inside a 25 M$_\odot$ star after the passage of the shock front. Both figures were adapted from Prantzos (2005).

and soft gamma-ray energies. The imaging Ge spectrometer SPI offers high spectral resolution of about 2.1 keV FWHM at 511 keV combined for the first time with moderate spatial resolution (FWHM about 3°). The imager IBIS offers excellent spatial resolution of about 12′ FWHM at moderate spectral resolution (FWHM about 38 keV at 511 keV).

2.1. ^{44}Ti

2.1.1. ^{44}Ti production sites and processes

The radioisotope ^{44}Ti is primarily produced in the so-called α-rich freeze-out of material initially in nuclear statistical equilibrium. The main site for α-rich freeze-out to occur is thought to be the innermost layers of core-collapse supernovae (ccSNe), although sub-Chandrasekhar mass white dwarf Type Ia SNe have also been proposed (Woosley & Weaver 1994).

The ^{44}Ti yield of ccSNe is notoriously difficult to calculate because it depends sensitively on the so-called mass cut, the explosion energy, and the (a)symmetry of the explosion. The mass cut, which has not yet been successfully calculated and is illustrated in the left panel of Fig. 1, is the notional surface separating material that is ejected from material that will fall back onto the compact remnant (neutron star or black hole) of the explosion. ^{44}Ti is believed to be produced in the deepest layers of the exploding star that may be ejected, depending on the precise location of the mass cut. The amount of synthesized ^{44}Ti also depends sensitively on the explosion energy and (a)symmetry. Theoretical calculations indicate that both increased explosion energy and increased asymmetry result in an increased ^{44}Ti yield.

Observationally, the presence of the radioisotope ^{44}Ti is revealed to the gamma-ray astronomer through the emission of three gamma-ray lines. The decay ^{44}Ti $\rightarrow$ ^{44}Sc ($\tau_{^{44}Ti} \sim 90$ y) gives rise to gamma rays at 67.9 keV and 78.4 keV; the subsequent decay ^{44}Sc $\rightarrow$ ^{44}Ca ($\tau_{^{44}Sc} \sim 5.7$ h) gives rise to a line at 1157.0 keV.

The astrophysical interest in ^{44}Ti is two-fold. Clearly, the amount and the velocity of ^{44}Ti is a very powerful probe of the explosion mechanism and dynamics of ccSNe, which are still poorly understood. In addition, the ^{44}Ti gamma-ray line emission is an ideal indicator of young SN remnants (SNRs). The lifetime is about 90 y, which roughly coincides with the expected recurrence time interval for ccSNe in our Galaxy. It is therefore

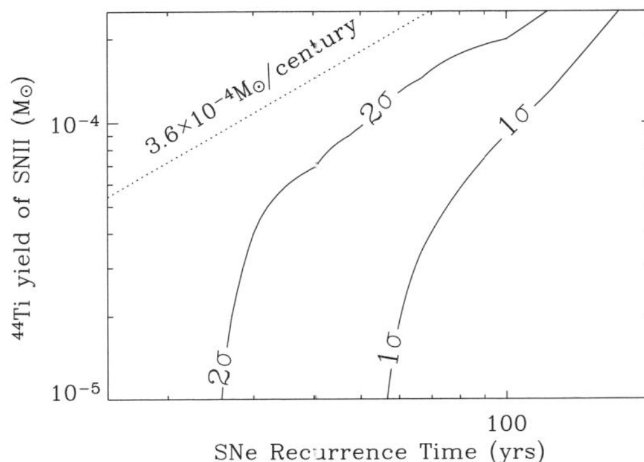

Figure 2. Contours denoting the probability that IBIS/ISGRI would have detected at least one ^{44}Ti point source as a function of the ^{44}Ti yield of Type II SNe and the Type II SN recurrence time (figure adapted from Renaud *et al.* 2004). The dotted line represents the present-day ^{44}Ti production rate that is required to explain the solar system abundance of ^{44}Ca, assuming that is solely produced as ^{44}Ti in Type II SNe.

expected that with a sufficiently sensitive instrument a few young SNRs should be visible in our Galaxy at the current epoch.

2.1.2. *Search for young SNRs in the inner Galaxy*

The most sensitive search to date for young SNRs at gamma-ray energies was performed by Renaud *et al.* (2004) who used the first year of INTEGRAL observations to search for 68 keV and 78 keV line emission in the inner Galaxy with the imager IBIS. This search addresses a long-standing puzzle linking the Galactic SN rate and Galactic chemical evolution: given current estimates of the present-day rates of thermonuclear and ccSNe and their yields, these events can only account for about 1/3 of the solar ^{44}Ca abundance based on chemical evolution models and assuming that all ^{44}Ca is formed as ^{44}Ti (Leising & Share 1994). At the same time, given these SN properties, combined with models for their Galactic distribution, past missions should have detected a few young SNRs even with their lower sensitivities – and detections were certainly expected for the unprecedented limiting point source sensitivity achieved with IBIS. However, as was the case in less sensitive previous searches, none have been found.

To assess the implications of the non-detection of young SNRs, Renaud *et al.* (2004) estimated the probability that at least one ^{44}Ti point source is detectable by generating Monte Carlo distributions using current estimates of the rates of thermonuclear and ccSNe, of their yields, and of their Galactic distribution; SN explosions were simulated as a function of the recurrence time and the ^{44}Ti yield of Type II SNe. A typical result is depicted in Fig. 2.

The Monte Carlo study rules out two obvious solutions to the ^{44}Ca puzzle. The estimated Galactic ^{44}Ti production rate of ccSNe could in principle be increased by postulating that the Galactic SN rate has been underestimated, or by postulating that the ^{44}Ti yields have been underestimated. Neither solution is viable, because in order to account for the solar ^{44}Ca abundance by SNe a present-day production rate of 3.6×10^{-4} M$_\odot$ century^{-1} is required – which is excluded at a greater than 95% confidence level for any acceptable combination of SN rates and ^{44}Ti yields. A third solution, compatible with existing data, is the existence of another, rare but high yield, Galactic ^{44}Ti source such as sub-Chandrasekhar mass white dwarf Type Ia SNe. The latest search for young SNRs with IBIS therefore highlights important open questions concerning the chemical evolution of our Galaxy, the production of ^{44}Ti in SNe, and the Galactic SN rate (Renaud *et al.* 2004).

2.2. ^{26}Al and ^{60}Fe

2.2.1. ^{26}Al and ^{60}Fe production sites and processes

^{26}Al and ^{60}Fe are produced hydrostatically as well as explosively in massive stars. The production of ^{26}Al requires free protons, hence it is mainly produced in the H, Ne, and C burning layers; the production of ^{60}Fe requires free neutrons, hence it is mainly produced in C, Ne, and He burning layers, as indicated in the right panel of Fig. 1. Products of hydrostatic nucleosynthesis can escape the massive star through stellar winds, in particular during the Wolf-Rayet (WR) phase of the most massive stars. The ^{26}Al, being produced in H burning layers near the surface, can escape most easily. When the star explodes in a ccSN, about equal amounts of ^{26}Al and ^{60}Fe are ejected. The yields of ^{60}Fe and in particular of ^{26}Al are hard to predict because of the important role of stellar winds, which in turn depend e.g. on the metallicity of the star or its rotation.

Both radioisotopes can be observed by gamma-ray line spectroscopy. The decay of ^{26}Al ($\tau_{^{26}Al} \sim 1.1 \times 10^6$ y) gives rise to a single gamma-ray line at 1808.6 keV. The decay ^{60}Fe $\rightarrow$ ^{60}Co ($\tau_{^{60}Fe} \sim 2.2 \times 10^6$ y) is revealed by a line at 58.6 keV, the subsequent decay ^{60}Co $\rightarrow$ ^{60}Ni ($\tau_{^{60}Co} \sim 7.6$ y) is revealed by two lines at 1173.2 keV and 1332.5 keV.

Astrophysically, the subtle differences in the production zones of the radioisotopes ^{26}Al and ^{60}Fe render their combined study an important probe of massive star nucleosynthesis. Furthermore, since the lifetimes of these two radioisotopes are much shorter than Galactic evolution timescales but are large enough to accumulate in the ISM, ^{26}Al and ^{60}Fe are ideally suited to study the sites and the distribution of recent Galactic star formation.

2.2.2. ^{26}Al in the inner Galaxy

The sky distribution of Galactic ^{26}Al was first mapped using the COMPTEL instrument; it was found to be very similar to that of massive stars in our Galaxy, establishing the massive star origin of ^{26}Al (Knödlseder *et al.* 1999). However, the relative importance of WR star winds and of ccSNe for seeding the ISM remained unclear. Early high-resolution spectroscopy with the balloon borne GRIS instrument by Naya *et al.* (1996) yielded an unexpectedly large line width of about 5.4 keV FWHM and complicated rather than clarified the origin of ^{26}Al as such substantial line broadening could only be explained by rather unusual circumstances (Sturner & Naya 1999). Spatially resolved spectroscopy, as afforded by the imaging spectrometer SPI, is expected to resolve the issue, as it will make it possible to probe the imprint of the dynamics of the ISM local to the ^{26}Al sources and that of Galactic differential rotation on shape and centroid of the 1808.6 keV line (modelled e.g. by Kretschmer *et al.* 2003).

Using the first 1.5 years of observations with SPI, Diehl *et al.* (2005a,b) obtained a spectrum of the 1808.6 keV line from the inner Galaxy of unprecedented quality (see left panel of Fig. 3). The line centroid is unshifted; the line width is intrinsically narrow with a FHWM of about 1.2 keV (and a 2σ upper limit of 2.8 keV), ruling out the earlier GRIS result. These line parameters are consistent with expectations from the Kretschmer *et al.* (2003) model, which for the inner Galaxy predicts no line shift and a line width of about 1 keV due to Galactic differential rotation (leaving room for a small amount of additional broadening due to modest ISM turbulence around the ^{26}Al sources). The ^{26}Al line is detected with such high significance by SPI that spatially resolved spectroscopy in the inner Galaxy can be performed for the first time, with important implications for the amount and origin of ^{26}Al, and the Galactic star formation and SN rates as discussed by Diehl *et al.* (2005a).

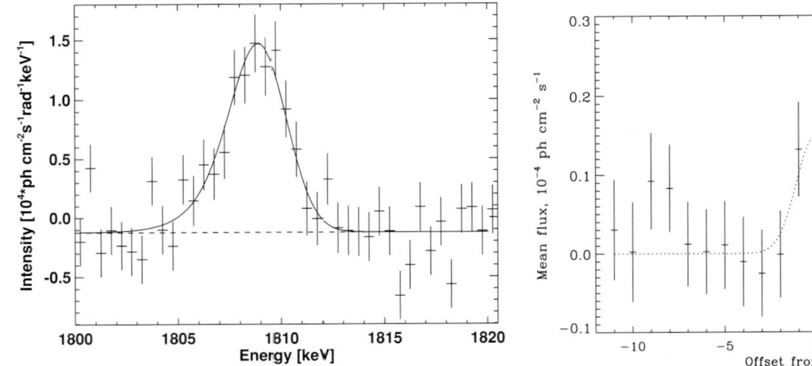

Figure 3. Left panel: the spectrum of the ^{26}Al gamma-ray line emission from the inner Galaxy (data points) and a Gaussian line fit (solid line) as obtained by Diehl *et al.* (2005b). Right panel: the overlayed and summed spectrum of the observed ^{60}Fe lines at 1173.2 keV and 1332.5 keV (data points) and a Gaussian line fit (dashed line) according to Harris *et al.* (2005).

2.2.3. ^{60}Fe and ^{26}Al from massive stars

One of the great successes for SPI was the first significant measurement of ^{60}Fe emission from our Galaxy by Harris *et al.* (2005) using the observations of the first year. The summed and overlayed spectrum of the two observed lines is depicted in the right panel of Fig. 3. The observed flux per line is $(3.7 \pm 1.1) \times 10^{-5}$ ph cm^{-2} s^{-1}, which translates into a line flux ratio of ^{60}Fe to ^{26}Al of 0.11 ± 0.03 per ^{60}Fe line.

The detection of ^{60}Fe has important implications for the nucleosynthesis in massive stars, as was first pointed out by Prantzos (2004, 2005). He integrated the predicted yields for Type II SNe over an initial mass function and found that if he only considered the lower mass range, where the WR phase is absent or weak, the ^{60}Fe to ^{26}Al line flux ratio is significantly overpredicted. This means that there must be an additional ^{26}Al source; Type II SNe do not produce enough ^{26}Al. This additional source could be WR stars, because when the integration is extended up to the highest masses, taking into account the ^{26}Al expelled in the WR winds, the observed ^{60}Fe to ^{26}Al line flux ratio can be reproduced. However, one should keep in mind that WR star yields are still uncertain, as they depend sensitively on e.g. metallicity and rotation.

2.3. Galactic positrons

2.3.1. Production processes and positron sources

The annihilation of positrons with electrons gives rise to two characteristic emissions at gamma-ray energies: the hallmark line at 511 keV, and the unique three-photon positronium (Ps) continuum emission (cf. Guessoum, Jean, & Gillard 2005). The detailed shape of the 511 keV annihilation line, as well as the positronium fraction (the fraction of positrons that annihilate through Ps formation), depend on the physical properties of the annihilation media; therefore detailed spectroscopy of the positron annihilation can provide unique information on the annihilation media and processes.

Positrons can be produced through a variety of processes and by a variety of objects in our Galaxy. The main production processes are β^+ decay of radioactive nuclei (formed in SNe, novae, hypernovae, WR stars, and AGB stars), the decay of π^+s resulting from interactions of energetic nuclei (in the ISM), pair (electron-positron) production in γ-γ interactions (close to compact objects and in gamma-ray bursts), and pair production by electrons in strong magnetic fields (around pulsars). Exotic processes, such as the

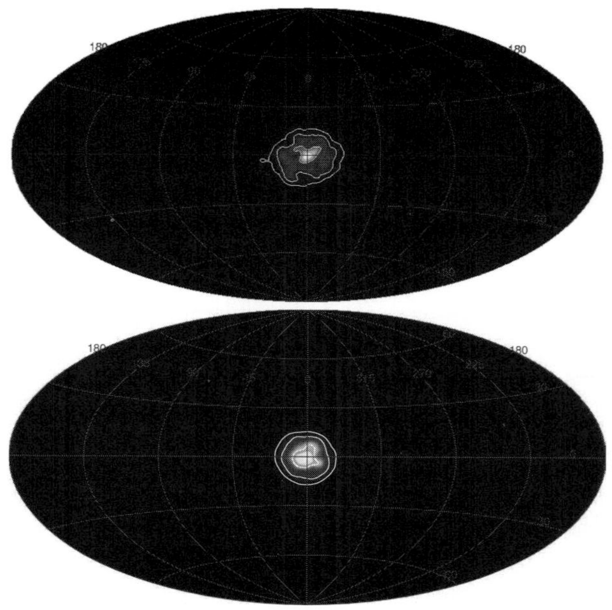

Figure 4. Sky maps in the 511 keV line (top panel, Knödlseder *et al.* 2005) and the Ps continuum emission (bottom panel, Weidenspointner *et al.* 2005). The maps were generated employing an implementation of the Richardson-Lucy algorithm. To reduce noise artifacts, the iterative corrections were smoothed during image reconstruction with a $5° \times 5°$ boxcar average. Further details are given in the text.

annihilation or decay of light dark matter, have also been proposed (Boehm *et al.* 2004). In spite of this well developed understanding of the fundamental physics of positron production, theoretical models for the positron yield of the various sources are still highly uncertain due to uncertainties concerning the astrophysical conditions in the different sources, and their Galactic distribution and duty cycle. Considering also the limited quality of existing observations, it is not surprising that the origin of the positrons is still poorly understood. At this stage improved measurements of the spatial distribution and the spectrum of the Galactic annihilation radiation are crucial for constraining better the many models.

2.3.2. *Annihilation radiation: Galatic distribution*

Investigations of the sky distribution of the annihilation radiation promise to provide clues to the identification of the source(s) of positrons in our Galaxy. First maps of the annihilation radiation, limited to the inner regions of our Galaxy, were obtained using the OSSE instrument on board the Compton Gamma-Ray Observatory in the 511 keV line and in Ps continuum emission (e.g. Purcel *et al.* 1997, Milne *et al.* 2001).

Improved mapping of the 511 keV line is feasible with the commissioning of SPI. During the first year of the mission most of the celestial sphere was observed. Using these observations, first sky maps in the 511 keV positron annihilation line (Knödlseder *et al.* 2005) and in the Ps continuum emission (Weidenspointner *et al.* 2005) have been obtained with the SPI spectrometer (Fig. 4, top and bottom). In both maps of positron annihilation radiation, the only prominent signal seen is that from the Galactic bulge region. The surface brightness of any emission from any other sky regions, in particular from the Galactic disk, is much fainter. The emission appears to be symmetric about the Galactic center (GC), and its centroid coincides well with the GC. The differences between the two maps, as well as any apparent small scale sub-structure in the maps, are not significant, as will be demonstrated below.

A more quantitative approach for studying the Galactic distribution of the observed extended emission is model fitting. From modeling the positron annihilation radiation

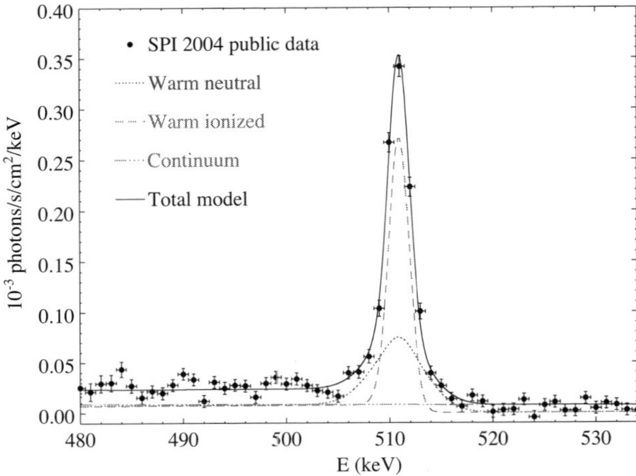

Figure 5. Best fit of the SPI spectrum of the Galactic bulge with the warm components of the ISM and the Galactic continuum emission (from Jean *et al.* 2005).

by an ellipsoidal distribution with a Gaussian radial profile it can be concluded that the annihilation radiation is spherically symmetric with a FWHM of about 8° and centered at the GC (Knödlseder *et al.* 2005, Weidenspointner *et al.* 2005). The annihilation line signal is strong enough to allow more detailed studies of its Galactic distribution. Knödlseder *et al.* (2005) fitted a variety of Galactic bulge, halo, and disk models to the data. Stellar bulge and halo models describe the annihilation line emission from the central region of our Galaxy equally well. On adding a Galactic disk component to the model fits, the faint annihilation radiation from the Galactic disk is detected at the 3-4σ level.

From these model fits a total 511 keV line flux from our Galaxy of $(1.5 - 2.9) \times 10^{-3}$ ph cm^{-2} s^{-1} is derived, with the bulge-to-disk flux ratio being about 1–3 and the bulge-to-disk annihilation luminosity ratio being even larger with a range of 3–9. Consequently, the model fits confirm the qualitative mapping result: SPI observations demonstrate for the first time that the annihilation radiation from our Galaxy is dominated by the bulge region (Knödlseder *et al.* 2005). Assuming a Ps fraction of 0.93 the total 511 keV line flux corresponds to a Galactic positron annihilation rate of $(1.6 - 2.8) \times 10^{43}$ s^{-1}.

Another approach to identifying the sources of Galactic positrons is to compare the sky distribution of the 511 keV line emission with all-sky intensity distributions observed at other wavelengths. The relatively best agreement is found at wavelengths dominated by emission from members of old stellar populations, but none of the tracer maps provides an acceptable fit to the data. The sky distribution of positron annihilation appears to be unique, as it is even more bulge dominated than potential old positron source populations such as Type Ia supernovae, novae, or low-mass X-ray binaries (Knödlseder *et al.* 2005).

Both Knödlseder *et al.* (2005) and Weidenspointner *et al.* (2005) searched for evidence for contributions from point sources on top of the extended emission, but none could be found in the 511 keV annihilation line and the Ps continuum emission. Similarly, De Cesare *et al.* (2004) did not find any evidence for point sources of 511 keV line emission in the GC region using observations by IBIS.

2.3.3. *Annihilation radiation: spectroscopy*

The most detailed spectroscopy to date of the annihilation radiation from the GC region has been performed by Jean *et al.* (2005). The line is found to be composed of a narrow and a broad component, consistent respectively with the expected line width in a warm ISM for Ps formation by radiative recombination and Ps formation in flight. This result is consistent with an earlier analysis by Churazov *et al.* (2004). When fitting the annihilation spectra that are expected for the five standard phases of the ISM to the SPI spectrum (see Fig. 5), it follows that about half of the positrons annihilate in each of the warm neutral and the warm ionized phases. Possible fitted contributions from annihilations in cold gas, molecular clouds, or hot gas are not significant so far. The importance of annihilations in the warm phases is consistent with our current understanding of positron propagation and annihilation and the gas distribution in the Galactic bulge (Jean *et al.* 2005). Jean *et al.* (2005) find a value for the Ps fraction of 0.97 ± 0.02, consistent with pre-INTEGRAL results, and with the SPI results obtained by Churazov *et al.* (2004) and Weidenspointner *et al.* (2005).

3. Prospects

The prospects for further nucleosynthesis studies with INTEGRAL are bright. As the mission continues, we expect to learn much more about the origin of positrons in our Galaxy from ever-improving mapping, spatially resolved spectroscopy, and dedicated observations of individual candidate sources. We also expect to study the Galactic distribution of ^{26}Al in detail by extracting distance information from line shifts, to investigate the Galactic distribution of ^{60}Fe which combined with studies of ^{26}Al will provide new insights into massive star nucleosynthesis as well as the Galactic star formation and SN rates, and to clarify the explosion physics of ccSNe by observing selected SNRs like Cas A.

References

Boehm, C., *et al.* 2004, *Phys. Rev. Lett.* 92, 101301

Churazov, E., *et al.* 2004, *MNRAS* 357, 1377

De Cesare, G., *et al.* 2004, *Advances in Space Research*, in press

Diehl, R., Prantzos, N. & von Ballmoos, P. 2005, *Nucl. Phys. A*, in press

Diehl, R., *et al.* 2005a, *Nature*, submitted

Diehl, R., *et al.* 2005b, *A&A*, submitted

Guessoum, N., Jean, P. & Gillard, W. 2005, *A&A* 436, 171

Harris, M.J., *et al.* 2005, *A&A* 433, L49

Jean, P., *et al.* 2005, *A&A* submitted

Knödlseder, J., *et al.* 1999, *A&A* 344, 68

Knödlseder, J., *et al.* 2005, *A&A*, in press

Kretschmar, P. 2005, *these proceedings*

Kretschmer, K., Diehl, R. & Hartmann, D.H. 2003, *A&A* 412, L47

Leising, M.D., & Share, G.H. 1994, *ApJ* 424, 200

Milne, P.A., *et al.* 2001, *Proc. of Gamma 2001* (AIP 587), 11

Naya, J.E., *et al.* 1996, *Nature* 384, 44

Prantzos, N. 2004, *A&A* 420, 1033

Prantzos, N. 2005, *Proc. 5th INTEGRAL Science Workshop* ESA SP-552, 15

Purcell, W.R., *et al.* 1997, *ApJ* 491, 725

Renaud, M., *et al.* 2004, *Proc. 5th INTEGRAL Science Workshop* ESA SP-552, 81

Sturner, S.J., & Naya, J.E. 1999, *ApJ* 526, 200

Weidenspointner, G., *et al.* 2005, *A&A*, submitted

Woosley, S.E. & Weaver, T.A. 1994, *ApJ* 423, 371

Discussion

GARCIA: Is 511 a point source?

WEIDENSPOINTNER: No: It is extended – must be at least 3 point sources or more.

VÖLK: Regarding ^{44}Ti sources you said that only Cas A has been detected by INTE-GRAL. Concerning the consistent claims of the past, concerning ^{44}Ti emission from the SNR Vela Jr., what is then your position?

WEIDENSPOINTNER: First INTEGRAL upper limits based on observations taken during Cycle 1 are still less constraining than previous COMPTEL results. By the end of Cycle 3 the exposure to Vela Jr. will have increased by more than a factor of 3 though, and then INTEGRAL should be able to provide a better answer to this intriguing question.

BOSCH-RAMON: Is the line e-e+ of the great annihilator completely ruled out?

WEIDENSPOINTNER: If the Great Annihilator is manifesting itself as a point source, or point-like source, of narrow 511 keV line emission, then it can indeed be completely ruled out.

ELVIS: 1. The ps + 511 keV maps appeared to show structure. Are these significant? 2. The ^{60}Fe line had similar sized features at slightly higher and lower energies, which suggests that ^{60}Fe may not be a significant feature. If treated as an upper limit does this solve the Fe/Al ratio problem?

WEIDENSPOINTNER: 1. No. We have not yet found any significant structure on angular scales of a few degrees or below.
2. One should keep in mind that the combined significance of the two high-energy ^{60}Fe lines is only 3.4σ. The feature above the overlayed ^{60}Fe lines is the residual of a well known background line from ^{69}Ge decay in the detectors, which our analysis reduces by about a factor of 100. The weak residual below the ^{60}Fe lines cannot be attributed to a single instrumental background feature. If the detected flux is treated as an upper limit, the need for a source of ^{26}Al in addition to ccSNe is even exacerbated.

Populations of High Energy Sources in Galaxies
Proceedings IAU Symposium No. 230, 2005
E. J. A. Meurs & G. Fabbiano, eds.

© 2006 International Astronomical Union
doi:10.1017/S1743921306008052

Diffuse X-ray Emission from Galaxies

Andrew M. Read[1]

[1]Dept. of Physics and Astronomy, University of Leicester, Leicester LE1 7RH, U.K.
email: amr30@star.le.ac.uk

Abstract. I review here the current ideas regarding the origin, evolution, and physical nature of hot diffuse gas in normal, starburst, interacting and merging galaxies, using recent X-ray observations with XMM-Newton and Chandra. Many types of diffuse X-ray structures, including winds, bubbles, halos, chimneys and fountains, can be formed in galaxies, and can enrich the intergalactic medium with mass, energy and metals. This has profound implications as regards galactic formation and evolution, and the enrichment and evolution of galaxy groups and clusters.

Keywords. ISM: general, ISM: jets and outflows, galaxies: general, galaxies: interactions, galaxies: ISM, galaxies: starburst, X-rays: galaxies, X-rays: ISM.

1. Introduction

Spiral galaxies undergoing moderate to high (starburst) levels of star-formation can become complex, multi-phased objects, and the diffuse thermal emission from 10^6–10^8 K gas is a very important probe of the conditions within the halos of these star-forming galaxies. If the star-formation is of a high enough level, then the most dramatic of diffuse X-ray emission structures can be formed − superwinds.

In the standard model for galactic superwinds driven by multiple ($\sim 10^6$) supernovae, the merged, metal-enriched SN ejecta are expected to have a central starburst temperature of $\sim 10^8$ K (and are hence a source of faint hard thermal X-rays). The warm neutral and ionized gas in superwinds is seen to have velocities of between $200-1000$ km s^{-1} (Heckman *et al.* 2000), and this material is believed to be embedded within or at the boundaries of a hot metal-enriched, wind fluid, with an even higher velocity (Strickland & Stevens 2000). Various processes, such as strong shocks ($T \sim 3.5 \times 10^6 [v_{Shock}/500\,\mathrm{km\,s^{-1}}]^2$) or thermal conduction at interfaces between the warm gas and the hot wind fluid, are able to create soft ($E < 2$ keV) thermal X-ray emission.

Diffuse thermal X-ray emission from galaxies is a mixture of continuum (bremsstrahlung and recombination) and line emission. For low-temperature thermal plasmas, the net emission is dominated by line emission for $Z > Z_\odot$. Because newly-synthesized metals are believed to enter the hot phases of the ISM, X-ray observations are critical in seeing the processes of galactic chemical evolution and IGM enrichment at work.

In the following sections, I shall review the recent Chandra and XMM-Newton X-ray observations of the diffuse gas in normal, starburst, and interacting/merging galaxies, i.e. attaining higher and higher levels of star-formation activity.

2. Normal Galaxies

Normal spiral galaxies, like our own, typically exhibit only low levels of diffuse emission. Typically, the emission can be described using two very soft low-metallicity thermal plasmas. Only in the most star-forming normal galaxies is diffuse emission seen to rise significantly above the galactic plane, in the form of hot coronae − very few significant

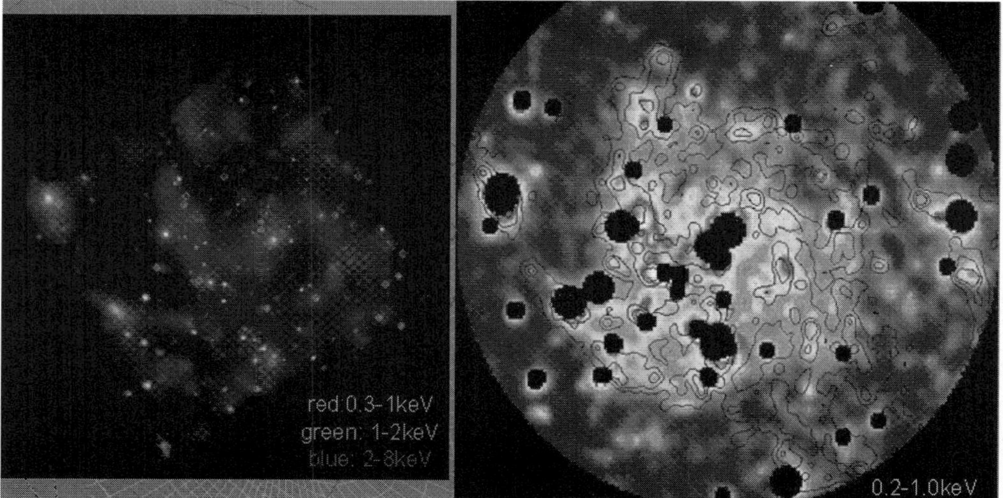

Figure 1. Diffuse X-rays from M101: (left) Chandra: Red 0.3−1 keV, green 1−2 keV, blue 2−8 keV (di Stefano & Kong 2004) (right) XMM-Newton: 0.2−1 keV with contours of GALEX UV emission (Warwick *et al.* 2005).

outflows, winds or superwinds are seen, and there is very little mass, energy or metal losses from these systems via hot diffuse gas.

Our nearest large neighbour, M31, is seen by XMM-Newton to have what appears to be genuinely diffuse gas only at the very centre of the galaxy (Shirey *et al.* 2001). Chandra is able to resolve out some of this emission into discrete sources (Garcia *et al.* 2001), but a small amount of genuine diffuse hot gas remains. Much more diffuse emission is seen in the more active normal galaxies. M101 for instance shows a lot of diffuse emission associated with the star-forming regions in its spiral arms (see Fig. 1). The correlation between the diffuse X-ray emission and the spiral arms is rather striking.

It is worth stressing at this point that the spectral analysis of soft diffuse X-ray features that cover a significant portion of any X-ray detector is prone to difficulties, in that it is essential that the background must be modelled as correctly as possible. There are many methods to do this, and some may be better than others, depending on the instrument, the type of data, the source, and the science attempted, to name but a few factors. Essentially though the problem is that extracted spectra contain vignetted photon background and non-vignetted particle background. Hence the source (diffuse) spectrum will have a rather different photon-to-particle background mix than the actual extracted background spectrum (taken from a very different part of the X-ray detector, away from the diffuse emission source). A way to correct for this is to use pure particle spectra, extracted from regions of the detector out of the field of view, and use the areas of the various extraction regions, as performed on the M101 EPIC data in Warwick *et al.* (2005). Here, using such a method, the spectral parameters for M101's diffuse emission are seen to be identical to those obtained by Chandra (Kuntz *et al.* 2003). The emission is described by two very soft low-metallicity thermal plasmas ($kT = 0.2$ & 0.65 keV), and such a result is typical of many normal galaxies (e.g. NGC300, Carpano *et al.* 2005).

3. Starburst Galaxies

Though there is a continuous distribution in galaxy star-formation activity, starburst galaxies are here defined using the widely-used IRAS $60\mu m$ to $100\mu m$ flux ratio

Figure 2. 8 Galaxies from a Chandra survey of superwinds (Strickland *et al.* 2004a,b): Green – optical R-band (stars), red – optical Hα emission (gas at T $\sim 10^4$ K), blue – Chandra 0.3−2 keV X-ray emission (gas at T $\sim 5 \times 10^6$ K), heavily smoothed.

$f_{60}/f_{100} > 0.4$. These more active, star-forming galaxies exhibit greater levels of diffuse emission than normal galaxies. The gas is hotter and exists not only in the form of coronae and halos, but also as outflows − superwinds. Consequently starburst galaxies can lose significant amounts of mass, energy and metals to the IGM.

A superwind starts in the nucleus of a galaxy some $\sim$40 Myr after a strong star-forming episode, when the high-pressure, high-temperature gas of some $\sim 10^6$ merged SN ejecta expands, and breaks out along the poles of the galaxy into the halo at $v \sim$ 1000's of km s^{-1}. On the way it entrains and accelerates cool dense, ambient gas from the galactic disk and halo, and it is this shocked gas that is believed to give rise to the soft X-rays, and the optical Hα emission. Possibly the wind fluid and the entrained gas is then able to escape the galaxy halo into the IGM. Superwinds are very important as regards galactic evolution and chemical evolution, mass-loss from galaxies, the metal enrichment of the IGM in clusters of galaxies (this is perhaps very important at high redshift from small galaxies) and energy injection into galaxy clusters.

Because of their small mass, dwarf galaxies are more susceptible to 'blow-outs' of the ISM, such as superwinds. Having said this however, it is not just the galaxy mass that comes into play here − the disk and halo parameters are poorly known and could play a very fundamental role (Strickland *et al.* 2004b). X-ray observations of dwarf galaxies (e.g. NGC1569; Martin *et al.* 2002) show low-temperature (0.3 keV), enhanced O/Fe abundance (2−4 times solar) outflows, carrying nearly all the metals ejected by the starburst. Most of the X-ray emitting material is thought to be entrained ISM.

Extra-planar soft diffuse thermal emission is seen in many starburst galaxies (Strickland *et al.* 2004a,b), as can be seen in Fig. 2. Thought to be *the* prototypical starburst galaxy, M82 has been observed several times in X-rays. It is close (D=3.6 Mpc), very IR bright and its starburst is triggered by the interaction with its neighbour, M81. Its superwind is visible in Hα and in X-rays (Stevens *et al.* 2003), extending some $\sim$14 kpc out of the plane of the galaxy. Various features are seen in the XMM-Newton view of the superwind, including a Hα 'cap' at the edge of the wind (thought to be an interaction of the superwind with an ionized cloud), an X-ray bridge connecting to this, and bright and dark lanes within the wind. The central region of the wind is hot, fitted with

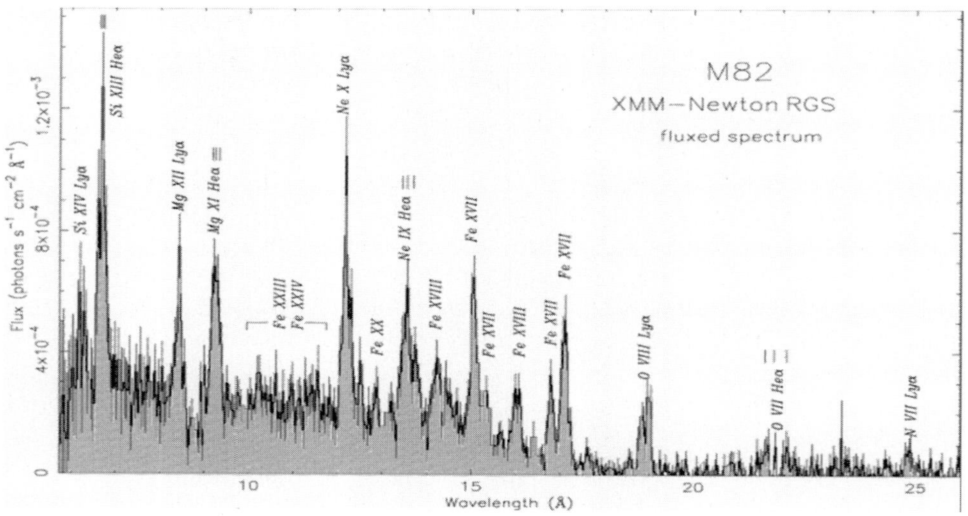

Figure 3. The fluxed RGS spectrum (RGS 1 & 2, order 1) of the M82 starburst. Bright emission lines (both hydrogen-like and helium-like triples) are indicated (Read & Stevens 2002).

a two-temperature (0.54 & 0.9 keV) model, while further out, the wind is cooler (0.37 & 0.54 keV). Furthermore there appears to be a HI hole associated with the superwind breakout, and an excess of HI associated with the dark lanes in the superwind (Stevens *et al.* 2003).

The XMM-Newton RGS observation of the M82 starburst (Read & Stevens 2002) yields the deepest high-resolution X-ray spectrum of a starburst galaxy (Fig. 3). Here, line broadening is seen due to the extended nature of the starburst wind. A complex array of lines is seen, including Lyman α emission from low-Z elements (N, O, Ne, Mg, Si), plus helium-like ions and the Fe-L series. The OVII line triplet can be resolved and is seen to be consistent with hot gas in collisional ionization equilibrium. A differential emission measure model fit to the data results in gas over a range of temperatures (0.2−1.6 keV).

M82 is also the galaxy in which we believe we have seen the first direct evidence of hard diffuse X-ray emission from a starburst galaxy. Old claims of diffuse hard X-rays from galaxies have now been proved to be incorrect, the emission just being due to unresolved point sources. New Chandra observations however show a few starbursts with possible genuine diffuse hard X-ray emission in the nuclear starburst region; M82, NGC 253, NGC 2146 (Strickland *et al.* 2003). Likely causes of this emission could be fluorescence or scattered AGN light from a LLAGN, or non-thermal (e.g. inverse Compton) or thermal (merged SN ejecta − the superwind) emission from the starburst. As M82 has no LLAGN, the first meaningful detection of diffuse hard (2−8 keV) emission in this starburst, using the damaged ACIS-I (Griffiths *et al.* 2000), may be the first direct detection of the wind fluid itself.

In their Chandra survey of superwinds, Strickland *et al.* (2004a,b) analysed many edge-on starburst and normal galaxies, including M82, NGC1482, NGC253, NGC3628, NGC3079, NGC4945, NGC4631, NGC891 and NGC6503. They found extended extraplanar emission in 80% of their sample, and the X-ray surface brightness is seen to be best fit with an exponential cf scale height ∼2–4 kpc. Further, it is seen that the extent of the diffuse X-ray emission scales with the size of the host galaxy. The Chandra spectra are best fit using two-temperature models with enhanced α-to-iron abundance ratios. The diffuse X-ray luminosity is proportional to the mechanical energy input from the stellar component, but the mass and energy estimates are strongly dependent on the (poorly constrained) X-ray filling factor (the fraction of the volume that is occupied by emitting gas). There is also evidence of a threshold in star-formation rate per unit surface area, before blow-out from the disk can occur, and evidence of nuclear processed material being swept out of these galaxies, and the entrained ISM being driven out as well.

As regards soft X-ray emission from superwinds, the current synopsis is that genuine spatial structure is seen in the X-ray, on scales similar to the structure in Hα. The best interpretation is that the X-rays are due to limb-brightening on small and large scales: There are kpc-scale nuclear outflow cones observed, where the soft X-rays and Hα are correlated on $10-20$ pc scales, and there is 10 kpc-scale extraplanar emission, where there is correlation on $\sim$1 kpc scales. Furthermore the soft X-rays and the Hα are well correlated in flux. The high alpha-element/iron abundance ratios indicate that the X-ray emitting material is either metal-enriched or dusty. It is now thought (Strickland *et al.* 2004a,b) that a superwind is not a volume-filling wind of the kind modelled by Suchkov *et al.* (1996). The majority of the emission must come from relatively low filling factor $(0.01-0.3)$ material. This low volume filling factor indicates that the visible X-rays are not representative of the full energy and volume of the wind. The soft X-ray emission arises in fact from the interaction of the wind with the galaxy ISM (disk and halo), and possible physical origins include: high-velocity shocks in clouds, or ambient disk or halo gas along the wind walls; conductively-evaporated layers around the clouds or walls; stand-off (reverse) shocks in the wind, i.e. compressed, enriched wind-fluid.

4. Interacting and Merging Galaxies

Interacting and merging galaxies, due to their rapidly changing environments, can reach very large star-formation rates, and very powerful starbursts and starburst-driven superwinds are observed. These galaxies can attain levels of diffuse emission greater even than in the classic starburst galaxies. Very large outflows and extensions are seen — both superwinds and possibly other more exotic structures — and it appears that very significant mass, energy and metal losses may be occurring.

The Mice, for instance, are a pair of spiral galaxies at an early stage of merging. Chandra observations (Read 2003) show starburst-driven galactic winds, similar to, though not yet of the size of the M82 wind, outflowing along the minor axes of both galaxies. That such a phenomenon can occur in such a rapidly evolving and turbulent system is surprising, and this is first time that the very beginning of starburst-driven hot gaseous outflow in a full-blown disk-disk merger has been seen. Other interacting galaxies show more peculiar phenomena. NGC520 for instance (Read 2005), shows what appears to be a starburst-driven galactic wind outflowing perpendicularly to only one of the nuclei — nothing is seen emanating from the second nucleus. This is thought due to the fact that NGC520 is believed to be the result of an encounter between a gas-rich and a gas-poor disk.

It is when the interacting galaxies come close to merging that very violent starbursts take place, and very large and unusual diffuse emission structures are seen. In the Antennae, one of the closest and best known very active merging pairs, a long 411 ks Chandra observation (Fabbiano *et al.* 2004) has revealed the presence of both extensive hot diffuse X-ray emission associated with the galactic disks, and what appears to be large (10 kpc) off-galaxy diffuse loops to the south (see Fig. 4). The diffuse emission within the galactic disks is seen to vary very considerably from region to region — some diffuse emission regions are best described by low-temperature, sub-solar abundance plasmas, while others require higher abundances, or additional strong power-law components, or very high abundances indeed. Results consistent with ASCA (where $\sim$0.2 solar abundances were

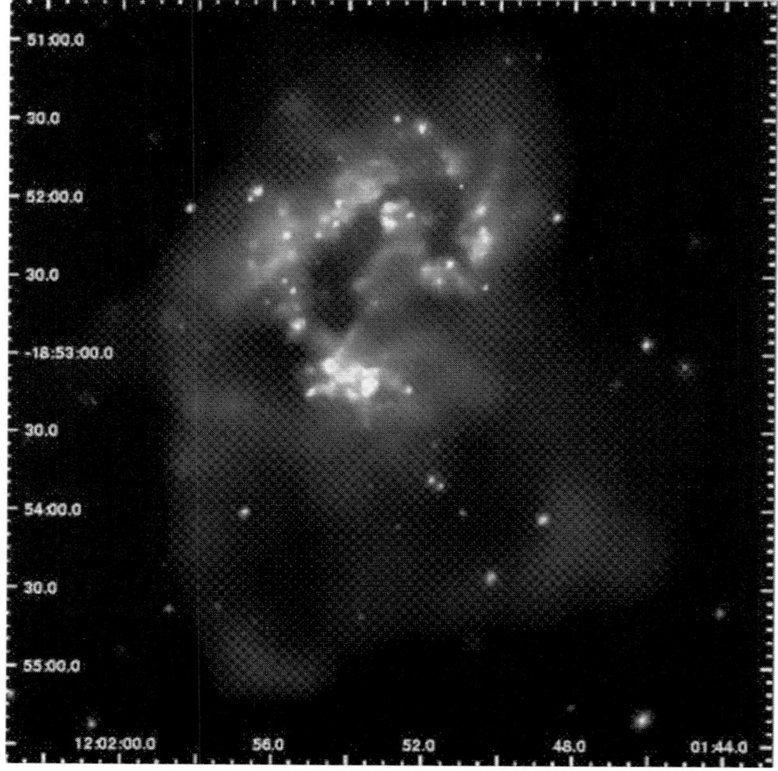

Figure 4. Chandra diffuse X-ray emission from the Antennae merging galaxy pair (Fabbiano *et al.* 2004): red 0.3−0.65 keV, green 0.65−1.5 keV, blue 1.5−6 keV. Two large loop/bubbles are seen to the south, lying outside of the optical disks and the HI tails.

found; Sansom *et al.* 1996) can be obtained if the entire integrated emission is fitted. Such apparent low abundances had been a common and puzzling feature of starburst X-ray spectra. In the Antennae at least, and by extension, perhaps in other starbursts as well, this is evidently a spatial resolution problem, and strong and variable line emission exists within different regions of the hot ISM, even after thorough point-source subtraction.

The large 10 kpc diffuse loops visible to the south of the Antennae are especially interesting. Not due to an artifact of any adaptive smoothing techniques, they lie outside of the optical disks and outside of the huge HI tidal tails, and resemble giant loops or bubbles. A radial profile of the soft band emission shows that there is significant hot gas extending even beyond these X-ray loop/bubbles. While this very extended gas is seen to be at $\sim$0.2 keV, the loops/bubbles are seen to be at $\sim$0.3–0.35 keV (both inside of, and in the rim of, each bubble). Can they be starburst-blown bubbles? Each bubble appears to have $L_X \sim 5 \times 10^{39}$ erg s^{-1} and contains $\sim$10^8 M$_\odot$ of gas. The cooling time of the gas ($\sim$1 Gyr) is some ten times longer than the merger timescale, and the thermal energy contained within each bubble is around 1.5×10^{56} erg, equivalent to $\sim$15,000 supernovae. As the supernova rate is $\sim$0.3 yr^{-1}, then there is certainly plenty of energy that is potentially available to power these bubbles. However, since the two bubbles are of the same size, they might seem to require two simultaneous events. Furthermore, the entropy ($\propto T/n_{\mathrm{e}}^{2/3}$) of the gas into which the bubbles are expanding appears to be *higher* than that in the bubble rims. This does not sit well with the idea that the emission at the bubble rims is due to shocks sweeping out into the surrounding halo gas (Ponman *et al.* 2005).

As a final thought regarding merging galaxies, it is very intriguing that similar bubble features are seen in Arp220 (McDowell *et al.* 2003), an extremely bright, active merging galaxy pair, whose nuclei are the point of colliding (i.e. at an evolutionary stage slightly later than the Antennae). It may even be that two double-bubbles are seen in diffuse X-rays within this system, and that therefore something rather unexpected may be occurring in merging galaxies. The evolution of the diffuse X-ray emission in merging galaxies is investigated more fully in Brassington *et al.* (2005).

5. Final Remarks

The best examples of normal galaxies, of dwarf and disk starbursts, and of interacting and merging galaxies have now been observed with XMM-Newton and Chandra, and we are steadily building up larger and large samples of each of these galaxy types. Although not discussed in this review, the same is also true of elliptical galaxies, of which there are now many examples. In order to study the diffuse emission correctly in external galaxies, it certainly seems the case that detailed studies can still only be performed on the closest, brightest objects. High spatial resolution is needed to disentangle the point sources from the diffuse emission in galaxies. Of great importance also, especially considering the relatively low temperature ($0.3-1.0\,$keV) of diffuse gas in galaxies, is to model and subtract the background as correctly as possible, and complex methods can be necessary to do this. Though we have gained many insights into the feedback processes from these X-ray observations, uncertainties do remain. It may have been that Astro-E2 would have been helpful for some of the bright starbursts — indeed, it would have perhaps been able to probe the velocities of the X-ray emitting gas, were it not for the XRS failure. Though a dark age may loom, in that XEUS et al. are perhaps still a long way off, their time is getting closer, and both Chandra and XMM-Newton have performed splendidly so far, and hopefully, will do for years to come.

Acknowledgements

I would like to thank the organizers for such a stimulating and well-run symposium. I would also like to thank my collaborators, in particular Pepi Fabbiano, Trevor Ponman, Ian Stevens, Dave Strickland & Bob Warwick.

References

Brassington, N.J., Read, A.M., & Ponman, T.J. 2005 *MNRAS* in prep.
Carpano, S., Wilms, J., Schirmer, M., & Kendziorra, E. 2005, *A&A* in press
di Stefano, R. & Kong, A.K.H. 2004 *ApJ* 609, 710
Fabbiano G., Baldi A., King, A.R., Ponman T.J., Raymond, J., Read, A.M., Rots, A., Schweizer, F., & Zezas, A. 2004 *ApJ* 605, 21
Garcia, M.R., Murray, S.S., Primini, F.A., Forman, W.R., Jones, C., & McClintock, J. E. 2001 *IAUS* 205, 38
Griffiths, R.E., Ptak, A., Feigelson, E.D., Garmire, G., Townsley, L., Brandt, W.N., Sambruna, R., & Bregman, J.N. 2000 *Science* 290, 1325
Heckman, T.M., Lehnert, M.D., Strickland, D.K., & Armus, L., 2000, *ApJS*, 129, 493
Kuntz, K.D., Snowden, S.L., Pence, W.D., & Mukai, K. 2003, *ApJ* 588, 264
Martin, C.L., Kobulnicky, H.A., & Heckman, T.M. 2002 *ApJ* 574, 663
McDowell, J.C., Clements, D.L., Lamb, S.A., Shaked, S., Hearn, N.C., Colina, L., Mundell, C., Borne, K., Baker, A.C., & Arribas, S. 2003 *ApJ* 591, 154

Ponman, T.J., *et al.* 2005 *ApJ* in prep.

Read, A.M. 2003 *MNRAS* 342, 715

Read, A.M. 2005 *MNRAS* 359, 455

Read, A.M. & Stevens, I.R. 2002, *MNRAS* 335, L36

Sansom, A.E., Dotani, T., Okada, K., Yamashita, A., & Fabbiano, G. 1996 *MNRAS* 281, 48

Shirey, R., *et al.* 2001 *A&A* 365, 195

Stevens, I.R., Read, A.M., & Bravo-Guerrero, J. 2003 *MNRAS* 343, L47

Strickland, D.K., Heckman, T.M., Colbert, E.J.M., Hoopes, C.G., & Weaver, K.A. 2004a *ApJS*, 151, 153

Strickland, D.K., Heckman, T.M., Colbert, E.J.M., Hoopes, C.G., & Weaver, K.A. 2004b *ApJ*, 606, 829

Strickland, D.K. & Stevens, I.R. 2000, *MNRAS* 314, 511

Suchkov, A.A., Berman, V.G., Heckman, T.M., & Balsara, D.S. 1996 *ApJ* 463, 528

Warwick, R.S., Jenkins, L.E., Read, A.M., & Roberts T.R. 2005 *MNRAS* in prep.

Discussion

WARD: What is the mass of the hot gas in the superwinds of M82 and NGC253, and what is the ratio of this to the HI?

READ: For M82, the EPIC observations yield values of 1.5 and $1.7 \times 10^7 \eta^{1/2} \, M_\odot$ for the hot gas mass in the nuclear and inner wind regions respectively. Approximate estimates of the hot gas masses in the entire X-ray superwinds range from 2.5 (NGC253) to 5.5 (M82) $\times 10^8 \eta^{1/2} \, M_\odot$. This is $\sim$10% of the HI mass.

VÖLK: Regarding outflow velocities from starburst galaxies, your problem with the soft X-ray gas not being representative might be resolved by comparing the synchrotron morphology with wind models to obtain the flow velocity. For NGC253, this gives $\sim$900 km s^{-1} (see arXiv:astro-ph/0509248). Do you agree?

READ: Certainly, it is believed from theory and simulations that the flow velocity of the hot wind fluid is somewhere close to this value, or possibly higher. As synchrotron halos have now been discovered in a number of starburst galaxies (e.g. NGC253, NGC4631), this new insight on starburst winds is very much worth pursuing further.

COURVOISIER: How do your results bear on the question of the emission of the iron that is observed in the intracluster gas?

READ: Results so far suggest that, for the outflowing or off-galaxy components of the diffuse emission, the metallicity appears to be very low ($\sim$0.2 solar). We may however be seeing a similar effect to that discussed in the context of the brighter central diffuse regions of the Antennae, whereby, only when we have sufficient spatial resolution and effective area are we able to detect high-metallicity regions within these low surface brightness diffuse emission features.

MEURS: The harder part of the X-ray emission from galaxies is often being used as a rather safe indicator of the SFR, since the diffuse emission is not expected to influence the results very much. You mentioned a few cases of diffuse emission extending into the hard X-ray region. Could this affect such SFR estimates?

READ: Not really, as, even in the very few galaxies where hard diffuse emission is perhaps detected, the hard X-ray emission is still very much dominated by the point source (compact object) population (XRBs LLAGN etc.).

Populations of High Energy Sources in Galaxies
Proceedings IAU Symposium No. 230, 2005
E. J. A. Meurs & G. Fabbiano, eds.

© 2006 International Astronomical Union
doi:10.1017/S1743921306008064

The X-ray Evolution of Merging Galaxies

Nicola J. Brassington,[1] **Andrew M. Read**[2]
and Trevor J. Ponman[1]

[1]School of Physics and Astronomy, University of Birmingham, Edgbaston, Birmingham B15
2TT, UK
email: njb@star.sr.bham.ac.uk

[2]Department of Physics and Astronomy, University of Leicester, University Road, Leicester
LE1 7RH, UK

Abstract. From a *Chandra* survey of nine interacting galaxy systems the evolution of X-ray emission during the merger process has been investigated. From comparing L_X/L_K and L_{FIR}/L_B it is found that the X-ray luminosity peaks ~ 300 Myr before nuclear coalescence, even though we know that rapid and increasing star formation is still taking place at this time. It is likely that this drop in X-ray luminosity is a consequence of outflows breaking out of the galactic discs of these systems. At a time ~ 1 Gyr after coalescence, the merger-remnants in our sample are X-ray dim when compared to typical X-ray luminosities of mature elliptical galaxies. However, we do see evidence that these systems will start to resemble typical elliptical galaxies at a greater dynamical age, given the properties of the 3 Gyr system within our sample, indicating that halo regeneration will take place within low L_X merger-remnants.

1. Introduction

One of the most important factors in galaxy evolution is the effect of galaxy collisions and mergers. It is likely that most galaxies have been shaped or created through an interaction or merger with another galaxy. In particular, many astronomers believe that the formation of many elliptical galaxies is the product of the merger of two spiral galaxies. This was first proposed by Toomre in 1977, illustrated by the 'Toomre' sequence (Toomre (1977)). This shows 11 examples from the New General Catalogue (NGC) of Nebulae and Clusters of Stars, from approaching disc systems to near-elliptical remnants. The systems within this sequence have been well studied over a range of wavelengths, including X-ray.

X-ray observations are of particular importance when observing interacting galaxies as they are able to probe the dusty nucleus of the system, enabling the nature of the point source population to be established. Imaging of the the soft X-ray emission allows us to map out the diffuse gas and observe galactic-winds outflowing from the system, enabling constraints to be placed on the energetics of the outflows. From observing a selection of these interacting systems, at different stages of evolution, the processes involved in the merging of galaxy pairs can be characterised. Here we briefly describe the behaviour of the evolutionary sequence for our selected sample and discuss the mechanisms involved at key stages in this merger process.

2. The Sample

Our sample contains nine merging systems, selected using the following criteria;
(*a*) All systems have been observed with *Chandra*.
(*b*) Systems comprise of, or, originate from, two similar mass spiral galaxies.

Table 1. Key properties of the nine interacting systems.

Galaxy System	Distance (Mpc)	Merger Age (Gyr)	L_X ($\times 10^{40}$erg s^{-1})	%L$_{diff}$	Paper reference
Arp 270	28	-0.65	2.95	28	Brassington *et al.* (2005a)
The Mice	88	-0.5	7.43	25	Read (2003)
The Antennae	19	-0.4	8.40	45	Fabbiano *et al.* (2004)
Mkn 266	115	-0.3	73.18	51	Brassington *et al.* (2005b)
NGC 3256	56	-0.2	90.00	59	Lira *et al.* (2002)
Arp 220	76	0	26.70	49	McDowell *et al.* (2003)
NGC 7252	63	1.0	6.91	30	Nolan *et al.* (2004)
Arp 222	23	1.2	1.46	45	Brassington *et al.* (2005b)
NGC 1700	54	3.0	17.58	83	Diehl & Statler (2005)

(*c*) Multi-wavelength information is available for each system.

(*d*) The absorbing column is low, maximising the sensitivity to soft X-ray emission.

(*e*) A wide chronological sequence is covered; from detached pairs to merger-remnants.

The main problem when working with a chronological study is assigning each system a merger age. This problem has been addressed through a combination of N-body simulations, dynamical age estimates and stellar population synthesis models. From this we have established an evolutionary sequence for the nine systems in our sample, this can be seen in Table 1 where: column (1) gives the system name, column (2) the distance (assuming $H_0 = 75$ km s^{-1} Mpc^{-1}), column (3) the merger age of the system (assigning 0 Myr at the time of nuclear coalescence), column (4) the total (0.3$-$6.0 keV) X-ray luminosity (L_X) of the system, column (5) the percentage of luminosity arising from the diffuse gas and column (6) the paper reference of a detailed *Chandra* study of the system.

3. The Evolution Plot

We have obtained the B-band, K-band and FIR luminosities for each system, in addition to their L_X value, as shown in Table 1. With this information we have compared the activity levels, indicated by L_{FIR}/L_B (a proxy for star-formation normalised by the galaxy mass) and L_X, scaled by galaxy mass for each system. For the normalisation of L_X we have used both the B-band and the K-band luminosities, L_X/L_B and L_X/L_K. This is to observe how L_B varies from L_K, as it is known that L_B is affected by star-formation.

These ratios are shown in Figure 1, where we have plotted, not only the luminosity ratios, but also the percentage of luminosity arising from diffuse gas (%L_{diff}), as a function of merger age. L_{FIR}/L_B (solid line), L_X/L_B (dot-dash line) and L_X/L_K (dashed line) have been normalised by the typical spiral galaxy, NGC 2403. Whilst %L_{diff} (dotted line) is plotted on the right y-axis of the image, as an absolute value, where %L_{diff} for NGC 2403 is 12%. The horizontal lines to the right of the plot indicate L_{FIR}/L_B, L_X/L_B, L_X/L_K and %L_{diff} for NGC 2434, a typical elliptical galaxy (Diehl & Statler (2005)).

From this figure, it can be seen that the activity increases with merger age up to the point of nuclear coalescence. At this time the activity peaks, showing that this point of evolution is the most violent stage in the merger process. After coalescence, the activity decreases for $\sim$1.2 Gyr and then flattens. At this point it exhibits a value of activity similar to that observed in typical elliptical galaxies, as indicated by NGC 2434.

Both of the normalised X-ray luminosity ratios, L_X/L_K and L_X/L_B, exhibit broadly the same trends, with values increasing until $\sim$300 Myr before nuclear coalescence takes place, and then dropping until $\sim$1.2 Gyr after the nuclei have coalesced. Both of these values then begin to rise again, and, from the indication of the trend lines, seem likely

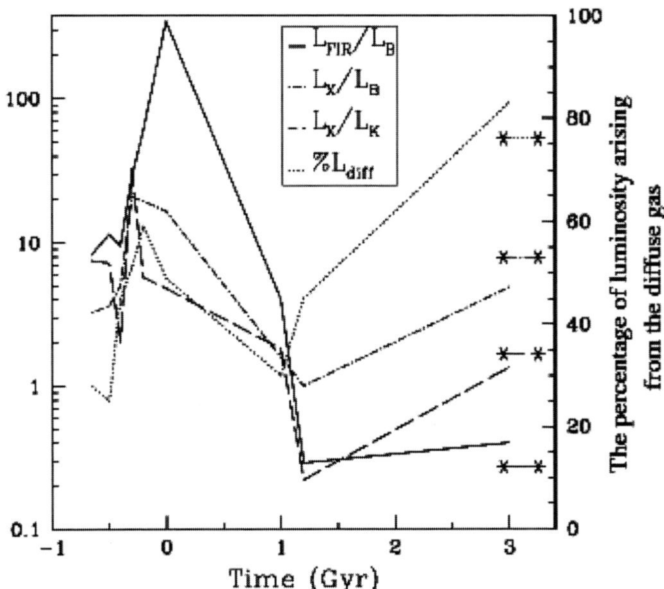

Figure 1. The evolution of X-ray luminosity in merging galaxies. Shown are $L_{\mathrm{FIR}}/L_{\mathrm{B}}$ (solid line), $L_{\mathrm{X}}/L_{\mathrm{B}}$ (dot-dash line), $L_{\mathrm{X}}/L_{\mathrm{K}}$ (dashed line) and $\%L_{\mathrm{diff}}$ (dotted line), plotted as a function of merger age, where 0 age is defined to be the time of nuclear coalescence. All luminosity ratios ($L_{\mathrm{FIR}}/L_{\mathrm{B}}$, $L_{\mathrm{X}}/L_{\mathrm{B}}$ and $L_{\mathrm{X}}/L_{\mathrm{K}}$) are normalised by the spiral galaxy NGC 2403. $\%L_{\mathrm{diff}}$ is plotted on a linear scale, shown on the right y-axis of the image and is an absolute value. $\%L_{\mathrm{diff}}$ for NGC 2403 is 12%. The horizontal lines to the right of the plot indicate $L_{\mathrm{FIR}}/L_{\mathrm{B}}$, $L_{\mathrm{X}}/L_{\mathrm{B}}$, $L_{\mathrm{X}}/L_{\mathrm{K}}$ and $\%L_{\mathrm{diff}}$ for NGC 2434, a typical elliptical galaxy.

to resemble an elliptical galaxy, at a later time, although this exceeds the evolution time within our sample. $\%L_{\mathrm{diff}}$ exhibits a similar trend, with the value peaking at a time before coalescence (although the peak is at $\sim$200 Myr) and then dropping, before rising once more, exhibiting a similar value to that of NGC 2434.

The results from this study are initially surprising, given that a similar study carried out with *ROSAT* observations (Read & Ponman (1998)), albeit a study that does not include all the same systems as ours, found that the peak in L_{X} was coincident with the peak in $L_{\mathrm{FIR}}/L_{\mathrm{B}}$ and nuclear coalescence. It is likely that this difference is due to the inclusion of the merging galaxy NGC 520 (a system which has been described as an 'anomalous half-merger' (Read (2005))), in the Read & Ponman (1998) study. We suggest that the reason that L_{X} peaks at a time before nuclear coalescence is a consequence of large-scale diffuse outflows being driven out of the interacting systems by the starbursts taking place within these galaxies.

Starburst-driven winds are responsible for the transport of gas and energy out of star forming galaxies. The energetics of these galactic winds were investigated in Strickland & Stevens (2000), via two-dimensional hydro-dynamical simulations. From modelling L_{X} they found that, once superbubbles had broken out of the galactic disc and formed galactic winds, L_{X} dropped rapidly ($\sim$10 Myr). It was also found that the soft X-ray emission from these winds arises from the cooler, denser, low volume filling factor gas ($\eta = 0.01-2\%$) which contains only $\sim$10% of the energy of the wind. Because of this, *Chandra* can only provide lower limits on the mass and energy content of the galactic winds. Consequently, more evolved, pre-merger systems that have wide-spread galactic winds exhibit lower L_X than starburst systems that have yet to experience large-scale diffuse outflows.

Another interesting result from this survey is the increase in L_X in the older merger-remnants. In previous studies post-merger systems have been found to be X-ray dim when compared to elliptical galaxies. In this study we have extended the merger sequence to include a 3 Gyr system, and, in doing so, have observed that these under-luminous systems seem likely to have increased L_X at a greater dynamical age. Given that these merger-remnants have been shown to be quiescent, we know that this increase in L_X is not due to any starburst activity within the system. Coupling this with the increase in $\%L_{\mathrm{diff}}$ indicates that diffuse X-ray gas is being produced, leading to the creation of X-ray haloes, as observed in mature elliptical galaxies. O'Sullivan et $al.$ (2001) investigated the relationship between L_X/L_B and spectroscopic age in post-merger ellipticals and found that there was a long term trend ($\sim$10 Gyr) for L_X to increase with time. The mechanism by which they explain the regeneration of hot gas haloes in these galaxies is one in which an outflowing wind to hydrostatic halo phase is driven by a declining SNIa rate. They argue that a scenario in which gas, driven out during the starburst, infalls onto the existing halo is not the dominant mechanism in generating X-ray haloes as this mechanism would only take $\sim$1$-$2 Gyr and would therefore not produce the long-term trend they observe. With the current information from our sample, neither scenario can be ruled out.

4. Conclusions

From our sample of nine interacting galaxy systems we have compared L_X/L_K and L_{FIR}/L_B, and shown that there is a peak in L_X $\sim$300 Myr before nuclear coalescence takes place. We suggest that the reason for this decrease in L_X at a time before coalescence is due to hot gaseous outflows breaking out of the galactic discs of these systems. We have also shown that the X-ray dim merger-remnant systems seem likely to evolve into more luminous systems at a greater dynamical age, given the properties observed in the 3 Gyr system within our sample. This indicates that halo regeneration will take place within low luminosity merger-remnant systems. Greater detail of this work can be found in Brassington et $al.$ (2005b).

References

Brassingon, N. J., Read, A. M., & Poman, T. J. 2005a, $MNRAS$ 360, 801

Brassingon, N. J., Read, A. M., & Poman, T. J. 2005b, in preparation

Diehl, S. & Statler, T. S. 2005, in preparation

Fabbiano, G., Baldi, A., King, A. R., Ponman, T. J., Raymond, J., Read, A., Rots, A., Schweizer, F., & Zezas, A. 2004, ApJ (Letters) 605, L21

Lira, P., Ward, M., Zezas, A., Alonso-Herrero, A., & Ueno, S. 2002, $MNRAS$ 330, 259

McDowell, J. C., Clements, D. L., Lamb, S. A., Shaked, S., Hearn, N. C., Colina, L., Mundell, C., Borne, K., Baker, A. C., & Arribas, S. 2003, ApJ 591, 154

Nolan, L. A., Ponman, T. J., Read, A. M., & Schweizer, F. 2004, $MNRAS$ 353, 221

O'Sullivan, E., Forbes, D. A., & Ponman, T. J. 2001, $MNRAS$ 324, 420

Read, A. M. 2003, $MNRAS$ 342, 715

Read, A. M. 2005, $MNRAS$ 359, 455

Read, A. M. & Ponman, T. J. 1998, $MNRAS$ 297, 143

Strickland, D. K. & Stevens, I. R. 2000, $MNRAS$ 314, 511

Toomre, A. 1977, in: Tinsley B. M. & Larson T. B. eds $Evolution$ of $Galaxies$ and $Stellar$ $Populations$, Yale University Observatory, New Haven. p. 401

Discussion

HORNSCHEMEIER: Would you comment on the uncertainties in the merger age estimates and your ∼300 Myr result in particular?

BRASSINGTON: These should be viewed as estimates but the merger sequence does work out very well. The ages of the systems prior to nuclear coalescence are determined with N-body simulations – so better, more detailed N-body simulations are needed to improve the age estimates.

FABBIANO: Are you developing or looking at hydro-dynamical simulations to try to explore detailed hot gas merger features such as the big loops of the Antennae?

BRASSINGTON: No, we're not currently carrying out any hydro-dynamical simulations. But they are definitely something we'd like to look into, along with improving current N-body simulation.

The internet corner.

Session 3

Detailed population studies in the nearer galaxies

Marina Kaufman Bernadó (left) and Rosanne DiStefano talking over coffee.

Populations of High Energy Sources in Galaxies
Proceedings IAU Symposium No. 230, 2005
E. J. A. Meurs & G. Fabbiano, eds.

© 2006 International Astronomical Union
doi:10.1017/S1743921306008076

The X-ray source populations in M31 and M33

Wolfgang Pietsch

Max-Planck-Institut für extraterrestrische Physik, Giessenbachstraße, 85741 Garching, Germany
email: wnp@mpe.mpg.de

Abstract. First, the X-ray source populations of M31 and M33 as known from *Einstein* and ROSAT observations are presented. Then, *Chandra* results on the galaxies are shortly summarized which not only spatially resolved the centre areas but also supernova remnants (SNRs) in both galaxies, and led to source catalogues of restricted areas with high astrometric accuracy. Also luminosity function studies and studies of individual sources based on *Chandra* and XMM-Newton observations led to a better knowledge of the X-ray source populations. After that I will concentrate on XMM-Newton surveys, in which more than 850 and 400 X-ray sources were detected in M31 and M33, respectively, down to a 0.2–4.5 keV luminosity of less than 10^{35} erg s^{-1}. EPIC hardness ratios as well as informations from earlier X-ray, optical, and radio catalogues were used to distinguish between different source classes (SNRs, supersoft sources (SSSs), X-ray binaries (XRBs), globular cluster sources within the galaxies, and foreground stars, and objects in the background). However, many sources could only be classified as hard. They may either be XRBs or Crab-like SNRs in the galaxies or background sources. Within M31, two globular cluster sources could be identified as low mass XRBs as they showed type I X-ray bursts. Many of the SSSs in both galaxies were identified as optical novae. Due to the high frequency of outbursts in the bulge area of M31 many novae can be monitored at the same time which makes the investigation of class properties much easier compared to novae in the Milky Way or the Magellanic Clouds.

Keywords. galaxies: individual (M31, M33), novae, cataclysmic variables, supernova remnants, X-rays: galaxies, X-rays: binaries, X-rays: bursts

1. Introduction

The Andromeda galaxy M31 is a massive early-type galaxy similar to the Milky Way with a large stellar bulge which is located in the Local Group at a distance of 780 kpc (Holland 1998; Stanek & Garnavich 1998; i.e. 1" corresponds to 3.8 pc) and is seen under an inclination of 78°. The optical extent of the SA(s)b galaxy can be approximated by an inclination-corrected D_{25} ellipse with a large diameter of 153.3' and axis ratio of 3.09 (de Vaucouleurs, de Vaucouleurs, Corwin *et al.* 1991; Tully 1998). In contrast to M31 and the Milky Way, M33 is a late-type spiral galaxy, type Sc, and does not have a stellar bulge. It is the third largest galaxy in the Local Group located at a distance of 795 kpc (van den Bergh 1991), is seen under a relatively low inclination of 56° (Zaritsky, Elston & Hill 1989) and the optical extent can be approximated by an inclination-corrected D_{25} ellipse with large diameter of 64.4' and axes ratio of 1.66. With their moderate Galactic foreground absorption ($N_{\rm H} = 6-7 \times 10^{20}$ cm^{-2}; Stark, Gammie, Wilson, *et al.* 1992) both galaxies are well suited to study the X-ray source population and diffuse emission. M31 and M33 have been the target of most imaging X-ray missions. In the following I will discuss results from the *Einstein*, ROSAT, *Chandra*, and XMM-Newton observatories.

2. Einstein and ROSAT observations of M31 and M33

In the M31 field, the *Einstein* X-ray observatory detected 108 individual X-ray sources brighter than $\sim5\times10^{36}$ erg s^{-1}, 16 of which were found to vary between *Einstein* observations (van Speybroeck, Epstein, Forman, *et al.* 1979; Collura, Reale & Peres 1990; Trinchieri & Fabbiano 1991). The sources were identified with young stellar associations, globular clusters (i.e. LMXBs) and SNRs (see e.g. Blair, Kirshner & Chevalier 1981; Crampton, Hutchings, Cowley, *et al.* 1984). With the ROSAT HRI, Primini, Forman & Jones (1993) reported 86 sources brighter than $\sim10^{36}$ erg s^{-1} in the central area of M31, nearly half of which were found to vary when compared to previous *Einstein* observations. With a separation of about one year between the M31 surveys, the ROSAT PSPC covered the entire galaxy twice and detected 560 X-ray sources down to a limit of $\sim5\times10^{35}$ erg s^{-1} (Supper, Hasinger, Pietsch, *et al.* 1997; Supper, Hasinger, Lewin, *et al.* 2001). The intensity of 34 of the sources varied significantly between the observations, and SSSs were established as a new class of M31 X-ray sources (see also Kahabka 1999). Many of the bright sources were identified as globular cluster sources.

In the M33 field, the *Einstein* X-ray Observatory detected diffuse emission from hot gas and 17 unresolved sources (Long, Dodorico, Charles, *et al.* 1981; Markert & Rallis 1983; Trinchieri, Fabbiano & Peres 1988). First ROSAT HRI and PSPC observations revealed 57 sources and confirmed the detection of diffuse X-ray emission which may trace the spiral arms within a 10' radius around the nucleus (Schulman & Bregman 1995; Long, Charles, Blair, *et al.* 1996). Combining all archival ROSAT observations of the field, Haberl & Pietsch (2001) found 184 X-ray sources within a 50' radius around the nucleus, identified some of the sources by correlations with previous X-ray, optical and radio catalogues, and in addition classified sources according to their X-ray properties. They found candidates for SSSs, XRBs, SNRs, foreground stars, and active galactic nuclei (AGN).

Two M33 sources are known for their outstanding X-ray properties (Peres, Reale, Collura, *et al.* 1989). The brightest source (X–8 in the nomenclature of Long, Dodorico, Charles, *et al.* (1981), luminosity of 10^{39-40} erg s^{-1}) is the most luminous X-ray source in the Local Group of galaxies and coincides with the optical center of M33. Its time variability (Dubus, Charles, Long, *et al.* 1997) and its point-like nature seen by ROSAT HRI (Pietsch & Haberl 2000) point towards a black hole XRB. A possible periodicity of 106 days was not confirmed in later observations (see Parmar, Sidoli, Oosterbroek, *et al.* 2001). The second source (X−7) is an eclipsing high mass XRB (HMXB) with a binary period of 3.45 d and possible 0.31 s pulsations, discovered in ROSAT PSPC and HRI observations (Schulman, Bregman, Collura, *et al.* 1993; Dubus, Charles, Long, *et al.* 1997; Dubus, Charles, Long, *et al.* 1999; Larson & Schulman 1997).

3. Chandra and XMM-Newton observations of M31 and M33

With the new generation of X-ray observatories, *Chandra* and XMM-Newton, only parts of M31 were surveyed. *Chandra* ACIS-I and HRC observations of the central region (covering areas of 0.08 and 0.27 square degree) resolved 204 and 142 X-ray sources, respectively (Garcia, Murray, Primini, *et al.* 2000; Kong, Garcia, Primini, *et al.* 2002a; Kaaret 2002). Three M31 disk fields, spanning a range of stellar populations, were covered with short *Chandra* ACIS-S observations to compare their point source luminosity functions to that of the galaxy's bulge (Kong, DiStefano, Garcia, *et al.* 2003a). A synoptic study of M31 with the *Chandra* HRC covered "most" of the disk (0.9 square degree) in 17 epochs using short observations, and resulted in mean fluxes and long-term light curves for the 166 objects detected (Williams, Garcia, Kong, *et al.* 2004). In these observations, several M31 SNRs were spatially resolved (Kong, Garcia, Primini, *et al.* 2002b; Kong, Sjouwerman, Williams, *et al.* 2003b: Williams, Sjouwerman, Kong, *et al.* 2004) and bright XRBs in globular clusters and SSSs and quasisoft sources (QSSs) could be characterized (Di Stefano, Kong, Garcia, *et al.* 2002; Di Stefano, Kong, Greiner, *et al.* 2004; Greiner, Di Stefano, Kong, *et al.* 2004). During the XMM-Newton guaranteed time program there were four observations of the central area of M31 and three aimed at the northern and two at the southern disk. These observations were used to investigate the bright and variable sources and diffuse emission (Shirey, Soria, Borozdin, *et al.* 2001; Osborne, Borozdin, Trudolyubov, *et al.* 2001; Trudolyubov, Borozdin & Priedhorsky 2001; Trudolyubov, Kotov, Priedhorsky, *et al.* 2004), and to derive source luminosity distributions (Trudolyubov, Borozdin, Priedhorsky, *et al.* 2002a). In addition, the time variability and spectra of several individual XRBs have been studied in detail (e.g. Trudolyubov, Borozdin, Priedhorsky, *et al.* 2002b; Barnard, Kolb & Osborne 2003; Barnard, Osborne, Kolb, *et al.* 2003; Barnard, Kolb & Osborne 2004; Mangano, Israel & Stella 2004; Barnard, Foulkes, Haswell, *et al.* 2005).

The full extent of M33 was covered by a homogeneous XMM-Newton survey with a sensitivity of 10^{35} erg s^{-1} in the 0.5–10 keV band, a factor of ten deeper than previous surveys. The survey consisted of a raster of 15 pointings of about 10 ks each. The directions were selected in a way that each area within the M33 optical D$_{25}$ extent was covered by the XMM-Newton EPIC detectors at least three times with the medium filter. First results of this survey were presented in Pietsch, Ehle, Haberl, *et al.* (2003). Three *Chandra* ACIS-S and ACIS-I observations were pointed at the M33 nucleus and the star-forming region NGC 604. Grimm, McDowell, Zezas, *et al.* (2005) present a source list of 261 sources in an area of $\sim$0.2 square degree down to a luminosity of $\sim$2 $\times$ 10^{34} erg s^{-1}. They determine luminosity functions consistens with those of other star-forming galaxies. Based on the same observations, Ghavamian, Blair, Long, *et al.* (2005) report the detection of 22 optically known SNRs in M33 of which at least four ar spacially extended, and propose the identification of one non-radiative SNR (based on the spectral index of the radio counterpart) and new optical counterparts for two soft X-ray SNR candidates from the XMM-Newton list of Pietsch, Misanovic, Haberl, *et al.* (2004).

Chandra observations confirm the point-like nature of M33 X–8 (Dubus & Rutledge 2002). The X-ray spectrum of the source can best be described by an absorbed power law plus disk blackbody model and the source flux can double within 5 ks (e.g. Ehle, Pietsch & Haberl 2001; La Parola, Damiani, Fabbiano, *et al.* 2003; Foschini, Rodriguez, Fuchs, *et al.* 2004; Dubus, Charles & Long 2004). The combination of the properties establishes the source as the nearest example of the ultra-luminous X-ray sources that have been uncovered in other nearby galaxies.

Pietsch, Mochejska, Misanovic, *et al.* (2004) analyzed XMM-Newton and *Chandra* data of M33 X–7 and determined improved binary parameters, however, they could not confirm the proposed pulsations. In a special analysis of DIRECT data, they identified an O7I star of 18.89 mag in V, as the optical counterpart, which shows the ellipsoidal heating light curve of a HMXB with the X–7 binary ephemeris. X–7 was the most distant eclipsing XRB until the detection of another such source in XMM-Newton and *Chandra* observations of the starburst galaxy NGC 253, which is located at more than three times the distance of M33 (Pietsch, Haberl & Vogler 2003).

4. XMM-Newton surveys of M31 and M33

Pietsch, Misanovic, Haberl, *et al.* (2004) have created merged medium and thin filter images for the three EPIC instruments, in five energy bands (0.2–0.5 keV, 0.5–1.0 keV,

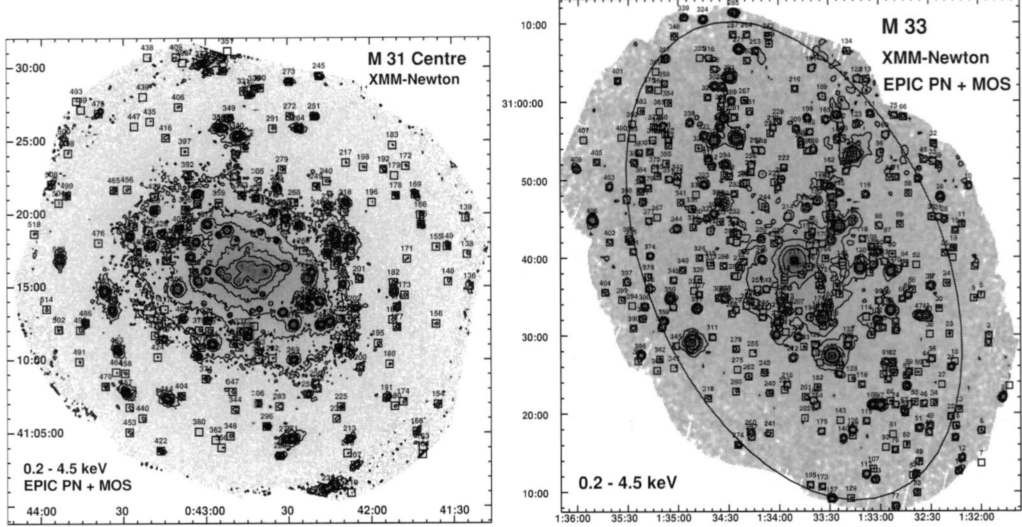

Figure 1. Combined XMM-Newton EPIC images in the 0.2–4.5 keV band. (left): M31 centre observations smoothed with a 5" FWHM Gaussian (Fig. 1 from Pietsch, Freyberg & Haberl (2005)). (right): M33 raster observations smoothed with a 20" FWHM Gaussian (Fig. 2 from Pietsch, Misanovic, Haberl, *et al.* (2004)).

1.0–2.0 keV, 2.0–4.5 keV, and 4.5–12 keV), using only times of low background from the XMM-Newton survey of M33†. A source search in these merged images – simultaneously using 5×3 images (5 energy bands and PN, MOS1 and MOS2 camera) – yielded a catalogue of 408 sources in an area of 0.80 square degree down to a luminosity in the 0.2–4.5 keV band of 10^{35} erg s^{-1}, more than a factor of 10 deeper than earlier ROSAT observations (see Fig. 1 for an overlay of the sources on a smoothed 0.2–4.5 keV all EPIC image). Hardness ratios were calculated only for sources for which at least one of the two band count rates had a significance greater than 2σ. In search for identifications, the X-ray source positions were correlated with sources in the SIMBAD and NED archives and within several catalogues. The catalogued X-ray sources are "identified" or "classified" based on properties in X-rays (hardness ratios (HR), variability, extent) and of correlated objects in other wavelength regimes. A source is counted as identified, if at least two criteria secure the identification. Otherwise, it is only counted as classified.

For the source catalogue and source population study of the archival observations of M31, Pietsch, Freyberg & Haberl (2005) analyzed the individual pointings and the merged data of the central area simultaneously in five energy bands in a similar way as described in their M33 analysis. The covered area of 1.24 square degree and limiting sensitivity is a significant improvement compared to the *Chandra* surveys, but only covers about 2/3 of the optical M31 extent (D_{25} ellipse) with a rather inhomogeneous exposure. The left panel of Fig. 1 shows an overlay of detected sources on an smoothed image of the M31 centre area. Fig. 2 shows colour/colour diagrams based on the HRs. To identify areas of specific source classes in the plots, colours of sources are over-plotted, that were derived from measured XMM-Newton spectra and model simulations.

---

† An analysis of the individual M33 pointings with less stringent background rejection and using all filters is presented in two posters by Misanovic *et al.* at this conference, and resulted in a source catalogue with improved positions (taking care of individual pointing offsets), that also provide information on source variability and spectral characteristics.

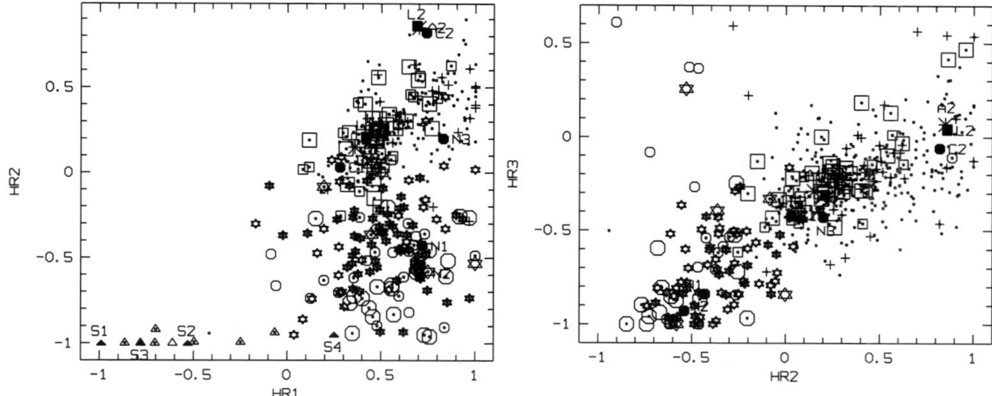

Figure 2. Hardness ratios detected by XMM-Newton EPIC. Shown as dots are only sources with HR errors smaller than 0.2 on both HRi and $HRi+1$. Foreground stars and candidates are marked as big and small stars, AGN and candidates as big and small crosses, SSS candidates as triangles, SNR and candidates as big and small hexagons, GlCs and XRBs and candidates as big and small squares. In addition, we mark positions derived from measured XMM-Newton EPIC spectra and models for SSSs (S1 to S4) as filled triangles, low mass XRBs (L1 and L2) as filled squares, SNRs (N132D as N1, 1E 0102.2–7219 as N2, N157B as N3, Crab spectra as C1 and C2) as filled hexagons, AGN (A1 and A2) as asterisk. (Extracted from Fig. 5 from Pietsch, Freyberg & Haberl (2005).)

Table 1. Summary of identifications and classifications of X-ray sources in M31 and M33.

		fgStar	AGN	GAL	GAL Cl	SSS	SNR	GlC	XRB	hard
M31	identified	6	1	1	1		21	27	7	
	classified	90	36		1	18	23	10	9	567
M33	identified	5		1			21 + 2		2	
	classified	30	12	1		5	23 − 2			267

Table 1 summarizes identifications and classifications according to the XMM-Newton catalogues. For the SNRs in M33 the Ghavamian, Blair, Long, *et al.* 2005 identifications are indicated. Detection of strong time variability in follow up analysis will certainly move many objects from the "hard" to "XRB" classification.

Based on the catalogue identifications, Pietsch & Haberl (2005) searched for X-ray bursts in XMM-Newton archival data of X-ray sources in M31 globular clusters (GCs) and GC candidates. Two bursts were detected simultaneously in EPIC pn and MOS detectors and some more candidates in EPIC pn. The energy distribution of the burst photons and the intrinsic luminosity during the peak of the bursts indicate that at least the strongest burst was a type I radius expansion burst (Fig. 3). The bursts identify the sources as neutron star low mass X-ray binaries in M31. The type I X-ray bursts in M31 are the first detected outside the Milky Way and show that with the help of XMM-Newton X-ray bursts can be used to classify neutron star low mass X-ray binaries in Local Group galaxies.

5. Novae as major class of SSS in M31

Pietsch, Fliri, Freyberg, *et al.* (2005) searched for X-ray counterparts to optical novae detected in M31 and M33. They combined an optical nova catalogue from the WeCAPP survey with optical novae reported in the literature and correlated them with the most

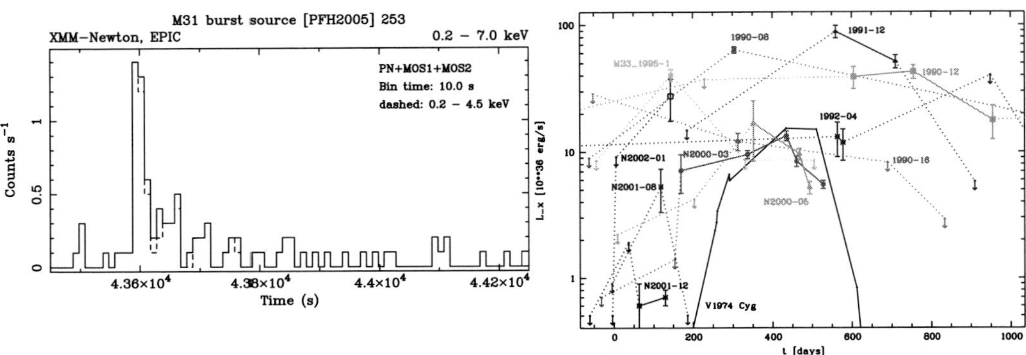

Figure 3. (left): Combinded XMM-Newton EPIC light curve of a type I X-ray burst of source [PFH2005] 253 in M31 (Fig. 3 from Pietsch & Haberl (2005)). (right): Light curves for M31 and M33 novae that were detected within 1000 d after outburst. Detections of individual novae are connected by solid lines, and connections to upper limits are marked by dashed lines (Fig. 3 from Pietsch, Fliri, Freyberg, *et al.* (2005)).

recent X-ray catalogues from ROSAT, XMM-Newton and *Chandra*, and – in addition – searched for nova correlations in archival data. They report 21 X-ray counterparts for novae in M31 (mostly SSS) and two in M33. This sample more than triples the number of known optical novae with supersoft phase. Most of the counterparts are covered in several observations which allows to constrain their X-ray light curves (see Fig. 3). Selected brighter sources were classified by their XMM-Newton EPIC spectra. Six counterparts are only detected in *Chandra* HRC I (3) or ROSAT HRI (3) observations, i.e. X-ray detectors with no energy resolution, and therefore can not be classified as supersoft. From the well-determined start time of the SSS state in two novae one can estimate the hydrogen mass ejected in the outburst to $\sim 10^{-5} M_\odot$ and $\sim 10^{-6} M_\odot$, respectively. The supersoft X-ray phase of at least 15% of the novae starts within a year. At least one of the novae shows a SSS state lasting 6.1 years after the optical outburst. Six of the SSSs turned on between 3 and 9 years after the optical discovery of the outburst and may be interpreted as recurrent novae. If confirmed, the detection of a delayed SSS phase turn-on may be used as a new method to classify novae as recurrent. At the moment, the new method yields a ratio of recurrent novae to classical novae of 0.3.

6. Conclusions

The sensitivity of XMM-Newton and *Chandra* observations of M31 and M33 combined with the wealth of multi-wavelength data for these galaxies allows a detailed study of the point source population. Many more interesting results can be expected from the approved *Chandra* ACIS-I very large proposal on M33 (sensitive mainly above 0.5 keV) and, hopefully, further monitoring of M31 and M33 with XMM-Newton and *Chandra* also in the energy band below 0.5 keV. The first results of the monitoring the SSS state of optical novae proved that these kind of studies can be more efficiently achieved by observing many candidates at the same time in one field in M31 or M33 than by monitoring individual sources in the Milky Way or the Magellanic Clouds. The results of the *Chandra* and XMM-Newton observations of M31 and M33 demonstrate the importance of arcsec spatial resolution, broad energy coverage, good energy resolution, and high collecting power – used together with deep images and catalogues at other wavelengths – also for future X-ray source population studies in nearby galaxies.

References

Barnard, R., Osborne, J. P., Kolb, U., & Borozdin, K. N. 2003, *A&A* 405, 505

Barnard, R., Kolb, U., & Osborne, J. P. 2003, *A&A* 411, 553

Barnard, R., Kolb, U., & Osborne, J. P. 2004, *A&A* 423, 147

Barnard, R., Foulkes, S. B., Haswell, C. A., Kolb, U., Osborne, J. P., & Murray, J. R. 2005, *MNRAS* submitted (astro-ph/0503259)

Blair, W. P., Kirshner, R. P., & Chevalier, R. A. 1981, *ApJ* 247, 879

Collura, A., Reale, F., & Peres, G. 1990, *ApJ* 356, 119

Crampton, D., Hutchings, J. B., Cowley, A. P., Schade, D. J., & van Speybroeck, L. P. 1984, *ApJ* 284, 663

de Vaucouleurs, G., de Vaucouleurs, A., Corwin, H. G., Buta, R. J., Paturel, G., & Fouque, P. 1991, *Third Reference Catalogue of Bright Galaxies* (vol. 1–3, Springer-Verlag Berlin Heidelberg New York)

Di Stefano, R., Kong, A. K. H., Garcia, M. R., Barmby, P., Greiner, J., Murray, S. S., & Primini, F. A. 2002, *ApJ* 570, 618

Di Stefano, R., Kong, A. K. H., Greiner, J., Primini, F. A., Garcia, M. R., Barmby, P., Massey, P., Hodge, P. W., Williams, B. F., Murray, S. S., Curry, S., Russo, T. A., *et al.* 2004, *ApJ* 610, 247

Dubus, G., Charles, P. A., Long, K. S., & Hakala, P. J. 1997, *ApJ* (Letters) 490, L47

Dubus, G., Charles, P. A., Long, K. S., Hakala, P. J., & Kuulkers, E. 1999, *MNRAS* 302, 731

Dubus, G. & Rutledge, R. E. 2002, *MNRAS* 336, 901

Dubus, G., Charles, P. A., & Long, K. S. 2004, *A&A* 425, 95

Ehle, M., Pietsch, W., & Haberl, F. 2001, in ASP Conf. Ser. 251: *New Century of X-ray Astronomy*, 300

Foschini, L., Rodriguez, J., Fuchs, Y., Ho, L. C., Dadina, M., Di Cocco, G., Courvoisier, T. J.-L., & Malaguti, G. 2004, *A&A* 416, 529

Garcia, M. R., Murray, S. S., Primini, F. A., Forman, W. R., McClintock, J. E., & Jones, C. 2000, *ApJ* (Letters) 537, L23

Ghavamian, P., Blair, W. P., Long, K. S., Sasaki, M., Gaetz, T. J., & Plucinsky, P. P. 2005, *AJ* 130, 539

Greiner, J., Di Stefano, R., Kong, A., & Primini, F. 2004, *ApJ* 610, 261

Grimm, H.-J., McDowell, J., Zezas, A., Kim, D.-W., & Fabbiano, G. 2005, *ApJS* accepted (astro-ph/0506353)

Haberl, F. & Pietsch, W. 2001, *A&A* 373, 438 [HP01]

Holland, S. 1998, *AJ* 115, 1916

Kaaret, P. 2002, textit*ApJ* 578, 114

Kahabka, P. 1999, *A&A* 344, 459

Kong, A. K. H., Garcia, M. R., Primini, F. A., Murray, S. S., Di Stefano, R., & McClintock, J. E. 2002a, textit*ApJ* 577, 738

Kong, A. K. H., Garcia, M. R., Primini, F. A., & Murray, S. S. 2002b, *ApJ* (Letters) 580, L125

Kong, A. K. H., DiStefano, R., Garcia, M. R., & Greiner, J. 2003a, *ApJ* 585, 298

Kong, A. K. H., Sjouwerman, L. O., Williams, B. F., Garcia, M. R., & Dickel, J. R. 2003b, *ApJ* (Letters) 590, L21

La Parola, V., Damiani, F., Fabbiano, G., & Peres, G. 2003, *ApJ* 583, 758

Larson, D. T. & Schulman, E. 1997, *AJ* 113, 618

Long, K. S., Dodorico, S., Charles, P. A., & Dopita, M. A. 1981, *ApJ* (Letters) 246, L61

Long, K. S., Charles, P. A., Blair, W. P., & Gordon, S. M. 1996, *ApJ* 466, 750

Mangano, V., Israel, G. L., & Stella, L. 2004, *A&A* 419, 1045

Markert, T. H. & Rallis, A. D. 1983, *ApJ* 275, 571

Osborne, J. P., Borozdin, K. N., Trudolyubov, S. P., Priedhorsky, W. C., Soria, R., Shirey, R., Hayter, C., La Palombara, N., Mason, K., Molendi, S., Paerels, F., Pietsch, W., Read, A. M., Tiengo, A., Watson, M. G., & West, R. G. 2001, *A&A* 378, 800

Parmar, A. N., Sidoli, L., Oosterbroek, T., Charles, P. A., Dubus, G., Guainazzi, M., Hakala, P., Pietsch, W., & Trinchieri, G. 2001, *A&A* 368, 420

Peres, G., Reale, F., Collura, A., & Fabbiano, G. 1989, *ApJ* 336, 140

Pietsch, W. & Haberl, F. 2000, in *The interstellar medium in M31 and M33*. Proceedings 232. WE-Heraeus Seminar, 22–25 May 2000, Bad Honnef, Germany. Edited by Elly M. Berkhuijsen, Rainer Beck & Rene A. M. Walterbos. Shaker, Aachen, 2000, p. 149

Pietsch, W., Haberl, F., & Vogler, A. 2003, *A&A* 402, 457

Pietsch, W., Ehle, M., Haberl, F., Misanovic, Z., & Trinchieri, G. 2003, *Astronomische Nachrichten* 324, 85

Pietsch, W., Mochejska, B. J., Misanovic, Z., Haberl, F., Ehle, M., & Trinchieri, G. 2004, *A&A* 413, 879

Pietsch, W., Misanovic, Z., Haberl, F., Hatzidimitriou, D., Ehle, M., & Trinchieri, G. 2004, *A&A* 426, 11

Pietsch, W. & Haberl, F. 2005, *A&A* 430, L45

Pietsch, W., Freyberg, M., & Haberl, F. 2005, *A&A* 434, 483

Pietsch, W., Fliri, J., Freyberg, M. J., Greiner, J., Haberl, F., Riffeser, A., & Sala, G. 2005, *A&A* in press (astro-ph/0504321)

Primini, F. A., Forman, W., & Jones, C. 1993, *ApJ* 410, 615

Schulman, E., Bregman, J. N., Collura, A., Reale, F., & Peres, G. 1993, *ApJ* (Letters) 418, L67

Schulman, E. & Bregman, J. N. 1995, *ApJ* 441, 568

Shirey, R., Soria, R., Borozdin, K., Osborne, J. P., Tiengo, A., Guainazzi, M., Hayter, C., La Palombara, N., Mason, K., Molendi, S., Paerels, F., Pietsch, W., Priedhorsky, W., Read, A. M., Watson, M. G., & West, R. G. 2001, *A&A* 365, L195

Stanek, K. Z. & Garnavich, P. M. 1998, *ApJ* (Letters) 503, L131

Stark, A. A., Gammie, C. F., Wilson, R. W., Bally, J., Linke, R. A., Heiles, C., & Hurwitz, M. 1992, *ApJS* 79, 77

Supper, R., Hasinger, G., Lewin, W. H. G., Magnier, E. A., van Paradijs, J., Pietsch, W., Read, A. M., & Trümper, J. 2001, *A&A* 373, 63

Supper, R., Hasinger, G., Pietsch, W., Trümper, J., Jain, A., Magnier, E. A., Lewin, W. H. G., & van Paradijs, J. 1997, *A&A* 317, 328

Trinchieri, G., Fabbiano, G., & Peres, G. 1988, *ApJ* 329, 1037

Trinchieri, G. & Fabbiano, G. 1991, *ApJ* 382, 82

Trudolyubov, S. P., Borozdin, K. N., & Priedhorsky, W. C. 2001, *ApJ* (Letters) 563, L119

Trudolyubov, S. P., Borozdin, K. N., Priedhorsky, W. C., Mason, K. O., & Cordova, F. A. 2002a, *ApJ* (Letters) 571, L17

Trudolyubov, S. P., Borozdin, K. N., Priedhorsky, W. C., Osborne, J. P., Watson, M. G., Mason, K. O., & Cordova. F. A. 2002b, *ApJ* (Letters) 581, L27

Trudolyubov, S., Kotov, O., Priedhorsky, W., Cordova, F., & Mason, K. 2004, *ApJ* submitted (astro-ph/0401227)

Tully, R. B. 1988, *Nearby galaxies catalog* (Cambridge and New York, Cambridge University Press)

van den Bergh, S. 1991, *PASP* 103, 609

van Speybroeck, L., Epstein, A., Forman, W., Giacconi, R., Jones, C., Liller, W., & Smarr, L. 1979, *ApJ* (Letters) 234, L45

Williams, B. F., Garcia, M. R., Kong, A. K. H., Primini, F. A., King, A. R., Di Stefano, R., & Murray, S. S. 2004, *ApJ* 609, 735

Williams, B. F., Sjouwerman, L. O., Kong, A. K. H., Gelfand, J. D., Garcia, M. R., & Murray, S. S. 2004, *ApJ* 615, 720

Zaritsky, D., Elston, R., & Hill, J. M. 1989, *AJ* 97, 97

Discussion

ERACLEOUS: Did you get any useful information from the optical monitor that can help with the source identification?

PIETSCH: Not all fields have OM observations. We have not looked into that yet, but in principle the UV information can help.

KIM: How many background AGNs do you expect in M31, M33 field of view and only in the bulge where the majority of the hard X-ray sources would be LMXBs? Did you make the luminosity function and compare with the Chandra results?

PIETSCH: In M33, about half of the detected sources are expected to be background AGN. In M31, we did not make an estimate. However, as the observations are more concentrated to the denser areas of the galaxy, the fraction should be less, specifically in the bulge area. We did not create luminosity functions yet.

DISTEFANO: This is a comment on the connection you have discussed between SSSs and novae. Among SSSs in the Galaxy and Magellanic Clouds (the so-called "classical" SSSs), novae comprise only a small fraction of SSSs. In more distant spirals, such as M101, M83, and M51 a large fraction of SSSs are found in the spiral arms, not consistent with a population that is previously old. This is true for even the softest sources in the galaxies.

PIETSCH: I made clear in the talk that if not all but most of the SSS in M31 correlate with novae. Many of the QSS in more distant galaxies are not SSS in the strong spectral selection. However, even if they were SSS, it would be difficult to identify with optical novae as there is no nova monitoring in these galaxies.

Populations of High Energy Sources in Galaxies
Proceedings IAU Symposium No. 230, 2005
E. J. A. Meurs & G. Fabbiano, eds.

© 2006 International Astronomical Union
doi:10.1017/S1743921306008088

The Next Nearest Black Holes: Chandra and HST Observations of X-ray sources in M31

M. Garcia

Harvard-Smithsonian Center for Astrophysics, Cambridge, Mass., USA
garcia@cfa.harvard.edu

Abstract. From October 1999 to the present Chandra has taken nearly monthly snap-shot observations of M31. The first 3 years of this dataset in combination with deeper but less frequent XMM observations has allowed the detection of 45 X-ray transients within M31. By analogy to our Galaxy, many of these transients are likely black hole candidates. We have indentified a few optical counterparts of these possible black holes via simultaneous HST imaging. The census allows a study of the endpoints of stellar evolution in our nearest neighbor galaxy. When stacked, the observations also allow a deep study of the M31 X-ray source population, which turns up a few surprises. The supermassive black hole in the center of M31 is the next nearest one after Sgr A*. Chandra and HST observations allow the detection of a weak X-ray source at the position of this SMBH. These observations provide some of the most secure and severe limitations on accretion in SMBH.

Discussion

ERACLEOUS: With repeated observations you should be able to construct the luminosity function every time and study its temporal variations. Have you done that? It would be a very useful piece of information to compare with models.

GARCIA: No, we have not done that. Our sensitivity in each visit is $\sim 10^{36}$ erg/s.

ERACLEOUS: That would still be very useful.

UBERTINI: I am puzzled by the ratio between BHC/NS seen by Chandra. The BEP-POSAX and INTEGRAL data set seems to indicate a lower number of BHC. Can you comment on the possible lower threshold for weak transients observed with your Chandra observation? It is very interesting to understand if there is an intrinsic difference in the BHC/NS ratio within M31 and our Galaxy.

GARCIA: The BEPPOSAX/INTEGRAL transients which are NS may be faint and missed due to our sensitivity limit. These NS transients are also fast – we do have a few, but only a few, transients which are fast, and could be the analogue of the BEPPOSAX NS transients.

MACCARONE: 1. Does the P_{orb}/M_V correlation differ for BHs and NSs?
2. Have you looked for quiescent optical emission from longest orbital period systems?

GARCIA: 1. It probably should, but the present correlation mixes the source categories together. It would be a good idea, now that we know more about which sources are black holes and which are neutron stars, to try to refine that correlation.
2. We have seen emission as they were fading but we haven't yet looked in true quiescence.

Populations of High Energy Sources in Galaxies
Proceedings IAU Symposium No. 230, 2005
E. J. A. Meurs & G. Fabbiano, eds.

© 2006 International Astronomical Union
doi:10.1017/S174392130600809X

XMM-Newton reveals ~100 new LMXBs in M31 from variability studies

R. Barnard[1], L. Shaw Greening[1], C. Tonkin[1], U. Kolb[1] and J.P. Osborne[2]

[1]The Open University, Walton Hall, Milton Keynes, MK7 6AA, UK
email: R.Barnard@open.ac.uk, L.Shaw-Greening@open.ac.uk., U.C.Kolb@open.a.c.uk

[2]University of Leicester, University Road, Leicester LE1 7RH, UK
email: julo@star.le.ac.uk

Abstract. We have conducted a survey of X-ray sources in XMM-Newton observations of M31, examining their power density spectra (PDS) and spectral energy distributions (SEDs). Our automated source detection yielded 535 good X-ray sources; to date, we have studied 225 of them. In particular, we examined the PDS because low mass X-ray binaries (LMXBs) exhibit two distinctive types of PDS. At low accretion rates, the PDS is characterised by a broken power law, with the spectral index changing from ~0 to ~1 at some frequency in the range ~0.01–1 Hz; we refer to such PDS as Type A. At higher accretion rates, the PDS is described by a simple power law; we call these PDS Type B. Of the 225 sources studied to date, 75 exhibit Type A variability, and are almost certainly LMXBs, while 6 show Type B but not Type A, and are likely LMXBs. Of these 81 candidate LMXBs, 71 are newly identified in this survey; furthermore, they are mostly found near the centre of M31. Furthermore, most of the X-ray population in the disc are associated with the spiral arms, making them likely high mass X-ray binaries (HMXBs). In general these HMXBs do not exhibit Type A variability, while many central X-ray sources (LMXBs) in the same luminosity range do. Hence the PDS may distinguish between LMXBs and HMXBs in this luminosity range.

Keywords. X-rays: binaries, galaxies: individual (M31), accretion discs, black hole physics.

1. Introduction

We present results from a survey of X-ray point sources in XMM-Newton observations of the Andromeda Galaxy (M31). M31 is the nearest spiral galaxy to our own, at 760 kpc (van den Bergh 2000), and its X-ray population is dominated by X-ray binaries.

Our work differs from previous surveys of M31 in two important ways. Firstly, we made power density spectra (PDS) from combined EPIC, background subtracted lightcurves of each source. This was made possible for the first time by the unprecedented sensitivity of XMM-Newton, allowing us to detect variability in M31 X-ray sources on ~100 s timescales. Secondly, we obtained the 0.3–10 keV luminosity of each source from best fit models to their spectral energy distribution (SED); previous works have used an assumed model to convert from count rate to luminosity in all but the brightest few sources.

We have examined four XMM-Newton observations of the central region of M31, along with one observation each of the North 1, North 2, North 3, South 1 and South 2 fields (see e.g. Pietsch *et al.* 2005). We performed automated source detection on these regions, and found 535 sources that were observed by all three EPIC detectors (see Shaw Greening *et al.* 2005, for details). However, we have not yet studied these sources in a uniform manner.

Table 1. Classifications of the X-ray population of M31 by power density spectrum (PDS) and spectral energy distribution (SED). The SED is classified by the photon index of the power law component, Γ, or by the temperature of the thermal component, kT_{BB}.

PDS	SED	Classification
Type A	$\Gamma \sim 1.5$–2.1	NS or BH LMXB in low/hard state
Type A	$\Gamma \sim 2.4$–3.0	BH LMXB in very high/ intermediate (steep power law) state
Type B	BB dominated $kT_{\mathrm{BB}} \sim 0.7$–$2$ keV	BH LMXB in high/soft state
Type B	Not BB dominated	NS LMXB in high state

2. Combining PDS and SED information

X-ray binaries exhibit many phenomena on time-scales spanning milliseconds to years. Low mass X-ray binaries (LMXBs) are particularly interesting because they exhibit very distinctive power density spectra that depend more on the accretion rate than on the nature of the primary. Both neutron star and black hole LMXBs exhibit PDS that are characterised by a broken power law at low accretion rates; the spectral index changes from ~0 to ~1 at a break frequency in the range ~0.01–1 Hz (van der Klis 1994). We call this Type A variability (Barnard *et al.* 2004). At higher accretion rates, the PDS is described by a simple power law over the some frequency range; we refer to this as Type B variability (Barnard *et al.* 2004). Van der Klis (1994) suggested that the transition from Type A to Type B variability occurs at some constant accretion rate, and suggested a transition at $\sim1\%$ Eddington. We have empirical evidence that suggests that the transition from Type A to Type B variability occurs at $\sim10\%$ Eddington in the 0.3–10 keV band (Barnard *et al.* 2004, 2005). Since the Eddington limit is proportional to the mass of the primary, we expect black hole LMXBs to exhibit Type A variability at higher luminosities than neutron star LMXBs. We assume a maximum neutron star mass of 3.1 $M_\odot$ and classify as a possible black hole LMXB any X-ray source that exhibits Type A variability at a 0.3–10 keV luminosity $>4 \times 10^{37}$ erg s^{-1}.

We also obtained EPIC-pn SEDs for each source in the 0.3–10 keV band, and an equivalent background SED. Response matrices and ancillary response files were generated for each source. The resulting SEDs were grouped for a minimum of 50 count bin^{-1} from sources with >500 counts, and for a minimum of 20 count bin^{-1} for SEDs with <500 counts. We fitted several spectral models to each SED and obtained the 0.3–10 keV luminosity from the best fit. If the source+background spectrum contained <150 counts, and/or the source spectrum contained <20 counts, then the luminosity was estimated from the 0.3–10 keV pn intensity; we assumed an absorbed power law model with line of sight absorption equivalent to 10^{21} H atom cm^{-2} and a photon index of 1.7.

By combining PDS and SED information we were able to classify the X-ray population of M31 to an unprecedented degree; Table 1 summarises our classification scheme for LMXBs (see e.g. McClintock & Remillard 2004; Barnard *et al.* 2005). We do not yet have such a scheme for HMXBs. For luminosity functions of the different regions of M31, see Shaw Greening *et al.* (2005).

3. Results

The initial survey covered the four observations of the 63 brightest sources in the central region of M31 (Barnard *et al.* 2005); we have yet to study the ~130 remaining sources. Similarly, we have only surveyed 36 out of 87 sources in South 1 region; this

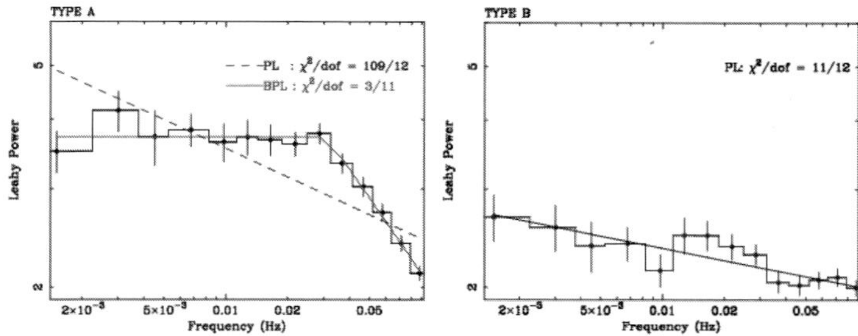

Figure 1. Power density spectra from the 2002 January XMM-Newton observation of two sources in the central region of M31. The PDS are averaged over 96 intervals of 128 bins, lasting 666 s, and grouped geometrically.

survey also predates the automated source detection. However, we have studied all 83 sources in North 1 and all 43 sources in North 2. We have not yet studied any of the 88 sources in North 3 nor the 41 sources in South 2. All the remaining sources will be surveyed over the next few months.

We have examined the four observations of the 63 X-ray sources in the core in most detail (Barnard *et al.* 2005); we were able to classify 107 out of 252 PDS as either Type A, Type B or flat; flat PDS had no significant power above 2, which is the expected power for Poisson noise when using Leahy normalisation. Figure 1 shows examples of Type A and Type B PDS exhibited by M31 X-ray sources during the ~60 ks XMM-Newton observation on 2002, January 6. Furthermore, we found that the Type B and flat PDS were both consistent with high accretion rate LMXBs in M31 (Barnard *et al.* 2005).

In the core, 40 out of 63 sources exhibited Type A variability in at least one observation; these are almost certainly X-ray binaries. A further 6 sources exhibited Type B but not Type A variability, making them likely LMXBs. In the North 1 and North 2 fields, 25 out of 83 and 1 out of 43 sources respectively exhibit Type A variability; in the South 1 field, 9 out of 36 sources exhibited Type A variability. Of the 40 sources that exhibited Type A variability in the core, 13 did so at luminosities $>4 \times 10^{37}$ erg s^{-1}, and were classed as black hole candidates (see Barnard *et al.* 2005).

Pietsch *et al.* (2005) surveyed these same XMM-Newton observations, finding 856 point sources. They catalogued 7 LMXBs and 9 likely LMXBs, as well as 27 certain and 10 likely associations of X-ray sources with globular clusters. These 53 sources are all likely LMXBs. We have surveyed 28 of these sources to date, and find that 10 of them exhibit Type A variability. Hence, 71 of the 81 X-ray binary candidates were newly identified by this work. In Fig. 2 we present a mosaic of GALEX images of M31 in the far ultraviolet (see e.g Thilker *et al.* 2005, and http://www.galex.caltec.edu), with the X-ray sources that we have already studied superposed. Black dots represent X-ray sources that exhibited Type A variability in at least one observation, while white dots represent sources that have not. We note that the central region is dominated by black dots; 63% of the surveyed sources exhibited Type A variability in the core. Meanwhile, 25–30% of sources in the North 1 and South 1 regions, and only 2% of sources in North 2, exhibited Type A variability (see Fig. 2).

The majority of sources in the North 1, North 2 and South 1 regions have 0.3–10 keV luminosities $<10^{37}$ erg s^{-1}. Almost all the core X-ray sources that we have surveyed in this regime with classified PDS show Type A variability; hence we would expect all LMXBs in the M31 disc to exhibit Type A variability also. However, we only detect Type A variability in ~20% of the disc sources. Furthermore, we see that the disc population of X-ray sources largely follows the spiral arms, suggesting that they are HMXBs. We

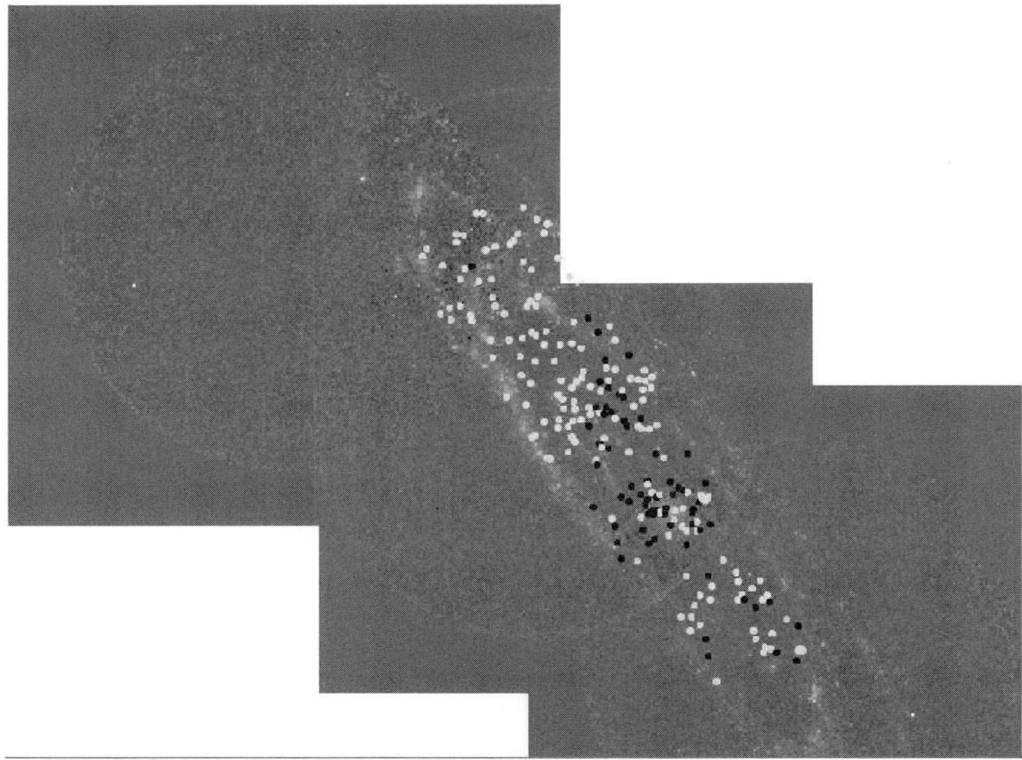

Figure 2. X-ray sources that we have already surveyed, superposed on a mosaic of GALEX FUV observations of M31. Black sources exhibit Type A variability, and are hence likely LMXBs. White sources do not exhibit Type A variability. Out of the 225 sources surveyed to date, 75 exhibit Type A variability.

therefore suggest that it may be possible to distinguish between LMXBs and HMXBs, based on their PDS at 0.3–10 keV luminosities $<10^{37}$ erg s^{-1}; the HMXBs are not expected to exhibit Type A variability as they are not disc fed. We expect most of the $\sim$130 faint sources in the core to exhibit Type A variability.

Acknowledgements

For this work, we used GALEX data that is publicly available from the archive at http://www.galex.stsci.edu/GR1. RB is supported by PPARC.

References

Barnard, R., Kolb, U., & Osborne, J. P. 2004, A&A, 423, 147

Barnard, R., Osborne, J. P., Kolb, U., & Haswell, C. A. 2004, To appear in "Interacting Binaries: Accretion, Evolution and Outcomes", Eds. L. Angellini *et al.*, astro-ph/0409122

Barnard, R., Kolb, U., & Osborne, J. P. 2005, A&A, submitted, astro-ph/0508284

McClintock, J. E. & Remillard, R. A. 2004, Compact Stellar X-ray Sources (Cambridge University Press), in press. astro-ph/0306213

Pietsch, W., Freyberg, M., & Haberl, F. 2005, A&A, 434, 483

Shaw Greening, L., Tonkin C., Barnard, R., Kolb, U., & Osborne, J. P. 2005, these proceedings

Thilker. D. A., Hoopes, C. G., Bianchi, L., *et al.* 2005, ApJL, 619, L67

van der Klis, M. 1994, ApJS, 92, 511

van den Bergh, S. 2000, The galaxies of the Local Group, Cambridge University Press, Cambridge Astrophysics Series Series, vol no: 35

Discussion

GHOSH: Would you say that the fraction of your LMXB candidates in the disk of M31 exceeds that in our galaxy? If so, what would this imply?

BARNARD: We have found that 3 black hole candidates that exhibit Type A (low accretion rate) variability are associated with a spiral arm, and are likely HMXBS, like CygX-1.

The one X-ray binary in North 2 is also in a spiral arm and may be a black hole HMXB. Otherwise, I dont know enough about the Galactic population to comment with expertise.

Populations of High Energy Sources in Galaxies
Proceedings IAU Symposium No. 230, 2005
E. J. A. Meurs & G. Fabbiano, eds.

© 2006 International Astronomical Union
doi:10.1017/S1743921306008106

XMM-Newton survey of the Local Group galaxy M 33 – catalogue results and global properties

Z. Misanovic[1], W. Pietsch[1], F. Haberl[1], M. Ehle[2], D. Hatzidimitriou[3] and G. Trinchieri[4]

[1]Max-Planck-Institut für extraterrestrische Physik, Giessenbachstraße, 85741 Garching, Germany
email: zdenka@mpe.mpg.de, wnp@mpe.mpg.de, fwh@mpe.mpg.de

[2]XMM-Newton Science Operations Centre, ESAC, ESA, P.O. Box 50727, 28080 Madrid, Spain
email: Matthias.Ehle@sciops.esa.int

[3]Department of Physics, University of Crete, P/O. Box 2208, 71003 Heraklion, Crete, Greece
email: dh@physics.uoc.gr

[4]Osservatorio Astronomico di Brera, via Brera 28, 20121 Milano, Italy
email: ginevra@brera.mi.astro.it

Abstract. Using 24 overlapping XMM-Newton observations of the Local Group spiral galaxy M 33, we have detected 447 sources in each individual pointing and in deep combined images. A total of 61 sources exhibit significant flux variations by a factor of up to 144, on time scales of hours to months or years. The detected variability, together with the hardness ratio (HR) method and optical identification (when available), is used to classify the sources as X-ray binaries (XRBs), supernova remnants (SNRs) and super-soft sources (SSS) in M 33, as well as background AGN and foreground stars in the field of view. The majority of sources can only be classified as 'hard', according to their HRs. We find that the luminosity distribution of the detected SNRs and SNR candidates in M 33 is similar to M 31, and slightly steeper than that of the LMC.

Keywords. X-rays: galaxies, X-rays: binaries, (ISM:) supernova remnants.

---

1. Catalogue results

Using deep combined XMM-Newton images of M 33, Pietsch *et al.* (2004) detected 408 X-ray sources. We have analysed all observations individually to study long-term X-ray flux variability, and compiled an additional catalogue with 350 objects. These two catalogues contain source positions, X-ray fluxes in 5 energy bands, HRs and cross-correlations with optical, infrared and radio data. The detected X-ray properties (HRs and variability) and cross-correlations have been used to classify sources as: XRBs, SNRs and SSS as intrinsic M 33 sources, and also foreground stars and background AGN. The majority of sources could only be classified as 'hard' according to the HRs. The summary of all sources and source classes is presented in the left panel of Fig. 1, while on the right we show their spatial distribution.

2. SNR population of M 33

Fig. 2 (left) shows the luminosity distributions of the SNR populations in M 33, M 31 and the LMC. Although M 31 has fewer detected and classified SNRs and SNR candidates per surveyed area, the distribution of this class of sources seems to be almost the same

160

Class	Cat(1)	Cat(2)	Total*/Common
XRB	2	10	10/2
SNR	44	25	39/25
SSS	5	11	12/4
fgStar	35	29	37/21
AGN & Gal	14	13	15/12
'hard'	267	206	271/187
Unclassified	41	56	69/28
Total	408	350	447/311

Cat(1): combined catalogue, Pietsch et al. 2004
Cat(2): individual catalogue, Misanovic et al. 2005
*Some sources have been re-classified in Cat(2)

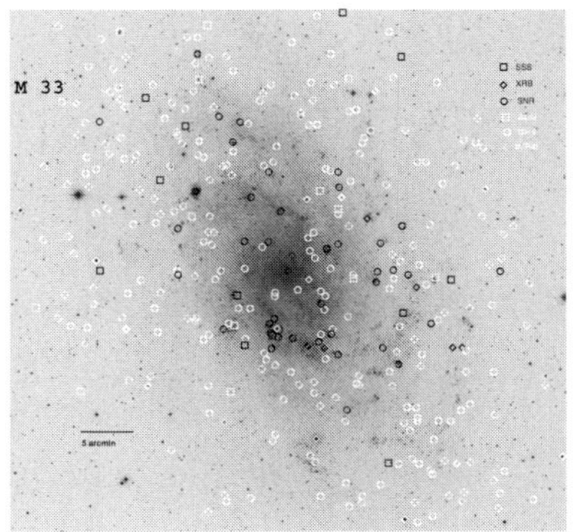

Figure 1. Left panel: The detected and classified sources in the combined and individual observation catalogues. **Right panel:** Spatial distribution of all X-ray sources classified in these two catalogues.

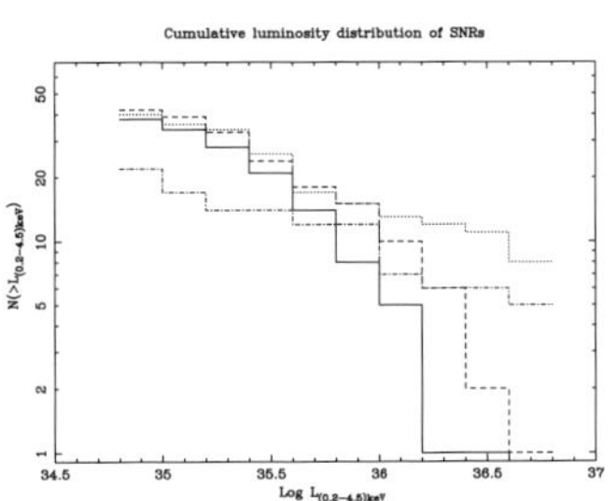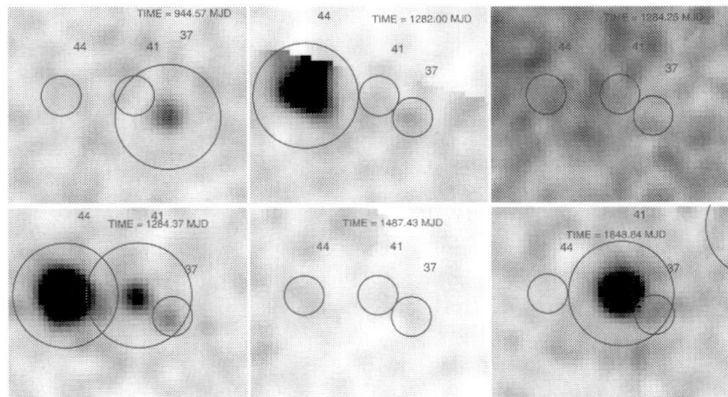

Figure 2. Left panel: Luminosity distribution of SNRs in M 33 (solid line), M 31 (dashed line) and the LMC (detected by ROSAT; dotted-dashed and dotted lines). For detailes see Misanovic *et al.* (2005). **Right panel:** Broad band XMM-Newton images of the region around two sources demonstrating extreme flux variability between observations on a time scale of several hours. The sources are classified as XRB candidates.

in both spiral galaxies. The SNRs in the LMC have a slightly flatter distribution, which was attributed to relatively higher metallicity of the LMC by Haberl & Pietsch (2001), and also recently confirmed by Ghavamian *et al.* (2005).

3. Variability

There are 61 significantly variable sources in our catalogue, with flux variability amplitudes in the range from 0.3 to more than 144. The detected variability was used as an efficient tool to classify sources. In particular, the extreme variability on relatively short time-scale was used to select several new XRB candidates in M 33 (e.g. the two sources shown at the right panel in Fig. 2).

References

Ghavamian P., Blair W.P., Long K.S., *et al.* 2005, AJ, 130, 539
Haberl F. & Pietsch W., 2001, A&A, 373, 438
Misanovic Z., Pietsch W., Haberl F., *et al.* 2005, A&A, in press
Pietsch W., Misanovic Z., Haberl F., *et al.* 2004, A&A, 426, 11

Populations of High Energy Sources in Galaxies
Proceedings IAU Symposium No. 230, 2005
E. J. A. Meurs & G. Fabbiano, eds.

© 2006 International Astronomical Union
doi:10.1017/S1743921306008118

XMM-Newton survey of the Local Group galaxy M 33 – bright individual sources

Z. Misanovic[1], W. Pietsch[1], F. Haberl[1], G. Trinchieri[2], M. Ehle[3], and D. Hatzidimitriou[4]

[1]Max-Planck-Institut für extraterrestrische Physik, Giessenbachstraße, 85741 Garching, Germany
email: zdenka@mpe.mpg.de, wnp@mpe.mpg.de, fwh@mpe.mpg.de

[2]Osservatorio Astronomico di Brera, via Brera 28, 20121 Milano, Italy
email: ginevra@brera.mi.astro.it

[3]XMM-Newton Science Operations Centre, ESAC, ESA, P.O. Box 50727, 28080 Madrid, Spain
email: Matthias.Ehle@sciops.esa.int

[4]Department of Physics, University of Crete, P/O. Box 2208, 71003 Heraklion, Crete, Greece
email: dh@physics.uoc.gr

Abstract. As shown in our first poster, in a recent survey of M 33 with XMM-Newton we detected the X-ray source population of this nearby spiral galaxy down to the (0.2-4.5) keV luminosity of 10^{35} erg s^{-1}, a factor of 10 deeper than in previous observations. The majority of the detected sources was classified using, in many cases, only their X-ray properties. In particular, 8 new X-ray binary (XRB) candidates were selected, based on their long-term X-ray light curves. We also classified supernova remnants (SNRs), super-soft sources (SSS), AGN, foreground stars and a population of 'hard' sources using the hardness ratio (HR) method. A detailed spectral and timing analysis of the brightest sources is in progress. We present a few examples of spectra for particular source classes. We find that bright 'hard' sources can be divided into two broad families: one best modelled by a powerlaw with photon index in the range of 1.0–2.0, and the other displaying disk blackbody spectra with kT of 0.8 to 1.5 keV.

Keywords. X-rays: galaxies, X-rays: binaries.

1. Introduction

As demonstrated by Pietsch *et al.* (2004a), we can use HRs to distinguish very soft, soft and hard sources. SNRs and foreground stars exhibit soft spectra with the emission mainly below 1.0 keV, while SSS have extremely soft spectra (below 0.5 keV). XRBs, Crab-like SNRs and background AGN have significantly harder spectra. We perform detailed spectral and timing analysis of the brightest sources in all individual XMM-Newton observations of M 33. Below we show several examples of typical source spectra.

2. Source spectra

The first source in Fig. 1 is a SNR, which we fitted by a two-temperature plasma (MEKAL) model and an absorption column of 6×10^{20} cm^{-2}. The variable source 253 also requires a two component model (MEKAL+POWERLAW). This source is identified as a late G-type star in optical follow up observations (Hatzidimitriou *et al.* 2005), and its relatively hard spectrum can, most probably, be attributed to flaring, although we do not have enough counts in any of the individual observations to confirm this. We fit an extremely variable SSS by an absorbed black body model with a kT of 62 eV.

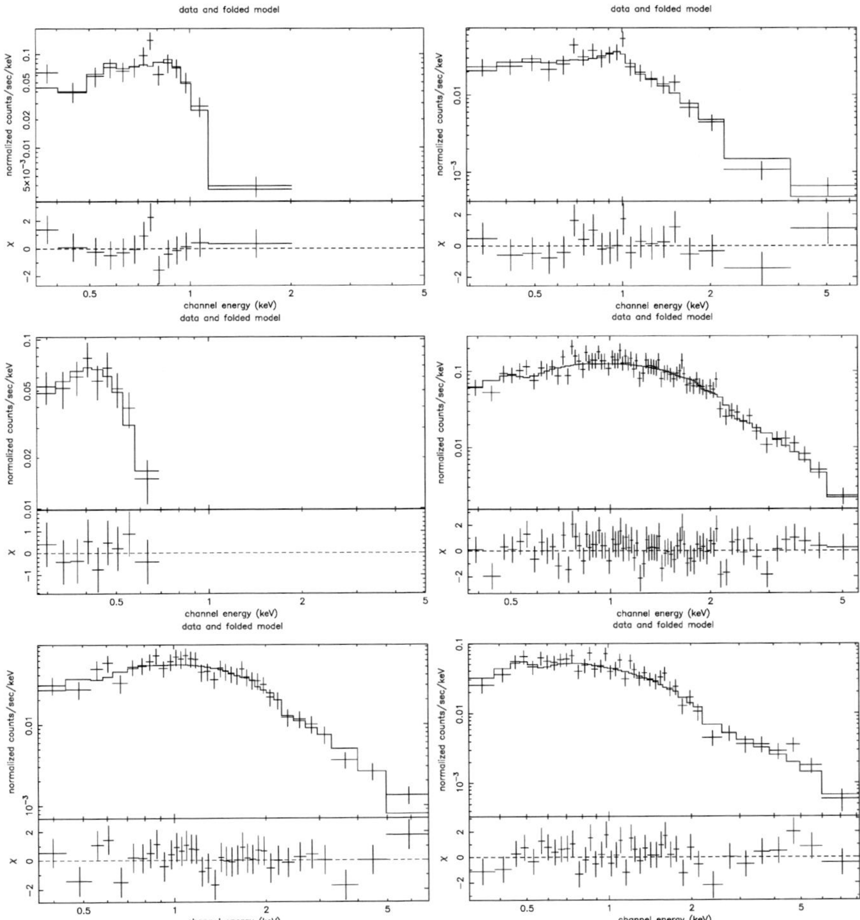

Figure 1. Spectra of bright X-ray sources detected in XMM-Newton survey of M 33. From upper left to lower right: Src 108 (SNR), 253 (fgStar), 207 (SSS), 150 (XRB M 33 X-7), 124 ('hard') and 248 (AGN). The source number and classification are from Misanovic *et al.* (2005).

The remaining sources from our selection display hard spectra. For M 33 X-7, a known eclipsing high mass XRB (see e.g. Pietsch *et al.* 2004b), we extract a spectrum in the high state and fit it by an absorbed disk blackbody with an intrinsic absorption column of about 3×10^{20} cm^{-2} and a kT of 0.89 keV. A 'hard' source (Src 124) has a similar spectral shape – almost the same absorption and a temperature of 1.03 keV. The powerlaw model also gave acceptable fits for these two sources, however, requiring significantly higher absorption columns and spectral indices in the range 2.6–3.0.

We fit a source classified as an AGN and a 'hard' source (SRC 131) with the powerlaw models with spectral indices of 2.0 and 1.6 respectively.

References

Hatzidimitriou, D., Pietsch, W., Misanovic, Z., Rieg, P. & Haberl, F. 2005, A&A, in preparation
Misanovic, Z., Pietsch, W., Haberl, F., *et al.* 2005, A&A, in press
Pietsch, W., Misanovic, Z., Haberl, F., *et al.*, 2004a, A&A, 426, 11
Pietsch, W., Mochejska, B.J., Misanovic, Z., *et al.* 2004b, A&A, 413, 871

Populations of High Energy Sources in Galaxies
Proceedings IAU Symposium No. 230, 2005
E. J. A. Meurs & G. Fabbiano, eds.

© 2006 International Astronomical Union
doi:10.1017/S174392130600812X

Dwarf Galaxies of the Local Group

Rosanne Di Stefano[1,2], Roberto Soria[1], F.A. Primini[1], and Albert Kong[3]

[1]Harvard-Smithsonian Center for Astrophysics, 60 Garden St., Cambridge, MA 02138, USA
email: rd@cfa.harvard.edu
[2]Department of Physics and Astronomy, Tufts University, Medford, MA 02155
[3]Center for Space Research, MIT, Cambridge, MA 02139, USA

Abstract. *XMM-Newton* and *Chandra* have ushered in a new era for the study of dwarf galaxies in the Local Group. We provide an overview of the opportunities, challenges, and some early results. The large number of background sources relative to galaxy sources is a major theme. Despite this challenge, the identification of counterparts has been possible, providing hints that the same mechanisms producing X-ray sources in larger galaxies are active in dwarf galaxies. A supersoft X-ray source within $2''$ of the supermassive black hole in M32 may be a remnant of the tidal disruption of a giant, although other explanations cannot be ruled out.

1. Introduction

Dwarf galaxies constitute the largest number of galaxies residing in groups and clusters. Dwarf galaxies also inhabit the field. They are the building blocks of galaxy formation. Therefore, in the young universe and throughout vast volumes of the present-day universe, dwarf galaxies provide the typical environments for the formation and evolution of X-ray sources (XRSs).

A vast array of multiwavelength observations of Local Group dwarf galaxies allow the detailed star formation histories of individual galaxies to be derived. This is important, because theoretical considerations and observations of larger, more distant galaxies suggest that there are correlations between star formation history and the formation of XRSs (see, e.g., Gilfanov 2004, Gilfanov, Grimm, & Sunyaev 2004, and references therein). There is, e.g., a strong correlation between star formation rates and the formation of high-mass X-ray binaries (HMXBs) and supernova remnants (SNRs). There is also a correlation between low-mass X-ray binaries (LMXBs) and the total galaxy mass; the LMXB population in a large galaxy provides a measure of the average star formation history. Detailed information about individual dwarf galaxies can therefore help to both test and refine theory. The Local Group dwarf galaxies in effect form a set of individual laboratories in which the foundations of X-ray astronomy can be studied.

We can best study X-rays from those dwarf galaxies in the Local Group, where our observations can discover sources of relatively low luminosity, and where we can determine source positions well enough to identify counterparts at other wavelengths. *Chandra* and *XMM-Newton* provide new tools. Each can resolve individual systems in the dwarf galaxies of the Local Group, where the X-ray populations are generally not spatially concentrated. The large effective area of *XMM-Newton* collects enough photons to study many spectral and timing characteristics without prohibitively long exposure times. The superb angular resolution of *Chandra* allows more reliable detection of counterparts, especially in combination with high resolution optical or radio observations; in some cases, structures in extended sources, such as SNRs, may be resolved.

Table 1. Source Counts for 4 Galaxies Observed with *XMM-Newton*

Galaxy	$n_{\rm H}$ $(10^{20}\ \mathrm{cm}^{-2})$	Distance (Mpc)	pn exp. time (ks)	MOS exp. time (ks)	Number of sources in EPIC field
NGC 205	6.7	0.72	13.1	14.6	60
Sextans B	2.7	1.3	14.4	16.1	56
NGC 147	12.0	0.75	10.8	12.4	53
WLM	2.0	1.0	6.7	9.2	38

2. Expected Numbers of Sources: The Background Dominates

The Local Group consists of the Milky Way, M31, M33, and more than 30 dwarf galaxies (see Mateo 1998; van den Bergh 2000). Dwarf galaxy masses range from $<10^8 M_\odot$ to $\sim 10^{10} M_\odot$. Morphological types include dwarf spheroidals, irregular galaxies of various shapes, and one elliptical. Central surface densities range from <11.6 to >26 in mag arcsec^{-2}, with most values lying between 20 and 26. Some dwarf galaxies are close enough to either M31 (e.g., M32 and NGC 205) or the Milky Way (e.g., the Sagittarius dwarf galaxy) to experience tidal interactions. Because of their proximity to Earth, the Magellanic Clouds have been well-studied at X-ray as well as at other wavelengths. In this presentation we will focus on the *other* dwarf galaxies. These are on the frontiers of X-ray astronomy, and present challenges that can be met by observations with the present generation of X-ray telescopes and with the next generation.

Despite differences among the galaxies, we can predict the number of XRSs expected if the formation of XRSs proceeds in dwarfs much as it does in larger galaxies. Because it is not clear that the correlations we use are correct for masses as small as those of individual dwarf galaxies, we make predictions only for a set of galaxies: M32, NGC 205, IC-10, WLM, NGC 147, Sextans B, the Sagittarius dwarf galaxy. These 7 galaxies represent a range of morphologies, star-formation rates, and star-formation histories. Some are near one of the large Local Group galaxies, and some are relatively isolated. In this set of galaxies we expect to find 2–10 globular cluster XRSs; 6–30 LMXBs, 2–10 HMXBs and SNRs. That is, we predict a total of 10–50 sources with X-ray luminosities near or above 10^{36} erg s^{-1}. (Details will be presented in Di Stefano *et al.* 2005.)

Foreground and Background Sources: Based on deep field studies, we expect there to be roughly 400 XRSs per square degree with flux (0.5–2 keV) greater than 2×10^{-15} erg s^{-1} cm^{-2} (see, e.g., Giacconi *et al.* 2002 and references therein). One circular field of radius $13'$, which roughly mimics *XMM-Newton* coverage, should therefore contain approximately 60 background sources.

Comparisons: Table 1 shows the results of *XMM-Newton* observations for 4 galaxies. In each case, the total number of XRSs detected is comparable to the number of background sources expected. The impression that the background plays the dominant role is reinforced by the fact that the overall spatial distributions appear to be more-or-less uniform, with little or no clustering obviously associated with galaxy features. This is illustrated in Figure 1, which shows the sources detected by *XMM-Newton* in the vicinity of NGC 205, a dwarf galaxy very close to M31. A small fraction of the detected XRSs are associated with the galaxy or its environment. Note, e.g., that one of the sources in Figure 1 is associated with a globular cluster (GC). To determine which other sources may also be associated with the NCG 205 or its environs, IDs with objects at other wavelengths are crucial. The same is true for other dwarf galaxies.

3. Individual Sources

3.1. *M32: Skirting a Supermassive Black Hole*

M32 is the only Local Group dwarf galaxy known to house a massive ($\sim 2 \times 10^6 M_\odot$) black hole (BH). There is a clear excess of sources in the vicinity of the BH: 3 within a few arcseconds of the nucleus, while the average density is roughly 1 per 10 square arcminutes (Ho *et al.* 2003). It is highly likely that the sources near the center are there because of some activity related to the presence of the BH. Indeed, stellar interaction in the dense region around the BH can produce XRSs.

It is especially interesting that one of the sources within $2''$ of the nucleus is a supersoft source (SSS). SSSs have luminosities above 10^{36} erg s^{-1} and emit little, if any radiation above $1-1.5$ keV. Nine SSSs located in the Galaxy or in Magellanic Clouds are associated with hot white dwarfs (WDs), and accreting WD models provide promising explanations for the other local SSSs. But interactions in dense stellar environments do not appear to favor the formation of SSS WD binaries based on theoretical arguments and on observations of GCs. Outside of GCs in M31 (which contain no SSSs), SSSs constitute roughly 10% of all observed XRSs (Di Stefano *et al.* 2004).

The only other galaxy in the Local Group in which close associations between a super-massive BH and SSSs can be detected is in M31, where there is also a SSS within $2''$ of the central BH. That observation inspired the conjecture the tidal stripping of giants by a massive BH can leave behind a core which will be bright in soft X-rays for an interval of 10^3–10^6 years (Di Stefano *et al.* 2001). The M32 observations provide an independent venue in which to test this conjecture and other possibilities as well.

3.2. *The Globular Cluster XRSs of NGC 205 and Sagittarius*

The X-ray luminosity function for Galactic GC XRSs is bimodal, consisting of "bright" sources ($L_x > 10^{36}$ erg s^{-1}) and "dim" sources ($L_x < 10^{34}$ erg s^{-1}). The bright sources are LMXBs with neutron star accretors. Even relatively short observations (>10 ksec) with *XMM-Newton* or *Chandra* will detect such sources in most of the Local Group dwarf galaxies. The dim sources are either CVs or LMXBs in quiescence. They can only be detected in nearby dwarfs, such as the Sagittarius dwarf galaxy. The results sketched below indicate that the luminosity distribution of XRSs in dwarf galaxy GCs is likely to be similar to that already studied in the Galaxy. The discovery of additional XRSs in the GCs of dwarf galaxies will allow more detailed comparisons to be made. **NGC 205:** Like M32, NGC 205 is close to M31. Soria *et al.* (2005) have discovered an XRS associated with one of the GCs, B024, in the field of the galaxy. (See Figure 1; Galleti et al 2004). **The Sagittarius Dwarf Galaxy** is the nearest dwarf galaxy to us. Ramsay & Wu (2005) have discovered 7 dim XRSs within the half-mass radius of M54, one of the GCs that is near Sagittarius and that also has a radial velocity consistent with membership in the Sagittarius dwarf galaxy system.

3.3. *Dwarf Galaxies with Active Star Formation*

HMXBs and X-ray active SNRs are expected only in galaxies which have experienced star formation within the past 10^6–10^8 years. IC 10 and NGC 6822 both have active star formation. Below we focus on two XRSs for which there is a wealth of supporting evidence linking each to its galaxy. **NGC 6822** contains an XRS near its center, long suspected to be associated with a SNR. The combination of *Chandra* observations, emission line studies by the Local Group Survey (LGS), and radio observations have made this identification secure. In fact, the SNR is resolved (see Figure 4), and the structures detected in X-ray, radio, and optical wavelengths coincide with each other. (See Kong *et al.* 2004 for details and additional references.) **IC 10** contains an XRS with $L_X(0.3$–$8\text{keV}) = 1.2 \times 10^{38}$ erg s^{-1}.

This source has been associated with a Wolf Rayet star, indicating that it is a high-mass X-ray binary. Variations by a factor of as much as 6 have been observed to occur over time intervals as short as 10^4 seconds. The spectrum has been fit with a multicolor disk model with $T_{in} = 1.1$ keV. (See Wang, Whittaker, & Williams [2005] for details; see also Bauer & Brandt [2004] and references therein, and Brandt *et al.* [1997].) These timing and spectral properties are consistent, but are not unique to what we expect from an accreting BH. This interesting system is clearly worthy of further study.

4. What We Have Learned, What We Can Learn

Most XRSs found in the vicinity of Local Group dwarf galaxies are associated with background objects. Nevertheless, previous and ongoing studies have discovered several intriguing bright sources with clearly associated counterparts in dwarf galaxies. Additional work should be able to identify galaxy counterparts for several times as many XRSs. Statistical tests can also be helpful. Although they cannot identify individual galaxy sources, they can help to quantify the fraction of observed sources associated with the galaxies by quantifying the level of deviations from spatial uniformity, or deviations from the luminosity function expected for the background.

The primary result derived so far is that the mechanisms that produce XRSs in dwarf galaxies appear to be those already well-studied in other, larger galaxies. This is important to know if we are to estimate the X-ray source contribution from dwarf galaxies in other parts of the Universe.

Refined predictions of the XRS population for individual dwarf galaxies, based on the star formation history inferred from multiwavelength studies, are needed. X-ray observations can then test the predictions, and possibly discover new effects. One effect already discovered in both M31 and M32 is the proximity of an SSSs to a massive BH. These SSSs could be remnants of giants that have been tidally stripped by the BH. Additional work is needed to determine is this is likely, or if other scenarios are preferred.

Together, *XMM-Newton* and *Chandra* make a good team for conducting studies of Local Group dwarf galaxies. The immediate scientific returns are large. In addition, ongoing observations can play an important role in preparing the way for the next generation of X-ray telescopes. The new telescopes will have significantly larger effective areas, while spatial resolution may not be quite as good as presently available. A modest decline in spatial resolution will not hobble future studies of Local Group dwarf galaxies, because the distribution of most sources is not spatially concentrated. On the other hand, larger effective area will provide spectral and timing information that can help us to better understand the physical nature of the sources.

References

Bauer, F. E. & Brandt, W. N. 2004, ApJL, 601, L67
Brandt, W. N., Ward, M. J., Fabian, A. C., & Hodge, P. W. 1997, MNRAS, 291, 709
Di Stefano, R., *et al.* 2004, ApJ, 610, 247
Di Stefano, R., Greiner, J., Murray, S., & Garcia, M. 2001, ApJL, 551, L37
Galleti, S., Federici, L., Bellazzini, M., Fusi Pecci, F., & Macrina, S. 2004, A&A, 416, 917
Giacconi, R., *et al.* 2002, ApJS, 139, 369
Gilfanov, M., Grimm, H.-J., & Sunyaev, R. 2004, MNRAS, 347, L57
Gilfanov, M. 2004, MNRAS, 349, 146
Ho, L. C., Terashima, Y., & Ulvestad, J. S. 2003, ApJ, 589, 783
Kong, A. K. H., Sjouwerman, L. O., & Williams, B. F. 2004, AJ, 128, 2783
Mateo, M. L. 1998, ARA&A, 36, 435
Ramsay, G. & Wu, K. 2005 (private communication).
van den Bergh, S. 2000, The galaxies of the Local Group, ISBN: 0521651816.
Wang, Q. D., Whitaker, K. E., & Williams, R. 2005, MNRAS, 362, 1065

Figure 1. NGC205 XRSs (yellow open circles) detected by *XMM-Newton*, overlaid on an optical image. The position of one XRS coincides with the position of a globular cluster.

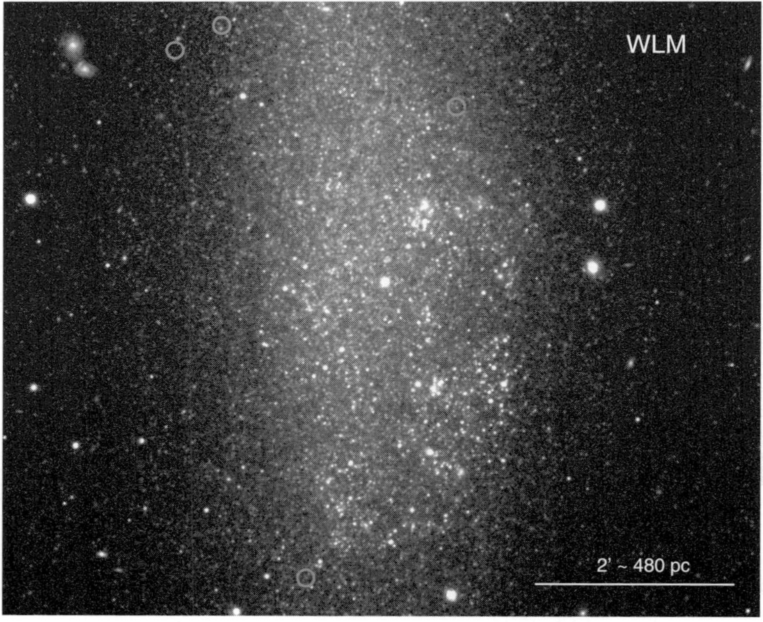

Figure 2. XRSs (green open circles) detected by *XMM-Newton*, overlaid on an optical image of WLM from the Local Group Survey.

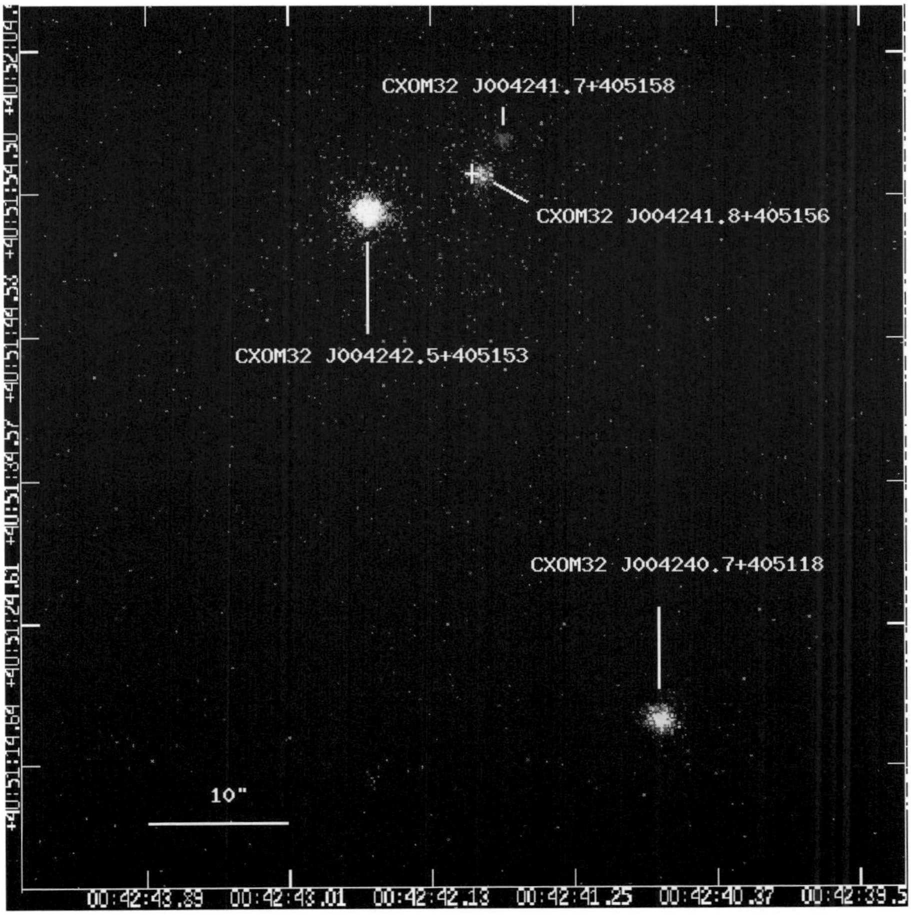

Figure 3. M32: The white cross marks the galaxy center; the red source is supersoft (the NW-most source, J004241.7+405158).

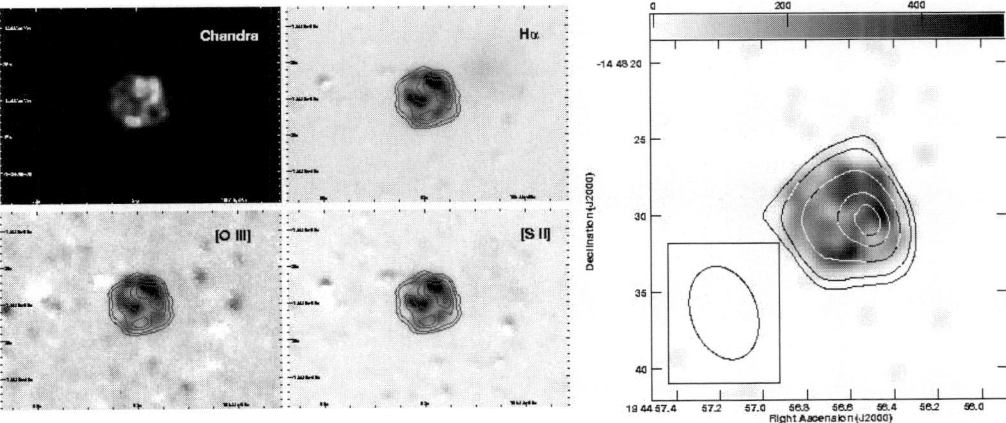

Figure 4. SNR near the center of NGC 6822. Right: radio contours superposed on the Chandra image. Left: upper left panel shows the Chandra image. Proceeding clockwise, the Chandra contours are superposed on the H$_\alpha$, [S II], and [O III] images.

Populations of High Energy Sources in Galaxies
Proceedings IAU Symposium No. 230, 2005
E. J. A. Meurs & G. Fabbiano, eds.

© 2006 International Astronomical Union
doi:10.1017/S1743921306008131

X-ray Pulsars in the Small Magellanic Cloud

R. H. D. Corbet[1,2], M. J. Coe[3], J. Galache[3] S. Laycock[4],
C. B. Markwardt[1,5] and F. E. Marshall[1]

[1]X-ray Astrophysics Laboratory, Code 662,
NASA/Goddard Space Flight Center, Greenbelt, MD 20771, USA

[2]Universities Space Research Association

[3]Department of Astronomy, University of Southampton,
Southampton, SO17 1BJ, UK

[4]Center for Astrophysics, 60 Garden Street, Cambridge, MA 02138, USA

[5]University of Maryland

Abstract. The SMC is now known to contain many more transient X-ray pulsars than would be expected based on a simple scaling of the number of such sources in the Galaxy by the relative mass of the SMC. We have been conducting regular monitoring observations of the SMC with the Proportional Counter Array on the Rossi X-ray Timing Explorer since 1997. This has resulted in the discovery of many of these X-ray pulsars and also provided orbital period measurements from detections of regular outbursts. We can now investigate the differences and similarities of the Galactic and SMC X-ray pulsar populations and consider the origin of the huge SMC X-ray pulsar over-abundance.

Keywords. galaxies: Magellanic Clouds, X-rays: binaries, stars: neutron.

1. Introduction

The first known X-ray pulsar in the SMC was the persistent supergiant system SMC X-1. Two luminous transients (SMC X-2, SMC X-3) were discovered with SAS-3 (Clark *et al.* (1978)). These were thought to be transient Be/neutron star systems although pulsations were not detected due to the low sensitivity of SAS-3. It was hypothesized that SMC pulsars were exceptionally luminous, possibly related to the low metallicity of the SMC (Westerlund (1990)). This was later to be disproved and an alternative explanation found for the high luminosity of the first few SMC X-ray pulsars to be discovered. Over subsequent years a few pulsars were also found with satellites such as ROSAT (Hughes (1994)).

2. SMC X-ray pulsars with RXTE

Serendipitous RXTE slew observations in 1997 showed a possible outburst from the vicinity of SMC X-3. Follow up target of opportunity pointed RXTE observation showed a complicated power spectrum with several peaks that were not all harmonically related to each other. Imaging ASCA observations were next made which showed two separate pulsars, however neither was found to be located at the position of SMC X-3. A more detailed look at the RXTE power spectrum showed that in fact *three* pulsars were simultaneously active (Corbet *et al.* (1998)). These observations were the first sign of the existence of a very large SMC X-ray pulsar population.

3. The RXTE monitoring program

RXTE has been regularly monitoring the SMC since 1997 using the Proportional Counter Array (PCA). We have discovered very many transient X-ray pulsars. For those sources where optical counterparts have been identified they are all found to be Be stars. We primarily make weekly observations of one particularly active region near SMC X-3. Other SMC regions have been monitored monthly depending on the amount of time awarded in a particular observing cycle. We use power spectra to extract pulsed flux from any pulsars. In this way, although the PCA is not an imaging instrument, the pulsed flux from multiple sources can be monitored independently. When new sources are identified we use cross scans in Right Ascension and Declination to localize the position of the new sources. For new sources we thus obtain at least minimal positional information. However, position determination can be problematic if too many pulsars are simultaneously active. A log of known X-ray pulsars in the SMC is currently maintained at http://lheawww.gsfc.nasa.gov/~corbet/pulsars/. About 45 sources are currently known.

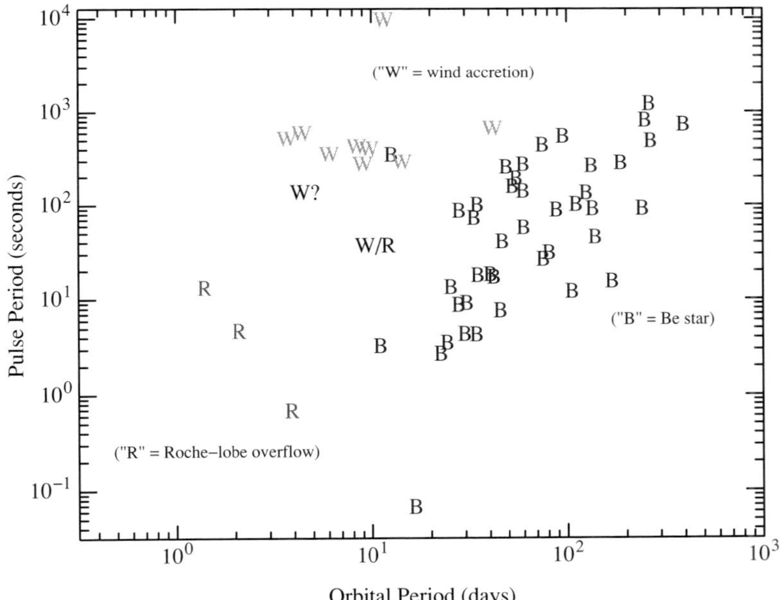

Figure 1. Pulse and orbital periods of HMXB pulsars. This plot includes systems in the Milky Way, SMC, and LMC.

4. Measuring Orbital Periods

From our RXTE monitoring program we have been able to determine the orbital period of a number of systems from the detection of regular outbursts. These period measurements are augmented by optical period determinations using OGLE and MACHO light curves (e.g. Schmidtke & Cowley (2005), Edge *et al.* (2005)).

5. The Be/Neutron Star Spin/Orbital Period Relationship

When Be HMXBs are in an active state (i.e. the Be star is surrounded by a sufficiently extensive disk) they often show regular X-ray outbursts which reveal the orbital period of the system. In addition, the X-ray light curves also show pulsations on the rotation period of the neutron star. It was first noted over twenty years ago that, for these systems, there is a correlation between pulse period and orbital period (Corbet (1984)). This correlation has been interpreted as arising from the spin period of the neutron star being determined by a balance of spin-up and spin-down torques. The spin period is forced to remain close to a time-averaged value where the corotation and magnetospheric radii are equal (Corbet (1984), Corbet (1986), Waters & van Kerkwijk (1989)). In this way the spin/orbital period relationship probes the circumstellar envelope around Be stars.

In order to characterize the P_s vs. P_{orb} relationship a simple linear fit was made to the log of these values. As the spin period is believed to be determined by the average local circumstellar environment, which depends on the size of the orbit, the orbital period may be regarded as the independent variable.

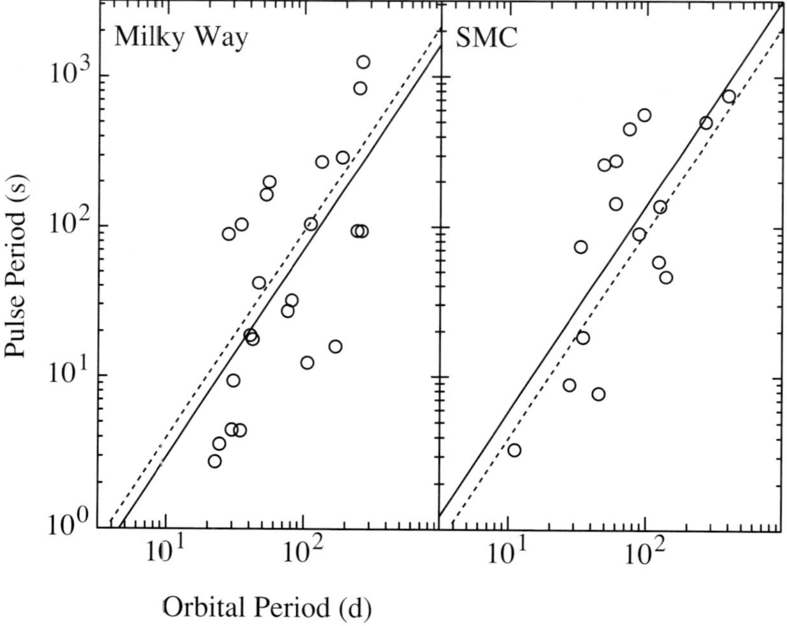

Figure 2. Pulse and orbital periods of Be/neutron star binaries in the Milky Way and SMC plotted separately. In each panel the dashed lines show the fit to the combined Milky Way/SMC data set and the solid lines show the fits to just the indicated subset.

6. Discussion

The relationship between pulse and orbital periods is found to be identical, within the errors, for both Galactic and SMC Be/neutron star binaries. This was not completely expected before hand as there is a number of differences between Galactic and SMC Be stars: (i) Maeder *et al.* (1999) find that the Be star phenomenon is more common in the SMC, (ii) the metallicity of the SMC is significantly lower than in the Galaxy with an overall metallicity of 1/5 solar, and even young clusters have low metallicities (e.g. Hill & Spite (1999)), (iii) Coe *et al.* (2005) find from optical photometry that the spectral types of primaries in Be/neutron star systems in the SMC may span a larger range than Galactic systems where the primaries are B2 or earlier. Because of these numerous differences it is somewhat surprising that we see no difference in the overall spin/orbital relationship. For early type supergiant stars where mass loss in winds depends on absorption of UV lines by metals there does appear to be a difference between the SMC and the Milky Way. Theoretically, radiatively driven mass loss is expected to depend on metallicity (Kudritzki *et al.* (1987), Vink *et al.* (2001)) as:

$$\dot{M} \propto Z^{0.5 \rightarrow 0.7} \tag{6.1}$$

The Be star mass loss mechanism appears not to be strongly dependent on metallicity and hence, by implication, is not strongly affected by radiation pressure.

References

Clark, G., Doxsey, R., Li, F., Jernigan, G., & van Paradijs, J. 1978, *ApJ* 221, L37
Coe, M., Edge, W., Galache, J., & McBride, V. 2005, *MNRAS*, 356, 502
Corbet, R. 1984, *A&A*, 141, 91
Corbet, R. 1986, *MNRAS*, 220, 1047
Corbet, R., Marshall, F., Lochner, J., Ozaki, M., & Ueda, Y. 1998, *IAUC*, 220
Edge, W., Coe, M., Galache, J., McBride, V., Corbet, R., Okazaki, A., Laycock, S., Markwardt, C., Marshall, F., & Udalski, A. 2005, *MNRAS*, 361, 743
Hill, V. & Spite, M. 1999, *Ap & Space Sc.*, 265, 469
Hughes, J. 1994, *ApJ*, 427, L25
Kudritzki, R., Pauldrach, A., & Puls, J. 1987, *A&Ap*, 173, 293
Laycock, S., Corbet, R., Coe, M., Marshall, F., Markwardt, C., & Lochner, J. 2005, *ApJ* in press
Maeder, A., Grebel, E., & Mermilliod, J.-C. 1999, *ApJ*, 346, 459
Schmidtke, P.C. & Cowley, A.P. 2005, *AJ*, 130, 2220
Vink, J., de Koter, A., & Lamers, H. 2001, *A&Ap*, 369, 574
Westerlund, B. 1990, *A&ARv*, 2, 29
Waters, L. & van Kerkwijk, M. 1989, *A&Ap*, 223, 196

Discussion

VAN DEN HEUVEL: Could it not be that we are missing many Be/X-ray systems in our Galaxy? They tend to be off for long periods, so we observe only a fraction of them. An over-abundance by a factor 50 for the SMC seems very extreme to me. I could imagine an over-abundance with respect to the Milky Way by a factor a few (perhaps as much as 5) due to a starburst, but not 50.

CORBET: Unless systems in the SMC are active for a larger fraction of time than Galactic systems then the effects of sources being off should affect estimates of source populations in both the SMC & Milky Way in a similar way. Searches for new Be sources in the Milky Way with both the RXTE ASM & Galactic plane scans with the PCA have not revealed very large numbers of new Be star sources. (However, absorption in the plane of the Galaxy does of course affect the sensitivity of these searches.) We also observed a single position in the LMC with the PCA, already known to contain a few pulsars, once per month for one year. No new pulsars were found, indicating a substantial difference between the LMC & SMC.

DISTEFANO: Is it possible that we are missing X-ray pulsars in the Galaxy?

174   Corbet *et al.*

CORBET: Please see my reply to the similar question from Van den Heuvel.

ZEZAS: Just a comment: Studies of the star-formation history of the SMC show that there was a recent burst of star-formation $\sim$10–30Myr ago, which may explain the observed excess of Be-X-ray binary pulsars.

CORBET: I agree. Such a recent burst of star formation is reported by Harris & Zaritsky (2004). Star Formation at that time might explain the surplus of Be star X-ray binaries but lack of excess of the shorter-lived supergiant wind accretors.

GHOSH: The discrepancy in the OB HMXB/Be-star binary ratio between SMC & Milky Way is most interesting and needs to be understood. If we believe that the massive companions from the same generation are in both types, somehow the parts of the mass-range or P_{orb} range relevant for OB companions are absent in SMC.
Also a comment: If the P_s–P_{orb} correlation is dominated by Be-star binaries, it is not surprising that SMC & Milky Way are very similar in this respect.

CORBET: In looking at the P_s–P_{orb} correlation, I only considered sources known to be Be star sources, or strongly suspected to be so because they are transient. The SMC metallicity is a factor $\sim$5 lower than the Galaxy. If the Be star phenomenon, e.g. mass loss rate or structure of the Be star envelope, depends on metallicity that would affect the correlation.

Populations of High Energy Sources in Galaxies
Proceedings IAU Symposium No. 230, 2005
E. J. A. Meurs & G. Fabbiano, eds.

© 2006 International Astronomical Union
doi:10.1017/S1743921306008143

The discovery of X-ray binaries in the Sculptor Dwarf Spheroidal Galaxy

Thomas J. Maccarone[1], Arunav Kundu[2], Stephen E. Zepf[2], Anthony L. Piro[3] and Lars Bildsten[3,4]

[1]School of Physics and Astronomy, University of Southampton, Southampton, SO17 1BJ, UK
email: tjm@phys.soton.ac.uk

[2]Department of Physiscs and Astronomy, Michigan State University, East Lansing, MI, USA

[3]Department of Physics, University of California at Santa Barbara, Santa Barbara, CA, USA

[4]Kavli Institute for Theoretical Physics,University of California at Santa Barbara, Santa Barbara, CA, USA

Abstract. We report the results of a deep Chandra survey of the Sculptor dwarf spheroidal galaxy. We find five X-ray sources with L_X of at least 6×10^{33} ergs/sec with optical counterparts establishing them as members of Sculptor. These X-ray luminosities indicate that these sources are X-ray binaries, as no other known class of Galactic point sources can reach 0.5-8 keV luminosities this high. Finding these systems proves definitively that such objects can exist in an old stellar population without stellar collisions. Three of these objects have highly evolved optical counterparts (giants or horizontal branch stars), as do three other sources whose X-ray luminosities are in the range which includes both quiescent low mass X-ray binaries and the brightest magnetic cataclysmic variables. We predict that large area surveys of the Milky Way should also turn up large numbers of quiescent X-ray binaries.

Keywords. X-rays:binaries – X-rays:galaxies – stars:binaries:close – galaxies:individual:Sculptor dwarf spheroidal – stellar dynamics – stars:Population II.

1. Introduction

It is quite difficult for field star populations to produce low mass X-ray binaries (LMXBs), especially those with neutron star primaries. Supernova explosions which eject more than half the mass of a system normally leave the systems unbound. Therefore the only ways to produce LMXBs through binary stellar evolution are through common envelope evolution which ejects much of the mass of the black hole's or neutron star's progenitor before the supernova occurs (Paczynski 1976; Kalogera & Webbink 1998), or through a finely-tuned asymmetric velocity kick which occurs at the birth of the neutron star or black hole (Brandt & Podsiadlowski 1995; Kalogera 1998). Intermediate mass X-ray binaries (i.e. X-ray binaries where the donor star is 1–8 $M_\odot$) can lose large amounts of mass and then evolve into LMXBs, which may help to solve the problem of formation rates of such LMXBs, since these systems will be more likely to survive the supernova explosions (see e.g. Podsiadlowski, Rappaport & Pfahl 2002). Because of the uncertainties in how to keep LMXBs bound, and other uncertainties regarding, e.g., the correlation between the initial masses of the stars in a binary system, theoretical rates of X-ray binary formation are highly uncertain.

The difficulties in producing LMXBs are even larger for old stellar populations. The mass accretion rate required to power a persistent bright LMXB is about $10^{-8} M_\odot \mathrm{yr}^{-1}$, meaning that the lifetime of such a system can be only about 100 Myrs from when accretion starts. It is generally believed that most LMXBs will begin their accretion phases only a few Gyrs after the supernova which creates the compact object (White & Ghosh 1998). Several possibilities still exist for producing X-ray binaries in old populations such as elliptical galaxies – these LMXBs we see in old stellar populations may be very low duty cycle transients (meaning that they will take a long time to accrete their entire mass donors) which begin accretion only after their donor stars evolve off the main sequence (Piro & Bildsten 2002), they may be ultracompact X-ray binaries (Bildsten & Deloye 2004), or they may be normal X-ray binaries which were produced through dynamical encounters in globular clusters, and then released into the field (e.g. White, Sarazin & Kulkarni 2002; see also Grindlay 1988).

If the bulk of the elliptical galaxy field sources are low duty cycle transients, then there must exist an underlying population of quiescent X-ray binaries which is much larger than the fraction which is flaring at any given time. Therefore, we have identified a large sample of old stars which is nearby, and does not contain any globular clusters, the Sculptor dwarf spheroidal galaxy. By finding five X-ray binaries in this galaxy, four of which have highly evolved donor stars, we have verified that the mechanism of Piro & Bildsten (2002) is, at the very least, a substantial contributor to the X-ray binary populations seen in old field star populations.

2. Data

We have obtained 21 exposures of 6-kiloseconds each with the Chandra X-ray observatory over the time period from 26 April 2004 to 10 January 2005. The monitoring, rather than a single deep observation, was performed in order to search for bright transient sources, but none were found. We then stacked the data to make a single image of the data on ACIS-S3, and ran WAVDETECT, finding 74 sources in the 0.5–8.0 keV band. Based on previous deep field measurements (e.g. Hornschemeier *et al.* 2001), we estimate that about 50 of these sources should be background AGN and that it is unlikely that more than one is a foreground star. In order to determine whether there are supersoft X-ray sources, we have also extracted an image from 0.1–0.4 keV, but this image contained no bright sources (i.e. with more than 10 photons) that were not found in the 0.5–8.0 keV image.

We have compared the positions of the 9 sources with at least 100 detected counts (giving accurate positions, and making them bright enough to rule out the possibility of a white dwarf accretor) with the positions of bright optical stars ($V < 20.5$) in the Sculptor galaxy from Schweitzer *et al.* (1995). The stars in the Schweitzer et al. (1995) catalog are all red giants, asymptotic giant branch stars, or horizontal branch stars. We have also checked whether the single brightest source has an optical counterpart in the deep optical photometry of Hurley-Keller, Mateo & Grebel (1999).

Allowing for an 0.6" boresight correction we find four matches within 0.4" of an optical star in Schweitzer *et al.* (1995). The chance superposition probability is 0.04 for these objects, and all have proper motion confirmations that they are members of Sculptor, rather than foreground or background sources. The fifth match is the brightest X-ray source in our sample, which is found at RA=1h00m13.9s, Dec=-33d44m42.5s. While there is no strong evidence for it to be variable in our monitoring campaign, it was not detected in the ROSAT bright source all-sky survey, so it must either be variable at the level of a factor of a few, or it must have a very hard spectrum at soft X-rays. Its optical counterpart is $R = 23.68, B - R = 0.95$ (D. Hurley-Keller private communication – see Hurley-Keller et al. 1999 for a description of how the photometry was done), giving it a luminosity and color roughly consistent with a solar-type star in the low metallicity environment of the Sculptor galaxy, and it appears to be near the turnoff of the main sequence in the Sculptor galaxy.

3. Discussion

3.1. *The nature of the matches*

It has been suggested that the field populations of elliptical galaxies are low duty cycle transients, with red giant donors (Piro & Bildsten 2002). Obviously the finding that several of these systems have red giant donors provides support for this hypothesis. The number of bright (i.e. $L_X > 10^{37}$ ergs/sec) LMXBs per unit stellar mass seems to be roughly constant across a sample of giant elliptical galaxies (Gilfanov 2004), with a typical value of about one per 10^8–$10^9 M_\odot$ of stars. We would then expect about $\sim.02$ such systems in Sculptor, given its mass of about $2 \times 10^6 M_\odot$ (Mateo 1998). Given that the duty cycles of these transients are likely to be less than about $1/200$ in the scenario of Piro & Bildsten (2002), one would expect that the number of quiescent transient X-ray binaries in the Sculptor galaxy would be at least a few, and possibly much higher if the duty cycles were even smaller than $1/200$. Our discovery of one clear case of an X-ray binary with a red giant counterpart and a few more candidates is thus in reasonable agreement with the picture suggested by Piro & Bildsten (2002).

3.2. *Are there other quiescent X-ray binaries in Sculptor?*

It is likely that there are actually many more quiescent X-ray binaries in the Sculptor galaxy than the five strong matches we have presented here and the three additional tentative matches. Our detection limits in this data set are at a luminosity of a little bit above 10^{32} ergs/sec, while quiescent LMXBs have been seen to be as faint as 2×10^{30} ergs/sec, with most of the faintest sources having black hole accretors (Garcia *et al.* 2001). Therefore, some quiescent X-ray binaries could be below our sensitivity limits, although most neutron star accretors are brighter than these limits in quiescence, and there do seem to be higher quiescent luminosities for long period transients, even for black hole accretors. There may also be other quiescent X-ray binaries which are detected in our observations, but which have faint optical counterparts. These could be, for example, ultracompact X-ray binaries, which have also been suggested to be systems which should be present in old stellar populations (Bildsten & Deloye 2004), or even X-ray binaries with lower main sequence donors which simply took longer than the typical few Gyrs (White & Ghosh 1998) to come into contact. We plan to investigate these possibilities in future work with deeper optical photometry and with spectroscopic follow-ups.

3.3. *Implications for X-ray binaries in other stellar populations*

These results clearly indicate that the canonical factor of 100 enhancement in X-ray binaries per unit stellar mass in globular clusters applies only to bright LMXBs. Comparing this galaxy with NGC 6440, which shows the highest number density of quiescent LMXBs (Heinke *et al.* 2003), we find that the globular cluster density is enhanced by a factor of only about 10, even assuming that we have found all the quiescent LMXBs in Sculptor already. Since other clusters are less rich in LMXBs than NGC 6440, the enhancement factor is even lower on the whole. This is probably because very wide binaries such as those with red giant donors do not survive unperturbed for a Hubble time in dense globular clusters.

Finally, we consider the implications of these results for the Milky Way's bulge's X-ray binary population. The bulge of the Milky Way contains about 5×10^2 times as many stars as the Sculptor dwarf spheroidal galaxy, meaning that it should have about 2,500

X-ray binaries if the number of binaries scales linearly with mass. In fact, since X-ray binary production is enhanced in metal rich systems (at least for globular clusters – see Kundu, Maccarone & Zepf 2003 for evidence of this effect and Maccarone, Kundu & Zepf 2004 for a discussion of a binary evolution model for this effect which should work equally well in field populations as in globular clusters) the Milky Way might contain a few times more than this number of X-ray binaries. Our estimate is comparable to numbers from theoretical predictions (Iben *et al.* 1997; Belczynski & Taam 2004).

Finally, we note that with only about 500 optical stars in the Schweitzer *et al.* (1995) catalog, at least $\approx 1\%$ of the giant branch/horizontal branch stars in Sculptor are in systems with accreting neutron stars or black holes. Based on this finding, one would expect that a large number of similar systems are likely to be found in surveys that include hundreds of red giant stars. This hypothesis should be testable as the results from the ChaMPlane survey (Grindlay *et al.* 2003) begin to come in, or in a large area survey of the Galactic Bulge.

Acknowledgements

We are grateful to Carine Babusiaux, Kyle Cudworth, Peter Jonker, Christian Knigge, Mike Muno, Eric Pfahl, Andrea Schweitzer, Jeno Sokoloski, Kyle Westfall and Rudy Wijnands for useful discussions. We thank Denise Hurley-Keller, Eva Grebel and Mario Mateo for the optical photometric measurement of the brightest X-ray source. AK and SEZ acknowledge support from NASA grants SAO G04-5091X and LTSA NAG-12975. LB acknowledges support from the NSF via grant PHY99-07949.

References

Belczynski, K. & Taam, R.E., 2004, ApJ, 616, 1159

Bildsten, L. & Deloye, C.J., 2004, ApJ, 607L, 119

Brandt, N. & Podsiadlowski, P., 1995, MNRAS, 274, 461

Garcia, M.R., McClintock, J.E., Narayan, R., Callanan, P., Barret, D., & Murray, S.S., 2001, ApJ, 553L, 47

Gilfanov, M., 2004, MNRAS, 349, 146

Grindlay, J.E., 1988, IAU Symp. 126, The Harlow-Shapely symposium on Globular Cluster Systems in Galaxies (Dordrecht: Kluwer), 347

Grindlay, J., *et al.*, 2003, AN, 324, 57

Heinke, C.O., Grindlay, J.E., Lugger, P.M., Cohn, H.N., Edmonds, R.D., Lloyd, D.A., & Cool, A.M., 2003, ApJ, 598, 501

Heinke, C.O., Grindlay, J.E., Edmonds, P.D., Cohn, H.N., Lugger, P.M., Camilo, F., Bogdanov, S., & Freire, P.C., 2005, ApJ, 625, 796

Hurley-Keller, D., Mateo, M., & Grebel, E.K., 1999, ApJ, 523L, 25

Iben, I., Tutukov, A.V., & Fedorova, A.V., 1997, ApJ, 486, 955

Kalogera, V., 1998, ApJ, 493, 368

Kalogera, V. & Webbink, R.F., 1998, ApJ, 493, 351

Kundu, A., Maccarone, T.J., & Zepf, S.E., 2003, ApJ, 574, L5

Maccarone, T.J., Kundu, A., & Zepf, S.E., 2004, ApJ, 606, 430

Mateo, M., 1998, ARA&A, 36, 435

Paczynski, B., 1976, IAU Symp. 73, Structure and Evolution of Close Binary Systems, ed. P. P. Eggleton, S. Milton, & J. Whelan (Dordrecht: Reidel), 75

Piro, A.L. & Bildsten, L., 2002, ApJ, 571L, 103

Podsiadlowski, P., Rappaport, S., & Pfahl, E.D., 2002, ApJ, 565, 1107

Schweitzer, A.E., Cudworth, K.M., Majewski, S.R., & Suntzeff, N.B., 1995, AJ, 110, 2747

White, N.E. & Ghosh, P., 1998, ApJ, 504, L31

White, R.E. III, Sarazin, C.D., & Kulkarni, S., 2002, ApJ, 571, L23

Discussion

DiStefano: For many reasons, I think your conclusions are likely to be correct. But, to play devil's advocate, I'd like to ask about possible degeneracies in the stellar models for the donor stars?

Maccarone: We have proper motion confirmations that these stars are Sculptor galaxy members from the work of Schweitzer *et al.* (1995).

Sarazin: What is the probability of chance overlaps of the X-ray error circles and optical stars in Sculptor?

Maccarone: For the good matches, the error circles are all less than 0.5", so the chance probability of superposition is about 0.02, if memory serves me correctly. For the fainter sources, because there are more sources and we used larger error circles, we got 4 more matches with a false match probability of about 0.4, so there's a reasonably large chance that at least one of those is a chance superposition.

Sarazin: Just a clarification – when you say that GC LMXB production has been over-estimated, I assume you are referring only to fainter sources.

Maccarone: Correct. Another way to look at this is that the duty cycles for the globular cluster sources are strongly enhanced compared to the field sources, and that this, rather than the number density enhancement is the dominant part of the reason why GCs show $\sim$100 times as many bright X-ray binaries per unit stellar mass as the field stars do.

Ivanova: It's a comment: You should be a bit more careful making the link between metal-poor dwarf galaxies and metal-rich giant ellipticals, as both the mass transfer rates and formation rates of X-ray binaries with a MS or a RG donors are not the same.

Maccarone: I agree that these are important considerations, and that my extra polations to larger galaxies should be taken as very rough estimates.

Populations of High Energy Sources in Galaxies
Proceedings IAU Symposium No. 230, 2005
E. J. A. Meurs & G. Fabbiano, eds.

© 2006 International Astronomical Union
doi:10.1017/S1743921306008155

X-ray Populations in Galaxies

G. Fabbiano

Harvard-Smithsonian Center for Astrophysics, 60 Garden St., Cambridge MA 02138, USA

Abstract. Today's sensitive, high-resolution Chandra X-ray observations allow the study of many populations of X-ray sources. The traditional astronomical tools of photometric diagrams and luminosity functions are now applied to these populations, and provide the means for classifying the X-ray sources and probing their evolution. While overall stellar mass drives the amount of X-ray binaries in old stellar populations, the amount of sources in star forming galaxies is related to the star formation rate. Short-lived, luminous, high mass binaries (HMXBs) dominate these young X-ray populations.

Keywords.

1. *Chandra* observations of X-ray binary (XRB) populations

It is well known that the Milky Way hosts both old and young X-ray source populations, reflecting its general stellar make up. In 1978, the *Einstein Observatory*, the first imaging X-ray telescope, opened up the systematic study of the X-ray emission of normal galaxies, and revealed populations of X-ray sources, at least in nearby spiral galaxies (Fabbiano 1989). With *Chandra*'s sub-arcsecond angular resolution, combined with CCD photometric capabilities (Weisskopf *et al.* 2000), the study of normal galaxies in X-rays has taken a revolutionary leap: populations of individual X-ray sources, with luminosities comparable to those of the Galactic X-ray binaries, can be detected at the distance of the Virgo Cluster and beyond.

We can now study these X-ray populations in galaxies of all morphological types, down to typical limiting luminosities in the 10^{37} ergs s^{-1} range. At these luminosities, the old population X-ray sources are accreting neutron star or black-hole binaries with a low-mass stellar companion, the LMXBs (life-times $\sim 10^{8-9}$ yrs). The young population X-ray sources, in the same luminosity range, are dominated by neutron star or black hole binaries with a massive stellar companion, the HMXBs (life-times $\sim 10^{6-7}$ yrs; see Verbunt & van den Heuvel 1995 for a review on the formation and evolution of X-ray binaries), although a few young supernova remnants (SNRs) may also be expected. At lower luminosities, reachable with *Chandra* in Local Group galaxies, Galactic sources include accreting white dwarfs and more evolved SNRs. Fig. 1 shows two typical observations of galaxies with *Chandra*: the spiral M83 (Soria & Wu 2003) and the elliptical NGC4697 (Sarazin, Irwin & Bregman 2000), both observed with the ACIS CCD detector. The images are color coded to indicate the energy of the detected photons (red 0.3–1 keV, green 1–2 keV and blue 2–8 keV). Populations of point-like sources are easily detected above a generally cooler diffuse emission from the hot interstellar medium. Note that luminous X-ray sources are relatively sparse by comparison with the underlying stellar population, and can be detected individually with the *Chandra* sub-arcsecond resolution, with the exception of those in crowded circum-nuclear regions.

To analyze this wealth of data two principal approaches have been taken: (1) a photometric approach, consisting of X-ray color-color diagrams and color-luminosity diagrams, and (2) X-ray luminosity functions (XLFs). Whenever the data allow it, time and spectral

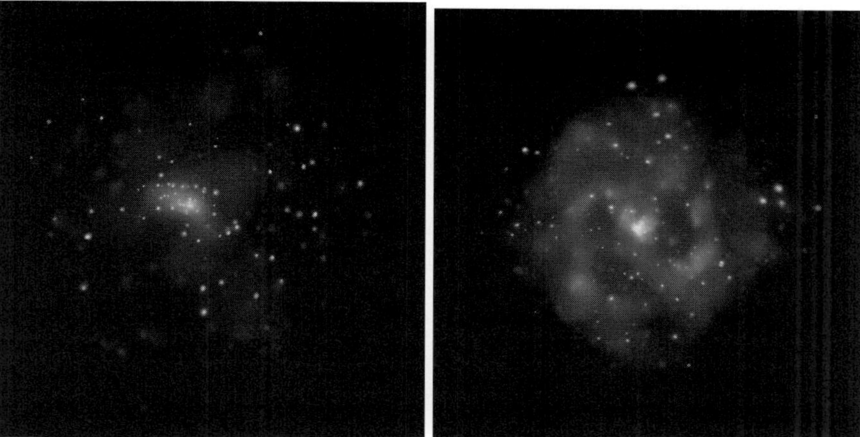

Figure 1. *Chandra* ACIS images of NGC4697 (left, box is 8.64 × 8.88 arcmin) and M83 (right, box is 8.57 × 8.86 arcmin). See text for details. Both images are from the *Chandra* web page http://chandra.harvard.edu/photo/; credit NASA/CXC

variability studies have been pursued. Optical and radio identifications of X-ray sources and association of their position with different galaxian components are also being increasingly undertaken.

2. X-ray colors

The use of X-ray colors to classify X-ray sources is not new. For example, White & Marshall (1984) used this approach to classify Galactic XRBs, and Kim, Fabbiano & Trinchieri (1992) used *Einstein* X-ray colors to study the integrated X-ray emission of galaxies. Unfortunately, given the lack of standard X-ray photometry to date, different definitions of X-ray colors have been used in different works; in the absence of instrument corrections, these colors can only be used for comparing data obtained with the same observational set up. Colors, however, have the advantage of providing a spectral classification tool when a limited number of photons are detected from a given source, which is certainly the case for most X-ray population studies in galaxies. Also, compared with the traditional derivation of spectral parameters via model fitting, color-color diagrams provide a relatively assumption-free comparison tool. *Chandra*-based examples of this approach can be found in Zezas *et al.* (2002a, b) and Prestwich *et al.* (2003), among others. The X-ray color-color diagram of Prestwich *et al.* (2003, Fig. 2) illustrates how colors offer a way to discriminate among different types of possible X-ray sources.

3. XLFs and parent stellar populations

Luminosity functions are well known tools in observational astrophysics. XLFs have been used to characterize different X-ray binary populations in the Milky Way (e.g., Grimm, Gilfanov & Sunyaev 2002), but these studies have always required a model of the spatial distribution of the sources, so to estimate their luminosities, which is inherently a source of uncertainty. External galaxies, instead, provide clean source samples, all at the same distance. Moreover, the detection of X-ray source populations in a wide range of different galaxies allows us to explore global population differences that may be connected with the age and/or metallicity of the parent stellar populations. XLFs establish the observational basis of X-ray population synthesis (Belczinsky *et al.* 2004).

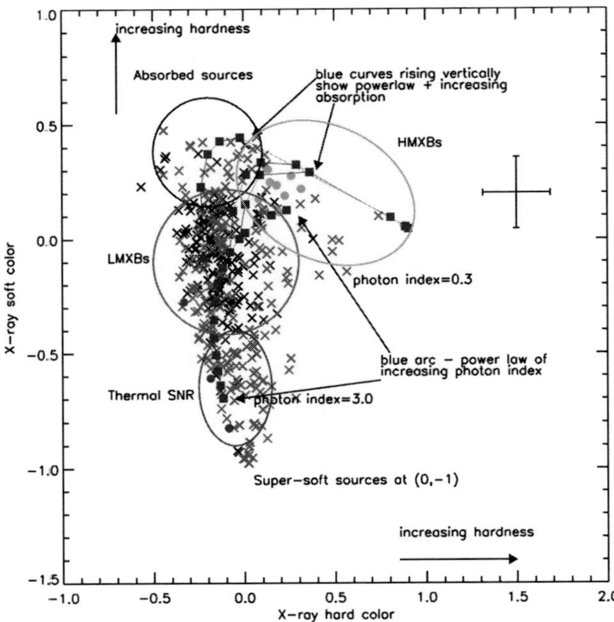

Figure 2. Chandra color-color diagram from Prestwich *et al.* (2003).

The XLFs have been fitted with power laws or broken power laws. The main parameters are: power-law slope (giving the relative luminosity distribution of X-ray sources), normalization (the total number of sources) and eventual breaks, pointing to changes in the X-ray source population (for example, different breaks are seen in the XLFs of the inner and outer bulge of M31, Kong *et al.* 2002). *Chandra* and *XMM-Newton* studies of M31 have revealed a variety of XLFs, connected with the different stellar populations of the field in question (see review of Fabbiano & White 2005 and references therein; Kong *et al.* 2003). In M81, the XLF of the spiral arm stellar population is flatter than that of the inter-arm and bulge regions, consistent with the prevalence of short-lived luminous HMXBs in younger stellar populations (Tennant *et al.* 2001, Fig. 3; Swartz *et al.* 2002).

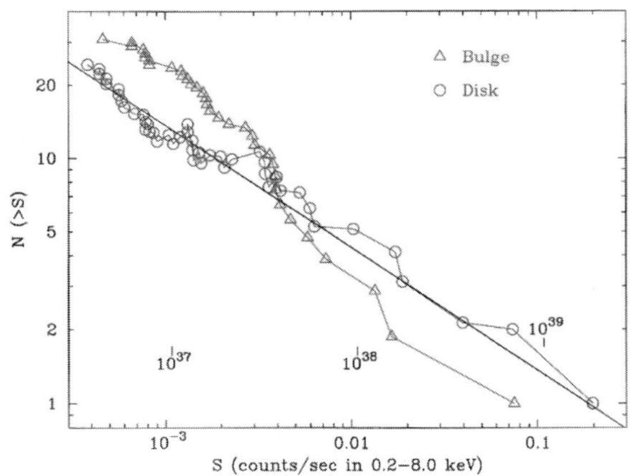

Figure 3. Bulge and disk XLFs of M81 (Tennant *et al.* 2001).

In general, flatter XLFs are found in more actively star-forming galaxies (e.g., a cumulative slope of ~-0.5 is found in the actively star-forming merger system, the Antennae;

Zezas & Fabbiano 2002). The early comparisons of XLFs of different types of galaxies (Zezas & Fabbiano 2002; Kilgard *et al.* 2002) also suggested that the normalizations are related to either the star formation rate (SFR) or the mass of the parent galaxy. Grimm, Gilfanov & Sunyaev (2003) took these ideas a step further, suggesting that HMXB XLFs follow a universal -0.6 cumulative power-law, with normalization proportional to the SRF. Gilfanov (2004) suggests that the normalization of LMXB XLFs is driven by the stellar mass of the galaxy (see also Kim & Fabbiano 2004).

In E and S0 galaxies the shape of the XLF has also been parameterized with models consisting of power-laws or broken power-laws. The overall shape (in a single power-law approximation in the observed range of $\sim 7 \times 10^{37}$ to a few 10^{39} ergs s^{-1}) is fairly steep (cumulative slopes -1 or steeper), i.e. with a relative dearth of high luminosity sources, when compared with the XLFs of star-forming galaxies. A lot of discussion has focused on a reported break at ~ 2–5×10^{38} ergs s^{-1}, near the Eddington limit of an accreting neutron star (Sarazin, Irwin & Bregman 2000 in NGC4697), which may be related to the transition between neutron star and black hole binaries in the population. Although some reported breaks are the result of incompleteness at the low luminosities (Kim & Fabbiano 2003), and typically breaks are not found in the completeness corrected XLFs of individual galaxies, Kim & Fabbiano (2004; see also Gilfanov 2004) report a break at $(5 \pm 1.6) \times 10^{38}$ ergs s^{-1} in the co-added corrected luminosity function for a sample of 14 E and S0 galaxies (Fig. 4).

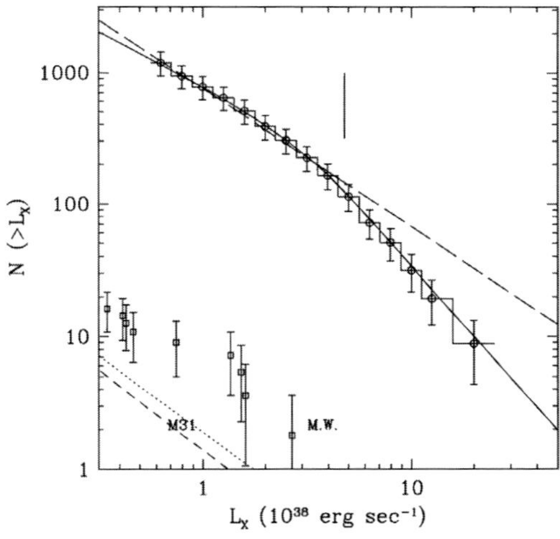

Figure 4. Co-added, completeness corrected XLF of 14 E and S0 galaxies, compared with the LMXB XLF of the Galaxy and M31 (Kim & Fabbiano 2004).

The paucity of very luminous X-ray sources in galaxies makes uncertain the definition of the high luminosity XLF, which may be better approached by co-adding 'consistent' samples of X-ray sources (e.g., Kim & Fabbiano 2004), but still uncertainties persist. Interestingly, the evaluation of the total X-ray luminosity of a galaxy may be significantly affected by statistics when a relatively small number of X-ray sources is detected (Gilfanov, Grimm & Sunyaev 2004).

Compact X-ray sources are notorious for their variability and this variability could in principle also affect the XLF, which is typically derived from a snapshot of a given galaxy. However, repeated *Chandra* observations in the case of both M33 (Grimm *et al.* 2005) and the Antennae galaxies (Zezas *et al.* 2005, in preparation) demonstrate that the XLF is remarkably steady, even when high luminosity sources clearly vary.

4. Conclusions

The results discussed in this talk represent only the beginning of what I hope will be a very fruitful field of investigation for many years to come. The tools that are being developed for characterizing and understanding the X-ray source populations of nearby galaxies lay the foundation of future work in X-ray population synthesis (see Belczynski *et al.* 2004). This approach, and future more sensitive high resolution X-ray observations, such as those anticipated from a possible *Gen-X* mission now in preliminary study by NASA, will allow the study of the X-ray evolution of galaxies. Given the strong link between young X-ray sources and star formation, these studies will also provide a direct probe of the process of galaxy formation and evolution.

Acknowledgements

This work was partially supported under NASA contract NAS8-39073 (CXC). This material was also covered in a review talk delivered at the COSPAR Colloquium Spectra and Timing of Compact X-ray Binaries, held in Mumbai (India), January 17–20, 2005.

References

Belczynski, K., Kalogera, V., Zezas, A., & Fabbiano, G. 2004, ApJ, 601, L147

Fabbiano, G. 1989, ARA&A, 27, 87

Fabbiano, G. & White, N. E. 2005, in Compact Stellar X-ray Sources in Normal Galaxies, eds. W. H. G. Lewin & M. van der Klis, Cambridge: Cambridge University Press, in press

Fabbiano, G., Zezas, A.. King, A. R., Ponman, T. J., Rots, A., & Schweizer, F. 2003a, ApJ, 584, L5

Gilfanov, M. 2004, MNRAS, 349, 146

Gilfanov, M., Grimm, H.-J., & Sunyaev, R. 2004, MNRAS, 351, 1365

Grimm, H.-J., Gilfanov. & M., Sunyaev, R. 2002, A&A, 391, 923

Grimm, H.-J., Gilfanov. & M., Sunyaev, R. 2003, MNRAS, 339, 793

Grimm, H.-J., McDowell, J., Zezas, A., Kim, D.-W., & Fabbiano, G. 2005, ApJ, in press.

Kilgard, R. E., Kaaret, P., Krauss, M. I., Prestwich, A. H., Raley, M. T., & Zezas, A. 2002, ApJ, 573, 138

Kim, D.-W. & Fabbiano, G. 2003, ApJ, 586, 826

Kim, D.-W. & Fabbiano, G. 2004, ApJ, 611, 846

Kim, D.-W., Fabbiano, G., & Trinchieri, G. 1992, ApJ, 393, 134

Kong, A. K. H., Di Stefano, R., Garcia, M. R., & Greiner, J. 2003, ApJ, 585, 298

Kong, A. K. K., Garcia, M. R., Primini, F. A., Murray, S. S., Di Stefano, R., & McClintock, J. E. 2002, ApJ, 577, 738

Prestwich, A. H., Irwin, J. A., Kilgard, R. E., Krauss, M. I., Zezas, A., Primini, F., Kaaret, P., & Boroson, B. 2003, ApJ, 595, 719 Sarazin, C. L., Irwin, J. A., & Bregman, J. N. 2000, ApJ, 544, L101

Shapley, A., Fabbiano, G., & Eskridge, P. B. 2001, ApJS, 137, 139

Soria, R. & Wu, K. 2003, A&A, 410, 53

Swartz, D. A., Ghosh, K. K., Suleimanov, V., Tennant, A. F., & Wu, K. 2002, ApJ, 574, 382

Tennant, A. F., Wu, K.. Ghosh, K. K., Kolodziejczak, J. J., & Swartz, D. A. 2001, ApJ, 549, L43

Verbunt, F. & van den Heuvel, E. P. J. 1995 in X-Ray Binaries, eds. W. H. G. Lewin, J. van Paradijs & E. P. J. van den Heuvel, Cambridge: Cambridge University Press, p.457

Weisskopf, M., Tananbaum, H., Van Speybroeck, L., & ODell, S. 2000, Proc. SPIE, 4012, 2

White, N. E. & Marshall, F. E. 1984, ApJ, 281, 354

Zezas, A. & Fabbiano, G. 2002, ApJ, 577, 726

Zezas, A., Fabbiano, G., Rots, A. H., & Murray, S. S. 2002a, ApJS, 142, 239

Zezas, A., Fabbiano, G., Rots, A. H., & Murray, S. S. 2002b, ApJ, 577, 710

Populations of High Energy Sources in Galaxies
Proceedings IAU Symposium No. 230, 2005
E. J. A. Meurs & G. Fabbiano, eds.

© 2006 International Astronomical Union
doi:10.1017/S1743921306008167

The X-Ray Population of NGC 300

S. Carpano[1], J. Wilms[2], E. Kendziorra[1] and M. Schirmer[3]

[1]Institut für Astronomie und Astrophysik, Abteilung Astronomie, Universität Tübingen, Sand 1, 72076 Tübingen, Germany
email: stefania.carpano@uni-tuebingen.de

[2]Department of Physics, University of Warwick, Coventry, CV4 7AL, United Kingdom

[3]Isaac Newton Group of Telescopes, 38700 Santa Cruz de La Palma, Spain

Abstract. We present X-ray properties of NGC 300 point sources, extracted from 66 ksec of *XMM-Newton* data taken in 2000 December and 2001 January. A total of 163 sources was detected in the energy range of 0.3–6 keV. We report on the global properties of the sources detected inside the D_{25} optical disk, such as the hardness ratio and X-ray fluxes, and spectral fitting of the brightest sources. We also present some properties of their optical counterparts found in B, V, and R images from the 2.2 m MPG/ESO telescope. Furthermore, we cross-correlate the X-ray sources with SIMBAD, the USNO-A2.0 catalog, and radio catalogues.

Keywords. Galaxies: individual: NGC 300 – X-rays: galaxies.

1. Introduction

NGC 300 is a normal dwarf galaxy of type SA(s)d belonging to the Sculptor galaxy group. Due to its small distance ($\sim$2.02 Mpc; Freedman *et al.* 2001), its low Galactic column density ($N_\mathrm{H} = 3.6 \times 10^{20}$ cm^{-2}; Dickey & Lockman 1990) and its face-on orientation, this galaxy is an ideal target for the study of the entire X-ray population of a typical normal quiescent spiral galaxy. The major axes of the D_{25} optical disk are 13.3 kpc and 9.4 kpc ($22' \times 15'$; de Vaucouleurs *et al.* 1991).

NGC 300 has already been observed by *ROSAT* five times between 1991 and 1997. Read & Pietsch (2001) discovered a total of 29 sources within the D_{25} disk, the brightest being a black hole candidate with $L_\mathrm{X} = 2.2 \times 10^{38}$ erg s^{-1} in the 0.1–2.4 keV band. More recently, NGC 300 was observed with *XMM-Newton* on 2000 December 26 during *XMM-Newton*'s revolution 192 and 6 days later during revolution 195. The luminous supersoft X-ray source XMMU J005510.7−373855 in the center of NGC 300 has been analysed by Kong & Di Stefano (2003). In addition to these X-ray data, observations with the 2.2 m MPG/ESO telescope in La Silla were performed. We use archival images in the broad band B, V, and R filters for this work.

Here, we present some X-ray properties of NGC 300 point sources, extracted from 66 ksec of *XMM-Newton* data and spectral fitting of the brightest sources. Some previous results of these observations were presented by Kendziorra *et al.* (2001) and Carpano *et al.* (2004), a full description of our results can be found in Carpano *et al.* (2005).

2. Results

We reduced the data using the standard *XMM-Newton* Science Analysis System (SAS), version 6.1.0. Spectra, images, and lightcurves were extracted using `evselect`.

Event and attitude file of each instrument were first merged for both orbits 192 and 195 using the `merge` task. Point source detection was then performed on the three cameras using the maximum likelihood approach of the `edetect_chain` task. After removing

Carpano *et al.*

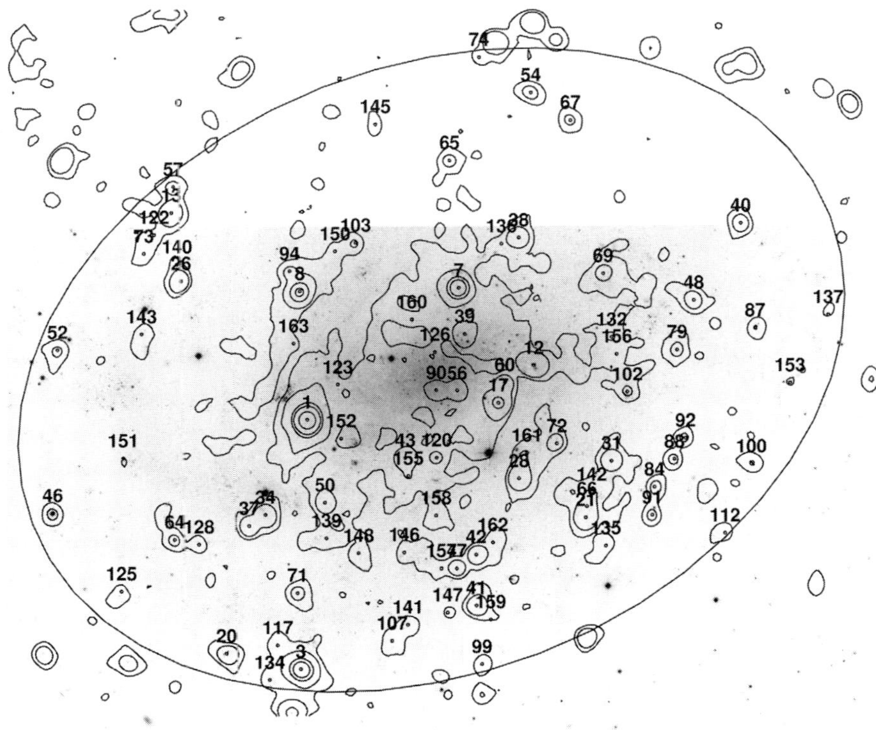

Figure 1. Optical image of NGC 300 in the visible band overlaid by a contour map of the merged 0.3–6.0 keV raw X-ray image from all three EPIC cameras and from both orbits. The D_{25} optical disk and the sources detected inside the disk are also shown.

sources associated with the galaxy cluster CL 0053−37, a total of 163 sources was found above a maximum likelihood threshold of 10 in the 0.3–6.0 keV band, 86 sources of which are within the D_{25} optical disk. This increases the X-ray inventory of NGC 300 by a factor of $\sim$3.

Fig. 1 shows the V band optical image of NGC 300 and the contour map of the merged X-ray raw image from both orbits and all three EPIC cameras in the 0.3–6.0 keV energy band. The D_{25} optical disk and the sources detected inside the disk, which are numbered in order of decreasing X-ray count rate as determined by the `edetect_chain`, are also shown.

Our detection limit in the 0.3–6.0 keV energy band is $F_{0.3-6} \sim 7 \times 10^{-16}\,\mathrm{erg\,cm^{-2}\,s^{-1}}$ for sources inside the optical disk. We computed the hardness ratio of the X-ray sources using a soft, medium, and hard energy band of 0.3–1.0, 1.0–2.0, 2.0–6.0 keV respectively. Comparing the color-color diagram for sources inside the D_{25} optical disk and having more than 20 net counts, with empirical color-color diagrams, we were able to determine the shape of the X-ray spectrum for each source individually and to estimate source fluxes. These are between $\sim$3.5 $\times 10^{-13}\,\mathrm{erg\,cm^{-2}\,s^{-1}}$ for the brightest one to $\sim$7 $\times 10^{-16}\,\mathrm{erg\,cm^{-2}\,s^{-1}}$.

After correcting the X-ray positions for the systematic shift between optical and X-ray coordinates, we searched for all possible optical counterparts in the merged BVR optical image and then calculate their fluxes in each of these three optical bands. Results are tabulated in Carpano *et al.* (2005). For the four brightest X-ray sources within the D_{25} ellipse, Fig. 2 shows the resulting optical counterparts in the merged optical image. Furthermore, we cross-correlated the X-ray sources with SIMBAD, the USNO-A2.0 catalog, and radio catalogues: 14 of our X-ray sources detected inside the optical disk have already

Figure 2. $15'' \times 15''$ optical images of the region centered on the corrected position of the four brightest X-ray sources inside the D_{25} optical disk. The circle indicates the 2σ positional error of the X-ray coordinates, the thicker circles show the optical sources found within the X-ray error circle. Catalog sources having a distance of less than $5''$ from the X-ray position are shown with a box.

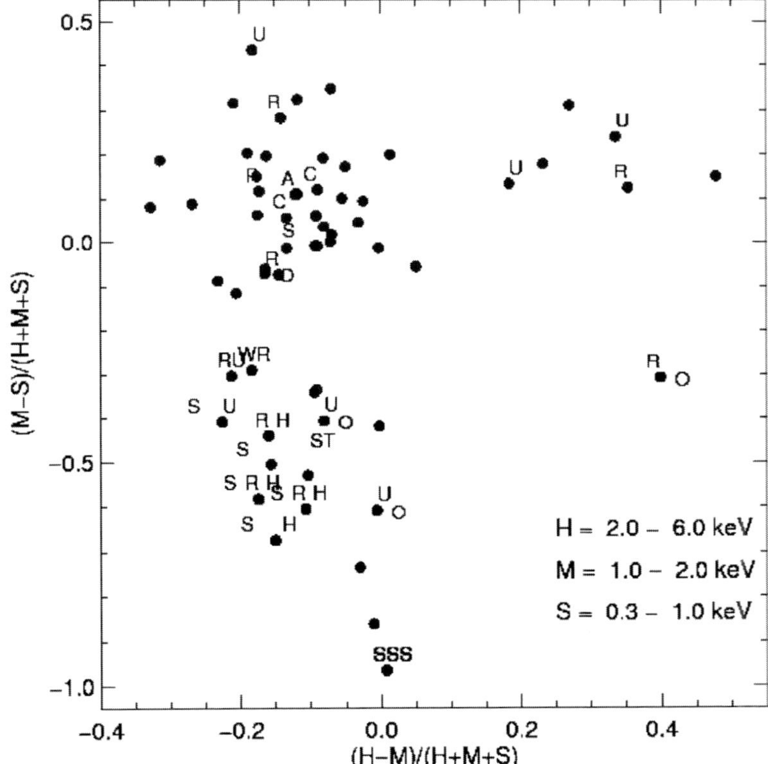

Figure 3. Color-color diagram for sources inside the optical disk. SNR are labeled with a 'S', radio sources with a 'R', H II regions with an 'H', Cepheid stars with a 'C', associations of stars with an 'A', stars in NGC 300 with a 'ST', Wolf-Rayet stars with a 'WR' and sources from the USNO catalog with a 'U'. 'SSS' refers to the supersoft source and 'O' are possible foreground stars.

been observed in the X-rays, there are 9 (suspected) Supernova Remnants (SNRs), 11 radio sources, from which 3 are associated with SNRs and 8 are possible AGNs. Other sources are matching with an association of stars, with H II (ionized) regions, with regions close to Cepheid variable stars, or with stars.

Fig. 3 shows the color-color diagram for the sources detected inside the optical disk and, when available, a label indicating the counterparts found in SIMBAD, USNO and radio catalogs. Following Pietsch *et al.* (2004) and Pietsch *et al.* (2005), soft sources with $\log(F_{\mathrm{X}}/F_{\mathrm{vis}}) = \log(F_{\mathrm{X}}) + m_{\mathrm{V}}/2.5 + 5.37 < -1$, where m_{V} is the visual magnitude, are identified with foreground stars.

Table 1. Best fit parameters for the four brightest sources.

source label	model	$N_{\mathrm{H}}(\times 10^{22}\,\mathrm{cm}^2)$	T (keV)	Γ	χ^2	dof	χ^2_ν
1 (revol. 192)	power+brems	$0.08^{+0.05}_{-0.04}$	$0.28^{+0.04}_{-0.06}$	$1.78^{+0.23}_{-0.23}$	566.53	495	1.14
(revol. 195)		$0.10^{+0.03}_{-0.02}$	$0.43^{+0.11}_{-0.12}$	$2.25^{+0.20}_{-0.19}$			
3	power+brems	$0.18^{+0.03}_{-0.07}$	$0.28^{+0.15}_{-0.06}$	$2.24^{+0.34}_{-0.45}$	125.82	116	1.08
7	power	$0.35^{+0.05}_{-0.03}$		$1.91^{+0.12}_{-0.13}$	154.64	110	1.41
8	bbody	$0.11^{+0.04}_{-0.04}$	$0.061^{+0.004}_{-0.003}$		73.29	53	1.38

Table 2. 0.3–2.4 keV source luminosity in units of $\times 10^{38}\,\mathrm{erg\,s^{-1}}$ of the four brightest sources.

source label	revol. 192	revol. 195	ROSAT ($L_{0.1-2.4}\,\mathrm{keV}$)
1 (pn)	$0.79^{+0.10}_{-0.11}$	$1.74^{+0.24}_{-0.49}$	2.06 ± 0.07
3 (pn)	$0.35^{+0.06}_{-0.07}$	$0.45^{+0.06}_{-0.06}$	0.64 ± 0.04
7 (pn)	$0.28^{+0.02}_{-0.02}$	$0.21^{+0.01}_{-0.01}$	0.12 ± 0.02
8 (MOS1)	$0.44^{+0.13}_{-0.13}$	$0.22^{+0.04}_{-0.04}$	0.48 ± 0.04

Except for one source, all SNR are located in a same region of the color-color diagram. The brightest source, having the WR star as optical counterpart, and which is a good candidate for a black hole X-ray binary, is slightly above the SNR group. The SSS is at the very bottom of the diagram and all harder sources, mainly hard X-ray binaries and background AGN, are spread on a large region in the upper part of the diagram.

We also performed spectral fitting for the four brightest X-ray sources inside the D_{25} optical disk. Results of the best spectral fits are shown in Table 1.

Fits were performed using data from the pn, MOS1, and MOS2, and both revolutions. The brightest source is the only one for which the hardness ratio has changed significantly between the two observations. Except for this source, we constrain the parameters of the spectral fitting to be equal for both revolutions.

For each of the sources, the pn (MOS1 for source 8) 0.3–2.4 keV luminosities are tabulated (Table 2) and compared to the value given by ROSAT (Read & Pietsch 2001), after that their luminosities have been recalculated considering a distance to NGC 300 of 2.02 Mpc (Freedman *et al.* 2001). All sources seem to be variable and are candidates for X-ray binaries.

References

Carpano, S., Wilms, J., Schirmer, M., & Kendziorra, E. 2004, Mem. Soc. Astron. Italiana, 75, 486

Carpano, S., Wilms, J., Schirmer, M., & Kendziorra, E. 2005, A&A, accepted

de Vaucouleurs, G., de Vaucouleurs, A., Corwin, jr., H., *et al.* 1991, Third Catalogue of Bright Galaxies (New York: Springer)

Dickey, J. M. & Lockman, F. J. 1990, ARA&A, 28, 215

Freedman, W. L., Madore, B. F., Gibson, B. K., *et al.* 2001, ApJ, 553, 47

Kendziorra, E., Wilms, J., Lamer, G., & Staubert, R. 2001, in New Visions of the X-ray Universe in the XMM-Newton and Chandra Era, ed. F. Jansen *et al.*, ESA SP-488, ESA Publications, Noordwijk, in press

Kong, A. K. H. & Di Stefano, R. 2003, ApJ, 590, L13

Pietsch, W., Misanovic, Z., Haberl, F., Hatzidimitriou, D., Ehle, M., & Trinchieri, G. 2004, A&A, 426, 11

Pietsch, W., Freyberg, M., & Haberl, F. 2005, A&A, 434, 483

Read, A. M. & Pietsch, W. 2001, A&A, 373, 473

Populations of High Energy Sources in Galaxies
Proceedings IAU Symposium No. 230, 2005
E. J. A. Meurs & G. Fabbiano, eds.

© 2006 International Astronomical Union
doi:10.1017/S1743921306008179

Optical Environments of M51 X-ray Sources

Roy E. Kilgard[1,3]**, Andrea H. Prestwich**[1]**, Martin J. Ward**[2]
and Timothy P. Roberts[3]

[1]Harvard-Smithsonian Center for Astrophysics, Cambridge, MA 02138, USA
email: rkilgard@cfa.harvard.edu

[2]Department of Physics, University of Durham, Durham, DH1 3LE, UK

[3]Department of Physics, University of Leicester, Leicester, LE1 7RH, UK

Abstract. The X-ray populations of spiral galaxies consist almost entirely of accreting X-ray binaries and supernova remnants. For the most luminous sources, it is possible to use X-ray spectroscopy and variability studies to gain insights into the nature of the sources. However, without unambiguously identified optical counterparts, it is impossible to definitively classify sources as, e.g. high-mass or low-mass X-ray binaries. The nearby interacting galaxy M51 is one of the best-studied galaxies across all wavelengths. At a distance of around 8 Mpc, it is possible to resolve features on scales of a few parsecs both in the X-ray and the optical. Recently, M51 was observed with the Advanced Camera for Surveys on the Hubble Space Telescope as part of a Hubble Legacy program. M51 has also been observed 3 times with the Chandra X-ray observatory. Combining these two datasets, we present initial results on the optical environments of M51 X-ray sources as the first part of a truly multi-wavelength study of X-ray sources in nearby galaxies.

Keywords. X-rays: galaxies, X-rays: binaries, galaxies: spiral.

---

1. Introduction

The discrete X-ray source populations of spiral galaxies consist almost entirely of accreting X-ray binaries and supernova remnants (Fabbiano (1989)). Prior to the launch of the *Chandra* X-ray Observatory in 1999, it was possible only to study these sources in detail in the most nearby galaxies: the Milky Way, the Magellanic Clouds, and, to a lesser extent, M31. However, with the large collecting area and sub-arcsecond spatial resolution of *Chandra*, it is possible to study X-ray source populations in galaxies well beyond 10 Mpc. This has led to much understanding about the relation between X-ray binary luminosity functions and star-formation (Kilgard *et al.* (2002), Grimm, Gilfanov, & Sunyaev (2003)) and the relation between X-ray color and source classification (Prestwich *et al.* (2003)).

In order to study the X-ray source population of spiral galaxies in detail, a survey of 11 galaxies has been performed with *Chandra*, with detailed results presented in Kilgard *et al.* (2005). This work characterizes the X-ray population based entirely on the X-ray spectral and temporal properties. However, it is impossible to truly understand the nature of an X-ray source without detecting an unambiguous counterpart in at least one other wavelength. As such, we have begun an extensive archival analysis project to collect as much multiwavelength data as possible for our sample of 11 galaxies, beginning with M51. The goals of the project are: to completely classify all X-ray sources in nearby galaxies; to determine the binary fraction within nearby galaxies and compare with the star formation rate and history as determined from observations at other wavelengths; to compare X-ray populations of different environments within individual galaxies, e.g. bulge vs. disk, arm vs. inter-arm, tidal tails, halos, etc.

The nearby (8.4 Mpc, Feldmeier, Ciardullo & Jacoby (1997)) interacting galaxy M51 has been studied extensively in almost all wavelengths. It is the closest example of an interacting galaxy pair, with massive star formation along the tidal tail and in the northern arm (Larsen (2000)). In early 2005, the *Hubble* Space Telescope observed M51 with the Advanced Camera for Surveys (ACS) in 3 wide optical filters and in a narrow $H\alpha$ filter (the F435W, F555W, F814W and F658N filters) as part of a *Hubble* Legacy program to celebrate the 15th anniversary of the telescope. The data was combined and cleaned at the STScI, producing science quality images of 8,600 × 12,200 pixels at 0.05" per pixel, and reaching limiting magnitudes of 27.3, 26.5 and 25.8 in B, V and I, respectively. As such, the data provides an excellent counterpart to the high-resolution X-ray data provided by *Chandra*. M51 has been observed 3 times by *Chandra* for a total of around 100 ks. We detect around 120 discrete X-ray sources plus diffuse emission to a limiting luminosity of 3×10^{36} erg s^{-1} in the 0.3–8 keV band (results are presented in Terashima & Wilson (2004) and Kilgard *et al.* (2005)).

2. Data Analysis

In order to correctly locate potential optical counterparts to the X-ray sources, it is necessary to register all the images to a common world coordinate system. We utilize the USNO B1.0 catalog, with astrometric accuracy of around 0.2" (Monet *et al.* (2003)). In performing this correction, it is necessary to select the catalog stars by hand, as an analysis of the ACS data shows that many of the objects identified as stars in the catalog are actually not discrete sources: some are more than one point source with very small separation, but many are extended galaxies. Once these have been eliminated, the coordinates of the corrected optical and X-ray images are good to around 0.4", slightly better than the *Chandra* ACIS pixel size of 0.5". At the distance of M51, this corresponds to an absolute accuracy of around 16 parsecs.

At the distance of M51, the limiting V magnitude of 26.5 corresponds to an absolute magnitude of $M_v = -3.1$, which allows us to directly detect most O and B stars. In addition, the astrometric accuracy will allow us to place sources within individual star clusters or HII regions. Of the 116 X-ray sources in M51, we exclude 16 sources: 5 because they are not covered by the ACS field-of-view, and 11 that are found in the extremely dusty nuclear region of M51a.

3. First Results

We shall consider source coincidences of 4 types: apparently unambiguous matches, matches with star clusters, matches with $H\alpha$ structures, and X-ray sources with no optical matches.

- *X-ray sources with unambiguous matches:* Of the 100 X-ray sources considered, 25 have discrete optical counterparts: 4 are background AGN, 3 are background galaxies, 1 is a foreground star, and one is the nucleus of M51b. The remaining 16 have stellar counterparts that appear to be associated with M51. 11 appear to be O or B type stars, while the remaining 5 are in heavily obscured dust lanes where it is hard to obtain accurate photometry. All 16 of the sources have X-ray colors that are consistent with the sources being soft X-ray binaries, likely HMXBs in the high/soft state.
- *X-ray sources in star clusters:* 9 of the sources are coincident with young star clusters. Of these, 4 are extremely faint in X-ray, one has extremely soft X-ray color (either a SNR or supersoft source), and 4 have colors consistent with X-ray binaries. Of the 4

Figure 1. *Left:* The combined optical image of M51 red is $H\alpha$, green is V band and blue is B band. *Right:* The adaptively smoothed, combined X-ray image; red is 0.3–1 keV, green is 1–2 keV, and blue is 2–8 keV.

X-ray binary candidates, 3 are ultraluminous X-ray sources, with luminosities in excess of 10^{39} erg s^{-1}.

- *X-ray sources and $H\alpha$ features:* A large number of sources (24/100) is coincident with features in the $H\alpha$. Most of these features are either shell-like or circular, and a few pixels across in the optical, corresponding to sizes of a few parsecs (each pixel in the optical images is around 2 parsecs across). The X-ray colors are consistently soft: 9 have SNR colors, 3 have colors that fall on the SNR/XRB cusp, and 9 have too few counts to classify, but appear to have soft colors. Only 3 have colors typical of X-ray binaries, and it is worth noting that, as one adds absorption, the colors of sources in the SNR category will move into the XRB category in the color classification scheme used (Kilgard *et al.* (2005)). 7 of the 24 sources are also coincident with 20cm sources (L. Maddox, private communication), thus suggesting that X-ray observations may be sensitive to SNR of different ages.

Amongst these sources, there appears a clear divide in luminosity space. 7 of the sources have luminosities greater than 6×10^{37} erg s^{-1}: the 3 SNR/XRB cusp sources, the 3 XRB-candidates, and one source with SNR colors. This source is a ULX with a cool disk blackbody spectrum and is a good candidate for an intermediate-mass black hole (Miller, Fabian, & Miller (2004)).

- *No optical matches:* Of the remaining sources, 19 have more than one O or B star in the X-ray error ellipse. The remaining sources are largely located within dust lanes, making it difficult to detect optical counterparts. The X-ray colors of these sources are typical of harder X-ray binaries: either LMXBs or HMXBs with heavy absorption. We also expect about 5-10 of these sources to be due to background contamination.

4. Conclusions

In an analysis of X-ray and optical data of M51, it is possible to classify around half of the X-ray source population as either HMXBs or SNR, using a combination of the X-ray colors of the sources and coincidence with optical star clusters, individual stars, or extended features in the $H\alpha$ that appear to be SNR. We can also remove a substantial portion of the foreground and background contamination by identifying direct optical counterparts to the sources. In addition, for those sources with no detected optical counterpart, we are able to make inferences as to their classification. For example, the absence of optical counterparts to any M51b X-ray sources except the nucleus indicates that the population is dominated by LMXBs. With this combination of X-ray color analysis and identification of optical counterparts, it is now possible to study separately the HMXB, LMXB, and SNR populations of spiral galaxies beyond the Milky Way.

Acknowledgements

Based on observations made with the NASA/ESA Hubble Space Telescope, obtained via the Hubble Legacy Program, at the Space Telescope Science Institute, which is operated by the Association of Universities for Research in Astronomy, Inc., under NASA contract NAS 5-26555. These observations are associated with program HST-AR-10669.01. A. Prestwich would also like to acknowledge support from the Chandra X-ray Center, operated under NASA contract NAS 8-39073.

References

Fabbiano, G. 1989, *ARA&A* 27, 87

Feldmeier, J.J., Ciardullo, R., & Jacoby, G.H. 1997, *ApJ* 479, 231

Grimm, H.-J., Gifanov, M., & Sunyaev, R. 2003, *MNRAS* 339, 793

Kilgard, R.E., Kaaret, P., Krauss, M.I., Prestwich, A.H., Raley, M.T., & Zezas, A. 2002, *ApJ* 573, 138

Kilgard, R.E., Cowan, J.J., Garcia, M.R., Kaaret, P., Krauss, M.I., McDowell, J.C., Prestwich, A.H., Primini, F.A., Stockdale, C.J., Trinchieri, G., Ward., M.J., & Zezas, A. 2005, *ApJS* 159, 214

Larsen, S.S. 2000, *MNRAS* 319, 893

Miller, J.M., Fabian, A.C., & Miller, M.C. 2004, *ApJ*, 614, 117

Monet, D.G., Levine, S.E., Canzian, B., Ables, H.D., Bird, A.R., Dahn, C.C., Guetter, H.H., Harris, H.C., *et al.* 2003, *AJ* 125, 984

Prestwich, A.H., Irwin, J.A., Kilgard, R.E., Krauss, M.I., Zezas, A., Primini, F., Kaaret, P., & Boroson, B. 2003, *ApJ* 595, 719

Terashima, Y. & Wilson., A.S. 2004, *ApJ* 601, 735

Discussion

GARCIA: Tell us more about the ULX/IMBH candidate.

KILGARD: It is a fuzzy blob on $H\alpha$, and detected on B and V.

DI STEFANO: Are any of the ULXs coincident with features in the $H\alpha$?

KILGARD: Yes, one of the ULXs is coincident with a shell-like feature in the $H\alpha$. It has a soft spectrum coincident with being an X-ray binary in the high-soft state.

Populations of High Energy Sources in Galaxies
Proceedings IAU Symposium No. 230, 2005
E. J. A. Meurs & G. Fabbiano, eds.

© 2006 International Astronomical Union
doi:10.1017/S1743921306008180

An Unprecedented View of a Galaxy's Dynamic X-ray Source Population

David Pooley

Astronomy Department, University of California at Berkeley, 601 Campbell Hall, Berkeley,
CA 94720, USA
email: dave@astron.berkeley.edu

Abstract. In May/June 2005, *Chandra* observed M81 fifteen times, roughly once every three days, as part of our proposal to explore the days to weeks timescale of variability for extragalactic point sources. Each observation reached a sensitivity of 5×10^{36} erg/s. Because these observations probe the timescale on which X-ray binaries typically evolve, we can now compare extragalactic sources to Galactic X-ray binaries on a more equal level than has been possible in the past. In addition, we can measure and quantify any possible time variability of M81's X-ray luminosity function and investigate alternative methods to characterize a galaxy's dynamic X-ray source population. We present here preliminary results of the observations.

Keywords. galaxies: individual (M81), X-rays: binaries.

---

1. Motivation

X-ray luminosity functions (XLFs) are almost universally used as a tool to compare different galaxies or to investigate sub-populations within a single galaxy. In fact, in a recent review of X-ray sources in normal galaxies, Fabbiano & White discuss XLFs on 10 of the 11 pages in the section on the X-ray binary populations in spiral galaxies (Fabbiano & White 2003); there are eight figures in that section — three are *Chandra* images (M31, M83, and the Antennae) and the other five are XLFs. However, despite the ubiquity of XLFs in studying other galaxies, the variability of the XLF of a galaxy has never been properly addressed before (see Maccacaro *et al.* 1987 for the variability of the XLF of a set of random fields observed multiple times with *Einstein*). An analysis of five observations of NGC 3877 (Gonzales 2004) reveals a strongly variable XLF (see Fig. 1 which shows the XLF of two different observations as well as an XLF constructed by considering the maximum $L_{\rm X}$ achieved by each source). Above $\sim 3 \times 10^{38}$ erg s^{-1}, the XLF of Obs. 4 indicates only 1 source while the XLF of Obs. 5 indicates 7 sources, and the maximum-XLF shows 10 sources. This order of magnitude difference is very interesting, to say the least.

Clearly, the XLF of a single observation may not be a good representation of the galaxy, especially since we have almost no information about the full range of XLF variability. NGC 3877 is one of only a few galaxies that have been observed more than once or twice (note: the observations were performed to follow the evolution of supernova 1998S; Pooley *et al.* 2002). Our observations of M81 are intended to provide much needed information in this area. *Chandra* observations over the past five years have resulted in the discovery of an astounding number (many thousands) of X-ray sources in other galaxies. As mentioned above, XLF properties (such as slope, maximum $L_{\rm X}$, a break at a certain $L_{\rm X}$, etc.) have been almost universally used to investigate these sources and compare galaxy populations, yet our findings on NGC 3877 could cast doubt on their reliability as an accurate characterization of a galaxy. As long as such doubts remain, the power

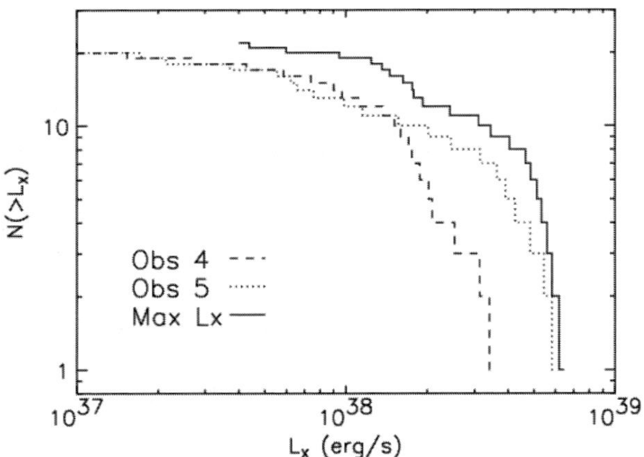

Figure 1. Three X-ray luminosity functions of the Sc galaxy NGC 3877 (Gonzales 2004). Two are of individual observations (notice the large difference) and one is based on the maximum observed luminosity of each source (SN 1998S was not included).

of these galaxy to galaxy comparisons is greatly weakened. To truly utilize the *Chandra* data, we must first understand and calibrate the XLF tool.

2. Observations

To explore XLF variability, we used *Chandra* ACIS-S in imaging mode to observe M81 fifteen times over the course of about a month and a half during the summer of 2005. Each observation lasted 12 ksec and reached a limiting luminosity (near the aimpoint) of 5×10^{36} erg/s at the distance of M81 (3.6 Mpc; Freedman *et al.* 2004). The observations log is listed in Table 1, and a composite image is shown in Figure 2.

3. Preliminary Results

We constructed XLFs of the inner $5' \times 5'$ region of M81, shown in Figure 3, and performed pairwise Kolmogorov-Smirnov tests to compute the probability that the observed XLFs came from the same parent distribution. The most discrepant pair of XLFs had only a 4% chance of being from the same parent distribution, but, considering the number of trials performed, this is not significant. Therefore, the bulge population of M81 shows no evidence for a varying XLF. We have not yet performed the same test on the spiral-arm population. It has been shown that the XLF of the M81 bulge X-ray sources is different than that of the disk X-ray sources (Tennant *et al.* 2001).

Individual sources showed a remarkable amount of variability, and we present preliminary results of the two most luminous. The ultraluminous source X-6 (Fabbiano 1988) has had

Table 1. Observation Log

Date	Good time	Date	Good time	Date	Good time
2005-05-26	11.0 ks	2005-06-09	12.0 ks	2005-06-24	11.6 ks
2005-05-28	11.4 ks	2005-06-11	11.8 ks	2005-06-26	12.0 ks
2005-06-01	12.0 ks	2005-06-15	11.9 ks	2005-06-29	10.7 ks
2005-06-03	11.8 ks	2005-06-18	12.0 ks	2005-06-03	12.0 ks
2005-06-06	11.8 ks	2005-06-21	11.8 ks	2005-07-06	12.0 ks

Figure 2. Images of the inner $6' \times 8'$ of M81 taken with *Spitzer* (left) and *Chandra* (right). The *Spitzer* image is courtesy NASA/JPL-Caltech/K. Gordon (University of Arizona) & S. Willner (Harvard-Smithsonian Center for Astrophysics). The *Chandra* image is the composite of the observations listed in Table 1. In the *Spitzer* image, red represents 24 μm emission, green represents 1.2–2.5 μm, and blue represents 2.5–6 μm. In the *Chandra* image, red represents 0.5–1.2 keV, green represents 1.2–2.5 keV, and blue represents 2.5–6 keV.

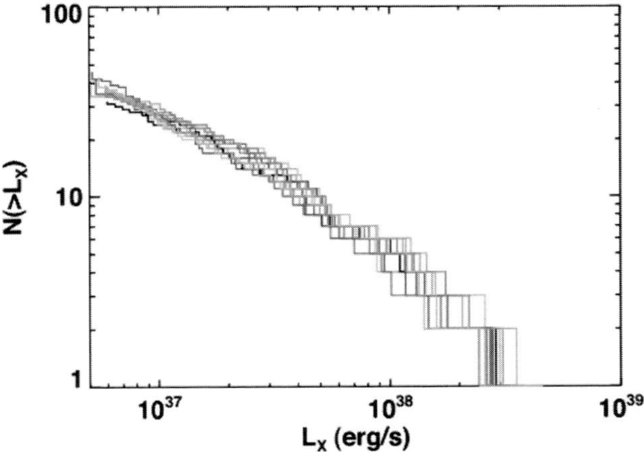

Figure 3. Individual X-ray luminosity functions of the inner $5'$ of M81. There is no evidence for variation among them.

conflicting reports of its variability. Immler & Wang reported a factor of two variability over six days (Immler & Wang 2001), but Swartz *et al.* reanalyzed those data, as well as a previous *Chandra* observation, and found no evidence for variability on short or long times scales (Swartz *et al.* 2003). We find variations of $\sim$30% on a timescale of days (see Figure 4).

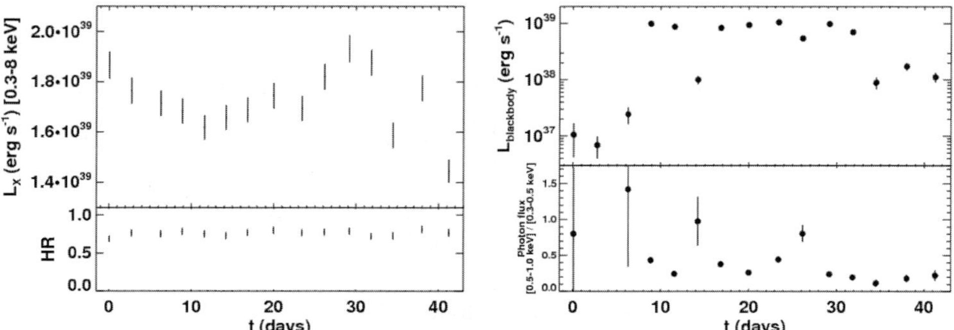

Figure 4. *Left:* X-ray luminosity and hardness ratio (HR) of the ultraluminous X-ray source X-6 in M81. The average absorbed 0.3–8 keV luminosity of each observation is plotted. HR is defined as the ratio of photon flux in the 2–8 keV band to that in the 0.5–2 keV band. *Right:* The bolometric luminosity of a very luminous soft source in M81. The source specrum is well described by a blackbody of $kT = 80$ eV. There is roughly a factor of two correction going from the detected in-band flux to the unabsorbed in-band flux, and another factor of two going from in-band to bolometric.

We also observe dramatic variability of a very soft source in M81, as shown in Figure 4. The source is very soft, with almost no emission detected above 1.5 keV and a spectrum well described by an absorbed blackbody of temperature $kT = 80$ eV. The source flux was observed to rise by two orders of magnitude over the course of a about a week. The very soft spectrum and extreme luminosity of this source have been interpreted in terms of an intermediate-mass black hole of $\sim$1200 $M_\odot$ (Swartz *et al.* 2002).

4. Conclusions

As *Chandra* and *XMM* will no doubt continue to probe nearby galaxies very deeply, it will be beneficial to split up the observations into series of closely-spaced short observations to probe this timescale of variability that is virtually unexplored outside the Milky Way.

Acknowledgements

DP gratefully acknowledges support provided by NASA through Chandra Postdoctoral Fellowship grant number PF4-50035 awarded by the Chandra X-ray Center, which is operated by the Smithsonian Astrophysical Observatory for NASA under contract NAS8-03060.

References

Fabbiano, G. 1988, ApJ, 325, 544

Fabbiano, G. & White, N. 2003, astro-ph/0307077

Freedman, W., *et al.* 1994, ApJ, 427, 628

Gonzales, E. 2004, undergraduate thesis, Massachusetts Institute of Technology

Immler, S. & Wang, Q. D. 2001, ApJ, 554, 202

Maccacaro, T., *et al.* 1987, AJ, 93, 1484

Pooley, D., *et al.* 2002, ApJ, 572, 932

Swartz, D., *et al.* 2002, ApJ, 574, 382

Swartz, D., *et al.* 2003, ApJS, 144, 213

Tennant, A.F., *et al.* 2001, ApJ, 549, L43

Populations of High Energy Sources in Galaxies
Proceedings IAU Symposium No. 230, 2005
E. J. A. Meurs & G. Fabbiano, eds.

© 2006 International Astronomical Union
doi:10.1017/S1743921306008192

A *Chandra* Observation of the Nearby Sculptor Group Sd Galaxy NGC 7793

Thomas G. Pannuti[1], Eric M. Schlegel[2] and Christina K. Lacey[3]

[1]*Spitzer* Science Center, California Institute of Technology, Mailstop 220-6,
Pasadena, CA 91101, USA
email: tpannuti@ipac.caltech.edu

[2]Harvard-Smithsonian Center for Astrophysics, 60 Garden Street, Cambridge, MA 02138 USA
email: eschlegel@cfa.harvard.edu

[3]Department of Physics, University of South Carolina, Columbia, SC 29208 USA
email: lacey@sc.edu

Abstract. We present the results of a *Chandra* observation made of the nearby spiral galaxy NGC 7793: the effective exposure time of this observation was 49094 seconds. Twenty-two discrete sources were identified at a minimum of a 3σ level to an estimated limiting luminosity of $\sim 2 \times 10^{36}$ ergs sec^{-1}. We have performed a spectral analysis of the known ultraluminous X-ray source (ULX) in this galaxy: statistically-acceptable fits to the spectrum can be obtained with either a power law model, a bremsstrahlung model or a DISKBB model. We have also searched for counterparts at multiple wavelengths to these sources: based on this search, we have classified two supernova remnants, one HII region and two foreground stars.

Keywords. X-rays: galaxies, galaxies: individual (NGC 7793).

---

1. Introduction

The Sd galaxy NGC 7793 is a member of the nearby Sculptor Group of galaxies and lies at a distance of approximately 3.38 Mpc (Puche 1988). Based on X-ray observations made with *Einstein* (Fabbiano *et al.* 1992) and the *Röntgensatellit* (*ROSAT*) (Read & Pietsch 1999), seven discrete X-ray sources – including an ultraluminous X-ray source (ULX) located along the southern edge of the galaxy – had been identified within the optical extent of NGC 7793. The supernova remnant (SNR) population in NGC 7793 is one of the best studied of any external galaxy: based on analysis of X-ray, optical and radio observations (Blair & Long 1997, Pannuti *et al.* 2002) a total of 31 SNRs have been identified in this galaxy. We have observed NGC 7793 with *Chandra* as part of our study of X-ray emission from SNRs in nearby galaxies.

2. Observations

We used the Advanced Charge-Coupled Device (CCD) Imaging Spectrometer (ACIS) aboard *Chandra* in Very Faint Mode to observe NGC 7793: virtually all of the galaxy was sampled with the ACIS-S3 chip and the effective exposure time of the observation was 49094 seconds. We re-filtered the Level 1 data, correcting for the induced charge-transfer inefficiency. Point sources were detected using the "wavdetect" algorithm (Freeman *et al.* 2002) at 1", 2 " and 4" scales.

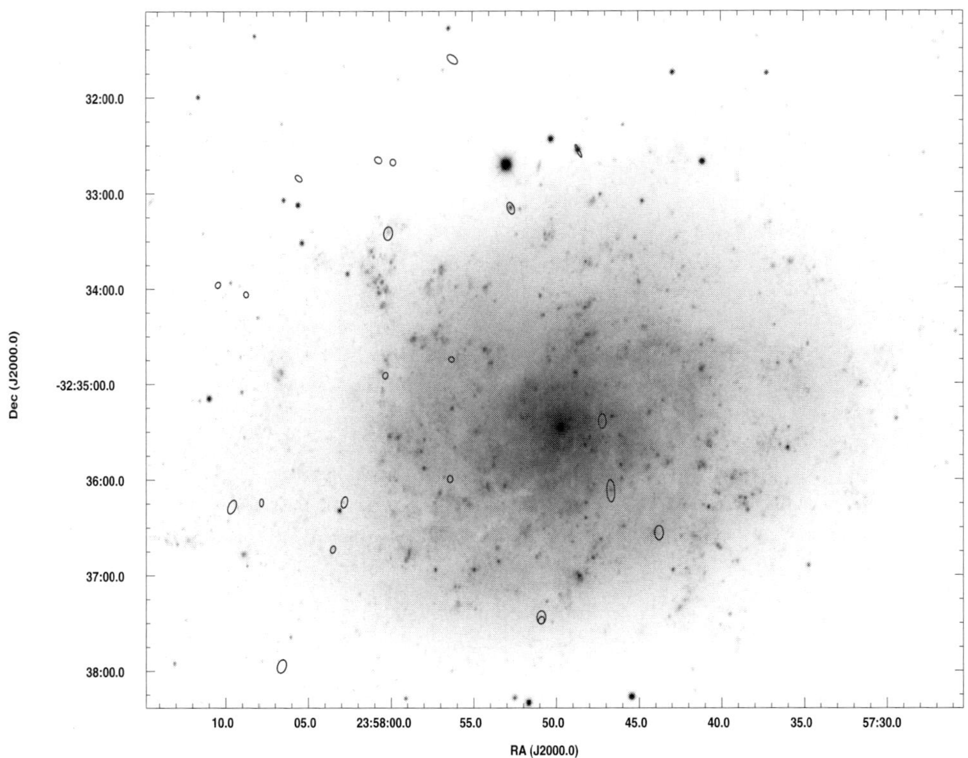

Figure 1. An *R*-band image of NGC 7793 with the positions of the 22 discrete X-ray sources detected by *Chandra* indicated.

3. Results

A total of 22 discrete sources were detected at a minimum of a 3σ level, corresponding to a limiting unabsorbed luminosity of approximately 3×10^{36} ergs sec^{-1}. The positions of the detected sources are plotted on an *R*-band image of NGC 7793 in Figure 1. Statistically-acceptable fits to the spectrum of the ULX can be obtained with either a power law model (with a photon index $\Gamma = 1.4 \pm 0.1$), a bremsstrahlung model ($kT = 25^{+20}_{-9}$ keV) or a DISKBB model ($kT_{\mathrm{in}} = 1.94^{+0.16}_{-0.14}$). We have searched for counterparts at other wavelengths for the detected X-ray sources: and identify X-ray counterparts to two SNRs, one HII region and two foreground stars. The remaining X-ray sources are most likely resident X-ray binaries in NGC 7793, or background galaxies.

Acknowledgements

T.G.P. acknowledges support for this work from *Chandra* Grant GO3-4104Z.

References

Blair, W. P. & Long, K. S. 1997, *ApJS*, 108, 261

Fabbiano, G., Kim, D.-W., & Trinchieri, G. 1992, *ApJS*, 80, 531

Freeman, P. E., Kashyap, V., Rosner, R., & Lamb, D. Q. 2002, *ApJS*, 138, 185

Pannuti, T. G., Duric, N., Lacey, C. K., Ferguson, A. M. N., Magnor, M. A., & Mendelowitz, C. 2002, *ApJ* 565, 956

Puche, D. & Carignan, C. 1988, *AJ*, 95, 1025

Read, A. M. & Pietsch, W. 1999, *A&A*, 341, 8

Populations of High Energy Sources in Galaxies
Proceedings IAU Symposium No. 230, 2005
E. J. A. Meurs & G. Fabbiano, eds.

© 2006 International Astronomical Union
doi:10.1017/S1743921306008209

A Survey of Massive Star Clusters in the X-ray Emission from Spiral Galaxies

E. M. Schlegel

High Energy Astrophysics Division, Smithsonian Astrophysical Observatory,
Cambridge MA, USA
email: eschlegel@cfa.harvard.edu

Abstract. The massive star clusters identified by S. Larsen are compared to the available Chandra observations of face-on spiral galaxies. In each galaxy, a few percent of the Larsen-identified clusters match X-ray-emitting point sources. An additional few match knots of emission in the diffuse emission. The cluster properties are examined to ascertain whether massive star clusters are X-ray sources.

Populations of High Energy Sources in Galaxies
Proceedings IAU Symposium No. 230, 2005
E. J. A. Meurs & G. Fabbiano, eds.

© 2006 International Astronomical Union
doi:10.1017/S1743921306008210

Low Mass X-ray Binaries and Globular Clusters in Early-Type Galaxies

Craig L. Sarazin

Department of Astronomy, University of Virginia, Charlottesville, VA 22903-0818, USA
email: sarazin@virginia.edu

Abstract. Chandra observations have allowed the detection of a large number of low mass X-ray binaries (LMXBs) in early-type galaxies. Comparison to catalogs of globular clusters (GCs) from Hubble Space Telescope observations have shown that a high fraction of the LMXBs in early-type galaxies are associated with GCs. The fraction of LMXBs associated with globular clusters increases along the Hubble sequence from spiral bulges to S0s to Es to cDs. On the other hand, the fraction of globular clusters which contain X-ray sources appears to be roughly constant ($\sim$4% for $L_X \gtrsim 10^{38}$ ergs/s, $\sim$10% for $L_X \gtrsim 10^{37}$ ergs/s). There is a strong tendency for the X-ray sources to be associated with the optically more luminous GCs. There is a trend for the X-ray sources to be found preferentially in redder, more metal-rich GCs, which is independent of optical luminosity correlation.

The relative role of formation of LMXBs in GCs and in situ formation in the field is uncertain. One of the best ways to study this is to compare the spatial distribution of GC-LMXBs, field LMXBs, GCs, and optical light in the galaxies. Theoretical models and results of fits to the observed distributions are presented.

Keywords. binaries: close, galaxies: elliptical and lenticular, galaxies: star clusters, galaxies: individual (NGC 4365), globular clusters: general, X-rays: binaries, X-rays: galaxies.

1. Introduction

Chandra observations have resolved most of the X-ray emission in X-ray-faint early-type galaxies into individual point-like sources (e.g., Sarazin, Irwin, & Bregman 2000). Given their properties and the stellar populations in these galaxies, these X-ray sources are assumed to be Low Mass X-ray Binaries (LMXBs). A significant fraction ($\sim$20–70%) of the LMXBs are associated with globular clusters in the host galaxies (Sarazin *et al.* 2001; Angelini *et al.* 2001). The fraction of LMXBs located in GCs is much higher than the fraction of optical light, which indicates that stars in GCs are much more likely (by a factor of $\sim$500) to be donor stars in X-ray binaries than field stars. As has been known for a number of years, a similar result applies to our own Galaxy and to the bulge of M31 (e.g., Hertz & Grindlay 1983). This is generally believed to result from stellar dynamical interactions in globular clusters, which can produce compact binary systems.

X-ray observations with *ASCA* indicated that the total luminosity of LMXBs in early-type galaxies correlated better with the number of GCs than with the optical luminosity of the galaxy (White *et al.* 2002). This is somewhat surprising, as a nontrivial fraction ($\sim$50%) of the LMXBs in most of the early-type galaxies observed so far with *Chandra* are not identified with GCs. This suggests that most (perhaps all?) of the LMXBs in early-type galaxies were made in GCs (Grindlay 1984; Sarazin *et al.* 2001; White *et al.* 2002). The field LMXBs might have been ejected from globular clusters individually by stellar dynamical processes (or possibly by supernova kicks), or emerged when globular clusters were destroyed by tidal effects. Alternatively, the field LMXBs may have been made in situ from primordial binaries.

200

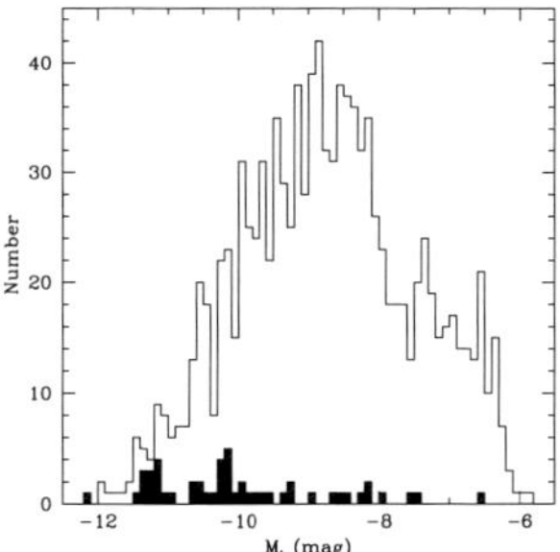 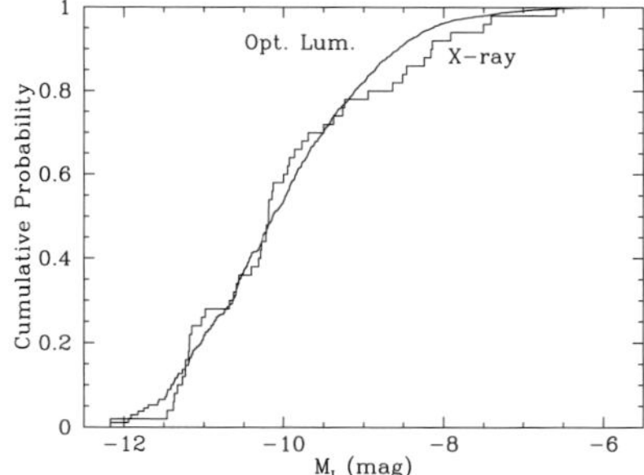

Figure 1. (*a*) Histograms of the number of globular clusters versus their absolute magnitude M_I in a sample of galaxies with Chandra data (Sarazin *et al.* 2003). The upper histogram is for all of the GCs in the galaxies. The lower shaded histogram shows the GCs which contain identified LMXBs. (*b*) Cumulative distribution functions for the probability that a GC contains an X-ray source ("X-ray") and for the optical luminosity of GCs ("Opt. Lum.").

2. Statistics of LMXBs and GC Populations

The fraction of LMXBs in a galaxy which are associated with GCs increases along a Hubble sequence from spiral bulges ($\sim$10%) to S0s ($\sim$20%) to giant ellipticals ($\sim$50%) to cD galaxies ($\sim$70%) (Sarazin *et al.* 2003). There is a well-established trend for the specific frequency of GCs in galaxies (S_N, the number of GCs per optical luminosity) to increase along the same Hubble sequence (e.g., Harris 1991). The detailed correlation of the fraction of LMXBs in GCs with S_N is more consistent with most of the field LMXBs being made in situ in the field (Juett 2005; see also Sarazin *et al.* 2003; Maccarone *et al.* 2003) On the other hand, Irwin (2005) argued recently that a significant portion of the field LMXBs in S0 galaxies may have come from disrupted GCs.

The fraction of globular clusters which contain X-ray sources appears to be roughly constant from galaxy to galaxy. For samples of LMXBs with a high limiting X-ray luminosity, $L_X \gtrsim 10^{38}$ ergs/s, the fraction is $\sim$4% (Kundu *et al.* 2002; Sarazin *et al.* 2003). For NGC 4697, a nearby elliptical with deep X-ray and GC observations, the fraction reaches $\sim$10% for $L_X \gtrsim 10^{37}$ ergs/s (Sivakoff *et al.* 2005).

3. Properties of GCs Containing LMXBs

Figure 1(*a*) shows histograms of the absolute I magnitude, M_I, of the total GC sample (upper histogram) and of the GCs containing LMXBs (shaded histogram). The LMXBs seem to be associated preferentially with the more optically luminous GCs (Angelini *et al.* 2001; Kundu *et al.* 2002; Sarazin *et al.* 2003). For example, the median value of M_I for non-X-ray GCs is -8.7, while the corresponding value for the X-ray GCs is -10.2. Using the Wilcoxon or equivalent Mann-Whitney rank-sum tests, the distribution of X-ray and non-X-ray GC luminosities are found to disagree at more than the 6σ level.

Of course, a correlation between optical luminosity and the probability of having an X-ray source is not unexpected. LMXBs contain normal stars, and globular clusters which have higher luminosities have more stars as potential donors in LMXBs. Thus, it is interesting to test the hypothesis that the probability that a GC contains a LMXB is proportional to its optical luminosity. Figure 1(*b*) compares the cumulative probability distribution of LMXBs versus the cumulative distribution of the optical luminosity in GCs. The two cumulative distribution functions track one another fairly well. For example, half of the optical luminosity comes from GCs brighter than $M_I = -10.1$, while the medium absolute magnitude of GCs with LMXBs is -10.2. The KS two-sample test was used to compare the two distributions; they are not significantly different. Thus, the current data indicate that optically bright GCs are much more likely to contain LMXBs than faint GCs, but the distribution is consistent with a constant probability per unit optical luminosity (Kundu *et al.* 2002; Sarazin *et al.* 2003). Recently, Jordán *et al.* (2004) found a correlation between the density of stars in M87 GCs and the occurrence of LMXBs. This would be consistent with the formation of LMXBs by dynamical collision processes in GCs, although the detailed form of the correlation found by Jordán *et al.* (2004) was also nearly equivalent to a simple dependence on the number of stars.

Figure 2 shows histograms of the $V - I$ colors for the total GC sample (upper histogram) and for the GCs containing LMXBs (shaded histogram). Because the sample contains a number of different galaxies, the overall color distribution may be less obviously bimodal than that seen in some individual galaxies. The LMXBs appear to be associated preferentially with the redder GCs (larger values of $V - I$) (Angelini *et al.* 2001; Kundu *et al.* 2002; Sarazin *et al.* 2003). The median color of the non-X-ray GCs is $V - I = 1.07$, while the corresponding median for the X-ray GCs is 1.14. Using the Wilcoxon or Mann-Whitney rank-sum tests or the KS test, the probabilities that the two color distributions where drawn from the same distribution are <0.2%. Roughly, red GCs are three times as likely to harbor a LMXB as blue GCs.

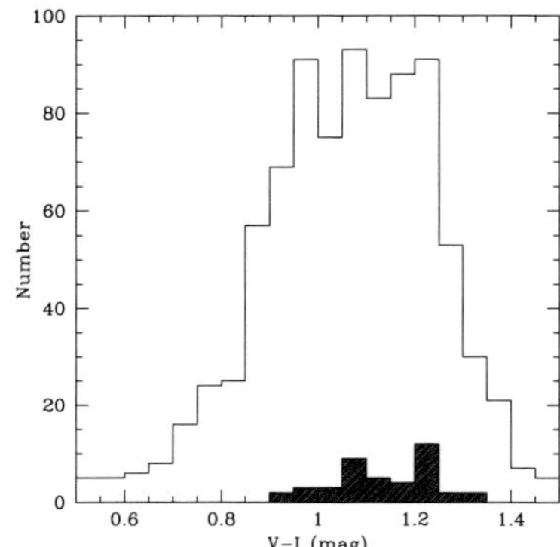

Figure 2. Histograms of the number of GCs versus their optical color, $V - I$. The upper histogram is for all of the GCs, while the lower shaded histogram shows the GCs which contain LMXBs.

4. Spatial Distributions of Stars, GCs, and LMXBs

The radial distribution of GCs in elliptical galaxies is more extended than that of the field stars; in particular, the optical light profiles of ellipticals typically show a central cusp, whereas the spatial distribution of GCs has a constant surface density core. This may indicate that GCs never formed in the central regions, or that the GCs which were initially formed there were disrupted by tidal effects. One way to test this would be to search the central regions of ellipticals for a stellar population which is characteristic of GCs. This is difficult for optical stars; however, as noted above, LMXBs are preferentially produced in GCs. At the same time, we would like to know what fraction of the field LMXBs were made in GCs. Some of the GC LMXBs might have escaped individually due to stellar dynamical interactions, or they may have been released when their host GC was disrupted by tidal effects. In either case, these field LMXBs would have a spatial distribution which reflected the initial spatial distribution of the GCs. Thus, by studying the spatial distributions of the optical light, GCs, and field and GC LMXBs in galaxies, we can constrain models for the formation and destruction of GCs and the origin of LMXBs.

Figure 3 shows models for the spatial distribution of field and GC LMXBs in an elliptical galaxy (Sarazin 2005). The stellar and GC distribution were based on observations of NGC 4365, and the X-ray sources are from our earlier Chandra observation of this galaxy. Here, I show only three extreme models. In Model 1, all field LMXBs were made in situ. This model probably provides an adequate fit to the existing data. Note the general result that the observed LMXB distribution is broader than that of the stars, reflecting the contribution of GC LMXBs. In Model 2, all LMXBs are made in GCs, and the field LMXBs were individually ejected from GCs. In this model, the predicted

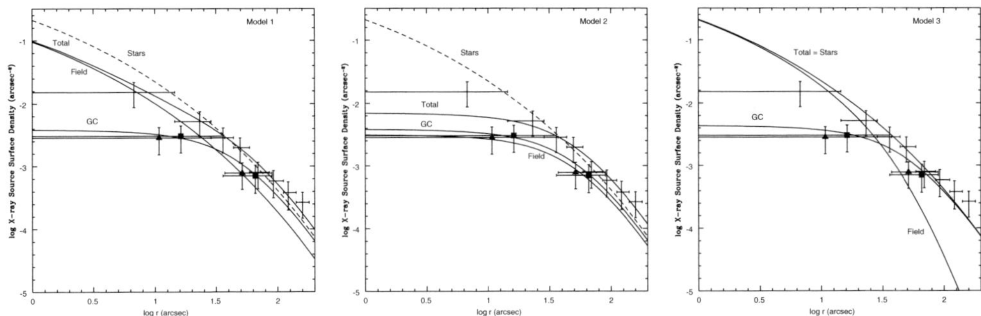

Figure 3. Predicted surface density distributions of X-ray sources in elliptical galaxies (Sarazin 2005). The solid curves labeled GC, Field, and Total show the surface densities of LMXBs in GCs, in the field, and in total, respectively. The dashed curve shows the total X-ray distribution if the X-ray sources followed the distribution of field stars. The data points are from NGC 4365. The error bars with triangles are the field LMXBs, those with squares are the GC sources, and the plain error bars are the total source distribution. Model 1 assumes that no GC LMXBs are lost from GCs either due to individual ejection or GC destruction. In Model 2, all LMXBs are made in GCs, no GCs are destroyed, but all of the field LMXBs were individually ejected from GCs. In Model 3, all LMXBs are made in GCs, and all field LMXBs result from the tidal disruption of GCs.

distribution of the LMXBs is broader than that observed. Finally, in Model 3, all LMXBs are also made in GCs, but the only mechanism for the release of the field LMXBs is the tidal disruption of GCs. In this model, the predicted distribution of field LMXBs is more strongly peaked than that observed due to the high rates of GC destruction at the center of the galaxy. Although better data are needed, the comparison of the models with the present data indicate that at least half of the field LMXBs were made in situ.

Acknowledgements

I am very grateful to Greg Sivakoff and Adrienne Juett for helpful comments. This work was supported by the National Aeronautics and Space Administration through Chandra Award Numbers GO3-4099X, AR4-5008X, GO4-5093X, GO5-6081X, and GO5-6086X, and Hubble Space Telescope Award Number HST-GO-10003.01-A.

References

Angelini, L., Loewenstein, M., & Mushotzky, R. F. 2001, *ApJ* 557, L35
Grindlay, J. E. 1984, *AdSpR* 3, 19
Harris, W. E. 1991, *ARA&A* 29, 543
Hertz, P. & Grindlay, J. E. 1983, *ApJ* 275, 105
Irwin, J. A. 2005, *ApJ* 631, in press
Jordán, A., Côté, P., Ferrarese, L., Blakeslee, J. P., Mei, S., Merritt, D., Milosavljević, M., Peng, E. W., Tonry, J. L., & West, M. J. 2004, *ApJ* 613, 279
Juett, A. M. 2005, *ApJ* 621, L25
Kundu, A., Maccarone, T. J., & Zepf, S. E. 2002, *ApJ* 574, L5
Maccarone, T. J., Kundu, A., & Zepf, S. E. 2003, *ApJ* 586, 814
Sarazin, C. L. 2005, *ApJ* submitted
Sarazin, C. L., Irwin, J. A., & Bregman, J. N. 2000, *ApJ* 544, L101
—. 2001, *ApJ* 556, 533
Sarazin, C. L., Kundu, A., Irwin, J. A., Sivakoff, G. R., Blanton, E. L., & Randall, S. W. 2003, *ApJ* 595, 743
Sivakoff, G. R., *et al.* 2005, *ApJ* submitted
White, R. E., Sarazin, C. L., & Kulkarni, S. R. 2002, *ApJ* 571, L23

Discussion

GHOSH: Do we always assume that each GC contains a single X-ray binary? In other words, is the total luminosity associated with an X-ray source coincident with a globular cluster always thought to be that of a single binary? In particular, at the high-L end ($\sim 10^{39}$ erg s^{-1}), beyond the famous break in the XLF, is it possible that a GC has multiple unresolved sources?

SARAZIN: In general, we assume that all X-ray sources, including those in GCs, are single. Based on the X-ray luminosity function and fraction of GCs containing LMXBs, it would be unlikely that a GC would contain 2 bright LMXBs. (However, note that one Galactic GC does contain 2 active LMXBs). So, the sources are probably single, unless some subset of GC has a particularly high probability of having LMXBs.

KIM: In the case of the GC disruption for the formation of field LMXBs, the test of the SN proportionality may not work, because SN does not include those GC's disrupted.

SARAZIN: Yes. A worry with that test is that one has to assume that the original number of GCs is proportional to the present number. If different fractions of GCs were destroyed in different galaxies, this might be wrong. This is an advantage of the spatial distribution test I described.

Chairman Vladimir Lipunov (left) following Craig Sarazin via a secret monitor.

Populations of High Energy Sources in Galaxies
Proceedings IAU Symposium No. 230, 2005
E. J. A. Meurs & G. Fabbiano, eds.

© 2006 International Astronomical Union
doi:10.1017/S1743921306008222

The trail of discrete X-ray sources in the early-type galaxy NGC 4261: anisotropy in the globular cluster distribution

Ginevra Trinchieri[1], Lea Giordano[2], Luca Cortese[2], Anna Wolter[1], Monica Colpi[2], Giuseppe Gavazzi[2] and Lucio Mayer[3]

[1]INAF-Osservatorio Astronomico di Brera, Via Brera 28, 20121 Milano, Italy
[2]Università degli Studi di Milano Bicocca, Piazza della Scienza 3, 20126 Milano, Italy
[3]Inst. of Theoretical Physics, University of Zürich, Winterurestr. 190, 8057 Zürich, Switzerland

Abstract. The recent evidence of a peculiar distribution of X-ray sources in the elliptical galaxy NGC 4261 reported by Zezas *et al.* has prompted us to study this galaxy combining archive X-ray and optical observations, from Chandra, INT, and HST. We find that a sizable fraction of the X-ray sources has a globular cluster as optical counterpart. This together with the shape of the luminosity function of the X-ray sources suggest that they are accreting low-mass binaries. We further show a remarkable similarity in the anisotropy of the projected spatial distributions of the optical and X-ray sources, which leads us to conclude that the spatial anisotropy of the X-ray sources in NGC 4261 is due to the anisotropy of the globular cluster population.

Keywords. X-rays: galaxies, galaxies: individual (NGC 4261).

1. NGC 4261

NGC 4261 is a massive early-type galaxy in the Virgo W Cloud. The optical spectrum and light distribution indicate an old unperturbed galaxy (Gavazzi *et al.* 2002, Schweizer & Seitzer 1992, Colbert *et al.* 2001), in a group that that does not show prominent signs of gravitational interactions (Garcia 1993, Noltenius 1993). The only evidence of activity is in the nuclear region that hosts a FRI radio source (3C 270).

Chandra observations of NGC4261 have been reported in the literature, mostly for the nuclear region (Chiaberge *et al.* 2003, Sambruna *et al.* 2003, Gliozzi *et al.* 2003). Zezas *et al.* (2003) have concentrated on the galaxy proper and have discovered the presence of about sixty bright off-nuclear X-ray sources in NGC 4261, which stand out for their large-scale spatial anisotropy. Zezas *et al.* (2003) interpret this peculiar distribution as evidence of an association with a young stellar population, possibly formed in a recent episode of star formation dynamically triggered along tidal tails. Since most to all X-ray sources in an early type galaxy are expected to be Low Mass X-ray binaries (LMXB), this association would make NGC 4261 a remarkable exception worth of further study.

2. Our contribution

We have used archival data from *Chandra*, the Isaac Newton Telescope (INT), and the Hubble Space Telescope (*HST*), to better understand the nature of the X-ray sources, to search for their optical counterparts, and to study their relation with the GC population of NGC 4261. XMM-Newton data were also considered but could not add useful information for this project.

The optical data at our disposal is not ideal: the HST dataset does not cover the whole galaxy and is not homogeneous in time and filters (Fig. 1); INT covers the whole optical

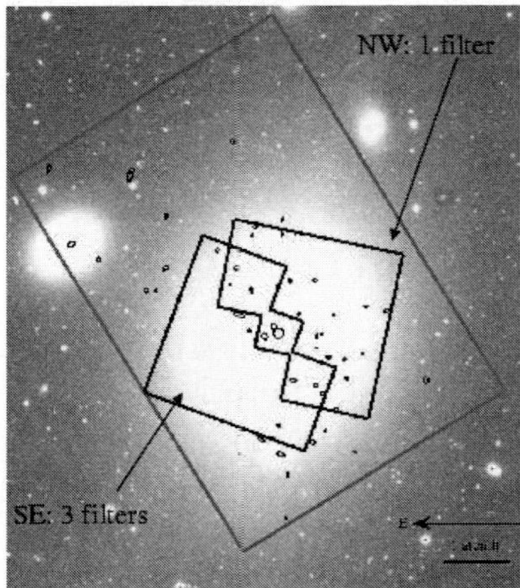

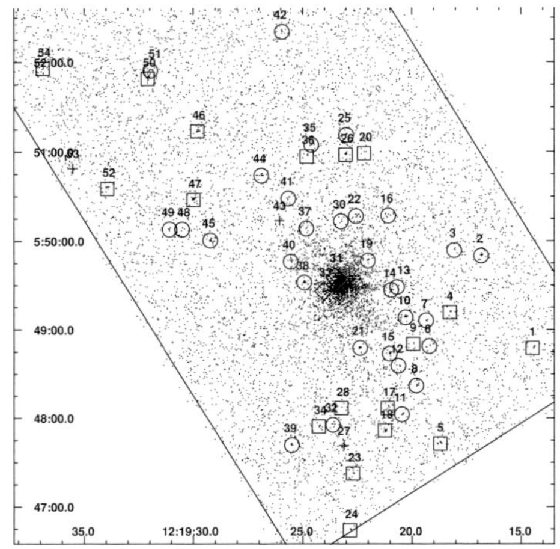

Figure 1. LEFT: R band image of NGC 4261 from the INT. The active field of the *Chandra* ACIS-S3 CCD and *HST* WFPC2 fields are shown. Only the SE field was observed in three filters, allowing us to select GCs on the basis of their colors as well. Ellipses indicate the positions of the X-ray sources. RIGHT: Positions of the X-ray sources identified with GC (open squares); with no id. (open circles); for which no optical counterpart was searched for (model subtraction is too noisy or outside the galaxy radius - diamonds); central AGN (cross); back-/foreground galaxies (+).

galaxy but in a single filter and at a lower ($\sim1''$) resolution (see details in Giordano *et al.* 2005). Nonetheless we have combined all information and produced a valid list of candidate globular clusters (GC) in NGC 4261. We have first selected GCs in the SE HST field, using the color information. The 325 objects selected span a V-I range from ~0.5 to ~2, typical of the globular cluster populations observed in other massive early-type galaxies, and indicative that both "red" and "blue" GCs are represented. We have then used this list to "calibrate" the results from the NE HST field and INT data, for which we have no color information. The cross-correlation between the three lists indicates that $\sim97\%$ (37/38) and $\sim83\%$ (49/59) of the sources in the common area between *HST*-SE and *HST*-NW and INT field, respectively, coincide with a GC in the *HST*-SE field. We have then used the INT "GC candidate" list, that provides an homogeneous set of optical sources over the whole galaxy, to determine the optical identification of X-ray sources obtained from the *Chandra* data.

We find that 50% of the X-ray sources can be associated with a GC cluster, preferentially belonging to the "metal rich" bright population (when the color information is available). Therefore it appears that the properties of the sources in NGC 4621 are like those in other early type galaxies (see for example Kundu *et al.* 2002, Maccarone *et al.* 2003), except for their spatial distribution (a rather big "except"!).

This is also confirmed by the shape of the luminosity distribution in the X-ray sources (Fig. 2): unquestionably the functional form appropriate for low mass X-ray binaries is a significantly better representation than that for High Mass systems.

We therefore conclude that NGC 4621 has a population of LMXB associated with GCs, like other early type galaxies studied. But unlike any other, the spatial distribution of these sources is peculiar, and totally at odds with the distribution of the optical light.

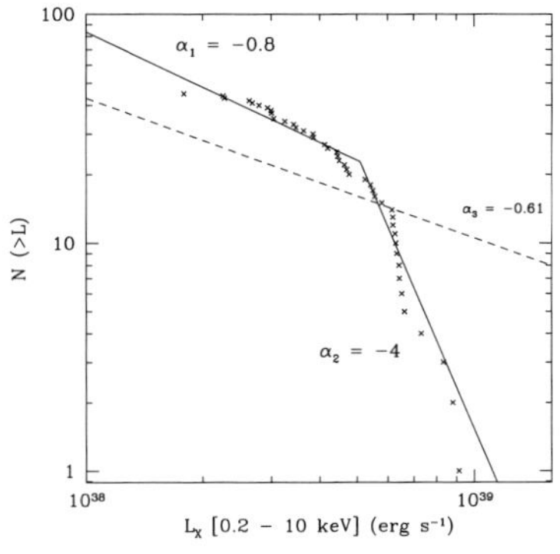

The integral luminosity function in the 0.2-10 keV band of the X-ray sources in NGC 4261. The solid line gives a broken power-law with fixed slopes, from Gilfanov (2004), renormalized to our data. We overlay the single power-law (dashed line) with a slope $\alpha_3 = -0.61$ as suggested by Grimm *et al.* (2003) to describe a population of HMXBs.

Figure 2.

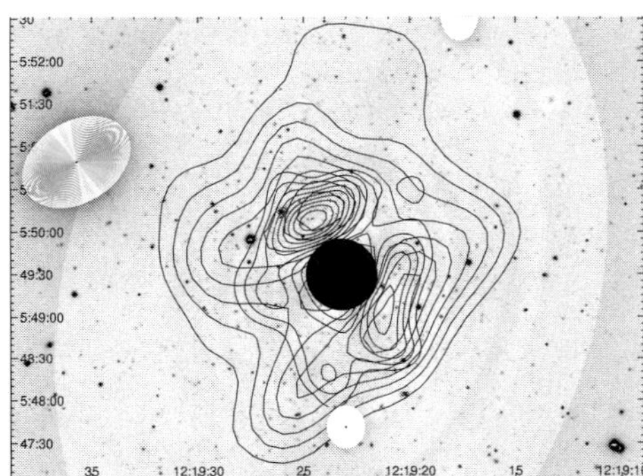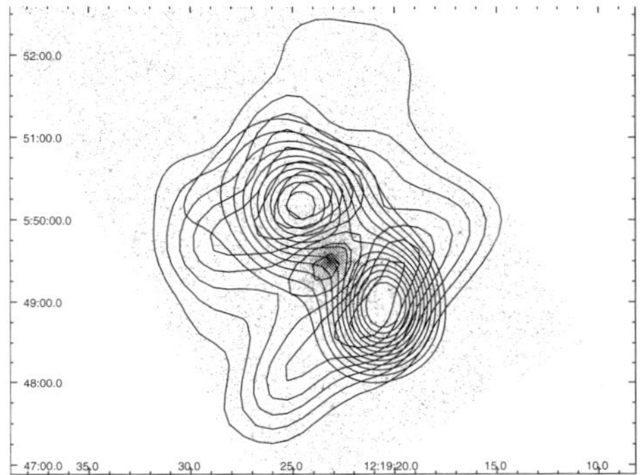

Figure 3. Iso-Density contours from the kernel density analysis. Left: comparison between the distributions of GCs in the INT (thin line) and HST (thick line). Right: comparison between the distributions of GCs in the INT (thin line) and of the *Chandra* X-ray sources (thick line).

3. Why is the X-ray source distribution so peculiar?

Since the spatial distribution of the sources is not that of the stellar light, we have looked at a correlation with the distribution of the GCs to which they are associated. The comparison between the sky projected distributions of the INT and *HST* GC candidates, estimated with the aid of an adaptive kernel density analysis (Silverman 1986) indicates a peculiar distribution of the GC population, with two main concentrations NE and SW of the nuclear region (Fig. 3-left). The same analysis applied to the distribution of the X-ray sources (Fig. 3-right) also shows two peaks. We have also simply computed the surface number density of sources for both GCs and X-ray sources in several regions, chosen to maximize the differences in the X-ray source population. The plots in Fig. 3 and Fig. 4 (left panel) show a remarkable similarity between the two distributions. We have also looked at the surface brightness distribution of hard photons (expected from individual sources of all luminosities) and found that the anisotropy might not be confined to the brightest sources (Fig. 4-center), since it is visible also when the detected sources are removed. This suggestion is also intriguing and worth of further investigation.

This therefore indicates that the anisotropy in the LMXBs is but a reflection of the peculiar spatial distribution of the GCs. It has been instrumental in discovering the large scale anisotropy of the globular cluster distribution as well, which to our knowledge is not observed in other systems.

Unfortunately this only implies that we have moved the focus onto the GC system, without uncovering the reason for this anisotropy. While we would like a confirmation of our GC list with better optical data, these results are already important for several reasons:

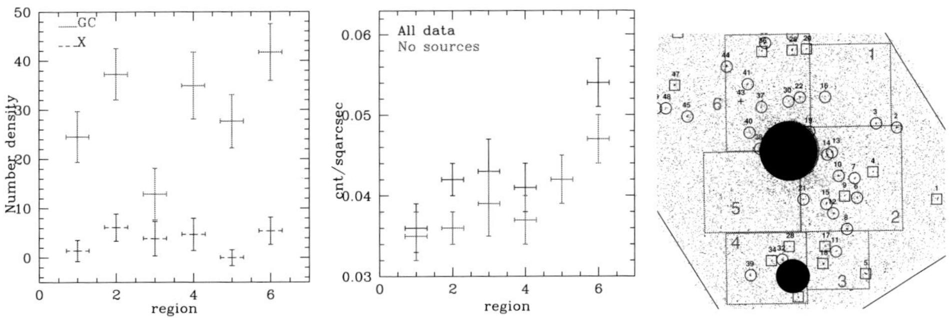

Figure 4. Distribution of the surface number density of GC candidates and of X-ray sources (Left) in the regions (shown on the Right) chosen to maximize the contrast in the X-ray sources. Central: photon surface brightness in 2-5 keV in the same area. Symbols indicate all data (black) and contribution of the (brighter) detected sources excluded (red: the lower symbol in each bin).

- The spatial correlation between the LMXB and the GC must be taken into account when studying the LMXB formation: all X-ray sources are concentrated in the same peculiar region of the galaxy, suggesting a strong link between the formations of "all" LMXB in GC.

- The spatial anisotropy of GC system of NGC 4261 may reflect a peculiar history of formation of this galaxy. Major mergers are thought to redistribute the original GC systems of the progenitor galaxies and trigger the formation of new ones in the center as well as along the tidal tails of the object arising from the merger. Whether this structure is the signature of such a phenomenon is unclear, given that all other optical parameters suggest a relaxed system. However, NGC 4261 could be an excellent target for further investigations, and in fact we are planning to follow this up with both improved data and a theoretical study with high-resolution simulations of galaxy mergers.

- Whether this is a unique system or whether this is the first discovered thanks to the X-ray image remains to be seen. Detailed X-ray data are available for only a limited number of nearby early type galaxies, so we could yet discover more examples of peculiar distributions of LMXB/GC in other, perhaps more distant systems.

References

Chiaberge, M., Gilli, R., Macchetto, F. D., Sparks, W. B., & Capetti, A., 2003, Ap.J. 582, 645

Colbert, J. W., Mulchaey, J. S., & Zabludoff, A. I., 2001, A.J. 121, 808

Garcia, A. M., 1993, A&A Suppl. 100, 47

Gavazzi, G., Bonfanti, C., Sanvito, G., Boselli, A., & Scodeggio, M., 2002, Ap.J. 576, 135

Gilfanov, M., 2004, MNRAS 349, 146

Giordano, L., Cortese, L., Trinchieri, G., Wolter, A., Colpi, M., Gavazzi, G., & Mayer, L. 2005 ApJ, in press (astro-ph/0508189)

Gliozzi, M., Sambruna, R. M., & Brandt, W. N., 2003, A&A 408, 949

Grimm, H. J., Gilfanov, M., & Sunyaev, R., 2003, MNRAS 339, 793

Kundu, A., Maccarone, T. J., & Zepf, S. E., 2002, Ap.J. 574, L5

Maccarone, T. J., Kundu, A., & Zepf, S. E., 2003, Ap.J. 586, 814

Nolthenius, R., 1993, Ap.J.Suppl. 85, 1

Sambruna, R. M., Gliozzi, M., Eracleous, M., Brandt, W. N., & Mushotzky, R., 2003, Ap. J. 586, L37

Schweizer, F. & Seitzer, P., 1992, A.J., 104, 1039

Silverman, B. W., 1986, Density estimation for statistics and data analysis, Monographs on Statistics and Applied Probability, London: Chapman and Hall, 1986

Zezas, A., Hernquist, L., Fabbiano, G., & Miller, J., 2003, Ap.J. 599, L73

Discussion

KIM: Are GCs bimodal in their optical colours? If so, is there any difference between blue & red GCs in their spatial distributions?

TRINCHIERI: The colour distribution is such that both red and blue globular clusters are included, although they don't seem to show a clear bimodal distribution as other galaxies. Unfortunately colours are available for the HST field only (where there are few X-ray sources!) and the blending between the two populations does not allow us a clear cut between red and blue to do a very good comparison of the spatial distributions. But we will look again.

Populations of High Energy Sources in Galaxies
Proceedings IAU Symposium No. 230, 2005
E. J. A. Meurs & G. Fabbiano, eds.

© 2006 International Astronomical Union
doi:10.1017/S1743921306008234

Multi-epoch Observations of LMXBs in Early-type Galaxies

Gregory R. Sivakoff[1]**, Andrés Jordán**[2,3]**, Adrienne M. Juett**[1]**,
Craig L. Sarazin**[1]**, and Jimmy A. Irwin**[4]

[1]Department of Astronomy, University of Virginia, P. O. Box 3818,
Charlottesville, VA 22903-0818, USA
email: grs8g@virginia.edu, ajuett@virginia.edu, sarazin@virginia.edu

[2]European Southern Observatory, Karl-Schwarzschild-Str. 2
85748 Garching bei München, Germany;
email: ajordan@eso.org

[3]Astrophysics, Denys Wilkinson Building, University of Oxford,
1 Keble Road, Oxford, OX1 3RH, UK

[4]Department of Astronomy, 909 Dennison Building, University of Michigan,
Ann Arbor, MI 48109-1042, USA;
email: jairwin@umich.edu

Abstract. *Chandra* observations of early-type galaxies have resolved large populations of low-mass X-ray binaries (LMXBs) in early-type galaxies. The majority of these observations have been snapshots on the order of a day or less. In our own Galaxy, LMXBs are known to exhibit a range of luminosity and spectral variability. Multi-epoch observations of early-type galaxies are just beginning to explore the regime of variability on timescales of days to years. We present results for NGC 4365 and NGC 4697, and compare them to the Milky Way.

Keywords. binaries: close, galaxies: elliptical and lenticular, galaxies: individual (NGC 4365), galaxies: individual (NGC 4697), X-rays: binaries, X-rays: galaxies.

---

1. Introduction

Milky Way (MW) low mass X-ray binaries (LMXBs) have been studied since the birth of X-ray astronomy (Giacconi *et al.* 1962). However, it is only with the sub-arcsecond resolution of the *Chandra X-ray Observatory* that the nature of the X-ray emission in early-type galaxies (E/S0s) can be studied in detail and that the very presence of LMXBs, suggested from spectral results by Kim, Fabbiano, & Trinchieri (1992), could be verified. Starting with the *Chandra* observation of NGC 4697 (Sarazin, Irwin, & Bregman 2000, 2001), the majority of X-ray emission in X-ray faint E/S0s has been resolved into sources whose properties are consistent with LMXBs.

Studies of LMXBs in the MW and in E/S0s are very complementary. Galactic LMXBs can be studied in great detail during both active ($L_x \gtrsim 10^{36}$ ergs s^{-1}) and quiescent ($\lesssim 10^{34}$ ergs s^{-1}) states across all wavelengths. From this, binary properties (e.g., donor type, compact object mass, orbital period, jet presence) can be determined, allowing for a better understanding of LMXB formation and evolution. However, there are several limitations in studying Galactic LMXBs: source distances are known for only a small subset, it is difficult to observe the whole Galaxy at once, absorption columns vary from source to source, and the size of the observed sample is limited.

For E/S0s, typical observations can only detect bright ($\gtrsim 10^{37}$ ergs s^{-1}), active LMXBs and such LMXBs cannot be studied in as great detail as in the MW. However, there

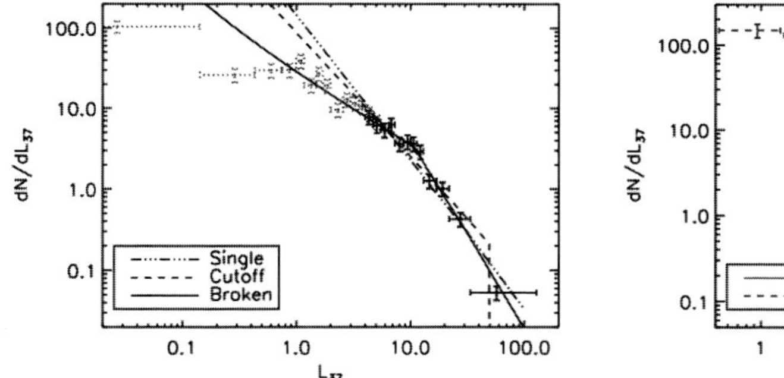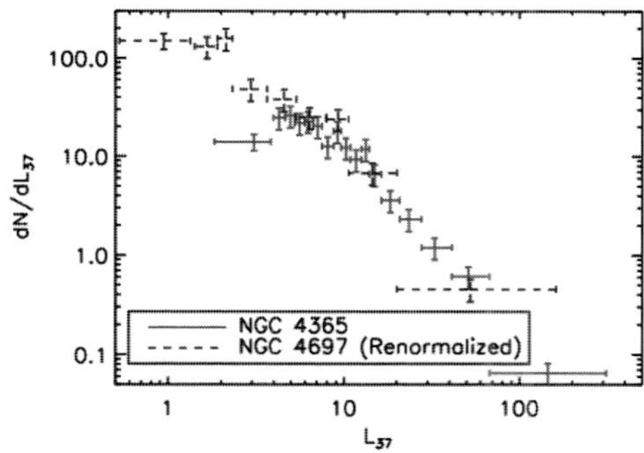

Figure 1. (Left) Completeness-corrected, instantaneous, differential luminosity function of five observations of NGC 4697. Only bins with solid lines (see text) were fit with the three power-laws displayed. (Right) A comparison of the differential luminosity function of the sum of all the observations of NGC 4365 and NGC 4697 (renormalized).

are distinct advantages to E/S0 observations. Since all LMXBs in an E/S0 share a common absorption column and distance, and most LMXBs in a nearby E/S0 fit in the field-of-view of *Chandra*, instantaneous luminosity functions (LFs) can be easily determined. Furthermore, E/S0s often have 50–200 sources brighter than $\sim 5 \times 10^{37}\,\mathrm{ergs\,s^{-1}}$, as opposed to the few bright sources in the Galaxy.

2. *Chandra* & *Hubble* Observations of NGC 4697 & NGC 4365

With this complementary nature in mind, we have performed multi-epoch observations of NGC 4697 and NGC 4365. Four new observations per galaxy increase the total exposure time on these galaxies from $\sim 40\,\mathrm{ks}$ for NGC 4697 (Sarazin, Irwin, & Bregman 2000, 2001) and NGC 4365 (Sivakoff, Sarazin, & Irwin 2003) to $\sim 200\,\mathrm{ks}$ each. These observations were designed to not only detect fainter LMXBs, but also open up the regime of LMXB variability studies in E/S0s through year timescales. The completed multi-epoch study of NGC 4697 is discussed in detail in Sivakoff *et al.* (2005, in preparation). The fifth and final observation of NGC 4365 is scheduled for 2005 November. Hubble observations of the centers of these galaxies (Côté *et al.* 2004; Jordán *et al.* 2004, 2005 in preparation), reveal the globular clusters (GCs), which are known to harbor large fractions of LMXBs (e.g., Angelini, Loewenstein, & Mushotzky 2001; Sarazin *et al.* 2003). Flanking fields of both galaxies will be observed in HST Cycle 14.

3. LMXB Luminosity Functions

Standardized reduction and flare removal were performed on all observations. With the increased exposures of the combined observations, 185 ks (NGC 4697) and 157 ks (NGC 4365), we detected 158 (NGC 4697) & 284 (NGC 4365) sources using CIAO WAVDE-TECT. For NGC 4697, we determined the count rates from PSF-scaled extraction source regions and the local background. With these photometric count rates, we determined the luminosities (assuming a 9.09 keV bremsstrahlung spectrum and correcting for vignetting, the PSF, and QE degradation) in individual observations. Since detection was performed in the combined observation of NGC 4697, we determined completeness from the combined counts of the five observations. We created the completeness-corrected, instantaneous, differential LF (Fig. 1 Left) and fit the sum of power-law models, $dN/dL \propto L^{-\alpha}$, and the background LF from Kim *et al.* (2004), for $4.0 \times 10^{37} < L_X < 1.3 \times 10^{39}$ ergs s^{-1}. A single power-law in that luminosity range was rejected at 99.5%, which also rejects cutoff power-laws with cutoffs beyond the upper luminosity bin. A cutoff power-law with a cutoff at $4.9 \pm 0.6 \times 10^{38}$ ergs s^{-1} and $\alpha = 1.5 \pm 0.1$ was only marginally rejected (88%). We note that Kim & Fabbiano (2004) found evidence for a possible break in the LF at this luminosity. A broken power-law was an acceptable fit with a break at $1.1 \pm 0.2 \times 10^{38}$ ergs s^{-1}, $\alpha_l = 0.8 \pm 0.3$, and $\alpha_h = 2.4 \pm 0.2$. We note that the broken power-law also goes through many of the low luminosity data points that we did not fit.

Since the observations of NGC 4365 are unfinished, we only roughly compare its LF to that of NGC 4697. In Fig. 1 (Right), we display the differential LF of the combined observations without completeness corrections, renormalizing the NGC 4697 LF by matching the number of sources with $L_X > 10^{38}$ ergs s^{-1} with that of NGC 4365. The rough shapes of the LFs match. Since NGC 4365 has almost twice as many sources as NGC 4697, we should be able to probe the LF of NGC 4365 in greater detail than that of NGC 4697.

4. GC-LMXB Connection

One emerging pattern of the GC-LMXB connection in E/S0s is that above 10^{38} ergs s^{-1}, $\approx 4\%$ of GCs contain an LMXB (Sarazin *et al.* 2003). With deeper observations we can explore this connection at lower luminosities. For instance, in NGC 4697, at our 3σ detection limit of $\approx 1.4 \times 10^{37}$ ergs s^{-1}, $\approx 8 \pm 2\%$ of GCs contain an LMXB within $1''$ after correcting for random matches. Among all detected sources ($L_X > 0.6 \times 10^{37}$ ergs s^{-1}), this rises to $\approx 10 \pm 2\%$. Since active LMXBs can have $L_X \gtrsim 10^{36}$ ergs s^{-1}, it is likely that the percentage of GCs with an active LMXB is even higher.

Over the history of X-ray astronomy, 12/150 MW-GCs have had an active LMXB (Liu, van Paradijs, & van den Heuvel 2001; Harris 1996). This corresponds to $\approx 8^{+3}_{-2}\%$ of GCs having contained an active LMXB. Since some LMXBs in Galactic GCs are transient, this is an upper limit to the instantaneous percentage of GCs containing active LMXBs.

It is likely that the percentage of GCs with an active LMXB in NGC 4697 will be higher than that of the MW. Given that NGC 4697 has a higher fraction of red GCs to blue GCs than the MW (Jordán et al. 2005) and that red GCs appear to be more likely to contain an LMXB (e.g., Sarazin *et al.* 2003), this result is not unexpected.

5. Variability Results

Milky Way LMXBs exhibit a wide range of variability, including transient outbursts, Type I bursts, and milder fluctuations. Such behavior is expected from LMXBs in E/S0s.

In Sivakoff, Sarazin, & Jordán (2005), we discussed the detection of flaring sources in NGC 4697 using a new technique. Two sources show ~ 1000 s flares with $L_{\rm bol} > 4 \times 10^{38}$ ergs s^{-1}. Although the timescale of the flares is similar to superbursts, the luminosity is higher than expected for a neutron star (NS). Furthermore, the recurrence timescale of flares ($\lesssim$ days) is much shorter than the approximately year-long timescale of Galactic superbursts. An even more perplexing source with recurrent ~ 100 s flares was also found (e.g., Fig. 2 left). The flaring in this source is most like that of LMC X-4. However, the flaring mechanism in LMC X-4 requires a very strong magnetic field (Moon, Eikenberry, & Wasserman 2003). Such an interpretation would require that accretion is the source of magnetic field decay and that the NGC 4697 source recently began accretion. The source is also similar to V4641 Sgr, a HMXB with a radio jet inclined only $\sim 15°$ to the line-of-sight. The flaring in the source in NGC 4697 could indicate that it is a micro-blazar.

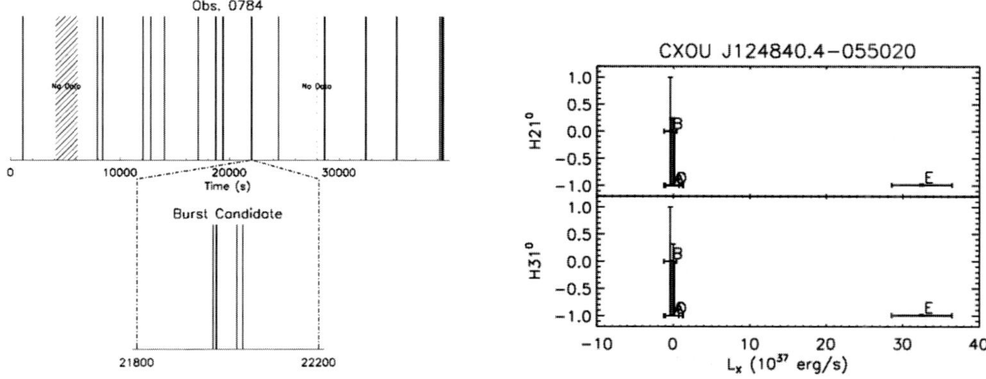

Figure 2. (Left) Impulse diagram indicating the time of arrival of photons in flaring source CXOU J124839.0-054750. (Right) Luminosity versus hardness ratio of a supersoft transient.

A variety of Galactic LMXBs exhibit long term variability (LTV), including Z/Atoll NS-LMXBs, soft X-ray transients, long-term transients like GRS1915+105, and black hole (BH) state changes. We detect LTV for 26/124 sources in NGC 4697. Of these sources, 11/26 are transient candidates. Assuming these transients are long-term transients, we calculate a mean outburst time of 105 yrs. One candidate is clearly a supersoft transient (Fig. 2 right). It goes from being undetected to $L_{\rm bol} \approx 8 \times 10^{38}$ ergs s^{-1} with a disk blackbody temperature of $\approx$0.14 keV. From its temperature, such a source could be interpreted as an intermediate-mass BH; however, this would require the source to be accreting well below its Eddington limit. We have also found evidence for a source showing a luminosity/spectral state transition. The source goes from being soft and faint to hard and bright, opposite of the conventional relationship in BH state transitions. It is possible that this discrepancy results from the softer bands used to calculate the hardness with *Chandra*; the *Chandra* soft state might result from the same power law spectrum as the *RXTE* hard state (Juett *et al.* 2005, in preparation).

Acknowledgements

Support for this work was provided by NASA through HST Award GO-10003.01-A and *Chandra* Awards GO5-6086X, GO4-5093X, AR3-4005X, GO3-4099X, and AR4-5008X.

References

Angelini, L., Loewenstein, M., & Mushotzky, R. F. 2001, *ApJ* 557, L35
Côté, P., *et al.* 2004, *ApJS*, 153, 223
Giacconi, R., Gursky, H., Paolini, F. R., & Rossi, B. B. 1962, *Phys Rev Lett* 9 439
Harris, W. E. 1996, *AJ*, 112, 1487
Jordán, A., *et al.* 2004, *ApJS*, 154, 509
Liu, Q. Z., van Paradijs, J., & van den Heuvel, E. P. J. 2001, *A&A*, 368, 1021
Kim, D.-W., *et al.* 2004, *ApJ*, 600, 59
Kim, D.-W. & Fabbiano, G. 2004, *ApJ*, 611, 846
Kim, D.-W., Fabbiano, G., & Trinchieri, G. 1992, *ApJ* 393, 134
Moon, D.-S., Eikenberry, & Wasserman, I. M. 2003, *Apj* 586, 1280
Sarazin, C. L., Irwin, J. A., & Bregman, J. N. 2000, *ApJ* 544, L101
Sarazin, C. L., Irwin, J. A., & Bregman, J. N. 2001, *ApJ* 556, 533
Sarazin, C. L., Kundu, A., Irwin, J. A., Sivakoff, G. R., Blanton, E. L., & Randall, S. W. 2003, *ApJ* 595, 743
Sivakoff, G. R., Sarazin, C. L., & Jordán, A. 2005, *ApJ* 624, L17
Sivakoff, G. R., Sarazin, C. L., & Irwin, J. A. 2003, *ApJ* 599, 218

After the 2nd Session. Gregory Sivakoff (in the middle) has just presented his contribution.

Conference Dinner – at the dessert. Notice the forest of glasses. From left to right: Kevin Nolan, Ines Brott, Brian McBreen, Laura Norci, Martin Elvis, Pepi Fabbiano, George Miley.

Session 4

Source classes that emerge from sampling over galaxies

Coffee break. Ronan McSwiney, Thierry Courvoisier and Andrey Bykov (left to right) in line
for coffee and refills.

Populations of High Energy Sources in Galaxies
Proceedings IAU Symposium No. 230, 2005
E. J. A. Meurs & G. Fabbiano, eds.

© 2006 International Astronomical Union
doi:10.1017/S1743921306008246

Nuclear sources in galaxies

M. Elvis

Harvard-Smithsonian Center for Astrophysics, Cambridge, Mass., USA
email: elvis@cfa.harvard.edu

Abstract. In the local Universe most massive black holes at the centers of galaxies are not luminous quasars. Is this because (1) they are starved of gas, (2) they accrete without emitting radiation, (3) they refuse to eat, ejecting the incoming material, or (4) they are storing up matter in an accretion disk to feast later?

With Chandra ACIS we have imaged a pilot sample of 6 nearby (D < 30 Mpc) elliptical galaxies chosen to be especially quiescent based on the careful optical spectroscopy of Ho, measured black hole masses (Mbh > 10(7)Msol), and with existing X-ray upper limits (Lx < 10(40)erg/s) implying far sub-Eddington accretion. In these galaxies we can measure, or limit, the diffuse hot interstellar medium, and so constrain the Bondi accretion rate.

Faint X-ray emission is detected at or around the nucleus in each galaxy. The morphology of these weak X-ray sources is complex. The X-ray colors of the sources can be determined, and a moderate quality spectrum for one was obtained. We discuss these results against the possible explanations of black hole quiescence.

On the other hand, a few percent of all galaxies shows evidence for nuclear activity and a brief review of the high energy emission from Active Galactic Nuclei is given.

Discussion

LIPUNOV: The stationary Bondi solution exist only in Keplerian gravitational potential. Did you include the contribution of the mass of central Globular Cluster? This is important for low massive Super Massive Black Holes (MBH < 10^6).

ELVIS: We almost resolve the Bondi radius with Chandra and do resolve it with HST so our estimates are quite robust.

COURVOISIER: 1. Where do you put the nucleus of our Galaxy in your discussion?
2. Do you consider the X-ray luminosity of the star population that is the source of gas to be accreted?

ELVIS: 1. Sgr A* is on our plots and is consistent with ADAF like solutions. I don't know if a comparable stellar mass loss rate has been calculated for the Milky Way but it would be a good thing to attempt.
2. The Stellar population is old and will have a negligible X-ray luminosity.

CHERNYAKOVA: Can the model of the accretion with low angular momentum of Beloborodov & Illarionov explain the observed AGN?

ELVIS: Low angular momentum accretion can always be invoked to accrete without radiation and should be investigated, however the mass loss from stars will show the angular momentum of the stellar population, so I suspect that this mechanism will not apply to this cold ISM component.

VÖLK: How would you estimate the possibility of short-term interruptions of accretion (like probably in our own Galactic Center) by SN explosions of the surrounding massive stars? You have not mentioned it in your list.

ELVIS: That is certainly possible in principle, but the galaxies we studied have only old stellar populations in their cores and so have no Supernovae; moreover this is unlikely to be the main mechanism keeping SMBH inactive since it would have to be effective 99% of the time.

MACCARONE: Have you considered whether the accreted gas in AGN might pile up in a disk and undergo occasional outbursts, instead of being blown out in winds, and are there any AGN with luminosities well in excess of what their gas supplies suggest they should be?

ELVIS: I think the bright AGN will make it difficult to measure the gas supply in those systems, so it is difficult to say whether those outbursting systems exist from observations. As for whether it is likely theoretically, I find the idea and the analogy to X-ray binaries intriguing, and have done some work with Aneta Siemiginowska on unstable disks. However, it is probably difficult to get duty cycles so small that 99% of galactic nuclei will be quiescent.

Populations of High Energy Sources in Galaxies
Proceedings IAU Symposium No. 230, 2005
E. J. A. Meurs & G. Fabbiano, eds.

© 2006 International Astronomical Union
doi:10.1017/S1743921306008258

Core X-ray sources in the Local Group galaxies

E. J. A. Meurs

Dunsink Observatory, Dublin Institute for Advanced Studies, Castleknock, Dublin 15, IRL
email: ejam@dunsink.dias.ie

Abstract. As the nearest galaxies around us, the Local Group systems offer especially good opportunities for observations of their nuclear X-ray radiation. Certain or possible nuclear X-ray sources in the Local Group suggest a minimum luminosity for activity to become manifest.

Keywords. galaxies: active galactic nuclei, Local Group, X-rays: galaxies.

1. Introduction: prevalence of active cores in galaxies

Among the many individual X-ray sources that nowadays are seen in neighbouring galaxies, the nuclear cores take a special position. Not only do they tend to be the centralmost source, but all other individual sources are products of stellar processes while the nuclear cores constitute the (usually) one and only source in a galaxy that is instead related to the supermassive black holes that power QSOs on a cosmic scale. There may incidentally be a kind of intermediate class of sources (that is, between stellar source and core source), the UltraLuminous X-ray sources that have also been discussed extensively during this Symposium, but their precise nature is still not conclusively established. There is still another type of source, which is mostly the result of stellar processes, diffuse X-ray emission, but we will not consider such emission here, as not being individual sources.

Nearby galaxies are usually no QSOs, fortunately enough, and from the nearest galaxies to the luminous QSOs far away spans a large range in level of nuclear activity. Upon closer examination, some form of nuclear activity can often be recognized in the centres of many galaxies around us. It is usually thought that remnant supermassive black holes, left behind at a period of higher nuclear activity some time in the past, make their presence known in these instances, when they receive fuel in the form of gas or stars.

The recognition of weak forms of nuclear activity leads naturally to questions like: how frequent is nuclear activity present among galaxies; do all types of galaxies exhibit nuclear activity; is a minimum size (i.e., mass) required for galaxies in order to harbour a nuclear core; till what low level of energy output can active cores be recognized. Obviously, the closer the galaxies, the easier to address these questions. Therefore, in this contribution, we restrict ourselves to the very nearest galaxies, those of the Local Group. Their small distances yield the best spatial resolution, notably important when low flux levels are reached at which cores would be comparable to all kinds of stellar objects.

Using X-rays is efficient in the search for low-level cores. Nuclear cores emit notably at high energies, as is known from e.g. Seyfert and Liner galaxies. Other objects in the galaxies are generally less likely to emit at X-rays, thus compared with all stellar objects in a galaxy there are less competitors for our attention.

2. The Local Group sample

The Local Group (LG) contains at least some 36 galaxies, from the big spirals (M31; our own Galaxy) down to small and loose aggregations of stars (e.g. the Draco dwarf galaxy). The number of LG members may vary from author to author. Important for our purposes is that galaxies of nearly all sorts are present: spirals (our Galaxy; M31; M33), ellipticals (e.g. NGC147 and NGC185) and irregulars (e.g. NGC6822); but no lenticular. Also, within these classes they show a range of sizes, although notably there is no normally sized elliptical. Below we will exclude the two Magellanic Clouds, which are very extended objects without clear centres in their stellar distribution.

3. Existing X-ray surveys of the Local Group

(a) Einstein

Results were obtained for about six LG members, this work has been described by Helfand (1984) and Fabbiano (1989). The presence of a strong central source in M33 was noted.

(b) ROSAT

A good number of the LG galaxies was observed with ROSAT. This was the basis for an X-ray survey of their cores (Zang & Meurs 2001; Meurs 2003).

(c) Chandra and XMM-Newton

The two currently flying X-ray satellites have so far yielded results for some of the LG galaxies, and several more are in the process of being observed, being analysed, or being published.

In the sense of providing an assessment of the cores of LG galaxies, the most comprehensive overview to date is by Zang & Meurs (2001), employing the ROSAT database. For some of these galaxies, Chandra and/or XMM-Newton data improve on the ROSAT results. In the next Section, therefore, we review the currently available X-ray data on LG galaxy cores. In the Section following after that, a general discussion of the LG cores is given, based on Zang & Meurs with updates where relevant.

4. Cores of individual Local Group galaxies

(i) Galactic Centre

Using the ROSAT PSPC, Predehl & Trümper (1994) found a source coincident with Sgr A* within 10 arcsec ($\Delta_{XO} = 7.8$ arcsec), visible only above 1.2 keV. With the greater spatial resolution of Chandra, Baganoff *et al.* (2003) resolve the source seen with ROSAT into a number of components, the strongest of which is only 0.27 arcsec from the radio position as determined interferometrically by Reid *et al.* (1999). The flux in the 0.1–2.4 keV band is about two orders of magnitude less than was found with ROSAT.

A flare of limited duration and magnitude may have occurred, but this does not seem to be a very regular phenomenon (see also Baganoff *et al.* 2001). Insights on the activity history of the Galactic Centre are found in Koyama (this volume).

(ii) M31

One of the sources discerned with the ROSAT HRI was coincident with the nucleus of M31 (Primini *et al.* 1993). This source could be resolved into 4 or 5 sources by Chandra (Garcia *et al.* 2000). Chandra and HST images have subsequently been registered to an accuracy of 0.1 arcsec (Garcia *et al.* 2005), resulting in a 2.5σ detection of a source directly next (West) to the northernmost of the nuclear Chandra sources in Garcia *et al.* (2000). The coincidence with the nuclear radio source (Crane *et al.* 1992) is within 1 arcsec, the X-ray luminosity in 0.1–2.4 keV is down to $\log L_X = 35.93$.

Table 1. Core sources in LG galaxies

Galaxy	log L$_X$ (0.1–2.4 keV)	Δ_{XO} (arcsec)	M$_{SMBH}$ (M$_\odot$)
M33	39.01	<1	<1.5 10^3
M32	36.45	0.7	2.5 10^6
M31	35.93	<1	3 10^7
GC	34.66	0.27	2.6 10^6

(iii) M33

The bright X-ray source in the nucleus of M33 has been known since the Einstein observations (e.g. Fabbiano 1989). It is the most luminous source in the LG (M33 X-8) and qualifies as an UltraLuminous X-ray source. Analysis of ROSAT observations led to a suggested long binary period (Dubus *et al.* 1997), which in later work could not be confirmed (Parmar *et al.* 2001, using BeppoSAX). The currently most popular explanation probably is a microquasar-like X-ray binary (Dubus & Rutledge 2002), supporting evidence has been derived from spectral analyses and short term variability at X-rays (La Parola *et al.* 2003; Foschini *et al.* 2004; Dubus *et al.* 2004). The close coincidence of M33 X-8 with the nucleus of M33, to within 0.6 arcsec (Dubus & Rutledge 2002), is nevertheless very noticeable (Schulman & Bregman 1995). A putative central black hole may have a mass in such an uncommon range (< 1500 M$_\odot$, Gebhardt *et al.* 2001) that spectral models pursued so far are not entirely appropriate. Therefore we include, for the time being, the M33 central source as a possible nuclear core, as in Zang & Meurs (2001).

(iv) M32

Based on analysis of ROSAT PSPC and HRI observations, Zang & Meurs (1999) advanced reasons that the bright central source in this small elliptical galaxy could conceivably be a low level active nucleus. Using Chandra data, Ho *et al.* (2003) report a weak source at 0.7 arcsec from the 2MASS position of M32. Adopting the latter result, the luminosity in the 0.1–2.4 keV band is log L$_X$ = 36.45.

(v) NGC6822

Zang & Meurs (2001) found an intriguing, weak source in a ROSAT HRI observation, only $\sim$7 arcsec away from the nominal optical position of the galaxy. NGC6822 is however an irregular galaxy, without well-defined centre, thus it is difficult to attribute this source to a core. We glanced at a later Chandra observation of this galaxy (J. Hartwell, private communication), but this has not very good statistics.

(vi) WLM

This is also an irregular galaxy, posing the same problems as for NGC6822. ROSAT HRI data indicate a source only $\sim$8 arcsec from the nominal optical position (Zang & Meurs 2001).

Thus, X-ray detection of core sources in LG galaxies include the Galactic Centre, M31 and M32, and possibly M33. The central sources in NGC6822 and WLM require more extensive observational studies than available to date.

5. Discussion: core X-ray sources in the Local Group

The recent advances in high resolution X-ray imaging of LG galaxies as quoted in the previous Section leave the main conclusions of Zang & Meurs (2001) unaffected. The current status of core X-ray sources in the LG is summarized in Table 1; the last column gives the estimated mass of a central supermassive black hole.

The most important results on galaxy cores in the LG are:

(1) Sgr A* in the Galactic Centre is still the weakest detected nuclear core within the LG.

(2) The nucleus of M33 remains the strongest nuclear source within the LG – if indeed it is one.

(3) All three spiral galaxies in the LG have a nuclear X-ray source (when including M33).

(4) Only the most luminous (dwarf) elliptical, M32, has a nuclear source.

(5) None of the irregular galaxies features a recognizable core source.

(6) Galaxies seem to need $L_B > 10^8$ $L_{B,\odot}$ for having a core source.

The LG conclusions are based on X-ray observation of 25 out of (ca.) 36 galaxies. Apart from M33, the detected core sources exhibit luminosities as for stellar X-ray sources. It will be interesting to obtain similarly good coverage (as to number of members observed) for next following nearby groups. The Chandra survey of nearby galaxies by Ho *et al.* (2001) emphasizes AGN candidates. Another way to recognize non-AGN cores, albeit very incidentally, is to observe stellar disruption flares that temporarily highlight otherwise dormant supermassive black holes in galaxy centres (e.g. Cunniffe 2003).

References

Baganoff, F.K., Maeda, Y., Morris, M., *et al.* 2003, ApJ 591, 891

Crane, P.C., Dickel, J.R., & Cowan, J.J. 1992, ApJ 390, L9

Cunniffe, J. 2003, PhD Thesis, Trinity College Dublin, Ireland

Dubus, G., *et al.* 1997, ApJ 490, L47

Dubus, G. & Rutledge, R.E. 2002, MN 336, 901

Dubus, G., Charles, P.A., & Long, K.S. 2004, A&A 425, 95

Fabbiano, G. 1989, ARAA 27, 87

Foschini, L., Rodriguez, J., Fuchs, Y., *et al.* 2004, A&A 416, 529

Garcia, M.R., Murray, S.S., Primini, F.A., *et al.* 2000, ApJ 537, L23

Garcia, M.R., Williams, B.F., Yuan, F., *et al.* 2005, ApJ 632, 1042

Gebhardt, K., Lauer, T.L., Kormendy, J., *et al.* 2001, AJ 122, 2469

Helfand, D. 1984, PASP 96, 913

Ho, L.C., Feigelson, E.D., Townsley, L.K., *et al.* 2001, ApJ 549, L51

Ho, L.C., Terashima, Y., & Ulvestad, J.S. 2003, ApJ 589, 783

La Parola, V., *et al.* 2003, ApJ 583, 758

Meurs, E.J.A. 2003, A&A 408, 95

Parmar, A.N., Sidoli, L., Oosterbroek, T., *et al.* 2001, A&A 368, 420

Predehl, P. & Trümper, J. 1994, A&A 290, L29

Primini, F.A., Forman, W., & Jones, C. 1993, ApJ 410, 615

Reid, M.J., *et al.* 1999, ApJ 524, 816

Schulman, E. & Bregman, J.N. 1995, ApJ 441, 568

Zang, Z. & Meurs, E.J.A. 1999, NewA 4, 521

Zang, Z. & Meurs, E.J.A. 2001, ApJ 556, 24

Discussion

LIPUNOV: I remember that before the Chandra era there was some excess of the BH candidate systems in the central part of our Galaxy. What is the situation now?

MEURS: The number of candidate supermassive black holes in the Galactic Centre is essentially down to one candidate, for which there are indications at a variety of wavelengths.

Populations of High Energy Sources in Galaxies
Proceedings IAU Symposium No. 230, 2005
E. J. A. Meurs & G. Fabbiano, eds.

© 2006 International Astronomical Union
doi:10.1017/S174392130600826X

X-ray nuclear emission of a sample of LINER Galaxies

O. González-Martín[1]†, J. Masegosa[1] & I. Márquez[1]

[1]Instituto de Astrofísica de Andalucía (CSIC), Apdo 3004, 10080 Granada, Spain

Abstract. We report the results from an homogeneous analysis of the X-ray (ACIS-S/Chandra) data available for a sample of 52 LINER galaxies. The X-ray morphology has been classified attending to their nuclear compactness in the hard band (4.5–8.0 keV), into 2 categories: AGN-like nuclei (with a clearly identified unresolved nuclear source) and Starburst-like nuclei (without a clear nuclear source). 60% of the total sample are classified as AGNs, with a median luminosity of $L_X(2.0–10.0\ \mathrm{keV}) = 2.5 \times 10^{40}$ erg s^{-1}, which is an order of magnitude higher than that for SB-like nuclei. The spectral fitting allows to conclude that most of the objects need a non-negligible thermal contribution. When no spectral fitting can be performed (low signal-to-noise ratio), the Color-Color diagrams allow us to compute physical parameters such as density column, temperature of the thermal model or spectral index and therefore to analyze the origin of the X-ray emission. All X-ray morphology, spectral fitting and Color-Color diagrams allow conclude that a high percentage of LINER galaxies host AGN nuclei.

Keywords. galaxies, AGN, LINER, X-ray, Chandra.

---

1. Introduction & Sample

LINERs are very common in the nearby universe. Pioneering works already estimate that at least 1/3 of all the spiral galaxies are LINERs (Heckman *et al.* 1980). More than two decades after they were clasified, there is still an ongoing strong debate on the origin of the energy source in LINERs, with two main alternatives for the ionizing source being explored: either it is a low luminosity AGN (Filippenko & Halpern, 1984), or it is of thermal origin from massive star formation (Filippenko & Terlevich, 1992) and/or from shock heating mechanisms resulting from the massive stars evolution (Fosbury *et al.* 1978 and Dopita 1976). The search for a compact X-Ray nucleus in LINERs is indeed one of the most convincing evidences about their AGN nature. The excellent resolution of Chandra allows an investigation of the X-Ray nuclear properties of these galaxies.

All the 476 LINER galaxies in the compilation by Carrillo *et al.* (1999) have been searched by coordinates in Chandra archives, with 137 galaxies having public ACIS data. To minimize any bias due to incompleteness, only the 65 objects with high enough exposure times and therefore high S/N ratio have been selected (t > 10000 seconds). The final sample, after optical re-identification as LINER nuclei, comprises 52 galaxies.

2. Data process

The data products were analyzed in an uniform, self-consistent, manner using *CXC Chandra Interactive Analysis of Observations* (CIAO‡) software version 3.1. The spectral analysis was done with *XSPEC¶* (version 11.3.2). Level 2 event data from ACIS

---

† e-mail address: omaira@iaa.es
‡ See http://asc.harvard.edu/ciao
¶ See http://cxc.heasarc.gsfc.nasa.gov/docs/xanadu/xspec/

instrument have been extracted from Chandra archive. Time intervals with high background levels have been excluded, using task *lc_clean.sl*†† in source-free sky regions of the same observation.

3. Spectral fitting

For the source selection, we made use of the nuclear positions from NVSS and 2MASS data base. Nuclear spectra were extracted using regions defined to include as many photons coming from the source as possible, but at the same time minimizing contamination from nearby sources and background. The background region was defined as either a source-free circular annulus or several circles surrounding each source, in order to take into account the spatial variations of the diffuse emission and to minimize effects related to the spatial variation of the CCD response. In order to use the χ^2 statistic, the data were grouped to include at least 20 counts per spectral bin, before background substraction. For the spectral fitting any events with energies above 10.0 keV or below 0.5 keV are excluded.

The spectra in the 0.5–10.0 keV passband were modelled with a single [MEKAL (ME), Raymond-Smith (RS) or Power Law (PL)] component first and second with a two component [ME+PL or RS+PL] model. Single models are representative for sources dominated either by thermal emission (such as Supernova Remnants and galactic bubles), or non-thermal emission (such as accreting compact objects), and two component models correspond to composite objects. Note that number counts were insufficient to employ detailed spectral fitting in 28 out of 52 objects.

Attending to the spectral fittings the use of two components models better reproduce the X-ray energy distribution of most of these 24 objects, indicating a non-negligible non-thermal contribution in our sample.

4. Luminosities

The luminosities and fluxes of the individual nuclear sources have been computed based on the best-fit model for the 24 galaxies above‡‡.

We have done an empirical calibration from these 24 objects with high signal-to-noise ratio from flux estimation using PIMMS¶¶ assuming an intrinsic power law slope of 1.8, corrected for Galactic absorption ($N_H = 3 \times 10^{20}$ cm^{-2}). In Fig. 1 (left) our estimated luminosity is plotted versus the value obtained from the direct integration of the spectral energy distribution. The luminosities are well correlated within a factor of three of the real one. The 2.0–10.0 keV luminosities are therefore provided for the whole sample, using the spectral energy distribution fit, when it is available, and from this calibration otherwise.

5. Morphological Classification

Since we focus our attention in the nuclear sources, no attempt has been made to fully characterize the flux and the spectral properties (when possible) of the extranuclear sources whose study is out of the scope of the paper. We have classified the nuclear morphology attending to the compactness in the hard band (4.5 to 8.0 keV). Therefore the sample has been grouped into 2 main categories:

†† See http://cxc.harvard.edu/ciao/download/scripts/
‡‡ Throughout this paper we use a cosmological constant of $H_o = 75$ km s^{-1} Mpc^{-1}
¶¶ See http://heasarc.gsfc.nasa.gov/Tools/w3pimms.html

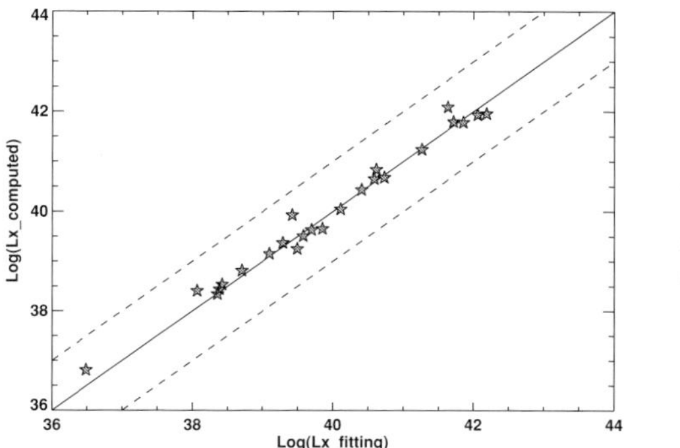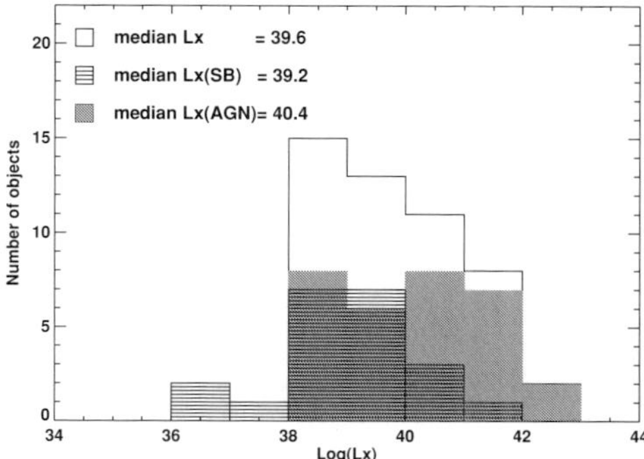

Figure 1. (left): Luminosity estimation versus real luminosity. Dashed-lines are 10 times above and below real value, respectively and continuous line indicates the best luminosity estimation. (right): Luminosity histogram for the whole sample (empty histogram), Objects clasified as AGN-like (grey histogram) and Starburst -like objects (dashed histogram).

- **AGN-like nuclei:** We include all the galaxies with a clearly identified unresolved nuclear source in the hard band. 59.6% (31/52) has been clasified as AGN-like nuclei in our sample and their median luminosity is $L_X(2\text{--}10\ \mathrm{keV}) = 2.5 \times 10^{40}\mathrm{erg\ s}^{-1}$ (Fig. 1, right).

- **Starburst-like nuclei (SBs):** Here we include all the objects without a clear nuclear source in the hard band. 40.4% of the sample of LINERs fall in this classification and their median luminosity is $L_X(2\text{--}10\ \mathrm{keV}) = 1.0 \times 10^{39}\ \mathrm{erg\ s}^{-1}$ (Fig. 1, right).

6. Color-Color Diagrams

The X-ray colors have also been studied for all the nuclear sources in our sample. Colors have been defined as the ratio of counts observed in the following energy bands: 0.6 to 0.9, 0.9 to 1.2, 1.2 to 1.6, 1.6 to 2.0, 2.0 to 4.5, and 4.5 to 8.0 keV. The bands were chosen in order to maximize the detection as well as to obtain a good characterization of the spectra. In the last energy band, the range from 6.0 to 7.0 keV has been excluded to avoid the possible contamination due to the FeK emission line. Therefore, 3 colors has been defined (Q_A, Q_B and Q_C) as Q=(Hard-Soft)/(Hard+Soft). Synthetic colors have been computed for PL, RS and PL+RS models by using PIMMS (see Fig. 2).

We have only considered the data with error less than 30%. In the less energetic diagram (Q_A versus Q_B, Fig. 2, left) there are only a few objects highly obscured. Note that these objects has been clasified as AGN-like objects. Most of the objects are in the combination grid with low column density, where AGN and SB like nuclei appear together.

A handful of objects lie quite far of any grid; most of these objects have been studied in the literature as hosting a radio-jet, an additional contribution that has not been taken into a account in our fitting. In the Q_B versus Q_C plot (Fig. 2, centre) we can see that Q_C is a good AGN activity estimator. However, not only a few objects classified as SB-like have a high Q_C, but also most of the objects classified as SB are not in the thermal model grid. This might be indicative that an additional non-thermal mechanism is needed to fit the spectral energy distribution.

Therefore, the use of Color-Color diagrams allows to also analyze the properties of the nuclear sources for which the spectral fitting is not possible.

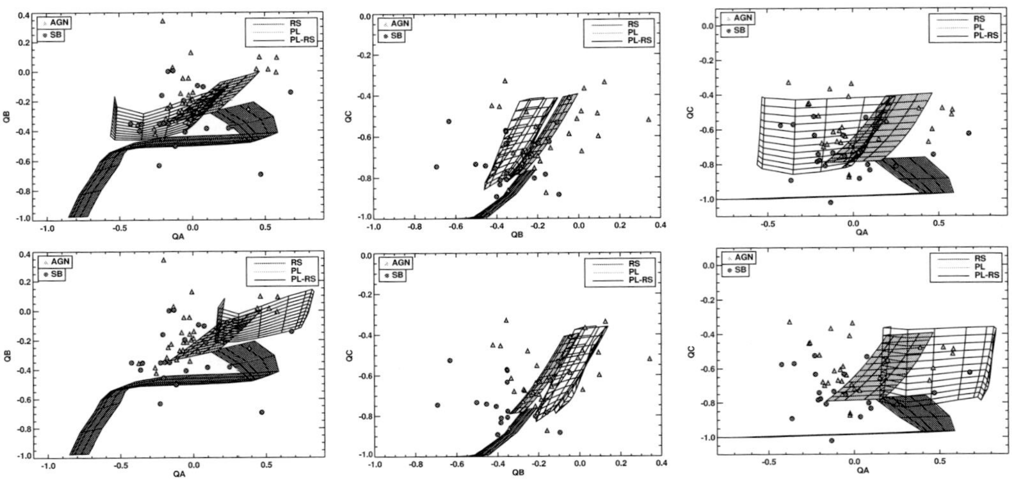

Figure 2. Color-color diagrams for a RS model (dark grey filled grid), PL model (light grey filled grid) and combinated model (empty grid) for $N_H = 10^{20}$ cm^{-2} (top) and $N_H = 10^{22}$ cm^{-2} (bottom). QB versus QA (left), QC versus QB (centre) and QC versus QA (right). Light-grey triangles are AGN-like objects and dark-grey circles are SB-like objects. The grids were calculated for $\Gamma = 0.4$ to 2.6 for PL model, for $kT = 0.4$ to 4.0 keV for the RS model and $N_H = (0.5 - 30.) \times 10^{20}$ cm^{-2} in the single model case.

7. Conclusions

In this paper we have described the detailed analysis of the X-ray spectral properties of the nuclear sources in an optically classified LINER galaxy sample, observed with Chandra ACIS-S. Our findings are summarized below:

- Morphologically, 50% of LINERs have been classified as AGN-like candidates, with median luminosity 10 times higher than that of SB-like objects.

- Both thermal and non-thermal contributions are required for the spectral fitting of most of the objects. Color-Color diagrams have confirmed this result.

- An empirical calibration for estimating X-ray luminosities ($L_X(2.0$–10.0 keV$)$) has been done based on total counts, which allows a reliable estimation when the spectral fitting is not possible.

- Color-Color diagrams are a valid tool to estimate physical parameters, specially interesting to be used when the spectral fit is not possible.

References

Carrillo, R., Masegosa, J., Dultzin-Hacyan, D., & Ordoñez, 1999, Rev. Mex. A&A, 35, 18

Dopita, M. A., 1976 ApJ, 209, 395

Filippenko, A. V. & Halpern, J. P. 1984 ApJ, 285, 458

Filippenko, A. V. & Terlevich, R. 1992 ApJ, 397, L79

Fosbury, R. A. E., Mebold, U., Goss, W. M., & Dopita, M. A. 1978 MNRAS, 183, 549

Heckman, T. M., Crane, P. C., & Balick, B., 1980, A&AS, 40, 295

Ho, L. C., Feigelson, E. D., Townsley, L. K., Sambruna, R. M., Garmire, G. P., Brandt, W. N., Filippenko, A. V., Griffiths, R. E., Ptak, A. F., & Sargent W. L. W., 1985, ApJ, 549, L51

Discussion

ELVIS: Is your definition of starburst based just on seeing extended X-ray emission? If so it is important to note that some extended X-ray emission is related to the nucleus as e.g. in NGC4151.

GONZALEZ-MARTIN: Our starburst definition is only based on the non existence of a point-like source in the hard band (4.5-8.0 KeV).

MEURS: Some of your LINER sample may also have been investigated with ISO for the IR-based AGN versus starburst separation. If there is such overlap, to what degree do your results agree with the IR diagnostics?

GONZALEZ-MARTIN: It is difficult to draw conclusion based in the present sample. Since our ACIS sample has been taken from the archive, when you compare with ISO data few objects have been observed. Therefore any conclusion based in the comparison between ISO and Chandra data need to be taken very carefully due to the bias from the way they have been selected.

MACCARONE: Have you looked at the radio properties of these objects to see if e.g. the radio to X-ray flux ratios or radio spectral indices are systematically different between AGN dominated LINERs and starburst dominated LINERs?

GONZALEZ-MARTIN: We haven't yet had a chance to look at the radio data.

Populations of High Energy Sources in Galaxies
Proceedings IAU Symposium No. 230, 2005
E. J. A. Meurs & G. Fabbiano, eds.

© 2006 International Astronomical Union
doi:10.1017/S1743921306008271

Very High Energy Observations of BL Lacs

John Quinn, Peter Cogan, Michael K. Daniel, David J. Fegan, Stephen Gammell and Andrew McCann

School of Physics, University College Dublin, Belfield, Dublin 4, Ireland
email: john.quinn@ucd.ie

Abstract. The status of Very High Energy (VHE, $E > 50$ GeV) gamma-ray observations of BL Lacertae objects is presented. The catalogue of well-established BL Lacertae objects detected at VHE energies contains seven members, and there have been recent reports of the detection of another four. All are nearby, X-ray bright sources. The temporal, spectral and broadband multi-wavelength properties of the sources are reviewed and possible implications for the gamma-ray production mechanism discussed. The most recent detections provide more stringent constraints on the cosmic extragalactic background light level and imply that the Universe is more transparent to VHE gamma radiation than previously thought.

Keywords. BL Lacertae objects: individual, gamma rays: observations.

1. Introduction

The maturing of the Imaging Atmospheric Cherenkov Technique (IACT) has opened a new astronomical window on the cosmos with observations in the 50 GeV – 50 TeV now routinely being made with ground-based instruments. Large optical reflectors are used to image the Cherenkov light emitted by extensive air showers in the atmosphere and off-line image analysis techniques distinguish between gamma-ray and charged cosmic-ray events. The technique has been used to detect in excess of 30 sources of VHE gamma rays, including 11 BL Lacs. For an introduction to the field of ground-based gamma-ray astronomy see Catanese and Weekes (1999) and Cogan *et al.* (2005).

BL Lac objects are a subset of the blazar class of Active Galactic Nuclei, and are characterised by a featureless optical continuum while exhibiting rapid variability. The blazar class, which also contains Flat-Spectrum Radio Quasars (FSRQs), are believed to be powerful radio galaxies which have their jets oriented towards the Earth. Blazars are very prominent in the gamma-ray sky; the EGRET detector on board the Compton Gamma-Ray Observatory detected 50 such objects above 100 MeV, (almost 40% of the entire EGRET third catalogue), 14 of which are BL Lacs with the rest being FSRQs. Thus, the EGRET sources were strong candidates to be VHE emitters and were amongst the first objects to be investigated with IACT instruments.

2. VHE Observations

2.1. *The VHE Catalogue*

There are seven well-established and well-studied BL Lacs in the VHE band. These are listed in Table 1. In addition, at the recent International Cosmic-Ray Conference in India an additional four detections (Table 2) were announced. All belong to the sub classification called HBL (High Frequency BL Lacs) which refers to the x-ray emission from these objects showing a peak at keV energies. The first objects detected were the closest (Markarian 421 and Markarian 501) but as more extensive surveys were carried

Table 1. Established VHE BL Lacs

Catalogue Name	Source	Date	Group	Type	Redshift
TeV 1104+3813	Markarian 421	1992	Whipple	HBL	0.031
TeV 1429+4240	H1426+428	2002	Whipple	HBL	0.129
TeV 1654+3946	Markarian 501	1995	Whipple	HBL	0.033
TeV 2000+6509	1ES1959+650	1999	Telescope Array	HBL	0.048
TeV 2005-489	PKS 2005-498	2005	HESS	HBL	0.071
TeV 2159-3014	PKS 2155-304	1999	Durham	HBL	0.116
TeV 2347-5142	1ES2344+514	1997	Whipple	HBL	0.044

Table 2. Recently announced VHE BL Lacs

Source	Date	Group	Type	Redshift
PKS 2005-489	2005	HESS	HBL	0.071
H 235-309	2005	HESS	HBL	0.165
1ES 1101-232	2005	HESS	HBL	0.186
1ES 1218+304	2005	Magic	HBL	0.182

out and increasingly sensitive instruments came on-line the horizon was extended. The most distant BL Lac detected at VHE energies is 1ES 1101-232 at $z = 0.186$.

2.2. *Flux Variability*

As at other wavelengths, the BL Lac emission at VHE energies is characterised by variability. The time scales for significant flux change range from years to less than one hour in some cases. Figure 1 (top) illustrates this for two sources; Markarian 421 and Markarian 501. In 1996 Markarian 421 was observed to undergo a series of flares which were remarkable for their amplitudes as well as their rapidity; on May 15 Markarian 421 went from being virtually undetectable to become the brightest known object at VHE energies in the space of less than an hour, and then rapidly faded to a low flux state. In 1997 Markarian 501, after a few years of relative inactivity, underwent a series of rapid flaring with dramatic month- and day-scale flaring with evidence for variability on a time-scale of hours. Monitoring of all the other established VHE BL Lacs also revealed significant variability with little evidence of any steady base-line emission.

2.3. *Spectral Properties*

The VHE energy spectra can all be fit either by power laws or power laws with exponential cutoffs:

$$\frac{\mathrm{d}N}{\mathrm{d}E} \propto E^{-\Gamma}\, e^{-E/E_0} \tag{2.1}$$

Measured spectral indices of the established BL Lacs range from $\Gamma = 1.9$ to $\Gamma = 3.5$ with a tendency for a harder spectrum with increasing distance (Schroedter, 2005). The cut-off energy for Markarian 421 has been measured to be ≈ 4.3 TeV (Krennrich *et al.*, 2002), while for Markarian 501 has been found to be ≈ 6 TeV (Aharonian *et al.*, 1999). In addition, Markarian 421 was monitored at a variety of different flux levels and spectral analysis revealed that there was a correlation between spectral index and flux level, with harder indices associated with higher flux levels (Krennrich *et al.*, 2002). Also, it was found that there was no change in the cut-off energy with flux.

Broadband multiwavelength observations have been carried out on several of the established VHE BL Lacs. The general trend appears to be that there is some correlation between the x-ray and gamma-ray emission. However, some 'orphan' flares have been

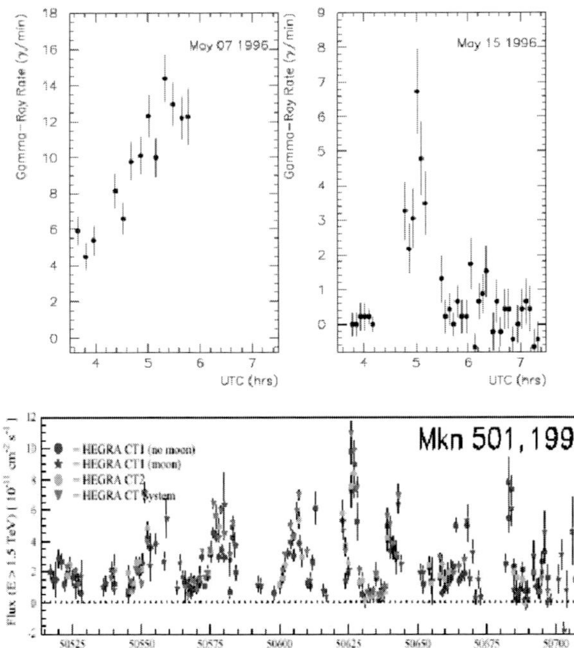

Figure 1. Top: Two remarkable flaring episodes from Markarian 421 in 1996 (Gaidos *et al.*, 1996). Bottom: daily fluxes from Markarian 501 in 1997 (Kranich *et al.*, 2001).

observed (e.g. Krawczynski *et al.*, 2004), where a flare is seen in either the X-ray or gamma-ray band but not both simultaneously. In general the broadband spectral energy distribution of BL Lacs consists of a double-peaked structure, as illustrated in Figure 2.

3. Implications of VHE BL Lac Detections

3.1. *Gamma-ray production mechanism*

There are two main classes of models for gamma-ray production in BL Lac jets, distinguished by the dominant particle species in the jet. In one class it is electrons that are believed to produce the gamma radiation. It is generally accepted that the first peak is due to beamed incoherent synchrotron radiation from electrons. These electrons can then inverse-Compton scatter photons to gamma-ray energies. In the Synchrotron Self-Compton model the electrons scatter the synchrotron photons that they produced while in the External Inverse-Compton model the seed photons originate from outside the jet. Conversely, in the hadronic family of models protons are responsible for the gamma-ray emission, either through direct synchrotron radiation or through the decay of pions which occur when the protons collide with target material. As yet, no class of models have been ruled out but the multi-wavelength observations indicate that the same particles which produce the synchrotron photons may also produce the gamma-ray photons, hence favouring electron models.

3.2. *Cosmic Extragalactic Background Light*

VHE gamma rays interact with low energy photons to produce electron-positron pairs. For a 1 TeV gamma ray the cross section is a maximum when the wavelength of the soft photon is 1.33 μm. Thus VHE photons will be attenuated on their journey from extragalactic sources and the spectrum of VHE photons modified. Whilst this presents

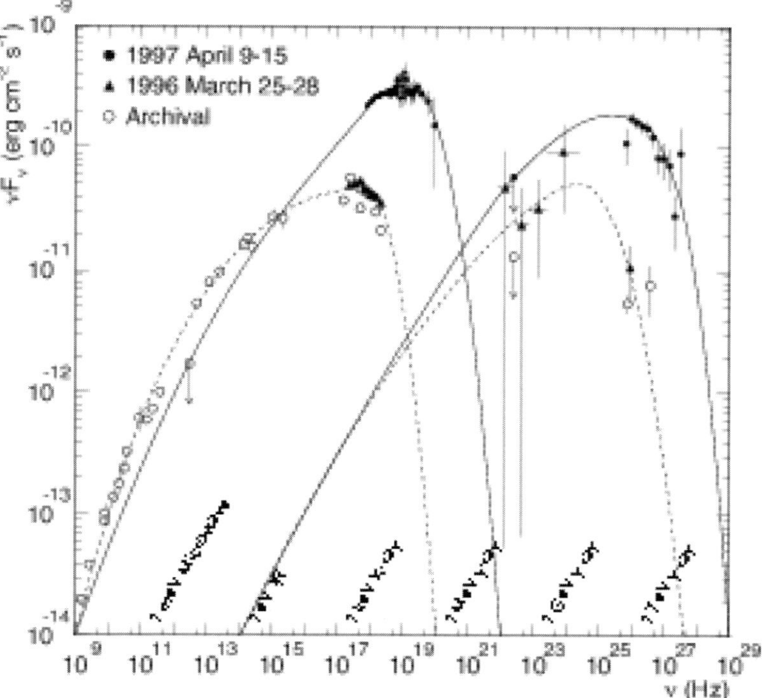

Figure 2. Broadband Spectral Energy Distribution of Markarian 501, from Catanese and Weekes (1999).

a difficulty for extragalactic VHE astronomy, as distant sources are difficult to detect, it also provides a mechanism to probe the cosmic EBL density, of which direct measurement is extremely difficult. The most recent results by HESS (Aharonian *et al.*, 2005) indicate that the EBL must be quite low, at the limit of that expected from galaxy source counts, and that the Universe is more transparent to VHE gamma rays than previously thought.

4. Conclusions and future prospects

The field of BL Lac astrophysics has been reinvigorated by the results from VHE instruments. There are now claimed detections of 11 objects, with 7 having been detected by multiple experiments. Third-generation IACT experiments, currently either online or coming coming online, are expected to dramatically expand our catalogue of gamma-ray emitting BL Lacs and allow more detailed studies of individual objects.

References

Aharonian, F., *et al.*, 1999, *A& A*, 349, 11

Aharonian, F., *et al.*, 2005, *submitted to Nature, astro-ph/0508073*

Catanese, M. and Weekes, T.C., 1999, *PASP*, 111, 1193

Cogan, P., *et al.*, *these proceedings*

Gaidos, J., *et al.*, 1996, *Nature*, 383, 319

Kranich, D., *et al.*, 2001, *Proc. 27th ICRC (Hamburg)*

Krawczynski, H. *et al.*, 2004, *ApJ*, 601, 151

Krennrich, F., *et al.*, 2002, *ApJ*, 575, L9

Schroedter, M., 2005, *ApJ.*, 628, 617S

232   Quinn, J. *et al.*

Discussion

MIRABEL: Can you explain the blazer phenomenology with bulk Lorentz factors of a few or you need longer than AGN?

QUINN: In general higher Lorentz factors ($\sim$50–100) are needed to fit one spectrum and variability observed $\sim$TeV blazers.

COURVOISIER: Do you take the IR radiation of the stars of the host galaxies when deducing the intergalactic IR background?

QUINN: This could explain why certain types of AGN are not seen at TeV easier. However, if absorption is occurring at the source then our estimates of the IR background level are underestimated.

MEURS: You showed extensive TeV lightcurve data of Mark 501. Have there been any attempts to perform a power spectrum analysis of these data to search for any periodicities?

QUINN: Some individuals analysed the data and claimed that there was evidence for a 23-day periodicity. Given that Cherenkov telescopes cannot operate in moonlight a 20-something day periodicity may be expected. The HEGRA collaboration themselves do not claim that there is any periodicity in the data.

John Quinn indicates a black hole size.

Populations of High Energy Sources in Galaxies
Proceedings IAU Symposium No. 230, 2005
E. J. A. Meurs & G. Fabbiano, eds.

© 2006 International Astronomical Union
doi:10.1017/S1743921306008283

High energy emission from flat-spectrum radio sources with $\sim$ kpc-scale structure

Pedro Augusto[1]

[1]Departamento de Matemática e Engenharias, Universidade da Madeira,
Caminho da Penteada, 9000-390 Funchal, Portugal
email: augusto@uma.pt

Abstract. Active Galactic Nuclei emit a substantial portion of their bolometric luminosities in X-rays. For example, the knots in radio jets are prominent sources of synchrotron X-rays while the hotspots of the brightest FRIIs emit self-synchrotron or Inverse Compton radiation. Most high-energy studies on flat-spectrum radio sources have been conducted for blazars which are dominant at γ-rays.

Augusto *et al.* (1998) have built a sample of 55 flat-spectrum radio sources dominated by structures (knots, hotspots, etc.) $\sim$0.1–2 kpc away from the nucleus. Seventeen (31%) of these are detected in X-rays (they tend to be the radio strongest) evenly splitting, morphologically, *both* at optical (radio) bands: nine QSO/BLLac (core-jets) on one-side; eight Galaxy/Sy2 (CSO/MSO/FRII) on the other. We have identified five confirmed compact/medium symmetric objects (CSO/MSOs) as X-ray emitters. A comparable type of source to CSO/MSOs is the physically similar (1–15 kpc) compact steep spectrum source (CSS), 28/129 (22%) of which are detected in X-rays, from a literature-selected sample (the percentage is smaller than for the 55-source sample due to a lower $<S_{4.85}>$). A 95% conf. level relation is found for CSSs: $S_X \propto (S_{4.85})^{0.6}$ and we found undistinguishable radio/X-ray properties for both the 55-source and CSS samples: clearly, their similar morphologies (e.g. knots in jets) stand up stronger than their radical radio spectrum differences.

Only two sources among the 55 (4%) have γ-ray detections and they seem quite abnormal (in $\alpha_{x\gamma}$ values, at least) – one of them is in a Sy2, not in a blazar.

Keywords. galaxies: active, (galaxies:) BL Lacertae objects: general, galaxies: jets, (galaxies:) quasars: general, galaxies: Seyfert, radio continuum: general, X-rays: galaxies, X-rays: general.

Augusto *et al.* (1998) have selected a sample of 55 flat-spectrum ($S_{8.4\,\text{GHz}} > 100$ mJy; $\alpha_{1.40}^{4.85} < 0.50$) radio sources dominated by structures (knots, hotspots, etc.) typically 0.1–2 kpc away from the nucleus. Using the NASA Extragalactic Database (NED) and the High Energy Missions Catalogue we have looked for any high-energy information (X-ray and γ-ray) for each of the 55 sources: 17 sources (31%) have it in X-rays while 38 are X-ray and γ-ray "quiet". Why? The hypothesis of similarity of the 4.85 GHz distributions is rejected at the 98% confidence level (χ^2-test), implying that the 17 sources tend to be the radio strongest in the 55-source sample.

There are 13 sources (75%) of the X-ray loud sub-sample that are bright enough to have X-ray spectral information, although for four of these we are limited to ROSAT hardness ratios (e.g. Voges *et al.* (1999)). About one-third of AGN in surveys shows a soft X-ray excess (Mushotzky *et al.* 1993, Gambill *et al.* 2003) which seems to be of thermal origin – e.g. Colbert *et al.* (1998). Soft excesses are seen in only two of our 17 (12%) X-ray loud sources (in a BLLac and in a QSO).

Out of the 23 compact/medium symmetric object (CSO/MSO) candidates of Augusto *et al.* (1998), seven (30%) are X-ray loud. Of these, five are already confirmed as CSO/MSOs (Augusto *et al.* 2006). Since the 55 sources were radio-morphologically selected, the only comparable type of source is the physically similar (1–15 kpc) compact steep spectrum source (CSS; $\alpha_{1.40}^{4.85} > 0.50$). We combined the O'Dea (1998) and Fanti *et al.* (2001) samples getting a total of 129 CSSs of which only 28 (22%) have X-ray information available. We compared the radio and X-ray properties ($S_{4.85}$, S_X, α_r and α_{rx}) between our 17 X-ray loud flat-spectrum sources and the morphologically similar 28 X-ray loud CSSs: χ^2 tests cannot rule out similarity for any of the parameters. Thus, the homogeneous population of CSSs and the heterogeneous population of flat-spectrum radio sources have similar radio/X-ray properties. Clearly, their similar morphologies (e.g. knots in jets) stand up stronger than the radical radio spectrum differences. As regards the samples individually, we reject "no correlation" at the $> 95\%$ level and derive approximate regression line fits for the following parameters: $\alpha_{rx} = 0.4\alpha_r + 0.7$ (55-source sample); $\log S_X = 0.6\log S_{4.85} - 6.8$ (CSS sample).

The 30 core-jets (CJs) identified in Augusto *et al.* (1998) split into 14 bent-jet and 16 straight-jet sources, the same splitting remaining for the X-ray loud subsample: three bent-jets vs. six straight ones. So, there is no apparent preference for selecting CJs in the X-ray loud subsample (possibly because the hot spots in CSO/MSOs are strong competitors for the knots in jets, or simply due to our poor statistics).

Flat-spectrum jets with pc-scale bends, are not likely to have γ-ray emission (von Montigny *et al.* 1995, Tingay *et al.* 1998); this is confirmed by the only bent-jet of the 55-source sample, X-ray and γ-ray quiet, which has VLBI data showing bending from pc- to kpc-scales – Augusto *et al.* (1998). Virtually all γ-ray emitting AGN are blazars (Sowards-Emmerd *et al.* 2005). However, there are two γ-ray detected sources in the 55-source sample (one BLLac) one of which is a Sy2, *not* a blazar! They both have abnormal $\alpha_{x\gamma}$ values: the BLLac value is too flat (0.42 vs. 0.83 $\pm$ 0.18 – Comastri *et al.* (1997)) while the one for the Sy2 is steeper than any known AGN value ($\alpha_{x\gamma} = 1.87$), although the soft γ-ray (100 keV) energy used in the calculation differs a lot from the usual mid γ-rays (100 MeV).

Acknowledgements

The author acknowledges the travel grants from the International Astronomical Union and the Fundação Calouste Gulbenkian.

References

Augusto, P., Wilkinson, P.N. & Browne, I.W.A. 1998, *MNRAS* 299, 1159

Augusto, P., Gonzalez-Serrano, J.I., Perez-Fournon, I. & Wilkinson, P.N. 2006, *MNRAS* submitted

Colbert, E.J.M., Baum, S.A., O'Dea, C.P. & Veilleux, S. 1998, *ApJ* 496, 786

Comastri, A., Fossati, G., Ghisellini, G. & Molendi, S. 1997, *ApJ* 480, 534

Fanti, C., Pozzi, F., Dallacasa, D., Fanti, R., Gregorini, L., Stanghellini, C. & Vigotti, M. 2001, *A&A* 369, 380

Gambill, J.K., Sambruna, R.M., Chartas, G., Cheung, C.C., Maraschi, L., Tavecchio, F., Urry, C.M. & Pesce, J. E. 2003, *A&A* 401, 505

Mushotzky, R.F., Done, C. & Pounds, K. A. 1993, *ARA&A* 31, 717

O'Dea, C.P. 1998, *PASP* 110, 493

Sowards-Emmerd, D., Romani, R.W., Michelson, P.F., Healey, S.E. & Nolan, P.L. 2005, *ApJ* 626, 95

Tingay, S.J., Murphy, D.W. & Edwards, P.G. 1998, *ApJ* 500, 673

Voges, W., *et al.* 1999, *A&A* 349, 389

von Montigny, C., *et al.* 1995, *A&A* 299, 680

Populations of High Energy Sources in Galaxies
Proceedings IAU Symposium No. 230, 2005
E. J. A. Meurs & G. Fabbiano, eds.

© 2006 International Astronomical Union
doi:10.1017/S1743921306008295

The Central Regions of Galaxies Hosting LINERs as Viewed by Chandra

H. Flohic[1], M. Eracleous[1], G. Chartas[1], J. Shields[2] and E. Moran[3]

[1]Dept. of Astronomy & Astrophysics, Penn. State University, University Park, PA 16802
[2]Dept. of Physics & Astronomy, Ohio University, Athens, OH 45701
[3]Astronomy Dept., Van Vleck Observatory, Wesleyan University, Middletiwn, CT 06459

Abstract. We have used Chandra archival observations of 19 galaxies hosting LINERs to explore the morphology and source population of their inner kiloparsec. Our goal was, in general, to determine the power source behind their nuclear X-ray emission and, in particular, to investigate the presence of an AGN. We find an AGN in 12 of the 19 galaxies in the sample. We also find that diffuse, thermal emission is common with properties very similar to what is found in normal galaxies. In 10 out of the 19 galaxies, the diffuce emission dominates the nuclear X-ray power. The X-ray point-source populations were studied by producing cumulative luminosity functions and their properties are also similar to what is found in normal galaxies.

1. Introduction

Low Ionization Nuclear Emission-Line Regions (LINERs) are defined based on their oxygen emission-lines ratios (Heckman 1980). They are common in nearby galaxies and are found in 30% of all galaxies, especially early types (Ho *et al.* 1997). A few mechanisms have been proposed to explain their line ratios (shocked gas, dense gas photoionized by hot stars or by X-rays from a LLAGN), but it is still unclear which mechanism is most relevant in LINERS. The goal of our study is to determine the power source of LINERs based on the X-ray properties of their central region; in particular, we have looked for indications for the presence of an AGN and/or unusual X-ray point source content.

2. Sample and Analysis

We selected all the LINERs in the Chandra Archives that were nearby ($d < 25$ Mpc) and observed for more than 15 ks to ensure high spatial resolution and S/N, obtaining a sample of 19 objects. The host galaxies in this sample span a wide range of morphology and the relative numbers of LINER types are similar to those found in optical surveys.

For each galaxy we produced an image of the central kiloparsec in both the soft (0.5–2 keV) and hard (2–10 keV) bands. We extracted the spectrum of all the detected sources using the package ACIS Extract, checked for the presence of a central point source, and fitted the spectrum of the brightest sources to determine their nature. We also extracted the spectrum of the diffuse emission in the central kiloparsec of each galaxy. Since the diffuse emission of individual galaxies did not produce enough counts to perform a meaningful fit, we stacked the spectra of spiral and elliptical galaxies separately.

3. Results

The galaxies in the sample have, on average, three X-ray point sources in their inner kiloparsec with a maximum of 12 sources. In 12 of the 19 galaxies we find a central X-ray

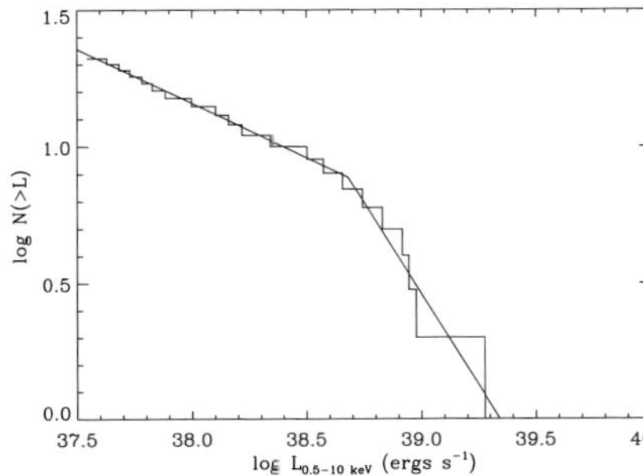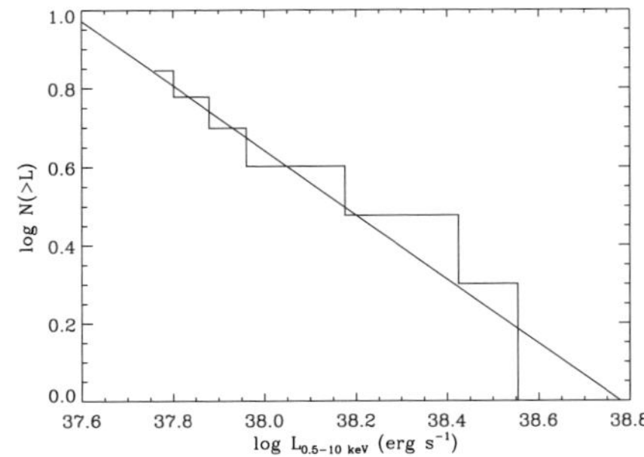

Figure 1. Cumulative luminosity function for the points sources in the elliptical galaxies (left, 21 sources) and for the spiral galaxies(right, 7 sources). The continuous lines show single or broken power-law fits to the data.

point source, which is often hard and always radio-loud. The typical X-ray luminosity of these AGN candidates is $L_{0.5-10\,\mathrm{keV}} = 10^{38} - 10^{39}\mathrm{erg\ s}^{-1}$. We compared the luminosity of the candidate AGNs to that of a $2''5$ diameter circle (similar to the area used in optical spectroscopic surveys) and found that the contribution of the AGN candidate to the X-ray luminosity of this region can vary between 30 and 100%. We calculate that L_{Bol}/L_{Edd} is between 10^{-7} and 10^{-4} for these AGNs (assuming a black hole mass of $10^7 M_\odot$).

Diffuse emission is common in the central region of galaxies hosting LINERs (it is found in 70% of the sample). We created stacked spectra for the spiral and elliptical galaxies separately, which are well fitted by a thermal plasma model ($\mathrm{kT} \sim 0.6$ keV) plus a power-law (photon index 1.3–1.5). This is very similar to the spectrum of diffuse emission found in normal galaxies (Sarazin *et al.* 2001, Sivakoff *et al.* 2004).

In Figure 1 we show separate cumulative luminosity function for all the non-nuclear sources found in the central region of the spiral and elliptical galaxies. These are well fitted by power-laws (broken for the elliptical galaxies) with indices very similar to those found for normal galaxies (Sarazin *et al.* 2001, Colbert *et al.* 2004).

4. Conclusions

LINERs are a heterogeneous population. 60% of the LINERs in our sample possibly harbor an AGN. In 70% we find significant diffuse emission in the inner kiloparsec. Some LINERs have significant contributions from both the diffuse emission and the possible AGN to the X-ray luminosity of the innermost $2''5$. The X-ray point-source populations found in the central region of galaxies hosting LINERs do not differ from those found in normal galaxies. Thus, even though a significant fraction of the X-ray power is produced by stellar processes, a recent starburst does not seem to be involved.

Acknowledgements: This work was supported by NASA through grant AR4-5010A from the Smithsonian Astrophysical Observatory.

References

Colbert, E. *et al.* 2004, ApJ, 602, 231
Heckman, T. M 1980, A&A, 87, 152
Ho, L. C., Filippenko, A. V. & Sargent, W. L. W. 1997c, ApJ, 487, 568
Sarazin, C. L., Irwin, J. A. & Bregman, J. N. 2001, ApJ, 556,533
Sivakoff, G. R., Sarazin, C. L. & Irwin, J. A. 2003, ApJ, 599, 218

Populations of High Energy Sources in Galaxies
Proceedings IAU Symposium No. 230, 2005
E. J. A. Meurs & G. Fabbiano, eds.

© 2006 International Astronomical Union
doi:10.1017/S1743921306008301

Blazars – INTEGRAL and Supermassive Black Hole Binaries

R. Hudec[1], F. Munz[1], M. Basta[1] and P. Kubanek[1,2]

[1]Astronomical Institute, Academy of Sciences of the Czech Republic, CZ–251 65 Ondrejov,
Czech Republic
email: rhudec@asu.cas.cz

[2]ISDC, Versoix, Switzerland

Abstract. We refer on analysis of the ESA INTEGRAL satellite data for blazars, promising sources to be observed during their active states. We further refer on searches for and analysis of supermassive binary black holes requiring very long time intervals (50 years and more) provided by digitised astronomical plates.

Keywords. Active Galactic Nuclei, Blazars, INTEGRAL.

1. Introduction

Blazars represent the most extreme class of active galaxies. They are observed in all wavelength bands – from radio through VHE gamma frequencies, with maximum spectral output and largest variability often at gamma ray energies. It is obvious that blazars represent suitable targets for INTEGRAL satellite (Winkler *et al.*, 2003) especially during active states (flares).

2. INTEGRAL observations

The INTEGRAL observations are divided into the following categories: (i) AO-1,2,3 Program (allocated pointed observations), (ii) Core Program CP (Galactic Plane Scans, Galactic Center Deep Exposure,...), and (iii) Objects inside FOV of AO-1,2,3 observations. Blazars in the INTEGRAL Galactic Plane Scans (GPS) represent a promising group of objects for the study within the INTEGRAL CP. The GPS zone is usually neglected by extragalactic astronomers due to heavy obscuration: in optical, $\sim$20% of the sky is obscured by our Galaxy, while the gamma–ray telescopes on board INTEGRAL allow detectability of up to few mCrabs in the most exposed GPS regions. Seven optically bright (with V$\leqslant$17 mag, to be detected by the INTEGRAL OMC camera) blazars were identified within galactic scans of INTEGRAL, namely: 1ES 0647+250, PKS 0823-223 (no gamma from EGRET, grav. lensing candidate), 1ES 2344+514 (TeV gamma ray source, very close), 8C 0149+710 (BL Lac candidate?), 4C 47.08, 87GB 02109+5130 (poorly understood blazar, TeV candidate), and BL Lac (the prototype). While the prototype object BL Lac is well studied, most of the INTEGRAL GPS blazars are poorly investigated and poorly understood so far. The study with Sonneberg Observatory Archival Plates reveals that most of these objects are optically variable, hence a gamma ray variability can be expected. Below the detection limit of the INTEGRAL OMC on board camera is blazar NRAO530 (1730-130), which is an example of blazar with violent optical activity (4 mag within 1 month). In flare, the object is expected to be much brighter also in gamma. This strengthens the role of optical monitoring and ToO program – the flare can be recognized by optical monitoring with small (D $\sim$50 cm) telescopes. All the above mentioned blazars in INTEGRAL GPS have been investigated with INTEGRAL CP data (IBIS and JEM-X telescopes). We have no positive detection by high energy instruments on board INTEGRAL yet (except marginal detection of 1ES 0647+250). The targets' quiet level is still below the sensitivity threshold of the instruments. However, the positive detection may be possible in the future as (i) there will be more cumulative time available and (ii) the probability to see a blazar during a flare (and hence much brighter) will also increase with time. Additional blazars have been identified in the fields represented by the AO-1 and AO-2 observations of other scheduled targets, covered by up to 400 ksec cumulative exposure time. The analysis of these objects is in progress.

Regarding the pointed observations of blazars by INTEGRAL, the AO-2 ToO blazar observation No. 220049 by Pian *et al.* (2005) has provided promising results. This collaborative proposal was based on extended optical and/or X-ray monitoring (RXTE ASM and others) of flaring activity of a large list of blazars and, alternatively, on soft gamma-ray monitoring by INTEGRAL itself (serendipitous detection of a flaring blazar in the IBIS FOV). Then ToO INTEGRAL observation was activated meeting the trigger criteria (major flaring event). Blazar S5 0716+714 was the target of this ToO observation. This is a BL Lac object, intensively monitored at radio and optical wavelengths. The ToO was triggered by optical activity – 2 outbursts up to the extreme level of R = 12.1 mag (historical maximum, light increase by 1 mag in 2 weeks and 2 magnitudes in 4 months) and, consequently, the INTEGRAL ToO observation was performed in the time interval 2004 April 2–7 (Pian *et al.*, 2005). Very recently, an INTEGRAL AO-3 ToO observation of 3C454.3 (z=0.859) was performed, with preliminary results given by L. Foschini *et al.* (2005, PI E. Pian with a large collaboration). This ToO was triggered by high optical (T. Balonek, VSNET alert) and X-ray (BAT Swift) activity of the source. The INTEGRAL observation started 2005 May 15, at 18:40 UT, with exposure of 200 ksec. The source was clearly detected by IBIS/ISGRI in the 20–40 and 40–100 keV energy bands, with a significance of 20 and 15 sigma.

In the second project, we have gathered data from the literature and observational campaigns in order to establish long-term optical light curves of the selected blazar binary black hole candidates – to study periodic behavior in their light curves and other interesting features (intense outbursts, flares, quiescent level behavior). However, there are several crucial data gaps that disable to confirm periodicity or a BBH model (that has already been built up for several of these blazars). Therefore we intend to go to databases of astronomical plates (e.g. Sonneberg Observatory, Germany (about 280,000 plates), Harvard College Observatory, USA (about 500,000 plates), UKSTU plate collection ROE Edinburgh, UK (18,000 very deep plates), and Leiden Observatory, NL (40,000 plates)) to fill in these gaps. Within this project, we intend to reach the following results: (i) improve historical light curves of candidates, (ii) periodicity and light curve analysis, (iii) confront the new light curve of the selected blazars with the corresponding theories, (iv) establish a detailed model at least of one of the candidates, (v) draw statistical conclusions, (vi) provide the data to wide scientific community.

Acknowledgements. This study is partly linked to the grant A3003206 by the Grant Agency of the Academy of Sciences of the Czech Republic.

References

Foschini, L. *et al.*, 2005, ATEL 497

Pian, E. *et al.*, 2005, A&A 429, 427

Winkler, C. *et al.*, 2003, A&A, 411, 7

Populations of High Energy Sources in Galaxies
Proceedings IAU Symposium No. 230, 2005
E. J. A. Meurs & G. Fabbiano, eds.

© 2006 International Astronomical Union
doi:10.1017/S1743921306008313

Supermassive binary black holes in blazars

Frank M. Rieger

Department of Mathematical Physics, University College Dublin, Belfield, Dublin 4, Ireland

Abstract. Low-frequency ($f \leqslant 10^{-6}$ Hz) quasi-periodic variability observed from radioloud, jet-emitting Active Galactic Nuclei (AGNs) can provide substantial inductive support for the presence of close ($d \lesssim 0.1$ pc) supermassive binary black holes in their centers. Such periodic variability may arise, for example, due to differential Doppler boosting along helical jet paths driven by the orbital motion or jet precession. If the jet path is non-ballistic, travel time effects can lead to a strong reduction of the observable period by up to a factor of γ_b^{-2}. Here we suggest a binary model where the optical periodicity with timescale of several years is related to accretion disk interactions, radio periodicity to Newtonian jet precession and periodicities in the high energy bands to the orbital motion of the jet. We analyze the explanatory potential of such a framework and comment on the possible origin of periodicities in AO 0235+16.

Keywords. black hole physics – galaxies: active – galaxies: jets.

Supermassive binary black holes (SBBHs) are generally expected to form during the cosmological evolution of galaxies. In hierarchical clustering scenarios, for example, giant elliptical galaxies such as the host galaxies of radioloud AGNs are thought to be the product of mergers between spiral galaxies each containing its own black hole (BH). Interacting galaxies are indeed observationally well-known and give strong reasons for such an evolutionary picture. In particular, direct observational support has been recently provided by the Chandra discovery of two wide activity centers (separation $d \sim 1$ kpc) in the merging galaxy NGC 6240 (Komossa *et al.* 2003). While the existence of wide SBBHs seems thus well grounded, the existence of close ($d \sim 0.003 - 0.1$ pc) SBBHs appears much more ambiguous. Dynamical friction and slingshot interactions with stars normally ensure that the binary BHs quickly get closer (e.g., Begelman *et al.* 1980). However, due to a 'loss cone depletion' it is still an open question today whether a substantial fraction of SBBHs can really coalesce within a Hubble time.

The presence of close SBBHs has been repeatedly invoked as plausible source for a number of observational findings in blazars (e.g., misalignment and precession of jets, apparent helical trajectories of knots or periodic variability; cf. Rieger 2005 for a review). Interactions of the companion with the accretion disk around the primary, for example, can provide a natural explanation for the longterm optical periodicity with periods $P_{\mathrm{obs}}^{\mathrm{opt}}$ of several years as observed in a number of blazars. It seems likely that the combined effects of dynamical friction, accretion disk interactions and gravitational radiation produce orbits that are nearly circular. Assuming that the observed periodicity is caused by the secondary BH crossing the disk around the primary BH on a slightly non-coplanar orbit, we can derive an upper limit for the real Keplerian orbital period $P_k \simeq 2/(1+z)\, P_{\mathrm{obs}}^{\mathrm{opt}}$. The observations of helical jet paths in many blazar sources suggest that periodic variability (especially in the energy range dominated by the jet, e.g., radio, X- and γ-ray) can also arise due to differential Doppler boosting for a periodically changing viewing angle. For non-ballistic motion, travel time effects will lead to a reduction of the observable period P_{obs} with respect to the real driving period P such that $P_{\mathrm{obs}} \simeq (1 + z)\, P/\gamma_b^2$, where $\gamma_b \sim (5-15)$ is the bulk flow Lorentz factor (Rieger 2004). The orbital motion of the jet-emitting BH and (Newtonian) precession of the jet generally represent the most obvious

Table 1. Properties of a sample of blazar SBBH candidates, cf. Rieger (2005) for details.

name	redshift z	periods P_{obs}	$(m+M)/10^8\,M_\odot$	P_k [yr]	$d/10^{16}$ cm
Mkn 501	0.034	23.6 d (X-ray)	(2–7)	(6–14)	(2.5–6)
		$\sim$23 d (TeV)			
BL Lac	0.069	13.97 yr (optical)	(2–4)	(13–26.1)	(4.8–9.7)
		$\sim$4 yr (radio)			
3C 273	0.158	13.65 yr (optical)	(6–10)	(11.8–23.5)	(6.5–12)
		8.55 yr (radio)			
OJ 287	0.306	11.86 yr (optical)	6.2	(9.1–18.2)	(5.5–8.8)
		$\sim$12 yr (infrared)			
		$\sim$1.66 yr (radio)			
0235+16	0.940	2.95 yr (optical)?	$\geqslant 1$	(1.5–3.1)	$\geqslant 0.95$
		5.7 yr (radio)			

driving sources for helical jet paths. Whereas the high energy emission is usually produced on small jet scales and thus likely to be modulated by the orbital motion of the SBBHs so that $P_{\mathrm{obs}} \sim 30\,(P_{\mathrm{obs}}^{\mathrm{opt}}/10\,\mathrm{yr})\,(15/\gamma_b)^2$ d, the main part of the radio jet emission usually originates from larger scales where the jet motion is likely to be dominated by Newtonian precession with a period which is (at least) an order of magnitude higher than P_k, i.e., $P = P_p \gtrsim 10\,P_k$ (Rieger 2004), so that one finds $P_{\mathrm{obs}}^{\mathrm{radio}} \gtrsim 20 P_{\mathrm{obs}}^{\mathrm{opt}}/\gamma_b^2$. Hence, if P_p is rather small (say $P_p \sim 10\,P_k$), moderate bulk Lorentz factors can be sufficient to account for $P_{\mathrm{obs}}^{\mathrm{radio}} < P_{\mathrm{obs}}^{\mathrm{opt}}$. Table 1 shows properties of a sample of blazar SBBH candidates, where the observed periodicities have been used to estimate the last two columns.

The results above may be particularly relevant for an appropriate understanding of the nature of the BL Lac AO 0235+16 (cf. Table 1), where two different SBBH models have been proposed recently (cf. also Rieger 2005): (i) In Romero *et al.* (2003) the observed optical periodicity has been related to accretion disk interactions (implying $P_k \simeq 2 \times 2.95/[1+z] \sim 3$ yr) and the radio one to Newtonian jet precession. (ii) In Ostorero *et al.* (2004) the radio and optical periodicity has been related to a helically bent inhomogeneous jet, driven by the orbital motion (using the radio period as direct tracer). Scenario (i) requires that the jet fluid motion is non-ballistic with $\gamma_b > 3$ as otherwise P_p would be too short. Moreover, the orbital motion should lead to some quasi-periodic modulation in the high energy bands, with observable periods ranging from $\sim$7 months (for $\gamma_b \sim 3$) to $\sim$20 days (for $\gamma_b \sim 10$). Scenario (ii), on the other hand, seems less plausible if $P_{\mathrm{obs}}^{\mathrm{opt}} \neq P_{\mathrm{obs}}^{\mathrm{radio}}$ (cf. Table 1), as is indeed confirmed by further studies. It is likely, however, that the real situation is much more complex, e.g., the radio jet flow may repeatedly approach the line-of-sight, resulting in a maximization of beaming effects ("radio knots") and an apparent short Keplerian period, e.g., $P_{\mathrm{obs}}^{\mathrm{radio}} \simeq 5.7$ yr may in reality imply $P_k \simeq 26$ yr (for $\gamma_b = 3$), and the radio lightcurves may have pronounced peaks separated by $(1+z)\,P_k$, with intermediate peaks occurring every 5.7 yr.

References

Begelman, M. C., Blandford, R. D. & Rees, M. J. 1980, *Nature* 287, 307

Komossa, S., Burwitz, V., Hasinger, G. *et al.* 2003, *ApJ* (Letters) 582, L15

Ostorero, L., Villata, M. & Raiteri, C.M. 2004, *A&A* 419, 913

Rieger, F.M. 2004, *ApJ* (Letters) 615, L5

Rieger, F.M. 2005, *Proc. 22nd Texas Symposium on Relativistic Astrophysics*, Stanford, eds. P. Chen *et al.* (eConf:C041213), 1601

Romero, G.E., Fan, J. & Nuza, S. E. 2003, *ChJAA* 3, 513

Populations of High Energy Sources in Galaxies
Proceedings IAU Symposium No. 230, 2005
E. J. A. Meurs & G. Fabbiano, eds.

© 2006 International Astronomical Union
doi:10.1017/S1743921306008325

The X-ray monitoring of quasar 3C273

M. Stuhlinger

XMM-Newton SOC, ESA, Madrid, Spain
email: Martin.Stuhlinger@sciops.esa.int

Abstract. XMM-Newton has observed the quasar 3C273 several times between 2000 June and 2005 July. In addition data of NASA's Rossi X-ray Timing Explorer is used. We present the results of this monitoring campaign.

The X-ray spectrum of 3C273, in the energy regime of 0.2–100 keV, is believed to be composed of three components: a power law component above about 3 keV, a soft excess below about 2 keV, and an Fe-line. The Fe-line, detected in only a few observations of 3C273, is not detected significantly in individual XMM-observations. Only the 2000 June observations show evidence for a presence of a broad line feature. The high energy component above 3 keV is well described by a single power law model with varying indices and fluxes. The steeper spectra show significantly higher flux levels. The soft excess component below 2 keV also shows variations in flux and spectral slope. Different models are tested to identify the soft excess to be thermal or non-thermal emission.

Populations of High Energy Sources in Galaxies
Proceedings IAU Symposium No. 230, 2005
E. J. A. Meurs & G. Fabbiano, eds.

© 2006 International Astronomical Union
doi:10.1017/S1743921306008337

GRBs as rare stages of stellar evolution

E. P. J. van den Heuvel

Sterrenkundig Instituut "Anton Pannekoek", University of Amsterdam,
The Netherlands
email: edvdh@science.uva.nl

Abstract. The association of "long" Gamma-Ray Bursts (durations > 2 seconds) with peculiar Type Ic supernovae suggests strongly that this type of GRB is produced by the collapse of the rapidly rotating core of an initially very massive star to a black hole. At the time of collapse the star has lost its hydrogen-rich envelope and the GRB is thought to be produced by a collimated relativistic jet of matter ejected along the star's rotation axis. The angular momentum constraints for producing such a "collapsar" or "hypernova" suggest that the GRB-producing core collapses constitute only a small fraction of all core collapses of massive stars. As to the short-duration GRBs (< 2 seconds), which make up about one third of all GRBs, the most favoured model is that of the coalescence of a double neutron star or of a neutron star-black hole binary. Also these events are expected to be very rare, having a frequency of at most one event per hundred thousand years for a galaxy like our own. Due to the collimation of the relativistically ejected matter, the observable frequency of GRB events will, like in the case of the "long" bursts, be at least a factor hundred smaller.

Discussion

COURVOISIER: Is what we know of NS population and binary evolution compatible with the observed rate of short GRBs?

VAN DEN HEUVEL: Berger *et al.* (2005), who discovered the short GRB 050724, argue in their paper that the observed incidence of short bursts, corrected for beaming (which makes that we see only 1 per cent of the GRBs produced by the mergers), is at least one order of magnitude lower than the expected merger rate of double NSs in galaxies. So, it appears that only a small fraction of the mergers succeeds in making a (short) GRB.

KRYVDYK: Very strong magnetic field is present in SN and BH. Is this fact taken into account in GRB models?

VAN DEN HEUVEL: Yes, this might be another parameter that could determine whether or not a NS + NS or NS + BH merger makes a Gamma Ray Burst.

MIRABEL: Do GRBs require merger of the two massive stars? Could BH binaries in our Galaxy be fossils of GRB Sources?

VAN DEN HEUVEL: Langer has suggested that only in the case of a merger of two helium cores one will have sufficient angular momentum (rotation) to produce a GRB when the core collapses to a black hole. This would be then a very rare event. I personally think that a helium star produced by mass transfer in a close binary could still be kept in sufficiently rapid rotation (by tidal forces) to produce a GRB. The helium-star (Wolf-Rayet) binary should then be very close (like V444 Cygni) for this to work.

Populations of High Energy Sources in Galaxies
Proceedings IAU Symposium No. 230, 2005
E. J. A. Meurs & G. Fabbiano, eds.

© 2006 International Astronomical Union
doi:10.1017/S1743921306008349

GRB observations with INTEGRAL and XMM

B. McBreen

Department of Experimental Physics†, University College Dublin, Dublin, Ireland
email: Brian.McBreen@ucd.ie

Abstract. The results obtained during the first three years of operation of INTEGRAL will be reviewed. Over 30 GRBs have been localised by the INTEGRAL burst alert system (IBAS) and the coordinates distributed. The follow-up observations with XMM-Newton and large telescopes have led to many interesting results. A comparison will also be made between the sensitivities of INTEGRAL and SWIFT.

Brian McBreen pointing out room safety regulations.

† Now: UCD School of Physics, Science Centre North, UCD.

Populations of High Energy Sources in Galaxies
Proceedings IAU Symposium No. 230, 2005
E. J. A. Meurs & G. Fabbiano, eds.

© 2006 International Astronomical Union
doi:10.1017/S1743921306008350

Collapsing Stars as Sources of High Energy Particles and Radiation

Volodymyr Kryvdyk†

Dept. Astronomy, Faculty of Physics, Kyiv National University
email: kryvdyk@univ.kiev.ua

The particle acceleration and the non-thermal radiation in the magnetospheres of collapsing stars are considered. The collapsing stars can be powerful sources of the high-energy charged particles and the non-thermal radiation bursts. These bursts can be observed by means of modern astronomical instruments.

Collapse begins when the mass of stellar core exceeds the Chandrasekhar bound and the star becomes dynamically unstable. The star compresses and its radius decreases. After that, the stars can evolve by several ways. The first is that the star explodes and loses mass. The massive stars will collapse to neutron stars or black holes. The star can loss its mass and we will observe this phenomenon as supernovae. The stars with the smaller mass will collapse to white dwarfs (Shapiro & Teukolsky (1985); Zeldovich & Novikov (1977)). The second possibility will realize when stars collapse without the loss of mass. In this case it is very difficult to observe the collapse, and we not have the astronomical date for the conformation of this stage.

The stars will emit electromagnetic radiation under the collapse (Kryvdyk (1999) and Kryvdyk (2001)). In this paper the non-thermal radiation from the collapsing stars with the initial dipole magnetic field on the non-relativistic stage is investigated, and the method for the search of collapsing stars is proposed. The magnetic field will increase during the collapse. The charged particles will be accelerated to relativistic energy in the magnetospheres of collapsing stars. These particles will emit electromagnetic waves from radio waves to gamma rays. The radiation flux increases during the collapse in the millions times and more compared with the initial flux. This radiation will be observed as the bursts in all frequency ranges. The intensity of these bursts is very high. The radiation flux from collapsing stars exceeds of the initial flux in millions on the final stage of collapse. Fig. 1 portrays the increase of the particles energy and the particles density during collapse. On the Fig. 2 is shown the polar jets arising by collapse. As follows from obtaining results, the collapsing stars can be the powerful sources of the non-thermal radiation. Where these impulses can be observed? First of all in middle of powerful gamma bursts and X-ray bursts which are not periodical and can be connected with the collapsing stars. These impulses can be observed also from the pre-supernovae. The stars go to this stage when the star begins to compress under the influence of the gravitational field. The powerful sources of the non-thermal radiation can be also the white dwarfs in double systems on the stage of the accretion-induced collapse. The periodical impulse non-thermal radiation can be generated also by the pulsating of the stars with magnetic field, since in this case the charged particles will accelerate and the non-thermal emission will generate. What difficulties can arise for the observation of collapsing stars? First, we can not point at the location of collapsing stars with enough accuracy. Above we indicated only at the types of stars that can be collapsed. But the theory not enables to make the detailed chronology of collapse, therefore we can not

† Present address: Dept. Astronomy, Faculty of Physics, Kyiv National University.

indicate where and when exactly the collapsing star can arise. This fact is a principal problem for the observational program of the search of collapsing stars. The next problem is how to choose the radiation bursts from collapsing stars from the great number of bursts with unknown origin.

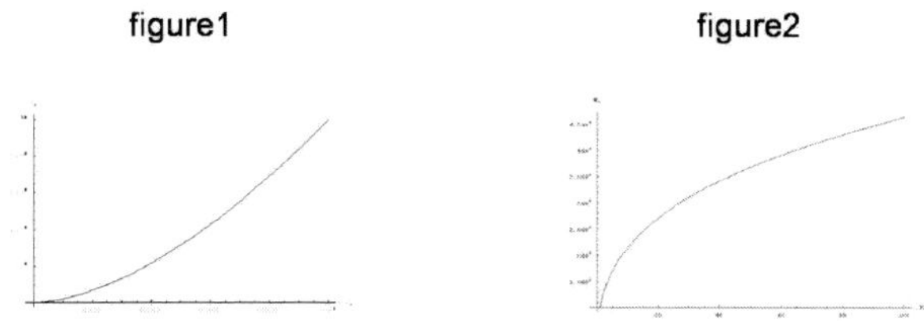

Figure 1. Figure Left: An increase of the particles energy during collapse; Figure right: An increase of the particles density during collapse.

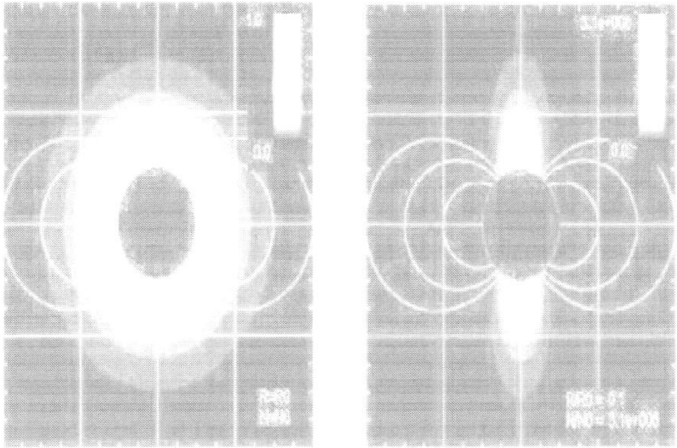

Figure 2. Polar jets in the magnetosphere of collapsing stars.

References

Shapiro, S.L. & Teukolsky, S.A. 1985, *Black Holes, White Dwarfs, and Neutron Stars*. Nauka: Moscow
Zeldovich, J.B. & Novikov, I.D. 1977, *Theory of gravity and stellar evolution*
Kryvdyk, V. 1999, *MNRAS*, 309, 593
Kryvdyk, V. 2001, *AdSpR*, 28, 463

Populations of High Energy Sources in Galaxies
Proceedings IAU Symposium No. 230, 2005
E. J. A. Meurs & G. Fabbiano, eds.

© 2006 International Astronomical Union
doi:10.1017/S1743921306008362

Using jet breaks to estimate GRB distances

P. Ward, E. J. A. Meurs and C. del Burgo

Dunsink Observatory, Castleknock, Dublin 15, Ireland

Abstract. Recent observations have suggested that the true energy release of GRBs is potentially far less than previously thought. This is due to beaming, a signature of which is a broadband break in the power-law decay of the afterglow emission. Taking these results we have constructed a basic distance estimator, which may be useful as a diagnostic tool for the large amount of GRBs without a spectroscopically measured redshift.

Keywords. gamma rays: bursts, cosmology: distance scale.

The value for the isotropic equivalent energy output of a GRB, E_{iso}, in a given bandpass is found by $E_{iso} = S_\gamma \frac{4\pi D_l^2}{(1+z)} k$ where S_γ is the fluence received in the observed band pass and D_l is the luminosity distance at redshift z. The quantity k is the k-correction. However, it has been shown that true energy release, E_γ, of a GRB is in fact much less than this when the burst is beamed into a collimated jet of half opening angle θ_j. E_γ will therefore be less than E_{iso} by a factor $(1 - \cos\theta_j)$. The beaming fraction can also be described as the ratio of the true energy release and isotropic equivalent energy; $\frac{E_\gamma}{E_{iso}} \approx \frac{\theta_j^2}{2}$.

One signature of such a jet is a broadband break in the power-law decay of the afterglow emission which occurs at a time t_j when the bulk Lorentz factor of the blast wave (Γ) has slowed down to $\Gamma < \theta_j^{-1}$. According to the formulation made by Sari *et al.* (1999) the spherical adiabatic evolution of the Lorentz factor is $\gamma(t) \approx 6(\frac{E_{iso}}{n_1})^{1/8} t_j^{-3/8}$. Subsequently if the break occurs when $\gamma \approx \theta_j^{-1}$, we find that $(\frac{2E_\gamma}{E_{iso}})^{-1/2} \approx 6(\frac{E_{iso}}{n_1})^{1/8} t_j^{-3/8}$ which can be rearranged to give $E_{iso} = 119(2E_\gamma)^{4/3}(n)^{-1/3}(t_j)^{-1}$, giving us an alternative approach for determining E_{iso}.

If GRBs are in fact standard candles the value for E_γ is a constant. Using this presumption we can use the two equations for E_{iso} to construct the following relationship between intrinsic burst parameters and the redshift,

$$\mathbf{D_T(z) = Q_T(t_j, S_\gamma, n, k) E_\gamma^{4/3}}$$

We can separate the relationship into a distance quantity D_T, which is a function of redshift z, and a burst quantity Q_T, which is a function of the burst properties t_j, S_γ, n and k; we define Q_T and D_T as $Q_T \equiv \frac{238.7}{t_j n^{(1/3)} S_\gamma k}$ and $D_T \equiv [\frac{2c}{H_0}(1 + z - \sqrt{(1+z)})]^2 \times \frac{1}{1+z} pc$.

We can now test this relationship using existing data for 12 bursts which have well established values for z, n, t_j, S_γ and k. Subsequently, for each burst, we derive a value for Q_T (Table 1). Figure 1 shows a plot of Q_T vs. z (logarithmic scale). The dotted line represents the observed trend; the value for Q_T appears to increase with redshift. Amati *et al.* (2002) previously noticed a trend of E_{iso} to increase with z.

The derived relationship is $\mathbf{Q_T = A \times z^\beta}$, where A $\sim 0.1\,\mathrm{cm}^3\,\mathrm{erg}^{-1}\,\mathrm{s}^{-1}$ and β=1.5 (see Figure 1). We therefore have a method for determining z*, the estimated redshift, according to the relationship derived from the plot in Figure 1: $\mathbf{z* = (\frac{Q_T}{0.1})^{2/3}}$.

Table 1. The parameters used for the 12 bursts in our sample

GRB	z	n $[cm^{-3}]$	t_j [days]	S_γ $[10^{-6}$ erg $cm^{-2}]$	k	Q_T $[10cm^3\ erg^{-1}\ s^{-1}]$
GRB970508	0.8349 ± 0.0003	1 ± 0.5	25 ± 5	1.8 ± 0.3	1.55 ± 0.08	0.3420 ± 0.3140
GRB980329	2.95 ± 0.95	29 ± 10	0.29 ± 0.2	65 ± 5	0.97 ± 0.09	0.4250 ± 0.5120
GRB980703	0.9662 ± 0.0002	28 ± 10	3.4 ± 0.5	22.6 ± 2.26	0.94 ± 0.08	0.1090 ± 0.0751
GRB990510	1.6187 ± 0.0015	0.29 ± 0.15	1.57 ± 0.03	19 ± 2	1.29 ± 0.03	0.9370 ± 0.6548
GRB991208	0.7055 ± 0.0	18 ± 22	<2.1	100 ± 10	1.09 ± 0.03	0.0398 ± 0.0537
GRB991216	1.02 ± 0.02	4.7 ± 6.8	1.2 ± 0.4	194 ± 19.4	0.88 ± 0.09	0.0696 ± 0.1379
GRB000301	2.0335 ± 0.0003	26 ± 12	7.3 ± 0.5	2 ± 0.6	1.37 ± 0.36	0.4028 ± 0.4402
GRB000418	1.1182 ± 0.0001	27 ± 256	25.7 ± 5.1	20.00 ± 2	1.00 ± 0.02	0.0155 ± 0.1519
GRB000926	2.0369 ± 0.0007	27 ± 3	1.8 ± 0.1	6.20 ± 0.62	3.91 ± 1.33	0.1823 ± 0.0548
GRB010222	1.4769	1.7 ± 0.85	0.93 ± 0.15	120.00 ± 3	1.03 ± 0.04	0.1740 ± 0.1261
GRB021004	2.3351	30 ± 270	6.5 ± 0.2	2.55 ± 0.69	1.04 ± 0.06	0.4456 ± 4.1707
GRB030329	0.1685	5.5 ± 2.75	0.48 ± 0.03	163.00 ± 1.4	1.01 ± 0.03	0.1711 ± 0.1028

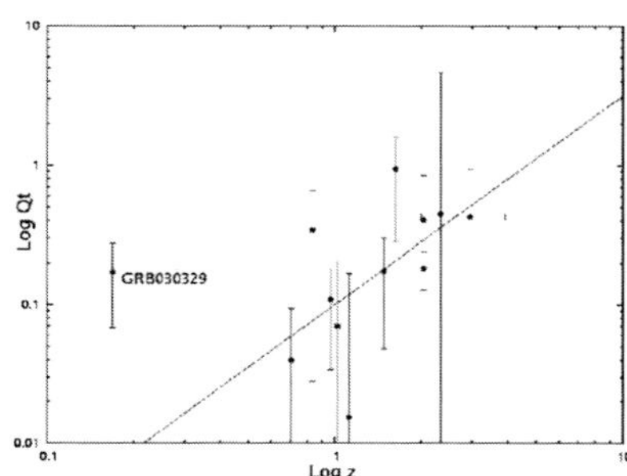

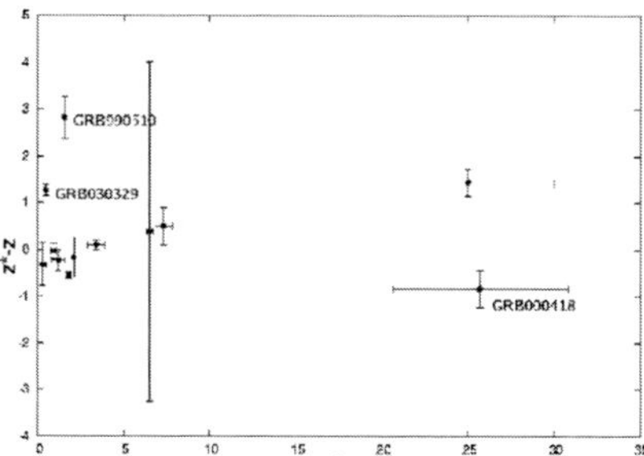

Figure 1. (left) A plot of $logQ_T$ vs. $logz$. We observe one main outlier to the relationship, GRB030329. This is the only GRB in our sample associated with a supernova, it has the highest fluence and is also the closest at z=0.1685. (right) The dispersion of t_j around z*-z.

If the observed trends are indeed due to some intrinsic characteristic of the bursts, then the above method would be extremely useful not only as a redshift estimator but would also be a useful tool to fill in the many gaps in the current set of GRB results. It is clear however that more data is required to understand fully both the nature of GRBs and their potential as probes of the high redshift universe. Continued research should include data mining of bursts with well established values for S_γ, n, t_j, and z. This would better provide a costraint to the relationship found from Figure 1. Also a more complete treatment involving D_T is necessary. If this relationship is true then a plot of D_T vs. Q_T should give a linear relationship with a slope equal to the value for $E_\gamma^{4/3}$.

References

Amati, *et al.* 2002, *A&A* 390, 81

Berger, Kulkarni and Frail 2003, *ApJ* 590, 379

Bloom, *et al.* 2001, *AJ* 121, 2879

Bloom, *et al.* 2003, *ApJ* 594, 674

Frail, D. A., Waxman, E., and Kulkarni, S. R. 2000, *ApJ* 537, 191

Frail, D. A. *et al.* 2001, *ApJ* 562, L55

Ghirlanda, *et al.* 2004, *ApJ* 616, 331

Rees and Mezsaros 1994, *ApJ* 430, 93

Rhoads, J. E. 1999, *ApJ* 525, 737

Sari and Piran 1999, *A&A* 138, 537

Sari, R. 1999, *ApJ* 524, L43

Sari, R., Piran, T., and Halpern, J. P. 1999, *ApJ* 519, L17

Wei and Lu 1997, *A&A* 323, 312

Populations of High Energy Sources in Galaxies
Proceedings IAU Symposium No. 230, 2005
E. J. A. Meurs & G. Fabbiano, eds.

© 2006 International Astronomical Union
doi:10.1017/S1743921306008374

Observing High-Energy Sources with REM: A Facility for Fast GRB Follow-up

P. Ward and E.J.A. Meurs (on behalf of the REM collaboration)

Dunsink Observatory, Castleknock, Dublin 15, Ireland

Abstract. REM is a fast slewing automatic telescope dedicated to the prompt observation of GRB afterglows. The telescope automatically reacts to GCN alerts, beginning follow-up observations in both infra-red and optical wavelengths. Recent observations of GRBs have shown the capabilities of REM to begin follow up observation in the order of 10s of seconds after the burst.

Keywords. gamma rays: bursts, instrumentation: miscellaneous.

REM is a fully robotic fast slewing telescope hosting a near infra-red imaging camera covering the 0.95–2.3 μm range and optical imager/spectrograph covering the range 0.45–0.95 μm. With such instrumentation REM is an extremely useful facility for a broad range of applications. It is REMs objective to observe GRBs simultaneously in both the optical and IR. For about half of the bursts with an X-ray afterglow an optical afterglow has been detected as well. In these cases the monochromatic flux decreases in time as a power law $F_\nu(t) \sim t^\delta$ with δ in the range 0.8–2. The first detection of an afterglow made the community realise that a significant amount of information about GRB physics is contained in the early afterglow.

Below are the initial results of GRB observations made by the REM telescope. In some cases the burst was not visible for a number of hours from the telescope site.

GRB050607 – Observations were performed in fully automated mode simultaneously in the near infrared and in the optical, starting approximately **47** seconds after the alert. Two R-band 3s exposures and five H-band 10s exposures confirm a non-detection, from which upper limits at 3 sigma level of R=17.1 and H=16 can be derived. Magnitudes have been calibrated with a set of stars around the position of the optical detection (GCN 3527) extracted from the 2MASS catalog for the H filter and from the USNO B1.0 catalog for the R filter.

GRB050509c – Observations of GRB 050509 were performed approximately **56** seconds after the burst (**32** seconds after the reception of the burst alert). At 2.5 hours after the burst no object was found at the position of the candidate down to a limiting magnitude of R=19 and H=15.5 (3-sigma level). We note that our R limit is consistent with the light curve power-law index observed by Gorosabel *et al.* (GCN 3425).

GRB050408 – The field was imaged with both REM instruments in V, R, I, J, H and Ks filters starting at 00:31 UT (approximately 8.2 hours after the burst) for a total integration time of 200 seconds for each filter. No sources are detected within SWIFT XRT error circle (Wells *et al.*, GCN 3191) and at the position of the Optical Transient (Ugarte *et al.*, GCN 3192; Huang *et al.*, GCN 3196) down to a limiting magnitude of 18.5, 18.3, 17.9, 16.8, 17.4, 18.4 (5-sigma upper limit) for V, R, I, J, H, and K filters.

GRB050306 – The field was imaged in R, J, H and Ks filters starting at 8:28 UT (approximately 28.9 hours after the burst). Total integration time was 32, 84, 84 and 94 seconds respectively. A quicker observation could not be performed before due to the

Moon proximity. Comparison with USNO B1.0 and 2MASS catalogues did not reveal new sources down to a limiting magnitude of R > 17.0, J > 15.4, H > 15.2 and Ks > 15.0 (3-sigma upper limit).

GRB050209 – Observations were performed in the V, R and I filters (REM IR camera was in maintenance). Observation started on 2005 Feb 09.163 UT, approximately 2.4 h after the GRB. The field was imaged in each filter for a total exposure time of 400s, under good seeing conditions. No new sources were detected within the HETE error circle by comparison with the Digitized Sky Survey, down to limiting magnitudes V=19.1, R=19.1, I=18.5 (5-sigma limits).

GRB050128 – Optical observations were carried out in V, R, I filters starting on 2005 Jan 28, at 7:10 UT, and ending at 9:45 UT (approximately 3 hours after the burst). Observations have been partly affected by the presence of the bright Moon close to the field, so only frames with short exposure time can be used. The field was imaged in each filter for a total exposure time of 50s. No new sources were detected within the XRT error circle (Antonelli *et al.*, GCN 2991) down to limiting magnitudes of V > 18.2, R > 18.2 and I > 17.9.

GRB041224 – Observations were carried out in the J, H and K filters, starting on 2004 Dec 25, at 1:06 UT, and ending at 1:39 UT; approximately 5 hours after the burst. Visual inspection of the NIR frames didn't reveal any new object when compared with the 2MASS catalog.

GRB041223 – Observations were carried out on 2004 Dec 24 from 02:37 UT to 03:11 UT; about 12 hours after the burst. The comparison with the 2MASS catalog did not reveal new infrared sources with S/N > 5 at the limit of the catalog. In particular no source is present at the position of the X-ray afterglow detected by Swift-XRT (Burrows *et al.* GCN 2901) and of the optical transient reported by Berger *et al.* (GCN 2902) and Malesani *et al.* (GCN 2903). The optical observations did not show any convincing candidate either. The 3-sigma upper limit in the R band is R ∼ 18.

Initial results with REM, since commissioning phase, have shown the telescope's ability to begin automated observations within 10s of seconds of the burst trigger. In the case of the most recent bursts, GRB050607 and GRB050509c, observations were performed simultaneously in the near infrared and optical starting approximately **47** and **32** seconds, respectively, after the alert.

References

Antonelli, *et al.*, *GCN* 2991
Berger, *et al.*, *GCN* 2902
Burrows, *et al.*, *GCN* 2901
Gorosabel, *et al.*, *GCN* 3425
http://www.merate.mi.astro.it/docM/reports/ann2002/annuario/rem/node3.html
http://gcn.gsfc.nasa.gov/gcn3_archive.html
Malesani, *et al.*, *GCN* 2903
Rhoads, *et al.*, *GCN* 3527
Wells, *et al.*, *GCN* 3191

Populations of High Energy Sources in Galaxies
Proceedings IAU Symposium No. 230, 2005
E. J. A. Meurs & G. Fabbiano, eds.

© 2006 International Astronomical Union
doi:10.1017/S1743921306008386

INTEGRAL and *XMM-Newton* observations of GRB 040223

S. McBreen[1] and S. McGlynn[2]

[1]Astrophysics Mission Division, RSSD of ESA, ESTEC, Noordwijk, the Netherlands
email: smcbreen@rssd.esa.int
[2]Department of Experimental Physics, University College Dublin, Dublin 4, Ireland
email: smcglynn@bermuda.ucd.ie

Abstract. We present gamma–ray and X-ray analysis of GRB040223 observed by *INTEGRAL* and *XMM-Newton*. GRB 040223 has a peak flux of $(1.6 \pm 0.1) \times 10^{-8}$ ergs cm^{-2} s^{-1}, a fluence of $(4.4 \pm 0.4) \times 10^{-7}$ ergs cm^{-2} and a steep photon power law index of -2.3 ± 0.2, in the energy range 20–200 keV. The steep spectrum implies that it is an X-ray rich GRB with emission up to 200 keV and $E_{\mathrm{peak}} < 20$ keV. The luminosity-lag relationship was used to obtain a redshift $z = 0.10^{+0.04}_{-0.02}$. The isotropic energy radiated in γ-rays and X-ray luminosity after 10 hours are both orders of magnitude less than classical GRBs.

Keywords. gamma rays: bursts, gamma rays: observations.

1. Introduction

The prompt emission from GRBs and the afterglow give valuable information on the radiation processes and the environment. There seems to be a continuum of spectral properties for X-ray flashes (XRF), X-ray rich GRBs and classical GRBs and it is probable that they have a similar origin (Sakamoto *et al.* 2004).

ESA's International Gamma-Ray Astrophysics Laboratory *INTEGRAL* (Winkler *et al.* 2003) is composed of two main coded-mask telescopes; an imager IBIS (Ubertini *et al.* 2003) and a spectrometer SPI (Vedrenne *et al.* 2003) with a combined energy range of 15 keV to 8 MeV. *INTEGRAL* has detected and localised 30 GRBs so far. The EPIC cameras on *XMM-Newton* (Turner *et al.* 2001) have been used to obtain X-ray afterglows of 6 of those GRBs. We report here observations of the prompt and afterglow emission of GRB 040223 and a detailed account is given elsewhere (McGlynn *et al.* 2005).

2. Data Analysis & Results

GRB 040223 was detected by the INTEGRAL burst alert system IBAS (Mereghetti *et al.* 2003). The IBIS light curve is given in Fig. 1a. GRB 040223 is in the long duration class with a well resolved pulse. The IBIS light curve was denoised with a wavelet analysis (Quilligan *et al.* 2002) and the risetime, fall time and FWHM of the pulse are 19 s, 22 s, and 13 s respectively. The IBIS data was divided into two energy channels i.e. 25–50 keV and 100–300 keV. The cross-correlation analysis (Norris 2002, Schaefer 2004) was performed between the two channels and the lag was determined to be 2.2 ± 0.3 s.

The IBIS spectral analysis was performed using the standard method (Moran *et al.* 2005). The IBIS data (20–200 keV) is well fit by a single power law with photon index -2.3 ± 0.2 with a reduced χ^2 of 1.01 for 20 degrees of freedom (dof) with errors at the 90% confidence level (Fig. 1b). The peak flux is $(1.6 \pm 0.1) \times 10^{-8}$ ergs cm^{-2} s^{-1} over the brightest second and the fluence is $(4.4 \pm 0.4) \times 10^{-7}$ ergs cm^{-2}.

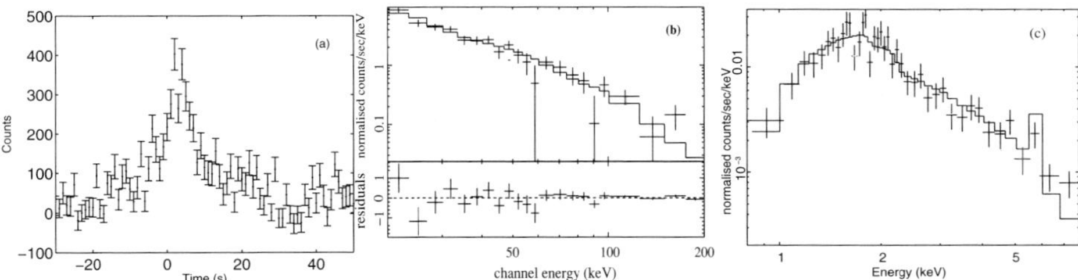

Figure 1. a) IBIS lightcurve of GRB 040223 in the energy range 15–200 keV; zero time is the IBAS trigger at 13:28:10 UTC. **b)** IBIS spectrum of GRB 040223 fit by a power law model from 20–200 keV. **c)** EPIC-PN spectrum of the GRB 040223 afterglow and its best fit absorbed power law model.

XMM-Newton observed the location of the GRB for 42 ks starting 18 ks after the burst where a fading X-ray source was detected. The temporal decay of the X-ray afterglow ($F_\nu(t) \propto t^{-\delta}$) was fit by a power law with index $\delta = -0.7 \pm 0.25$ by Gendre *et al.* (2004). Our analysis is consistent with this result. We obtained 3 afterglow spectra from the PN and MOS Cameras (0.2–10 keV) after standard data screening. The spectra were well fit by a power law $F_\nu \propto \nu^{-\beta_x}$ where the spectral index $\beta_x = 1.7 \pm 0.2$ with reduced χ^2 of 1.29 for 111 dof (Fig. 1c). The absorption column density has a high value of $N_H = 1.8 \times 10^{22}$ cm^{-2}, exceeding the high galactic value in this direction of 6×10^{21} cm^{-2}.

There are no direct measurements of the redshift to GRB 040223 so model dependent distance indicators were used. The luminosity-lag relationship (Norris 2002) was used to calculate the peak luminosity of $3.8^{+3.8}_{-1.7} \times 10^{47}$ ergs s^{-1} (McGlynn *et al.* 2005). The redshift to the source is z $= 0.10^{+0.04}_{-0.02}$ when the peak flux of $1.6 \pm 0.1 \times 10^{-8}$ ergs cm^{-2} s^{-1} is combined with the peak luminosity. The fluence gives a total isotropic γ–ray luminosity (E_{ISO}) of approximately 10^{49} ergs which is about three orders of magnitude less than classical GRBs. GRB 040223 is sub-luminous in γ–rays by a large factor.

The X-ray flux after 10 hours is $2.4 \pm 0.4 \times 10^{-13}$ ergs cm^{-2} s^{-1} in the 2–10 keV region. The X-ray luminosity of GRB 040223 is 6×10^{42} ergs s^{-1} and is orders of magnitude fainter than observed from classical GRBs (Bloom *et al.* 2003). The X-ray and γ-ray luminosities of GRB 040223 and XRF 030723 are very comparable.

References

Bloom, J. S., Frail, D. A., & Kulkarni, S. R. 2003, ApJ, 594, 674

Gendre B., Piro L., De Pasquale M.[astro-ph/0412302]

McGlynn, S., McBreen, S., Hanlon, L., *et al.* 2005, [astro-ph/0505349]

Mereghetti, S., Götz, D., Borkowski, J., Walter, R., & Pedersen, H. 2003, A&A, 411, L291

Moran, L., Mereghetti, S., Götz, D., *et al.* 2005, A&A, 432, 467

Norris, J. P. 2002, ApJ, 579, 386

Quilligan, F., McBreen, B., Hanlon, L., *et al.* 2002, A&A, 385, 377

Sakamoto, T., Lamb, D. Q., Graziani, C., *et al.* 2004, ApJ, 602, 875

Schaefer, B. E. 2004, ApJ, 602, 306

Turner, M. J. L., Abbey, A., Arnaud, M., *et al.* 2001, A&A, 365, L27

Ubertini, P., Lebrun, F., Di Cocco, G., *et al.* 2003, A&A, 411, L131

Vedrenne, G., Roques J. P., Schönfelder V. *et al.* 2003, A&A, 441, L63

Winkler, C., Courvoisier, T. J.-L., Di Cocco, G., *et al.* 2003, A&A, 411, L1

Populations of High Energy Sources in Galaxies
Proceedings IAU Symposium No. 230, 2005
E. J. A. Meurs & G. Fabbiano, eds.

© 2006 International Astronomical Union
doi:10.1017/S1743921306008398

Young X-ray-Emitting Supernovae in Galaxies

Eric M. Schlegel[1,2]†

[1]High Energy Astrophysics Division, Smithsonian Astrophysical Observatory, Cambridge, MA
02138 USA
email: eschlegel@cfa.harvard.edu

[2]Department of Physics and Astronomy, University of Texas-San Antonio,
San Antonio, TX 78249 USA
email: eric_schlegel@utsa.edu

Abstract. This talk reviews the observations of the high-energy emission of young supernovae, providing an update from previous reviews in 1995 and 2003. A summary plot shows the number distribution of X-ray luminosities, currently totalling 25 supernovae, from which it is clear that SN IIP are weak X-ray sources, SN IIn are very luminous sources, and SN Ib/c cover a broad range in luminosity.

Keywords. supernovae; supernovae: individual; X-rays: individual; supernova remnants.

1. Introduction

The X-ray emission of supernovae (SNe) provides insight into the late phases of stellar evolution by probing the explosion and subsequent interaction of the outgoing shock with the progenitor's circumstellar medium. Studies of this X-ray emission commenced with the discovery of SN1980K in NGC 6946 by Canizares *et al.* (1982) and accelerated with the launches of *ROSAT* and *ASCA* in the 1990s. The use of both satellites yielded studies of three important objects, SN1978K in NGC 1313, SN1993J in NGC 3031 (M81), and SN1987A (Schlegel *et al.* (1999), Schlegel, Petre, & Colbert (1996), Ryder *et al.* (1993); Immler *et al.* (2001), Zimmermann *et al.* (1994), Suzuki & Nomoto (1995), Kohmura *et al.* (1994), Leising *et al.* (1994); and Hasinger, Aschenbach, & Truemper(1996) and references therein, respectively). In addition, observations of SN1998bw using *BeppoSAX* provided the first link between the X-ray afterglow of a GRB and a SN (Pian *et al.* (1999)).

The observations have been reviewed in Immler & Lewin (2003) and Schlegel (1995). Here, I update these reviews to the current epoch (*circa* July 2005).

2. Emission Mechanisms

The emission mechanisms may be divided into two groups based on the temporal evolution.

The first group of mechanisms may be labelled as 'prompt' or 'early'. The mechanisms include Compton scattering of γ-rays, inverse Compton scattering of optical or radio photons, and the prompt thermal burst of photons associated with the break-out of the shock through the progenitor's atmosphere. Compton scattering predicts zero X-ray flux at energies below $\sim$16–20 keV because of the photoelectric absorption. Inverse Compton scattering may produce X-rays but is expected to be confined to the earliest phases; the process is temporally limited as the expansion causes the debris field to become

† This research was supported by NASA Contract NAS8-39073 to SAO for the *Chandra Observatory*.

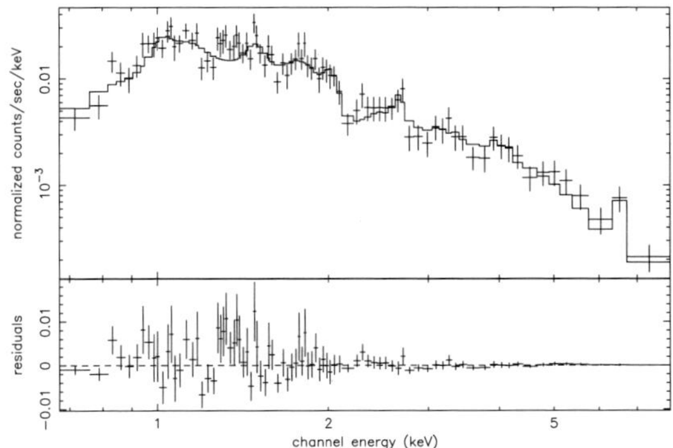

Figure 1. SN1986J as observed by *XMM-Newton*. A feature at $\sim$1 keV is consistent with Ne emission; Fe K emission is also reported.

increasingly optically thin. The prompt thermal burst is expected to produce $\sim$100 keV photons, but the burst should be brief, on the order of seconds. For a review, please see Schlegel (1995) and references therein.

The second group consists of the 'sustained' or 'late' mechanisms and include X-rays arising from pulsar input and circumstellar (CS) interactions. The pulsar input mechanism is expected to produce a maximum luminosity of $\sim$5$\times$10^{38} erg s^{-1} with an efficiency of $\sim$1.5% as described in Chevalier & Fransson (1992). The CS interaction has been described by Chevalier & Fransson (1994). In this process, the outgoing shock probes the CS medium and is expected to generate hard X-rays ($>$10 keV) for the first $\sim$100 days. As the outgoing shock plows up material, a reverse shock is eventually established that probes the progenitor 'atmosphere'. The spectrum of the reverse shock softens as the optical depth decreases. Measurements of the decline of the X-ray light curves in a soft and a hard band in principle provide measures of the density exponents of the progenitor and CS matter.

3. Recent Observations

Here, I provide a brief summary of observations obtained over the past few years.

SN1970G in M101 has been recovered in a subset of the 1 Msec *Chandra* observation of M101 and is described in Immler & Kuntz (2005). A light curve was obtained by re-analyzing a set of *ROSAT* observations once the *Chandra* observation showed that the *ROSAT* sources were consistent with SN1970G. The detected decline rate of -1.7 ± 0.6 is typical of the late emission of SNe.

P. Chandra and co-workers using *Chandra* observed SN1995N and reported in Chandra *et al.* (2005). An 'uptick' in the *ROSAT+ASCA* light curve, reported by Fox *et al.* (2000), was shown to be the emission of blended sources. Interpolating the trend between the *ROSAT* and *Chandra* data points at the *ASCA* epoch provided an approximate measure of the SN emission. Re-scaling of the *Chandra* flux for SN1995N and convolving it plus the surrounding field with the *ASCA* point spread function showed that the original *ASCA* extracted flux and the convolved flux were consistent to within 20–25%. Blended point sources at the *ASCA* epoch provide a natural explanation for the apparent increase. The extracted *Chandra* spectrum of SN1995N shows evidence for enhanced Ne emission at $\sim$1 keV.

An *XMM-Newton* GTO observation of SN1986J in NGC 891 has been described by Temple *et al.* (2005) (Figure 1). The spectrum shows evidence for enhanced Ne, S, and possibly Mg as well as the presence of an emission line at Fe Kα. The slope of the light

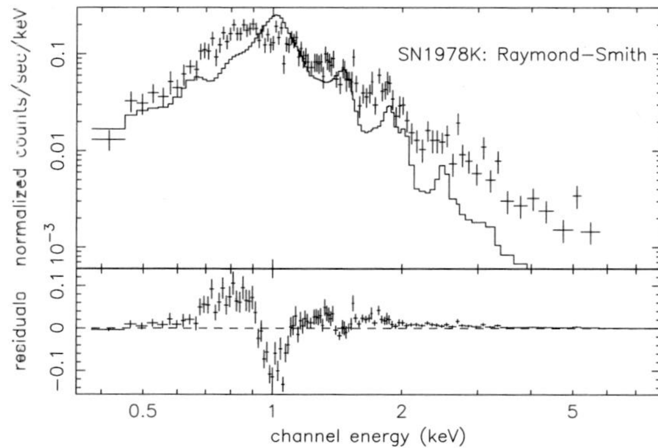

Figure 2. The spectrum from a *Chandra* observation of SN1978K in NGC 1313 with the best-fit *ASCA* spectrum overlaid. Clear spectral evolution has occurred.

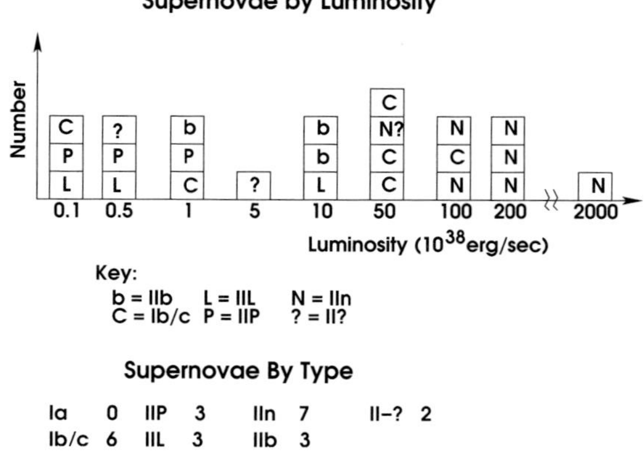

Figure 3. The distribution of X-ray emitting SNe by luminosity and type. For this figure, *no* adjustment has been made to generate a uniform energy band from which the flux and luminosity were calculated. As a result, the figure illustrates general trends, but may be incorrect in detail.

curve is -2.99 ± 0.45 whereas free-free emission or line emission produce slopes of -1 and -1.7, respectively. As the slope is determined by several *ROSAT* data points, it is possible that the point spread function of the *ROSAT* mirrors provides extra contamination however, the authors took care to account for the resolution. The slope thus remains a puzzle for future observations of SNe.

SN1988Z is still present in a *Chandra* observation as described by Schlegel & Petre (2005). The quantile cclor (Hong, Schlegel, & Grindlay (2004)) is consistent with a temperature of ~ 2 keV. The light curve shows a decline that is slower than that implied by considering only the *ROSAT* data points. In contrast to SN1986J, for SN1988Z, spatial resolution is not an issue as the *Chandra* field shows the sources are well-separated.

Finally, Schlegel *et al.* (2004) report that SN1978K, the SN-equivalent of a particular well-known advertising bunny, is still generating $\sim 10^{39}$ erg s^{-1}. Spectral evolution is definitely present as the best-fit model from the *ASCA* epoch no longer is a good fit (Figure 2). A two-temperature model is required with soft (0.6 keV) and hard (3.2 keV) components. The soft band light curve remains consistent with a constant while the hard band light curve shows a drop by $\sim 50\%$ from the *ASCA* epoch. This likely signals the start of the decline of SN1978K.

Table 1. Band Passes for X-ray Luminosities of SNe

Band	Number	Band	Number	Band	Number
0.3–2	2	0.1–2.4	6	0.2–2	1
0.5–2	1	2–10	4	0.5–10	3
0.3–5	1	0.2–10	1	0.1–10	1
0.4–8	1	0.5–8	2		

Note: energies are in keV; 'Number' represents # SNe luminosities reported in that band.

New SNe have been detected, but generally have been observed only once (atrocious!) or twice (bad!) each. These include 2001em (Ib/c), 2001gd (IIb), 2001ig (IIb), 2002hh (II), 2002hi (IIn), 2003bg (Ic), 2003L (Ic), and 2004dj (IIP).

We can then update the numbers: Schlegel (1995) reviewed 5 SNe (78K, 80K, 86J, 87A, 93J); Immler & Lewin (2003) described an additional 10 SNe (79C, 88Z, 94I, 94W, 95N, 98S, 98bw, 99em, 99gi, and 02ap). Currently, 24 SNe are present on S. Immler's list† of SNe. The number should be 25: SN1968D was recovered as described in Holt *et al.* (2003). This list leads to Figure 3 showing the distribution of SNe by luminosity (note that the luminosity axis is not to scale).

4. X-ray Emission by SN Type

4.1. *SN Ia*

Type Ia SNe have been predicted to be strong X-ray sources. Unfortunately, to date, none have been detected. The best upper limit for a SN Ia prior to the launches of *Chandra* and *XMM-Newton* was that of Schlegel & Petre (1993) for SN1992A from a *ROSAT* observation obtained at an age of +15 days. The estimated limit was 1.5×10^{38} erg s^{-1} in the 0.5–2 keV band. The value was correctly criticized for not considering circumstellar absorption. It still represents a real upper limit, however, it does not provide a limit on the amount of circumstellar absorption. The newest upper limit, by Hughes *et al.* (2005) from a *Chandra* observation of SN2002bo at an age of +2 days, is 5×10^{38} erg s^{-1} in the 2-8 keV band. As the host galaxy is NGC 3190 (D $\sim$ 19–22 Mpc), there remains room for improvement if Nature cooperates.

4.2. *SN IIP*

In 2001, Schlegel (2001) hypothesized that SN IIP have very little CS matter and will never be strong X-ray sources. That hypothesis was based upon *Chandra* observations of SN1999em and SN1999gi. Since that time, the only additional IIP object is SN2004dj (NGC 2403). The detected luminosity does not counter the hypothesis, being in the 10^{36-37} erg s^{-1} range.

4.3. *SN Ib/c*

The association of GRBs and SN Ib/c was firmly established with the observations of GRB 030329 and SN2003dh by Stanek *et al.* (2003). The key question is whether every GRB produces an SN Ib/c and vice versa. The available evidence implies the answer is 'no': Soderberg *et al.* (2005) argue that radio emission from a sub-relativistic jet would spread into our line-of-sight regardless of the pointing direction of the original jet. Their radio survey shows <17% of Ib/c SNe show radio emission, suggesting a small fraction of jets are produced. We clearly have more to learn about the association of GRBs and SNe.

† Available at lheawww.gsfc.nasa.gov/users/immler/supernovae_list.html

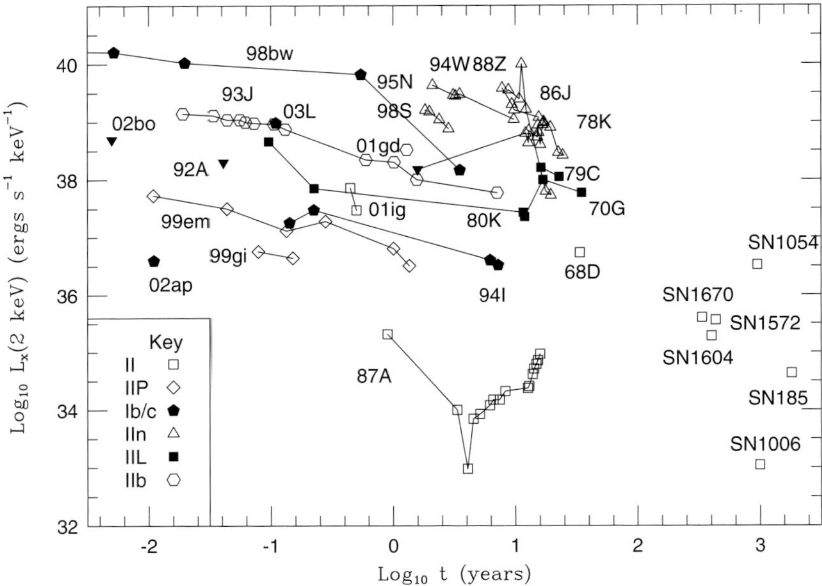

Figure 4. Plot of most of the available observations of the X-ray emission of SNe at 2 keV. The "2 keV" flux was chosen because it is common to all bandpasses of the various X-ray detectors represented. Missing from the figures are data points for 2004dj, 2003bg, 2002hi, 2002hh, and 2001em. Of these, all are single points except 2004dj. The key symbols apply to the SNe and not the remnants in the lower right corner. A variety of remnants are presented to show the eventual 'end point' of the SNe light curves.

Finally, we may aggregate all (nearly all!) of the light curve data to date by estimating the flux at 2 keV. Why 2 keV? To date, this energy is common to every satellite that has been used to study the X-ray emission of SNe (Table 1). Figure 4 shows the resulting light curves. The light curves have been keyed to the SN type and include the two upper limits on the SNe Ia. On the lower right side of the figure are shown the 2 keV luminosities of several well-known SN remnants, the eventual destination of each of the light curves. From this figure, we may infer that all of the SN types are represented with detections except for SN of Type Ia, that the least luminous objects appear to be Type IIP (and highlight the distinct difference of SN1987A), that the most luminous objects are SN IIn, and that the SN Ib/c show a very broad range in luminosity (a factor of several 1000 separate SN2002ap and SN1998bw). This figure also highlights the rather spotty temporal coverage as fully 1/3 of the objects are present with one or two data points, and an additional 1/3-1/2 are represented by 3–4 data points, leaving about 1/5 (ie, ~4) with sufficient temporal coverage that a study of trends could be possible (78K, 87A, 93J, and 99em). This illustrates the need for better temporal coverage of a range of SN types.

5. Summary

The X-ray emission of SNe is hard at early times, then softens. The observations to date provide a mix of slow and fast fading, suggesting that a higher rate of coverage could provide detailed information on the late stages of the progenitor's mass loss. As a result of the outward propagation of the shock at velocities of 1 to a few $\times 10^4$ km s^{-1}, we view the mass loss history at a considerably increased rate.

The data collected to date also provides some observational goals. First, the observation of a nearby SN Ia (D $\lesssim$5–8 Mpc) would either detect or set a critical upper limit on the presence of a circumstellar medium. To avoid possible interpretation problems, the detector should cover a hard X-ray band. Second, the detection and evolution of emission lines in a SN IIP would provide details for the hypothesized tenuous circumstellar medium surrounding these objects. Third, SN Ic and SN IIn objects, preferably more than one of each, need excellent temporal coverage to follow their light curves and spectral evolution.

Acknowledgements

I thank the conference organizers for the invitation to speak. The research and travel were supported by the Chandra X-ray Center through NASA contract number NAS8-39073 to the Smithsonian Astrophysical Observatory for the operation of the *Chandra Observatory*.

References

Canizares, C., Kriss, G., & Feigelson, E. 1982, ApJ, 253, L17

Chandra, P., *et al.* 2005, ApJ, 629, 933

Chevalier, R. & Fransson, C. 1992, ApJ, 395, 540

Chevalier, R. & Fransson, C. 1994, ApJ, 420, 268

Fox, D. W., *et al.* 2000, MNRAS, 319, 1154

Hasinger, G., Aschenbach, B., & Truemper, J. 1996, A&A, 312, L9

Holt, S., *et al.* 2003, ApJ, 588, 792

Hong, J. S., Schlegel, E. M., & J. E. Grindlay, 2004, ApJ, 614, 508

Hughes, *et al.* 2005, in preparation

Immler, S. & Kuntz, K. 2005, astro-ph/0506023

Immler, S. & Lewin, W. 2003, in Supernovae and Gamma-Ray Bursters, ed. K. Weiler (New York: Springer), 91

Immler, S., Aschenbach, B., & Wang, Q. D. 2001, ApJ, 561, L107

Kohmura, Y., *et al.* 1994, PASJ, 46, L157

Leising, M. D., *et al.* 1994, ApJ, 431, L95

Pian, E., *et al.* 1999, A&AS, 138, 463

Ryder, S., *et al.* 1993, ApJ, 416, 167

Schlegel, E. M. 2001, ApJ, 556, L25

Schlegel, E. M. 1995, Rprt Prog Phys, 58, 1375

Schlegel, E. M. & Petre, R. 2005, in preparation

Schlegel, E. M., *et al.* 2004, ApJ, 603, 644

Schlegel, E. M., *et al.* 1999, AJ, 118, 2689

Schlegel, E. M., Petre, R., & Colbert, E. J. M. 1996, ApJ, 456, 187

Schlegel, E. M. & Petre, R. 1993, ApJ, 412, L29

Soderberg, A., *et al.* 2005, astro-ph/0507147

Stanek, K., *et al.* 2003, ApJ, 591, L17

Suzuki, T. & Nomoto, K. 1995, ApJ, 455, 658

Temple, R., Raychaudhury, S., & Stevens, I. 2005, astro-ph/0506657

Zimmermann, H.-U., *et al.* 1994, Nature, 367, 621

Discussion

ZEZAS: The possibility that some ULXs could be SNRs is valid if we have single snapshot observation. However, for many galaxies we now have multiple exposures showing that many ULXs are variable associating them with X-ray binaries.

SCHLEGEL: I agree.

ERACLEOUS: Can you please summarize the best estimate of the duration of the "early hard phase" and the best estimate of the 2–10 KeV luminosity during this phase.

SCHLEGEL: SN 1993 is probably the best case for answering this question. The duration is about 100 days and luminosity is about 10^{39} erg/s, although theoretically it could reach 10^{40} erg/s.

Populations of High Energy Sources in Galaxies
Proceedings IAU Symposium No. 230, 2005
E. J. A. Meurs & G. Fabbiano, eds.

© 2006 International Astronomical Union
doi:10.1017/S1743921306008404

Impact of Historical Chinese astronomical records on high-energy sources in our Galaxy

Zhenru Wang

Department of Astronomy, Nanjing University, Nanjing, 210093, China
email: zrwang@nju.edu.cn

Abstract. Owing to the rapid progress of space sciences and the high-energy astrophysics in the past 2–3 decades, historical supernova records have made important contributions to modern astrophysics and will continue to do so in the future. The main topics here are the earliest supernova observed by human beings and the AD393 guest star as well as our new concept of "Po stars"

Keywords. Supernova, Historical Supernova; ISM, Supernova Remnant.

1. Introduction

Before renaissance, most historical SN (supernova) records are from oriental countries, mainly from China. Biot (1846) and Houmbolds (1850) were the pioneer western astronomers who firstly introduced historical Chinese astronomical records to the West. Since the 20th century more and more astronomers have worked on this field.

Historical SN records are very important because they include the historical clues to the origin and evolution of SNR (supernova remnant) and NS (neutron star) and can provide new information to the active fields of astrophysics and physics. The best example is the famous Tian-Guan guest star with its remnant Crab Nebula (Wang (1987a)). The main topics here are the earliest SN observed and recorded by human beings and AD 393 guest star as well as the new concept of the "Po stars".

2. The Great New Star

Which SN is the earliest SN observed and recorded by human beings? My answer is the Great New Star that occurred in the 14th century BC, observed by ancient Chinese and recorded in an oracle bone (see Fig. 1 Left) with the following inscription: "On the 7th day, a Ji-Si night, a Great New Star appeared side by side with the Antares (α Sco)."— From 2nd part of 9th chapter of Yin-Xu-Shu-Qi-Hou-Bian. Antares was an important star to the agriculture and calendar at the ancient time in China.

The Great New Star was considered as a SN by Dong (1945), Xi (1955), Needham (1959) and Xi & Bo (1965). But they did not consider what its remnant was. In 1987, I proposed that the remnant of the Great New Star be the contemporary γ-ray source 2CG 353+16 discovered by the European γ-ray satellite COSB. And 2CG 353+16 had been identified as the dark molecular cloud near ρ Oph (Mayer-Hasselwander *et al.* 1980), but it cannot offer enough γ-rays to explain the observed γ-ray flux (Black & Fazio 1973; Hiller 1984). A good idea is that a SN – the Great New Star – had explored inside there. It can resolve the above difficulty and so 2CG 353 +16 as the compact remnant (Wang 1987b; Wang 1987c) and diffuse remnant (Xu Wang & Qu 1992) of the Great New Star are still there. Narike (1999) more exactly determined the age of the Great New Star by use of the 5 lunar eclipses in the reign period of King Wu-ding.

After searching for the oracle bones, we can find more pieces of them probably related to the Great New Star as shown in Fig. 1 Middle & Fig. 1 Right a, b, c (taken from Guo 1980; Yao 1988). Combined Fig. 1 Left and Fig. 1 Right, the Great New Star was visible at least 40 days.

If we search the 3rd EGRET Catalogue (Hartman *et al.* 1999), the Great New Star seems to correspond to 3EG J1627-2419. And there is a note as '...the scale of the variations is much smaller than the pixel size in the EGRET maps. This may lead to apparent sources.' It seems to be helpful for further identification in the future.

3. AD393 – RX J1713.7-3946 (G347.3-0.5)

SNR RX J1713.7-3946 was discovered by the ROSAT All Sky Survey (Pfeffermann & Aschenbach 1996). We suggested it to be the remnant of AD393 guest star (Wang *et al.* 1997) as the principal basis for four-dimension identification (Wang *et al.* 1986). The historical record of AD393 guest star is in Fig. 1 (see Wang *et al.* 1997) and translated as follows : "A guest star appeared within the asterism Wei in the 2nd lunar month and disappeared in the 9th lunar month in the year of AD393 (the 18th year of the Tai-Yuan reign period)".

Both RX J1713.7-3946 and AD393 guest star are in good positional coincidence. The morphology of RX J1713.7-3946 from ROSAT observation seems to be central-ring-brighten expected to have originated from a SN evolving in a stellar wind bubble (Wang *et al.* 1997). The internal ring is confirmed by XMM-Newton observation (Cassam-Chenie *et al.* 2004) and the low density feature in the center of the SNR for the stellar wind bubble is obtained as less than $0.02/cm^3$ by XMM observation(Cassam-Chenaie *et al.* 2004) and as $0.01/cm^3$ by 2.6mm observation by Fukui *et al.* (2003).

The distance of RX J1713.7-3946 was in controversy at the beginning around its discovery. The distance of 1kpc for it was first obtained from ROSAT (Pfeffermann & Aschenbach 1996). Wang, Qu & Chen (1997) adopted it as 1kpc. Koyama *et al.* (1997) also got it as 1kpc from ASCA observation. But Slane *et al.* (1999) preferred 6kpc as its distance.

Recently, Fukui *et al.* (2003) used 4m radio telescope equipped with a sensitive super-conducting mixer receiver to observe 1.1 million positions in CO emission. They discovered the interacting molecular gas toward the TeV γ-ray peak of the SNR. According to this identification and the measurement for the mean velocity of the non-acceleration component, they determine the distance of the SNR to be $\sim$1kpc (Moriguchi *et al.* 2005; Fukui *et al.* 2003). The column density of HI and CO observation also supports it to be at 1kpc (Koo *et al.* 2004). From its XMM observation combined with the CO and HI survey (Cassam-Chenaie *et al.* 2004), it is pointed out that the CO and HI absorption column densities are in excellent agreement with their XMM measurements in different places of the remnant only if the SNR is at a 1.3±0.4kpc and not at 6kpc. Lazendic *et al.* (2004) suggested that its observed X-ray flux appears problematic to the total energy budget of RX J1713.7-3946 unless the source distance is a factor of 5 smaller than that they preferred before. Hence it seems that a lot of recent observations support it at 1kpc. It will be important for the further investigation of its specific properties.

Now RX J1713.7-3946 is a hot spot in the high-energy astrophysics. Besides it is one of the three shell type SNRs with nonthermal X-ray emission (Koyama *et al.* 1997; Slane *et al.* 1997), the TeV γ-rays were detected from it by CANGAROO (Enomoto *et al.* 2002) and H.E.S.S. experiment (Aharoniam *et al.* 2004). It is important whether the TeV γ-rays originate from the acceleration of protons and the decay of pions (Aharonian & Atoyan 1999; Enomoto *et al.* 2002). If it is the case, it would give the first evidence for the acceleration of protons in SNRs and then the neutrino emission would be further expected from it (Costantini & Vissani 2005). It would be very exciting for the development of the origin of cosmic rays and the high-energy astrophysics.

Although there are some different opinion on the emission mechanism of the TeV γ-rays by hadronic hypothesis right now, such as Reimer & Pohl (2002), it can stimulate to study RX J1713.7-3946 in more details both in observation and theoretically. Hopefully GLAST will offer more sensitive and good resolution data of spectra in the band up to Gev and subTeV for clearing up some divergences.

4. The new concept of "Po Star"

The bright suspected new star visible for longer than 6 months as a necessary condition (Clark & Stephenson (1997)) for SNe is certainly applicable only for nearby SNe. We expect distant SNe to be faint and shortly visible. There are many appellations for the ancient suspected new stars, such as guest star, strange star, broom star, and "Po star"etc. The most complicated one is "Po star". "Po star" is a term that was usually used to describe a faint comet or a comet just during its faint phase. So a "Po star" just looked like a very weak comet in visual magnitude, usually without a tail if no other additional notations. Our forbears could discover new faint point source never seen before, but they could not distinguish the nature of them. This is why the concept of a "Po star" came about (Wang Li & Zhao 2002).

Today a "Po star" can be considered as a faint point source around the visual magnitude of 5. There are about one hundred Po stars recorded from 2320BC to the beginning of 20th century. Xi & Bo (1965) suggested 7 criteria to distinguish SNe or Novae from them. The main standard is that the object was nonmoving and without tails. 20-30 Po Stars are suggested to be the candidates of SNe or novae (Xi & Bo 1965; Chen 1986). Based on our new concept of Po star the SGR 1900+14 (Hurley 1999) and the youngest PSR J1846-0258 (Gotthelf *et al.* 2000) are suggested to be the remnant of BC 4 Po star and AD1523 Po star respectively (Wang Li & Zhao 2002; Wang 2001).

Acknowledgements

Thanks to Zhao, Y. and Narike for their helpful discussion on my research, to Li, M. for his help on the manuscript. This work is supported by NNSF of China and the IAU Grant.

References

Aharonian, F. A. & Atoyan, A. M. 1999, *A&A*, 351, 330
Aharonian, F. A., *et al.* (H.E.S.S. Collaboration) 2004, *Nature*, 432, 75
Biot, E. 1846, *Connaissance des Temps* Additions, p. 67
Black, J. H. & Fazio, G. G.1973, *ApJ*, 185, L7
Cassam-Chenai, G., Decourchelle, A., Ballet, J., Sauvageot, J.-L., & Dubner, G. 2004, *A&A*, 427, 199
Chen, Z. G. 1986, *Chinese Astronomical History*, Vol. 3 (Taiwen: Ming Wen)
Clark, D. & Stephenson, F. 1997, *The Historical Supernovae* (Pergamon, Oxford)
Costantini, M. & Vissani F. 2005, *Astroparticle Physics*, 23, 477
Dong, Z. B. 1945, *Yin-Li-Pu (A study on the calendar and astronomy of the Shang Dynasty)*, Sichuan, China
Enomoto, R., *et al.* (CANGROO Collaboration) 2002, *Nature*, 416, 823
Fukui, Y., *et al.* 2003, *PASJ*, 55, L61
Gotthelf, R., *et al.* 2000, *ApJ*, 542, L37

Guo, M. R. 1980, *The Compilation of the Inscription of Tortoise Shells and Bones*

Hartman, R. C., *et al.* 1999, *APJS* 123, 79–202

Hiller, R. 1984, *Gamma Ray Astronomy*, Clarendon Press , Oxford

Houmboldt, K. 1850, *Kosmos, III*, p. 220

Hurley, K. 1999 *ApJ*, 510, L111

Koo, B., *et al.* 2004, *J. Korean Astron. Soc.*, 37, 61; IAU Sym. 218, Camilo & Gaensler eds. 73

Koyama, K. *et al.* 1997, *PASJ*, 49, L7

Lazendic, J., *et al.* 2004, *ApJ*, 602, 271

Mayer-Hasselwander, H. A., *et al.* 1980, *Proc. Texas Symp. 9th Relativ. Astroph.*, Ann. NY Acad. Sci., 336, 211

Moriguchi, Y., *et al.* 2005, *in X-Ray and Radio Connections* (eds. L. O. Sjouwerman and K. K Dyer)

Narike, T. 1999, *AAPPS Bulletin*, 9, No. 2, 10

Needham, J. 1959, *Science and Civilisation in China*, Vol. 3, Cambridge Univ. Press

Pfeffermann, E. & Aschenbach, B. 1996, *in Roentgenstrahlung from the Universe*, International Conf. on X-ray Astron. & Astroph., MPE report 263 (Zimmermann, Trumper & Yorke ed.), p. 267

Reimer, O. & Pohl, M. 2002, *A&A*, 390, L43

Slane, P., *et al.* 1997, *AAS*, 29, 1368

Slane, P., *et al.* 1999, *ApJ*, 525, 3

Wang, Z. R., Liu, J. Y., Gorenstein, P., & Zombeck, M. V. 1986, *in Highlights of Astronomy*, Vol 7, J. P. Swings ed., p. 583

Wang, Z. R. 1987a, *in The Origin and Evolution of Neutron Stars*, IAU Symp. 125, Helfand & Huang ed. (Reidel, Dordrecht), p. 305

Wang, Z. R. 1987b, *Science*, 235, 1485

Wang, Z. R. 1987c, *La Recherche*, 18, 1416

Wang, Z. R, Qu, Q. Y., & Chen, Y. 1997, *A&A*, 318, L59

Wang, Z. R. 2001, *in New Century of X-ray Astron.*, ASP Conf. Ser. 251, Inoue & Kunieda ed., 284

Wang Z. R., Li, Z. Y., & Zhao, Y. 2002, *ApJ*, 544, L49

Xi, Z. Z. 1955, *Acta Astronomica Sinica*, 3, 183

Xi, Z. Z. & Bo, S. R. 1965, *Acta Astronomica Sinica*, 13, 1

Xu, J. J., Wang, Z. R., & Qu, Q. Y. 1992, *A&A*, 256.

Yao, X. S. 1988, *The compilation of the explanation for facsimile copy of Yin-Xu oracle bones* (Beijing-Zhonghua Book Co.)

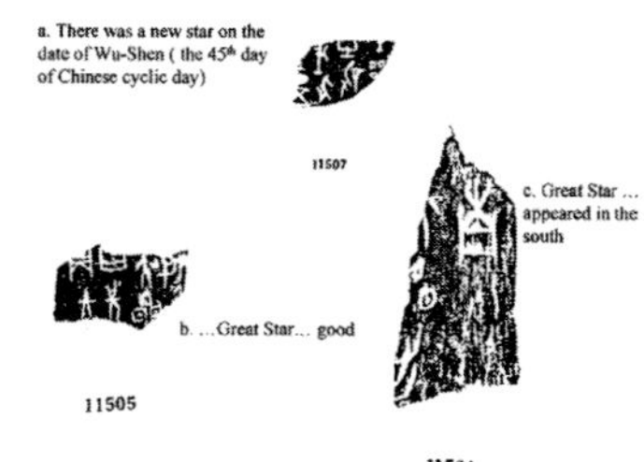

Figure 1. Chinese oracle bone records in the 14th century BC. **Left:** On the 7th day, a Ji-Si night, a Great New Star appeared side by side with the Antares (α Sco). **Middle:** There was a new star on the date of Xin-Wei (the 8th day of Chinese cyclic day) in the Heavenly Stems and Earthly Branches. **Right:** a. There was a new star on the date of Wu-Shen (the 45th day of Chinese cyclic day) b. ...Great Star...good c. Great Star...appeared in the south.

Discussion

MILEY: From a comparison of multiple reports of the same guest star in ancient records what was the typical positional accuracy of the reports?

WANG: It is hard to give a typical positional accuracy for guest stars. In the ancient records, their positions are usually recorded as near a certain famous star, related to an asterism or a constellation. So the positional accuracy is different for various cases. It depends on the stellar environment nearby the guest star.

Populations of High Energy Sources in Galaxies
Proceedings IAU Symposium No. 230, 2005
E. J. A. Meurs & G. Fabbiano, eds.

© 2006 International Astronomical Union
doi:10.1017/S1743921306008416

Supernovae astrophysics from Middle Age documents

V. F. Polcaro[1] and A. Martocchia[2]†

[1]IASF-INAF, Rome, Italy
email: polcaro@rm.iasf.cnr.it

[2]OAS, Strasbourg, France
email: martok@quasar.u-strasbg.fr

Abstract. The supernova explosion of 1054 AD, which originated the Crab Nebula and Pulsar, is probably the astronomical event which has been most deeply studied by means of historical sources. However, many mysteries and inconsistencies, both among the different sources and between what is deduced by the historical records and the present day astronomical data, are demanding extraordinary efforts by theoretical astrophysicists in order to put all the data in a meaningful framework. An accurate analysis of the historical sources, like the one we are presenting here, may contribute to solve some of these problems.

Galactic Supernovae are rare events and their testimonies are extremely important for astrophysics. To date, seven astronomical events documented by historical texts are believed to have been galactic supernovae (see Table 1). Information gathered from the historical sources concerning all of these events have been used in some way in astrophysical studies, though it is not always easy to extract quantitative data from ancient measurements.

Before the Cepheid distance scale was extended by HST to include the host galaxies of many SNe of Type Ia with good photometry, historical Type Ia SNe had been used to compute the Hubble constant by combining their brightness at maximum with the modern distances to their remnant in order to estimate the peak absolute magnitude. For instance, Schaefer (1996), from a careful reconstruction of the Type Ia SN 1572 light curve, obtained a peak magnitude of $V = -4.53 \pm 0.18$, corresponding to an absolute magnitude of $V_o = -18.64 \pm 0.31$, deriving $H_o = 66 \pm 12$ km s^{-1} Mpc^{-1}: this is an astonishing precise result, considering that it was obtained by using data gathered by naked eye more than 400 years ago!

However, it may be risky to use historical astronomical data for astrophysics if they have not been carefully checked from a historical point of view. The most indicative case is the one of SN 185, which was used for calibrating the brightness of Type Ia SNe and hence for deriving the Hubble constant. But, as has been shown by Schaefer (1995), most likely this event was *not* a supernova, but a transit of comet P/Swift-Tuttle – indeed the derived value of H_o was the unrealistic one of $H_o \simeq 150$ km s^{-1} Mpc^{-1}.

The importance of a careful historical analysis of the historical sources before using them for astrophysical purposes is further illustrated by the case of SN 1054.

The Song Empire sources were the first to be suggested as witnesses of the birth of the Crab Nebula by Hubble (1928) and Mayall (1939). It was found that they report a date when the Emperor was notified by the astronomer Yang Weide about the 1054 "guest star" appearance (*4th* July), the length of the period in which this star was visible in daylight (23 days), the date of the last sighting (*17th* April 1056), and the star position

† Present address: CESR, Toulouse, France.

Table 1. Possible historical Supernovae

Year	Date	mag	Remnant	SN Type	Source(s)
185 AD	Dec 7(?)	-2(?)	RWC 86(?)	Ia(?) Not a SN(?)	Chinese
393	Feb 27-28	-3(?)	RX J1713.7-3946(?)		Chinese
1006	Apr 30	-7.5	SNR 1006	Ia	Arabic, Chinese, Japanese, European
1054	Apr 11	-4(?)	M1 (Crab)	II(?) Ib(?)	Chinese, North American(?), Arabic, Japanese, European
1181	Aug 6	-1(?)	3C 58(?)	II(?) Ib(?)	Chinese, Japanese, European
1572	Nov 6	-4	Tycho SNR	Ia	Tycho Brahe, etc.
1604	Oct 9	-3	Kepler SNR	Ib(?)	Johannes Kepler, etc.

Table 2. The historical records of SN1054 (revised version of Tab.1 from Collins *et al.*, 1999)

Date	MJD	Ref.	location	appearance as...	likely m_V	Notes
04/11/1054	6126	a.	Fustat	*star*		Ibn Butlan
04/11/1054	6126	b.	Flanders	*bright disk at noon*	~-7	
04/24/1054	6139	c.	Ireland	*fiery pillar*		
late April 1054		d.	Rome	*bright light*	<-3.5	
05/10/1054	6155	e.	Liao Kingdom	*star*		Sun eclipse
05/14/1054	6159	f.	Armenia	*star*		
late May 1054(?)		g.	Italy	*very bright star*		
late May 1054		h.	Japan	*new star... as Jupiter*	~-4.5	
June 1054		h.	Japan	*star*		
$\sim$06/20/1054				Crab in conjunction with Sun		not visible
07/04/1054 ?	6210 ?	i.	Song Empire	*star... like Venus*	~-3.5	for 23 days (*)
08/27/1054 ?	6264 ?	i.	Song Empire	*star... like Venus*	~-3.5	for 23 days (*)
1055		a.	Constantinople	*star*		
04/17/1056	6863	i.	Song Empire	*no more visible*	$>+6$	

JD=MJD+2100000

References: a. *Diary of Ibn Butlan* b. *Tractatus de ecclesia* c. *Irish Annals* d. *De Obitu Sancti Leonis* e. *Sung-shih hsin-pien* f. *Etum Patmich* g. *Rampona Chronicles* h. *Mei Getsuki* i. *Sung hui-yao*.
(*) Datation may have been falsified (or may just refer to the communication to the Emperor), thus to be possibly shifted before.

in the sky. These data made it possible to prove, though with some problems, the link between the historical event and the explosion of the precursor of the Crab Nebula (Mayall & Oort, 1942; Duyvendak, 1942). Nevertheless, another surely independent Chinese source, the *K'i-tan-kuo-chih*, the history of the Kingdom of Liao written about 1350 AD, refers only briefly to the "guest star", stating that the event occurred near the time of (or "during", following a more recent translation: Collins *et al.*, 1999) a total eclipse of the Sun. Duyvendak (1942) has shown that the only important eclipse of that period occurred on 10*th* May 1054. This evidence of an earlier date of the SN 1054 explosion has been often neglected because of the contradiction with respect to the date reported by the sources of the more developed Song Empire. On the other hand, the latter texts give us just two photometric points: on 4*th* July 1054, the star is defined "as luminous as Venus" and thus with V $= -3.5 \pm 1$; on 17*th* April 1056, it is declared "no more visible" and thus having V $\geqslant 5.5$. However, as shown by Collins *et al.* (1999),

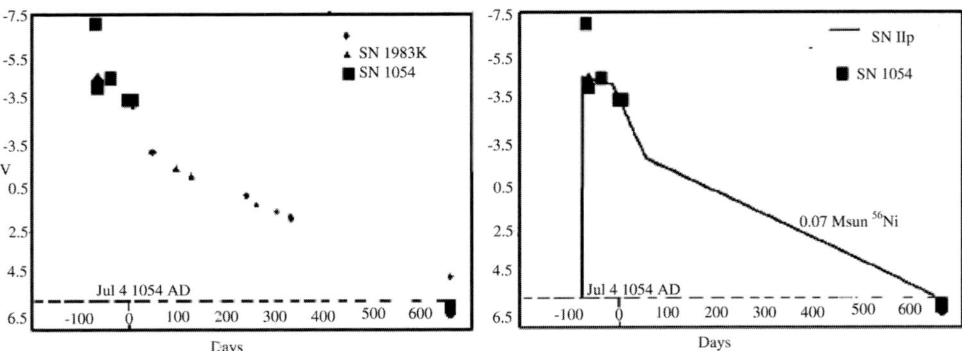

Figure 1. *Left:* The historical light curve of the SN 1054 overlapped to the photometric points of two modern type II SNe, reduced to the distance of the Crab Nebula. *Right:* the same light curve compared to a simple model of SN-IIp, assuming a production of 0.07 $M_\odot$ of ^{56}Ni.

these two photometric points are not compatible with the average light curve of any kind of core-collapse supernova to have exploded on *4th* July 1054: they could be only marginally fitted by the average light curve of a Type Ia SN or by a Type II-L SN with its maximum on *20th* June. While the first hypothesis is ruled out by the presence of the Crab Pulsar, both hypotheses are actually in contradiction with the Song sources themselves, stating the appearance of the star on *4th* July. A problem is that Type II SNe may have significantly different luminosity at peak (although their light curves have a very homogeneous behaviour when the radioactive nuclides decay becomes the dominant energy source: see e.g. Cappellaro & Turatto, 2001, for a general review of SN types); however, one cannot but assume that there is something wrong in the official Song Court report. And, indeed, we have nowadays the possibility to consider the whole set of data concerning the SN 1054 event. It is witnessed, to our knowledge, by at least 13 primary historical sources all over the world (see Table 2). They are of course much different from each other, both in clarity and in style; however, from the appearance of the reported phenomena, it is often possible to guess the corresponding visual magnitude of the SN 1054. We may therefore try reconstructing a rough light curve, by using the formulas which give the naked eye visibility of astronomical objects (see e.g. Schaefer, 1993). If we overlap to this curve the observed magnitudes of a few modern Type II SNe, scaled to the distance of Crab Nebula (Fig. 1, left), and a simple model of Type IIp SN with production of 0.07 $M_\odot$ of ^{56}Ni (Collins *et al.*, 1999; Sollerman *et al.*, 2001; see Fig. 1, right), we can notice that the fit is fairly good in both cases, though the historical data crowd in the high-luminosity segment of the curve, as obvious, making an actual quantitative "best fit" procedure impossible.

We notice that a single point is out from the fit: that is the one related to the very first appearance of the SN, obtained from the *Tractatus de Ecclesia S. Petri Aldenburgensi,* the chronicle of the Church of St. Peter in Oudembourg (in present day Belgium) written by an anonymous monk or clerk some 20 years after the reported events. This is the text translation, in full (for the latin original see: Guidoboni, Marmo & Polcaro, 1994):

And the most blessed Pope Leo, after the beginning of the construction of the aforementioned church of St. Peter, in the following year, on the 18th day before the first of May, a Monday, around midday, happily departed this world. And at the same hour as his leaving of the flesh, not only in Rome, where his body lies, but also all over the world there appeared to men a circle in the sky of extraordinary brightness which lasted for about half an hour. Perhaps the Lord wished to say that he [the Pope] was worthy to receive a crown in Heaven between those who love Him.

A deep textual analysis of this reference, one of the very few written by an actual eye-witness of the SN 1054, allowed the reconstruction of the exact date of the reported phenomenon as the 11*th* April 1054, in agreement with the date that can be deduced from the reports of the Arabic scholar Ibn Butlan (Guidoboni *et al.*, 1994). Furthermore, we can observe that in this document, the author describes the phenomenon in neutral terms, unaffected by any set of beliefs: the disk-like shape, the intense brightness and the duration of the phenomenon are all elements common to very different cultures. The author separates the description of the phenomenon from his cautious symbolic interpretation, showing a clear awareness of the different levels of discourse. The Flemish chronicler saw a bright point source at $\sim 30°$ from the horizon: in the foggy sky of Flanders in Spring, this would appear exactly like a disk. Such a short optical transient, in the very first 30 minutes following the collapse of a SN-II precursor, could very easily escape the detection even with the nowadays observatories; therefore, if confirmed by further evidences, this testimony could be very important for the understanding of the physics of the core-collapse supernovae.

Thus, if the SN 1054 exploded on 11*th* April 1054 (as Guidoboni, Marmo & Polcaro, 1992, first suggested and Collins *et al.*, 1999, demonstrated) it was surely still visible, and very impressive, near to the zenith of the Song capital Kaifeng during the Sun eclipse of 10*th* May 1054. From this sky configuration and by using the standards of the Chinese astrology, which are perfectly documented, it is easy to reconstruct the *omen* that must have been deduced at the time: the Sun represents the Emperor (actually, the Emperor *was* the Sun); the eclipse is a danger for the Emperor's life; moreover, the contemporary presence of the "guest star" indicates the loss of the Heavenly support; therefore the danger is unavoidable: *the Emperor will die*. It is not surprising that such a terrible omen must have been subject to some form of censorship: what the astronomer Yang Weide did – most probably following the wish of Emperor Renzong himself (Polcaro, 2005) – was to censor all the references to the "guest star" preceding its conjunction with the Sun, which occurred at the end of June.

On the other hand, the Chief Astronomer of the Liao kingdom, whose king actually died one year or so after the eclipse, had no reasons to maintain the secret concerning the "guest star" visibility during the eclipse and the related omen.

This is a clear example of the need of taking the historical and cultural context into account, to derive meaningful scientific results from the study of ancient reports.

References

Cappellaro, E. & Turatto, M. 2001, in: Proc. of the Meeting *The influence of binaries on stellar population studies*, Dordrecht: Kluwer Academic Publishers, 2001, Astrophysics and Space Science Library (ASSL), 264, 199

Collins, G.W., Claspy, W.P., & Martin, J.C. 1999, *PASP*, 111, 871

Duyvendak, J.J.L. 1942, *PASP*, 4541, 645.

Guidoboni, E., Marmo, C., & Polcaro, V. F. 1992, *Le Scienze*, 292, 24

Guidoboni, E., Marmo, C., & Polcaro, V.F. 1994, *Mem. SAIt*, 65, 623

Hubble, E. 1928, *PASP Leaflet*, 1, 14

Mayall, N.U. 1939, *PASP Leaflet* 3, 119, 145

Mayall, N.U. & Oort, J.H. 1942, *PASP*, 54, 95

Polcaro, V.F. 2005, in: *Proc. of SEAC 2005*, Isili, Jun 28–Jul 3, 2005

Schaefer, B.E. 1993, *Vistas in Astronomy*, 36, 311

Schaefer, B.E. 1995, *AJ*, 110, 1973

Schaefer, B.E. 1996, *ApJ*, 459, 438

Sollerman, J., Kozma, C., & Lundqvist, P. 2001, *A&A*, 366, 1971

Discussion

VAN DEN HEUVEL: a) Where was Oudembourg located?
b) I saw in Sky and Telescope once an article about a coin of Byzantium of 1054, where the emperor had a star next to him, while on coins from other years there is no star. Could you comment on this?

POLCARO: Oudembourg is located in the present-day Belgium, not far from Bruges, at a perfect latitude to have the Crab at the right height in the centre of the field of view of a man standing and looking to the flat horizon of Flanders.
Concerning the Byzantium coin, its connection with the SN1054 has been called into discussion, since there is a number of other Byzantine coins, of very different epochs, showing a star near to the Emperors' images. However I think that this particular coin should be actually connected with the Supernova: we have many cases of artistic representations of impressive astronomical events, for instance the transit of the Halley Comet on the "Cappella degli Scrovegni" fresco by Giotto, possibly the SN1181 in the Abbey of San Pietro in Valle Fresco, and many other ones. And, for sure, the SN1054 was a very impressive event.

Thoughtful after an Irish lunch. Left to right: Laura Norci, Brian McBreen and Francesco Polcaro, who came first in the Albert Einstein look-alike competition.

Populations of High Energy Sources in Galaxies
Proceedings IAU Symposium No. 230, 2005
E. J. A. Meurs & G. Fabbiano, eds.

© 2006 International Astronomical Union
doi:10.1017/S1743921306008428

X-ray emission from Young Supernovae in Starbursts

L. Norci[1] and E. J. A. Meurs[2]

[1]School of Physics, Dublin City University, Dublin, Ireland
email: lno@physics.dcu.ie

[2]Dunsink Observatory, Dublin Institute for Advanced Studies, Castleknock, Dublin 15, Ireland

Abstract. We have collected from the literature X-ray fluxes of Young Supernovae, measured with various instruments. After converting the data to one energy range, we have compared the X-ray light curves of these objects. The X-ray luminosities of early Supernovae show coherent trends with Supernova type and provide significant, though short-lived contributions to the X-ray luminosity of Starbursts.

Young Supernovae are strong soft X-ray emitters. Produced by a massive progenitor, they have luminosities that span a large energy range, from about 10^{41} erg s^{-1} down to 10^{34} erg s^{-1}. Imaging X-ray satellite observatories have been successful in detecting a number of recent extragalactic Supernovae.

We have collected from the literature and from the Web page of S. Immler† the pertinent X-ray measurements of these Young Supernovae and have reduced their lightcurve fluxes to the 0.1–2.4 keV spectral window, with appropriate assumptions for spectral shapes (see next paragraph). After conversion to luminosities, the resulting data points are presented homogeneously in one diagram (Figure 1).

In Figure 1 we have plotted these X-ray luminosities versus Supernova age for almost all known X-ray detected Young Supernovae. Objects for which Supernova type or date since outburst are uncertain have been excluded from this sample. Fluxes have been converted to the chosen energy range by using the spectral parameters provided in the original papers (see Immler†; Sutaria *et al.* 2003 (SN2002ap)). Noticeable is the case of SN 1993J for which it is believed that two components are present of which the hard component is dominant at early times (Immler, private communication). For this case we have assumed the hard spectrum up to day 100 and a purely soft component after that.

If we compare the X-ray luminosities belonging to various kinds of Supernova events in Figure 1 and the distribution of Supernova types with mass and metallicity in Figure 2 of Heger *et al.* (2003), we can identify relatively narrow luminosity ranges in which certain Supernova types occur. We notice that the two SN IIp types have very similar luminosities of the order of 10^{38} erg s^{-1} at peak. These Supernovae are supposed to have progenitors of masses between 10 and 25–40 M$_\odot$ and a wide range of metallicities. In a similarly well defined energy band one can find the II L types with 10^{39} erg s^{-1} at peak. In the same area one finds the only example of a IIb type. Both IIL and IIb types are thought to be produced at higher metallicities, in a very narrow range of masses between approximately 20–40 M$_\odot$. The Supernovae IIn occupy a well defined band at the top of the luminosity range above approximately $10^{39.5}$ erg s^{-1} at peak. A progenitor is not explicitly identified by Heger *et al.* for these Supernovae. There are indications in the

† http://www.astro.umass.edu/astpage/xray/heag.html

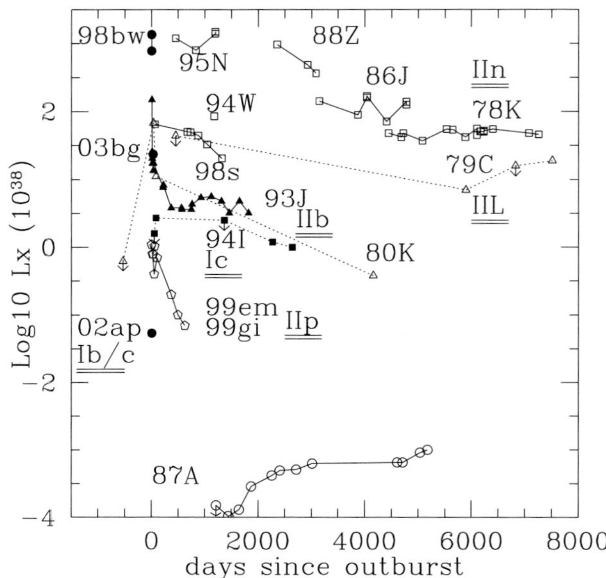

Figure 1. X-ray light curves in the energy range 0.1–2.4 keV versus Supernova age for all the known X-ray Supernovae for which reliable data are available. Different Supernova types are indicated with different colours and symbols. The 1998bw Supernova is believed to be associated with a γ-ray burst.

literature that their high luminosity depends on special conditions in the circumstellar medium, such as high density or clumpiness of the medium.

The X-ray luminosities of Type Ib/c Supernovae are actually varying widely over an energy range from a little over 10^{37} erg s^{-1} for SN 2002ap and around 10^{41} erg s^{-1} for SN 1998bw. On the other hand, three out of four of these Supernovae are fairly recent and only one X-ray point could be determined for each case. These Supernovae are thought to be formed from high mass progenitors ($>40 M_\odot$) and for high metallicity, conditions that are required for the star to lose effectively the outer H envelope. In low metallicity environments, binary evolution may be required in order to lose the H envelope. If to this we add that SN 1998bw has been associated with a γ-ray burst it becomes evident that there maybe more than one kind of progenitor associated to these Supernova events, which would fit with them showing a wide range of X-ray luminosities.

The Supernova 1987A stands out for the very low X-ray luminosity. On the other hand this is a peculiar Supernova in many respects, possibly because the progenitor was in a region of low metallicity. Also the X-ray light curve is still increasing and therefore a peak X-ray luminosity is not easily established.

Most of the light curves lie above 10^{38} erg s^{-1} for the first ca 50 years. Thus, Young Supernovae provide significant, though short-lived contributions to the X-ray luminosity of Starbursts.

References

Heger, A., Fryer, C.L., Woosley, S.E., *et al.* 2003, ApJ 591, 288
Sutaria, F.K., Chandra., P., Bhatnagar, S., & Ray, A. 2003, A&A 397, 1011

Populations of High Energy Sources in Galaxies
Proceedings IAU Symposium No. 230, 2005
E. J. A. Meurs & G. Fabbiano, eds.

© 2006 International Astronomical Union
doi:10.1017/S174392130600843X

Ultra-luminous X-ray Sources

Martin Ward

Department of Physics, University of Durham, Durham, UK
email: m.ward@durham.ac.uk

Abstract. ULXs have been studied for 20 years, but only recently have detailed X-ray spectral/timing studies, and investigations of their statistical properties and their environments and counterparts been undertaken. I will review progress in these areas.

Keywords. X-rays: Ultra-luminous sources, starformation.

1. Introduction

Ultra-luminous X-ray sources (ULXs) are a sub-set of X-ray sources which emit $L_X > 10^{39}$ erg s^{-1}, and can reach luminosities of up to 10^{41} erg s^{-1}. They are so-called because these luminosities significantly exceed those of known discrete X-ray sources within our Galaxy and M31. Note: the X-ray luminosities quoted for ULXs are always broad-band, but the actual energy range can differ from study to study, and so comparisons need to be made with some care, particularly if the range used extends to low energies, and for sources with soft excesses. Symmetric accretion onto a 1.4 M$_\odot$ compact object with $L_X > 10^{39}$ erg s^{-1} implies a substantially super-Eddington accretion rate. Or, if the compact object is emitting at the Eddington limit then the black hole must be 10's of solar masses. The situation of anisotropic emission will be considered later. Under the standard assumptions the highest mass black hole that can be formed in a galactic binary system is $\sim$20 M$_\odot$ (see Fryer & Kalogera 2001), which is close to that observationally determined in the case of a galactic micro-quasar (GMQ). For super-massive black holes (SMBHs) at the centres of galaxies, the lowest mass estimated is in the range $\sim$10^4 to 10^5 M$_\odot$, Filippenko and Ho (2003). From this is it clear that there are about three orders of magnitude between the highest mass black hole in a binary system, and the lowest mass black hole at the centre of a galaxy. It is natural therefore that studies of ULXs have become inextricably linked to the search for intermediate mass black holes (IMBHs). For completeness one should note that globular clusters have been proposed as hosts for IMBHs with masses of $\sim$1000's M$_\odot$, but previous claims to have detected such objects should be treated with caution (Gerssen *et al.* 2003). In this review I will discuss the multi-frequency, temporal and environmental properties of ULXs with particular attention to the two main models proposed to explain their nature; that they are IMBHs or alternatively that they are variations on the types of X-ray binary system that we have identified within our Galaxy.

1.1. *A brief history of ULXs*

ULXs have been known of observationally since the early 1980's, although they were not originally referred to using this acronym. In a study of M51 using the *Einstein Observatory* Palumbo *et al.* (1985) suggested that the brightest extra-nuclear sources have compact object masses of $\sim$10 M$_\odot$. In a comprehensive review, Fabbiano (1989) pointed out that such sources are quite common in X-ray studies of nearby galaxies. Later observations of a dwarf star forming galaxy, NGC5408, led Fabian and Ward (1993) to suggest a mass of the accreting object of $\sim$100 M$_\odot$. There followed statistical studies of galaxy samples, e.g. Roberts and Warwick (2000), who showed that in a *ROSAT* sample one in five galaxies contained at least one source with $L_X > 10^{39}$ erg s^{-1}. The *Chandra* era, with its sub-arcsecond imaging capability, drastically reduced the problem of source confusion, and contamination from extended diffuse emission. One of the early *Chandra* results was the identification of an off-nucleus highly variable X-ray source in M82, which reached a luminosity of more than 10^{41} erg s^{-1}; Kaaret *et al.* (2001), Matsumoto *et al.* (2001). The ULX in M82 is a single spectacular example, but the best studied ensemble of ULXs are those found in the Antennae system, which are the subject of a very detailed analysis, see Fabbiano *et al.* (2003) and references therein. In the remainder of this review I will focus on specific results that have bearing on the nature of ULXs.

2. Models of ULXs

If we adopt the most straightforward interpretation of the Eddington limit, then the highest luminosity ULXs require accretion onto a compact object of $\sim$1000 M$_\odot$, or much more if the accretion process is inefficient at radiating energy, i.e. an ADAF. At the other extreme it is possible that the emission is really super-Eddington, resulting from radiatively driven inhomogeneities in the accretion disc, as propsed by Begelman (2002). Alternatively, if the emission is anisotropic which could result from thermal timescale mass-transfer (King *et al.* 2001), this could explain the luminosity of an *average* ULX. More extreme beaming is present in galactic micro-quasars, and a link between ULXs and these objects was first proposed by Reynolds *et al.* (1997) for Dwingeloo X-1, and in more detail by Kording *et al.* (2002). It is these generic models that the studies described below aim to confront.

3. X-ray Properties of ULXs

Because the definition of ULXs is based on a single parameter, its luminosity, it is probable that the class will be heterogeneous, and in particular may contain examples of both high and low mass X-ray binaries, and some supernova remnants. In order to focus the discussion I will exclude ULXs that are positively identified with SNRs, and also those found in early type and elliptical galaxies. Irwin *et al.* (2004) have studied the properties of ULXs suggested to be associated with early type galaxies, and find their number consistent with that expected for background sources. This means the incidence of ULXs in ellipticals and S0s is consistent with zero. In addition, no ULX at the upper end of the luminosity distribution has so far been found in an elliptical. These facts point to any confirmed (non-background) ULX in early type galaxies being associated with the high luminosity end of the low-mass X-ray binary population. In the following sections I will discuss results for ULXs associated with spiral and starforming galaxies.

3.1. *X-ray Spectra*

In standard accretion discs the temperature is inversely related to the black hole mass as $L \propto M^{-1/4}$. This means that if IMBHs exist in ULXs their X-ray spectra should be fitted by cooler accretion discs than those found in stellar-mass black hole candidates. Based on good quality X-ray spectra of a number of ULXs at the higher end of the luminosity distribution $L_X > 10^{40}$ erg s^{-1}, Miller *et al.* (2004) show that their disc temperatures are 5–10 times lower than found in Galactic stellar mass black hole systems in their high states. This would imply black holes in these ULXs of up to a few 1000 M$_\odot$.

However, as the authors point out the spectral complexity at soft X-ray energies is not well understood in these sources, and if it does not arise from a disc as assumed, but rather from an optically thin thermal plasma, then the black hole mass determinations are incorrect. However, no emission lines characteristic of a thermal plasma have so far been detected, supporting the cool disc hypothesis. It is somewhat disappointing to see that even given good signal/noise X-ray spectra of some ULXs, e.g. NGC 5204 X-1, and HoII X-1, Roberts *et al.* (2005) and Goad *et al.* (2005), the models still result in ambiguity in distinguishing between stellar mass and intermediate mass black holes. It is likely that we need a more sophisticated understanding of the soft energy components and the role of jets and winds, as well as better data at hard energies to constrain power-law versus multi-colour disc models.

3.2. *X-ray Variability*

Perhaps the most obvious and secure use of high amplitude X-ray variability data for ULXs, is to rule out the presence of multiple unresolved sources, and an SNR origin of the X-rays.

There are many X-ray variability studies of ULXs, and I will highlight results for just a few of the best studied cases. The ULX in M82 is at the high end of the luminosity distribution with an X-ray luminosity up to 10^{41} erg s $^{-1}$. Strongly beamed X-rays are ruled out in this source by the presence of a QPO, thought to be related to disc oscillations, and the detection of iron K alpha emission, Strohmayer & Mushotzky (2003). This is because a significant beamed component would have diluted the QPO amplitude and K alpha emission, making these features difficult or impossible to detect. Unfortunately X-ray spectral fitting is not very useful for this source because of the column density and significant contamination from an extended diffuse plasma component. In another example, NGC4559 X-7, Cropper *et al.* (2004) have identified a break in the power density spectrum. In the case of SMBHs and stellar mass BHs there are often two breaks in the power density spectrum, and so the break in NGC4559 X-7 is consistent with either 38 $M_\odot$ or 1300 $M_\odot$. Another impressive case of a quasi period in a ULX is seen in M74(NGC628) X-1, see Liu *et al.* (2005), and Krauss *et al.* (2005). This source exhibits extreme variablity on timescales of a few 1000 seconds. Once again any mass estimate depends strongly on the chosen interpretation of the QPO.

So it seems at present that the detection of QPOs and power density breaks can be useful for constraining the effects of beaming, but not as a reliable direct mass determination.

4. The Environment of ULXs

The environments of ULXs may be conveniently separated into three topics; their global environment, i.e. the host galaxy, their local environment within the host galaxy, e.g. associated nebulae, star clusters etc. and finally direct association with an optical or compact radio counterpart.

4.1. *The Host Galaxy*

It is evident that the ULX phenomenon is associated with star formation. A spectacular example being the Antennae system which contains 9 ULXs. At higher redshift the Cartwheel starforming ring contains 20 ULXs, although without variablity data we cannot exclude the possiblity that some sources are multiple but not spatially resolved, Gao *et al.* (2003) and Wolter and Trinchieri (2004). It is proposed by Grimm *et al.* (2003), that there exists a universal X-ray luminosity function (LF) for high mass X-ray binaries, if one normalizes each separate LF by the appropriate star formation rate for the galaxy concerned. Wolter and Trinchieri find that the LF for the Cartwheel fits this universal function, but with a factor 5 higher cut-off in L_X. This may be connected with the fact that ULX occurrence is related to galaxy type (and/or metal abundance) since they are often found in nearby dwarf galaxies.

4.2. *The Local Environment*

The more luminous ULXs are frequently located near to sites of active star formation such as super star clusters containing 10^6 $M_\odot$ of stars within a volume only a few parsecs across. It is suggested that encounters within such a high stellar density environment could result in a high fraction of binaries. An interesting result found by Kaaret *et al.* (2004b), is that there is often a slight offset between the position of the ULX and a nearby star cluster. This could be interpreted as resulting from the ULX being ejected from the cluster via dynamical interactions. The closeness to the location of birth assumed to be within the cluster, means that the ULX phenomenon switches on soon after their formation. There are a number of ULX that are cospatial with emission nebulae, and spectroscopy and imaging of these nebulae have been the subject of intensive effort over the last few years, (see contribution from Manfred Pakull in these proceedings). Some of these nebulae are very large, e.g. 400pc across for NGC1313 X-2, Pakull and Mirioni (2002), and 200pc across for IC 342 X-1, Roberts *et al.* (2003). The energies required to produce the emission lines and to drive the expansion velocities, are higher than those found in a typical supernova, and so multiple supernova events have been suggested, or evidence of a hypernova event.

Generally the optical emission lines from these nebulae are consistent with those seen in SNRs. However, in the case of HoII the spectrum has a strong line from HeII, Pakull and Mirioni (2002). They used the spectrum to estimate the luminosity required to photoionise the nebula. Kaaret, Ward and Zezas (2004) observed HoII using the HST to provide images in several emission lines including HeII, and derived a photoionised luminosity for the nebula consistent with that of Pakull and Mirioni. This implies that the detected X-rays are not significantly beamed, and so the black hole should have a mass in the range $\sim$25 to 40 $M_\odot$.

4.3. *Optical Counterparts*

The search for optical counterparts to ULXs is hampered by their association with star forming regions. This often means that the fields are crowded, so that accurate astrometry is essential. By alignment of the position of an X-ray source in the field with a known counterpart, such as a SNR, it is possible to achieve relative astrometry between the X-ray and optical of a few tenths of arcseconds. Table 1 summarises the information on the five best candidates for optical counterparts of ULXs. They are all young stars, consistent with the connection between ULXs and starformation. These counterparts are all predictably faint, but in the case of Holmberg II, HST/STIS spectroscopy was obtained by Liu *et al.* (2004). The spectrum contained the high ionisation emission line of NV, strongly supporting this star as the X-ray source counterpart. Similarly Kuntz *et al.* (2005) detected spatially unresolved HeII and HeI emission lines at the position of the counterpart to M101 X-1, confirming the physical association with the X-ray source. No ULX has so far been observed to eclipse, but there is the possiblity of obtaining a direct measurement of the black hole mass function via determination of an orbital period and the orbital velocities. This would be extremely challenging currently for such faint sources, but could be viable in the era of extremely large telescopes.

Table 1. Optical Counterparts to ULXs

Source	Spectral Type	Reference
M81 X-11	O8V	Liu *et al.* (2002)
NGC1313	OV-O-BI	Zampieri *et al.* (2004)
Holmberg II	O4V-B3 Ib	Kaaret *et al.* (2004)
NGC5204	B0 Ib	Liu *et al.* (2004)
		Goad *et al.* (2002)
M101	mid-B supergiant	Kuntz *et al.* (2005)

4.4. *Radio Observations of ULXs*

Given that X-rays dominate the energy output of ULXs, it might not seem obvious that radio observations would prove a fruitful band to employ in the study of these objects. However, there are two main reasons why radio observations are potentially very useful. First, given the strong evidence that ULXs are related to starformation, they may be direcly or indirectly associated with supernovae and their remnants. This likelihood increases for sources with lower X-ray luminosities, soft X-ray spectra and with no evidence of X-ray variability. An example is SN1987K in NGC1313, Schlegel *et al.* (2000). In relatively rare cases some young SNR can reach X-ray luminosities in excess of 10^{40} erg s^{-1}, Fox *et al.* (2000), also see the contribution of Norci and Meurs in these proceedings. Such examples are clearly interesting in their own right, but they do not help us to address questions regarding intermediate mass black holes. There are now several examples of ULXs previously identified with radio bright SNRs, that based on their X-ray variability are now re-classified as compact binaries eg. the radio hypernova remnants in M101 (Snowden *et al.* 2001) and the source M16 in NGC6969 (Roberts and Colbert 2003) are now claimed to be X-ray binaries. Furthermore, the radio and optical emission line nebulae around some ULXs may arise from the SNR associated with the birth of the ULX, for example IC342 X-1, Roberts *et al.* (2003), and M81 X-9, Wang (2002). The second reason for radio studies is that if radio emission is detected it might be intrinsic to the ULX, and hence provide a direct probe of the physics of the ULX phenomena. The first suggestion of this is the radio detection of the ULX in NGC5408, for which it is claimed the emission may be beamed in a way analogous to that in the galactic micro-quasars (GMQs), Kaaret *et al.* (2003). However, a second epoch radio observation of NGC5408 reported by Kording *et al.* (2005), shows the emission to be consistent with a steady source, unlike the typical radio properties of GMQs. Radio emission has also been detected from the ULX in the dwarf galaxy Holmberg II, Miller *et al.* (2005). In this case it is extended by about 60x40pc, and is cospatial with the optical HeII line nebula. The optical and radio properties argue against the radio arising from an HII region or a SNR, and Miller *et al.* suggest that we may be seeing the same phenomena as in the W50 nebula surrounding the jet source in SS433, but with 10 times greater energy within the nebula.

To establish a link between radio emission from ULXs and galactic micro-quasars requires radio monitoring, since the beamed radio emission from GMQs is highly variable. Intriguingly one of the most luminous and best studied of all ULXs, that in M82, is spatially coincident with a transient radio source, detected only once (Kording *et al.* 2005). This could be interpreted as evidence in favour of a radio outburst in a beamed GMQ, but it is also consistent with a faded radio supernova, which might be related to the birth of the ULX. Clearly any future reappearance of the radio emission would be a critical test of the beaming model.

5. Conclusions

ULXs are undoubtably a heterogeneous class, but the variable high luminosity examples, $L_X > 10^{40}$ erg s $^{-1}$, may be more homogeneous, and are the likely candidates for IMBHs. The association between ULXs and starformation is proven, although their formation and duty cycle in different spectral states remain open questions. The issue of anistropic emission/beaming has been investigated by studies of their associated nebulae, but so far this has not resulted in mass estimates of more than a few $10's$ $M_\odot$. Improved understanding of both their soft and hard X-ray spectral components will help in deciding whether a cool accretion disc, hence high BH mass, is the correct interpretation.

References

Begelman, M. 2002, *ApJ* 568. L97

Cropper, M., *et al.* 2004, *MNRAS* 349, 39

Fabbiano, G. 1989, *A.Rev.Astron.Astrophys.* 27, 87

Fabbiano, G., *et al.* 2003, *ApJ* 584, L5

Fabian, A.C. & Ward, M.J. 1993, *MNRAS* 263, L51

Filippenko, A.V. & Ho, L.C. 2003, *ApJ* 588, L13

Fox, D.W., *et al.* 2000. *MNRAS* 319, 1154

Fryer, C.L. & Kalogera, V. 2001, *ApJ* 554, 548

Gao, Y., *et al.* 2003, *ApJ* 596, L171

Grimm, H.-J., Gilfanov, M., & Sunyaev, R. 2003, *MNRAS* 339, 793

Gerssen, J., van der Marel, R.P., Gebhardt, K., & Guhathakurta, P. 2003, *AJ* 125, 376

Goad, M,R., Roberts, T.P., Knigge, C., & Lira, P. 2002, *MNRAS* 335, L67

Goad, M.R., Roberts, T.P., Reeves, J., & Uttley, P. 2005, *astro-ph/* 0510185

Irwin, J.A., Bregman, J.N., & Athey, A.E. 2004, *ApJ* 601, L143

Kaaret, P., *et al.* 2001 , *MNRAS* 321, L29

Kaaret, P., *et al.* 2003, *Science* 299, 365

Kaaret, P., Ward, M.J., & Zezas, A. 2004, *MNRAS* 351, L83

Kaaret, P., *et al.* 2004b, *MNRAS* 348, L28

King, A., *et al.* 2001, *ApJ* 552, L109

Kording, E., Falcke, H., & Markoff, S. 2002, *Astron. & Astrophys* 382, L13

Kording, E., Colbert, E., & Falcke, H. 2005, *Astron. & Astrophys* 436, 427

Krauss, M.I., *et al.* 2005, *ApJ* 630, 228

Kuntz, K.D., *et al.* 2005, *ApJ* 620, L31

Liu, J-F., *et al.* 2002, *ApJ* 580, L31

Liu, J-F., Bregman, J.N., & Seitzer, P. 2004, *ApJ* 602, 249

Liu, J-F., *et al.* 2005, *ApJ* 621, L17

Miller, N.A., Mushotzky, R.F., & Neff, S.G. 2005, *ApJ* 623, L109

Miller, J.M., Fabian, A.C., & Miller, M.C. 2004, *ApJ* 607, 931

Matsumoto, H., *et al.* 2001, *ApJ* 547, L25

Pakull, M.W. & Mirioni, L. 2002, *ESA SP-488, astro-ph/* 0202488

Palumbo, G., *et al.* 1985, *ApJ* 298, 259

Reynolds, C.S., *et al.* 1997, *MNRAS* 286, 349

Roberts, T. P. & Warwick, R.S. 2000, *MNRAS* 315, 98

Roberts, T.P. & Colbert, E.J.M. 2003, *MNRAS* 341, L49

Roberts, T.P, Goad, M.R., Ward, M.J., & Warwick, R.S. 2003, *MNRAS* 342, 709

Roberts, T.P., *et al.* 2005, *MNRAS* 357, 1363

Schlegel, E.M., *et al.* 2000, *ApJ* 120,791

Strohmayer, T.E. & Mushotzky, R.F. 2003, *ApJ* 586, L61

Wang, D.Q. 2002, *MNRAS* 332, 764

Wolter, A. & Trinchieri, G 2004, *Astron. & Astrophys.* 426, 787

Zampieri, L., *et al.* 2004, *ApJ* 603, 523

Discussion

BOSCH-RAMON: Comment: A way for distinguishing ULXs with intermediate BH and strongly Doppler boosted μQ jets could be the presence of radio emission: lacking in intermediate BH ($L_R \propto L_X{}^{0.7}$) and probably present in strongly Doppler boosted μQ jets.

WARD: Radio observations are certainly a valuable tool in the multi-frequency study of ULX. One problem is ambiguity – if detected, is the radio related to a Supernovae event or a microquasar? Clearly spectral index and temporal monitoring will help answer this question.

Dr Who in the audience.

Populations of High Energy Sources in Galaxies
Proceedings IAU Symposium No. 230, 2005
E. J. A. Meurs & G. Fabbiano, eds.

© 2006 International Astronomical Union
doi:10.1017/S1743921306008441

Properties of SS433 and ultraluminous X-ray sources in external galaxies

S. Fabrika, S. Karpov, P. Abolmasov and O. Sholukhova

Special Astrophysical Observatory of the Russian AS, Nizhnij Arkhyz 369167, Russia
email: fabrika@sao.ru, karpov@sao.ru, pasha@sao.ru, olga@sao.ru

Abstract. We suggest that the ultraluminous X-ray sources located in external galaxies (ULXs) are supercritical accretion disks like that in SS433, observed close to the disk axis. We estimate parameters of the SS433 funnel, where the relativistic jets are formed. Emergent X-ray spectrum in the proposed model of the multicolor funnel (MCF) is calculated. The spectrum can be compared with those of ULXs. We predict a complex absorption-line spectrum with broad and shallow $K\alpha/Kc$ blends of the most abundant heavy elements and particular temporal variability. Another critical idea comes from observations of nebulae around the ULXs. We present results of 3D-spectroscopy of nebulae of two ULXs located in Holmberg II and NGC6946. In both cases the nebula is found to be powered by the central black hole. The nebulae are compared with SS433 nebula (W50).

Keywords. X-ray sources, accretion disks, jets, individual: SS433.

1. Introduction

The main properties of the ultraluminous X-ray sources (ULXs) – huge luminosities ($10^{39-41}\,erg/s$), diversity of X-ray spectra, strong variability, connection with star-forming regions, their surrounding nebulae, were reviewed by Ward (2005). ULXs may be supercritical accretion disks observed close to the disk axis in close binaries with a stellar mass black hole or microquasars (Fabrika & Mescheryakov (2001); King *et al.* (2001); Koerding *et al.* (2001)), or they may be intermediate-mass black holes with "normal" accretion disks (Colbert & Mushotzky (1999); Miller, Fabian & Miller (2004)). It is also possible that ULXs are not homogeneous class of objects. It was suggested originally by Katz (1987) that SS433 being observed close to the jet axis will be extremely bright X-ray source. Fabrika & Mescheryakov (2001) discussed observational properties of face-on SS433-like objects and concluded that they may appear as a new type of extragalactic X-ray sources. Here we continue to develop this idea.

2. The funnel in SS433 disk

The main difference between SS433 and other known X-ray binaries is highly supercritical and persistent mass accretion rate ($\sim 10^{-4}\,M_\odot/y$) onto the relativistic star (a probable black hole, $\sim 10 M_\odot$), which has led to the formation of a supercritical accretion disk and the relativistic jets. SS433 properties were reviewed recently by Fabrika (2004).

SS433 is an edge-on system, its total luminosity (mainly in UV) is $L_{bol} \sim 10^{40}\,erg/s$. A temperature of the source is $T = (5-7) \cdot 10^4\,K$, a mass loss rate in the wind is $\dot{M}_w \sim 10^{-4}\,M_\odot/y$. Both optical and X-ray jets are well collimated ($\sim 1°$). SS433 X-ray luminosity (the cooling X-ray jets) is $\sim 10^4$ times less than bolometric luminosity, however kinetic luminosity of the jets is very high, $L_k \sim 10^{39}\,erg/s$.

The jets have to be formed in a funnel in the disk and the disk wind. Supercritical accretion disks simulations (Eggum, Coroniti, Katz (1985); Okuda (2002); Okuda *et al.* (2005)) show that a wide funnel is formed (a full opening angle $\theta_f \approx 40°–50°$) close to the black hole. Convection is important in the inner accretion disk outside the funnel walls.

If one adopts for the funnel luminosity, that it is about the same as SS433 bolometric luminosity, $L_f \approx L_{bol}$, (Fabrika & Mescheryakov (2001); Fabrika & Karpov (2005)) and the funnel opening angle $\theta_f \approx 40°–50°$, one obtains "observed" face-on luminosity of SS433 $L_x \sim 10^{41}\, erg/s$ and expected frequency of such objects $\sim$0.1 per a galaxy like Milky Way. On the other hand a critical luminosity for a $10\,M_\odot$ black hole is $L_{edd} \sim 10^{39}\, erg/s$. At highly supercritical accretion rate $\dot{M}/\dot{M}_{cr} \sim 10^3$ it increases (the inner disk geometry) by a factor of $(1 + \ln(\dot{M}/\dot{M}_{cr})) \sim 10$. A doppler boosting factor is not large for SS433 ($\beta = V_j/c = 0.26$), it is $1/(1 - \beta)^{2+\alpha} \sim 2.5$. The third factor is the geometrical funneling ($\theta_f \approx 40°–50°$), it is $\Omega_f/2\pi \sim 10$. Thus, one may expect a face-on luminosity of such supercritical disk $L_{edd} \sim 2 \cdot 10^{41}\, erg/s$.

3. The multicolor funnel model

A size of SS433 wind photosphere is $r_{ph,wind} \sim 10^{12}\, cm$. An estimate of the jet photosphere (Fabrika & Karpov (2005)), which is a bottom of the wide funnel, gives $r_{ph,jet} \sim 4 \cdot 10^9\, cm$. We find that the jet is not transparent down to the black hole horizon. However, the gas supply to the funnel is very variable due to convection (Okuda *et al.* (2005)).

Fabrika & Karpov (2005) developed a simple model of the funnel (multicolor funnel, MCF) to estimate the emerging X-ray spectrum. They considered both gas pressure dominated, $T(r) \propto r^{-1}$, and radiation pressure dominated, $T(r) \propto r^{-1/2}$ cases. They based on the observed temperature of the wind photosphere in SS433 and found temperatures of the inner funnel walls (at a level of $r_{ph,jet}$), $T_{ph,wind}$ between $\sim 1.7 \cdot 10^7\, K$ and $\sim 1 \cdot 10^6\, K$. Below $r_{ph,jet}$ the walls can not be observed.

In Fig. 1 we present MCF energy spectra with different ratios of radiation to gas pressure ($\xi = aT_0^3/3k_b n_0$) at the deepest visible parts of the funnel's walls ($r_{ph,jet}$). The temperature $T_0 = 1\,keV$ was adopted. They are spectra with no radiation losses in the funnel ($\theta_f \approx 0$), and with radiation losses included ($\theta_f \approx 40°$). The multicolor disk (MCD) spectrum with the same inner temperature $T_{inn} = 1\,keV$ is shown for comparison. A low-temperature break in the MCF spectra appears due to finite length of the funnel.

The jets velocity ($0.26\,c$) in SS433 and its surprising stability suggest line-locking mechanism (Shapiro, Milgrom & Rees (1986)) to operate in the funnel. A photosphere of the funnel's inner walls may radiate absorption spectrum with Lc and Kc edges of H- and He-like ions. The MCF model predicts very complex absorption-line spectrum. Fig. 1 presents expected absorption line profiles of OVIII Lα and Lc transitions.

The absorption bands should belong to H- and He-like ions of the most abundant heavy elements and should extend from the Kc to the Kα energies of the corresponding ions. Variations of the gas parameters along the funnel – its velocity, density, temperature and volume filling factor (the collimation) – could make the absorption-line profiles appreciably more complex, necessitating the use of X-ray spectra with high signal/noise ratios in searches for these lines.

The supposition that ULXs are face-on SS433 objects leads to the following predictions: 1) We expect typical MCD-like spectra, however the MCF spectra may be more diverse depending on the funnel parameters. 2) Absorption spectrum with broad and shallow Lα/Lc, Kα/Kc blends of the most abundant heavy elements. 3) Temporal variability on

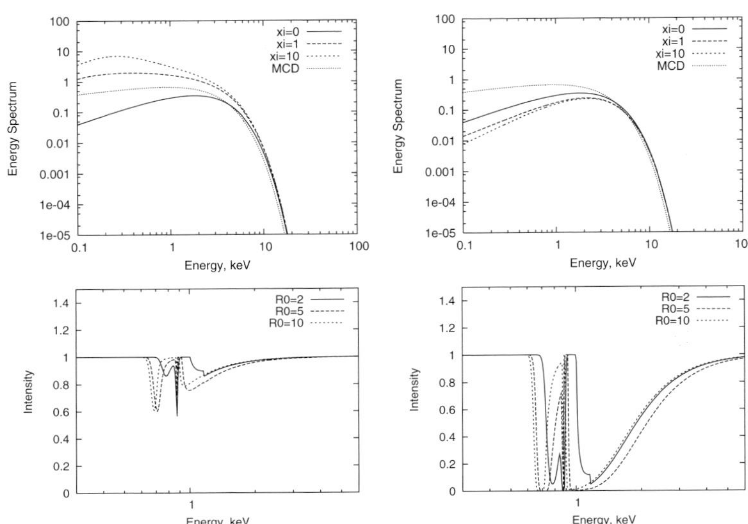

Figure 1. Top: energy spectra of the multicolor funnel (MCF) compared with the MCD spectrum with the same inner temperature $1\,keV$. Bottom: absorption line profiles of OVIII Lα and Lc transitions at different efficiencies of jet acceleration for optically thin (left) and for optically thick (right) regimes. The acceleration starts at $R_0 = 1$ (in units of the jet bottom photosphere, $r_{ph,jet}$), and ends at $R_0 = 2, 5, 10$. The line intensities are in arbitrary scale.

time scales $r_{ph,wind}/c \sim 30\,sec$ and $r_{ph,jet}/c \sim 0.1\,sec$. 4) A typical accretion disk power density spectrum at scales $>> 0.1\,sec$. 5) Very bright UV source ($L_{UV} \sim 10^{40}\,erg/s$, $T \geqslant 10^5\,K$) is less collimated than X-ray source. The predicted complexity of the dependence of the absorption-line profiles on the structure of the funnel and mechanisms of acceleration and collimation of the gas in the funnel presents excellent opportunities for direct probing of these structures in supercritical accretion disks and for studies of mechanisms of jet formation.

4. W50 and ULXs nebulae

The radio nebula W50 was produced (or distorted) by SS433 jets. We show W50 in Fig. 2 together with nebulae surrounding two ULXs in Holmberg II and NGC 6946 galaxies, which were studied recently by Integral-field spectroscopy method (Lehmann *et al.* (2005); Fabrika, Abolmasov & Sholukhova (2005)). W50 contains bipolar nebula with optical filaments located at $\pm 0.5°$ or $\pm 50\,pc$ from SS433 at places of jets termination. A total energy of the nebula is $E_k \sim 2 \cdot 10^{51}\,erg$ (Zealey *et al.* (1980)), which corresponds to the jet kinetic luminosity $L_k \sim 3 \cdot 10^{39}\,erg/s$ for 20000 years. The observed velocity dispersion in the filaments is $\sim 50\,km/s$, however [NII]/Hα line ratio corresponds to dispersion $\sim 300\,km/s$. SS433 is an edge-on system, $i = 79°$. If one takes into account this factor, the velocity dispersion may reach $250 - 300\,km/s$.

ULXs are located frequently in bubble-like nebulae. New data (Pakull (2005)) show that the nebulae are expanding with a velocity $\sim 80\,km/s$ (up to $\sim 250\,km/s$). In two well-studied ULXs nebulae radial velocity gradients in a high ionisation emission line He IIλ4686 were found. They are $\pm 50\,km/s$ on spatial scale $\sim \pm 30\,pc$ in Holmberg II ULX-1 (Lehmann *et al.* (2005)), and $\sim \pm 50\,km/s$ on a scale $\pm 20\,pc$ in NGC 6946 ULX-1 (Fabrika, Abolmasov & Sholukhova (2005)). In all cases line ratios indicate shock ionisation. This means that the nebulae are powered by the central source, probably via jets activity.

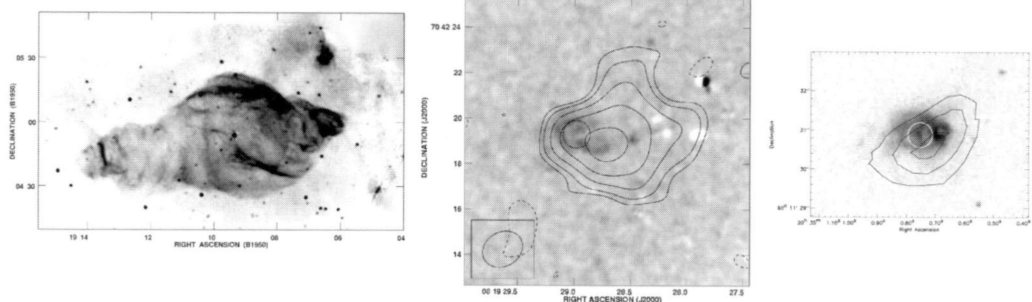

Figure 2. Three nebulae on the same scale in parsecs. VLA image of W50 (Dubner *et al.* (1998)) with SS433 in center, left; Holmberg II ULX-1 in HST HeII image with VLA isophotes, middle (Miller, Mushotzky & Neff (2005)) and NGC6946 ULX-1 in HST Hα + [SII] image with VLA isophotes, right (van Dyk *et al.* (1994)). Circles show X-ray Chandra positions.

The nebulae in Holmberg II and NGC6946 have circle-like features in the line-images. In both cases the radio sources are shifted to a brighter circle-like feature and in the both cases this part of the nebulae is approaching, the opposite part is receding (Lehmann *et al.* (2005); Fabrika, Abolmasov & Sholukhova (2005)). At some imagination one may conclude that the nebulae around these two ULXs are face-on versions ($i = 10° - 30°$) of the SS433 nebula.

Acknowledgements

The work is supported by RFBR under grants number 03-02-16341 and 04-02-16349.

References

Colbert, E.J.M. & Mushotzky, R.F. 1999, *ApJ* 519 89

Dubner, G.M., Holdaway, M., Goss, W.M., & Mirabel, I.F. 1998, *AJ* 116 1842

Eggum, G.E., Coroniti, F.V., & Katz, J.I. 1985. *ApJ* (Letters) 298 L41

Fabrika, S. 2004, *Astrophysics and Space Physics Reviews*, vol. 12, p. 1

Fabrika, S. & Mescheryakov, A. 2001, in: R.T. Schilizzi (ed.), *Galaxies and their Constituents at the Highest Angular Resolution*, IAU Symp. N205 (Manchester, UK), p. 268

Fabrika, S. & Karpov, S. 2005, in preparation

Fabrika, S., Abolmasov, P., & Sholukhova, O. 2005, in preparation

Katz, J.J. 1987, *ApJ* 317 264

King, A.R., Davies, M.B., Ward, M.J., Fabbiano, G., & Elvis, M. 2001, *ApJ* (Letters) 552 L109

Koerding, E., Falcke, H., Markoff, S., & Fender, R. 2001, *Astron. Gesells. Meet. Abstr.* 18 176

Lehmann, I., Becker, T., Fabrika, S., Roth, M., Miyaji, T., Afanasiev, V., Sholukhova, O., Sanchez, S.F., Greiner, J., Hasinger, G., Costantini, E., Surkov, A., & Burenkov, A. 2005, *A&A* 431 847

Miller, J.M., Fabian, A.C., & Miller, M.C. 2004, *ApJ* (Letters) 614 L117

Miller, N.A., Mushotzky, R.F., & Neff, S.G. 2005, *ApJ* (Letters) 623 L109

Okuda, T. 2002, *PASJ* 54 253

Okuda, T., Teresi, V., Toscano, E., & Molteni, D. 2005, *MNRAS* 357 295

Pakull, M., 2005, this conference

Shapiro, P.R., Milgrom, M., & Rees, M.J. 1986, *ApJSS* 60 393

van Dyk, S.D., Sramek, R.A., Weiler, K.W., Hyman, S.D., & Virden, R.E., 1994 *ApJ* (Letters) 425 L77

Ward, M.J., 2005, this conference

Zealey, W.J., Dopita, M.A., & Malin, D.F. 1980, *MNRAS* 192 731

Populations of High Energy Sources in Galaxies
Proceedings IAU Symposium No. 230, 2005
E. J. A. Meurs & G. Fabbiano, eds.

© 2006 International Astronomical Union
doi:10.1017/S1743921306008453

Ultra-luminous Supersoft X-ray Sources in Nearby Galaxies

Albert Kong[1] and Rosanne Di Stefano[2]

[1]MIT Kavli Institute for Astrophysics and Space Research, 77 Massachusetts Avenue,
Cambridge, MA 02139, U.S.A.
email: akong@space.mit.edu

[2]Harvard-Smithsonian Center for Astrophysics, 60 Garden Street, Cambridge,
MA 02138, U.S.A.
email: rd@cfa.harvard.edu

Abstract. *Chandra* and *XMM-Newton* improve our understanding of X-ray populations in galaxies. In particular, there exists a class of ultra-luminous X-ray sources (ULXs) for which the observed luminosity is greater than 10^{39} ergs s^{-1}. ULXs are of great interest since they represent a population of possible intermediate-mass black holes. While the spectra of majority of ULXs are similar to Galactic X-ray binaries, a few ULXs have very soft X-ray emission ($kT = 50$–100 eV) resembling supersoft X-ray sources (SSSs) discovered in the Milky Way and the Magellanic Clouds. We here report some recent multiwavelength observations of three ultra-luminous SSSs in M101, NGC 300, and the Antennae. They have shown many interesting behaviors such as state transitions, spectral changes, and time variabilities in different timescales. Unlike typical SSSs, ultra-luminous SSSs are unlikely associated with white dwarfs because of the high X-ray luminosities. We discuss some binary models involving stellar-mass and intermediate-mass black holes to explain the nature of the systems.

Keywords. accretion, black hole physics, X-ray: binaries.

1. Introduction

Recent high angular resolution X-ray observations reveal that there is a large number of ultra-luminous X-ray sources (ULXs) in many nearby galaxies. ULXs are luminous ($L_X > 10^{39}$ ergs s^{-1}) non-nuclear X-ray point sources with apparent X-ray luminosities above the Eddington limit for a $\sim 10 M_\odot$ black hole (BH). While some ULXs have been associated with supernovae, many are thought to be accreting objects with X-ray flux variability observed on timescales of hours to years. A natural possibility is that the compact object is an intermediate-mass black hole (IMBH) with mass of $\sim 10^{2-4} M_\odot$ (Miller & Colbert 2004). The origin of such objects remains uncertain. Some ULXs may have a stellar-mass BH with beamed emission (King *et al.* 2001; Körding *et al.* 2002). While current observations are inconclusive about the nature of ULXs, recent X-ray observations show that some ULXs have a cool accretion disk ($kT \sim 0.1$ keV), suggesting the presence of IMBHs (Miller *et al.* 2003, 2004; Wang *et al.* 2004).

While the majority of ULXs has X-ray emission from 0.1 to 10 keV and the high energy tail contributes significant emission, a few ULXs have very soft spectra with no X-ray emission above 1 keV (Fabbiano *et al.* 2003; Kong & Di Stefano 2003; Di Stefano & Kong 2003), similar to supersoft X-ray sources (SSSs) in the Milky Way and the Magellanic Clouds. The high luminosities of ultra-luminous SSSs are inconsistent with typical nuclear burning white dwarf models for Galactic SSSs. These ultra-luminous SSSs could, however, very well be IMBHs. Their luminosities and temperatures are consistent with what is predicted for accreting BHs with masses between roughly 100 and 1000 $M_\odot$.

Alternatively, outflows from stellar-mass BHs could also achieve such a high luminosity and low temperature (King & Pounds 2003). We here review 3 well studied ultra-luminous SSSs in nearby galaxies.

2. M101 ULX-1

M101 ULX-1 is one of the most luminous ULXs. It was discovered with *ROSAT* and was confirmed as a SSS with a blackbody temperature of about 100 eV, with *Chandra* (Pence *et al.* 2001; Mukai *et al.* 2003; Di Stefano & Kong 2003). During 2000 March, *Chandra* detected it at $L_X \sim 4 \times 10^{39}$ ergs s^{-1}, and then in 2000 October, its luminosity was around 10^{39} ergs s^{-1}. In 2004, *Chandra* conducted a monitoring program for M101. Figure 1 (left) shows the long-term X-ray lightcurve of M101 ULX-1 during 2000–2005. M101 ULX-1 was near the detection limit during 2004 January, March, and May; the X-ray spectra were harder with a power-law shape (see Figure 1 right), and the X-ray luminosity was about 2×10^{37} ergs s^{-1}, a factor of about 10^2 fainter than that during the outbursts in 2000 (Kong *et al.* 2004). The source was found to be in outburst during the July 5 observation, with an X-ray luminosity of about 7×10^{39} ergs s^{-1}. Data taken on 2004 July 6, 7, and 8 show that the source was in outburst with a peak bolometric luminosity (for assumed isotropic emission) of about 10^{41} ergs s^{-1} (Kong *et al.* 2004). In general, the X-ray spectra are best described with an absorbed blackbody model with temperatures of $\sim$50–100 eV (see Figure 1 right). In addition, we found absorption edges at 0.33, 0.57, 0.66, and 0.88 keV in two of the high state spectra. These features may signal the presence of highly ionized gas in the vicinity of the accretor (e.g., warm absorber). A DDT *XMM-Newton* observation was made on 2004 July 23 and the luminosity was about 6×10^{38} ergs s^{-1}. A harder X-ray spectrum with a power-law tail ($kT = 53$ eV and $\alpha = 0.72$) was seen up to 7 keV. More recently, *Chandra* observations made in 2004 September and November indicated that the source returned to the low state with a power-law spectrum and a luminosity of $\sim$2$\times 10^{37}$ ergs s^{-1}. In 2004 December, the source was in outburst again with very soft spectra ($kT = 40$–150 eV) and a peak bolometric luminosity of about 10^{40} ergs s^{-1}. During the rise of the outburst, the spectra were supersoft with blackbody temperatures of 40–70 eV. The peak of the outburst occurred on 2004 December 30. The source showed a cool accretion disk on 2005 January 1. The X-ray spectrum can be fitted with a disk blackbody model with $kT_{DBB} = 0.17$ keV which is significantly harder than previous supersoft X-ray spectra. The peak luminosity of the 2004 December outburst is about 2×10^{40} ergs s^{-1} which is very similar to the 2004 July outburst. A DDT *XMM-Newton* observation made on 2005 January 8 indicated that the source was slightly brighter than 2005 January 1 ($L_X = 6 \times 10^{39}$ ergs s^{-1}) and the X-ray spectrum returned to a supersoft state ($kT = 56$ eV; Kong 2005).

Based on the position provided by *Chandra*, an optical counterpart of M101 ULX-1 was found in archival *Hubble Space Telescope (HST)* data. The region of M101 ULX-1 was observed with *HST* using Wide Field Planetary Camera 2 and Advanced Camera for Surveys in 1994, 1995, and 2002. There is a blue object ($V = 23.8$, $B - V = -0.2$) within the $0.6''$ *Chandra* error circle (Kong *et al.* 2005; Kuntz *et al.* 2005). The optical counterpart is also detected with ground-based telescopes (Kong *et al.* 2005). At the distance of M101 ($d = 6.7$ Mpc), the absolute magnitude corresponds to $M_V = -5.3$. The magnitudes and colors of the blue star are consistent with an OB star. The source is also clearly located near star forming regions in a spiral arm from GALEX far UV image (Kong *et al.* 2005).

The IMBH model is the likely model for M101 ULX-1 since stellar-mass black hole has difficulty in explaining the state transitions, temperature changes, and the extremely

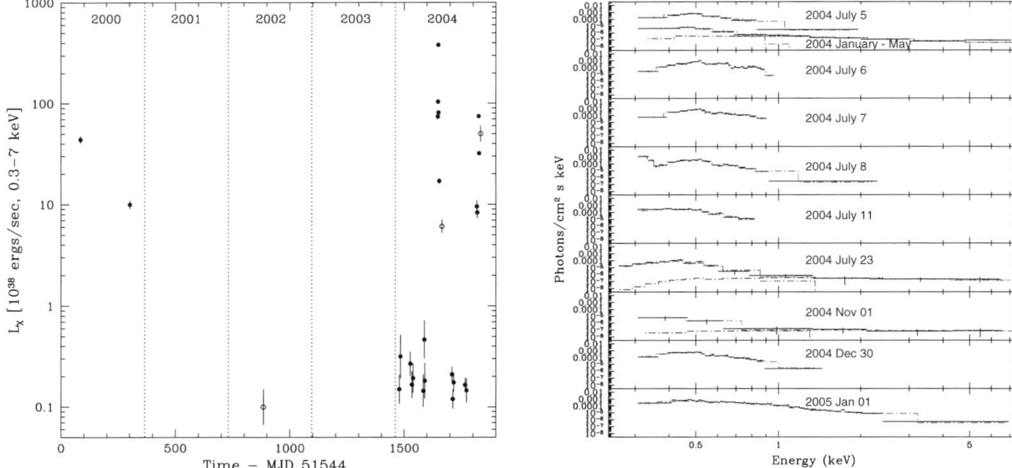

Figure 1. *Left:* Long-term X-ray light curve of M101 ULX-1 during 2000-2005. The 4 outbursts are clearly shown. The time between outbursts in 2000 was about 210 days, in 2004 this was about 200 days. *Right:* Unfolded spectra of M101 ULX-1. The total spectrum, blackbody component, and power-law component are shown in green, red, and blue, respectively. During the 2004 July outburst, the spectra can be fitted with blackbody/disk blackbody model with temperatures of 50–100 eV. On July 7, the peak bolometric luminosity reached 10^{41} ergs s^{-1}. The *XMM-Newton* spectrum taken on July 23 can be fitted with a blackbody ($kT = 53$ eV) and a power-law ($\alpha = 0.72$) model. The combined Jan-May spectrum and the Nov 1 spectrum are similar to the *XMM-Newton* spectrum, but with much lower luminosities ($\sim 10^{37}$ ergs s^{-1}). The second outburst in 2004 occurred in December. The December 30 spectrum can be fitted with a blackbody model with $kT = 70$ eV. On 2005 January 1, the source is hotter with $kT = 150$ eV.

high luminosities. The location of M101 ULX-1 near star formation regions also favors a high mass companion orbiting around an IMBH.

3. XMMU J005510.7-373855 in NGC 300

By using archival *XMM-Newton* observations, a very luminous SSS ($kT \approx 60$ eV) was found in NGC 300; the X-ray luminosity is 10^{39} ergs s^{-1} during the "high" state and 10^{38} ergs s^{-1} in the "low" state (Kong & Di Stefano 2003). *This SSS is the nearest ultra-luminous SSS and is one of the two recurrent ultra-luminous SSSs* (Kong *et al.* 2004). The source was previously seen with *ROSAT* in 1992 and fell below the detection limit in subsequent *ROSAT* observations (Fig. 2 left). It reappeared in *XMM-Newton* observations with bolometric luminosities between 10^{38} and 10^{39} ergs s^{-1}. More importantly, the source showed a 5.4-hr periodicity (Fig. 2 right). The high X-ray variability on both short timescales (hours) and long timescales (days to years) indicates that the SSS is an accreting system. If the 5.4-hr variability is the *orbital* period of the binary system, the X-ray emission can in principle be explained by accretion onto a white dwarf, a neutron star or a BH (see Kong & Di Stefano 2003 for details). However, the high luminosity in the "high" state makes accreting white dwarf model unlikely, while the supersoft spectra are difficult to be explained by a neutron star accretor. There is no *Chandra* observation of NGC 300 and it is difficult to locate the optical counterpart in the crowded region by using the *XMM-Newton* position. A planned *Chandra* observation in 2006 will help us to resolve this problem and provide important information about the nature of the source.

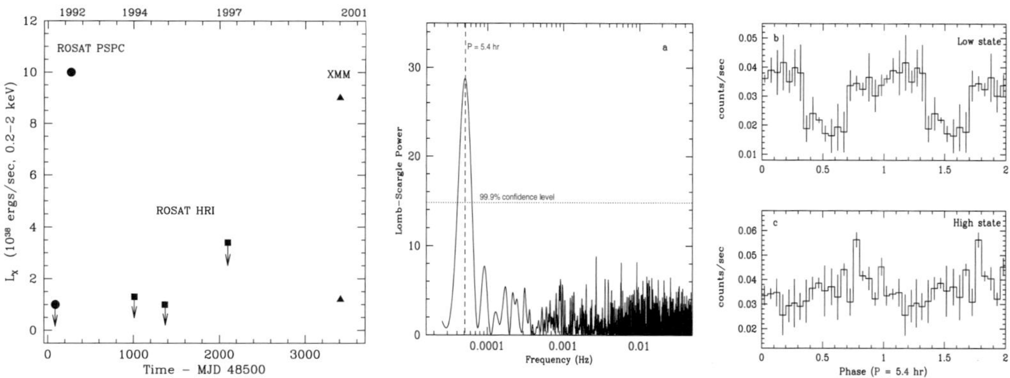

Figure 2. *Left:* Long-term light curve of the ultra-luminous SSS in NGC 300 from 1991 to 2001 (filled circles: *ROSAT* PSPC; filled squares: *ROSAT* HRI; filled triangles: *XMM-Newton*). *Right:* The Lomb-Scargle periodogram as obtained by *XMM-Newton* on 2001 January. The horizontal dotted line is the 99.9% confidence level. (b) Folded light curve of the low state (2001 January) data on a period of 5.4 hr. (c) Folded light curve of the high state (2000 December) data on a period of 5.4 hr. The T_0 of both light curves are set at the time of the first data point.

4. CXOAnt J120151.6-185231.9 in the Antennae

This SSS is one of the earliest ultra-luminous SSSs discovered by *Chandra* (Zezas *et al.* 2002). During a monitoring program between 1999 and 2002, the source was detected in all four *Chandra* observations. The luminosity varied between 10^{38} ergs s^{-1} and 10^{40} ergs s^{-1} and the temperature was between 90 eV and 150 eV (Fabbiano *et al.* 2003). There was no obvious optical counterpart from *HST* at the position of the SSS (Zezas *et al.* 2002). While IMBH model is a natural explanation if we assume an isotropic emission, stellar-mass BH model is feasible if the emission is mildly anisotropic resulting from accretion at close to the Eddington rate onto the BH. White dwarf is also a possible solution but it requires extreme beaming.

References

Di Stefano, R. & Kong, A. K. H. 2003, ApJ, 592, 884

Fabbiano, G., *et al.* 2003, ApJ, 591, 843

Kong, A. K. H. & Di Stefano, R. 2003, ApJ, 590, L13

Kong, A. K. H., Di Stefano, R., & Yuan, F. 2004, ApJ, 617, L49

Kong, A. K. H. 2005, ATel, 409

Kong, A. K. H., Rupen, M. P., Sjouwerman, L. O., & Di Stefano, R. 2005, in the proceedings of the XXII Texas Symposium on Relativistic Astrophysics, Stanford University (astro-ph/0503465)

Körding, E., Falcke, H., & Markoff, S. 2002, A&A, 382, L13

King, A. R., Davies, M. B., Ward, M. J., Fabbiano, G., & Elvis, M. 2001, ApJ, 552, L109

King, A. R. & Pounds, K. A. 2003, MNRAS, 345, 657

Kuntz, K. D., *et al.* 2005, ApJ, 620, L31

Miller, M. C. & Colbert, E. J. M. 2004, International Journal of Modern Physics D, 13, 1

Miller, J. M., Fabbiano, G., Miller, M. C., & Fabian, A. C 2003, ApJ, 585, L37

Miller, J. M., Fabian, A. C., & Miller, M. C. 2004, ApJ, 614, L117

Mukai, K., Pence, W. D., Snowden, S. L., & Kuntz, K. D. 2003, ApJ, 582, 184

Pence, W. D., Snowden, S. L., Mukai, K., & Kuntz, K. D. 2001, ApJ, 561, 189

Wang, Q. D., Yao, Y., Fukui, W., Zhang, S. N., & Williams, R. 2004, ApJ, 609, 113

Zezas, A., Fabbiano, G., Rots, A. H., & Murray, S. 2002, ApJS, 142, 239

Discussion

BARNARD: We have observed several supersoft sources with back body temperatures of 50 eV and apparent luminosities of 10^{40}–10^{41} erg/s. However, with other thermal models, e.g. NSA model, the luminosity can be 5-6 orders of magnitude lower.

KONG: I totally agree that we have to examine every possible spectral model to derive the best fit and the correct luminosity.

MIRABEL: Could SS433 be an Ultraluminous Supersoft source?

KONG: If SS433 satisfies the condition of ultraluminous supersoft X-ray source (i.e. the spectrum can be fit with a thermal model with $kT < 0.1$ keV and has a luminosity of $>10^{39}$ erg/s), then SS433 can be an ultraluminous SSS.

Populations of High Energy Sources in Galaxies
Proceedings IAU Symposium No. 230, 2005
E. J. A. Meurs & G. Fabbiano, eds.

© 2006 International Astronomical Union
doi:10.1017/S1743921306008465

Intermediate-Mass Black Hole Candidate ULXs

J. Miller

Harvard-Smithsonian Center for Astrophysics, Cambridge, Mass., USA
email: jmmiller@cfa.harvard.edu

Abstract. In a subset of the most luminous of the so-called "ultra-luminous" X-ray sources in nearby galaxies, there is evidence for black holes with masses considerably higher than found in Galactic binaries. Apart from extremely high X-ray luminosities, cool disks found in the X-ray spectra of these sources and X-ray timing measurements form the basis for present evidence for intermediate-mass black holes in these sources. New optical and radio measurements appear to support the X-ray evidence. I will review recent X-ray, optical, and radio observations of these ULXs, and discuss the strengths of the intermediate-mass black hole interpretation, arguments against this interpretation, and future prospects for revealing the nature of these ULXS more clearly.

Discussion

GHOSH: Regarding diagnostics from PDS, you described QPOs at some length. What about the possible diagnostic value of the "break" in PDS? Can you comment on the status of our understanding?

MILLER: The nature of breaks is not well-known. Breaks may trace a timescale which is fundamental (see correlation between stellar mass BHs and AGN). But, the modelling of PDS is ambiguous. Spectra are a better route to understanding ULXs.

MACCARONE: Could you comment on the fact that the spectral index of the power law is, for example for the NGC 1313 X-1 ULX, much harder than what is typically seen in Galactic very high state systems, but similar to low/hard states? Could this imply even higher masses than you've suggested?

MILLER: I think these might be similar to GX339-4 at its peak, where it's in the low/hard state, but very bright.

Populations of High Energy Sources in Galaxies
Proceedings IAU Symposium No. 230, 2005
E. J. A. Meurs & G. Fabbiano, eds.

© 2006 International Astronomical Union
doi:10.1017/S1743921306008477

New insights into ultraluminous X-ray sources from deep *XMM-Newton* observations†

T. P. Roberts[1], A.-M. Stobbart[1], M. R. Goad[1], R. S. Warwick[1], J. Wilms[2], P. Uttley[3] and J. N. Reeves[3,4]

[1]X-ray & Observational Astronomy Group, Dept. of Physics & Astronomy,
University of Leicester, University Road, Leicester LE1 7RH, UK
First author email: tro@star.le.ac.uk

[2]Astronomy & Astrophysics Group, Dept. of Physics, University of Warwick,
Coventry CV4 7AL, UK

[3]Exploration of the Universe Division, NASA Goddard Space Flight Center, Greenbelt Road,
Greenbelt, MD 20771, USA

[4]Dept. of Physics & Astronomy, Johns Hopkins University, 3400 N Charles Street, Baltimore,
MD 21218, USA

Abstract. The central controversy over whether or not ultraluminous X-ray sources (ULXs) contain a new *"intermediate-mass"* class of black holes (IMBHs) remains essentially unresolved. Indeed, whilst many recent X-ray spectroscopy results find evidence for a cool (100–200 eV) accretion disc – the expected signature of a $\sim$1000 $M_\odot$ IMBH – in ULX spectra, most of the circumstantial evidence (a combination of multiwavelength counterparts, theoretical modelling and the behaviour of accreting black holes in our own Galaxy) argues that the black holes underlying ULXs could be substantially less massive. I will present a new analysis of the deepest *XMM-Newton* observations of ULXs that directly addresses their underlying nature. This includes the results of a new 100-ks observation of the archetypal ULX Holmberg II X-1. Though a slight soft excess in its X-ray spectrum can be fitted by a cool accretion disc model, a rigorous analysis of the temporal data shows that the black hole cannot be larger than $\sim$100 $M_\odot$. Interestingly, we find evidence that the putative accretion disc corona is cool and optically thick in this source, unlike most Galactic binaries. We have also undertaken a detailed spectral analysis of the next 12 best ULX datasets in the *XMM-Newton* archive. Using physically self-consistent spectral modelling we show that whilst all the ULXs show possible cool accretion discs, the majority of these ULXs appear dominated by an optically-thick Comptonising medium. I will argue that this is evidence that most (though not necessarily all) ULXs contain black holes that are at most a few tens of solar masses in size.

Keywords. black hole physics – X-rays: binaries – X-rays: galaxies.

1. Introduction

The key recent evidence supportive of the presence of IMBHs in ULXs derives from fitting their X-ray spectra with the same empirical model as used for Galactic black hole X-ray binaries (BHXRBs). It has been found that a number of luminous ULXs are well fitted by this combination of a soft multi-colour disc blackbody (MCDBB) plus a hard power-law continuum model, with one crucial difference: the temperature of the accretion disc is a factor $\sim$10 lower, at $\sim$0.1–0.2 keV, than in Galactic systems. As the temperature

† Based on observations obtained with *XMM-Newton*, an ESA science mission with instruments and contributions directly funded by ESA Member States and NASA.

of the innermost edge of an accretion disc decreases as the mass of the compact object increases, this implies very massive black holes in these ULXs, at ~ 1000 M$_\odot$ (e.g. Miller *et al.* 2003; Miller *et al.* 2004). However, recent analyses of high spectral quality ULX data have shown that some ULXs are best described by a variant of this model where the power-law continuum dominates *at low energies* (e.g. Stobbart *et al.* 2004; Foschini *et al.* 2004). In at least one case there is a clear ambiguity, with both variants fitting the same data (Roberts *et al.* 2005). A serious challenge for this alternate model, though, is that the physical origin of the dominant soft power-law is unclear.

2. A sample of bright ULXs

We have therefore selected and uniformly reduced a sample of 13 ULXs, comprising the highest quality EPIC spectral data available from the *XMM-Newton* archive, to address the following questions: how easy is it to differentiate the two spectra? How common is each type of spectrum? And what are the physics underlying the alternate spectral model? Though this sample is small, the ULXs are representative of the full range of ULX luminosity ($\sim 10^{39} - 2 \times 10^{40}$ erg s^{-1}), and with a minimum of a few thousand counts per source represent the best defined X-ray spectra of ULXs to date.

The high definition and underlying complexity of the ULX spectra were highlighted by simple spectral fits, with none of the sources being well-fit (using a 95% probability of rejection criterion) by an absorbed MCDBB model, and only 5/13 being fit by a power-law continuum (including the three lowest quality datasets). The use of two-component models improved the fits greatly. In particular, the IMBH model (i.e. cool MCDBB + hard power-law) produced good fits to 8/13 datasets, with an inner-disc temperature $kT_{in} \sim 0.1$–0.25 keV and a power-law photon index $\Gamma \sim 1.6$–2.5†.

However, the alternate empirical model also produced a total of 8/13 good fits. 6 of these sources were also well-fit by the IMBH model, hence spectral ambiguity is present in 6/13 of the ULXs in our sample. Of the remaining sources, two apiece were uniquely well-fit by either the IMBH or alternate model, and the remaining three were well-fit by neither (though two favoured an IMBH fit, and one the alternate). As an attempt to find which model the ambiguous sources prefer, we note that the key observable distinction between the two models is that the alternate model should show curvature in the 2–10 keV range. Therefore we fit both power-law and broken power-law models to the 2–10 keV data from each source, and looked for a statistical improvement between the two fits. This approach was vindicated by demonstrating that those sources best fit by the alternate model showed strong curvature ($> 4\sigma$ improvements according to the F-test). Of the ambiguous sources, 3/6 also showed some evidence for curvature ($> 2\sigma$ significance) as, rather surprisingly, did two of the IMBH model sources. Hence we find at least marginal evidence for 2–10 keV curvature in $> 50\%$ of our spectra.

We next investigated the origin of this curvature using more physically-motivated models. The "slim disc" model of an accretion disc (e.g. Watarai *et al.* 2001; XSPEC parameterisation courtesy K. Ebisawa) was unsuccessful in fitting our spectra in 10/13 cases. Instead, we found a physically self-consistent accretion disc plus Comptonised corona model, using a `diskpn + eqpair` model in XSPEC (Gierlinski *et al.* 2001, Coppi 2000), gave good fits to 11/13 sources (plus another only marginally rejected at the 95% criterion). All the fits display a cool disc component, with temperatures ~ 0.1–0.3 keV, as would be expected from IMBHs. The majority of these fits also show a remarkable

---

† Though see Roberts *et al.* (2005) on the anomalous flatness of this photon index if the compact object is an IMBH.

characteristic; the spectral curvature originates in a coronal component that is optically thick, with typical optical depths of $\tau \sim 10$–40. This is very puzzling, because if ULXs are to be understood as higher-mass analogues of high-state BHXRBs, then their corona should be similarly optically thin ($\tau < 1$, which only appears to be the case in two sources). Hence this model demonstrates that many ULX spectra do not appear similar to the high spectral state of BHXRBs.

3. Holmberg II X-1

This source is regarded as the archetypal luminous ($L_X > 10^{40}$ erg s^{-1}) nearby ULX, and has been widely studied (e.g. Dewangan *et al.* 2004). We were awarded a 100-ks observation of this source in *XMM-Newton* AO-3 (Goad *et al.* 2005, submitted to MN-RAS). Though more than 60% of this observation was lost to space weather, we were still able to extract the first reasonable signal-to-noise RGS spectrum of an ULX. This showed a smooth continuum shape, with the exception of an excess of counts slightly above 0.5 keV. This could be fit by an O VII triplet, but a better solution was found by allowing the abundance of the absorbing material to drop to ~ 0.6 of the solar value.

Interestingly, this result strongly affects the EPIC data modelling. In particular, using a 0.6-solar abundance TBABS model in XSPEC greatly reduces the size of the apparent soft excess, and so the mass of the BH estimated from the IMBH model is reduced to $\sim 33\%$ of its value assuming a solar abundance absorber. However, the best fit to the data is found to be the physical accretion disc + corona model, with $kT_{in} \sim 0.2$ keV and $\tau \sim 4$–9 (i.e. a cool disc and optically thick corona).

The EPIC timing data showed Ho II X-1 to be remarkably invariant during the observation. A power spectral density (PSD) analysis was performed, finding that no power was evident (above the Poisson noise level) in the $\sim 10^{-4} - 6$ Hz range. This immediately ruled out Ho II X-1 being in a high BHXRB state. It does not rule out a state with a band-limited PSD, such as occurs in the low or very high states. However, the strong limits placed by the non-detection of power in the observed frequency interval implies any power must be present at higher frequencies. Assuming that BH timing properties scale linearly with mass (e.g. Uttley *et al.* 2002), we can place an upper limit on the mass of Ho II X-1 of 100 M$_\odot$ if it is in the low or very high state. Encouragingly, GRS 1915+105 shows very similar variability characteristics in its "χ-class" of behaviour, which is thought to be typical of the very high state.

4. Conclusions

Whilst it is evident from this work that optically-thick coronae may be common in ULXs, what is their origin? One possible explanation is offered by the model of Zhang *et al.* (2000), that explains accretion discs as a 2-layer system, with a cool (0.2–0.5 keV) interior seeding a warm, optically thick (1–1.5 keV, $\tau \sim 10$) Comptonising upper disc layer. This potentially explains both spectral components seen in our modelling. This model has also successfully been used to describe GRS 1915+105. We therefore suggest that this argues Ho II X-1 and many other ULXs may be analogues of GRS $1915 + 105$, probably with larger BH masses (10–100 M$_\odot$), accreting at around the Eddington limit.

Obviously, we cannot rule out the cool discs being the signature of an ~ 1000 M$_\odot$ IMBH. However, we note that similar spectral components – also modelled as cool discs – are seen in PG quasars. In these sources their temperature has been shown to be completely independent of the mass of the BH (Gierlinski & Done 2004). This could imply a radically different origin for the soft excesses, such as in an outflow, or perhaps even as atomic features on the accretion disc spectrum.

The bottom line from this work is that many ULXs do not appear to be in the expected high state for a $\sim$1000 $M_\odot$ IMBH. Instead, their unusual spectra may be more consistent with $<$100 $M_\odot$ BHs in a similar mode to GRS 1915 + 105 in the very high state. However, confirmation of this can only come from one source: a dynamical mass limit for the BH in an ULX derived from its orbital dynamics.

Acknowledgements

TPR, AMS and MRG gratefully acknowledge funding from PPARC.

References

Coppi, P.S., 2000, *HEAD*, 5.2311
Dewangan, G., Miyaji, T., Griffiths, R.E., & Lehmann, I., 2004, *ApJ* (Letters) 608, L57
Foschini, L., Rodriguez, J., Fuchs, Y., Ho, L.C., Dadina, M., Di Cocco, G., Courvoisier, T.J.-L., & Malaguti, G., 2004, *A&A* 416, 529
Gierliński, M. & Done, C., 2004, *MNRAS* 349, L7.
Gierliński, M., Maciolek-Niedzwiecki, A., & Ebisawa, K., 2001, *MNRAS* 325, 1253
Miller, J.M., Fabbiano, G., Miller, M.C., & Fabian, A.C., 2003, *ApJ* (Letters) 585, L37
Miller, J.M., Fabian, A.C., & Miller, M.C., 2004, *ApJ* 607, 931
Roberts, T.P., Warwick, R.S., Ward, M.J., Goad, M.R., & Jenkins, L.P., 2005, *MNRAS* 357, 1363
Stobbart, A., Roberts, T.P., & Warwick, R.S., 2004, *MNRAS* 351, 1063
Uttley, P., McHardy, I.M., & Papadakis, I.E., 2002, *MNRAS* 332, 231
Watarai, K., Mizuno, T., & Mineshige, S., 2001, *ApJ* (Letters) 549, L77
Zhang, S.N., Cui, W., Chen, W., Yao, Y., Zhang, X., Sun, X., Wu, X.-B., & Xu, H., 2000, *Science* 287, 1239

Discussion

MILLER: (Abridged) Aren't your timing results the wrong way round? Don't we expect no variability in the high state, and up to 30% RMS variation in the very high state? I also don't believe your spectroscopy results – by using specific physical models you are imprinting a limited set of assumptions on the data.

ROBERTS: I agree to an extent on your comment on spectroscopy, though note that our results included both optically-thin and -thick solutions so we imparted no strong bias on that parameter. The preference for optical thickness is consistent with the curvature we model empirically. On the subject of timing, the high state can have its intrinsic noise washed out if you look at the portion of the spectrum dominated by the accretion disc. We are looking predominantly at the supposed corona, and see no variation. Similarly, our point on the very high state is one would expect to see variability power at high frequencies for a stellar-mass black hole, and lower frequencies for larger black holes. As we see no variation at frequencies up to $\sim$5 Hz we can argue the black hole is no more massive than 100 $M_\odot$.

GHOSH: Do I understand you correctly that this high optical depth (τ greater or equal to 8) is one of the major diagnostics suggesting an unusual accretion mode? If so, one of the schematic model figures you showed is that of a "sandwich" accretion disk, wherein the top layer of the disk provides this optical depth. Such models are interesting, and their physical bases need to be understood better.

ROBERTS: Such high optical depths are not generally found in black hole coronae, so do suggest an unusual accretion state. I agree that a better physical understanding is very desirable!

Populations of High Energy Sources in Galaxies
Proceedings IAU Symposium No. 230, 2005
E. J. A. Meurs & G. Fabbiano, eds.

© 2006 International Astronomical Union
doi:10.1017/S1743921306008489

Ultraluminous X-ray Sources: Bubbles and Optical Counterparts

Manfred W. Pakull, Fabien Grisé and Christian Motch

Observatoire Astronomique de Strasbourg, France
email: pakull@astro.u-strasbg.fr

Abstract. Optical studies of ultraluminous X-ray sources (ULX) in nearby galaxies have turned out to be instrumental in discriminating between various models including the much advertised intermediate mass black hole hypothesis and various beaming scenarios. Here we report on ESO VLT and SUBARU observations of ULX that have revealed the parent stellar clusters with ages of some 60 million years in two cases. Thus we are able to derive upper limits of about 8 $M_\odot$ for the mass donors in these systems. The optical counterparts are dominated by X-ray heated accretion disks, and the discovery of the He II λ4686 emission line now allows to derive dynamical masses in these systems. Apparent radial velocity variations of 300 km/s have been detected in NGC 1313 X-2 which, if confirmed by further observations, would exclude the presence of IMBH in these systems.

Keywords. galaxies: individual (NGC 1313, Holmberg IX), ISM: bubbles, X-rays: galaxies, X-rays: binaries.

1. The enigma of ultraluminous X-ray sources

One of the most significant results from recent X-ray studies of nearby galaxies is the discovery of a number of non-nuclear point sources (Ultraluminous X-ray sources – ULX) with apparent (isotropic) X-ray luminosities of 10^{39}–10^{41} erg/s, a factor of 10–1000 times brighter than typical luminous X-ray binaries in our Galaxy.

From X-ray timing and spectral studies it is clear that ULXs are accreting compact objects. Therefore, their luminosity should not exceed the Eddington limit $L_E = 1.3 \times 10^{38}$ $M/M_\odot$ erg/s. One possible explanation is that the accreting object powering an ULX is an intermediate-mass black hole (IMBH), with M $\sim 10^2$–10^4 $M_\odot$ (Colbert & Mushotzky 1999, Makishima *et al.* 2000, Miller, Fabian & Miller 2004). An alternative explanation is that the emission is beamed along the observer's line of sight, either by geometrical effects (King *et al.* 2001), or due to relativistic jets like in microquasars (Fabrika & Mesheryakov 2001), so that the true total luminosity does not exceed the Eddington limit of a stellar-mass black hole. Yet another possibility is that the accretor is a stellar BH genuinely emitting above the classical Eddington limit (Begelman 2002).

At the same time, one has to explain the high accretion rate (up to $\sim 10^{-6}$ $M_\odot$/yr) required by the inferred luminosities. Wind accretion (typical for high-mass X-ray binaries) or Bondi-type accretion from the interstellar medium are clearly too inefficient. It is more likely that the mass transfer occurs via Roche-lobe overflow from a donor star to the black hole. Stable, high mass transfer can proceed on a nuclear time scale ($\leqslant$ several 10^7 yr), if the mass donor is massive – but not much more massive than the black hole (cf. Rappaport, Podsiadlowski & Pfahl 2005).

2. Bubble Nebulæ

An important piece of information for our understanding of ULX comes from the discovery of huge ionised bubble nebulæ around a significant fraction of unobscured ULX in nearby galaxies (Pakull & Mirioni 2002, 2003). It is even likely that all ULX sources are surrounded by such structures. Close to the X-ray source, we sometimes observe X-ray ionisation effects which in the case of Holmberg II X-1 has allowed to independently measure the (ionising) X-ray luminosity (Pakull & Mirioni 2002). The presence in the outer regions of strong [S II] and [O I] emission lines and of supersonic expansion speeds of 80-250 km/s derived from the width of Hα emission show that (at least the outer parts of) the bubbles are shock-excited rather then photoionised.

Among the best-studied large ULX bubbles are MH 9-11 around Holmberg IX X-1 (cf. Grisé *et al.*, this symposium) and NGC 1313 X-2 which both have diameters of about 500 pc, i.e. they are much larger than supernova remnants.

One of the many interesting aspects of ULX bubbles is that we might estimate the kinetic energy involved in the ULX phenomenon. Two possibilities for bubble formation have been discussed: either they were formed in an explosive event with kinetic energy E_0 (possibly the SN explosion that created the compact component in the ULX), or they are being inflated by ULX stellar wind/jet activity with mechanical energy L_w. Assuming for simplicity that energy is largely conserved (Sedov-Taylor solution), we have for the SNR case

$$E_0 \approx 1.9\ 10^{52}\ erg\ R_2^3\ v_2^2\ n \qquad t_6 \approx 0.4\ R_2\ v_2^{-1} \qquad (2.1)$$

and for the wind/jet case

$$L_w \approx 3.8\ 10^{39}\ erg\,s^{-1}\ R_2^2\ v_2^3\ n \qquad t_6 \approx 0.6\ R_2\ v_2^{-1} \qquad (2.2)$$

Here R_2 is the radius in units of 100 pc, and v_2 is the expansion velocity in units of 100 km/s of a bubble having an age of $10^6\ t_6$ yrs. The interstellar particle density, n, into which the bubble expands can be estimated from comparing the observed Hα emission with the intensity $I_\alpha = 10^{-6}\ I_{\alpha,-6}$ erg cm^{-2} s^{-1} sr^{-1} of a fully radiative shock (Dopita & Sutherland 1996):

$$n \approx 0.6\ cm^{-3}\ I_{\alpha,-6}\ v_2^{-2.4} \qquad (2.3)$$

Straightforward application of these equations to ULX bubbles yields typical ages of 10^6 yrs and densities in the range of 0.1–1.0 cm^{-3}. This results in very large energy/power requirements of $\sim 10^{52-53}$ erg for the SNR case, and $\sim 10^{39-40}$ erg s^{-1} for continuous inflation. However, a SNR is likely to expand into a region of very low density that has previously been excavated by the stellar winds from an evolving cluster (see below) before hitting the walls of that 'superbubble'. In this way, energy requirements could well be ten times lower, and more akin to typical SN energies of 10^{51} erg. Another complication arises from the clumpiness of the interstellar medium which results in smaller shock velocities in the dense optically emitting clouds as compared to a higher velocity of the main shock in the intercloud medium. Taking into account this effect, Blair *et al.* (1981) estimated the kinetic of SNR to be:

$$E_0 \approx 4\ 10^{50}\ erg\ R_2^3\ n(SII) \qquad (2.4)$$

where n(SII) is the density in the recombination zone behind the shock, as measured by the well-known forbidden [SII] line ratio. However, this method is not without severe problems either, as the energy calculated in this way appears to be positively correlated with the remnant diameter, an effect that may be related to the magnetic pressure in the dense clouds.

As outlined below, we favour the possibility of ongoing inflation by stellar winds/jets which would require for reasonable mass loss rates $\leqslant 10^{-6}$ M$_\odot$/yr mildly relativistic ejection velocities akin to the famous v=0.26 c jets in SS 433. However, inflation into a cloudy medium would also lower the power requirements somewhat.

3. Clusters

Most massive stars are born in clusters or associations. Searching for such birthplaces of ULX thus opens the possibility to learn more about their ages and possible formation scenarios. For the relatively isolated ULX in Holmberg IX and in NGC 1313 X-2 we have been able to identify small faint clusters of stars that in each case are clearly associated with the ULX (cf. Grisé *et al.*, this symposium). Multicolour photometry and isochrone

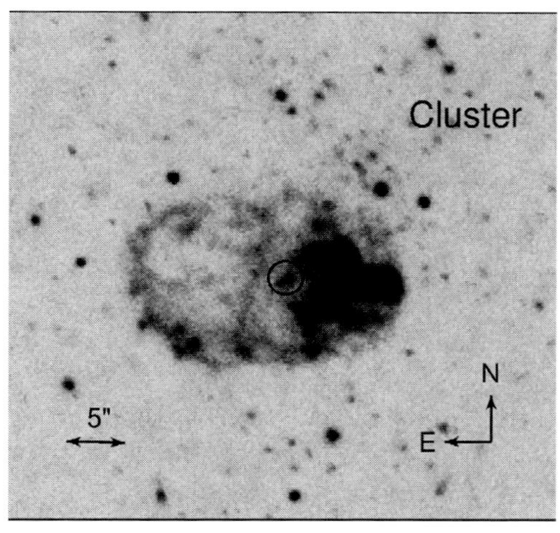

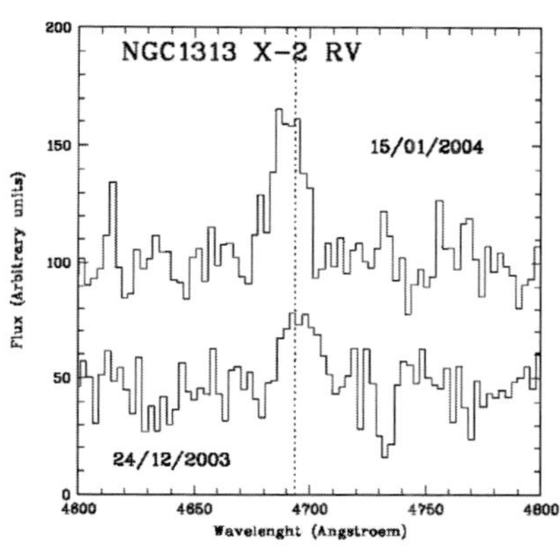

Figure 1.
Left: Multicolor image (including Hα) of the 400 pc diameter bubble around the ULX NGC 1313 X-2. The Chandra error circle includes a close pair of stars (star C in Zampieri *et al.* 2004). The N-W component is the optical counterpart.
Right: Blue range spectra of the counterpart taken 3 weeks apart. Most prominent is the He II λ4686 emission line which appears to have moved by 300 km/s.

fitting yield ages of some 40–70 Myrs and total cluster masses of some 10^3 M$_\odot$. Thus, all ionising O stars have already exploded a long time ago, implying again that the nebulæ cannot possibly be photoionized by stellar EUV continua. This has an important consequence for the mass donor component in the ULX: it cannot be more massive then about 8 M$_\odot$. Furthermore, if the bubble nebulæ indeed represent the remnants of the formation process of the ULX black holes, then these explosions (which took place about 1 Myr ago) have taken place in an advanced stage of cluster evolution, i.e. the progenitor stars of the ULX accretors would not have been much more massive then the current donors (i.e. $\leqslant$ 10 M$_\odot$). Such stars are of course not likely to become massive black holes.

We are therefore left with the hypothesis that the ULX nebulae represent Begelman *et al.* (1980) 'beambags' of jet inflation like the 'ears' in the (radio) nebula W 50 around SS 433.

4. Optical counterparts

The optical counterparts of ULXs Holmberg IX and NGC 1313 X-2 (see Fig. 1) have visual magnitudes of 22.9 and 23.4, respectively, and blue optical colors (B-V $\sim$ 0.0); at a distance of 3–4 Mpc this translates into $M_V \sim -5$. Optical spectra taken with the ESO VLT and with the SUBARU telescope reveal the presence of stellar He II $\lambda4686$ emission with equivalent widths of 10 and 18 Å, respectively. This high excitation line can be considered a hallmark of X-ray binaries, being formed in the X-ray heated disk around the compact component. In Galactic high mass X-ray binaries (such as the $L_X = 10^{38}$ erg/s systems Cen X-3, SMC X-1, LMC X-4, etc) the corresponding EWs are more than an order of magnitude smaller, attesting to their much smaller X-ray luminosities. Therefore, the presence of strong He II $\lambda4686$ emission in ULX might also be taken as strong evidence against the X-ray beaming scenarios mentioned earlier. We note by the way that a He II $\lambda4686$ emitting Wolf-Rayet star interpretation can be ruled out given the advanced age of the cluster.

Strong support for an accretion disk interpretation comes also from its position in the famous van Paradijs & McClintock (1994) $\Sigma - M_v$ diagram of low mass X-ray binaries (i.e. of X-ray ionised disks), at the very high-luminosity end of this relation. Here Σ is proportional to the optical light expected from an accretion disk that is heated by a given X-ray luminosity; we here have assumed a semi-detached binary of some 20 $M_\odot$ total mass.

The Right panel of Fig. 1 illustrates two observations of the NGC 1313 X-2 counterpart separated by about three weeks. Here the He II $\lambda4686$ emission appeared to have varied in radial velocity (RV) by about ~300 km/s around the velocity of the local H I gas in the galaxy (dotted line). *If* this variation reflects real RV changes (as opposed to possible profile changes of the line having intrinsic FWHM ~600 km/s), then the object in the center of the accretion disk cannot be very massive, i.e. one could rule out the presence of an IMBH.

We finally mention variations by ~0.2 mag amplitude over 9 nights in our B band photometry of the optical counterpart of NGC 1313 X-2. However, no stricly periodic signal is seen, such as expected from ellipsoidal variations, and we ascribe these changes to variable X-ray heating of the accretion disk.

Acknowledgements

We acknowledge collaboration with T.G. Tsuru, K. Sekigushi, A. Tajitsu and I. Smith.

References

Begelman, M.C., Sarazin, L., Hatchett, S.P., McKee, C.F., & Aarons, J. 1980, *ApJ* 238, 722

Begelman, M.C. 2002, *ApJ* 568, L97

Blair, W.P., Kirshner, R.P., & Chevalier, R.A. 1981, *ApJ* 247, 879

Colbert, E.J.M. & Mushotzky, R.F. 1999, *ApJ* 519, 89

Dopita, M.A. & Sutherland, R.S. 1996, *ApJ* 102, 161

Fabrika, S. & Mescheryakov, A. 2001, in High Angular Resolution in Astronomy, ed. R. Schlizzi *et al.* (ASP publication)

King, A.R., Davies, M.B., Ward, M.J., Fabbiano, G., & Elvis, M. 2001, *MNRAS* 55, L109

Makishima, K. *et al.* 2000, *ApJ* 535, 632

Miller, J.M., Fabian, A.C., & Miller, M.C. 2004 *ApJ* 614, L117

Pakull, M.W. & Mirioni, L. 2002, *astro-ph/0202488*

Pakull, M.W. & Mirioni, L. 2003, *Rev. Mexicana AyA* 15, 197

Rappaport, S.A., Podsiadlowski, Ph., & Pfahl, E. 2005, *MNRAS* 356, 401

van Paradijs, J. & McClintock, J.E. 1994, *AA* 290, 133

Zampieri, L. *et al.* 2004, *ApJ* 603, 523

Discussion

MACCARONE: The detection of moving emission lines is very exciting but I think one should be cautious about interpreting them as disk lines rather than as possible mixtures of disk lines and narrow Bowen fluorescence components from the irradiated surface of the donor star, as used e.g. by Hynes *et al.* to measure the rotation curve of GX 339-4.

PAKULL: Narrow components are fainter than broad components, in most or all cases, and trace the motion of the accretion disk in most or all cases.

Populations of High Energy Sources in Galaxies
Proceedings IAU Symposium No. 230, 2005
E. J. A. Meurs & G. Fabbiano, eds.

© 2006 International Astronomical Union
doi:10.1017/S1743921306008490

The recurrent ultra-luminous X-ray transient NGC 253 ULX1

M. Bauer and W. Pietsch

Max-Planck-Institut für extraterrestrische Physik, Giessenbachstraße, 85741 Garching, Germany
email: mbauer@mpe.mpg.de

Abstract. We present the results of *ROSAT* and XMM-Newton observations of the recurrent ultraluminous X-ray source (ULX) NGC 253 ULX1. This transient is one of the few ULXs that was detected during several outbursts. The luminosity reached 1.4×10^{39} erg s^{-1} and 0.5×10^{39} erg s^{-1} in the detections by *ROSAT* and XMM-Newton, respectively, indicating a black hole X-ray binary (BHXRB) with a mass of the compact object of >11 M$_\odot$. In the *ROSAT* detection NGC 253 ULX1 showed significant variability, whereas the luminosity was constant in the detection from XMM-Newton. The XMM-Newton EPIC spectra are well-fit by a bremsstrahlung model (kT= 2.24 keV, $N_H = 1.74 \times 10^{20}$ cm^{-2}), which can be used to describe a comptonized plasma. No counterpart was detected in the optical I, R, B, NUV and FUV bands to limits of 22.9, 24.2, 24.3, 22 and 23 mag, respectively, pointing at a XRB with a low mass companion.

Keywords. X-rays: binaries – X-rays: individuals: (NGC 253 ULX1) – black hole physics – accretion, accretion disks.

1. Introduction

ULXs are extra-nuclear compact X-ray sources with luminosities considerably exceeding the Eddington luminosity for stellar mass X-ray binaries of $\sim 2 \times 10^{38}$ erg/s (Makishima *et al.* 2000). ULXs obtain special attention as they may indicate the existence of intermediate mass black holes (100–1000 M$_\odot$, IMBHs). The ULX presented here, NGC 253 ULX1, was found by Liu & Bregman (2005) searching for ULXs in ROSAT HRI observations of 313 nearby galaxies. We analysed the source for variability (Fig. 1) and extracted the first spectrum from XMM-Newton data. For a more extensive report on NGC 253 ULX1 see Bauer & Pietsch (2005).

2. Properties of NGC 253 ULX1

NGC 253 ULX1 showed two outburst. The first was observed by ROSAT with $L_X = 1.4 \times 10^{39}$ erg/s (0.3–10keV), the second was detected by XMM-Newton with $L_X = 0.5 \times 10^{39}$ erg/s (0.3–10keV). From the first outburst we determined a lower mass limit of the compact object to 11 M$_\odot$. We were able to obtain the first spectrum of this source from the second outburst. It was best fitted with a bremsstrahlung model (kT $= 2.24$ keV, $N_H = 1.74 \times 10^{20}$ cm^{-2}, $\chi^2_{\mathrm{red}} = 0.961$). The brightness of the source varies by at least a factor of 500 ($L_{X,\mathrm{min}} = 0.003 \times 10^{39}$ erg/s). Its fastest change in luminosity ($L_{\mathrm{max}}/L_{\mathrm{min}}$) is greater than 71 in 120 days. In the ROSAT observation, NGC 253 ULX1 showed significant variability by at least a factor of 2 during 8 days. No short term variability was detected in the XMM-Newton observation.

3. Optical counterpart

We checked images taken with the Wide Field Imager (WFI) on the MPG-ESO 2.2m telescope at La Silla in the R-, I- and B-band (limiting magnitudes 24.2, 22.9 and 24.3,

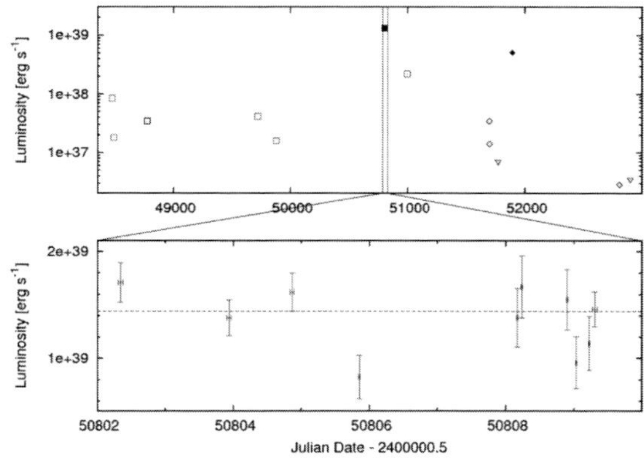

Figure 1. NGC 253 ULX1 light curves. Upper panel: solid symbols represent detections, open symbols 3σ upper limits of NGC 253 ULX1. Different symbols represent observing instruments: *ROSAT* (squares), XMM-Newton (diamonds) and *Chandra* (triangles). Lower panel: Luminosities of individual *ROSAT* HRI exposures from observation 601111h. In contrast to the upper panel, in the lower panel the plot is linear in luminosity.

respectively) and images taken with the Galaxy Evolution Explorer (GALEX, a space telescope from NASA observing in the ultraviolet) in the NUV and FUV (limiting magnitudes 22 and 23, respectively), but no counterpart could be detected.

4. The nature of NGC 253 ULX1

The best fitting bremsstrahlung spectrum indicates that NGC 253 ULX1 is a XRB system. We determined the lower mass limit of the compact object from its ROSAT luminosity to 11 $M_\odot$, typical for a stellar mass black hole. If we assume a multicolour disk blackbody model (XSPEC model *diskbb*, fitted with $N_H = 1.30 \times 10^{20}$ cm^{-2}, kT $= 0.62$ keV, $\chi^2_{\rm red} = 1.671$) then NGC 253 ULX1's position in the luminosity-disk temperature diagram (Fig. 2 in Miller *et al.* 2004) also indicates that NGC 253 ULX1 is a stellar mass black hole. The lack of an optical counterpart points at a low mass companion in the system. A high mass XRB should have been detectable in the WFI images (see Sect. 3) at about 22 to 24 mag (extrapolating V magnitudes from high mass XRBs in the Magellanic Clouds, Liu *et al.* 2000).

The recurrent outbursts exclude that the source is the luminous remnant of a recent supernova, like e.g. SN1993J in M81 (Zimmermann & Aschenbach 2003). We can also exclude that NGC 253 ULX1 is either a foreground object or a background active galactic nucleus (AGN) based on three arguments: (i) Its $\log(f_{\rm x}/f_{\rm opt})$ value of >3.2 exceeds that expected for galactic sources (-4.6 to -0.6) as well as AGNs (-1.2 to $+1.2$) (Maccacaro *et al.* 1988). (ii) The variability of NGC 253 ULX1 is by far larger than typically observed for AGNs ($\sim$10–60) and (iii) NGC 253 ULX1 shows a bremsstrahlung spectrum, whereas spectra of AGNs above 2 keV are typically fitted by a power law.

We conclude that NGC 253 ULX1 is a low mass XRB with a stellar black hole as compact companion.

References

Bauer, M. & Pietsch, W. 2005, astro-ph/0507248, A&A accepted

Liu, J. & Bregman, J. N. 2005, ApJS, 157, 59

Liu, Q. Z., van Paradijs, J., & van den Heuvel, E. P. J. 2000, A&AS, 147, 25

Maccacaro, T., Gioia, I. M., Wolter, A., Zamorani, G. & Stocke, J. T. 1988, ApJ, 326, 680

Makishima, K., Kubota, A., Mizuno, T., *et al.* 2000, ApJ, 535, 632

Miller, J. M., Fabian, A. C. & Miller, M. C. 2004, ApJ, 614, L117

Zimmermann, H.-U. & Aschenbach, B. 2003, A&A, 406, 969

Populations of High Energy Sources in Galaxies
Proceedings IAU Symposium No. 230, 2005
E. J. A. Meurs & G. Fabbiano, eds.

© 2006 International Astronomical Union
doi:10.1017/S1743921306008507

Irradiation models for ULXs and fits to HST observations of NGC 4559 X-7

Chris Copperwheat[1], Mark Cropper[1], Roberto Soria[1,2] and Kinwah Wu[1]

[1]Mullard Space Science Laboratory, University College London, Holmbury St. Mary, Dorking, Surrey, RH5 6NT, UK
[2]Harvard-Smithsonian Center for Astrophysics, 60 Garden Street, Cambridge, MA 02138, USA
email: cmc@mssl.ucl.ac.uk

Abstract. We detail a model we have created to describe the optical emission from a ULX in terms of an irradiated companion star and disk. We apply this model to optical observations of ULX X-7 in NGC 4559. We revise the parameters of the companion star in this system to be older, less massive and of a later spectral type than previously reported. We find the black hole to be of a few hundred solar masses at most.

Keywords. accretion, accretion disks, black hole physics, X-rays: galaxies, X-rays: stars.

1. Introduction

Ultra-luminous X-ray sources (ULXs) are point-like, non-nuclear sources with apparent isotropic luminosities greater than 10^{39} ergs s^{-1}. The brightest sources have X-ray luminosities in excess of 10^{40} ergs s^{-1}, implying bolometric luminosities of order 10^{41} ergs s^{-1}. This is significantly larger than the Eddington luminosity limit for an accreting stellar-mass black hole (BH). The nature of these objects is still unclear. The high luminosities may be a result of beaming towards the observer. Alternatively, the accreting object may be an intermediate mass black hole (IMBH) with mass $50 - 10000 M_\odot$. We have created a model describing the optical emission from ULXs. Our model is useful to identify optical counterparts. Our study has shown that optical observations are also powerful tools, as X-ray observations, to determine the properties and hence the nature of ULX.

2. Model

We consider a binary model, with the compact object accreting material from a companion star. We assume the X-ray emission is isotropic, and hence take the IMBH interpretation as a starting point. The brightest ULXs require an accretion rate greater than that which could be supplied by a stellar wind, so we assume the matter is transferred onto the compact object through Roche lobe overflow. We constrain the geometry of the system so that the companion star is filling its Roche lobe. This constraint necessitates a small binary separation and a large companion star. A large amount of X-ray flux will be incident on the surface of this star, and the optical/IR characteristics of this star will be modified. The irradiation will induce intensity and colour shifts compared to normal stars, which we use as a diagnostic. We assume the system is in a quasi-steady state, and the irradiated surfaces are in thermal, radiative and hydrostatic equilibrium. We consider the effects of radiative transport and radiative equilibrium in the irradiated surface of the star and an irradiated accretion disk. We consider a plane-parallel model and adopt the radiative transport formulation of Milne (1926) and Wu *et al.* (2001) to describe the

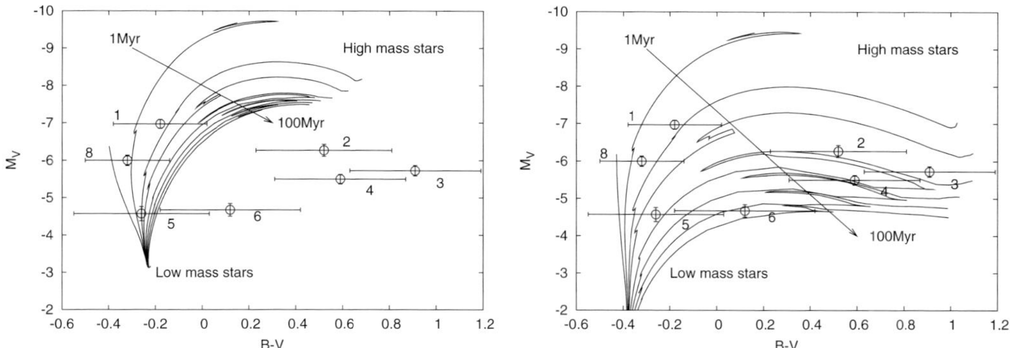

Figure 1. Colour – magnitude diagrams showing the Geneva isochrones modified to include our irradiation model. In the left panel we use an inclination of $\cos i = 0.5$ and a BH mass of $100 M_\odot$. In the right panel we use $\cos i = 0.0$ and a BH mass of $1000 M_\odot$. If we require candidates 2, 3 and 4 to fit with our model, then this combination of a low inclination and a high BH mass is necessary.

heated surface. We determine the total emergent radiation from a distorted, Roche lobe filling star numerically. We do the same for the disk, using a thin disk geometry.

The model is described more completely in Copperwheat *et al.* (2005).

3. Application to ULX X-7 in NGC 4559

A ULX with $L_x \simeq 10^{40}$ ergs s^{-1}was found in the late type spiral galaxy NGC 4559 (d=10Mpc). HST observations revealed eight candidates for the optical counterpart (Soria *et al.* 2005) We apply our irradiation model to stars evolved according to the Geneva isochrones of Lejeune & Schaerer (2001) and then compare these modified isochrones to the observations. We do this for a range of orientations, inclinations and BH masses (Figure 1). We find that some candidates are consistent with our model irradiated star and disk only when we use a very low inclination and a very high BH mass. If the inclination is such that the optical emission contains an appreciable disk component, we find that these candidates (2, 3 and 4) do not fit with any BH and star combination.

Soria *et al.* (2005) concluded that candidate 1 was the most likely counterpart to the ULX. The other candidates were found to be consistent with the unperturbed isochrones for stars with masses $10 - 15 M_\odot$ and ages of approximately 20Myr. If we assume the companion star is also this age and that candidate 1 is indeed the counterpart, then by comparing the 20Myr isochrone with the observed magnitudes we find the mass of the companion star to be $11 - 12 M_\odot$ and the radius to be $40 - 50 R_\odot$. We find also that these assumptions imply an upper bound on the BH mass range of a few hundred solar masses.

References

Copperwheat, C. M., Cropper, M., Soria, R. & Wu, K. 2005, MNRAS, 362, 79
Lejeune, T. & Schaerer, D., 2001, A&A, 366, 538
Milne, E. A. 1926, MNRAS, 87, 43
Soria, R., Cropper, M., Pakull, M., Mushotzky, R. & Wu, K., 2005, MNRAS, 356, 12
Wu, K., Soria, R., Hunstead, R. W. & Johnston, H.M. 2001, MNRAS, 320, 177

Populations of High Energy Sources in Galaxies
Proceedings IAU Symposium No. 230, 2005
E. J. A. Meurs & G. Fabbiano, eds.

© 2006 International Astronomical Union
doi:10.1017/S1743921306008519

The Ultraluminous X-ray Source in Holmberg IX and its Environment

**Fabien Grisé[1], Manfred W. Pakull[1]
and Christian Motch[1]**

[1]Observatoire Astronomique de Strasbourg, 13 rue de l'Université, FRANCE
email: grise@astro.u-strasbg.fr

Abstract. We present optical observations of an ultraluminous X-ray source (ULX) in Holmberg IX, a dwarf galaxy near M81. The ULX has an average X-ray luminosity of some 10^{40} erg/s. It is located in a huge (400pc × 300pc) ionized nebula being much larger than normal supernova remnants. From the observed emission lines (widths and ratios) we find that the structure is due to collisional excitation by shocks, rather than by photoionization.

We identify the optical counterpart to be a 22.8 mag blue star ($M_V = -5.0$) belonging to a small stellar cluster. From isochrone fitting of our multi-colour photometry we determine a cluster age of some 60 Myr. We also discovered strong stellar HeIIλ4686 emission (equivalent width of 10 Å) which proves the identification with the X-ray source, and which suggests the presence of an X-ray heated accretion disc around the putative black hole.

Keywords. galaxies: individual (IC 342), ISM: supernova remnants, X-rays: galaxies, X-rays: binaries

1. Introduction

The two main hypotheses to explain the high luminosity of ULXs are intermediate mass black holes (IMBHs) having 10^2 to 10^5 solar masses (Colbert & Mushotzky 1999) or non-isotropic emission beamed into our line-of-sight (King *et al.* 2001).

Here, we are interested in one of these objects, Holmberg IX X-1, located at a distance of 3.6 Mpc in a dwarf galaxy companion of M81. Miller (1995) discovered the nebula around the position of the X-ray source and proposed that this object was an extremely luminous supernova remnant, but the presence of X-ray variability (La Parola *et al.* 2001) has shown that it is a compact X-ray source.

Our optical observations were carried out in 2003 and 2004 with the 8.2 meter SUBARU telescope on Mauna Kea, Hawaii.

2. Results and discussion

In the huge supernova remnant-like complex, our images reveal that a "blue non-stellar object" seen in previous studies is in fact a star cluster where the brightest members are resolved (Figure 1). The most luminous object is a 22.8 mag object, located in the 1″ radius Chandra error circle; we note that in archive HST images this object has a fainter (by 1.9 mag) companion to the west.

One very interesting result is the discovery of the HeIIλ4686 emission in the brightest star (i.e. not in fainter companion) with an equivalent width (EW) of some 10 Å (Figure 1), proving that it is the optical counterpart of the X-ray source. This is a common feature in luminous massive X-ray binaries, but here with an EW ten to twenty times

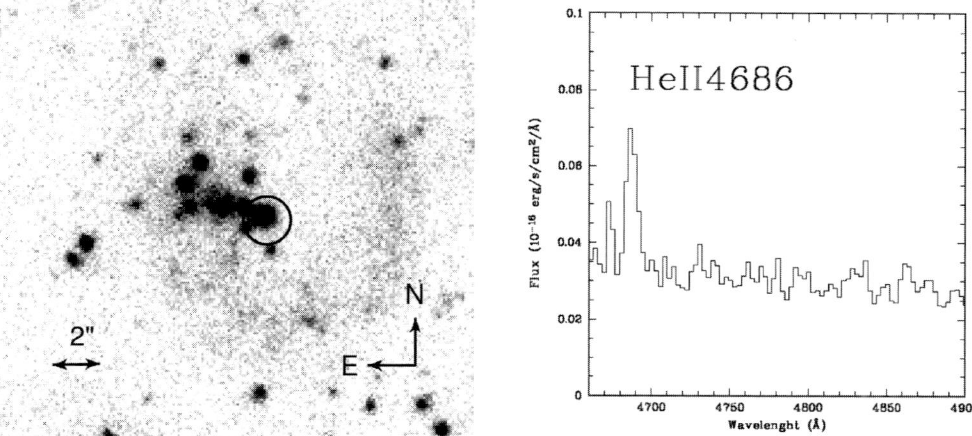

Figure 1.
Left : V-band image of the stellar cluster in the local environment of the ULX. The optical counterpart is the bright star in the Chandra error circle.
Right : 1D spectrum of the stellar counterpart of Holmberg IX X-1 displaying the HeIIλ4686 emission line.

higher. This strongly suggests the presence of an accretion disk that is heated by the very luminous X-ray source. This constitutes further evidence against beaming in ULX and opens the possibility to measure the binary orbit from radial velocity observations.

Isochrone fitting to our multi-colour photometry has permitted to constrain the age of the star cluster to which the ULX belongs to about 60 Myr, and to put an upper limit on the total cluster mass - some 10^3 M$_\odot$. This directly implies that the donor component in the ULX cannot be more massive than 8 M$_\odot$, because more massive cluster stars have already exploded as supernovae.

3. Perspectives

Optical observations of radial velocity variations of the HeIIλ4686 line in ULXs will hopefully allow to determine the masses of the components in these systems. This would be a decisive test on whether IMBH are present in ULXs or not. Optical observations of other ULX (in particular NGC 1313 X-2) show strikingly similar characteristics (see contribution at this symposium by Pakull *et al.*).

Future optical observations will be crucial to reveal nature and evolution of the exciting class of ultraluminous X-ray emitters.

References

Colbert, E. J. M. & Mushotzky, R. F. 1999, *ApJ* 519, 89
King, A.R. *et al.* 2001, *ApJ*, 552, L109
La Parola, V. *et al.* 2001, *ApJ*, 556, 47
Miller, B. W. *et al.* 1995, *ApJ*, 446, L75

Populations of High Energy Sources in Galaxies
Proceedings IAU Symposium No. 230, 2005
E. J. A. Meurs & G. Fabbiano, eds.

© 2006 International Astronomical Union
doi:10.1017/S1743921306008520

Ultra-Luminous X-ray Sources: Evidence for Very Efficient Formation of Population III Stars Contributing to the Cosmic Near-Infrared Background Excess?

Haruka Mii and Tomonori Totani

Department of Astronomy, Kyoto University
Sakyo-ku, Kyoto 606-8502, Japan

Abstract. Accumulating evidence indicates that some of ultra-luminous X-ray sources (ULXs) are intermediate mass black holes (IMBHs), but the formation process of IMBHs is unknown. One possibility is that they were formed as remnants of population III (Pop III) stars, but it has been thought that the probability of being an ULX is too low for IMBHs distributed in galactic haloes to acccunt for the observed number of ULXs. Here we show that the number of ULXs can be explained by such halo IMBHs passing through a dense molecular cloud, if Pop III star formation is very efficient as recently suggested by the excess of the cosmic near-infrared background radiation that cannot be accounted for by normal galaxy populations. We calculate the luminosity function of X-ray sources in our scenario and find that it is consistent with observed data. Our scenario can explain that ULXs are preferentially found at outskirts of large gas concentrations in star forming regions. A few important physical effects are pointed out and discussed, including gas dynamical friction, radiative efficiency of accretion flow, and radiative feedback to ambient medium. ULXs could last for $\sim 10^{5-6}$ yr to emit a total energy of $\sim 10^{53}$ erg, which is sufficient to power the ionized expanding nebulae found by optical observations.

Keywords. accretion, accretion disks, early universe, X-rays: stars.

1. Introduction

Ultra-Luminous X-ray sources (ULXs, Makishima *et al.* 2000 and references therein) are bright X-ray sources having luminosities greater than $\sim 3 \times 10^{39}$ erg/s found in off-nuclear regions of nearby galaxies. The luminosity exceeds the Eddington limit of a $\sim 20 M_\odot$ black hole (BH) that is the maximum mass expected from normal stellar evolutionary paths (Fryer & Kalogera 2001), and their origin is now a matter of hot debate. The high luminosity may also be explained by intermediate mass black holes (IMBHs) of $\gtrsim 100 M_\odot$, without violating the Eddington limit. Recently evidence for IMBHs has been accumulating for many ULXs, and it appears that at least a part of ULXs, especially the most luminous ones, are IMBHs.

However, the formation of IMBHs to become ULXs is a challenging problem. It is theoretically expected that IMBHs form as remnants of massive population III (Pop III) stars (e.g., Schneider *et al.* 2002), and they may become ULXs if they accrete sufficient gas in nearby galaxies. However, it is uncertain whether the number of Pop III IMBHs is sufficient to explain the observed number of ULXs by this scenario.

The efficiency of Pop III star formation is highly uncertain, but there are some observational hints. Independent groups (Cambrésy *et al.* 2001; Matsumoto *et al.* 2004) reported detections of the cosmic near-infrared background radiation (CNIB), which cannot be explained by normal galaxy populations (Totani *et al.* 2001). It has been shown that efficient formation of Pop III stars before $z \sim 8$ can explain this unaccounted excess of

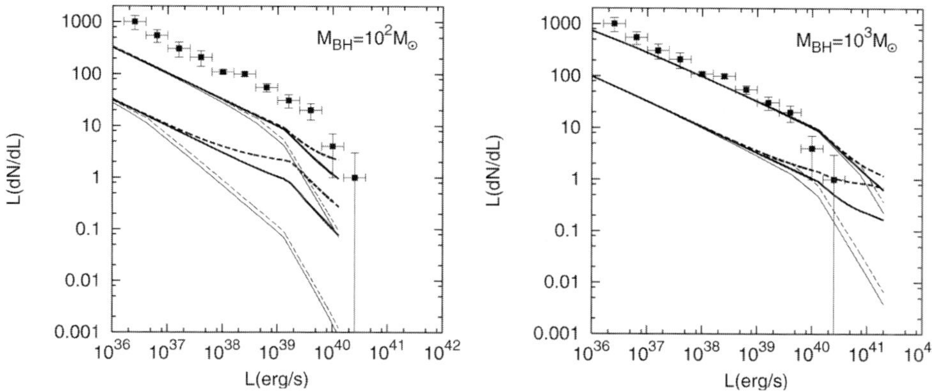

Figure 1. *Left*: The X-ray luminosity function of $M_{\mathrm{BH}} = 100\ M_\odot$ IMBHs normalized to $\mathrm{SFR}_{>5} = 50 M_\odot\ \mathrm{yr}^{-1}$. For each line marking, there are two model curves: the upper one for $\eta = 1$ and the lower one for $\eta = 0.1$. The sound speed is set as $c_s = 0.3$ and 10 km/s for each thick and thin curve, respectively. The dynamical friction parameter is set to be $\xi = 1$ and 10 for each solid and dashed curve, respectively. The data points show the observed universal X-ray luminosity function from Grimm *et al.* (2003). *Right*: Same as the left panel, but for $M_{\mathrm{BH}} = 10^3 M_\odot$.

CNIB (Santos, Bromm, & Kamionkowski 2002; Salvaterra & Ferrara 2003); the excess peak at $\sim 1.5 \mu$m corresponds to the redshifted Lyman α line emission.

The required amount of Pop III stars to explain the CNIB is somewhat extreme, but not entirely impossible; about $\sim 10\%$ of the cosmic baryons must be converted into Pop III stars. Most of Pop III stars could be massive enough to collapse into IMBHs without any metal ejection, meeting the metallicity constraints from quasar absorption line systems. The consequence of this scenario is that there should be IMBHs in halos of nearby galaxies, whose mass fraction in the baryonic matter is $f_{\mathrm{III}} \equiv \Omega_{\mathrm{IMBH}}/\Omega_B \sim 0.1$, where $\Omega_B \sim 0.05$ is the cosmic baryon density in the standard definition.

In this paper we show that, if this is the case, such IMBHs passing through dense molecular clouds of star forming regions should have a siginificant contribution to the observed ULX population, especially to the highest luminosity ones.

2. Predicted ULX Luminosity Function

In Fig. 1, we show our prediction of the ULX luminosity function. Both the absolute number and the shape of luminosity function are in agreement with the data. For the details, see the publised paper of Mii & Totani (2005).

References

Fryer, C. L. & Kalogera, V. 2001, ApJ, 554, 548
Makishima, K. *et al.* 2000, ApJ, 535, 632
Matsumoto, T. *et al.* 2004, ApJ, 626, 31
Mii, H. & Totani, T. 2005, ApJ, 628, 873
Salvaterra, R. & Ferrara, A. 2003, MNRAS, 339, 973
Santos, M. R., Bromm, V. & Kamionkowski, M. 2002, MNRAS, 336, 1082
Schneider, R., Ferrara, A., Natarajan, P. & Omukai, K. 2002, ApJ, 571, 30
Totani, T., Yoshii, Y., Iwamuro, F., Maihara, T. & Motohara, K. 2001, ApJ, 550, L137
Wright, E. L. & Reese, E. D. 2000, ApJ, 545, 43

Populations of High Energy Sources in Galaxies
Proceedings IAU Symposium No. 230, 2005
E. J. A. Meurs & G. Fabbiano, eds.

© 2006 International Astronomical Union
doi:10.1017/S1743921306008532

Keck Observations of Candidate Ultra-Luminous X-ray Sources

**Diane Sonya Wong, Ryan Chornock
and Alexei V. Filippenko**

Astronomy Department, 601 Campbell Hall, University of California, Berkeley, CA 94720 USA
email: dianew@astro.berkeley.edu

Abstract. We present results of optical follow-up observations of candidate ultra-luminous X-ray sources (ULXs). Using Keck optical spectroscopy, 17 of the candidates from the Colbert & Ptak (2002) catalog have been identified; this is one of the largest sets of optical identifications of such objects thus far. 15 are background active galactic nuclei (AGN); 2 are foreground stars in our Galaxy. These findings are compared with background and foreground object expectations, as derived from log N-log S relations. Also, the results are briefly discussed in terms of the spiral-galaxy/ULX connection.

Keywords. black hole physics, catalogs, galaxies: distances and redshifts, galaxies: individual (NGC 720, NGC 1316,NGC 1365, NGC 1399, NGC 2775, NGC 3266, IC 2597, NGC 3923, NGC 4151, NGC 4373, NGC 5128), quasars: general, X-rays: galaxies.

1. Observations

Ultra-luminous X-ray sources (ULXs; also referred to as intermediate-luminosity X-ray objects [IXOs]) are non-nuclear point X-ray sources with apparent (isotropic) X-ray luminosities $L_{\mathrm{X}} \approx 10^{39} - 10^{41}\,\mathrm{erg\,s^{-1}}$. Colbert & Ptak (2002, hereafter CP02) assembled the first complete catalog of candidate ULXs consisting of 87 objects listed in order of increasing right ascension.

Seventeen candidates (in 11 different galaxies) from the CP02 catalog were observed with the Keck 1 10-meter telescope on Mauna Kea, Hawaii, USA on 2003 February 28 and 2003 December 20-21 UT. Spectra were obtained using the Low Resolution Imaging Spectrometer (LRIS). The chosen setup yielded a wavelength coverage of $\sim$3150-9400 Å. A 1″ wide slit was used all three nights. Typical seeing on these nights was $\sim$ 1″. The resulting resolution was $\sim$6 Å.

Standard reduction was done with IRAF version 2.12 (flatfielding with dome flats, optimal extraction of the spectra, wavelength calibration) and IDL versions 5.5 & 6.1 (flux calibration, atmospheric band removal).

2. Discussion

We have found 15 of the candidate ULXs to be background objects with redshifts ranging from 0.239 to 2.784. The redshifts were found by fitting Gaussians to the line profiles, and comparison to a composite quasar spectrum. Additionally, 2 candidate ULXs are foreground objects. Of these, the counterpart of IXO 75 is actually a double M star. The results are shown in Figure 1.

The number of candidate ULXs reported by CP02 is the number found in a circle of *radius* D25† for each galaxy. Using the Hasinger *et al.* (1998) log N-log S relation, the

---

† D25 is the galaxy's size as defined by the apparent isophotal diameter measured at, or reduced to, the surface brightness level of 25.0 B mag arcsec^{-2}.

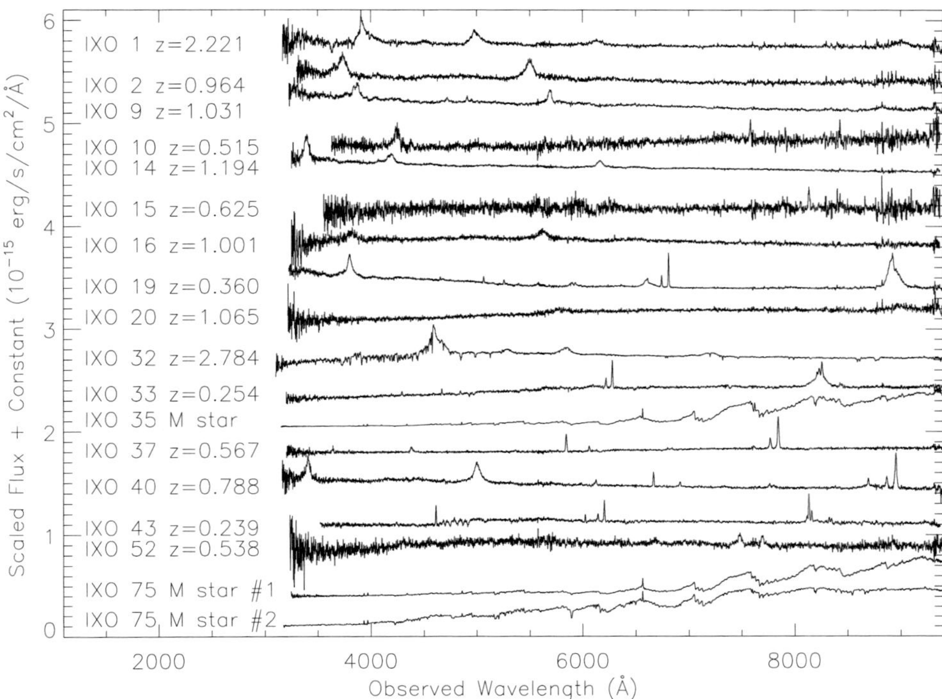

Figure 1. Candidate ULX Spectra. CP02 notation (e.g. "IXO 1") is used here. Most are AGN.

number of expected background and foreground sources ("fake" ULXs) was calculated for the same region. These two numbers were compared: the number of candidate ULXs reported by CP02 (24) is comparable to the expected number of "fake" ULXs (21) in these galaxies. Furthermore, the majority of the candidate ULXs we identified are outside of the D25 ellipse, so it is not surprising that they are not associated with the galaxies in whose fields they lie.

Of the remaining 70 candidate ULXs in the CP02 catalog, the majority (47) is in spiral galaxies. However, the mean of 1.3 candidates *per* spiral is comparable to (and slightly lower than) the mean of 2.3 per elliptical.

See the work by Gutiérrez & López-Corredoira (2005) for another set of candidate ULX optical identifications.

Acknowledgements

The authors acknowledge NSF grant AST-0307894.

References

Colbert, E. J. M. & Ptak, A. F. 2002, *ApJS*, 143, 25

Gutiérrez, C. M. & López-Corredoira, M. 2005, *ApJL*, 622, L89, and these proceedings

Hasinger, G., Burg, R., Giacconi, R., Schmidt, M., Trumper, J. & Zamorani, G. 1998, *A&A*, 329, 482

Populations of High Energy Sources in Galaxies
Proceedings IAU Symposium No. 230, 2005
E. J. A. Meurs & G. Fabbiano, eds.

© 2006 International Astronomical Union
doi:10.1017/S1743921306008544

Search for Serendipitous ULX Candidates in *XMM-Newton* Observations

Yueheng Xu[1], Mike Watson[1],
Timothy Roberts[1] and Weimin Yuan[2]

[1]Department of Physics & Astronomy, University of Leicester,
University Road, Leicester, LE1 7RH, UK
email: yx12@star.le.ac.uk, mgw@star.le.ac.uk, tro@star.le.ac.uk

[2]Yunnan Astronomical Observatory,
National Astronomical Observatories of China, Chinese Academy of Sciences,
Phoenix Hill, PO Box 110, Kunming, Yunnan, 650011, China P.R.
email: wmy@ynao.ac.cn

Abstract. We report the serendipitous discovery of several ULX candidates in *XMM-Newton* observations. Such discoveries suggest that ULXs are not a negligible component of the extragalactic X-ray source population.

Keywords. X-ray: galaxies

1. Introduction

Ultraluminous X-ray sources (ULXs) are extra-nuclear point-like X-ray sources in galaxies with luminosities only surpassed by active galactic nuclei (AGN). ULXs are observationally defined to have $L_x > 10^{39}$ erg s^{-1}, equal to the Eddington limit for accretion onto a $\sim 7 M_\odot$ compact object. The high apparent luminosities of many ULXs raise the question whether some ULXs may be powered by accretion onto "intermediate-mass" black holes (see Miller & Colbert *et al.* 2004 for a review). Alternatively ULXs may involve accretion onto stellar-mass black holes with the high luminosities originating either in truly super-Eddington radiation from inhomogeneous accretion discs, or through anisotropic radiation patterns and/or relativistic beaming. Most of the ULX candidates reported to date have been discovered in pointed observations of the specific galaxies. The sensitivity of typical *XMM-Newton* observations is such that ULX candidates can be detected to a distance of ~ 100 Mpc. This allows a search for serendipitous ULX candidates in *XMM-Newton* observations and thus provides the means to accumulate a valuable new and unbiased sample of ULX candidates in external galaxies. Here we present the preliminary results from just two relatively nearby galaxies.

2. The ULX candidate in UGC 03422

UGC 03422 is a nearby SAB(rs) galaxy ($z = 0.01355$, Falco *et al.* (1999), corresponding to a distance of 57 Mpc) lying at a distance of $\simeq 6.4'$ to the northwest of the Seyfert 2 galaxy Markarian 3. This field (including UGC 03422) has been observed 12 times by *XMM-Newton* with exposure times ranging in 3.5 ks to 58.5 ks. One of the serendipitous X-ray sources in the Markarian 3 field lies in the outer spiral arm of UGC 3422, $\simeq 43.4''$ southeast of the optical centre of the galaxy. Deep optical imaging of the Markarian 3 field obtained with Isaac Newton Telescope with the prime focus Wide Field Camera shows that the ULX candidate lies close to several prominent HII regions in the outer spiral arm of the galaxy, a comparable location to other confirmed ULXs in other galaxies.

The X-ray spectrum of the ULX candidate extracted from the longest *XMM-Newton* observation (ObsID: 0111220201) is relatively hard, broadly consistent with other ULXs. Assuming the source lies in the galaxy, its derived X-ray luminosity in the 0.2–10 keV band is $\approx 10^{40}$ erg s^{-1}. Its location in the galaxy, X-ray luminosity and X-ray spectrum make it a good ULX candidate. Further work is underway to investigate the long-term variability of this object and look for possible X-ray spectral changes.

3. ULX candidates in KUG 0214-057

Another example of the serendipitous detection of ULX candidates in an external galaxy is the discovery of at least three, and possibly four, ULX candidates in the barred spiral galaxy KUG 0214-057 as already reported by Watson *et al.* (2005).

KUG 0214-057 is a 15 mag. barred spiral galaxy at $z = 0.018$ (corresponding to a distance of 75 Mpc) covered by a relatively deep (50 ksec) *XMM-Newton* observation made as part of the Subaru *XMM-Newton* Deep Survey. The four ULX candidates lie in the prominent spiral arms of the galaxy. At least three, and possibly four, of them may be physically associated to the galaxy with X-ray luminosities from $\sim 5 \times 10^{39}$ erg s^{-1} to $\sim 10^{40}$ erg s^{-1} and X-ray colours consistent with what is expected for ULX source spectra. An interesting aspect of the discovery of these ULXs in KUG 0214-057 is that, using the ULX population as a metric, the implied star formation rate is very high for this relatively low-mass galaxy.

Acknowledgements

YX acknowledges financial support from a Dorothy Hodgkin Studentship.

References

Falco, E. E. *et al.* 1999, *Publ. Astron. Soc. Pac.* 111, 438 (The updated Zwicky catalog, UZC)
Miller, M. C. & Colbert, E. J. M. 2004, *Int. J. Mod. Phys. D*, 13, 1
Watson, M. G., Roberts, T. P., Akiyama, M., & Ueda, Y. 2005, *A&A*, 437, 899

Populations of High Energy Sources in Galaxies
Proceedings IAU Symposium No. 230, 2005
E. J. A. Meurs & G. Fabbiano, eds.

© 2006 International Astronomical Union
doi:10.1017/S1743921306008556

Identification of Optical Counterparts of ULX sources

Carlos M. Gutiérrez and Martín López-Corredoira

Instituto de Astrofísica de Canarias, 38200 la Laguna, Tenerife, SPAIN
email: cgc@ll.iac.es, martinlc@ll.iac.es

Abstract. We present the results of an on-going program for the identification and characterization of optical counterparts of Ultra-Luminous X-ray (ULX) sources. The targets have been selected from the catalogues by Colbert & Ptak (2002) and Swartz *et al.* (2004). A clear identification based on unambiguous optical spectral features was possible for 26 objects. A large number of objects result to be QSOs at higher redshift than the putative parent galaxy, and other ULXs seem to be associated to HII regions. In a few cases the optical counterpart results a foreground star in our galaxy. The observational program will continue to obtain a representative sample for statistical studies.

Keywords. X-rays: Ultra-luminous sources.

One of the most intriguing astrophysical objects are the ultra-luminous X-ray sources (ULXs) which have been discovered around nearby galaxies by the X-ray satellites Einstein, ROSAT, Chandra and XMM. Assuming that they are at the distance of their parent galaxies, their luminosities in the 0.1–2.4 Kev X-ray band are in the range $10^{39} - 10^{40}\,\mathrm{erg\,s^{-1}}$. Several explanations about the nature of these objects in terms of intermediate mass black holes associated with globular clusters, HII regions, supernova remnants, etc (Pakull & Mirioni 2002; Angelini *et al.* 2001; Gao *et al.* 2003; Roberts *et al.* 2003; Wang 2002), local QSOs (Burbidge *et al.* 2003), hypothetical supermassive stars or beamed emission (King *et al.* 2001; Kording *et al.* 2002) have been proposed. To disentangle the nature of these objects it is important but still scarce the identification of optical counterparts (e. g. Roberts *et al.* 2001; Wu *et al.* 2002; Maseti *et al.* 2003; Liu *et al.* 2004). This motivated us to start this study searching for such optical counterparts in the Digital Sky Survey (DSS) images, in the digitalized USNO catalogues, and in the released Sloan Digital Sky Survey (SDSS) images. For instance, we have identified possible optical counterparts, in an error circle compatible with the X-ray position, for ~50 % of ULXs listed in Colbert & Ptak (2002). The typical magnitudes of such objects are 18–20 in the b band and are therefore bright enough targets for spectroscopic observations with 2 to 4 m telescopes.

Previously we have presented the identification of 9 of such optical counterparts (Arp *et al.* 2004; Gutiérrez & López-Corredoira 2005) showing that 8 of these sources correspond to QSOs at higher redshift than their putative parent galaxy. Another object was associated with an HII region at the redshift of the parent galaxy (NGC 1073). Here, we present 17 new identifications of such optical counterparts based on long-slit spectroscopic observations obtained in several runs at the William Herschel Telescope (WHT) in the Observatory Roque de los Muchachos, La Palma, and at the 1.93 m telescope in the Haute Provence Observatory (OHP). The typical slit width was 2 arcsec, and the resolution ~$3 - 6$ Å. Exposure times were between 900 and 1800 sec for WHT observations, and between 1800 and 5400 sec for observations at OHP. The spectra were

ID	Obs.	Galaxy	z_{gal}	D_{gal} (Mpc)	$\log(L_X)$ (erg s^{-1})	Sep (')	z_{ULX}	Type
IXO 22	OHP Mar 2005	IC 342	0.0001	3.90	39.2	5.1	0.00	HII region
IXO 31	WHT Feb 2004	Holmberg II	0.0005	4.50	40.2	2.1	0.0009	HII region (or M star)
IXO 32	WHT Feb 2004	NGC 2775	0.0045	18.05	39.2	3.7	2.770	QSO
IXO 34	WHT Feb 2004	NGC 3031	-0.0001	3.55	40.0	12.5	0.001	HII region
IXO 35	WHT Feb 2004	NGC 3226	0.0038	17.63	39.2	2.4		M star
IXO 37	WHT Feb 2004	IC 2597	0.0076	40.09	40.9	1.9	0.567	ELG
IXO 40	WHT Feb 2004	NGC 3923	0.0058	22.24	39.9	4.4	0.787	QSO
IXO 43	OHP Mar 2005	NGC 4151	0.0033	20.30	39.3	5.0	0.24	ELG
IXO 54	OHP Mar 2005	NGC 4438	0.0002	16.80	39.9	3.5	0.66	QSO
IXO 57	OHP Mar 2005	NGC 4472	0.0033	16.80	39.8	8.1	0.85	QSO
IXO 65	OHP Mar 2005	NGC 4559	0.0027	9.70	39.9	2.0	0.0027	HII region
IXO 70	OHP Mar 2005	NGC 4649	0.0037	14.85	39.3	7.0	0.62	QSO
SW 38	OHP Mar 2005	NGC 3079	0.0037	15.51	39.5	1.9	0.004	HII region
SW 44	OHP Mar 2005	NGC 3556	0.0023	9.67	39.6	0.7		star
SW 45	OHP Mar 2005	NGC 3556	0.0023	9.67	39.2	0.1	0.0023	HII region
SW 90	OHP Mar 2005	NGC 4490	0.0019	7.80	39.4	1.2	0.0015	HII region
SW 134	OHP Mar 2005	NGC 5457	0.0008	7.24	38.9	6.2		M star

analyzed following a standard procedure using IRAF†. The Table presents a summary of our results (another four objects were observed but the optical spetra did not allow a clear characterization). The columns are: (1): name of the ULX (IXO or SW for objects from the Colbert & Ptak (2002) or Swartz *et al.* (2004) catalogues respectively). (2): telescope and epoch of observation. (3)–(5): name, redshift and distance of the parent galaxy. (6): X-ray luminosity assuming that the ULXs are at the distance of the parent galaxy. (7): projected distance between the ULXs and their parent galaxy. (8)–(9): redshift and nature of the optical counterpart of the ULXs.

References

Angelini, L., Loewenstein, M. & Mushotzky, R. F. 2001, *ApJ* 557, L35

Arp, H., Gutiérrez, C. M. & López-Corredoira, M. 2004, *A&A* 418, 877

Burbidge, G., Burbidge, E. M. & Arp, H. 2003, *A&A* 400, L17

Colbert, E. & Ptak, A. 2002, *ApJS* 143, 25

Gao, Y., Wang, Q. D., Appleton, P. N. & Lucas, R. A. 2003, *ApJ* 596, L171

Gutiérrez, C. M. & López-Corredoira, M. 2005, *ApJ* 622, L89

King, A. R., Davies, M. B., Ward, M. J., Fabbiano, G. & Elvis, M. 2001, *ApJ* 552, L109

Kording, E., Falcke, H. & Markoff, S. 2002, *A&A* 382, L13

Liu, J. F., Bregman, J. N. & Seitzer, P. 2004, *ApJ* 602, 249

Maseti, N. *et al.* 2003, *A&A* 406, L27

Pakull, M. W. & Mirioni, L. 2002, in *New Visions of the X-ray Universe in the XMM-Newton and Chandra Era*, ESTEC

Roberts, T. P. *et al.* 2001, *MNRAS* 325, L7

Roberts, T. P., Goad, M. R. & Warwick, R. S. 2003, *MNRAS* 342, 709

Swartz, D. A., Ghosh, K. K., Tennant, A. F. & Wu, K. 2004, *ApJS* 154, 519

Wang, Q. D. 2002, *MNRAS* 332, 764

Wu, H., Xue, S. J., Xia, X. Y., Deng, Z. G. & Mao, S. 2002, *ApJ* 576, 738

† IRAF is the Image Reduction and Analysis Facility, written and supported by the IRAF programming group at the National Optical Astronomy Observatory (NOAO) in Tucson, Arizona.

Francesco Polcaro tries to follow several Irish dialects. From left to right: Cóilín Ó Maoiléidigh, Brendan Jordan (hands + dessert only), Francesco Polcaro, Catherine Handley, Eamonn Cunningham, Carol Woods (who had avoided being photographed when the Conference Photo was taken), Hilary O'Donnell.

Session 5

Overall population characteristics

Hans-Jakob Grimm (right) and Laura Norci, overlooked by specimen of modern Irish art.

Populations of High Energy Sources in Galaxies
Proceedings IAU Symposium No. 230, 2005
E. J. A. Meurs & G. Fabbiano, eds.

© 2006 International Astronomical Union
doi:10.1017/S1743921306008568

The INTEGRAL Gamma-Ray Sky after 1000 days in orbit

P. Ubertini[1], A. Bazzano[1], A. J. Bird[2], L. Bassani[3] F. Lebrun[4], R. Walter[5] on behalf of the survey team

[1]INAF/Istituto di Astrofisica Spaziale e Fisica Cosmica, Area di Ricerca Tor Vergata, Via Fosso del Cavalire 100, 00133 Rome, Italy
email: pietro.ubertini@rm.isf.cnr.it
email: angela.bazzano@rm.iasf.cnr.it

[2]School of Physics and Astronomy, University of Southampton, Highfield, Southampton, SO17 1BJ, UK
email: ajb@astro.soton.ac.uk

[3]INAF/Istituto di Astrofisica Spaziale e Fisica Cosmica, Via P. Gobetti 101, 40129 Bologna, Italy
email: loredana.bassani@bo.iasf.cnr.it

[4]CEA Saclay Service d'Astrophysique 91191 Gif sur Yvette cedex France
email: lebrun@hep.saclay.cea.fr.

[5]ISDC ch. D'Ecogia 16 01290 O1290 Versoix CH
email: Roland.Walter@obs.unige.ch

Abstract. This paper will review the main astrophysical results obtained in the field of high energy Galactic sources with the INTEGRAL/IBIS Gamma-ray Imager (Ubertini *et al.* 2003) on-board INTEGRAL (Winkler *et al.* 2003), the ESA space Observatory successfully launched the 17th October 2002 from Baikonur with a Proton vehicle. In view of the high sensitivity of the two gamma ray instruments IBIS and SPI and their capability to provide at the same time image, spectra and time profiles of all the sources in their wide field of view, a key project was approved as "Core Programme" to deeply observe the Galactic Centre (GCDE) and to exploit regular scans of the whole Galaxy Plane. The major results obtained in terms of classes of high energy emitters are shortly outlined.

Keywords. gamma rays: observations, telescopes, surveys

---

1. Introduction

After more than 1000 days in orbit one of the main INTEGRAL/IBIS achievement is the detection of more than 200 high energy sources for E > 20 keV. Out of this unprecedented sample, $\sim$30 are detected above 150 keV, showing for the first time a populated "non thermal" sky. A substantial fraction of the whole sample of sources has been so far identified with already known objects, typically at X and IR wavelengths. Nevertheless, a few tens are still of unknown origin. Among them, we have recently discovered two different classes of objects:

- The slow highly absorbed pulsars (period between $\sim$100 to $\sim$1000 s) in high mass systems, populated by a young system composed by an evolved magnetised neutron star orbiting in the faint stellar environment of a "giant star"

- The INTEGRAL/IBIS soft gamma ray counterpart of the newly discovered TeV emitting objects. In fact, the first IBIS survey indicates a clear association for a number of TeV detected objects: AX J1838.0-0655, SGR A*, MSH 15-52/PSR1509-58, and Crab

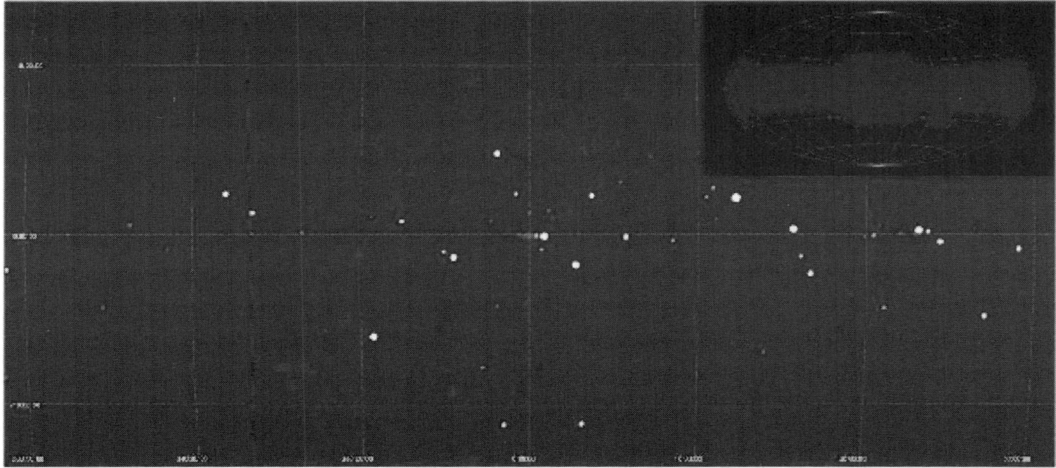

Figure 1. The 30-50 keV significance map for the 1st IBIS/ISGRI catalog

Nebula, while the second survey has discovered that IGR J18135-1751 is the soft gamma-ray counterpart of the TeV source HESS1813-178.

2. Surveys

From the outset, surveys were considered a key element of the INTEGRAL data results. The intention was to provide a temporally and spectrally resolved archival view of the gamma-ray sky, and this concept was included as part of the Core Programme for all instruments. Survey teams were established for each instrument, with the intention of providing *unbiased* surveys of the sky, while other teams were established to search for more specific source types.

2.1. *IBIS all-sky survey*

The *IBIS survey team* have so far conducted two surveys of the gamma-ray sky based on all available data. The 1st IBIS/ISGRI soft gamma-ray galactic plane survey catalogue (Bird *et al.* 2004), hereafter *the first catalogue*, used all data from revolutions 46-120, a total of 2529 pointings. As the catalog name suggests, the 5Ms exposure of this survey was biased heavily towards the Galactic Plane and specifically the central ∼40 degrees of the galactic sky covered by the Galactic Centre Deep Exposure. Construction of the first catalogue was optimised for the detection of weak, persistent sources, and followed this process:

• Image generation and cleaning for each pointing using the *OSA 3* software. This involved an iterative procedure to develop a catalogue of sources to be included for image cleaning.

• All-sky mosaic generation in two energy bands optimised for source detection: 30-50 keV and 20-40 keV. Figure 1 shows the 30-50 keV map.

• Determination of selection thresholds based on whole-image statistics to minimise false source detections. Conservative thresholds were chosen so as to limit the false detections to at most one source.

• Automatic source detection using the SExtractor software.

• Construction of final source list by visual inspection of each excess to remove systematic structures.

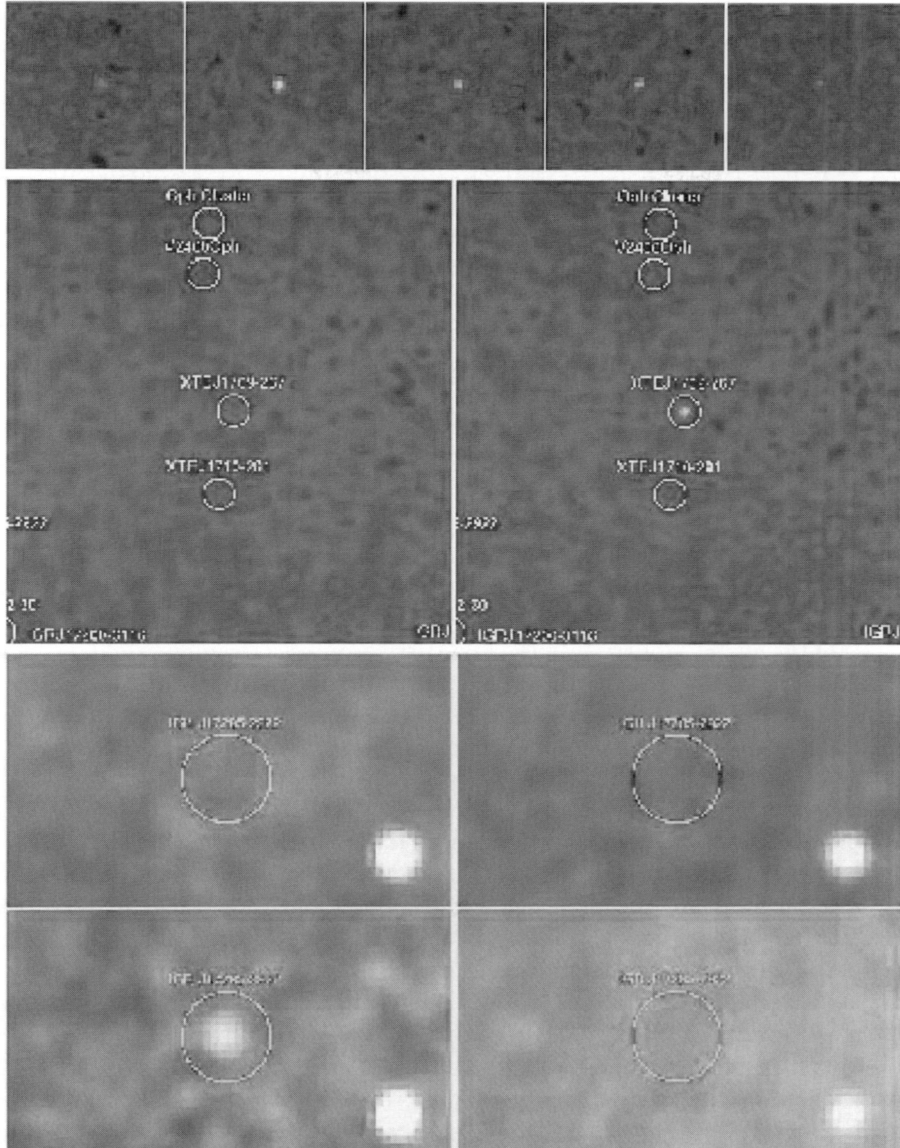

Figure 2. Examples of transient source detection on different timescales: (top) Short flare from XTE J1858+034 in March 2004 visible in a few science windows with duration ∼10000s; (middle) XTE J1709-267 visible during revolution 171; (lower) IGR J17285-2922 visible only during the Sept/Oct GCDE observation period (Sguera *et al.* 2005).

The first catalogue contained 123 sources dominated by galactic binary systems, at least partly because of the exposure bias. The detailed breakdown of this first catalog was 53 LMXB, 23 HMXB, 7 supernova products, 5 CVs, 5 AGN, 1 SGR, 1 galaxy cluster and 28 sources of unknown nature. Of these 28 sources, exactly half had been previously observed by previous missions, while for 14 sources, the INTEGRAL detection represented the first discovery. Each source in the first catalogue was localised with an accuracy of between $1'$ and $3'$ depending on source brightness, and the quoted sensitivity limit was ∼1mCrab in the regions of highest exposure. The fact that nearly a quarter of the first catalog consisted of sources of unknown origin has led to a large campaign of follow-up observations and detailed studies which have revealed the nature of several of these sources. Of the '28 unknown' sources, only 9 remain with no clue to their nature. The other 19 have some, at least tentative identification, including:

- IGR J16318-4848, in the Norma arm of the galaxy, was the first INTEGRAL-detected source to be identified by optical/NIR spectroscopy. It was found to be a supergiant B[e] star, and the spectra showed a very high level of line-of-sight obscuration ($N_H > 2 \times 10^{24}$). High levels of obscuration have been observed in many of the new IGR sources, interpreted as binary systems within a Compton-thick envelope (Walter *et al.* 2003).

- IGR J18027-2016 was identified, by a combination of XMM, SAX and IBIS data, to be an eclipsing HMXB (Hill *et al.* 2005) comprising a neutron star emitting X-rays through wind-fed accretion from an OB-supergiant. Again, a high hydrogen column density is observed in this source.

- IGR J16479-4514 has been identified as a fast transient system (Sguera *et al.* 2005), possibly in a LMXB system.

- IGR J18027-1455 and IGR J20247+5058 have been optically identified as Sy1 galaxies at redshifts of $z = 0.035$ and $z = 0.02$ respectively (Masetti *et al.* 2004). IGR J20247+5058 is also a radio galaxy, with an interesting core-jet morphology (Mantovani *et al.* 1982).

Analysis of the global properties of the first survey sources (Dean *et al.* 2005) concluded that INTEGRAL is detecting a previously un-noticed population of highly absorbed sources, probably binary systems with massive stars and a neutron star companion. Many of the IGR sources may fall into this class of object. A systematic correlation was carried out of the sources in the first catalog with those in the ROSAT bright source catalogue, a previous all-sky soft X-ray survey. This revealed a total of 75 IGR-ROSAT associations within a $3'$ error circle (Stephen *et al.* 2005a), while it was also shown that, on a statistical basis, less than 1 of these should be a random association. This study has paved the way for follow-ups of many IGR sources at optical and IR wavelengths, since the ROSAT association provides us with a much better source location for 10 of the IGR sources. The 2^{2nd} IBIS/ISGRI catalog (Bird *et al.* 2005) significantly increases both the breadth and depth of exposure, and is constructed from all data from revolutions 46-96, and all Core Programme data from revolutions 96-210, a total of more than 6500 individual pointings, more than doubling the exposure of the 1st catalog. In addition, several technical improvements were made over the first survey, both in terms of the core *OSA 4* software used and the survey-specific tools used for source detection and location. In particular, emphasis has been placed on the filtering of science window data to remove those pointings with poor image quality, usually due to inadequate background subtraction following bad 'space weather' such as solar flares. With the rapidly increasing overall exposure, one other significant change implemented for the second catalog has been the search for sources on different timescales. We cannot expect to detect a short-lived transient in a mosaic of many Ms of data, since detection is based on the average flux throughout the exposure on any given source. In order to counter this problem, detection is now carried out on: science window (typically 2000s) timescales; revolution (typically 200ks) timescales; revolution groups, typically encompassing one long exposure campaign such as a Galactic Centre Deep Exposure period; whole-exposure mosaics, as used for the first catalogue, covering a varying exposure depending on the location in the sky. Figure 2 illustrates sources becoming visible on several of these timescales (Sguera *et al.* 2005). The multi-timescale approach will become more important as the accumulated exposures become more incompatible with the detection of transient activity. The resulting second catalogue contains 209 sources, with a significantly different distribution of source types due to a reduction in the exposure bias towards the Galactic Centre. Significant parts of the extragalactic sky have now been viewed by INTEGRAL, and so the contribution of AGN towards the overall source population ($\sim$20%) is increasing dramatically. As in the first catalogue, a large fraction ($\sim$25%) of these sources are of unknown nature. A vigorous follow-up campaign is already underway to understand the detailed characteristics, and hence nature, of these sources. Again, correlation with the ROSAT Bright Source catalog has been fruitful (Stephen *et al.* 2005b), providing improved locations for a further nine sources. The IBIS survey team continues to enhance their techniques for detection of both persistent and transient sources.

2.2. *Other surveys*

Other surveys (of smaller fields) have been constructed. Revnivtsev *et al.* 2004 detected 60 sources in a 2 Ms private observation of the Galactic centre in Aug-Sept 2003, down to a sensitivity limit of $\sim$2mCrab. The spiral arm tangents have also been the subject of detailed study: Molkov *et al.* 2004 report 28 sources in a 0.8 Ms exposure of the Sagittarius Arm, while a more recent 2Ms exposure of the Crux arm tangent (Revnivtsev *et al.* 2005) has detected 46 sources down to a sensitivity limit of $\sim$1mCrab. Surveys of specific sources types are now beginning to be practical, an example of this being a survey of AGN detected (Bassani *et al.*, 2005, submitted) using similar techniques to *the IBIS survey*. This has detected 48 AGN, and a further 19 candidate AGN emitting in the 20-100 keV band. Work is also underway to compile a catalogue of sources visible at higher energies, above 100keV.

3. New absorbed persistent super-giant HMXB

The detection of new bright ($>$10 mCrab) persistent hard X-ray sources by INTE-GRAL has been a surprise as those have remained almost unnoticed in previous X-ray galactic surveys. XMM-Newton follow-up observations have been collected for seven of those sources. The X-ray counterparts (Figure 3) feature strong absorption ($>10^{23}\mathrm{cm}^{-2}$), intense fluorescence lines and soft X-ray excesses. Five of those sources appeared to be long spin period pulsars ($>$139 s) and a short orbital period ($<$10 days) was detected in two of them, thanks to eclipses (Figure 3). Long spin periods and persistency are characteristics of super-giant HMXB systems. A picture confirmed by the short orbital periods, the fraction of eclipsing systems and the early type near infrared counterparts. The 2MASS counterparts are not strongly reddened indicating that the X-ray absorption is indeed local to the sources. A persistent strong absorption is unusual among super-giant HMXB and may point out a peculiar stellar wind configuration. A number of the new, yet unidentified, galactic sources discovered by INTEGRAL could be new heavily absorbed HMXB. INTEGRAL therefore increases the number of know super-giant HMXB systems very significantly.

4. Active Galactic Nuclei Results

The extragalactic gamma-ray sky is still poorly explored with only two surveys performed above 10 keV in the past: the historical all-sky survey conducted in the 1980's (Levine *et al.* 1984) by the HEAO/A4 instrument in the 13-80 keV band and the more recent RXTE/PCA one made in the 8-20 keV band using slew observations (Revnivtsev *et al.* 2004, Sazanov and Revnivtsev, 2004). Both surveys lacked imaging capability, so that the association with an active galaxy was sometimes tentative. Despite being so rare, gamma-ray surveys are an efficient way to find AGN as they probe

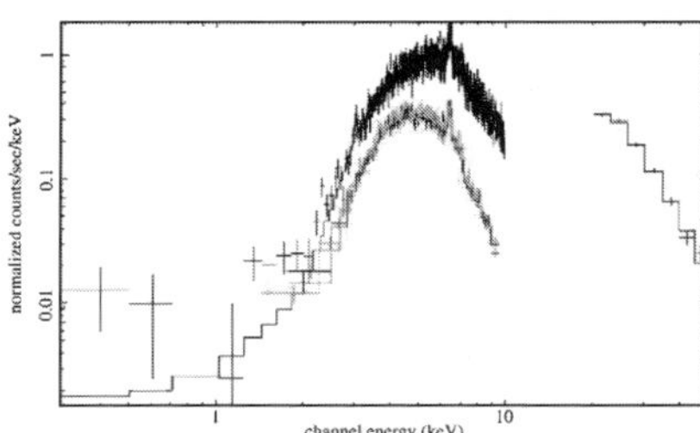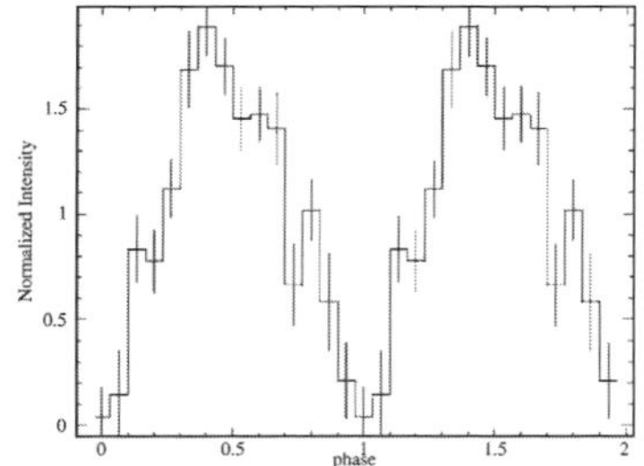

Figure 3. Left: XMM-Newton/INTEGRAL combined spectrum of IGR J17252-3616. The soft excess has not been modelled. Right: INTEGRAL orbital folded lightcurve of IGR J17252-3616. Phase 0 indicates the time of the eclipse.

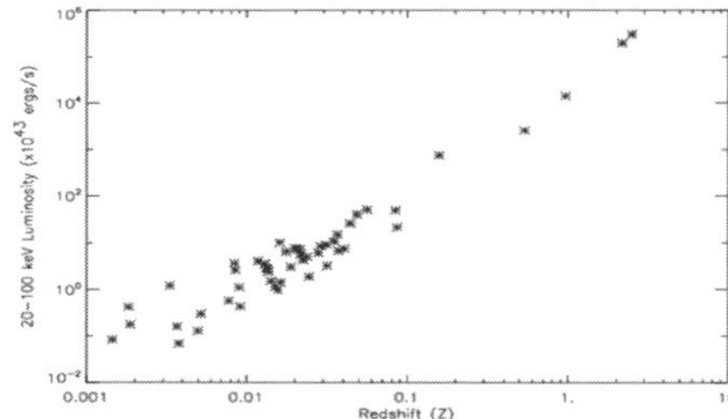

Figure 4. Redshift vs 20-100 keV luminosity for the INTEGRAL AGN sample.

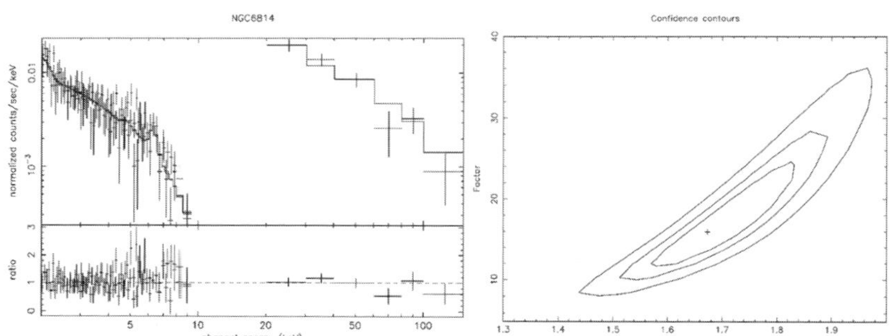

Figure 5. Left: ASCA/INTEGRAL combined spectrum of NGC6814. Right: Confidence levels of the cross-calibration constant $\frac{INT}{ASCA}$ vs photon index for NGC6814

heavily obscured regions/objects, i.e. those that could be missed in optical, UV, and even X-ray surveys. That absorbed AGN are common in the local Universe (Risaliti *et al.* 1999) is obvious from the fact that the 3 nearest (within 4 Mpc) active galaxies (NGC 4945, Centaurus A and Circinus galaxy) are all highly obscured with $N_H \geqslant 10^{23}$ atoms cm^{-2} (Matt *et al.* 2000). This situation has only recently been appreciated, perhaps due to all 3 objects lying close to the Galactic plane. The absorbing matter is often thick, since more than half of the Seyfert 2s in the local Universe have column densities in excess of, or equal, to 1.5×10^{24} atoms cm^{-2} (Risaliti *et al.* 1999). This result, coupled with the fact that type 2 objects are probably more numerous than type 1 sources, indicates the need for surveys free from biases against the Compton thick regime, i.e. above the 10-20 keV band. Quantifying the fraction of AGN missed by current surveys which are affected by selection due to absorption is necessary if we want to understand the accretion history of the Universe and study a population of objects so far poorly understood. Furthermore, the distribution of column densities is a key parameter for estimating the contribution of AGN to the X-ray cosmic diffuse background and for testing current unified theories. A step forward in this field is now provided by SWIFT/BAT and INTEGRAL/IBIS which are surveying a great fraction of the sky above 20 keV with a sensitivity better than a few mCrab and a point source location accuracy of the order of 1-3 arcminutes depending on the source strength. These two surveys are complementary to each other, as the first covers the high galactic latitude sky, while the second one concentrates on mapping mostly the galactic plane so that together they will provide the best yet sample of AGN selected in the gamma-ray band. The first sample of AGN detected by SWIFT/BAT lists 45 objects (Markwardt *et al.* 2005), all but 5 of which have archival X-ray spectra, enabling the estimation of the column density and other spectral properties. The column density distribution of these AGN is bimodal with 64% of the non blazar sources having absorption in excess of 10^{22} atoms cm^{-2}. None of the sources brighter than 3×10^{43} erg s^{-1} in the 14-195 keV band shows high column densities, while almost all those below this value are absorbed. The INTEGRAL/IBIS catalogue comprises 54 confirmed AGN and 13 candidates. Figure 1 shows a plot of the 20-100 keV luminosity (assuming $H_0 = 75$ Km/sec Mpc and $q_0 = 0$) against redshift when available for the INTEGRAL AGN sample. The complementarity of the SWIFT and INTEGRAL surveys is confirmed by the little overlap between the two, since only 9 objects are present in both catalogues. Within the sample of INTEGRAL AGN, 37 objects have an estimate of the intrinsic absorption. Within this subsample the fraction of absorbed $N_H \geqslant 10^{22}$ atoms cm^{-2} objects is 62%, much in agreement with the SWIFT result and also with the expectations of population synthesis models when Compton thick sources are taken into consideration (Comastri 2004). We also find that the ratio of Seyfert 1-1.5 to Seyfert 2 is $\sim$1:1, considerably different than the ratio found in optical or line selected surveys where the same ratio is 1:3 (Maiolino and Rieke, 1995, Ho, Filippenko and Sargent, 1999). We also find 6 Blazar type objects. Interestingly, in the sample of 17 Seyfert 2 with absorption measured, 75% are heavily absorbed ($N_H \geqslant 10^{23}$atoms cm^{-2}) with almost half of these being Compton thick. The fraction of Compton thick objects over the whole sample of INTEGRAL AGN having a measured column density is at the moment $\sim$30%, stressing again the need for considering Compton thick absorption in synthesis models of the X-ray background. A very preliminary LogN LogS distribution of our sample indicates substantial agreement with the HEAO/A4 and PDS fluxtuations analysis result (Fiore 1999, private communication), with typically 0.02 sources per square degrees above our sensitivity limit. It is important to note that IBIS can also provide detailed spectral information particularly if combined with lower energy data as obtained by past and current X-ray instruments also in view of the good cross calibration obtained between the various instruments. For example in the particular case of NGC6814, a source poorly studied in the past due to confusion with a nearby galactic source, IBIS in combination with ASCA has provided for the first time a broad band spectrum (Figure 5) as well as information on the extreme flux variability behaviour: a huge change in the cross calibration constant between the two instruments is required by the data, Figure 5 implying a flux change of a factor of 16 over a 10 year period (Molina *et al.* 2005). Source behaviour over time can also be checked through the light curves covering the entire INTEGRAL lifetime: while mainly developed for galactic objects this standardized analysis can also be used for extragalactic objects. Finally much is expected by IBIS on the study of absorbed objects. In Figure 6 we show the combined spectra (in the form of νF_ν) of IGR J07565-4139 (left panel) and NGC788 (right panel), observed respectively by INTEGRAL & Chandra and INTEGRAL & ASCA. IGR J07565-4139 has a Compton thick reflection dominated spectrum, with $N_H > 1.5 \times 10^{24}$ cm^{-2} but the pile-up fraction in Chandra data did not allow us to observe the expected (huge) iron line. NGC788 is characterized by a Compton thin model with $N_H \sim few 10^{23}$ cm^{-2}, and the ASCA data show the presence of the Kα iron line with EW = 600 eV, as expected with such strong absorption fraction.

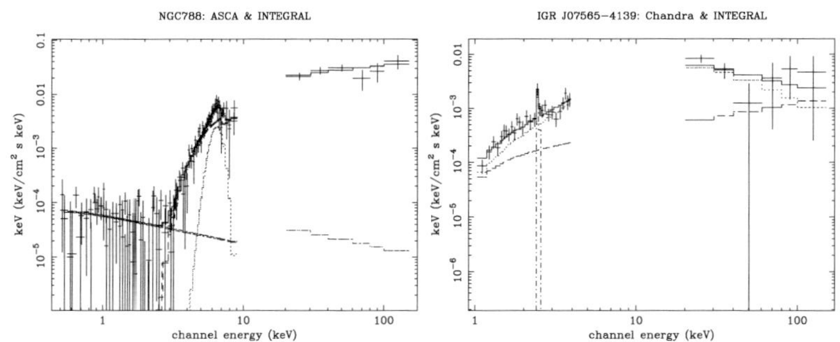

Figure 6. Left: ASCA/INTEGRAL combined spectrum of the Compton thin source NGC788. Right: Chandra/INTEGRAL combined spectrum of the Compton thick source IGRJ07565-4139.

Acknowledgements

The Italian co-authors acknowledge this research has been granted by the Italian Space Agency via contract n. I/R/046/04 ASI/IASF. We thank Alessandra De Rosa for the input on absorbed AGNs, Memmo Federici for data analysis system management and archive operability and Catia Spalletta for the editing of the paper.

References

Bassani, L. *et al.* 2005, *ApJ Letters*, submitted
Bird, A.J., Barlow, E.J., Bassani, L., *et al.* 2004, *ApJ*, 607, L33
Bird, A.J., Barlow, E.J., Bassani, L., *et al.* 2005, *ApJ*, in press
Comastri, A. 2004, *Supermassive Black Holes in the Distant Universe*, 245
Dean, A.J., Bazzano, A., Hill, A.B., *et al.* 2005, *A&A*, in press
Hill, A.B., Walter, R., Knigge, C., *et al.* 2005, *A&A*, 439, 255
Ho, L.C., Filippenko, A.V., Sargent, W.L.W. 1997, *ApJ Suppl.*, 112, 315
Levine, A., Lang, F., Lewin, W. *et al.* 1984, *ApJS*, 54, 581
Maiolino, R. & Rieke G.H. 1995, *ApJ*, 454, 95
Mantovani, F., Nanni, M., Salter, J.C., *et al.* 1982, *A&A*, 105, 176
Markwardt, C. B. *et al.* 2005, *ApJ Letters*, submitted, astro-ph. 05099860
Masetti, N., Palazzi, E., Bassani, L., *et al.* 2004, *A&A*, 426, 41
Matt G. *et al.* 2000, *MNRAS*, 318, 173
Molkov, S.V., *et al.* 2004, *Astron. Lett.*, 30, 534
Molina M. *et al.* 2005, *in preparation*
Revnivtsev, M.G., *et al.* 2004, *Astron. Lett.*, 30, 382
Revnivtsev, M.G., *et al.* 2004, *A&A*, 418, 927
Revnivtsev, M.G., *et al.* 2005, *Astron. Lett.*, in press
Risaliti, G., Maiolino, R. & Salvati, M. 1999, *Apj*, 522, 157
Sazanov and Revnivtsev 2004, *A&A*, 423, 469
Sguera, V., *et al.* 2005, *A&A*, in press
Stephen, J.B., Bassani, L., Molina, M., *et al.* 2005a, *A&A*, 432, L49
Stephen, J.B., Bassani, L., Malizia, A., *et al.* 2005b, *A&A*, in press
Ubertini, P., *et al.* 2003, *A&A*, 411, L131
Walter, R., *et al.* 2003, *A&A*, 411, L427
Winkler C., *et al.* 2003, *A&A*, 411, L1

Discussion

MACCARONE: Do you think that a large fraction of the X-ray background above 20 KeV is truly diffuse, and if so, what emission mechanism do you think causes this?

UBERTINI: Down to about 10^{-11} ergs/sec/cm^2, we certainly don't resolve much of it. With future work on the log N-logS distributions, we will be able to make better extrapolations of the effects of fainter sources, but it seems right now that a lot of the emission is truly diffuse. I don't want to speculate about what would cause this diffuse emission.

Populations of High Energy Sources in Galaxies
Proceedings IAU Symposium No. 230, 2005
E. J. A. Meurs & G. Fabbiano, eds.

© 2006 International Astronomical Union
doi:10.1017/S174392130600857X

The IBIS/ISGRI Survey of the Galactic Plane - Global Characteristics of the Gamma-Ray Sources

A. J. Dean[1], A. Bazzano[2], A. B. Hill[1], J. B. Stephen[3], L. Bassani[3], E. J. Barlow[1], A. J. Bird[1], F. Lebrun[4], V. Sguera[1], S. E. Shaw[1,5], P. Ubertini[2], R. Walter[5] and D. R. Willis[1]

[1]School of Physics and Astronomy, University of Southampton, Highfield, SO17 1BJ, UK.
[2]IASF-Rm, INAF, Via Fosso del Cavaliere 100, I-00133 Rome, Italy.
[3]IASF-Bo, INAF, Via Gobetti 101, I-40129 Bologna, Italy.
[4]CEA-Saclay, DAPNIA/Service d'Astrophysique, F-91191 Gif sur Yvette Cedex, France.
[5]Geneva Observatory, INTEGRAL Science Data Centre, Chemin d'Ecogia 16, 1291 Versoix, Switzerland.

Abstract. INTEGRAL is the first gamma-ray astronomy mission with a sufficient sensitivity and angular resolution combination appropriate to the detection and identification of considerable numbers of gamma-ray emitting sources. The large field of view enables INTEGRAL to survey the Galactic Plane on a regular ($\sim$weekly) basis as part of the core programme. The first source catalogue, based on the 1st year of core programme data ($\sim$5 Msec) has been completed and published (Bird *et al.* 2004). It contained 123 γ-ray sources (24 HMXB, 54 LMXB, 28 "unknown", plus 17 others) - sufficient numbers for a reasonable statistical analysis of their global properties. The detection of previously unknown γ-ray emitting sources generally exhibiting high intrinsic absorption, which do not have readily identifiable counterparts at other wavelengths, is intriguing. The substantial fraction of unclassified γ-ray sources suggests they must constitute a significant family of objects. In this paper we review the global characteristics of the known galactic sources as well as the unclassified objects.

Keywords. Gamma-rays: observations – X-rays: binaries – Galaxy: structure – Galaxy: stellar content.

1. Introduction

INTEGRAL is the first gamma-ray astronomy mission with a sufficient sensitivity and angular resolution combination appropriate to the detection and identification of considerable numbers of gamma-ray emitting sources (Winkler *et al.* 2003). With an observation time of roughly 1000 seconds required to detect a $1M_\odot$ neutron star emitting at the Eddington limit from a distance of 10kpc it is clear that a meaningful survey of discrete galactic gamma-ray emitting objects is possible in the lifetime of INTEGRAL.

A significant fraction of the INTEGRAL core programme is devoted to regular scans of the Galactic Plane and a deep exposure of the Galactic Centre. This is facilitated by the large field of view of the on-board telescopes ($9° \times 9°$ fully coded and $29° \times 29°$ zero response for IBIS, Ubertini *et al.* (2003)), which permits $\sim$100 ksec exposures of the Galactic Plane to be made $\sim$every 12 days). The 1st IBIS/ISGRI survey catalogue is constructed from $\sim$5 Msec of observations (all Core Programme observations between revolutions 46 and 120). The data is organised into short pointings (science windows) of $\sim$2 ksec. The $\sim$2500 individual science window images were then mosaiced using a custom tool to produce deep all-sky maps. An initial source list was obtained by searching all-sky mosaics using the *SExtractor* tool. It has since been discovered that the identification of IGR J17460-3047 was in fact a false detection. The typical point source location error range is $20''$–$3'$, with a 1 arc minute error circle for a source $>10\sigma$ (90% confidence) (Gros *et al.* 2003). The precision of the source locations allow for the clear association to sources seen by previous missions.

Of the 123 sources in the catalogue, 5 were unambiguously identified with AGN, 5 with white dwarfs, 4 with radio pulsars (free neutron stars), 3 supernova remnants and one cluster. Of the remainder, 54 were identified with known low mass X-ray binary (LMXB) systems, 24 with high mass X-ray binaries (HMXB) and the remaining 28 sources have no firm classification and are here after described as "unclassified". In this paper we have only considered the LMXBs, the HMXBs and the unclassified objects, discounting IGR J17460-3047 as a false detection.

2. The Log(N)-Log(S) Distributions

In this section we construct the number flux distribution for the various source types. We have assumed the usual power-law form of $N(>S) = KS^{-\alpha}$ for the relationship. As the sky coverage of the first survey is not complete, the Log(N)-Log(S) curves are only indicative of the true galactic distributions and must be corrected for both the area of sky covered and the depth of exposure (or the minimum detectable flux (MDF)) at each point. This is not straightforward due to the residual systematic structures in the sky maps, however it is possible by inspection of the error and exposure distributions to parameterise the variation in MDF as a function of exposure, thereby taking into account the effects of the residual systematic variations.

From this relationship and the general sky exposure map we can then correct the observed logN-logS resulting in the relationships shown in Figure 1.

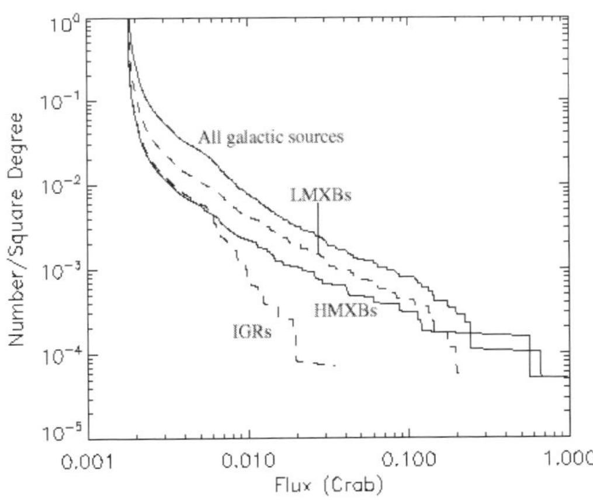

Figure 1. The Number-Flux relationships corrected for exposure and sky area observed.

The limiting detection threshold of 6 sigma is sufficiently high so that the maximum likelihood (ML) method (Murdoch et al. 1973) can be used to calculate the best-fit values of the slope of the number-flux relationship without the uncertainty in the correction factor for weaker sources dominating the correction itself. This has been performed for the corrected distributions and the value of the slope found in each case is shown below:

- All Galactic Sources, $\alpha = 0.91 \pm 0.09$
- LMXB, $\alpha = 0.95 \pm 0.13$
- HMXB, $\alpha = 0.81 \pm 0.15$
- Unclassified sources, $\alpha = 2.11 \pm 0.46$

It is of interest to consider and compare the Log(N)-Log(S) distributions of the various sub-groupings. The -0.91 slope of the power law for all the combined galactic sources is close to the -1 value expected for a uniform infinite plane distribution and lies between the value (-0.79 ± 0.07) measured for galactic sources by ASCA (Sugizaki *et al.* 2001) in the classic 2-10 keV X-ray band and the (-1.1) slope derived from the Einstein Galactic Plane survey (Hertz & Grindlay 1984). This all-source value is clearly dominated by the numerically superior LMXBs, which when detached provide a slope of -0.95 ± 0.13 as a sub-set. The value for LMXBs is thus consistent with -1, as qualitatively expected for a population with a larger scale height (Sugizaki *et al.* 2001). HMXBs are typically located within spiral arms and this would automatically lead to a value of α closer to 0.5. In this context, the -0.81 ± 0.13 slope measured for the HMXBs appears slightly flatter than the LMXBs and may reflect the likely location of these objects along the spiral arms. The steep slope of the unclassified sources may imply a different source population at low fluxes, this is far steeper than we may expect from even an extragalactic contamination. However, it must be remembered that there are relatively very few of these objects, and they are also amongst the weaker sources in the survey therefore care must be taken in any interpretation of this slope.

3. The Angular Distributions

The first year of INTEGRAL observations has been strongly biased towards the study of the Galactic Plane and in particular the Galactic Centre, hence the first IBIS/ISGRI catalogue is essentially a register of galactic sources. Figure 2 reveals the considerable differences in the longitudinal distribution of the HMXB and LMXB sources. Whereas the LMXB are concentrated in the Galactic Centre region, the HMXB are spread more extensively along the Galactic Plane, but in a non-uniform manner, with some evidence for bunching in the vicinity of the tangential points of the spiral arm structure. The locations of the spiral arm tangents were taken from Englmaier *et al.* (1999). Additionally, a number of HMXBs is seen at $l^{II} \sim \pm 90^o$, which is indicative of systems located within our particular spiral arm. This association had been previously noted through inspection of the Ginga data by Koyama *et al.* (1990) and in the RXTE/ASM data by Grimm *et al.* (2002).

The association with the spiral arms is entirely to be expected since the high mass binaries are young stellar systems and should be attached to regions where star formation has recently taken place, such as the spiral arms.

The unclassified sources appear to have a longitude distribution concentrated around the Galactic Centre, similar to that of the LMXBs. To some extent this is misleading as the Galactic Centre has had the most sky exposure and hence we are more likely to detect new systems in this region. However, upon closer examination, whilst some emulate the distribution of LMXB, it can be seen that the unclassified sources do not have a symmetrical distribution about the Galactic Centre and show some tendency to cluster around the locations of spiral arm tangents, specifically the Scutum, 3-kpc and Norma arms. Specifically the unclassified sources appear to concentrate in the 3-kpc and Norma spiral arms.

Examining the angular distribution off the Galactic Plane of the sources we find that the HMXBs exhibit a much narrower range of angular separations than the LMXBs, as expected if they are confined to the Galactic Disc. The much wider

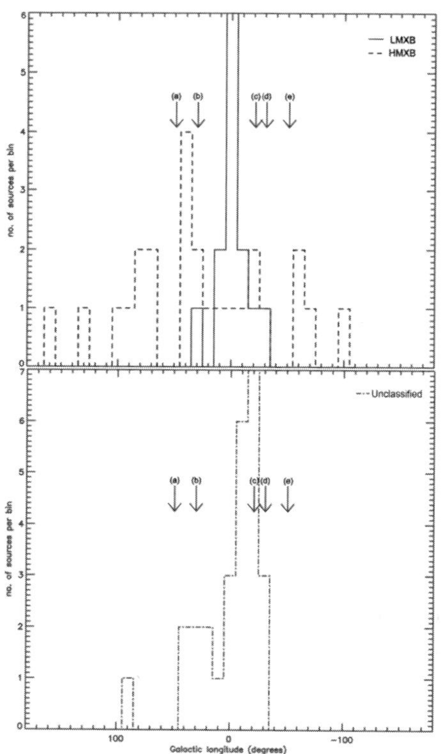

Figure 2. The Galactic longitude distribution of HMXBs, LMXBs and unclassified sources. Labelled are the locations of the spiral arm tangents: (a) - Sagittarius; (b) - Scutum; (c) - 3-kpc; (d) - Norma; (e) - Centaurus.

spread of the LMXBs is indicative of a population derived from the Galactic Bulge. The unclassified sources appear to have a sharp latitude cut-off similar to that of the HMXBs, this may indicate that they too are a primarily a population from the Galactic Disc. However, they do not precisely conform to either of the subgroups; this is not unexpected as they could consist of a mixture of both, or be a separated generic set.

4. Summary and Discussion

The INTEGRAL Galactic Plane survey and ensuing first catalogue has offered for the first time the possibility to investigate the global γ-ray characteristics of galactic sources on a reasonable statistical basis. In this paper we have studied the generalised observational features of γ-ray selected previously unclassified objects, and compared them with the corresponding parameters of known HMXB and LMXB systems. Although the sample of sources has been γ-ray selected the Log(N)-Log(S) and angular distributions of the known classes, as expected, emulate those of X-ray selected sets e.g. Grimm *et al.* (2002), there appear to be subtle differences. Of great interest are the 27 "unidentified" objects, which naturally have overhanging questions such as what are they and how do they fit into mainstream stellar evolution?

Clearly a vigorous programme of follow-up studies on individual objects, using such missions as XMM-Newton, Chandra, and RXTE for timing studies, is one means to pursue the answers to these questions, and such activities are already underway: Matt & Guainazzi (2003); Walter *et al.* (2003); Hill *et al.* (2005); Stephen *et al.* (2005). The complementary statistical approach employed here essentially comes to the same conclusion i.e. that a large fraction of the unclassified objects are obscured high mass X-ray binary systems. Detailed studies of the main distributional characteristics of the unclassified sources when compared to the equivalent distributions of known X-ray binaries all point in this direction.

It is hardly surprising that INTEGRAL should discover a population of previously unnoticed sources. The catalogue is γ-ray selected, and INTEGRAL operates above the energy threshold for which significant photoabsorption takes place in universal abundance matter, so that a strong emitter above $\sim$30keV can be rendered insignificantly weak in the classical X-ray band. As discussed by Lutovinov *et al.* (2005), the HMXB systems constitute the most likely candidates. The stellar wind accretion mechanism that dominates in HMXB systems, as opposed to the Roche lobe overflow associated with LMXBs, naturally provides a suitable dense and strongly absorbing circumstellar wind to veil the X-ray emission.

Exactly how the γ-ray selected highly absorbed systems fit into the overall picture of binary star evolution is currently unclear. This is principally due to the requirement of a suitable companion able to generate a suitably dense gas surrounding the compact object to stifle the X-ray emission. We need to understand why they are different to "normal" HMXB. Do these highly absorbed systems relate to the mass/giant nature of the primary star, or to the orbit configuration, or are they experiencing a phase the binary systems routinely pass through, but have not been exposed at other wavelengths? Do they need to be HMXB? Clearly a series of dedicated observations is required to solve this problem.

References

Bird, A.J., Barlow, E.J., Bassani, L., *et al.*, 2004, ApJ, 607L, 33B
Englmaier, P. & Gerhard, O., 1999, MNRAS, 304, 512
Grimm, H.-J., Gilfanov, M. & Sunyaev, R., 2002, A&A, 391, 923
Gros, A., Goldwurm, A., Cadolle-Bell, M., *et al.*, 2003, A&A, 411, L179
Hertz, P. & Grindlay, J., 1984, ApJ, 278, 137
Hill, A.B., Walter, R., Knigge, C., *et al.*, 2005, A&A, 439, 255
Koyama, K., Kamada, M., Kunieda, H., *et al.*, 1990, Nature, 343, 148
Lutovinov, A., Revnivtsev, M., Molkov, S., *et al.*, 2005, A&A, 430, 997
Matt, G. & Guainazzi, M., 2003, MNRAS, 341, L13
Murdoch, H. S., Crawford, D. F. & Jauncey, D. L., 1973, ApJ, 183, 1
Stephen, J.B., Bassani, L., Molina, M., *et al.*, 2005, A&A, 432, L49
Sugizaki, M., Mitsuda, K., Kaneda, H., *et al.*, 2001, ApJS, 134, 77
Ubertini, P., Lebrun, F., Di Cocco, G., *et al.*, 2003, A&A, 411, L131
Walter, R., Rodriguez, J., Foschini, L., *et al.*, 2003, A&A, 411, L427
Winkler, C., Courvoisier, T. J.-L., Di Cocco, G., *et al.*, 2003 A&A, 411, L1

Populations of High Energy Sources in Galaxies
Proceedings IAU Symposium No. 230, 2005
E. J. A. Meurs & G. Fabbiano, eds.

© 2006 International Astronomical Union
doi:10.1017/S1743921306008581

Very faint X-ray transients in our Galaxy

Rudy Wijnands[1]

[1]Astronomical Institute "Anton Pannekoek", University of Amsterdam, Kruislaan 403,
1098 SJ, Amsterdam, The Netherlands
email: rudy@science.uva.nl

Abstract. We present the first results of our X-ray monitoring campaign using the X-ray satellites *XMM-Newton* and *Chandra* on the 1.7 square degree region centered on Sgr A*. The purpose of this campaign is to monitor the X-ray behavior of X-ray sources (both persistent and transient ones) which are too faint to be detected by monitoring instruments aboard other satellites currently in orbit (e.g., *RXTE, Integral, Swift*). Our first observations were obtained on June 5, 2005, using the HRC-I detector aboard *Chandra*. Here we briefly discuss our results on the very faint X-ray transients we detected during these observations. We detected new outbursts (with peak luminosities close to 10^{36} erg s^{-1}) of the two very faint transients GRS 1741.9–2853 and XMM J174457-2850.3, as well as a potential new very faint X-ray transient with an outburst luminosity of a few times 10^{34} erg s^{-1}.

Keywords. accretion, accretion disks, X-rays: binaries, stars: neutron

1. Introduction

The X-ray luminosity of our Galaxy is dominated by accreting neutron stars and black holes in X-ray binaries. The majority of such systems is usually dormant but occasionally they exhibit outbursts during which their X-ray luminosity increases by a factor of >100 with peak luminosities of $>10^{37-39}$ ergs s^{-1}. The existence of these *bright X-ray transients*, mostly containing black holes, has been known since the early days of X-ray astronomy. Several years ago, Heise *et al.* (1999), using the *BeppoSAX* satellite, provided strong evidence for a class of *faint X-ray transients* with peak luminosities of of 10^{36-37} ergs s^{-1}. Interestingly, a large fraction of these faint systems harbor neutron stars instead of black holes, pointing to differences in formation histories (King(2000)).

Although transients with peak luminosities below 10^{36} erg s^{-1} have been sporadically reported in the past (e.g., Maeda *et al.* (1996)), it was only recently realized (e.g., Sakano *et al.* (2005)) that such *very faint X-ray transients* (VFXTs) likely form another distinct class of X-ray transients with peak luminosities of 10^{34-36} ergs s^{-1}, a factor >10 lower than the faint transients. All recently discovered VFXTs were found using the sensitive *Chandra* and *XMM-Newton* X-ray satellites (Sidoli & Mereghetti (2003), Hands *et al.* (2004), Porquet *et al.* (2005), Sakano *et al.* (2005), Muno *et al.* (2005a)). Some of them might be intrinsically bright transients at large distances; however, many are observed near the Galactic center, indicating source distances of $\sim$8 kpc and therefore very low peak luminosities. Some VFXTs might be intrinsically brighter than observed due to inclination effects (e.g., Muno *et al.* (2005b)), but only a small fraction of the VFXTs can be explained in this manner (see, e.g., the discussion in King & Wijnands (2005)).

The characteristics of the VFXTs (e.g., their spectra or outburst light curves; e.g., Sakano *et al.* (2005) and Muno *et al.* (2005a)) indicate that they are not a homogeneous class of sources but that different types of accreting neutron stars and black holes show themselves as VFXTs. Therefore, a variety of models are probably needed to explain all systems. However, it is likely that a significant fraction of the VFXTs are neutron stars

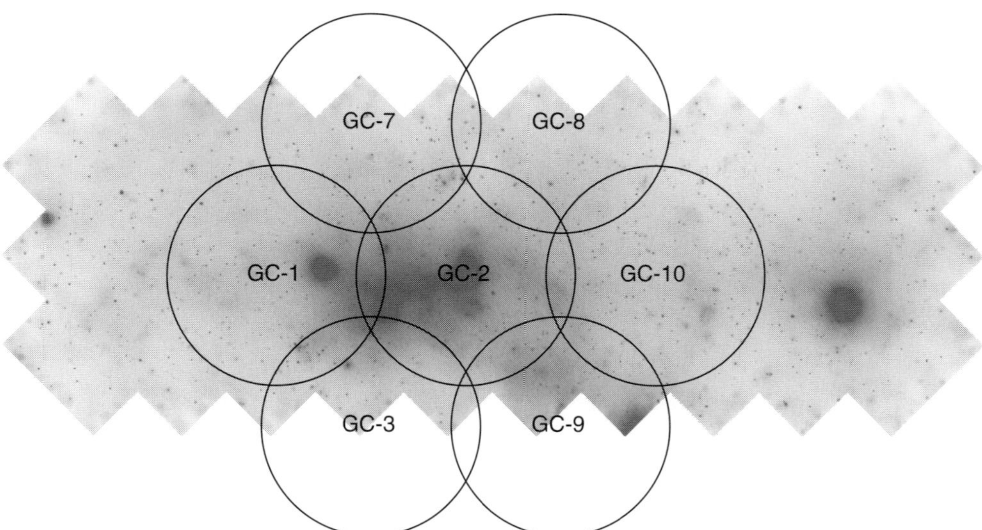

Figure 1. Field-of-view of the monitoring campaign in Galactic coordinates. The circles indicate the field-of-view of the *Chandra* and *XMM-Newton* observations. The image is the Wang *et al.* (2002) *Chandra* Galactic center survey data.

and black holes which accrete matter at a very low rate from a low-mass companion star. For example, at least two such systems have exhibited type-I bursts which have so far only been observed for accreting neutron stars in low-mass X-ray binaries. These low-mass X-ray binaries which accrete at very low rates are a challenge to our understanding of the evolution of such systems. If their time averaged accretion rates do not drop below $10^{-13} M_\odot$ yr^{-1} then these systems can be explained as low-mass X-ray binaries which have spent most of the age of the Galaxy reducing their companion stars to values $\sim 0.01 M_\odot$ (King & Wijnands (2005)). If future observations show that the time averaged accretion rates drop below this value then more exotic explanations may be needed, such as accretion from planetary companions or accretion onto intermediate mass black holes.

Our understanding of the nature of these enigmatic VFXTs is hampered by the lack of known systems, the limited information about their outburst duration, recurrence time and duty cycle (e.g., Muno *et al.* (2005a)), and the lack of optical/infrared counterparts. The VFXTs are usually too faint (if close to the Galactic center) to be detected by the current monitoring satellites (e.g., *RXTE*, *Integral* or *Swift*) and source confusion close to the Galactic center also inhibits their detection. Therefore, we have initiated a monitoring campaign using *XMM-Newton* and *Chandra* of the Galactic center region to discover new VFXTs and to detect new outbursts of the already known VFXTs.

2. Description of the program

We have secured a monitoring campaign using *XMM-Newton* and *Chandra* of the 1.7 square degree region centered around Sgr A* to search for new and recurrent VFXTs (Fig. 1). Our campaign consists of four observational epochs, two with *Chandra* (one performed on June 5 and the other scheduled for the week of 17 October 2005; using the HRC-I detector because of its relatively large field-of-view) and two with *XMM-Newton* (scheduled for the end of February and early April 2006). The first *Chandra*/HRC-I observations have been performed and we discuss our first results here.

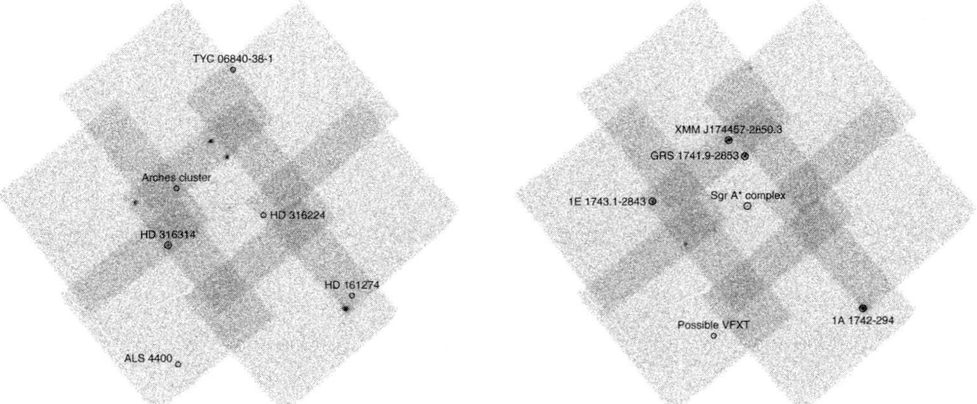

Figure 2. The merged image of the 7 HRC-I observations. Left panel: a sub-set of the detected stars is indicated; right panel: the detected X-ray binaries as well as the Sgr A* region.

3. Results

The combined image of the seven *Chandra*/HRC-I observations is shown in Figure 2. Preliminary searches for point sources resulted in 21 detected sources. Two of these sources (the Sgr A* region and the Arches cluster) are known to consist of a complex of point sources in combination with strong diffuse emission. Sgr A* itself was not detected during our observations. The majority of the remaining sources can be identified with known stars (e.g., HD 316314, HD 316224, TYC 06840-38-1) or have clear counterparts in the DSS images. Also, two persistent X-ray binaries (1E 1743.1–2843 and 1A 1742–294) are detected. Here we will not discuss those stars and persistent X-ray binaries; instead we will focus on the VFXTs detected.

Two known VFXTs were found to be in outburst again during our observations: GRS 1741.9–2853 and XMM J174457–2850.3 (Fig. 2 right panel; Wijnands *et al.* (2005)). Their luminosities are somewhat uncertain since the HRC-I detector does not have energy resolution. Therefore, we converted the observed count rate into luminosity using PIMMS and the spectra previously found for these sources (e.g., Muno *et al.* (2003), Sakano *et al.* (2005)) and found that both sources have luminosities slightly below 10^{36} erg s^{-1}. No type-I bursts were seen for either source. After the discovery of the new outbursts of both sources we obtained an additional *Chandra* observation (using the ACIS-I detector) of the two sources on July 1, 2005. During this observation, we could detect GRS 1741.9–2853 still at a luminosity of $\sim 10^{35}$ erg s^{-1} but XMM J174457–2850.3 had decreased in luminosity to a few times 10^{33} erg s^{-1}. The long term light curves of both sources are shown in Figure 3. A possible new VFXT (Fig. 2) was also detected at a luminosity of a few times 10^{34} erg s^{-1}, which was not seen in archival *XMM-Newton* data.

4. Conclusions

During the first *Chandra* observations of our monitoring campaign we detected new outbursts of the known VFXTs GRS 1741.9–2853 and XMM J174457-2850.3. We also discovered a potential new VFXT at a peak luminosity of a few times 10^{34} erg s^{-1}. Using our scheduled monitoring observations in the fall of 2005 and the spring of 2006, we will discover several more VFXTs and additional outbursts of known VFXTs. By studying the properties and behavior of many VFXTs we will get a better understanding of the nature of this enigmatic class of X-ray transients. For the systems accreting from a low-mass

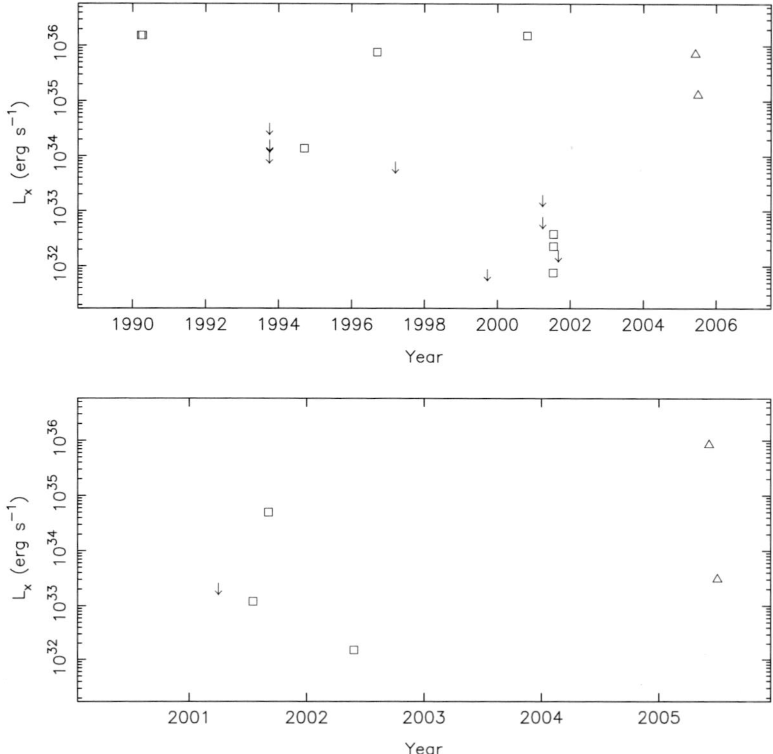

Figure 3. Light curves of GRS 1741.9–2853 (top) and XMM J174457–2850.3 (bottom). In both panels the triangles indicate our *Chandra* data. The squares and the upper limits are taken from, respectively, Muno *et al.* (2003) and Sakano *et al.* (2005) for the top and bottom panel.

companion star, our monitoring campaign will help to determine their time averaged accretion rates, needed to constrain theories about the formation of such systems (e.g., King & Wijnands (2005)).

References

Hands, A. D. P., Warwick, R. S., Watson, M. G., & Helfand, D. J. 2004, MNRAS, 351, 31

Heise, J., in't Zand, M. J. J., Smith, S. M. J., Muller, M. J., Ubertini, P., Bazzano, A., Cocchi, M., & Natalucci, L. 1999, Astrophysical Letters Communications, 38, 297

King, A. R. 2000, MNRAS, 315, L33

King, A.R. & Wijnands, R. 2005, MNRAS, submitted

Maeda, Y., Koyama, K., Sakano, M., Takeshima, T., & Yamauchi, S. 1996, PASJ, 48, 417

Muno, M. P., Baganoff, F. K., & Arabadjis, J. S. 2003, ApJ, 598, 474

Muno, M. P., Pfahl, E., Baganoff, F. K., Brandt, W. N., Ghez, A., Lu, J., & Morris, M. R. 2005a, ApJ, 622, L113

Muno, M.P., Lu, J.R., Baganoff, F. K., Brandt, W.N., Garmire, G.P., Ghez, A.M., Hornstein, S.D., & Morris, M.R., 2005b, ApJ, submitted (astro-ph/0503572)

Porquet, D., Grosso, N., Burwitz, V., Andronov, I. L., Aschenbach, B., Predehl, P., & Warwick, R. S. 2005, A&A, 430, L9

Sakano, M., Warwick, R. S., Decourchelle, A., & Wang, Q. D. 2005, MNRAS, 357, 1211

Sidoli, L., & Mereghetti, S. 2003, The Astronomer's Telegram, 147, 1

Wang, Q. D., Gotthelf, E. V., & Lang, C. C. 2002, Nature, 415, 148

Wijnands, R., *et al.* 2005, The Astronomer's Telegram, 512, 1

Discussion

IVANOVA: What can you say about outburst's durations and recurrence intervals for VFXT?

WIJNANDS: Both the durations and recurrence times vary a lot from source to source. Some VFXTs are on for <1 month, others >1 year. Not much is known about the recurrence time but at least a few recur every 3-5 years.

CHERNYAKOVA: What is the impact of the newly resolved point sources in the Sgr A* region to the total X-ray emission of the region?

WIJNANDS: Close (<10') to Sgr A* the X-ray flux is consistent with what has been found before (e.g. by Muno et al). In the whole FOV of our MRC-I observations the true VFXTs detected contribute about $2 \ 10^{36}$ erg/s (2-10 KeV) to the X-ray luminosity in this region.

Rudy Wijnands, Tom Maccarone and Jacco Vink (from left to right) follow eagerly the demonstration of modern Irish dancing.

Populations of High Energy Sources in Galaxies
Proceedings IAU Symposium No. 230, 2005
E. J. A. Meurs & G. Fabbiano, eds.

© 2006 International Astronomical Union
doi:10.1017/S1743921306008593

Synchrotron X-Ray SNR Candidates Discovered in the *ASCA* Galactic Plane Survey

M. Ueno[1], S. Yamauchi[2], A. Bamba[3], H. Yamaguchi[4], K. Koyama[4], and K. Ebisawa[5]

[1]Department of Physics, Faculty of Science, Tokyo Institute of Technology 2-12-1, Oo-okayama, Meguro-ku, Tokyo 152-8551, Japan
email: masaru@hp.phys.titech.ac.jp

[2]Faculty of Humanities and Social Sciences, Iwate University, 3-18-34 Ueda, Morioka, Iwate 020-8550, Japan

[3]RIKEN (The Institute of Physical and Chemical Research) 2-1, Hirosawa, Wako, Saitama 351-0198, Japan

[4]Department of Physics, Graduate School of Science, Kyoto University, Sakyo-ku, Kyoto 606-8502, Japan

[5]The Institute of Space and Astronautical Science (ISAS), Japan Aerospace Exploration Agency(JAXA) 3-1-1, Yoshinodai, Sagamihara, Kanagawa 229-8510, Japan

Abstract. Shell-like supernova remnants (SNRs) are primary candidates for the origin of Galactic cosmic rays. However, among the known SNRs (about 220), only a small fraction has been known to exhibit the synchrotron X-ray spectrum, that is considered to be a piece of evidence for high energy particle acceleration. Synchrotron X-ray emitting SNRs are known to be systematically radio-quiet compared to the SNRs that do not emit synchrotron X-rays. Therefore, most synchrotron X-ray emitting SNR candidates may have escaped detection in the previous systematic radio surveys. On the other hand, hard X-ray surveys are effective to search for synchrotron X-ray emitting SNRs, because of its penetration power. Thus we have searched for SNRs in the *ASCA* Galactic Plane Survey data, the first Galactic imaging survey in >2 keV, and discovered 14 candidates. Deep follow-up observations with *ASCA*, *XMM*, or *Chandra* on 5 of them revealed 2 sources to be synchrotron X-ray emitting SNRs. Furthermore we confirmed non-thermal X-ray spectra from the other 3 sources, though the origin is yet unknown. We report the observational results and discuss the X-ray origin.

Keywords. supernova remnants; cosmic rays; Galaxy: structure.

---

1. Introduction

Since the discovery of synchrotron X-ray emission from SN1006 (Koyama *et al.* 1995), synchrotron X-rays and TeV γ-rays are detected in several SNRs (e.g., RX J1713.7−3946; Enomoto *et al.* 2002), which are evidence for high energy ($\geqslant 10$ TeV) electrons located in the shells of SNRs. However, synchrotron X-rays or TeV γ-rays has been detected in only a small number of SNRs ($\sim$10) compared with the total number of SNRs now known ($\sim$220 excepting for filled-center type SNRs; Green 2004). Therefore, we cannot draw conclusion as to whether particle acceleration is a universal activity of SNRs or not. As a result, it is still unknown whether particle acceleration in SNRs is significant for the Galactic cosmic rays.

From the discoveries of several SNRs in the *ROSAT* All Sky Survey (e.g., Pfeffermann & Aschenbach 1996), it is suggested that a part of SNRs have low radio surface brightness and have been escaping detection in the previous systematic radio surveys. However, due to the interstellar absorption, soft X-ray observations with *ROSAT* are not suited for detecting sources embedded in the Galactic plane. On the other hand, hard X-rays are transparent through the Galactic plane. In addition, synchrotron X-rays are dominant against thermal emission in the hard X-ray band. Therefore, a hard X-ray survey would be the best way to find new SNRs emitting synchrotron X-rays (synchrotron X-ray SNRs) located in the Galactic plane.

On this point of view, we reanalyzed the data set of the *ASCA* Galactic Plane Survey, which is the first imaging survey of the Galactic plane in the hard X-ray band (> 2 keV). Comprehensive studies of the survey data have been reported by Sugizaki *et al.* (2001). In this paper, we concentrate on new detection of unidentified diffuse hard X-ray emissions.

2. Observations and Data Reduction

The *ASCA* Galactic Plane Survey (AGPS) consists of 173 pointing observations, covering the area of the Galactic inner disk of $|l| \lesssim 45°$ and $|b| \lesssim 0.4°$. Since about 34% of the surveyed regions suffer from stray light from nearby bright sources, we excluded these regions in our analysis. Since GISs had 4 times larger field of views than SISs, we concentrated on the X-ray data from GIS in this SNR survey.

Exposure corrected images of the *ASCA* Galactic Plane Survey data in the 2.0–7.0 keV band were produced using night earth data for the background, and blank sky data for the efficiency map. We searched for unidentified extended sources in the resultant exposure-corrected 2.0–7.0 keV band images. As a result, we detected 14 new extended sources in the hard band images (2–7 keV). Among them are 5 SNR candidates, which we have already reported (G28.6−0.1, G11.0+0.0, G25.5+0.0, G26.6−0.1, G32.45+0.1; Bamba *et al.* 2001; Bamba *et al.* 2003; Ueno *et al.* 2003; Yamaguchi *et al.* 2004), and G337.2+0.1, which was reported by Combi *et al.* (2005) as an SNR candidate. The hard band images of the other 8 sources are shown in Figure 1. By making radial profiles, we verified these sources are significantly extended compared with the point spread function of GIS.

3. Discussion

3.1. *Radio Faintness of the SNR candidates*

Radio surface brightness (Σ) of SNRs tends to become small as their diameters (D) become large, which is generally known as the Σ-D relation (Figure 2; e.g., Green 2004). Since the new SNR candidates in this paper are discovered with X-rays but not with radio observations, it is inferred that they are radio faint compared with the other SNRs. According to the Σ-D relation, the diameters of the candidates should be large. In order to test this, the distances of the candidates are roughly estimated from the absorption column densities of the X-ray spectra, assuming the density of the ISM, $N_{\rm H} = 1$ cm^{-3}. Since the $N_{\rm H}$ of two candidates is consistent with 0 cm^{-2} and their distances can not be determined, they are excluded here. On the other hand, the upper limit of radio brightnesses of the new SNR candidates can be estimated from the sensitivities of the previous radio surveys. The $l > 0°$ and $l < 0°$ regions are covered with the NRAO VLA Sky Survey (Condon *et al.* 1998) and with the Molonglo Galactic Plane Survey (Green *et al.* 1999), respectively. Upper limits (3σ) are calculated to be 6×10^{-22} and $(1 - 2) \times 10^{-21}$ W m^{-2} s^{-1} Hz^{-1} sr^{-1} (spectral index of 0.5 is assumed). These results are added to the Σ-D relation in Figure 2. Although there are large uncertainties in diameters, it is seen that newly found SNR candidates are faint in radio not because they have large diameters.

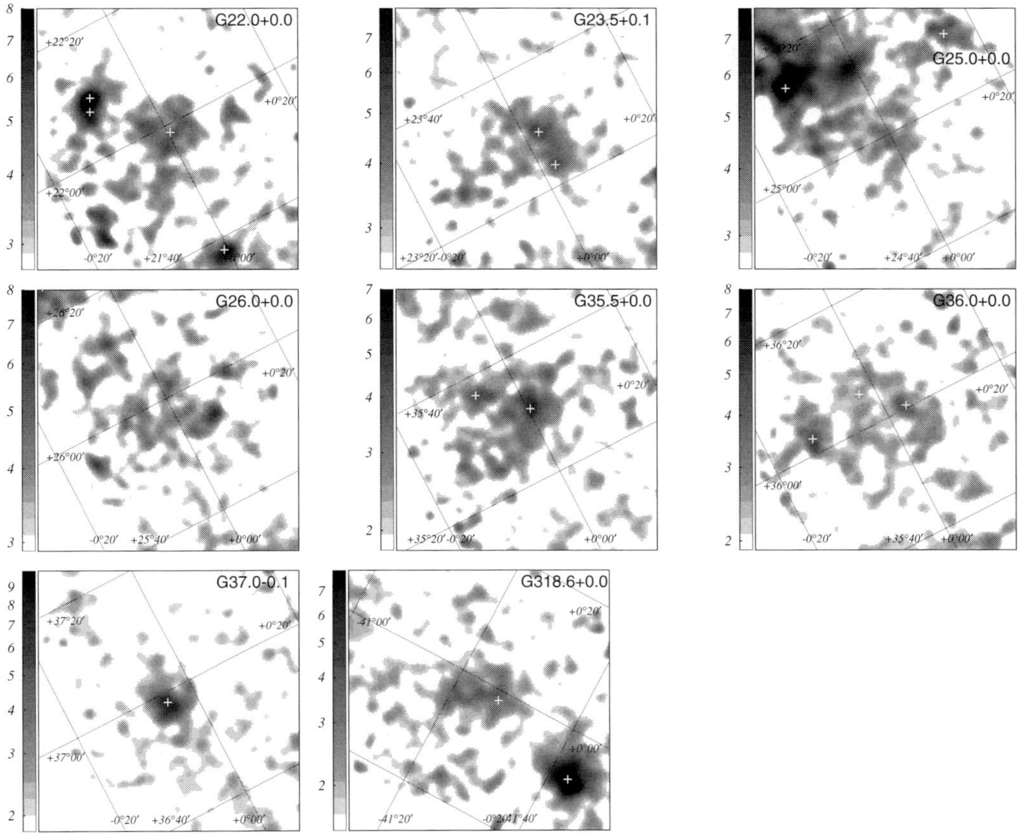

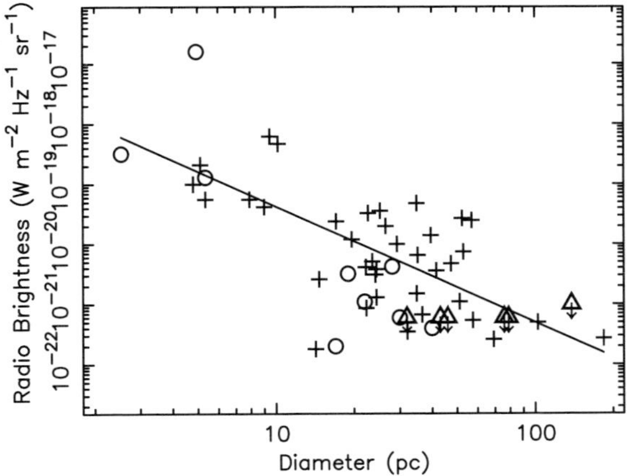

Figure 1. *ASCA* images of the new SNR candidates in the 2.0–7.0 keV band. Scales are logarithmic and the numbers next to the scale bars correspond to the surface brightness ($\times 10^{-6}$ counts cm^{-2} s^{-1} arcmin^{-2}). The orientation is equatorial (J2000), whereas the coordinates shown are galactic. The sources detected in Sugizaki *et al.* (2001) are designated with white crosses. The new SNR candidates are in the center of each image.

Figure 2. Radio surface brightness vs. diameter relation for shell SNRs (see Green 2004). SNR emitting synchrotron X-rays and those not are designated with circles and crosses, respectively. The new SNR candidates are shown with triangles, whose radio surface brightnesses shown are all upper limits.

Arbutina & Urošević (2005) suggested that SNRs located in high density media show high radio surface brightnesses, which is also theoretically deduced by Berezhko & Völk (2004). Since synchrotron X-ray SNRs tend to be faint in radio, they may be located in low density environment.

3.2. *Expected number of synchrotron X-ray SNRs in the Galactic plane*

If we assume the diameter (20 pc) and X-ray luminosity (2×10^{34} erg s^{-1} in 2-10 keV) of SN1006 for synchrotron X-ray SNRs expected in the Galactic plane, the deepness of the *ASCA* survey is $\sim$10 kpc for the synchrotron X-ray SNR survey, which is limited by the angular resolution. We assume that distribution of supernova remnants is the same as that of molecular clouds. In order to estimate coverage of the *ASCA* survey, we simply assume a finite disk model for the number density of SNRs η as;

$$\eta(r,z) = \begin{cases} \eta_0 \exp\left[-\frac{1}{2}\left(\frac{|z|}{Z_s}\right)^2\right] & (r \leqq R_d) \\ 0 & (r > R_d). \end{cases} \tag{3.1}$$

According to Sanders *et al.* (1984), the scale hight of CO (Zs) is $\sim$ 60 pc. We assume $R_d = 5$ kpc, at which radius the Galactocentric arms are located (Sanders *et al.* 1984). We found that 37% of the putative SNRs are located in the region covered by the *ASCA* survey ($|l| < 45°$, $|b| < 0.4°$, distance <10 kpc). However, 34% of the survey region suffers from stray light and could no be used for an extended-source search. Therefore, $1/0.37/(1-0.34) \sim$4 times the number of SNRs detected in the *ASCA* survey is expected in the Galactic plane. Since we discovered 14 candidates in the survey region and at least 2 sources, G28.6−0.1 and G32.45+0.1, are synchrotron X-ray SNRs, we expect 8$\sim$56 synchrotron X-ray SNRs in the Galactic plane.

References

Arbutina, B., & Urošević, D. 2005, MNRAS, 360, 76

Bamba, A., Ueno, M., Koyama, K., & Yamauchi, S. 2001, PASJ, 53, L21

Bamba, A., Ueno, M., Koyama, K., & Yamauchi, S. 2003, ApJ, 589, 253

Berezhko, E. G., Völk, H. J. 2004, A&A, 427, 525

Combi, J. A., Benaglia, P., Romero, G. E., & Sugizaki, M. 2005, A&A, 431, L9

Condon, J. J., Cotton, W. D., Greisen, E. W., Yin, Q. F., Perley, R. A., Taylor, G. B., & Broderick, J. J. 1998, AJ, 115, 1693

Enomoto, R., *et al.* 2002, Nature, 416, 823

Green, A. J., Cram, L. E., Large, M. I., & Ye, T. 1999, ApJS, 122, 207

Green, D. A. 2004, Bulletin of the Astronomical Society of India, 32, 335

Koyama, K., Petre, R., Gotthelf, E. V., Hwang, U., Matsuura, M., Ozaki, M., & Holt, S. S. 1995, Nature, 378, 255

Pfeffermann, E., & Aschenbach, B. 1996, Roentgenstrahlung from the Universe, 267

Sanders, D. B., Solomon, P. M., & Scoville, N. Z. 1984, ApJ, 276, 182

Sugizaki, M., Mitsuda, K., Kaneda, H., Matsuzaki, K., Yamauchi, S., & Koyama, K. 2001, ApJS, 134, 77

Ueno, M., Bamba, A., Koyama, K., & Ebisawa, K. 2003, ApJ, 588, 338

Yamaguchi, H., Ueno, M., Koyama, K., Bamba, A., & Yamauchi, S. 2004, PASJ, 56, 1059

Discussion

VINK: Wouldn't you think that the faint SNRs discovered by HESS are most likely to be TeV emitters because of Inverse Comptom emission from the same electrons that cause X-ray synchrotron emission, whereas pion decay is unlikely due to the low density?

UENO: I think that's reasonable.

Katsuji Koyama (right) with company before the Conference Dinner; next to him Masaru Ueno.

Populations of High Energy Sources in Galaxies
Proceedings IAU Symposium No. 230, 2005
E. J. A. Meurs & G. Fabbiano, eds.

© 2006 International Astronomical Union
doi:10.1017/S174392130600860X

Searching for continuum γ-ray emission from OB associations with INTEGRAL, some preliminary results

J.-C. Leyder[1,2] and G. Rauw[1]†

[1]Institut d'Astrophysique et Géophysique, 17 Allée du 6-Août, 4000 Liège, Belgium
email: leyder@astro.ulg.ac.be

[2]Integral Science Data Center, Versoix, Switzerland

Abstract. Recent studies suggested that there might be a correlation between unidentified γ-ray sources from the third EGRET catalogue and OB associations. Moreover, when extrapolating the fluxes measured by EGRET at energies above 100 MeV with a power-law down to the energy range of ISGRI, the expected count rates should be large enough to be detectable with INTEGRAL. Most of those OB associations being located in the Galactic plane, they are monitored by INTEGRAL as part of the Core Program during both the Galactic Plane Scans and the Galactic Center Deep Exposure. Combining public and CP data, we have performed a search for gamma-ray emission from OB associations and the first results are presented.

Keywords. gamma rays: observations, X-rays: stars, open clusters and associations: general.

1. OB Associations

OB associations harbour massive O and B-type stars that exhibit fast stellar winds ($v_\infty \simeq 2000$ km/s,) and large mass loss rates ($\dot{M} \simeq 10^{-5} M_\odot$/yr for O type stars).

The presence of massives stars implies that these associations are found in star forming regions such as the spiral arms of the Galaxy.

2. OB Associations in γ-rays

Recent studies on the possible association of unidentified EGRET sources with different types of galactic objects indicated a significant correlation of sources from the 3 EG catalogue with OB associations (Romero, Benaglia & Torres 1999). Most of these OB associations are located in the Galactic Plane and are therefore observed during the INTEGRAL Core Programme. Deriving (accurate) positions and fluxes of these sources with the IBIS instrument might allow us to identify their counterparts. Furthermore, it might also help us determine whether the observed continuum emission is coming: (a) from radio-quiet pulsars (as suggested by Romero *et al.* 1999); (b) from shock fronts created by the interactions of the stellar winds of massive stars with the ambient ISM (Manchanda *et al.* 1996); (c) from hydrodynamic shocks in the winds of individual stars (Chen & White 1991); or (d) from hydrodynamic shocks between the winds of massive binary systems (Eichler & Usov 1993, Mücke & Pohl 2001, Benaglia *et al.* 2001). The results of this study could thus help us to identify the so-far unknown counterparts of a large number of galactic γ-ray sources.

The sources listed by Romero *et al.* (1999) have photon indices near $\Gamma \sim 2$. Therefore the number of photons at a given energy is $n(E) = \alpha \times E^{-\Gamma}$. Using the EGRET fluxes

† Research Associate FNRS, Belgium.

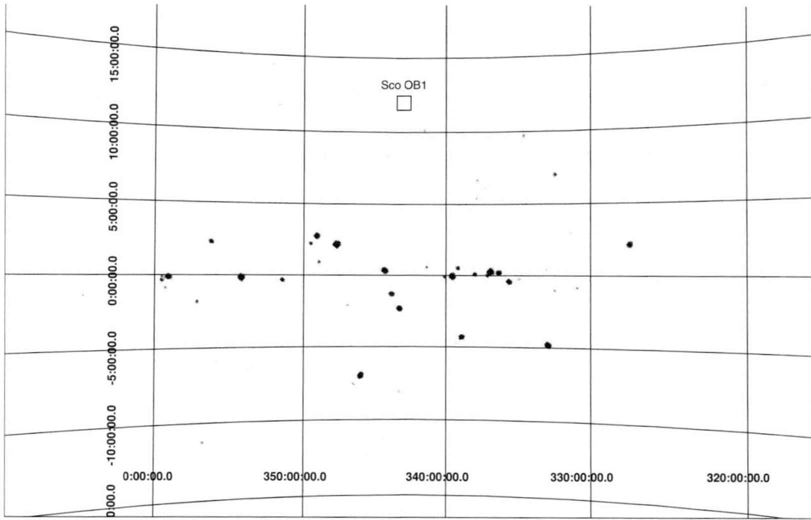

Figure 1. ISGRI significance map of the Sco OB 1 region in the 20 – 40 keV band, produced with OSA 5.0. The fainter sources detected on this image have fluxes of ∼4 mCrab.

(>100 MeV) given by Romero *et al.* and extrapolating them with a power-law spectrum to the IBIS energy range (15 keV – 10 MeV), we estimate for a typical OB association a continuum flux of $45\,10^{-9}\,\mathrm{ph\,cm^{-2}\,s^{-1}\,keV^{-1}}$ @ 1 MeV.

3. Analysis with INTEGRAL

Some preliminary results were first obtained with the Off-line Scientific Analysis (OSA) software, version 4.2. The analyses are being repeated with OSA 5.0 to take advantage of the better background reduction introduced with this new release.

The only instrument used up to now is ISGRI (Lebrun *et al.* 2003), a coded-mask imager offering a spatial resolution of 12′ over an energy range from approximately 23 to 1000 keV.

4. Preliminary results

The first tests carried out using OSA 4.2 did not permit to detect any signal for the sample of OB associations that were studied. The improvements in OSA 5.0, especially in the background correction, might help change this situation (see Figure 1).

References

Benaglia, P., Romero, G.E., Stevens, I.R., & Torres, D. F. 2001, *A&A* 366, 605
Chen, W. & White, R.L. 1991, *ApJ* 266, 512
Eichler, D. & Usov, V. 1993, *ApJ* 402, 271
Lebrun, F., Leray, J.P., Lavocat, P., *et al.* 2003, *A&A* 411, L141
Manchanda, R.K., Polcaro, V.F., Norci, L., Giovannelli, F., Brinkmann, W., Radecke, H.D., Manteiga, M., Persi, P., & Rossi, C. 1996, *A&A* 305, 457
Mücke, A. & Pohl, M. 2001, in: A.F.J. Moffat & N. St.-Louis (eds.), *Interacting Winds from Massive Stars*, ASP Conf. Series, vol. 260, p. 355
Romero, G.E., Benaglia, P., & Torres, D.F. 1999, *A&A* 348, 868

Populations of High Energy Sources in Galaxies
Proceedings IAU Symposium No. 230, 2005
E. J. A. Meurs & G. Fabbiano, eds.

© 2006 International Astronomical Union
doi:10.1017/S1743921306008611

High-Mass X-ray Binaries population in the Galaxy

A. A. Lutovinov[1,2], M. G. Revnivtsev[1,2], M. R. Gilfanov[1,2] and R. A. Sunyaev[1,2]

[1]Space Research Institute, Profsoyuznaya str. 84/32, Moscow 117997, Russia
email: lutovinov@hea.iki.rssi.ru

[2]Max-Planck Institut für Astrophysik, Karl-Schwarzschild Str. 1, D-85741 Garching, Germany

Abstract. We study high mass X-ray binaries (HMXBs) in the Galaxy using data of the IN-TEGRAL observatory. High sensitivity survey of the whole Galaxy with a possibility to detect absorbed sources significantly enlarged our sample of HMXBs in a comparison with the previous studies. Large fraction of detected high mass X-ray binaries is highly photoabsorbed. We investigated the HMXBs distribution along the Galactic plane and found their strong concentrations toward Galactic spiral arms, confirming previous results of Grimm *et al.* (2002) obtained using smaller sample of sources. We conclude that the mapping of Galactic HMXBs should be important tool to trace the star formation regions at the opposite side of the Galaxy.

Keywords. Galaxy: structure, X-rays: galaxies, X-rays: binaries.

1. Introduction

High- and low-mass binaries (LMXBs and HMXBs) harboring compact objects (neutron stars or black holes) are the brightest X-ray emitters in the Galaxy. At the moment more than 130 HMXBs and 200 LMXBs ever observed during last 35–40 years are known (Liu *et al.* 2000, 2001). Their positions in the Galaxy intimately connected with the origin of their optical star. The HMXBs being the younger X-ray population of the Galaxy (massive stars which are companion stars in HMXBs can not live longer than $\sim$10–100 Myr) should trace the star formation (SF) regions, while LMXBs (live time much more then $\sim$Gyr) should be more concentrated in the regions of high stellar mass density.

Using the RXTE/ASM survey of the sky (the 2–10 keV energy range) Grimm *et al.* (2002) found significant differences in the spatial distribution of HMXBs and LMXBs. The luminosity functions of LMXBs and HMXBs were found to be also quite different. Luminosity function of HMXBs essentially is a power law from $\sim10^{39}$ erg/s down to $\sim10^{33}$ erg/s where indications of some flattening were obtained by Shtykovskiy & Gilfanov (2005). The LMXB luminosity function is much flatter on lower luminosities end (Gilfanov 2004). It is interesting that the behavior of luminosity functions of LMXBs and HMXBs might be generally understood considering the mass trasfer mechanisms in these binary systems. It was shown that the shape of the HMXB luminosity function is governed by the properties of the massive star mass loss via stellar wind (Postnov 2003), while the luminosity function of low mass X-ray binaries (in which the Roche lobe overflow is the dominant way to supply the matter to the accretor) contains imprints of magnetic and gravitational braking of the binary system (Postnov & Kuranov 2005).

Below we describe some general properties of high and low mass X-ray binaries from a sample of sources obtained with latest INTEGRAL/IBIS survey of the Galaxy.

HMXB	LMXB
young population ($\times 10^7$ yr)	old population ($\times 10^9$ yr)
$M_c > 10 M_\odot$	$M_c \sim 1 M_\odot$
type: late $O - B[e]$ stars	type: $K - M$ stars
strong stellar wind	no stellar wind
intrinsic absorption	no intrinsic absorption
trace the SF regions	trace stellar mass
Vela X-1, 4U1700-37, GX301-2,	1E1740-294, GRS1915+105, X-ray
Cyg X-1, most X-ray pulsars	bursters and BHC

2. Inner part of the Galaxy

In the recent paper (Lutovinov *et al.* 2005) we focused on a sample of Galactic sources located in the Galactic plane in between the Norma and Sagittarius spiral arms (the inner part of the Galaxy). This Galactic region was selected because it had the best statistics of the available INTEGRAL data to that date. Using a $\sim$5 Msec exposure we constructed a flux limited sample of sources in this region and detected about 90 sources with a flux more than 1.5 mCrab in the 20–60 keV energy band. Most of them (49) were identified with LMXBs, but 23 were HMXBs. The usage of hard X-ray energy band (INTEGRAL/IBIS telescope) allowed one to reveal a considerable population of absorbed sources and significantly increase the number of known HMXBs in the inner part of the Galaxy. Most of detected HMXBs proved to be accreting X-ray pulsars with a strong intrinsic photoabsorption. As an example in Fig. 1 we present spectra of new heavily absorbed sources IGR J16318-4848 and IGR J16358-4726 obtained with INTEGRAL (>20 keV) and RXTE (3–20 keV) observatories.

3. Towards maping the whole Galaxy

Now (end of summer 2005) there is more than 24 Msec of publicly available data of INTEGRAL observations. These observations allow us to study large part of the sky including extragalactic (see Krivonos *et al.* 2005) and Galactic sources. At the moment more than 300 sources were detected in all data with a high statistical significance. About 60 from them are new sources discovered by INTEGRAL. X-ray binaries are concentrated towards the Galactic plane, however HMXBs and LMXBs have different vertical scale heights, reflecting the age of stellar companions of these sources: $\sim$150 pc for HMXBs

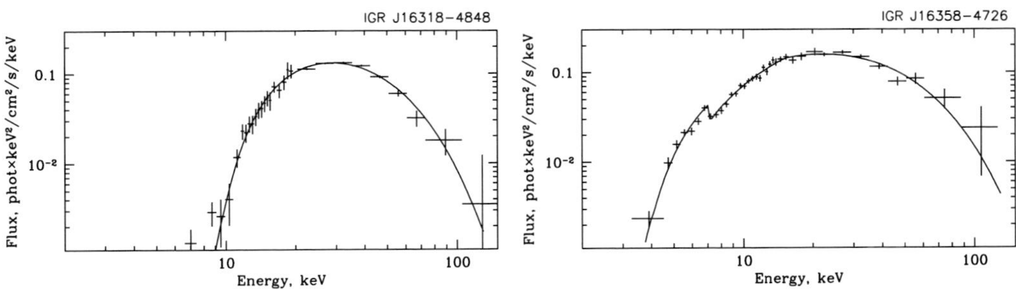

Figure 1. Broadband energy spectra of IGR J16318-4848 and IGR J16358-4726. The best-fit models are shown by solid lines.

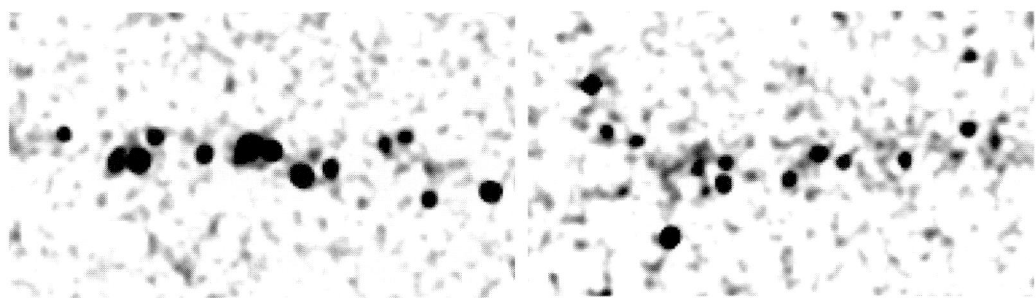

Figure 2. Maps of Norma (left) and Scutum-Sagittarius (right) spiral arm tangents obtained with IBIS/INTEGRAL.

and $\sim$400 pc for LMXBs (Grimm *et al.* 2002). At the Galactic Center distance from the Sun (assume 8.5 kpc) these scale heights correspond to angular scales of $\sim$1$°$and $\sim$2.7$°$, respectively. Below we present two current samples of sources with $|b| < 2°$ and $|b| < 5°$ from the Galactic plane.

	All sky
Total	300 sources
HMXB	49 sources
LMXB	68 sources
Galactic plane ($|b| < 5$, LMXB scale)	
Total	184 sources
HMXB	43 sources
LMXB	54 sources
Galactic plane ($|b| < 2$, HMXB scale)	
Total	115 sources
HMXB	36 sources
LMXB	33 sources

It is interesting to note that relative quantity of HMXBs and LMXBs changes considerably if we widen our selection region with the respect to the Galactic plane. Majority of absorbed sources, discovered by INTEGRAL, lies very close to the Galactic plane. On Fig. 2 one clearly sees a large number of sources near the Galactic plane and rapid drop of their surface density towards higher Galactic latitues.

Our Galaxy consists mostly of the disk and the bulge components (see e.g. Bahcall & Soneira 1980). In the Galactic disk there is a clear spiral structure which is believed to be a spiral density wave, initiating the intense star formation. Therefore we might naturally expect that increased number density of high mass X-ray binaries should be observed in spiral arms regions. In Fig. 3 we present distribution of surface density of LMXBs and HMXBs along the Galactic plane (note however, that the sample of sources considered here is not flux limited therefore some non-uniformity of exposure times should be imprinted in the observed sources distribution). It is obviously seen that there are concentrations of HMXB in the regions of tangents to the spiral arms, while it is much weaker in LMXB distribution.

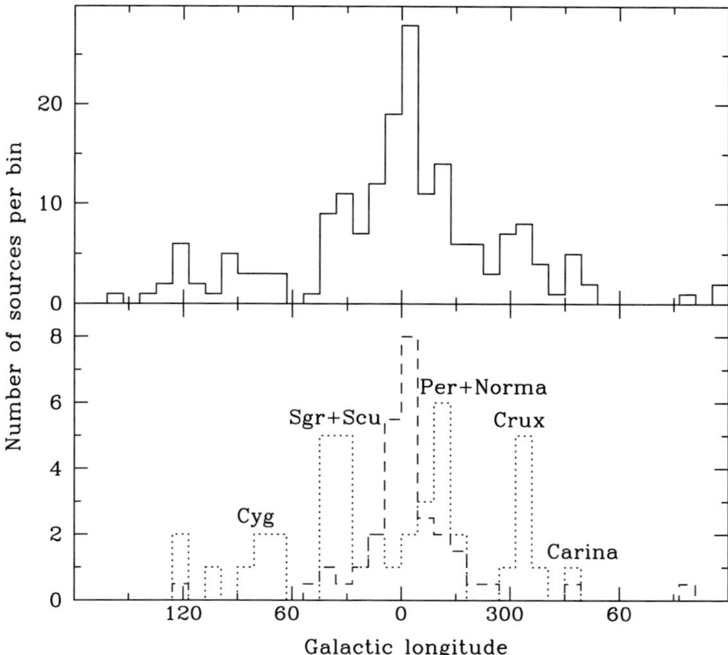

Figure 3. Angular distribution of all detected sources (upper panel), identified HMXBs and LMXBs (dotted and dashed lines in bottom panel, respectively). The number of LMXBs is divided by 2. The spiral arms are indicated by their names.

Note that the current sensitivity limit of the INTEGRAL/IBIS survey – $F_{\lim} \sim 10^{-11}$ erg/s/cm^2 in best places – corresponds to the detection sensitivity $L_{\lim} > 10^{35}$ erg/s till the end of the Galaxy. Therefore we might expect that with the INTEGRAL survey, which does not suffer from strong interstellar photoabsorption in the Galaxy, we should see all HMXB sources in the Galaxy brighter than the above limit. The position of newly discovered and will-be-discovered HMXBs should also fall into the region of intense star formation in the Galaxy. Consequently the mapping of HMXBs should be interesting tool to trace the star formation, especially at the opposite side of the Galaxy where most other methods to map the spiral structure fail (Vallee 1995).

Acknowledgements

We would like thank to E.Churazov for developing methods of analysis of the IBIS data and software. This work was partially supported by RFBR grant number 04-02-17276.

References

Bahcall, J. & Soneira, R. 1980, *ApJS* 44, 73
Gilfanov, M. 2004, *MNRAS* 349, 146
Grimm, H.-J., Gilfanov, & Sunyaev, R. 2002, *A&A* 391, 923
Krivonos, R., Revnivtsev, M., Churazov, E., & Sunyaev, R. 2005, *this issue*
Liu, Q., van Paradijs, J., & van den Heuvel, E. 2000, *A&AS* 147, 25
Liu, Q., van Paradijs, J., & van den Heuvel, E. 2001, *A&A* 368, 1021
Lutovinov, A., Revnivtsev, M., Gilfanov, M., *et al.* 2005, *A&A* in press, astro-ph/0411550
Postnov, K. 2003, *Astron. Lett.* 29, 372
Postnov, K. & Kuranov, A. 2005, *Astron. Lett.* 31, 7

Shtykovskiy, P. & Gilfanov, M. 2005, *A&A* 431, 597
Vallee, J. 1995, *ApJ* 454, 119

Discussion

ERACLEOUS: Can you tell from the data if the HMXBs are powered by accretion from the stellar wind of the companion or by Roche lobe overflow?

LUTOVINOV: It is possible that some are powered by Roche lobe overflow but the majority are accreting from a stellar wind.

GHOSH: Would you say that the discrepancy between the spatial distribution of HMXBs and the positions of the spiral arms is already statistically significant, or that more work needs to be done?

LUTOVINOV: Really, the significance of this discrepancy is not enough. We obtained at the moment only some indications for the possible displacement between HMXBs and spiral arms. More observations are needed to make final conclusions.

LIPUNOV: One remark. Most of X-Ray massive binary systems belong to MS + Be systems. Due to selection effect we see only 1% percent of them. The real luminosity function can be determined after X-ray observations of 100's of galaxies.

LUTOVINOV: Yes, most of high mass X-ray binary systems include Be-stars. But they usually have quite small duty cycles, therefore at any given moment the probability to see the bright Be-system is low. This is confirmed by our sample, that doesn't contain any known Be-system.

Populations of High Energy Sources in Galaxies
Proceedings IAU Symposium No. 230, 2005
E. J. A. Meurs & G. Fabbiano, eds.

© 2006 International Astronomical Union
doi:10.1017/S1743921306008623

IGR J16393−4643: a new heavily-obscured X-ray pulsar

A. Bodaghee[1,2]†, R. Walter[1,2], J. A. Zurita Heras[1,2], A. J. Bird[3], T. J.-L. Courvoisier[1,2], A. Malizia[4], R. Terrier[5,6], and P. Ubertini[7]

[1]INTEGRAL Science Data Centre, Chemin d'Ecogia 16, CH–1290 Versoix, Switzerland
email: Arash.Bodaghee@obs.unige.ch

[2]Observatoire de Genève, Chemin des Maillettes 51, CH–1290 Sauverny, Switzerland
[3]School of Physics and Astronomy, University of Southampton, Highfield, SO17 1BJ, UK
[4]IASF/CNR, Via Piero Gobetti 101, I–40129 Bologna, Italy
[5]CEA-Saclay, DAPNIA/Service d'Astrophysique, F–91191 Gif-sur-Yvette, France
[6]Fédération de recherche APC, Collège de France 11, place Marcelin Berthelot,
F–75231 Paris, France
[7]IASF/CNR, Via Fosso del Cavaliere 100, I–00133 Roma, Italy

Abstract. An analysis of the high-energy emission from IGR J16393−4643 is presented using data from *INTEGRAL* and *XMM-Newton*. The source is persistent in the 20–40 keV band at an average flux of 5.1×10^{-11} ergs cm^{-2} s^{-1}, with variations in intensity by at least an order of magnitude. A pulse period of 912 ± 3 s was discovered in the *ISGRI* and *EPIC* light curves. The source spectrum is a strongly-absorbed ($N_{\mathrm{H}} = (2.5 \pm 0.2) \times 10^{23}$ cm^{-2}) power law that features a high-energy cutoff above 15 keV. Two iron emission lines at 6.4 and 7.1 keV, an iron absorption edge $\gtrsim 7.1$ keV, and a soft excess emission of 7×10^{-15} ergs cm^{-2} s^{-1} between 0.5–2 keV, are detected in the *EPIC* spectrum. The shape of the spectrum does not change with the pulse. Its persistence, pulsation, and spectrum place IGR J16393−4643 among the class of heavily-absorbed HMXBs. The improved position from *EPIC* is R.A. (J2000)= $16^{\mathrm{h}}39^{\mathrm{m}}05.4^{\mathrm{s}}$ and Dec.= $-46°42'12''$ ($4''$ uncertainty) which is compatible with that of 2MASS J16390535−4642137.

Keywords. X-rays: binaries, X-rays: individual: IGR J16393−4643=AX J163904−4642.

1. Observations & Analysis

The *INTEGRAL* data consist of roughly 1500 core program and public pointings between revolutions 30–260. Version 4.2 of the *INTEGRAL* OSA software was used to reduce raw data into images. Intensity, significance, variance, and exposure mosaic images in various energy bands between 20 and 60 keV were constructed. From these mosaics, we extracted a spectrum and a source location of R.A. (J2000)= $16^{\mathrm{h}}39^{\mathrm{m}}05^{\mathrm{s}}$ and Dec.= $-46°42.3'$ ($26''$ uncertainty) which agrees with the *ISGRI* position of Bird *et al.* (2004) while improving it. The mean flux (20–60 keV) of the source is 0.73 ± 0.02 counts per second, or 4.9 mCrab, at a significance of 36σ.

XMM-Newton observed IGR J16393−4643 on March 21, 2004, for an effective exposure time of ~ 8 ks. We used the SAS v. 6.1.0 software to analyze the data and to extract the *EPIC* spectra. The refined X-ray position averaged from *MOS* and *PN* is R.A. (J2000)= $16^{\mathrm{h}}39^{\mathrm{m}}05.4^{\mathrm{s}}$ and Dec.= $-46°42'12''$ ($4''$ uncertainty). This position does not coincide with the counterparts proposed by Combi *et al.* (2004), but is about 1' away from the infrared source 2MASS J16390535−4642137.

A pulsation of 912 ± 3 s was detected in both the *ISGRI* light curve of revolution 37 (an epoch during which the source was particularly bright) and in the *PN* data (Fig. 1).

† Present address: ISDC, Chemin d'Ecogia 16, CH–1290 Versoix, Switzerland.

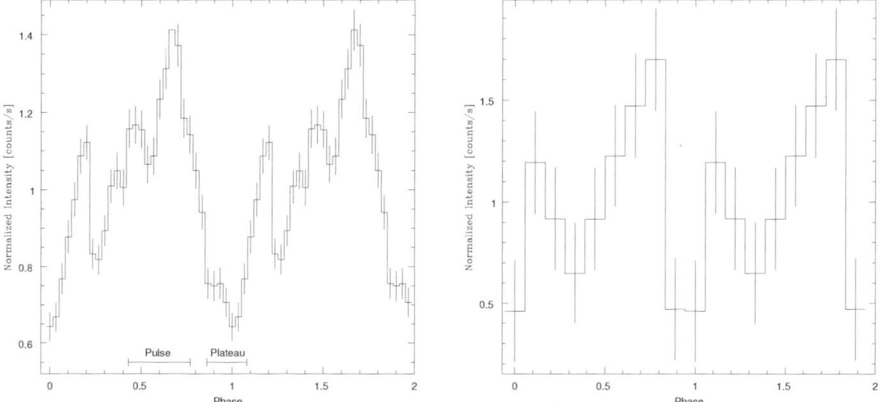

Figure 1. Light curves folded at 912 s for *PN* 2–10 keV (*left*) and *ISGRI* 15–40 keV (*right*).

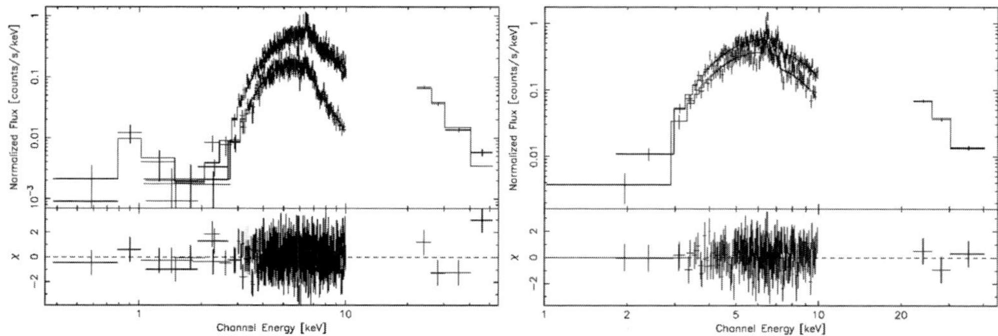

Figure 2. Average *PN*, *MOS*1–2, and *ISGRI* spectra (*left*), and the phase-resolved *PN* spectra constrained with the *ISGRI* spectrum (*right*).

The source spectrum can be fit with an absorbed power law plus an energy cutoff ($\Gamma_1 \sim 1$, $\Gamma_2 \sim 4$, $E_{\rm cut} > 15\,{\rm keV}$, $\chi^2_\nu = 0.96$) or with a Comptonized model ($kT_{\rm e} \sim 4.4\,{\rm keV}$, $\tau \sim 9$, $\chi^2_\nu = 0.95$, see Fig. 2 *left*). The spectrum features a large hydrogen column density of $N_{\rm H} \sim 25\,10^{22}\,{\rm cm}^{-2}$ which indicates that the absorption is intrinsic since it is an order of magnitude more than the galactic absorption. Iron lines at 6.4 keV (Fe Kα) and 7.1 keV (Fe Kβ) are detected, as is a soft excess emission of $7 \times 10^{-15}\,{\rm ergs\,cm}^{-2}\,{\rm s}^{-1}$ which requires a blackbody component in the model. The absorbed, integrated fluxes (in units of $10^{-11}\,{\rm ergs\,cm}^{-2}\,{\rm s}^{-1}$) are 4.4 in the 2–10 keV band, and 5.1 in the 20–60 keV band. The pulsation affects the normalizations but does not modify the shape of the spectra nor its parameters, specifically the $N_{\rm H}$, in any appreciable way (Fig. 2 *right*).

Details of the analysis can be found in Bodaghee *et al.*, 2005 (submitted to A&A).

References

Bird, A.J., Barlow, E.J., Bassani, L., *et al.*, 2004, *ApJ* 607, 33

Combi, J.A., Ribo, M., Mirabel, I.F., *et al.*, 2004, *A&A* 422, 1031

Cutri, R.M., Skrutskie, M.F., van Dyk, S., *et al.*, 2003, *2MASS All-Sky Catalog of Point Sources*

Sugizaki, M., Mitsuda, K., Kaneda, H., *et al.*, 2001, *ApJS* 134, 77

Walter, R., Rodriguez, J., Foschini, L., *et al.*, 2003, *A&A* 411, 427

Walter, R., Courvoisier, T.J.-L, Foschini, L., *et al.*, 2004, *in Exploring the Gamma-ray Universe, Proc. 4th INTEGRAL workshop*, ESA SP-459

Populations of High Energy Sources in Galaxies
Proceedings IAU Symposium No. 230, 2005
E. J. A. Meurs & G. Fabbiano, eds.

© 2006 International Astronomical Union
doi:10.1017/S1743921306008635

IGR J17252−3616: an accreting pulsar observed by *INTEGRAL* and *XMM-Newton*

J. A. Zurita Heras[1,2]†, G. de Cesare[3], R. Walter[1,2], A. Bodaghee[1,2], G. Bélanger[4], T. J.-L. Courvoisier[1,2], S. E. Shaw[5,1] and J. B. Stephen[6]

[1]INTEGRAL Science Data Centre, Versoix, Switzerland
email: Juan.Zurita@obs.unige.ch

[2]Observatoire de Genève, Sauverny, Switzerland
[3]IASF-INAF, Via Fosso del Cavaliere 100, 00133 Roma, Italy
[4]Service d'Astrophysique, DAPNIA/DSM/CEA, 91191 Gif-sur-Yvette, France
[5]School of Physics and Astronomy, University of Southampton, Highfield, SO171BJ, UK
[6]IASF/CNR, Via Piero Gobetti 101, 40129 Bologna, Italy

Abstract. IGR J17252−3616 is the hard X-ray counterpart of EXO 1722−363. The regular monitoring by *INTEGRAL* shows that IGR J17252−3616 is a persistent source with an average count rate of ∼6.4 mCrab in the 20–60 keV energy band. A follow-up observation with *XMM-Newton* showed that the source is located at R.A. (2000.0) $= 17^h 25^m 11.4^s$ and Dec. $= -36°16'58.6''$ with an uncertainty of $4''$.

The source is a binary X-ray pulsar with a spin period of 413.7 s. The spectral shape is typical for an accreting pulsar except that a huge intrinsic absorption and a cold iron fluorescence line are detected. The absorbing column density and cold iron line do not vary with the pulse period. The observations suggest that the source is a wind-fed accreting pulsar accompanied by a supergiant star.

Keywords. X-rays: binaries, X-rays: individual: IGR J17252−3616 = EXO 1722−363.

1. Introduction

EXO 1722−363 was discovered by EXOSAT in June 1984 (Warwick, Norton, Turner, *et al.* 1988). From Ginga observations in 1987 and 1988, Tawara, Yamauchi, Awaki, *et al.* (1989) and Takeuchi, Koyama & Warwick (1990) detected a pulsation of 413.9 s, important variations of the intensity in X-rays, a hard spectrum with important low-energy absorption and an emission line at 6.2 keV. Corbet, Markwardt & Swank (2005) resolved the orbital period of 9.741 days and detected a varying high column density with RXTE data. These investigations conclude that the system is a high mass X-ray binary (HMXB).

2. Observations & Analysis

INTEGRAL is a hard X-ray and γ-ray observatory of the European Space Agency (ESA). A total exposure of 6.5 Ms was accumulated between MJD 52671 and 53294. A follow-up observation with *XMM-Newton* was performed on March 21, 2004, for three hours (MJD 53085.542–53085.667).

We focused our work on *INTEGRAL* IBIS/ISGRI and *XMM-Newton* EPIC instruments. The data were reduced with OSA 4.2 and SAS 6.1.0. Images, light curves and spectra were generated. For ISGRI, a detailed timing and spectral analysis was performed

† Present address: ISDC, ch. d'Ecogia 16, 1290 Versoix, Switzerland.

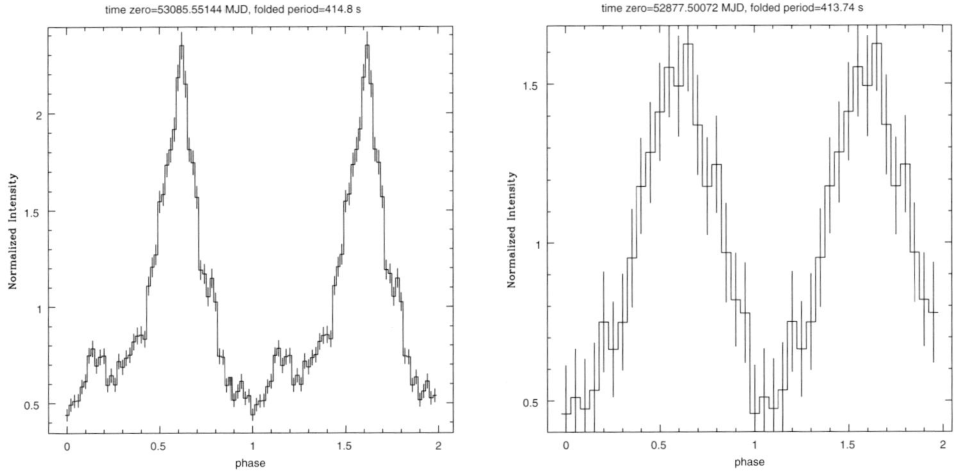

Figure 1. Folded light curves. *left:* pn, 0.5–10 keV; *right:* ISGRI, 20–60 keV.

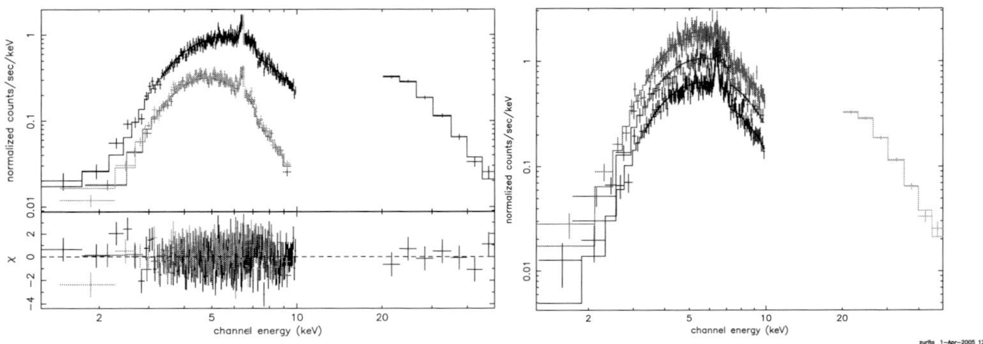

Figure 2. Combined EPIC+ISGRI spectra. The ISGRI spectrum is in the hard X-ray band. *left:* Average pn and MOS[12] spectra; *right:* Phase-resolved spectra for pn.

on revolution 106 (MJD 52877.4–52880.4) where the source reached its maximum flux. A pulsation has been detected in both pn and ISGRI data of 415 ± 5 s and 413.7 ± 0.4 s, respectively (see Fig. 1). The spectrum can be fitted with a flat power law plus an energy cutoff ($\Gamma \sim 0.02, E_c \sim 8.2$ keV, χ^2/d.o.f.=401/376) or with a comptonized model ($kT_e \sim 5.5$ keV, $\tau \sim 7.8$, χ^2/d.o.f.=401/376, see Fig. 2 *left*). The spectrum also indicates a large hydrogen column density of $N_H \sim 15\,10^{22}$ atoms cm^{-2} suggesting an intrinsic absorption. The Fe Kα line at 6.4 keV is clearly detected. Phase-resolved spectroscopy does not show any variation in the continuum except the total emitted flux (see Fig. 2 *right*). The absorption is constant along the pulse phase.

The observed features of IGR J17252−3616 clearly indicate that it is the hard X-ray counterpart of EXO 1722−363 even if the positions are not compatible.

References

Corbet, R. H. D., Markwardt, C. B., & Swank, J. H. 2005, *ApJ* in press

Takeuchi, Y., Koyama, K., & Warwick, R. S. 1990, *PASJ* 42, 287

Tawara, Y., Yamauchi, S., Awaki, H., Kii, T., Koyama, K., & Nagase, F. 1989, *PASJ* 41, 473

Warwick, R. S., Norton, A. J., Turner, M. J. L., Watson, M. G. & Willingale, R. 1988, *MNRAS* 232, 551

Populations of High Energy Sources in Galaxies
Proceedings IAU Symposium No. 230, 2005
E. J.A. Meurs & G. Fabbiano, eds.

© 2006 International Astronomical Union
doi:10.1017/S1743921306008647

IGR J16320−4751 as seen by simultaneous INTEGRAL and XMM observations

J. Rodriguez[1,2]†

[1]CE Saclay, DSM/DAPNIA/SAp (CNRS UMR AIM),
F-91191 Gif Sur Yvette Cedex, France

[2]ISDC, 16 Chemin d'Ecogia, CH-1290 Versoix, Switzerland

Abstract. We present the preliminary results of simultaneous XMM-Newton and INTEGRAL observations of the highly absorbed INTEGRAL source IGR J16320-4751. We refine the X-ray position with XMM-Newton, and then examine the spectral properties of the source using both satellites, separating two periods visible in the lightcurves, an initial flare and a more steady period. We show that the source spectrum and its behaviour are compatible with IGR J16320-4751 being a pulsar accreting from a high mass companion.

Keywords. Neutron stars, accretion, X-ray: observations, stars:individual IGR J16320−4751.

1. Introduction

IGR J16320-4751 was discovered during an INTEGRAL observation of 4U 1630-47 (Tomsick *et al.* 2003). Analyses of archival data showed IGR J16320-4751 is the hard X-ray counterpart to AX J16319-4752. A public ToO with XMM led to a first (arsec accuracy) X-ray position, leading to the identification of 2 infrared counterpart candidates (Rodriguez *et al.* 2003). Recently re-analysis of XMM+ASCA data led to discovery of 1300s pulsation (Lutovinov *et al.* 2005), indicating IGR J16320-4751 is probably a pulsar. We present here the first results of the analysis of strictly simultaneous XMM and INTEGRAL observations of this enigmatic source. Deep analysis of these data will be presented in Rodriguez *et al.* (2005, submitted to MNRAS).

2. Refining the position using old and new data

We produced PN images from our last observation (August 2004), and MOS images from the 2003 XMM ToO. Using ep_detect chain we can refine the X-ray position of the source to RA = 16^h 32^m 01.9^s, DEC = $-47°$ $52'$ $26.9''$ $(+/-3'')$. This new position supports more strongly the association of IGR J16320-4751 with the northern source in Fig. 1 of Rodriguez *et al.* (2003).

3. Spectral analysis

Two distinct "epochs" have to be separated for spectral analysis, "flare" and "non-flare", as shown in Fig. 2. We then fitted the spectra with simple model of (absorbed) power law and high energy cut-off plus Gaussian and iron edge. The data are well represented, but a soft excess is seen in the non-flare spectrum. The best fit parameters are reported in Fig. 2 for each of the two periods. In particular the main difference between the flare and non flare period seems related to variations of the absorption column

† On behalf of a larger collaboration.

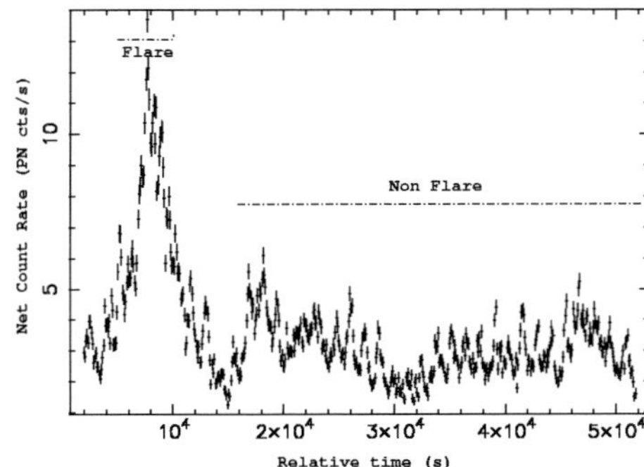

Figure 1. XMM 2–12 keV lightcurve showing the periods of flare and non-flare.

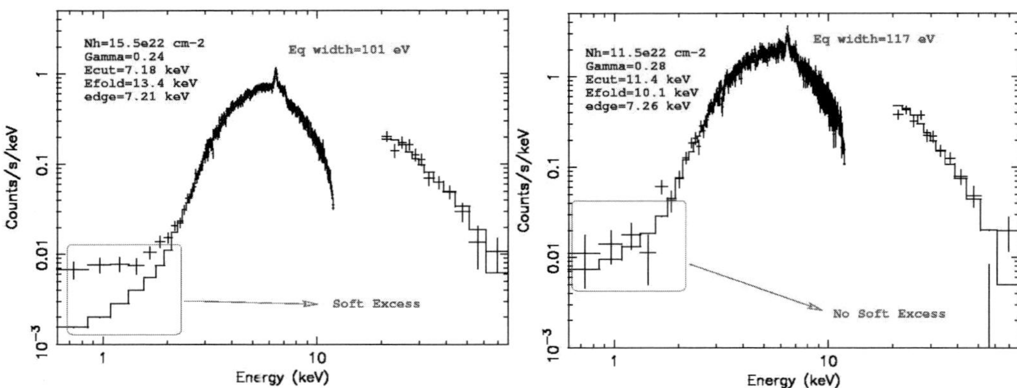

Figure 2. Left: Non-flare XMM/PN INTEGRAL/ISGRI spectrum. **Right:** Flare XMM/PN INTEGRAL/ISGRI spectrum. In both case the best model is superimposed as a line, while the main spectral parameters are reported in the panels.

density, although some evolution of the cut off energy and folding of the cut off is also visible.

4. Conclusion

The timing and spectral analysis reported here, although quite preliminsry, are strongly indicative of a pulsar primary given the presence of X-ray pulsations. In addition the spectral parameter of IGR J16320$-$4751 are clearly similar to those usually reported for systems containing a high mass star, therefore further suggesting it is a high mass X-ray binary (Rodriguez *et al.* 2003; Lutovinov *et al.* 2005).

Acknowledgements

I would like to acknowledge the great help of A. Bodaghee, M. Cadolle-Bel, A. Paizis & J. Zurita for various aspects of the work presented here.

References

Lutovinov, A., Rodriguez, J., Revnivtsev, M., & Shtykovskiy, P. 2005, *A&A*, 433, L41
Rodriguez, J. & Tosmick, J.A., Foschini, L., *et al.* 2003, *A&A*, 407, L41
Rodriguez, J., *et al.* 2005, submitted to *MNRAS*
Tomsick, J.A., Lingenfelter, R. Walter, R., Rodriguez, J., Goldwurm, A., Corbel, S., & Kaaret, P. 2003, *IAUC* 8076

Populations of High Energy Sources in Galaxies
Proceedings IAU Symposium No. 230, 2005
E. J. A. Meurs & G. Fabbiano, eds.

© 2006 International Astronomical Union
doi:10.1017/S1743921306008659

INTEGRAL discovers a population of persistent intrinsically absorbed super-giant High-Mass X-Ray Binaries

R. Walter[1]†, J. Zurita[1], L. Bassani[2], A. Bazzano[3], A. Bodaghee[1],
A. Dean[4], P. Dubath[1], A. Parmar[5], M. Renaud[6] and P. Ubertini[3]

[1]INTEGRAL Science Data Centre, Chemin d'Ecogia 16, 1290 Versoix, Switzerland
[2]IASF/CNR, Via Piero Gobetti 101, 40129 Bologna, Italy
[3]IASF-INAF, Via Fosso del Cavaliere 100, 00133 Roma, Italy
[4]School of Physics and Astronomy, University of Southampton, Highfield, S0171BJ, UK
[5]ESTEC, Postbus 299, NL-2200 AG Noordwijk, The Netherlands
[6]Service d'astrophysique, DAPNIA/DSM/CEA, 91191 Gif sur Yvette, France

Abstract. During the first year in operation, INTEGRAL has detected more than 28 new bright sources which emit the bulk of their emission above 10 keV. Follow-up observations of a subset of these sources in the X-ray band with XMM-Newton indicate that 80% of them are very strongly absorbed. More than half of these absorbed sources show strong pulsations with long periods ranging from 139 to 1300s, i.e., they are slow X-ray pulsars. Many of these new sources are super-giant high-mass X-ray binaries (HMXB) in which the stellar wind of the companion star is accreted onto the compact object. The large local absorption in these new sources can be understood if the compact objects are buried deep in their stellar winds. These new objects represent half of the population of active super-giant HMXB.

Keywords. X-rays: binaries, gamma rays: observations, Galaxy: stellar content,stars: mass loss,stars: neutron,stars: binaries: close.

---

1. A population of absorbed super-giant HMXB

The INTEGRAL survey of the galactic plane and central regions has revealed the presence of few hundred sources in the energy range 20–100 keV. Twenty-five percent of the objects in the first INTEGRAL soft γ-ray Galactic plane survey catalog had no known counterparts and so their nature was unclear. The majority of those sources were detected in the Galactic bulge and Norma arm tangent region.

XMM-Newton follow-up observations of the first source discovered by INTEGRAL, IGR J16318−4848, revealed a spectrum with intense fluorescence lines (up to keV equivalent width) of Fe Kα and Kβ and Ni Kα as well as strong ($2 \ 10^{24}$ cm^{-2}) low-energy absorption which makes the source extremely difficult to observe at energies below 4 keV. The strong absorption and unusual spectral properties led to the suggestion that IGR 16318−4848 is the first representative of a group of highly absorbed galactic binaries.

Here, we present the results of INTEGRAL, XMM-Newton and infrared observations of 10 among the new sources detected by INTEGRAL in the Galactic plane during the first year of the mission. Among those 10 sources we found.

- 7 intrinsically absorbed and persistent X-ray emitters. Five of them are pulsars featuring long (139–1300 sec) spin periods. Orbital periods (<10 days) are known for two objects in which eclipses could be detected. Five sources have early type stellar companions i.e. they are HMXB. The long spin periods and short orbital periods indicate

† email: Roland.Walter@obs.unige.ch

that the sources are very likely super-giant systems. Since the X-ray properties of all sources are very similar all those sources are likely persistent super-giant HMXB systems.

- 2 intrinsically absorbed transient super-giant HMXB.
- 1 unabsorbed persistent LMXB.

Detailed studies of the other new INTEGRAL sources are needed to confirm that a significant fraction of them are persistent absorbed super-giant systems. In addition, INTEGRAL has also discovered several new transient super-giant systems that should be slightly different objects. In any case, INTEGRAL has already almost doubled the number of active super-giant HMXB known in the Galaxy.

2. Source geometry

The observed spectra provide several clues about the source geometry:

- The continuum spectral shape is best characterized by a transmission geometry rather than by reflection. Fe line with keV apparent equivalent width are also not compatible with reflection. This indicates that the compact sources are embedded in dense material.
- The absence of reddening in excess to the galactic value in the infrared counterparts indicates that the absorption is local to the sources.
- The centroids of the Fe lines indicate low Iron ionization i.e. that the region where fluorescence takes place is larger than 10^7 km. The absence of modulation of the Fe line on the spin period indicate that the fluorescence region is larger than 10^8 km. The fluorescence/absorbing region is not associated to the accretion column nor with an accretion disk, its size is comparable to or larger than the orbital radius.
- The relation between the fluorescence line flux and the absorbing column density is as expected for a spherical geometry. The fluorescence region is not extremely patchy.
- Variable absorbing column densities or Iron line fluxes observed in few objects indicate that most of the absorbing/fluorescence region lies within 10^{8-9} km (IGR J16318-4848, IGR J16320-4751) or within a few orbital radii (IGR J17252-3616).

The source persistency and the strong absorption could be understood if the compact objects orbit continuously in a dense component of the stellar wind. Short orbital periods are expected and observed. High column densities could be related with slow dense stellar winds in the orbital plane. The lack of wind acceleration could be related to the photoionization of the X-ray sources.

3. Conclusion

INTEGRAL has unveiled a population of super-giant HMXB of long spin periods and short orbital periods in which the compact objects are orbiting within a dense component of the super-giant stellar wind. Such systems are characterized by persistent (but variable) hard X-ray emission.

References

Bodaghee, *et al.*, 2005, A&A submitted
Hill, *et al.*, A&A, A&A in press
Rodriguez, *et al.*, 2005, MNRAS submitted
Walter, *et al.*, 2003, A&A 411, 427
Walter, *et al.*, 2004, ESA-SP 552, 417
Walter, *et al.*, 2005, A&A submitted
Zurita, *et al.*, 2005, A&A submitted

Populations of High Energy Sources in Galaxies
Proceedings IAU Symposium No. 230, 2005
E. J. A. Meurs & G. Fabbiano, eds.

© 2006 International Astronomical Union
doi:10.1017/S1743921306008660

X-ray Luminosity Functions and Star Formation Rate

H.-J. Grimm[1]† M. Gilfanov[2] and R. Sunyaev[2]

[1]Harvard-Smithsonian Center for Astrophysics, 60 Garden Street, Cambridge, MA, USA
email: hgrimm@head.cfa.harvard.edu

[2]Max-Planck-Institut für Astrophysik, Karl-Schwarzschild-Strasse 1, 85471 Garching, Germany

Abstract. I will discuss the connection between the luminosity function of a population of high-mass X-ray binaries in a galaxy and the star formation rate in the this galaxy. The understanding of this connection provides on the one hand an independent measure of an important galaxy property, and on the other hand new insights into populations of high-mass X-ray binaries. In particular, observations with the Chandra X-ray telescope are uniquely suited to investigate the X-ray part of this connection and I will present examples of this.

Keywords. X-rays:binaries, X-rays:galaxies, galaxies:fundamental parameters.

1. Introduction

X-ray binaries can be separated into two different classes according to the mass of the secondary star. Low-mass X-ray binaries (LMXBs) contain black holes or neutron stars as the primary object, and a star of less than 2.5 $M_\odot$ as secondary. High-mass X-ray binaries (HMXBs) on the other hand have a secondary of more than 2.5 $M_\odot$. In fact, in the Milky Way most secondaries in LMXBs have masses less than 1 $M_\odot$, and most HMXBs have secondaries with more than 10 $M_\odot$. Only few objects are known to have secondaries in the gap between 1 $M_\odot$ and 10 $M_\odot$.

The difference in secondary masses results in two important physical distinctions between LMXBs and HMXBs. The first of these concerns the lifetime of the system. Lifetimes of main sequence stars are roughly proportional to M_*^{-3}. Thus, the massive stars in HMXBs have shorter lifetimes than the stars in LMXBs by a factor of ~ 1000 or more, i.e. 10^5-few times 10^7 years. Due to the short lifetime HMXBs are a good estimator of recent star formation rate. Moreover, the location of HMXBs is close to star formation regions because they live only shortly and also have smaller systemic velocities.

The second distinction is with regard to the mass loss (and thus accretion) mechanism of the secondary star. The stars in LMXBs lose mass to the primary exclusively through Roche lobe overflow. The mass is lost through the inner Lagrange point of the system and forms an accretion disk around the primary object. In HMXBs this mechanism exists as well. But because massive stars have strong stellar winds, the primary can also accrete mass from just the wind without the secondary filling its Roche lobe.

To investigate the relation between high-mass X-ray binaries and star formation rate we need to understand star formation measures. Star formation is measured by various different indicators in different spectral bands. The bands range from radio to UV, the most common being radio flux at 1.4 GHz, far-infrared flux, UV flux, and H_α. Despite the use of quite different energy bands almost all indicators measure the ionizing photon flux from massive stars ($M > 8 M_\odot$) in star forming regions. H_α and UV measure this

† Present address: Harvard-Smithsonian Center for Astrophysics, 60 Garden Street, Cambridge, MA, USA.

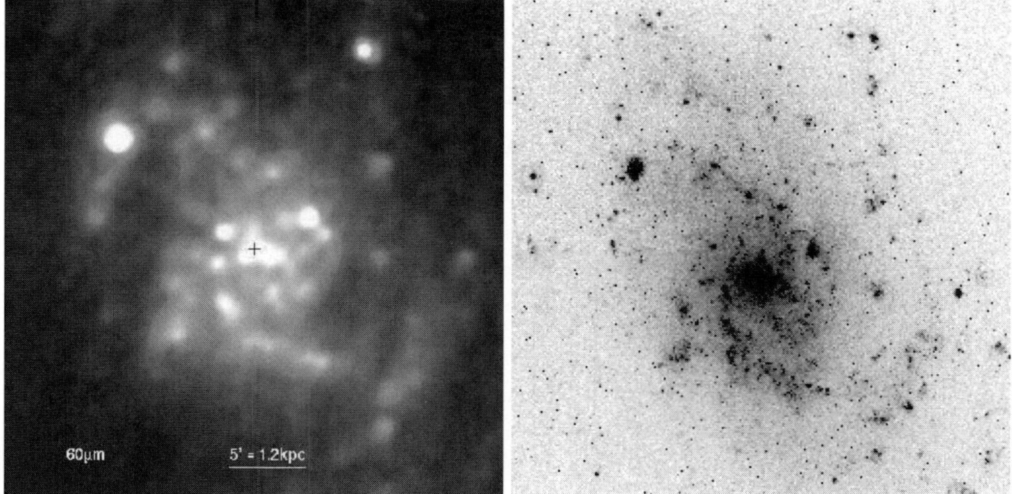

Figure 1. The panels show the galaxy M33 in different spectral bands, 60 micron far-infrared (left) and H_α (right). Both spectral bands are used to measure star formation rate. Despite the different bands (FIR versus UV), these images show the same structure. Both SFR indicators measure the ionizing photon flux from massive stars ($M > 8M_\odot$) in star forming regions. H_α measures this flux directly while FIR measures the reprocessed ionizing flux from dust. Pictures taken from the NED database. Sources: Hippelein *et al.* (2003) and Cheng *et al.* (1997).

flux directly while FIR measures the reprocessed ionizing flux from dust. The radio flux measures the thermal emission from HII star formation regions.

The main problem is the conversion from the measured fluxes to actual star formation rates. In measuring the thermal radio flux it is difficult to disentangle the thermal from non-thermal emission. Other indicators suffer from uncertainties in the escape fraction of ionizing photons, dust absorption, and contamination.

2. Luminosity functions

X-ray binary luminosity functions have become a diagnostic for populations with the advent of XMM-Newton and particularly Chandra. Observations of other galaxies have the advantage of uniform coverage and known distance for all sources. On the other hand only in the Milky Way, and the Magellanic Clouds, are unique optical counterparts available for identification. Also only in the Milky Way sources with luminosities below 10^{36} erg s^{-1} are observable. The difficulty with the X-ray binary population of the Milky Way is that the construction of the luminosity function requires knowledge of the mass distribution of the Milky Way to correct for the non-uniform X-ray luminosity coverage (Grimm *et al.* 2002). In other galaxies it is in general not possible to distinguish between LMXBs and HMXBs.

To investigate the connection between HMXBs and SFR we thus selected galaxies whose X-ray binary population is dominated by HMXBs, i.e. star formation (Grimm *et al.* 2003). This selection is based on the assumption that HMXBs are related to star formation rate whereas LMXBs are related to the mass of a galaxy. Galaxies that have a sufficiently high value of the ratio star formation rate versus galaxy mass were selected. All the galaxies are of late Hubble type. The luminosity functions cover a range of 4 orders of magnitude in luminosity and a factor of $\sim$40 in star formation rate. The observed luminosity functions are shown in the left panel of Fig. 2.

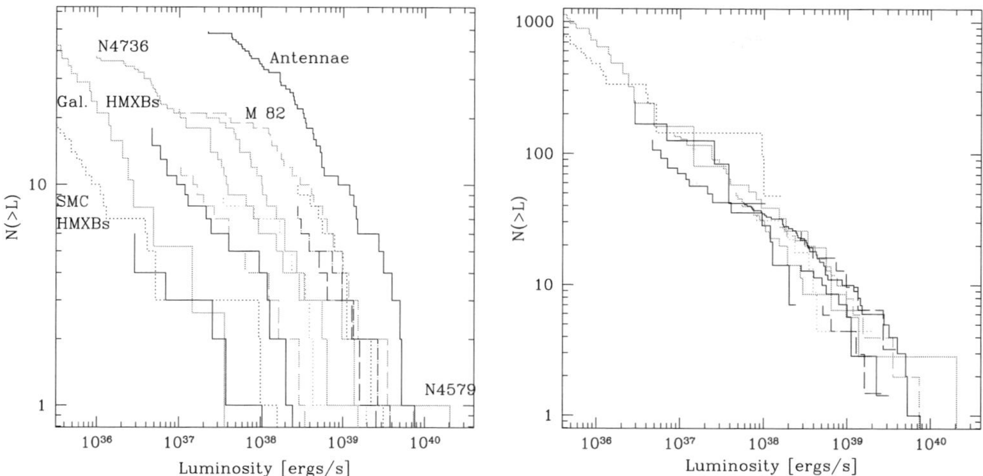

Figure 2. *Left:* Luminosity functions of various actively star forming galaxies observed with Chandra and of HMXBs in the Milky Way. Only very nearby galaxies have unique optical counterparts. The Milky Way luminosity function is dependent on knowledge of the mass distribution of the Milky Way. Note the wide spread in normalization of the luminosity functions observed by Chandra. *Right:* Chandra luminosity functions scaled by the ratio of their star formation rates to the star formation rate of the Antennae galaxies.

If one scales the luminosity functions by their respective star formation rates the large spread in normalizations disappears. The scaled luminosity functions fall into a relatively narrow range in the number–luminosity plane. Although there is still some spread at present this can be easily accounted for by uncertainties in the SFR measurement other measurement uncertainties. It is therefore possible to combine the scaled luminosity functions into a single "universal" luminosity function for HMXBs.

The universal luminosity function has the shape of a simple power law. There is an indication for a cutoff at $2 \cdot 10^{40}$ erg s^{-1}. From the luminosity function the evidence for a cutoff is not significant. However, there is independent evidence for the existence of a cutoff that will be presented below. The universal luminosity function is given by

$$\frac{dN}{dL_{38}} = 3.3 \cdot SFR \cdot L_{38}^{-1.61} \text{, if } L_{38} < 200 \tag{2.1}$$

with $L_{38} = 10^{38} L$ erg s^{-1}.

Surprisingly there are no strong deviations from this power law over the 5 orders of magnitude in luminosity where it has been determined. Although there are considerable uncertainties, especially with star formation rate measurements, the simple shape is unexpected because of the different source types contributing to the luminosity function. At low luminosities the main contributors are Be/X-ray binaries with neutron star primaries. At medium luminosities wind accreting systems with either black hole or neutron star primaries contribute mostly. At high luminosities back holes with supergiant secondaries are the main contributors. At the highest luminosities sources are defined as ULXs. As the exact nature of ULXs is still unknown, it is only possible to state that the smooth extension of the luminosity function to the ULX regime suggests that most of them are part of the X-ray binary population.

The upper end of the luminosity function and the objects that populate this luminosity range has been an area of intense research in recent years. Because of the excellent angular and spectral resolution of Chandra, and XMM, it is possible to identify ULXs in distant

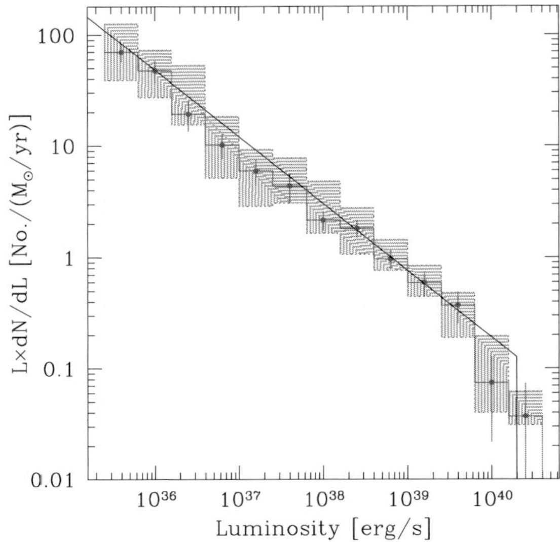

Figure 3. Universal luminosity function of HMXBs. There are no strong deviations from this power law over the 5 orders of magnitude in luminosity where it has been determined. The cutoff at high luminosities is not significant from the luminosity function alone but the relation of the total X-ray versus SFR requires the existence of a cutoff in the range of the indicated luminosities.

galaxies as individual objects that apparently violate the Eddington limit for known compact object masses. However, the bulk of the population of X-ray binaries are low luminosity sources that are difficult to observe in other galaxies. But these sources are important for our understanding of the formation of a whole population of X-ray binaries. To understand their formation processes and compare simulations to observations it is important to know how many sources there are in a galaxy. The shape of the observable luminosity functions, a relatively steep power law, indicates that there must be a cutoff or turnover at low luminosities.

Even with the sensitivity of Chandra and XMM the low luminosity end is only observable in the Milky Way and very nearby galaxies within the Local Group. Despite the observational difficulties there are indications that the lower end of the luminosity function is (almost) in the range of current observatories.

Shtykovskyi & Gilfanov (2005) have used XMM-Newton observations of HMXBs in the Large Magellanic Cloud. At the lower end of the observable luminosity range at $\sim 10^{34}$ erg s^{-1} the number of HMXBs and candidate HMXBs is below the expectation from the universal luminosity function for HMXBs. A possible explanation for this lack of HMXBs is the disappearance of HMXBs due to the "propeller effect". This effect was predicted by Illarionov & Sunyaev (1975) for accreting, young neutron stars. In these systems, which are only HMXBs due to the age requirement for the neutron star, the magnetic field in conjunction with the rapid rotation of the young neutron star is able to expel the matter that tries to accrete for certain parameter combinations of spin period and field strength. For realistic assumptions about spin periods and magnetic field strengths for HMXBs the "propeller effect" would prevent HMXBs from emitting X-rays in a luminosity range from $10^{32} - 10^{34}$ erg s^{-1}. At the higher luminosities the expected change in the luminosity function is only gradual. Therefore there are at present no definite conclusions possible.

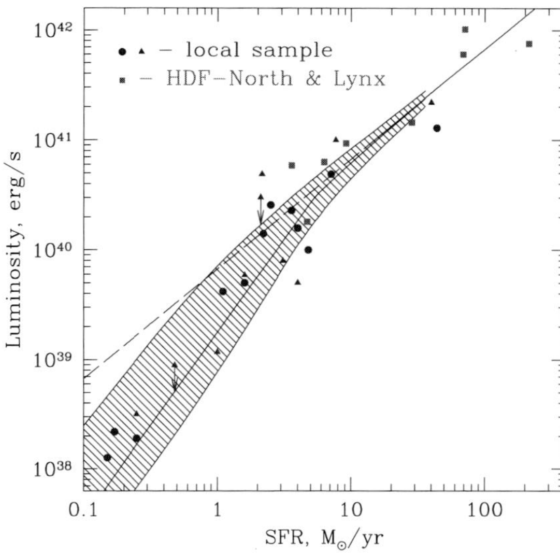

Figure 4. Total X-ray luminosity versus star formation rate. The filled circles and triangles are local galaxies, the filled circles galaxies with resolved X-ray binary populations. The squares are galaxies from Chandra observations of the Hubble Deep Field-North. The solid line is the predicted relation from the universal luminosity function for HMXBs. The shaded area shows the asymmetric 1 sigma uncertainty range.

The "propeller effect" applies only to young neutron stars, i.e. only some HMXBs. McClintock *et al.* (2004) have observed various transient X-ray binaries in quiescence. The observations reveal a clear distinction between neutron stars and black holes. Neutron stars have quiescent luminosities of 10^{32}–10^{34} erg s^{-1}, whereas black holes have quiescent luminosities of 10^{30}–10^{32} erg s^{-1}. The luminosity is correlated with the orbital period of the system. One of the questions raised by these observations is whether to include quiescent systems into the luminosity function at all.

3. Integrated properties

Despite the angular resolution achievable with current X-ray missions the vast majority of galaxies will always be unresolvable. However, it is possible to use the integrated measurements as well to study the relation between X-rays and star formation. The shape of the luminosity function results in a unique relation between the total X-ray flux and star formation rate. In this case some care has to be taken to ensure that the X-ray emission is not related to an AGN or other processes. For sufficiently distant galaxies redshifting ensures that the emission will not be related to supernova remnants and similar low energy emitters.

Fig. 4 shows the relation between star formation rate and X-ray emission for a sample of nearby galaxies where X-ray source populations are resolved and some distant galaxies observed in the Hubble Deep Field-North. There is clearly a non-linear relation at low SFR/X-ray luminosities, and only at a few 10^{40} erg s^{-1} the relation changes and becomes linear. The non-linear behaviour is the result of the shape of the luminosity function and its dependence on the star formation rate (Gilfanov *et al.* 2004). For the luminosity function it is interesting that the shape and break of the total X-ray luminosity versus star formation rate relation depend on parameters of the luminosity function, namely the slope and the cutoff luminosity. Obviously the normalization of the luminosity function is the main factor in the numerical constant of the relation. Thus all important parameters of the luminosity function can in principle be determined just from the total X-ray emission of a galaxy. Especially with respect to the cutoff this is important at the present stage. The cutoff is not strongly required from the observed luminosity functions alone. However, the existence of a linear regime in the relation between total X-ray luminosity and star formation rate requires the existence of such a cutoff. If the cutoff did not exist, the relation would be non-linear for all luminosities/star formation rates! The slope β of the non-linear part is defined by the slope α of the luminosity function,

$$\beta = \frac{1}{\alpha - 1}, \qquad (3.1)$$

and the break luminosity $L_{tot,break}$ is defined by the slope and the cutoff luminosity L_{cut},

$$L_{tot,break} \approx \begin{cases} \frac{L_{cut}}{2-\alpha} & \text{if } 1 < \alpha < 2 \\ \frac{L_{cut}}{\alpha-2} \times \left(\frac{L_{cut}}{L_1}\right)^{\alpha-2} & \text{if } \alpha > 2. \end{cases} \qquad (3.2)$$

Also the nature of ULXs could possibly be investigated with integrated properties of galaxies. One explanation for the ULX phenomenon is that these objects contain intermediate mass black holes with masses up to ~ 10000 M$_\odot$. Since these objects cannot be the result of stellar evolution processes like neutron stars and stellar mass black holes their luminosity function is different from the one for "normal" X-ray binaries. They are however correlated with star formation because most ULXs are observed in star forming galaxies. If intermediate mass black holes exist and form a different sub-population at high star formation rates, there might be a deviation from the linear regime in the X-ray luminosity/SFR diagram at very high star formation rates.

References

Cheng, K. P., Collins, N., Angione, R., *et al.* 1997, *UITVi; Vol. U*
Gilfanov, M., Grimm, H.-J., & Sunyaev, R. 2004, *MNRAS* 351, 1365
Grimm, H.-J., Gilfanov, M., & Sunyaev, R. 2003, *MNRAS* 339, 793
Grimm, H.-J., Gilfanov, M., & Sunyaev, R. 2002, *A&A* 391, 923
Hippelein, H., Haas, M., Tuffs, *et al.* 2003, *A&A* 407, 137
Illarionov, A. & Sunyaev, R. 1975, *A&A* 39, 185
McClintock, J., Narayan, R., & Rybicki, G. 2004, *ApJ* 615, 402
Shtykovskyi, P. & Gilfanov, M. 2005, *MNRAS* 362, 879

Discussion

PAKULL: Nearby dwarf galaxies tend to have either zero HMXBs or one, which is a ULX, and no lower luminosity sources (down to $\sim 10^{37}$ erg s^{-1}). The sample includes IZw18, the lowest metallicity galaxy there is. Don't you think that another parameter, possibly metallicity, plays an important role in the formation of ULXs, as mass loss rates from massive stars decrease with metallicity.

GRIMM: Metallicity is probably one of the more important parameters affecting primary masses. However, the low number of such sources and other sources of errors in the luminosity function – star formation relation do not allow to distinguish effects of metallicity on the luminosity function at present.

Populations of High Energy Sources in Galaxies
Proceedings IAU Symposium No. 230, 2005
E. J. A. Meurs & G. Fabbiano, eds.

© 2006 International Astronomical Union
doi:10.1017/S1743921306008672

Star-formation history, globular clusters and XRBs in early-type galaxies

L. A. Nolan[1]†, R. Kraft[2], T. Ponman[1], C. Jones[2] and S. Raychaudhury[1]

[1]School of Physics and Astronomy, University of Birmingham, Birmingham, B15 2TT, UK

[2]Harvard/Smithsonian Center for Astrophysics, 60, Garden St., MS-67, Cambridge, MA, 02138

Abstract. NGC 5102 has an unusually low number of XRBs. The deficit of LMXBs is even more striking because some of these sources may in fact be HMXBs, produced in one of the two recent bursts of star formation $\sim$15 and $\sim$300 Myr ago. Our UV-optical spectral synthesis analysis demonstrates that a significant fraction ($>$50%) of the stars in this galaxy are comparatively young ($<$3 Gyr). NGC 5102 also has an usually low number of globular clusters for its mass, luminosity and environment.

We discuss the relationship between the XRB population, the globular cluster population and the relative youth of the majority of the stars in this galaxy. We intend to extend our investigation of the relationship between XRB populations, star-formation history and globular clusters to a sample of ten early-type galaxies with a range of star-formation histories and investigate the implications for models of LMXB formation and evolution.

Keywords. X-rays: binaries, galaxies; galaxies: stellar content.

1. Introduction

The X-ray binary (XRB) populations of galaxies with a range of masses and morphological types have now been studied in detail, and it is clear that the variance in the number of XRBs per unit optical luminosity in early-type galaxies is greater than that expected based on counting statistics alone. This variance is likely to be related to the much-debated origin and evolution of XRBs.

Some have argued that most or all of the XRBS in early-type galaxies originate in globular clusters (GCs), where the stellar densities are sufficient for neutron stars to be captured in N-body interactions. Recent Chandra observations show that a significant fraction (40–60%) of the XRBs are spatially coincident with GCs, supporting this idea. Others have argued that that XRBs are related to the bulk of the stars in the galaxy, and that the variance in XRB populations is therefore related to the star-formation history (sfh) of the galaxy (White & Ghosh 1998, Wu 2001). The fact that the majority of LMXBs in the Galactic and M31 bulges do not appear to be associated with GCs supports this argument. However, recent analysis suggests both mechanisms may be important (Irwin & Bregman 2003).

We present the results of a 34ks Chandra/ACIS-S observation of the small ($M_B = -17.45$), blue, nearby (d = 3.1 $\pm$ 0.015 Mpc (McMillan, Ciardullo, & Jacoby 1994)), lenticular galaxy, NGC 5102. ROSAT and Einstein observations suggest it is under-X-ray luminous for its mass (Forman, Jones, & Tucker 1985, Irwin & Sarazin 1998, O'Sullivan, Forbes, & Ponman 2001). It is also unusually blue, with one or more recent ($<$0.1 Gyr ago) bursts of star-formation, which could have resulted in a high-mass X-ray binary (HMXB) population as well as the low-mass XRB (LMXB) population expected in early-types.

† email: lan@star.sr.bham.ac.uk

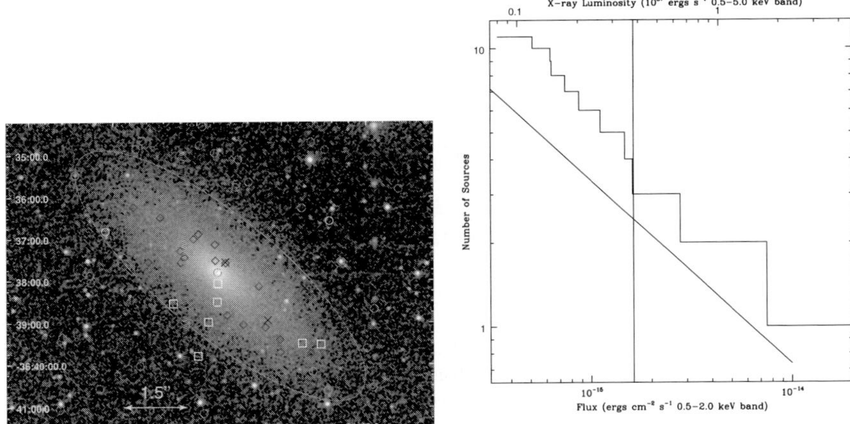

Figure 1. LHS: Positions of the X-ray point sources overlaid onto a 2MASS J band image of NGC 5102. The large green ellipse denotes the optical boundary (D_{25}) of the galaxy; the yellow and green circles the XPSs with $L_X \geqslant 10^{37}$ ergs^{-1} and $L_X < 10^{37}$ ergs^{-1} respectively. The red diamonds are planetary nebulae, and the white boxes are HII regions (both from McMillan, Ciardullo, & Jacoby 1994). The blue Xs represent GC candidates. RHS: Cumulative luminosity function (histogram) of X-ray point sources detected within the D_{25} ellipse. The log(N)-log(S) of the background AGN is plotted as the continuous curve (Tozzi *et al.* 2001). The corresponding X-ray luminosity of the sources in the 0.5-5.0 keV band (assuming the source to be at a distance of 3.1 Mpc) is shown across the top. The solid vertical line denotes the flux corresponding to 16 counts in our observation. Above this line, we are complete and unbiased.

Spectral synthesis modelling suggests that most or all of the stars are relatively young (<5 Gyr) (Pritchet 1979, Rocca-Volmerange & Guiderdoni 1987, Deharveng *et al.* 1997). Here, we study the sfh (via long-baseline spectral decomposition) and the GC population (using archival HST data) of NGC 5102, to investigate their relationship with the XRB population detected in the Chandra observation.

2. The X-ray point sources

The point sources in NGC 5102 were detected using the CIAO programme *wavdetect* in the 0.5–5.0 keV band. Here, we will consider only the X-ray point sources detected within the D_{25} ellipse which is shown on the left-hand side of Figure 1. The right-hand side of Figure 1 plots the cumulative X-ray luminosity function in this region. We find only 11 XRBs within the D_{25} ellipse, and only 2 with $L_X > 10^{37}$ ergs^{-1}. One of these sources is likely to be a background AGN (see Figure 1), so we are left with only 1 XRB in NGC 5102 which has $L_X > 10^{37}$ ergs^{-1}. We restrict ourselves to this energy range as we know that we are complete and unbiased here, and for ease of comparison to other galaxies. By comparison with other early-types, scaling with the B band luminosity, we predict that NGC 5102 should contain 6 LMXBs (5, if scaled with the J band luminosity). NGC 5102 clearly has a dearth of LMXBs, which is exacerbated by the possibility that this single source could be an HMXB associated with the recent star-formation.

3. The globular clusters

The archival HST/WFPC2 data, observed with the F569W filter, was searched for GCs using the image-finding algorithm SEXTRACTOR on the unsharp masked image. The positions of the detected GCs are shown in Figure 1. As the HST data only cover

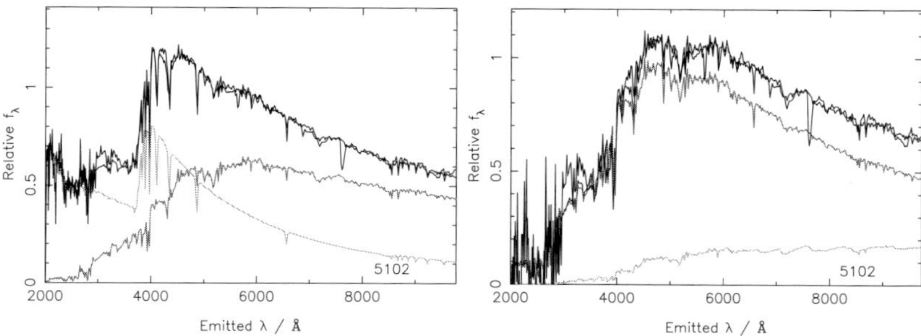

Figure 2. Left: The best-fitting two-component model (thin black line) superimposed over the spectrum of NGC 5102 (thick black line). The two component populations are also shown: the dominant population (3 Gyr, Z = 1.5 $Z_\odot$) is in red, and the lesser population (0.3 Gyr, Z = 0.2 $Z_\odot$) is in green. Right: The best-fitting two-component model (thin black line) superimposed over the residual of the spectrum of NGC 5102 following the subtraction of the 0.3 Gyr population (thick black line). The two component populations are also shown: the dominant population is in red (2 Gyr, Z = $Z_\odot$), and the lesser population (10 Gyr, Z = 2.5 $Z_\odot$), is in green. A faint population of older (10 Gyr) stars is not rejected by the fitting statistics.

$\sim$46% of the total light of the galaxy, the number of GCs was scaled by the galactic light fraction. We estimate that there are 7 GCs in NGC 5102. This gives a globular cluster specific frequency (GCSF) of 0.4, which is atypically low for a galaxy of this luminosity, morphology and environment. A more typical GCSF for an NGC 5102-like galaxy would be $\sim$1.5. However, if the number of GCs scales with the GCSF, then, by comparison with Cen A and NGC 4472, we would expect 1 XRB in NGC 5102 given 7 GCs.

4. The star-formation history

We have estimated the ages of the stellar populations in NGC 5102 by fitting a two-component stellar population evolutionary synthesis model to its UV-optical spectrum. We use the stellar population models of Jimenez *et al.* (2004). The age, metallicity and relative mass fraction of each single stellar population component is allowed to vary freely, and the best-fitting two-component model is found via χ^2 minimisation. The results are shown on the left-hand side of Figure 2. The dominant (93%) population has an age of 3 Gyr, and Z = 1.5 $Z_\odot$. The smaller population has age = 0.3 Gyr and Z = 0.2 $Z_\odot$. This is consistent with the results of other authors (Pritchet 1979, Rocca-Volmerange & Guiderdoni 1987, Bica 1988). The age of the dominant population is remarkably young, so, to check for the presence of an sub-population we subtracted the 0.3 Gyr model population from the observed spectrum, and apply the two-component model-fitting technique to the residual spectrum. The results are shown on the right-hand side of Figure 2. The addition of a third, older (10 Gyr) population is not ruled out by the fitting statistics. Even with the addition of this third population, the majority (>55% by mass) of the population is still young (2 Gyr). If the XRB population is directly linked to the stellar population, this would suggest that NGC 5102 has not had sufficient time to develop an XRB population appropriate to its stellar mass.

5. Discussion

We have demonstrated that NGC 5102 has a deficit of XRBs. We have also shown that more than half the stars in this unusually blue lenticular galaxy are less than 3 Gyr old, and that it has an atypically small number of GCs. Is the lack of XRBs related to the relative youth of the stellar population, to the lack of GCs, or to both? There seems little doubt that GCs play an important role in the formation of XRBs in early-types, so the question is, what role does the stellar population play? It is probable that both the sfh and the deficit of GCs play a role in the lack of XRBs in NGC 5102.

Even if $30-60\%$ of the XRBs in all early-type galaxies are contained within GCs, the other $40-60\%$ remains unaccounted for. It is possible that all XRBS in early types were formed in GCs and some have been ejected, although there is little evidence either way. On the other hand, the non-GC XRBs may be an entirely distinct population. Studies of the LMXB populations in the Galactic and M31 bulges suggest that these sources were likely formed via a stellar evolution scenario, where the LMXB begins life as a wide binary with a large mass ratio. The more massive star evolves rapidly, and explodes as a supernova. The binary orbit then decays, possibly through magnetic breaking, until Roche lobe overflow occurs, and the LMXB is born. If this is the case, our results suggest that there is a minimum timescale of a few Gyr for XRB populations to form. This timescale is roughly consistent with estimates of the timescale required for rotational breaking of the magnetic winds of companion stars sufficient to reduce the binary separation to the point where Roche lobe overflow occurs (Verbunt & Zwaan 1981). There would also be a minimum timescale for LMXBs to evolve if their companion stars were red giants, which have a characteristic timescale to evolve off the main sequence depending on their mass and metallicity.

There is clearly an important relationship between LMXBs and GC populations in massive ellipticals in rich environments with large GCSFs. However, sfh and binary evolution must also play an important part in the formation of LMXBs. We are currently investigating the relationship between XRBs, sfh and GCs in a sample of early-types.

These results are presented and discussed in detail in Kraft *et al.* (2005).

Acknowledgements

This work was funded by NASA grant GO2-3109X.

References

Bica, E., 1988, A&A, 195, 76

Deharveng, J.-M., Jedrzejewski, R., Crane, P., Disney, M. J., & Rocca-Volmerange, B., 1997, A&A, 326, 528

Forman, W., Jones, C., & Tucker, W., 1985, ApJ, 293, 102

Irwin, J. A. & Bregman, J. N., 2003, AAS, 203,

Irwin, J. A. & Sarazin, C. L., 1998, ApJ, 499, 650

Jimenez, R., MacDonald, J., Dunlop, J. S., Padoan, P., & Peacock, J. A., 2004, MNRAS, 349, 240

Kraft, R. P., Nolan, L. A., Ponman, T. J., Jones, C., & Raychaudhury, S., 2005, ApJ, 625, 785

McMillan, R., Ciardullo, R., & Jacoby, G. H., 1994, AJ, 108, 1610

Pritchet, C., 1979, BAAS, 11, 429

O'Sullivan, E., Forbes, D. A., & Ponman, T. J., 2001, MNRAS, 328, 461

Rocca-Volmerange, B., & Guiderdoni, B., 1987, A&A, 175, 15

Tozzi, P., *et al.*, 2001, ApJ, 562, 42

Verbunt, F. & Zwaan, C., 1981, A&A, 100, L7

White, N. E. & Ghosh P , 1998, ApJ, 504, L31

Wu, K., 2001, PASA, 18, 443

Discussion

SARAZIN: I just wanted to suggest that E + A elliptical galaxies are particularly good objects for studying the relation of LMXBs with star formation history.

NOLAN: Yes, that would be a good way to probe younger stellar populations, although care may have to be taken to avoid contamination of the population with HMXBs. I believe one of the galaxies in our sample B is in fact an E + A galaxy.

GHOSH: We're so used to thinking of early-type galaxies not having young, massive stars that NGC 5102 seems a remarkable case indeed. What's special about this lenticular galaxy that it had such recent bouts of star formation?

NOLAN: Early type galaxies are not the simple, mature stellar systems that they were once declared to be! They suffer from mergers (many are probably formed from major, recent ($z < 1$) mergers) both major & minor (e.g. capture of dwarfs) & interactions, which are likely to induce bursts of star formation. So we do expect to find a range of ages of the stellar populations in early-type galaxies. NGC 5102 is in a poor environment, and its recent star formation was probably as a result of the capture of a dwarf galaxy. It's unusual for its large mass fraction of young stars, but not for having a young stellar population.

Populations of High Energy Sources in Galaxies
Proceedings IAU Symposium No. 230, 2005
E. J. A. Meurs & G. Fabbiano, eds.

© 2006 International Astronomical Union
doi:10.1017/S1743921306008684

The Luminosity Function of X-ray Point Sources in Centaurus A

Rasmus Voss and Marat Gilfanov

MPA, Karl Schwarzschild-Strasse 1, 85741 Garching, Germany
email: voss@mpa-garching.mpg.de

Abstract. We studied the population of X-ray point sources in the elliptical galaxy Centaurus A, using archival Chandra data. Within a radius of 10 arcmin from the centre we detected 272 sources. Of these approximately half are LMXBs with the rest being CXB sources. The spatial distribution of the LMXBs is found to be consistent with the distribution of K-band light. We constrain the luminosity function (LF) of the X-ray sources down to $\sim 2 \times 10^{36}$ erg s^{-1}. The X-ray LF of the LMXBs flattens significantly below $L_X \sim 5 \times 10^{37}$ erg s^{-1} and follows the $dN/dL \propto L^{-1}$ law in agreement with results from the bulges of spiral galaxies.

Keywords. galaxies: individual (Centaurus A), X-rays: binaries, X-rays: galaxies.

1. Introduction

The faint end of the low-mass X-ray binary (LMXB) X-ray luminosity function (XLF) was studied by Gilfanov (2004), using observations of the bulges of two spiral galaxies, the Milky Way and M31. This study showed a broken powerlaw with a break at a few times 10^{37} erg s^{-1}, a power law index of ~ 1 below and ~ 1.8 above. It is important to extend the study of the low end of the XLF of LMXBs to also include the populations in elliptical galaxies as these have evolutionary histories that are different from the bulges of spiral galaxies.

Centaurus A (Cen A) is the primary candidate for such a study. It is massive enough to contain sufficient number of LMXBs and, on the other hand, is sufficiently nearby to reach luminosities below $\sim 10^{37}$ erg s^{-1} with moderate observing time. Details can be found in Voss & Gilfanov (2005). The analysis in this paper is based on 4 Chandra observations, two of them made with the ACIS-I array (OBS-ID 316 and 962), and the other two with the ACIS-S array (OBS-ID 2978 and 3965). Together these four observations cover most of Cen A within a 10 arcmin radius from the centre. The data preparation was done following the standard CIAO† threads (CIAO version 3.1; CALDB version 2.28), and limiting the energy range to 0.5–8.0 keV. The observations were merged and sources were detected in the merged image using `wavdetect`. To avoid crowding and strong diffuse emission, we excluded the area within a radius of 30 pixels (~ 15 arcsec) from the centre of the galaxy, as well as the region covering an X-ray jet emanating from the galaxy nucleus. Finally we end up with a sample of 272 point-like sources.

2. Optical identifications

For the identification of Chandra sources we used the results of the optical studies of the Cen A region by Peng *et al.* (2004), Minniti *et al.* (2003), Woodley *et al.* (2005) and Graham & Fasset (2002). In total we identified 6 X-ray sources as foreground stars,

† http://cxc.harvard.edu/ciao/

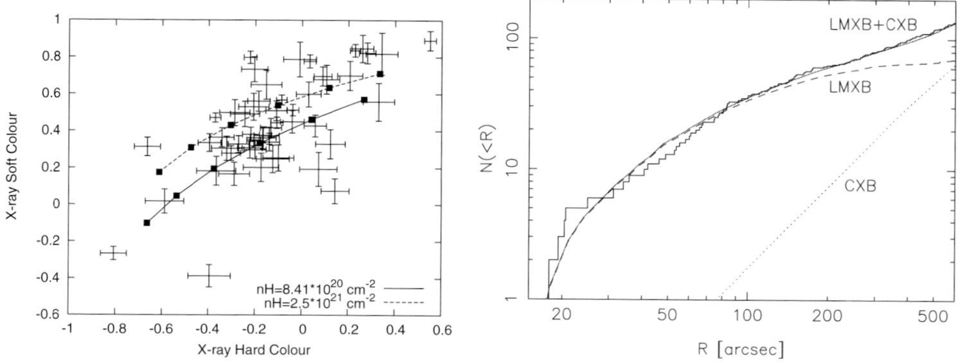

Figure 1. Left: The colour–colour diagram of the brightest, >200 counts, sources within 5 armin from the centre of Cen A. The sources coinciding with H_α-emitting regions are shown in bold. For reference, the two lines show the hardness ratios of power law spectra for two different values of absorption. Right: The radial distribution of observed sources (solid, black), compared to the best fit model (thick, grey) and the contributions of LMXBs (dashed) and CXBs (dotted).

leaving 266 sources of presumably extragalactic origin – either intrinsic Cen A sources or background AGNs. Of these, 37 were identified with the globular clusters in Cen A. Eight sources within 4 arcmin from the centre of Cen A coincide with H_α-emitting regions found in Minniti *et al.* (2003). The optical magnitudes of the H_α sources indicate that they may be young star clusters as well as individual X-ray binaries. In order to search for further indications of the high-mass X-ray binary (HMXB) nature of these sources we have compared their spectral properties with other sources and searched for periodic variability in their X-ray emission. The results of these attempts to find the nature of the objects were inconclusive, as no periodic variability was found, and the colour distribution does not deviate from the distribution of the rest of the sources (Fig. 1, left).

3. Populations of X-ray sources in the field of Centaurus A

LMXBs are related to the population of old stars, and there is therefore a correlation between their number and the stellar mass of a galaxy (Gilfanov 2004). In order to estimate the expected number and luminosity distribution of LMXBs we used a K-band image from 2MASS Large Galaxy Atlas (Jarret *et al.* 2003) and integrated the flux emitted in the parts of Cen A analyzed in this paper. Using the results of Gilfanov (2004) we predict $\approx$81 LMXBs with $L_X > 10^{37}$ erg s^{-1}.

Being young objects, HMXBs are associated with star formation. We used calibration of Grimm *et al.* (2003) to calculate the expected number of HMXBs (see comments in Shtykovskiy & Gilfanov 2005, regarding the normalization), from the FIR luminosity of Cen A (Marston & Dickens 1988). From this we get the expectation of $\approx$10 HMXBs brighter than 10^{37} erg s^{-1}.

To estimate the number of background sources we used results of the cosmic X-ray background (CXB) $\log(N) - \log(S)$ determination by Moretti *et al.* (2003). For the total area of our survey of 0.079 deg^2 we obtain from the source counts in the soft band $\approx$34 CXB sources above the flux corresponding to 10^{37} erg s^{-1}. From the hard band counts the predicted number is $\approx$47 sources.

We detected 136 sources with a luminosity above 10^{37} erg s^{-1}, fitting well the expectations above.

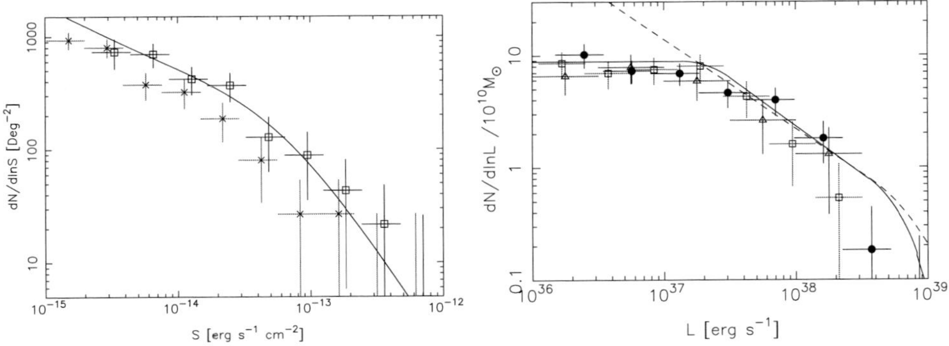

Figure 2. Left: The source counts (open squares) in the outer region (5 arcmin<r<10 arcmin). Thick solid line: CXB $\log(N) - \log(S)$ from Moretti *et al.* (2003) with best fit normalization from this paper. For comparison the source counts in the CDF-N Obs-ID 1671 are shown (crosses). Right: The LF of LMXBs in the inner $r \leqslant 5$ arcmin of Cen A (the CXB contribution subtracted) in comparison with LMXB XLFs in the Milky Way (triangles) and M31 (squares) in Gilfanov (2004). The latter two are multiplied by constant factors of 1.7 and 0.6, respectively. Solid line: average LMXB XLF in the nearby galaxies as determined by Gilfanov (2004). Dashed line: average LMXB XLF from Kim & Fabbiano (2004) and its extrapolation towards low luminosities.

4. Spatial distribution of point sources

The spatial distribution of the LMXBs has been shown (Gilfanov 2004) to follow, to the first order approximation, the distribution of the stellar mass. This can be represented by the distribution of the K-band light and was computed using the K-band image of Cen A from 2MASS. The density of the CXB sources can be assumed flat on the angular scales under consideration. Therefore the CXB growth curve is proportional to the enclosed solid angle. The (unknown) distribution of HMXBs has not been included as it is unlikely to exceed 10 per cent in the total number of sources.

The model has been compared with the observed distribution of sources more luminous than 10^{37} erg s^{-1}. The model adequately describes the data (Fig. 1, right) as confirmed with the KS-test, which gave the probability of 96 per cent. A Maximum Likelihood (M-L) fit to the radial distribution of the unbinned data gave 70.3 ± 10.0 LMXBs and 65.7 ± 9.8 CXB sources ($>10^{37}$ erg s^{-1}). The abundance of LMXBs is surprisingly close to the expectation value. The number of CXB sources, on the other hand, is higher than the expectation.

5. Source counts and the CXB source density

In analysing the LFs and $\log(N) - \log(S)$ distributions, we corrected for incompleteness effects as described in Voss & Gilfanov (2005). We estimated the normalization of the CXB $\log(N) - \log(S)$ distribution from the source counts in the outer region. This region is far enough from the inner parts of the Cen A to keep the number of sources related to the galaxy low, while close enough to the aimpoints of the observations to have a reasonable sensitivity. In this region the incompleteness corrected (Voss & Gilfanov 2005) number of sources with the 0.5–8 keV flux exceeding 2.7×10^{-15} erg s^{-1} cm^{-2} (luminosity 4.0×10^{36} erg s^{-1}) is 101.3 from which 13.4 are expected to be LMXBs. The implied number of CXB sources is therefore $\approx$88. This is consistent with the normalization obtained from the radial distribution of sources, and higher than the expectation from Moretti *et al.* (2003). To confirm that the difference is not due to different analysis methods, we analyzed one CDF-N observation Obs-ID 1671, using the same method as for Cen A. As can be seen from Fig. 2, left, the difference remains, although it is smaller $\sim$50%.

6. LMXB XLF

The LMXB XLF based on the combined data of $r \leqslant 5$ arcmin is plotted in Fig. 2, right, along with luminosity distributions of LMXBs in the Milky Way and M31. This is the first study to extend the LMXB XLF in elliptical galaxies below a few times 10^{37} erg s^{-1}. We find that the XLF of the LMXBs is consistent with the results from the bulges of spiral galaxies (Gilfanov 2004). The bright end is consistent with the slope found by Kim & Fabbiano (2004). The faint end is, however, inconsistent with an extrapolation of their XLF.

The plot illustrates qualitative and quantitative similarity of the LMXB luminosity distributions in Cen A and bulges of spiral galaxies. Spiral and elliptical galaxies have different evolutionary histories and could differ in the properties of their LMXB populations. As demonstrated here, the LFs nevertheless seem very similar.

7. Summary and conclusions

We have used archival data of Chandra observations to study statistical properties of the point source population of Cen A. About half of the detected sources are expected to be X-ray binaries in Cen A, mostly LMXBs, and the vast majority of the remaining sources are background galaxies constituting the resolved part of the CXB.

The spatial distribution of the detected sources can be well described by a sum of two components. One follows the K- band light, representing LMXBs in Cen A, while the other is uniform over the field, representing the resolved part of the CXB. The normalization of the LMXB component agrees well with the average value derived for the local galaxies by Gilfanov (2004). The normalization of the uniform component and source counts in the exteriors of the galaxy appears to indicate an overabundance of the CXB sources in the direction of Cen A by a factor of $\sim$1.5 or, possibly, more.

We were able to recover the the LMXB LF in the inner $r \leqslant 5$ arcmin down to $L_X \sim 2 \times 10^{36}$ erg s^{-1}. This is by a factor of $\sim$5–10 better than achieved previously for any elliptical galaxy (Kraft *et al.* 2001; Kim & Fabbiano 2004). The shape of the luminosity distribution is consistent with the average LMXB XLF in nearby galaxies derived by Gilfanov (2004) and of the bright end by Kim & Fabbiano (2004). In particular, we demonstrate that the LMXB XLF in Cen A flattens at the faint end and is inconsistent with extrapolation of the steep power law with differential slope of $\approx$1.8–1.9 observed above $\log(L_X) \sim 37.5$–38 in the previous studies of elliptical galaxies.

References

Gilfanov, M. 2004, *MNRAS* 349, 146
Graham, J. A. & Fasset, C. I. 2003, *ApJ* 575, 712
Grimm, H.-J., Gilfanov, M. R., & Sunyaev, R. A. 2003, *MNRAS* 339, 793
Jarret, T. H., Chester, T., Cutri, R., Schneider, S., & Huchra, J. P. 2003, *AJ* 125, 525
Kim, D.-W. & Fabbiano, G. 2004, *ApJ* 611, 846
Kraft, R. P., Kregenow, J. M., Forman, W. R., Jones, C., & Murray, S. S. 2001, *ApJ* 560, 675
Marston, A. P. & Dickens, R. J. 1988, *A&A* 193, 27
Minniti, D., Rejkuba, M., Funes, J. G., & Akiyama, S. 2003, *ApJ* 600, 716
Moretti, A., Campana, S., Lazzati, D., & Tagliaferri, G. 2003, *ApJ* 588, 696
Peng, E. W., Ford, H. C., & Freeman, K. C. 2003, *ApJS* 150, 367
Shtykovskiy, P. & Gilfanov, M., 2005, *A&A* 431, 597
Voss, R. & Gilfanov, M. 2005, *A&A* in press, astro-ph/0505250
Woodley, K. A., Harris, W. E., & Harris, G. L. H. 2005, *AJ* 129, 2654

Discussion

KIM: But Cen A is a complex system, with a large amount of different X-ray emission (with considerable sub-structures), dust lanes, jets etc. The complete correction is therefore very critical, likely more at the central region.

VOSS: Since Kim & Fabbiano (2004) results are based on X-ray sources with $L_X > \sim$ a few * 10^{37}, the XLF flattening which you have found is not really consistent.

MACCARONE: A comment: we would expect to see breaks in the luminosity function to state transitions between low/hard and high/soft states, since the bolometric corrections to the Chandra band change substantially there. These will be at a few percent of the Eddington limit – about 3 * 10^{37} erg/sec for BHs and 3 * 10^{36} erg/sec for NSs.

VOSS: That's an interesting point and I plan to look into it when I get home.

Populations of High Energy Sources in Galaxies
Proceedings IAU Symposium No. 230, 2005
E. J. A. Meurs & G. Fabbiano, eds.

© 2006 International Astronomical Union
doi:10.1017/S1743921306008696

X-ray source populations in star-forming galaxies

Andreas Zezas[1]

[1]Harvard-Smithsonian Center for Astrophysics, Cambridge, MA 02138, USA
email: azezas@cfa.harvard.edu

Abstract. We present a study of the variable X-ray source populations in the Antennae galaxies and a small sample of nearby star-forming galaxies. We find that source variability does not affect the shape of their X-ray luminosity functions. Several of the sources detected in the Antennae exhibit a wide range of spectral variability patterns. Finally, from our study of a small sample of nearby star-forming galaxies, we find a weak indication for variations of the shape of their luminosity functions.

Keywords. galaxies: starburst, X-rays: galaxies, X-rays: binaries.

1. Introduction

Star-forming galaxies provide a unique environment to study the High Mass X-ray binary populations (HMXBs). Recent studies which have focused on the X-ray luminosity functions (XLFs) of these X-ray source populations have suggested that there is a universal XLF for HMXBs which has a cumulative slope of $\alpha \sim 0.5$.† However, it is still unclear if and how these XLFs depend on the age of the stellar populations and the properties of their host galaxy. For example, Belczynski *et al.* (2004), based on theoretical modeling of HMXB populations, find that there is a non-monotonic dependence of the XLF slope on the age of the stellar populations.

An important component of the X-ray source populations found in star-forming galaxies are the Ultraluminous X-ray sources (ULXs), which are defined (rather arbitrarily) as sources with luminosities in excess of 10^{39} erg s^{-1}. Their nature, and most importantly their relation with the lower-luminosity X-ray source population, is unclear with possibilities including, Intermediate Mass black-holes (IMBHs), mechanical or relativistic beaming, super-Eddington accretion disks, or simply the upper tail of the regular X-ray binary populations (e.g. Zezas & Fabbiano, 2002; Miller & Colbert, 2004).

In order to investigate the nature of X-ray sources associated with young stellar populations, and more specifically address their connection with the stellar populations we embarked on a detailed study of nearby star-forming galaxies. Next we present results from our *Chandra* monitoring observations of the Antennae galaxies and an age sequence of nearby star-forming galaxies.

---

† Throughout this contribution the slopes of the XLFs will refer to their cumulative form: $N(> L) = K \times L^{-\alpha}$, where $N(> L)$ is the number of sources which have higher luminosity than L; K is the normalization of the XLF and α is the slope of the power-law function. However, all fits are performed on the differential form of the XLF.

2. The Antennae

2.1. *The monitoring program*

The Antennae (D = 19 Mpc), the prototypical merging system of galaxies, hosts one of the largest X-ray source populations observed in a star-forming galaxy, including a large number of ULXs (Zezas & Fabbiano 2002). In order to study the variability properties and probe the faint end of this population we obtained 8 monitoring observations of the Antennae over a span of two years (Fabbiano *et al.*, 2003).

We detect between 40 and 70 sources in the shortest and the longest exposures respectively, down to a typical luminosity of $\sim 7 \times 10^{37}$. In the coadded exposure (totaling 411 ksec) we detect 120 sources down to a luminosity of $2\text{–}5 \times 10^{37}$ erg s^{-1}, for low and high background regions respectively.

These monitoring observations allow us to investigate the spectral variability of the rich X-ray source population detected in the Antennae. We find that 43 sources are variable either in short or long timescales, and 21 sources show indications for spectral variability based on their hardness ratios. We find a wide range of spectral variability patterns: sources not showing any correlation between luminosity and spectral hardness; sources becoming softer while their intensity increases and sources showing the opposite behavior (i.e. exhibit hard spectra with increased intensity). We interpret the latter case as analog of the very high state observed in Galactic black-hole binaries and microquasars, where changes in the accretion rate affect the temperature of the accretion disk and its relative strength compared to the power-law component (e.g. Miller *et al.*, 2002). In the relatively narrow energy band of the *Chandra* spectra (0.3–8.0 keV) these transitions can be interpreted as changes in the power-law slope.

2.2. *Multiwavelength counterparts*

Correlation of the X-ray source catalog with star-cluster catalogs from the HST-WFPC2 observations of the Antennae (Whitmore & Schweizer 1995; Whitmore *et al.*, 1999), shows that the majority of the X-ray sources is associated with regions of recent star-formation. We note that in this comparison we took special care to match the relative astrometry of the HST and *Chandra* data using ground based wide field optical images. We find optical associations for 66 out of the 104 X-ray sources which fall within the field of view of the HST observations. Of those 64 are associated with young star clusters (younger than ~ 25 Myr) or are in regions of on-going star-formation, manifested by the detection of Hα emission. Only 2 X-ray sources are associated with older stellar populations, although 7 sources without optical counterparts are found in the bulge regions of the interacting galaxies. This indicates that the majority of the X-ray sources is most likely HMXBs. Even with the improved relative astrometric calibration we find that a significant number of sources (and 6 out of the 12 ULXs) is not associated with star-clusters but instead they lie in their vicinity. This suggests that these sources were either expelled from their parent clusters (Zezas & Fabbiano 2002) or that their parent clusters have dissolved in the meantime (Soria 2005).

3. XLF variability

These monitoring observations also provide a unique opportunity to investigate the effect of individual source variability on their XLF. In Fig. 1 we show the cumulative XLFs of the seven observations together with their associated errors (including Poisson uncertainties on the number of sources and the luminosity of each source, calculated as described in Zezas & Fabbiano 2002). All seven XLFs are well fitted by power-law models

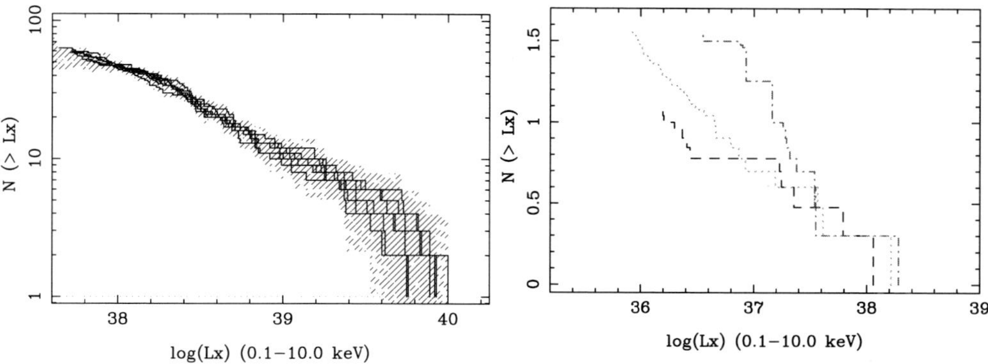

Figure 1. Left: Comparison of the cumulative XLFs from the 7 monitoring observations of the Antennae. The hatched area indicates the uncertainties on each luminosity function (see text). **Right:** Comparison between the X-ray luminosity functions of NGC 1569 (black, dashed line), the coadded observation of NGC 4214 (green, dotted line) and NGC 5253 (red dash-dot line). All XLFs are corrected for incompleteness.

with consistent slopes of $\alpha \sim 0.5$. A comparison using a Kolmogorov-Smirnov test also shows that they are consistent with a common parent population. This result indicates that X-ray source variability does not affect the shape of the XLFs, which allows us to use single observations of other galaxies in order to obtain a representative picture of their X-ray binary populations.

Since the individual XLFs are consistent with each-other, we fit the XLFs of the coadded observation. We find that to first order it is also fitted with a flat power-law ($\alpha \sim 0.5$) over three decades in luminosity. There is also indication for a bump at $\sim 3 \times 10^{38}$ erg s^{-1} and possibly a cutoff at high luminosities, but a preliminary statistical analysis shows that a broken power-law does not provide a better fit than a single power-law. The fact that there is no significant difference between the XLF shape of the ULXs ($L_X > 10^{39}$ erg s^{-1}) and the less luminous sources indicates that the ULXs are not associated with a strongly beamed fraction of the "normal" X-ray binary population.

4. A starburst age sequence

Although the Antennae offers a large population of X-ray sources, its complicated star-formation history results in X-ray binary populations from various stellar generations. This makes it difficult to address the connection between the observed X-ray source populations and their parent stellar populations. For this reason we embarked on a program to study a set of star-forming galaxies with well characterized star-formation histories dominated by a single burst of star-formation. A subset of this sample includes (from younger to older systems) NGC 1569, NGC 4241 and NGC 5253.

In the case of NGC 4241 we have exposures in three different epochs allowing us to look for the effect of source variability on the shape of the XLF. As with the Antennae we did not find any difference between the three XLFs.

In the right panel of Figure 1 we present the XLFs of these three galaxies, corrected for incompleteness. We find that the three XLFs are slightly different, although because of the small number statistics they are fitted with consistent power-law slopes ($\alpha \sim 0.5$–0.6). We also see a weak trend for NGC 5253, which has the oldest XRB population, to have the steepest XLF; forthcoming observations will allow us to extend the XLF to lower luminosities in order to test the significance of this trend.

5. Discussion and Conclusions

From our monitoring observations of the Antennae we find that:

• A large number of sources shows intensity and/or spectral variability. In some cases this spectral variability is consistent with spectral changes due to changes in the accretion rate, in the same way as is seen in Galactic sources.

• The majority of the sources is associated with recent star-formation indicating that they are HMXBs. However, a significant number of sources (including 6 ULXs) is not associated with any optically identified star-cluster, indicating either that they were ejected or that their parent star-clusters have dissolved.

• X-ray source variability does not affect the shape of the XLF, allowing us to perform population studies based on individual observations of galaxies.

• There is a weak indication for different XLFs associated with star-forming galaxies with stellar populations of different ages.

Acknowledgements

I would like to thank my collaborators in the different projects presented in this contribution: G. Fabbiano, F. Schweizer, A. Baldi, V. Kalogera, J. Gallagher, P. Kaaret, A. Prestwich. This work has been supported by NASA LTSA grant NAG5-13056 and *Chandra* grant G04-5098X.

References

Belczynski, K., Kalogera, V., Zezas, A., & Fabbiano, G. 2004, *ApJ Lett.*, 601, L147

Fabbiano, G., Zezas, A., King, A. R., Ponman, T. J., Rots, A., & Schweizer, F. 2003b, *ApJ Lett.*, 584, L5

Grimm, H.-J., Gilfanov, M., & Sunyaev, R. 2003, *MNRAS*, 339, 793

Kilgard, R. E., Kaaret, P., Krauss, M. I., Prestwich, A. H., Raley, M. T., & Zezas, A. 2002, *ApJ*, 573, 138

Miller, J. M., Wijnands, R., Rodriguez-Pascual, P. M., Ferrando, P., Gaensler, B. M., Goldwurm, A., Lewin, W. H. G., & Pooley, D. 2002, *ApJ*, 566, 358

Miller, M. C. & Colbert, E. J. M. 2004, International Journal of Modern Physics D, 13, 1

Whitmore, B. C., Zhang, Q., Leitherer, C., Fall, S. M., Schweizer, F., & Miller, B. W. 1999, *AJ*, 118, 1551

Whitmore, B. C. & Schweizer, F. 1995, *AJ*, 109, 960

Soria, R. 2005, in Proceedings of XXII Texas Symposium on Relativistic Astrophysics, Stanford University, 2004, (astro-ph/0503340)

Zezas, A. & Fabbiano 2002, *ApJ*, 577, 726

Discussion

McBreen: In the Antennae there are heavily obscured regions of star-formation as revealed in the mid-infrared with ISO. What X-ray sources are associated with these regions?

Zezas: There are a couple of sources associated with the heavily obscured overlap region between the two galaxies; these sources also show heavily absorbed X-ray spectra. From the same region we see hard diffuse emission which most probably is unresolved emission from X-ray binaries in this obscured, actively star-forming region. On the other hand we do not find many X-ray sources in the inner part of the spiral arm where there is also some dust obscuration. Also most of the ULXs are found in the outer edge of the spiral arm, rather than the highly obscured regions.

Populations of High Energy Sources in Galaxies
Proceedings IAU Symposium No. 230, 2005
E. J. A. Meurs & G. Fabbiano, eds.

© 2006 International Astronomical Union
doi:10.1017/S1743921306008702

Point X-ray Sources in Elliptical Galaxies

Natalia Ivanova[1] and Vicky Kalogera[1]

[1]Physics and Astronomy Department, Northwestern University, 2145 Sheridan Rd, Evanston
IL 60208, USA
email: nata@northwestern.edu

Abstract. We analyze the upper-end X-ray luminosity function (XLF) observed in elliptical galaxies for point sources. We propose that the observed XLF is dominated by transient BH systems in outburst and the XLF shape reflects the black hole (BH) mass spectrum among old X-ray transients. The BH mass spectrum – XLF connection depends on a weighting factor that is related to the transient duty cycle and depends on the host-galaxy age, the BH mass and the donor type (main sequence, red giant, or white dwarf). We argue that the assumption of a constant duty cycle for all systems leads to results inconsistent with current observations. The type of dominant donors in the upper-end XLF depends on what type of magnetic braking operates: in the case of "standard" magnetic braking, BH X-ray binaries with red-giant donors dominate, and in the case of weaker magnetic braking prescriptions main sequence donors dominate.

Keywords. galaxies:elliptical – binaries:close – methods: statistical – X-rays: binaries.

1. Transient and persistent Black Hole X-Ray Binaries in Ellipticals

With *Chandra* observations, elliptical galaxies out to the Virgo cluster have now been studied. These observations have revealed a large number of point X-ray sources and early on Sarazin *et al.* (2000) identified an XLF shape that required two power laws with a break at $\simeq 3.2 \times 10^{38}\,\mathrm{erg\,s^{-1}}$. Following longer exposures for more ellipticals (Gilfanov 2004, Kim & Fabbiano 2004) have confirmed that the combined sample of sources from all observed galaxies requires two power laws and a break at $5 \pm 1.6 \times 10^{38}\,\mathrm{erg\,s^{-1}}$ and the best-fit slope of the upper end to be $\alpha_{\mathrm{d}} = 2.8 \pm 0.6$.

We analyze the upper-end of XLF assuming: (i) the XLF above the break at $5 \times 10^{38}\,\mathrm{erg\,s^{-1}}$ is populated by X-ray binaries (XRBs) with black hole (BH) accretors (Sarazin *et al.* 2000); (ii) the vast majority of these BH-XRBs are part of the galactic-field stellar population in ellipticals; (iii) donor masses are lower than $\simeq 1\text{--}1.5\,\mathrm{M_\odot}$ (in accordance to the current estimates for the ages of stellar populations in ellipticals) and (iv) we adopt the current understanding for the origin of transient behavior in XRBs (King 2005). We consider typical mass transfer (MT) rate associated with each type of XRB donor ($\dot{M}_i$) and compare it to the critical MT rates for transient behavior ($\dot{M}_{\mathrm{crit}}$): if $\dot{M}_i < \dot{M}_{\mathrm{crit}}$, the binary system is assumed to be a transient X-ray source and we assume that during the outburst the X-ray luminosity is equal to the BH accretor Eddington luminosity.

Mass transfer in BH XRBs with low-mass main sequence (MS) donors is driven by angular momentum losses due to magnetic braking (MB) and gravitational radiation (GR). From detailed binary evolutionary calculations using the stellar evolution and MT code (Ivanova & Taam 2004), we find that if MB follows the Skumanich law (taken as in Rappaport *et al.* 1983), then XRBs with BHs $\leqslant 10 M_\odot$ are persistent as long as the donor masses are $> 0.3 M_\odot$. The MT rates are about 0.01–0.25 of the BHs' Eddington rate. As a result, the persistent X-ray luminosity for these systems is $\leqslant 10^{38}\,\mathrm{erg\,s^{-1}}$. For $M_{\mathrm{BH}} > 10\,M_\odot$, the outburst luminosity is in excess of $\simeq 1.5 \times 10^{39}\,\mathrm{erg\,s^{-1}}$. This limit is comparable to the highest luminosity seen currently in XLFs of ellipticals (Kim & Fabbiano 2004), and therefore these systems cannot contribute significantly to the observed XLFs. Outbursts are possible from transient BH-MS with $M_{\mathrm{BH}} < 10\,M_\odot$ and donors less massive than $\simeq 0.3\,M_\odot$; these low mass donors are out of thermal equilibrium. In the case of the MB prescription derived for fast rotating systems, which are BH-MS binaries (Ivanova & Taam 2003, IT), BH-MS systems are transient for all BHs masses $M_{\mathrm{BH}} > 3 M_\odot$ and for all low-mass MS donors. If a donor is a low-mass subgiant or red giant (RG), it has been shown (King 2005) that such XRBs are transient, regardless of the BH mass. The evolution of mass-transferring BH systems with a white dwarf (WD) very weakly depends on the BH mass. A typical life-time of such a system in the persistent stage is $\sim 10^7$ years and only during 10^6 years a BH-WD will have MT rates in excess of the Eddington limit. We conclude that all BH binaries that contribute to the current upper-end XLFs of ellipticals are expected to be transient sources.

2. Weighting Factor for the BH Mass Spectrum and the Duty Cycle

The transient BH XRBs (those in outburst) contributing to the upper-end XLF is a sub-set of the true population of BH XRBs in ellipticals determined by the duty cycle of BH transient binaries. For the general case of a transient duty cycle that is dependent on the BH accretor mass, the differential XLF $n(L)_{\mathrm{obs}}$ and the underlying BH mass distribution in XRBs $n(m_{\mathrm{BH}})$ are connected by:

$$n(L_X)_{\mathrm{obs}} = n(m_{\mathrm{BH}}) \times W(m_{\mathrm{BH}}), \qquad (2.1)$$

where $W(m_{\mathrm{BH}})$ is a weighting factor related to the dependence of the transient duty cycle on m_{BH}. The observed slope of the differential upper-end XLF is $\alpha_{\mathrm{d}} = 2.8 \pm 0.6$: $n(L_X)_{\mathrm{obs}} \propto L_X^{-\alpha_{\mathrm{d}}}$. Assuming that $n(m_{\mathrm{BH}}) \propto m_{\mathrm{BH}}^{-\beta}$ and $W(m_{\mathrm{BH}}) \propto m_{\mathrm{BH}}^{-\gamma}$, the slope characterizing the underlying BH mass distribution in XRBs is $\beta = \alpha_{\mathrm{d}} - \gamma$.

At present there are no strong constraints on the duty cycles either from observations or from theoretical considerations. Among known Galactic X-ray transients, typical duty cycles of a few % are favored for hydrogen donors (Tanaka & Shibazaki 1996) and there are no data on duty cycles for transients with a WD companion. For the standard assumption of a constant duty cycle, $\beta = \alpha_{\mathrm{d}} = 2.8 \pm 0.6$. However, according to binary population synthesis models for the Milky Way published so far, the number of formed BH-WD LMXBs exceeds the number of BH-MS LMXBs by a factor of 100 (Hurley et al. 2002). If the duty cycle were similar for both types of systems, we would observe a few hundreds BH-WD binaries; however none such binaries have been identified. We therefore investigate how plausible duty cycle assumptions affect the upper-end XLF shape, considering a variable (dependent on MT rates) duty cycle equal to $\eta = 0.1 \times \dot{M}_{\mathrm{d}}/\dot{M}_{\mathrm{crit}}$. This specific choice of the dependence is not solidly motivated, however, it implies a correlation of the duty cycle with how strong a transient the system is: the further away from the critical MT rate, the smaller the duty cycle.

The expression for η in the case of WD and RG donors can be found analytically. For RG we obtain (using MT rates for RGs from Webbink et al. 1983) $\dot{m}_{\mathrm{RG}}/\dot{m}_{\mathrm{crit}} \simeq 0.2 a^{-0.7} m_{\mathrm{d}}^{1.5}$. Therefore, for RG donors, $\gamma = 0$. Instead there is a significant dependence on the RG donor mass. Since in ellipticals the typical mass for RG donors is about the same as the mass of the turn-off of MS stars, the duty cycle for RG donors depends on the turn-off mass, and hence on the age T [Gyr] of the elliptical. Assuming a flat in the logarithm distribution of orbital separations before MT starts, we find:

$$W(T) \simeq 0.03 \times T^{-0.5}. \qquad (2.2)$$

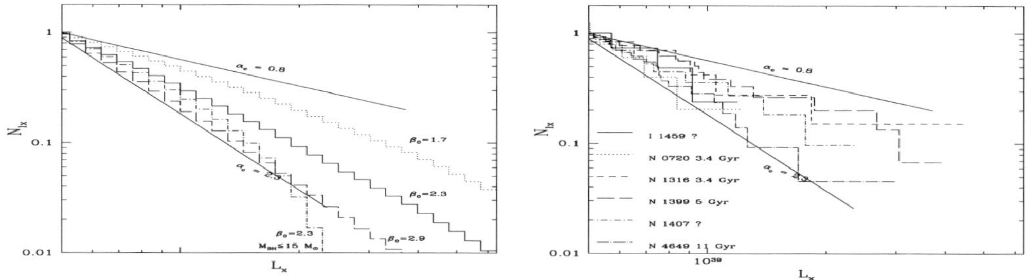

Figure 1. The left panel shows model XLFs for BH-MS binaries. Lines shows results for initial $\beta_0 = 1.7, 2.3, 2.9$ at an age of 10 Gyr (dotted, solid and dashed lines). The dash-dotted line is for a $15\,M_\odot$ BH mass cut-off with $\beta_0 = 2.3$. Thick solid lines corresponds to slopes $\alpha_c = 0.8$ and $\alpha_c = 2.3$. The right panel shows XLFs in observed early type galaxies (data are from Kim & Fabbiano 2004, the ages of the ellipticals are from Ryden *et al.* 2001 and Temi *et al.* 2005).

In order to find the probability that a BH-WD contributes to the upper-end XLF at an elliptical age, we assume that (i) all accreting BH-WD systems were formed within a short interval of elliptical ages several Gyrs ago; and (ii) the binary is a transient at time T. We further adopt a constant BH-WD formation rate during this interval. We obtain:

$$W(T; m_{\mathrm{BH}}) = 1.6 \times 10^{-4} m_{\mathrm{BH}}^{-0.5} t_1^{-7/4} \frac{1 - (t_2/t_1)^{-3/4}}{1 - (t_2/t_1)} \qquad (2.3)$$

Here t_1 and t_2 are the time in Gyrs from the current moment to the start and the end of BH-WD binaries formation. Therefore for WD donors $\gamma = 0.5$ and W depends on both the BH mass and the galaxy age.

As discussed previously, for IT MB, a BH-MS system is transient throughout the MT phase. It also can be seen that the MT dependent duty cycle is about few %, consistent with the observations. Prolonged mass accretion onto the BHs affects their mass spectrum. Since this effect cannot be included analytically, we have examined it quantitatively using simple Monte Carlo simulations. We set up the simulations assuming a flat BH-MS birth (MT onset) rate and a flat mass distribution for donors at the onset of the MT phase, without any restrictions on the mass ratio q_{BH}. For the MT evolution we took into account both IT MB and GR, and if the MT timescale is longer than the thermal timescale of the donor, the donor radius evolution is simply proportional to the mass lost due to MT. On the other hand, if the MT timescale is shorter than the donor's thermal timescale, the donor is out of the thermal equilibrium. In this case we modify the evolution of the donor radius using a prescription that is in acceptable agreement with our results from detailed MT calculations with the stellar evolution: $\delta r \sim \delta m \sqrt{\dot{m}_{\mathrm{TH}}/\dot{m}}$, where $\dot{m}_{\mathrm{TH}} = m_{\mathrm{d}}/t_{\mathrm{TH}}$ is the MT rate driven on the donor's thermal timescale t_{TH}.

Based on the results of our Monte Carlo simulations we find that: (i) due to accretion the BH mass spectrum slope increases by about 0.2, i.e., $\beta = \beta_0 + 0.2$, where β_0 is the BH mass slope at MT onset; (ii) the slope of the BH mass spectrum at the beginning of mass transfer best reproduces the observations with $\beta_0 = 2.3 \pm 0.6$ (see Fig. 1). We also find that the relation between β and β_0 is not sensitive to the age of the elliptical.

3. Conclusions

We conclude that all BH binaries contributing to the upper-end XLF of ellipticals are transient. A constant transient duty cycle independent of the donor type can be excluded unless BH-WD transients have very weak outbursts and can not be detected. The upper-end XLF is formed by an underlying mass spectrum of accreting BHs. The BH X-ray transients have a dominant donor type and an accreting BH mass spectrum slope β that depends on the strength of MB angular momentum loss. In particular, in the case of Skumanich-type MB, the XLF is dominated by BH-RG binaries and $\beta = 2.8 \pm 0.6$. In the case of MB for fast-rotators, only BH-MS transients significantly contribute to the upper-end XLF and $\beta = 2.5 \pm 0.6$ ($\beta_0 = 2.3 \pm 0.6$). If the relative fraction of BH-RG transients in ellipticals is larger than the observed relative fraction in our Galaxy, we expect that the BH-RG binaries contribution will lead to a time-dependence of XLF slopes, where younger ellipticals will have a slope predicted for BH-RG binaries, and older ellipticals a steeper slope predicted for BH-MS binaries.

Acknowledgements

This work is partially supported by a *Chandra* Theory Award to N. Ivanova and a Packard Foundation Fellowship in Science and Engineering to V. Kalogera.

References

Hurley, J. R., Tout, C. A., & Pols, O. R. 2002, MNRAS, 329, 897
Ivanova, N., & Taam, R. E. 2003, ApJ, 599, 516 (IT)
Ivanova, N., & Taam, R. E. 2004, ApJ, 601, 1058
Gilfanov, M. 2004, MNRAS, 349, 146
King, A., 2005, "Compact Stellar X-ray Sources," eds. W.H.G. Lewin and M. van der Klis, CUP
Kim, D., & Fabbiano, G. 2004, ApJ, 611, 846
Rappaport, S., Verbunt, F., & Joss, P. C. 1983, ApJ, 275, 713
Ryden, B. S., Forbes, D. A., & Terlevich, A. I. 2001, MNRAS, 326, 1141
Sarazin, C. L., Irwin, J. A., & Bregman, J. N. 2000, ApJL, 544, L101
Tanaka, Y., & Shibazaki, N. 1996, ARAA, 34, 607
Temi, P., Mathews, W. G., & Brighenti, F. 2005, ApJ, 622, 235
Webbink, R. F., Rappaport, S., & Savonije, G. J. 1983, ApJ, 270, 678

Discussion

LIPUNOV: The initial mass function can not be described as simple power law.

IVANOVA: It is not assumed or insisted. This is what the XLFs tell us about how the BHs mass spectrum looks like, for a specific type of donors. It does not represent the whole spectrum of all ever formed BHs, only for selected type of donors. We hope that this result will provide additional constrains for these scientific teams who work on the connection between pre-collapse stars and the resulting masses of formed compact objects.

KIM: Can you constrain the minimum mass of BH?

IVANOVA: No, we can not make it directly. We can only give constrains on the BH mass spectrum given by our understanding of their donors is correct. E.g., we can not rule out BH less massive than $3M_\odot$ as the possible lack of BH-MS binaries with BHs of such masses in XLFs can be the result of different selection effects originating in the binaries evolution, but not in BHs formation.

SIVAKOFF: You predict that all XRB in the high end of the XLF are transient. Can you comment on the time-scales expected for this transience? In particularly, are the time-scales observable on our life-times?

IVANOVA: We can only give estimates for the duty cycles, but not for the outbursts durations. The latter will have to come from observations.

Populations of High Energy Sources in Galaxies
Proceedings IAU Symposium No. 230, 2005
E. J. A. Meurs & G. Fabbiano, eds.

© 2006 International Astronomical Union
doi:10.1017/S1743921306008714

Cumulative luminosity functions of the X-ray point source population in M31

L. Shaw Greening,[1] C. Tonkin,[1] R. Barnard,[1] U. Kolb[1] and J. P. Osborne[2]

[1]Open University, Walton Hall, Milton Keynes, MK7 6AA, UK
email: L.Shaw-Greening@open.ac.uk

[2]University of Leicester, University Road, Leicester, LE1 7RH, UK

1. Introduction

We present preliminary results from a detailed analysis of the X-ray point sources in the XMM-Newton survey of M31 (e.g. Barnard *et al.* 2005; Pietsch *et al.* 2005). These sources are expected to be mostly X-ray binaries.

2. Source Detection

The left hand panel of Fig. 1 shows a mosaic of GALEX FUV images of M31 (Thilker *et al.* 2005) overlaid by the point sources detected by edetect_chain in SAS 6.0.0. Only sources which were on both the PN and MOS chips were analysed. The minimum detection likelihood was set to 10 and detections that overlapped at a radius of $20''$ were excluded. Lightcurves, energy spectra and power density spectra were then extracted for each region and analysed to determine source properties. Energy spectra were fit with power law, bremsstrahlung, black body and neutron star atmosphere models in xspec 11.3.1 with free parameters to determine the flux of the source. The unabsorbed luminosities were then calculated assuming a distance to M31 of 760 kpc (van den Bergh 2000).

3. Results

In the right hand panel of Fig. 1, we present the first cumulative luminosity functions of the XMM-Newton M31 fields to be created using individual fits for each source spectrum to obtain the unabsorbed luminosity for each source. The North 1 (solid) and 2 (dashed) field luminosity functions are very similar in shape despite having very different numbers of sources. The Core and South 1 luminosity functions have different shapes because these fields have not yet been fully surveyed down to the same luminosity limits as the North fields. The final luminosity limit for each field will also depend on the good-time interval of each observation. We will investigate the significance of these luminosity functions once the survey is complete.

We have also created a luminosity function of the North 1 and 2 fields combined in order to compare it to the luminosity function of these fields reported by Trudolyubov *et al.* (2002). Unlike in our study, these authors used individual spectral fits only for the brightest sources ($L_X > 5 \times 10^{36}$ erg s^{-1}); the faint sources had their luminosities estimated assuming an absorbed power law model with fixed power law index. Above $L_X = 10^{36}$ erg s^{-1} our luminosity function can be approximated by a power law $N(L > L_X) \sim L_X^{-\alpha}$ with slope $\alpha = 0.87$, inconsistent with $\alpha = 1.3$ found by Trudolyubov *et al.* (2002). This is probably a consequence of the fixed spectral model assumed by these authors.

Kong *et al.* (2002) used Chandra to study the core region of M31. They extracted energy spectra from the longest (8.8ks) observation and carried out spectral fitting only

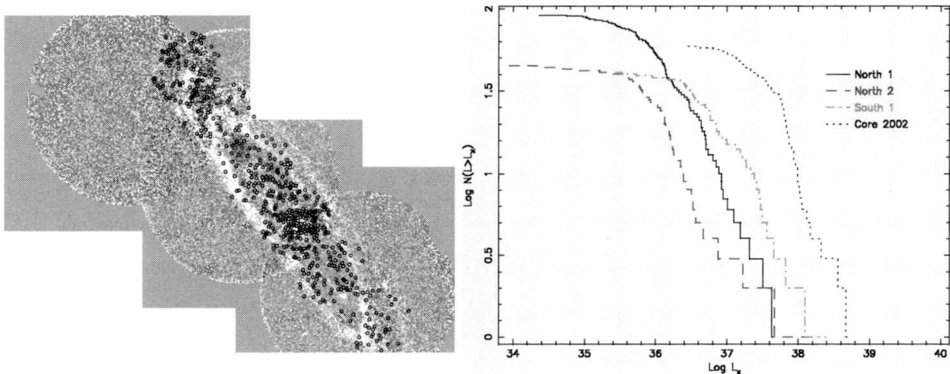

Figure 1. The left hand panel shows the X-ray sources detected, superimposed on a mosaic of GALEX FUV observations of M31. The right hand panel shows the luminosity functions of each of the fields we have studied.

on the brightest 20 of the 204 sources they found. All sources were well fit by both absorbed power law and Raymond-Smith models. For the other 184 sources, count rates were converted to luminosities assuming an absorbed power law model. They obtained a luminosity function described by $\alpha = 1.44$ above a break at 1.77×10^{37} erg s^{-1} for the whole core exposure. This is comparable with our $\alpha = 1.96$ for $37.7 < \log(\mathrm{L}_X$ erg s$^{-1}) < 38.2$ in the Core luminosity function.

Grimm *et al.* (2002) studied the Milky Way and created luminosity functions for the high and low mass binary populations separately. Their cumulative luminosity functions had $\alpha = 0.64$ for HMXBs and $\alpha = 0.26$ for LMXBs. Neither of these slopes is consistent with our luminosity functions of M31 but the binary population cannot be easily separated in M31 and both types of binary are present.

4. Future Work

We have so far studied 225 sources over 4 fields and our survey is continuing by looking at the unstudied sources we have detected in the 2 outermost disc fields, the 130 fainter core sources and the 51 fainter sources in the South 1 field. X-ray binaries will be identified by their energy spectrum and power density spectrum as outlined in Barnard *et al.* (2004, 2005). We will also continue to create luminosity functions for each field to allow us to compare populations. All the luminosity functions we create will also need to be corrected for foreground and background sources (Giacconi *et al.* 2001; Hasinger *et al.* 2001) and for incompleteness (see Kim & Fabbiano (2004))

References

Barnard, R., Kolb, U. & Osborne, J. P. 2004, A&A, 423, 147
—. 2005, astro-ph/0508284
Grimm, H.-J., Gilfanov, M. & Sunyaev, R. 2002, A&A, 391, 923
Giacconi, R., Rosati, P., Tozzi, P., *et al.* 2001, ApJ, 551, 624
Hasinger, G., Altieri, B., Arnaud, M., *et al.* 2001, A&A, 365, L45
Kim, D.-W. & Fabbiano, G. 2004, ApJ, 611, 846
Kong, A. K. H., Garcia, M. R., Primini, F. A., *et al.* 2002, ApJ, 577, 738
Pietsch, W., Freyberg, M. & Haberl, F. 2005 A&A 434, 483
Thilker, D. A., Hoopes, C. G., Bianchi, L., *et al.* 2005, ApJL, 619, L67
Trudolyubov, S. P., Borozdin, K. N., Priedhorsky, W. C., Mason, K. O. & Cordova, F. A. 2002, ApJL, 571, L17
van den Bergh, S. 2000, PASP, 112, 529

Populations of High Energy Sources in Galaxies
Proceedings IAU Symposium No. 230, 2005
E. J. A. Meurs & G. Fabbiano, eds.

© 2006 International Astronomical Union
doi:10.1017/S1743921306008726

X-ray perspective on young stellar populations in objects with nuclear activity - the case of our Galactic Center-

Katsuji Koyama, Atsushi Senda and Shin-Ichiro Takagi

Department of Physics, Kyoto University
email: koyama@cr.scphys.kyoto-u.ac.jp

Abstract. Although our Galactic Center harbors a black hole (Sgr A*) of a few million solar masses, it and its environments are very quiet at present. In X-rays however, the close vicinity of Sgr A* shows very unique and various phenomena mostly originated from young stellar populations. We report on the X-ray perspective on the young stellar populations which are related to our Galactic Center activities. The discussion is essentially based on the observational facts of new X-ray objects in the Galactic Center region in the $1° \times 2°$ area. They are;

1) Clusters of young high mass stars, which are Sgr B2, Arches, IRS 13 and Quintuplet.

2) X-ray reflections in the giant molecular clouds, such as Sgr B2, Sgr C, M0.01-0.09 and others.

3) New candidates of X-ray supernova remnants (SNRs), which are Sgr A East, G0.570-0.018 and G359.8-0.3.

4) Non-thermal Jets, Filaments and Shells, which are unique X-ray features in the GC region.

These X-ray features may be closely related with each other, hence may have common origins. A unified picture is presented for the X-ray activity of our Galactic Center comparing with the X-ray spectra from other type of galaxies such as;

5) Star burst galaxy (NGC 253), low luminosity AGN (M 81) and Seyfert 2 (NGC 1068).

Keywords. X-rays, The Galactic Center, High mass young stars, Young SNRs, Non-thermal jets and filaments, X-ray reflection nebula.

---

1. Introduction

High resolution infrared astrometry over a decade on the Galactic Center (GC) region detected orbital motions of a number of stars around Sgr A* (Schoedel *et al.* 2002). The orbital motions directly constrain the mass of Sgr A* to be about 3 millions of sun, hence support that the GC harbors a massive black hole. However no clear X-ray activity, commonly associated to a massive black hole, has been reported so far. A breakthrough is the discovery of highly ionized iron lines extending around the GC. Likely origin of this emission is a high temperature (10^7-10^8 keV) plasma due possibly to violent explosions in the GC region within the past 10^5 years (Koyama *et al.* 1989). The first hard X-ray imaging satellite, *ASCA* obtained a high quality spectrum of this plasma resolving the iron lines to H-like and He-like irons. The intensity ratio of these two highly ionized iron lines gives clear evidence for high temperature plasma of 5-10 keV. In addition, *ASCA* found strong neutral iron lines at 6.4 keV associated with giant molecular clouds (Koyama *et al.* 1996). The spectrum and morphology are explained to be fluorescence irradiated by strong X-rays (hence: X-ray reflection nebula) coming from the Sgr A* side. Recent *Chandra* observations on the GC have confirmed these early results and furthermore, with its unprecedented spatial resolution, have resolved

Figure 1. Left: The X-ray image of the Galactic Center region in 2-10 keV. Right: Same as the left, but in 6.4 keV (adopted from Koyama *et al.* 1996)

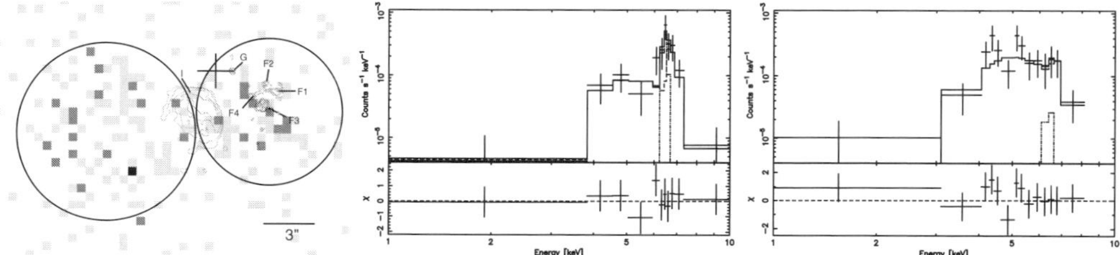

Figure 2. Left: The *Chandra* X-ray image of the ultra compact HII complex, Sgr B2 Main (contour). The X-ray bin size is $0.5'' \times 0.5''$. Middle: The X-ray spectrum and the best-fit model (solid histogram) of a thin thermal plasma with the 6.4 keV line (the dotted histogram) for No 10. Right: Same as No 10 but for No 13. (adopted from Takagi, Murakami & Koyama 2002)

a number of SNRs, clusters of high mass stars, non-thermal X-ray filaments as well as many other 6.4-keV clouds (Murakami *et al.* 2000, Baganoff *et al.* 2001, Murakami *et al.* 2001a, Murakami, Maeda & Koyama 2001b, Murakami 2001c, Wang *et al.* 2002a, Wang *et al.* 2002b, Baganoff *et al.* 2003, Sakano *et al.* 2003, Park *et al.* 2004, Muno *et al.* 2004, Senda 2004). This paper reviews all these observational facts and tries to draw a unified picture of our GC activities.

2. High Mass Star Forming Regions

2.1. *The Sgr B2 cloud*

The Sgr B2 cloud, one of the largest giant molecular clouds in our Galaxy, is located at a projected distance of 100 pc from Sgr A*. The radio continuum band observations have revealed nearly 60 ultra compact (UC) HII regions lying north-to-south (Gaume & Claussen 1990). Bipolar outflows are found from two of the HII complexes, Sgr B2 Main and North (Lis *et al.* 1993). Clusters of OH, H_2O and H_2CO masers are found near these HII regions (Mehringer, Goss, & Palmer 1994). Therefore the cores of Sgr B2 may harbor clusters of high mass stars in a zero-age main-sequence (ZAMS) phase.

Chandra observed Sgr B2 and found more than a dozen X-ray sources, some are at the positions of UCHII complexes (Takagi, Murakami & Koyama 2002). Figure 2 (Left) shows the *Chandra* X-ray image in the 2-10 keV band overlaid on the radio contours of the ultra compact (UC) HII complex, Sgr B2 Main. The UC HII complex comprises several components, F1–F4, G (De Pree, Goss & Gaume 1998) and I (Gaume & Claussen 1990). The cross (+) indicates the position and 1σ errors ($\sim1''$) of a mid-infrared source. The X-ray complex No 10 is extended along the UC HII "island" F3, F4 and G, hence would be comprised of several point sources, possibly high mass stars. In detail, not all the individual X-ray sources coincide to the UC HIIs. Therefore, these X-ray members would be younger than the phase to ionize the circumstellar medium, or ZAMS. Hence the ages of these stars would be well below 10^5 year.

No 13 is also a complex of X-ray sources lying in the east of the UC HII region I, but is associated with no UC HII region. This may indicate that all the high mass members in No 13 are in a pre-phase of UC HII formation, possibly younger than No 10. Still it exhibits X-rays with comparable luminosity to No 10.

Figure 2 (Middle and Right) shows the X-ray spectra, and the best-fit thin thermal models. The best-fit absorptions are 4×10^{23} Hcm^{-2}, one of the largest among the known X-ray emitting stars, indicating that they are really lying at/near the center of the cores of Sgr B2. This demonstrates that hard X-rays are very powerful to discover deeply embedded stars even if they suffer from a large optical extinction of $Av \sim 200$-300 *mag*. The absorption corrected luminosities in the 2-10 keV band for No 10 and No 13 are 8×10^{32} ergs s^{-1} and 13×10^{32} ergs s^{-1}, respectively. As already noted, these X-ray sources are likely to be complexes of several high mass stars, hence the luminosity of the individual high mass star would be in the order of 10^{32} ergs s^{-1}. This is similar value to the nearest high mass young stellar object(YSO), θ C Ori of spectral type O6-7 (Schulz *et al.* 2001) and larger than those of high mass young stars in the Monoceros R2 cloud (Kohno, Koyama & Hamaguchi 2002).

A notable feature of No 10 is strong lines at 6.7 keV and 6.4 keV. The former (6.7 keV line) is the K-shell transition line from He-like irons in thin hot plasma. The latter (6.4 keV line) may be due to neutral or low-ionization irons in a dense circum/inter stellar medium irradiated by the central stars. The iron abundance determined from both of the lines is more than 5 times solar, which is significantly larger than that from any other high mass star forming region (Kohno, Koyama & Hamaguchi 2002, Schulz *et al.* 2001, Yamauchi *et al.* 1996). The iron abundance of No 13, in contrast to No 10, is sub-solar.

2.2. *The Arches and Quintuplet Clusters*

The Arches Cluster is also a giant molecular cloud located near Sgr A* with young stars of ages about 2 Myr. The *Chandra* observations were made in 2001 and 2004, and found three bright X-ray sources at the position of the IR high mass stars No 21, 23 and 26 (Blum *et al.* 2001). No 23 and No 21 are also radio sources AR1 and AR4, hence would have large mass loss of 1.7×10^{-4} and 3.9×10^{-5} M$_\odot$, respectively (Lang, Goss & Rodríguez 2001). Figure 3 (Left) is the *Chandra* X-ray image made by the observation in 2004. The X-ray spectra of the three stars are fitted with a thin thermal model. They all show large X-ray absorptions of $\sim10^{23}$ H cm^{-2}, which is consistent with being the cluster members. The X-ray luminosities in the 2-10 keV band are 2–5×10^{33} ergs s^{-1}, one of the highest among any known young stars.

A unique feature is a diffuse emission elongated to the southeast of the cluster core. The X-ray spectrum shows a strong iron line at 6.4 keV. The morphology and spectral features indicate that the diffuse emission is due to particle or X-ray irradiation from the direction of the cluster core. If the source is X-ray, the required luminosity must be a few$\times10^{35}$ ergs s^{-1}, 10 times brighter than the present value. One possibility is that the

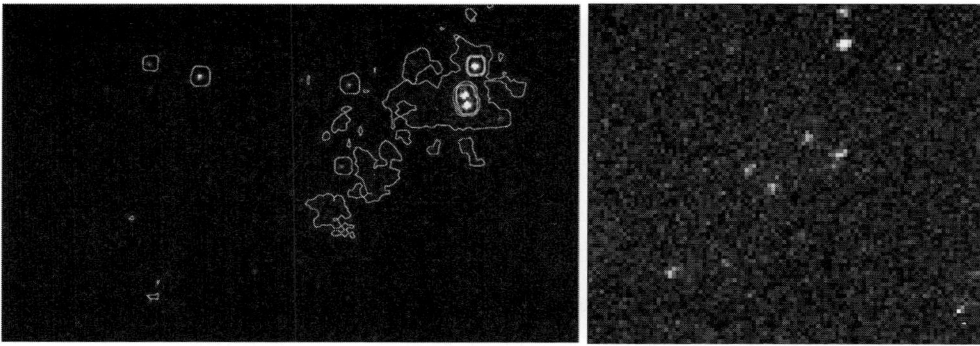

Figure 3. The 2-10 keV band X-ray image of the Arches Cluster (Left) and the Quintuplet Cluster (Right).

Arches cluster was brighter than the present value in the near past. It should be noted, however, no large flux variation is found between the two observations in 2001 and 2004. Another possibility is a jet of high energy particles emanating from the core, because the diffuse emission is collimated to one direction. We note that some young stars emit fairly high speed jets in a very early stage of the star evolution (e.g. Tsujimoto *et al.* 2004). The Arches young stars may have fur active jets, which may be related to extreme physical conditions (e.g., high magnetic field, high density, etc.) in the GC region.

The Quintuplet cluster is located nearly the same distance as the Arches cluster, but is more evolved with estimated age of about 4 Myr. This cluster also includes many bright IR sources, possibly high mass young stars. The *Chandra* X-ray image in Figure 3 (Right) shows several point sources, but only 2 sources are clearly associated to the IR stars (No. 231 and No. 211 in Figer, McLean & Morris 1999). The luminosity of the point sources is in the order of 10^{32} erg s^{-1}, which is not unusually high, but is typical to normal young high mass stars.

2.3. *The GC cluster: IRS 13E*

Many IR star clusters have been reported in the close vicinity of Sgr A*. Although *Chandra* found many moderately bright X-ray sources, possible X-ray association to the IR clusters is only IRS 13E (Baganoff *et al.* 2003). The X-ray luminosity is in the order of 10^{33} erg s^{-1}, which is typical to clusters of young high mass stars. Morris *et al.* (2002) pointed out that a jet-like structure is extending to the north, similar morphology to that of the Arches cluster. Unlike Arches however, the "jet" spectrum is featureless, non-thermal typical to relativistic jets from black holes, for example. Schoedel *et al.* (2005) argued that IRS 13E must include an intermediate mass black hole to be bounded against the large tidal force of Sgr A*. With the available data however, it is hard to conclude whether the jet-like structure is real jet emanating from IRS 13E, or only a series of X-ray sources accidentally aligned.

3. The X-ray Reflection Nebula

ASCA found that the narrow band X-ray image at 6.4 keV in the GC region is totally different from that of the wide band image (see Figure 1). Figure 1 (Right) shows the *ASCA* 6.4 keV band image. Clumpy structures on the giant molecular cloud Sgr B2, M0.01-0.09 and others are found. These 6.4 keV clumps are confirmed by *Chandra* with high resolution imaging. Since Sgr B2 is most deeply observed, we concentrate on Sgr B2 and discuss the origin.

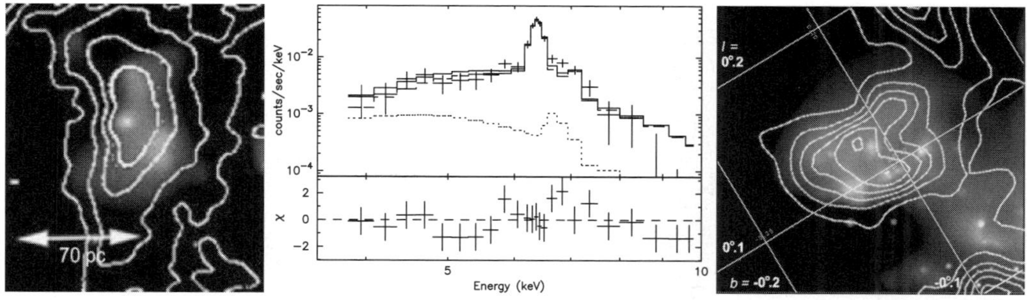

Figure 4. Examples of the X-ray reflection nebulae; Left: The 6.4 keV image of Sgr B2 overlaid on the radio contour. Middle: The spectrum of Sgr B2. Right: The 6.4 keV image of M0.01-0.09 overlaid on the radio contour (adopted from Murakami, Maeda & Koyama 2001b and Murakami 2001c).

Figure 4 (Left) is the *Chandra* image in 6.4 keV overlaid on the radio contours of Sgr B2. We see a crescent shape on the cloud in the near side of the GC and the convexed side pointing to the GC direction. The X-ray spectrum (Figure 4, Middle) shows a strong 6.4 keV line and a sharp edge at 7.1 keV. These structures can only be made by the irradiation of external X-rays on the neutral gas. The crescent morphology with the convexed side pointing to the GC direction also supports that the external X-rays are coming from the GC direction.

To demonstrate this scenario, we made a simulation. The simulated X-ray spectrum is shown by the solid line of Figure 4 (Middle). The morphology and spectrum can be well reproduced by the simulation, supporting the X-ray reflection scenario with external X-ray sources in the GC direction. Thus we call the 6.4 keV clump as the X-ray Reflection Nebula. Since no bright source is found near the GC region, we infer that the irradiation source is Sgr A*, which is presently inactive but was active about 300 years ago, the light traveling time from Sgr A* to Sgr B2.

Although not as clear as the case of the Sgr B2 cloud, we found many 6.4 keV clumps in the close vicinity of the GC (Bamba *et al.* 2002). If not all, these clumps may be also made by the past X-ray activity of Sgr A*. Using the reflection nebula scenario, and using the possible distance from Sgr A* to the reflection nebulae, we can infer that the past X-ray flux of Sgr A* has monotonously declined to the present value (Murakami 2001c).

4. Supernovae and the Remnants

Since the GC region is filled with the X-ray emitting plasma, X-rays due to possible SNRs are hidden in this high X-ray background. Still we discovered several SNR candidates with a thin thermal spectrum (Senda 2004). With *ASCA*, *Chandra* and *XMM-Newton*, we found a ring-like structure (G359.8-0.3) at the south of the GC. This source shows emission lines from highly ionized Si and S. The temperature is about 0.5 keV, hence may be a new SNR. Near the west of Sgr B2, we found a ring-and-tail structure (G0.570-0.018). The spectrum shows a 6.5-6.6 keV line, indicating very hot plasma (a few keV). This also may be a candidate young SNR (Senda 2004).

The most important fact is the confirmation of an energetic SNR associated to the radio shell Sgr A East (Maeda *et al.* 2002). The X-ray spectrum shows highly ionized Si, S, Ar and Fe lines, of which Fe is clearly over-abundant. The intensity ratio of He-like and H-like iron lines give the plasma temperature of about 4 keV (Sakano *et al.* 2003). Therefore, Sgr A East must be very young SNR with the age of about 1000 year.

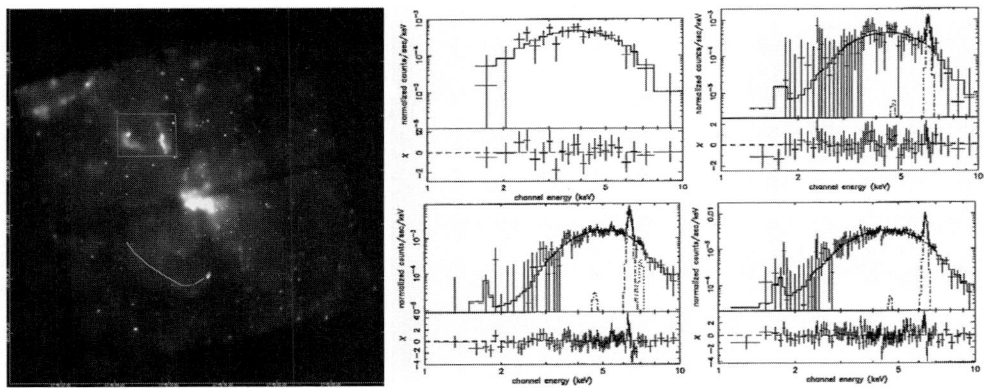

Figure 5. Left: The 3-color X-ray image near the GC region: red=0.5-2 keV, blue=2-6 keV. green=6.4 keV. Upper left is a cluster of X-ray reflection nebulae, while lower left is a non-thermal arc (the blue line, see text). Right: Examples of spectra from the X-ray Reflection Nebulae shown in the Left Figure (adopted from Senda 2004).

Interesting feature of this SNR is a bow-structure with a point source at the north head (Park *et al.* 2004). The spectrum of the point source is a power-law, which resembles to a pulsar. Although no pulsation is found, the point source would be a young neutron star kicked-off by the supernova explosion.

At the southeast from Sgr A*, we see an arc-structure (see Figure 5, Left). This arc may be a part of SNR or a relic of a possible outburst of the GC. Interestingly the spectrum is non-thermal like SN 1006 (Koyama *et al.* 1995). We will discuss the non-thermal X-rays in more detail in the following section.

5. Non-Thermal Knots, Jets and Filaments

In the GC region, there are many knots, or filaments with a non-thermal spectrum (see Figures 6 and 7, and also Morris *et al.* 2002). In Figure 6, we see three elliptical knots on a line going to the northwest from Sgr A*. The major axis of the ellipses are also aligned on this line. Thus the three knots are likely ejected from the GC in several 100 years ago. Most of the other knots and filaments are inside or near the boundary of the SNR Sgr A East, hence may be related to the SN explosion. However, it is not clear whether they are ejected from Sgr A East or Sgr A*, or due to other origins.

From the reflection nebulae, non-thermal jets and the supernova explosion of Sgr A East, we infer a unified model of our Galactic Center activity. As already pointed out a supernova explosion occurred at the center of Sgr A East about 1000 year ago. This made an expanding shocked shell. About 300 years ago, the shocked shell arrived at Sgr A*, the super massive black hole. Since the shocked shell has higher density than the GC environments, a larger mass accretion to Sgr A* had occurred. This produced bright X-rays and ejection of high speed jets.

6. What is our Galactic Center ?

As we have been discussing, many categories of X-ray objects are contributing to the GC activity. If our GC is at a long distance, for example 10-20 Mpc, we can not resolve individual objects in the GC, but can see it as a point or a small spot like the center of other galaxies. The GC spectrum, which includes all the relevant objects, is shown in Figure 8 (Left). A question is "who has a similar spectrum ?" We compare the X-ray

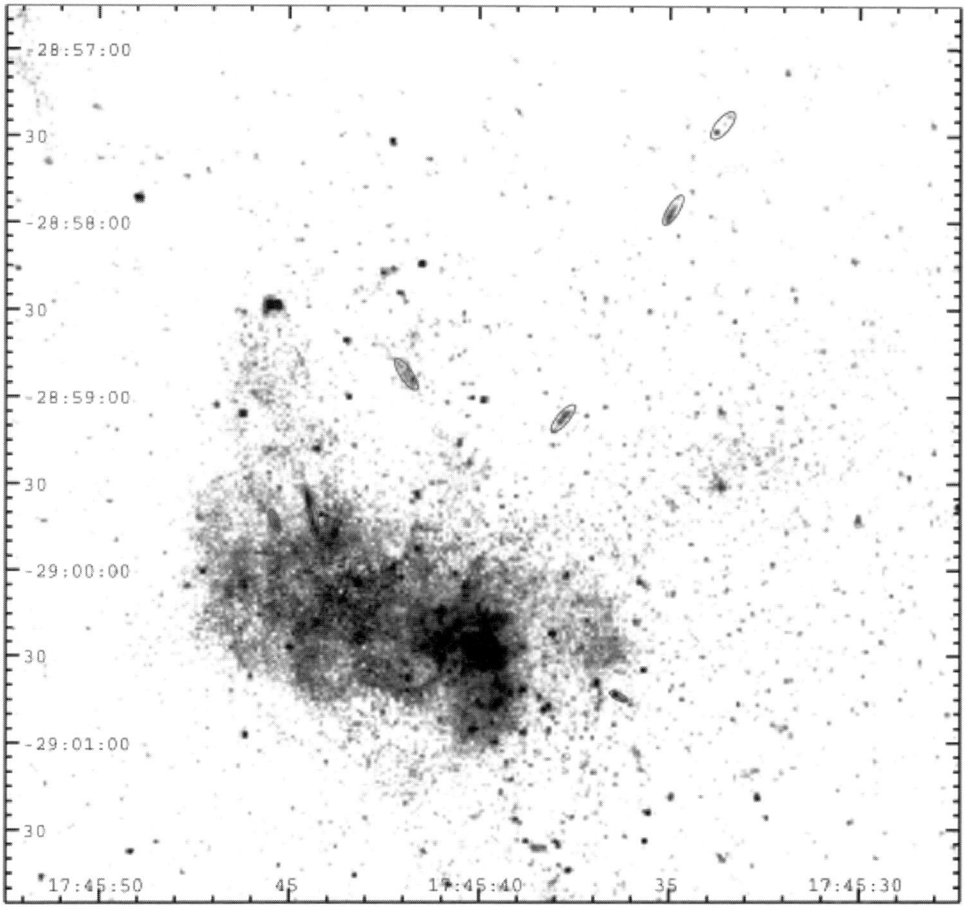

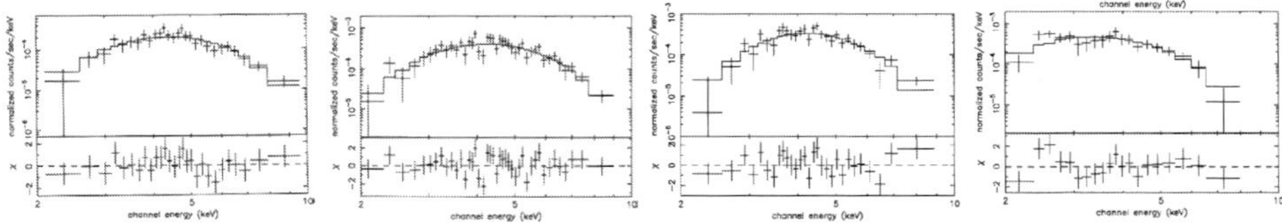

Figure 6. Non-thermal jets, arcs and filaments around Sgr A* (green ellipses). Three knots which are aligned on a line toward the northwest direction would be X-ray jets (adopted from Senda 2004).

Figure 7. Examples of non-thermal spectra for the four nearest knots from Sgr A* (adopted from Senda 2004).

spectra of the centers of different types of galaxies, M31 (twin of the Milky Way), M81 (low luminosity AGN), NGC 253 (starburst galaxy) and NGC 1068 (Seyfert2 galaxy) (Takahashi *et al.* 2004, Pietsch *et al.* 2001, Page *et al.* 2004, Weaver *et al.* 2002, Matt *et al.* 2004). Among them, the NGC 1068 spectrum is very similar to the GC, but its absolute luminosity is far larger. NGC 1068 has a very active nucleus, but is blocked by a dense torus, and what we observe is the reflected component. Also the center of NGC 1068 is currently in a very active star forming phase. Thus the spectrum of NGC 1068 is a combination of the reflected AGN and the starburst activity. Our Galactic Center must be a scale-down version of NGC 1068. The essential difference is that NGC 1068

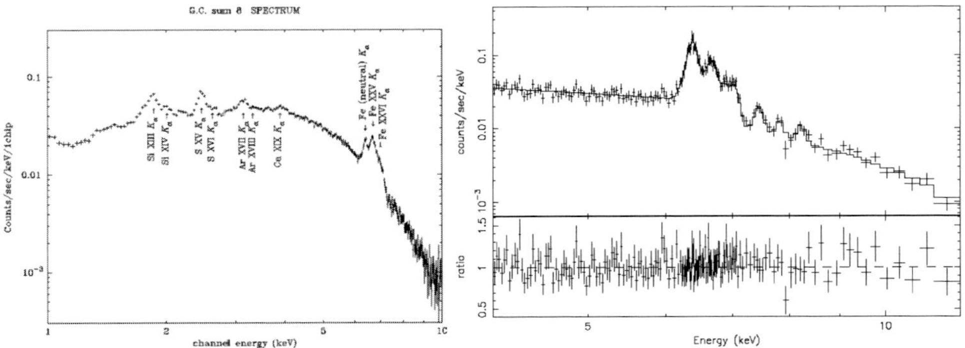

Figure 8. Comparison of the Galactic center spectra of the Milky Way (Left) and the Seyfert 2 galaxy NGC 1068 (Right) (adopted from Koyama *et al.* 1996 and Matt *et al.* 2004).

presently has an active nucleus but is hidden. The supper massive black hole in our Galactic Center is not active at present, but was active in the past and the reflected X-rays have just now arrived at the Earth.

Acknowledgements

We would like to acknowledge Drs Maeda and Murakami, who provide the *ASCA* and *Chandra* results. This work is supported by the Grant-in-Aid for the 21st Century COE "Center for Diversity and Universality in Physics" from the Ministry of Education, Culture, Sports, Science and Technology (MEXT) of Japan.

References

Baganoff, F. K., Bautz, M. W., Brandt, W. N., Chartas, G., *et al.* 2001, *Nature*, 413, 45

Baganoff, F. K., Maeda, Y., Morris, M., Bautz, M. W., *et al.* 2003, *ApJ*, 591, 891

Bamba, A., Murakami, H., Senda, A., Takagi, S. Yokogawa, J. & Koyama, K. 2002, *Proceedings of the Symposium "New Visions of the X-ray Universe in the XMM-Newton and Chandra era"*, Noordwijk, NL

Blum, R. D., Schaerer, D., Pasquali, A., Heydari-Malayeri, M., Conti, P. S. & Schmutz, W. 2001, *AJ*, 122, 1875

De Pree, C. G., Goss, W. M. & Gaume, R. A. 1998, *ApJ*, 500, 847

Figer, D. F., McLean, I. S. & Morris, M. 1999, *ApJ*, 514, 202

Gaume, R. A. & Claussen, M. J. 1990, *ApJ*, 351, 538

Kohno, M., Koyama, K.& Hamaguchi, K. 2002, *ApJ*, 567, 423. also *Erratum* in *ApJ*, 580, 626

Koyama, K., Awaki, H., Kunieda, H., Takano, S. & Tawara, Y. 1989, *Nature*, 339, 603

Koyama, K., Petre, R., Gotthelf, E. V., Hwang, U., Matsuura, M., Ozaki, M. & Holt, S. S. 1995, *Nature*, 378, 255

Koyama, K., Maeda, Y., Sonobe, T., Takeshima, T., Tanaka, Y. & Yamauchi, S. 1996, *Publ. Astron. Soc. Japan*, 48, 249

Lang, C. C., Goss, W. M. & Rodríguez, L. F. 2001, *ApJ* (Letter), 551, L143

Lis, D. C., Goldsmith, P. F., Carlstrom, J. E. & Scoville, N. Z. 1993, *ApJ* (Letter), 402, L238

Maeda, Y., Baganoff, F. K., Feigelson, E. D., Morris, M. *et al.* 2002, *ApJ*, 570, 671

Matt, G., Bianchi, S., Guainazzi, M. & Molendi, S. 2004, *A&A*, 414, 155

Mehringer, D. M., Goss, W. M. & Palmer, P. 1994, *ApJ*, 434, 237

Morris, M., Baganoff, F., Muno, M., Howard, C. *et al.* 2002, *Proceedings of the Galactic Center Workshop*, 167, Hawaii, USA

Muno, M. P., Baganoff, F. K.. Bautz, M. W., Feigelson, E. D. *et al.* 2004, *ApJ*, 613, 326

Murakami, H., Koyama, K., Sakano, M. Tsujimoto, M. & Maeda, Y. 2000, *ApJ*, 534, 283

Murakami, H., Koyama, K., Tsujimoto, M., Maeda, Y. & Sakano, M. 2001a, *ApJ*, 550, 297

Murakami, H., Koyama, K. & Maeda, Y. 2001b, *ApJ*, 558, 687

Murakami, H. 2001c, *PhD thesis, Kyoto University*

Page, M. J., Soria, R., Zane, S., Wu, K. & Starling R. L. C. 2004, *A&A*, 422 , 77

Park, S., Muno, M. P., Baganoff, F. K., Maeda, Y. *et al.* 2004, *ApJ*, 603, 548

Pietsch, W., Roberts, T. P., Sako, M. Freyberg, M. J. *et al.* 2001, *A&A*, 365, L174

Sakano, M., Warwick, R. S., Decourchelle, A. & Predehl, P. 2003, *Mon. Not. Roy. Astron. Soc.*, 340, 747

Sakano, M., Warwick, R. S., Decourchelle, A. & Predehl, P. 2004, *Mon. Not. Roy. Astron. Soc.*, 350, 129

Schoedel, R., Ott, T., Genzel, R., Hofmann, R. *et al.* 2002, *Nature*, 419, 694

Schoedel, R., Eckart, A., Iserlohe, C., Genzel, R. & Ott, T. 2005, *ApJ (Letter)*, 625, L111

Schulz, N. S., Canizares, C., Huenemoerder, D., Kastner, J. H., Taylor, S. C. & Bergtrom, E. J. 2001, *ApJ*, 549, 441

Senda, A. 2004, *PhD thesis, Kyoto University*

Takagi, S., Murakami. H. & Koyama, K. 2002, *ApJ*, 573, 275

Takahashi, H., Okada, Y., Kokubun, M., & Makishima, K. 2004, *ApJ*, 615, 242

Tsujimoto, M., Koyama, K., Kobayashi, N., Saito, M., Tsuboi, Y. & Chandler, C. J. 2004, *Publ. Astron. Soc. Japan*, 56, 341

Wang, Q. D., Gotthelf, E. V. & Lang, C. C. 2002a, *Nature*, 415, 148

Wang, Q. D., Lu, F. & Lang, C. C. 2002b, *ApJ*, 581, 1148

Weaver, K. A., Heckman, T. M., Strickland, D. K. & Dahlem, M. 2002, *ApJ (Letter)* , 576, L19

Yamauchi, S., Koyama, K., Sakano, M. & Okada, K. 1996, *Publ. Astron. Soc. Japan*, 48, 719

Discussion

PANNUTI: Have the three SNRs described in the talk been detected in the radio as well as the X-ray.

KOYAMA: G359 has been detected in the radio: The radio counterpart is shell-like. Another SNR, Sgr A East also has a well-known radio counterpart. The third SNR, G0.3, has not been detected in the radio.

A demonstration of modern Irish dancing at the end of the Dinner.

Session 6

High energy population synthesis

Pranab Ghosh during talk.

Populations of High Energy Sources in Galaxies
Proceedings IAU Symposium No. 230, 2005
E. J. A. Meurs & G. Fabbiano, eds.

© 2006 International Astronomical Union
doi:10.1017/S1743921306008738

Population synthesis of binary relativistic stars

V. M. Lipunov[1]†

Sternberg Astronomical Institute, Moscow State University, Russia
13, Universitetskij pr-t, Moscow, Russia
email: lipunov@sai.msu.ru

Abstract. Population synthesis of binary stars with relativistic component is presented.

Keywords. population synthesis, binary stars, relativistic stars, high energy sources.

1. Introduction

In the early 1980s, our understanding of binary star evolution based on the pioneer work by Paczynski (1971), Tutukov & Yungelson (1973), van den Heuvel & Heise (1972), allowed us to construct a general evolutionary scenario which successfully explained the genesis of well-studied normal stars and offered potential explanation for new X-ray sources discovered in space experiments. On the other hand, it was clear that dramatic new processes should occur after a compact star (white dwarf (WD), neutron star (NS) or black hole (BH)) has been formed in a binary system. Taking account of these processes was especially important in the cases of WD and NS as they can have strong magnetic field and rotate rapidly. Here, we come across a new phenomenon in stellar evolution - the evolution of gravitating magnetic compact stars (gravimagnetic rotators). The original idea goes back to pioneer work by Schwartzman (1971), Illarionov & Sunyaev (1975), Bisnovatyi-Kogan & Komberg (1975), Shakura (1975), Wickramasinghe & Whelan (1975), Lipunov & Shakura (1976), Savonije & van den Heuvel (1977), van den Heuvel (1977), and Lipunov (1982a). The astrophysical manifestations of the magnetized compact star are mainly determined by its interaction with the surrounding plasma by means of two types of physical fields: electromagnetic and gravitational, and the evolution itself represents a gradual change of the character of this interaction. The universality of such an approach is not only its ability to explain apparently such different objects as radiopulsars, X-ray pulsars, X-ray bursters, cataclysmic variables, polars, transient X-ray sources, etc., but also its ability to predict completely new and still undiscovered objects. Therefore, the realistic treatment of binary star evolution must include both types of evolution: the nuclear evolution for the normal stars, and the rotational evolution for the compact magnetized stars Lipunov (1982a). The last fact complicates the evolutionary tree to such a point that the need for a special numerical tool for studying binary evolution (the Scenario Machine), analysis of the observed picture and approval of the evolutionary scenarios becomes quite obvious (Kornilov & Lipunov (1983a), Kornilov & Lipunov (1983b)). Now it consists of a large numerical code that incorporates the crucial physical processes in binary systems and takes into account:

(a) mass exchange between binary components (van den Heuvel & Heise (1972), Tutukov & Yungelson (1973));

† Present address: Sternberg Astronomical Institute, 13, Universitetskij pr-t, Moscow, Russia.

(b) loss of the orbital momentum due to gravitational waves (Kraft (1965), Pachinski (1967), Tutukov & Yungelson (1979));

(c) loss of the orbital momentum due to magnetic wind (Skumanich (1972));

(d) common envelope (Paczynski (1976));

(e) spin evolution of magnetic compact stars (Schwartzman (1970), Lipunov (1982a));

(f) natal kick (Ozernoy (1965), Shklovskii (1970)).

The history of the Scenario Machine can be briefly summarized as follows:

a+e - Kornilov & Lipunov (1983a), Kornilov & Lipunov (1983b) Massive binaries

a+e +f - Kornilov & Lipunov (1984) Massive binaries

a+b + c+d+e - Lipunov & Postnov (1987b) Low Massive Binaries

a + d - Dewey & Cordes (1987) Massive Binaries + radiopulsar evolution

a+b+c+d+e+f - Scenario Machine Lipunov & Postnov (1987a), Lipunov & Postnov (1987b), Lipunov & Postnov (1988) All mass

a+b+c+d - Tutukov & Yungelson (1993) All mass

a+b+c - Kolb & Ritter (1992) Low massive binaries

a - Pols & Marinus (1994) Open Young Clusters

a+b+c+d+f -Portegies & Spreeuw (1996)

a+d+f -Arzoumanian *et al.* (1997) e - for Radiopulsars

a+d+f - Norci & Meurs (2001)

a+b+c+d+f -Kolb *et al.* (2000)

a+b+c+d+f - Tauris *et al.* (2000)

a+d+f - Kalogera & Belczynski (2001)

a+b+c+d+f - Pfahl *et al.* (2003)

The results obtained with the Scenario Machine include the following: 1) Prediction of nearly 100 new stages of binary systems with magnetized compact companions; 2) Prediction of millisecond X-ray accreting pulsars (see Kornilov & Lipunov (1983b)), later confirmed by discovery of A0538-66 (P = 0.067 s); 3)Explanation of some of the ultrasoft superluminous sources as superaccreting compact stars in binaries (Lipunov *et al.* (1993)), which was confirmed by the discovery of a transient X-ray pulsar RX J0059.2-7138 in the Small Magellanic Cloud (SMC) (Hughes (1994)); 4) Prediction of the strong dependence of the X-ray luminosity on star formation history in galaxies; 5) Prediction and estimation of the number of binary radiopulsars with OB-stars (1 per 700 visible galactic pulsars) (Kornilov & Lipunov (1984), Lipunov & Prokhorov (1987)), which was confirmed by discovery of PSR B1259-63 (Johnston *et al.* (1992)); 6) Estimation of the number of binary radiopulsars with black holes (Lipunov (1994)); 7) Calculation of the gravitational wave background formed by galactic and extragalactic binaries (Lipunov & Postnov (1987a), Lipunov & Postnov (1988); Lipunov *et al.* (1995a)); 8) Anisotropy of supernova explosions giving "kick"-velocities of young pulsars of 70-100 km/s (Kornilov & Lipunov (1984); similar estimates are given by Dewey & Cordes (1987)), a more detailed discussion will be given below; 9) A significant evolution of X-ray luminosity of the galaxies (discovered on RXTE, Chandra, INTEGRAL) as indicator of star formation rate history (Tatarintzeva *et al.* (1989)); 10) A significant evolution of supernova rates and binary NS merging rates at cosmological distances (Lipunov & Postnov (1988); Lipunov *et al.* (1995b)).

2. Full classification of NSs

The property that these objects have in common is that their astrophysical manifestations are determined by interaction with the surrounding matter. This interaction is provided by two kinds of physical forces, gravitational and electromagnetic. The importance of this fact was initially understood for NS (Schwartzman (1970), Schwartzman (1971)). In the early 1980s, this approach led to the creation of a complete classification scheme involving various regimes of interaction between neutron stars and their environment, as well as to the first Monte Carlo simulation of the NS evolution (see Lipunov (1982a) and Kornilov & Lipunov (1983a)).

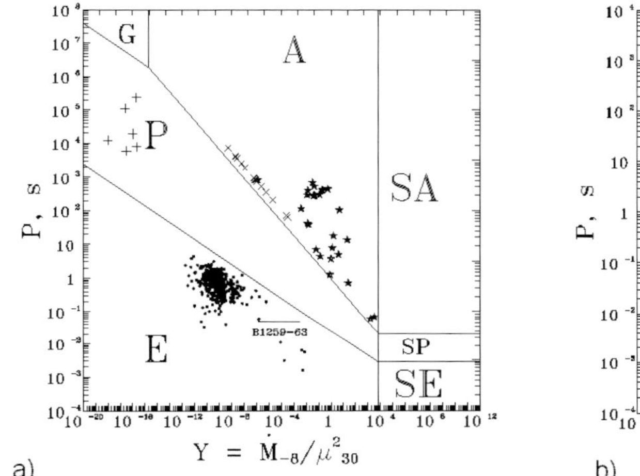

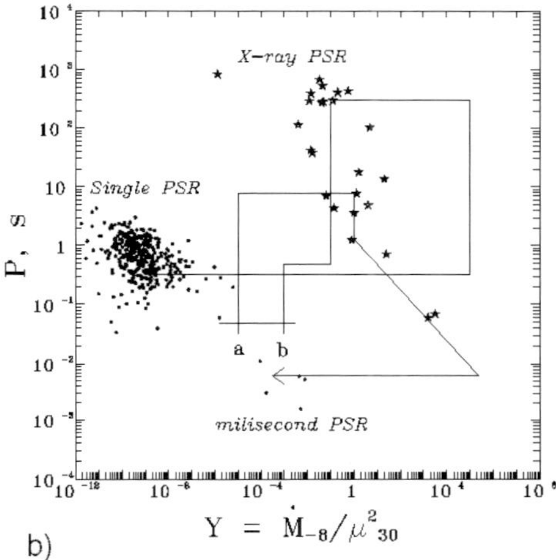

Figure 1. a)The observed magnetic rotators on the universal period - gravimagnetic parameter diagram: "+" - isolated WD × - intermediate polars; ∗ - accreting NS; • - radiopulsars. For isolated rotators, the accretion rate $\dot{M} = 10^{-16} M_\odot/yr$ is assumed. The horizontal bar shows the orbital eccentricity-induced accretion rate change in binary pulsar PSR B1259-63. **A** - Accretor, **P** - Propeller, **E** - Ejector, **G** - Georotator, **SA** - SuperAccretor, **SP** - SuperPropeller, **SE** - SuperEjector (Lipunov *et al.* (1996a)). b)The period-gravimagnetic parameter diagram for NS in binary systems. (a) with NS magnetic field decay (the oblique part of the track corresponds to "movement" of the accreting NS along the so-called "spin-up" line), (b) a typical track of a NS without field decay in a massive binary system (Lipunov *et al.* (1996a)).

The interaction of a magnetic rotator with the surrounding plasma to a large extent depends on the relation between the four characteristic radii: the stopping radius, R_{st}, the gravitational capture radius, R_G, the light cylinder radius, R_l, and the corotation radius, R_c. One can note that the magnetic dipole moment μ and the accretion rate $\dot{M}_c$ always appear in the same combination, $y = \frac{\dot{M}_c}{\mu^2}$, as was first noticed by Davies & Pringle (1981). Analysis of the nature of interaction of a magnetized star with the surrounding plasma allows us to write an approximate evolution equation for the angular momentum of a magnetic rotator in the general form (Lipunov (1982a)) $\frac{dI\omega}{dt} = \dot{M}k_{su} - \kappa_t \frac{\mu^2}{R^3_t}$, where k_{su} is a specific angular momentum applied by the accretion matter to the rotator and R_t is a radius of the interaction.

The evolution of NS in binaries must be studied in conjunction with the evolution of normal stars. For more detail see monographs (Lipunov (1992), Lipunov *et al.* (1996a)).

3. The Effect of Kick Velocity on Binary Pulsar Populations

Let us consider what fraction of different types of binary radiopulsars can be obtained within the framework of the modern evolutionary scenario of binary stars if a phenomenological kick velocity caused by the collapse anisotropy is added. Several attempts of this kind have been made over the last 20 years (Kornilov & Lipunov (1984); Tutukov et al. (1984); Dewey & Cordes (1987)). The observational data existing at that time convincingly pointed to the presence of a small kick velocity of about 70-100 km/s. However, the statistics of binary radio-pulsars at that time was very poor. Of course, all such studies are restricted considering the evolution of an ensemble of stars that initially originated from binaries and were not tidally captured, as occurs in globular clusters; so the numerous binary pulsars observed in globular clusters and whose evolution is not yet fully understood must be excluded from consideration (see Kuranov & Postnov (2004)).

If we make comparison for a number of different binary species, we write this criterion in the form:

$$COOC = \frac{\sum_i w_i (C/C + O/C)_i}{\sum_i w_i}, i = 1, 2...$$

where the sum is taken for each i-th species, and w_i are corresponding weights. Obviously, this function reaches a minimum value of 2 once C=O for all species. Lipunov et al. (1997) compare the calculated ("C") and observed ("O") numbers of binary radio-pulsars with NS (PSR+NS), WD (PSR+WD) and normal OB-stars (PSR+OB).

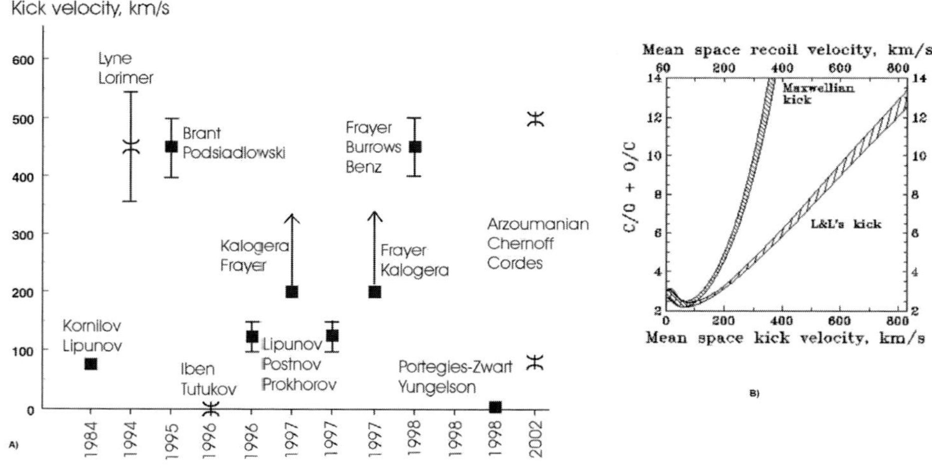

Figure 2. a)History of the kick estimations. b)The COOC criterion for binary radio-pulsar species as a function of the mean kick velocity (Lipunov et al. (1997)).

The calculations presented above clearly demonstrate a strong decrease in number of binary pulsars with diverse secondary components on the assumed kick velocity the NS acquires at birth. Putting aside a thorough analysis of the observational data (which could well be influenced by a number of selection effects leading to an underestimation

Table 1. The observational examples of Binary Stars with relativistic companion. Examples with references are the discoveries predicted by Kornilov & Lipunov (1984).

Optical & compact star's stage	Main Sequence	Giant	Roche Lobe Overflow	WR	Compact
E	PSRB 1259-63 Johnston *et al.* (1992)	?	?	?	PSR 1913+16
P	X0331+53 Corbet (1996)	?	?	?	?
A		Vela X-1	Her X-1	?	XTEJ0929-314 Galloway *et al.* (2002)
SE	No	No	?	No	?
SP	No	No	?	No	?
SA	No	No	SSS?	No	?
BH,	?	CygX-1		Cyg X-3	?
SBH			SS433?UULXs,SSS?		

of low-velocity pulsars; e.g. Tutukov *et al.* (1984)). Recently, Arzoumanian *et al.* (2002) practically confirmed it.

4. X-ray Binaries and Black Holes in Binaries

The first version of the Scenario Machine predicted a lot of different stages of binary stars with relativistic companion, which were discovered after 1983 (see Table 1).

Two different types of stellar BH formation exist: (1) direct core collapse of a massive star and (2) AIC of a NS. Recently calculation of the Black Holes Mass Function (Bogomazov *et al.* (2005)) predicted the Observable Number of Low Massive AIC Black Holes (1 per 20-40 usual Black Holes).

The discovery of binary radio-pulsars with massive unseen companions ($>3 - 5M_\odot$) would be of great importance to fundamental physics and modern theory of stellar evolution, providing a compelling evidence for the existence of BH in nature. The formation of binaries consisting of a BH in a pair with a radio-pulsar (PSR) has been predicted by Kornilov & Lipunov (1983b) and discussed previously by Narayan *et al.* (1991), Lipunov (1994). Lipunov *et al.* (2005) predicted observable number of such systems: (BH+PSR)/PSR $\sim 1/1500$.

5. X-ray luminosity function and Star Formation Rate

The total X-ray luminosity from different types of objects is traced separately at any instant. The X-ray luminosity evolution after an instantaneous star formation burst for $t > 2 \times 10^9$yr (long time scale) can be rather well fitted by a power law (Tatarintzeva *et al.* (1989)): $L(t) \approx 3 \times 10^{40}(\frac{N}{10^{12}})(\frac{t}{10^9 yr})^{-1.56} erg/s$. N is the total number of stars in the galaxy.

Lipunov *et al.* (1996b) studied the evolution of stellar populations on the time-scale of 10 Myr for starbursts. We estimate the expected number of X-ray transient sources (NS + MS star), super accreting BH, and BH + supergiant binaries, all as functions of time elapsed after the burst.

The numerous point-like extragalactic X-ray sources were discovered in last years due to Chandra and XMM-Newton missions. Some authors (Grimm *et al.* (2002), Gilfanov (2004)) report about power law X-ray luminosity function: $\frac{dN}{dL} \sim L^{-\alpha} x SFR$, $\alpha \sim 1.5$., which discussed by Postnov (2003) from theoretical point of view. My critical points concern both observational and theoretical aspect. There is no observed universal luminosity function because: a) The number of the bright X-ray binaries is very small per galaxy; b) All people build the cumulative function with very small number of examples (almost all statistically small population show power low cumulative declination in Log). We must test the differential luminosity distribution; c) We do not know real X-ray luminosity due to high variability binary X-ray sources (on scale from sec. to 100 years)

There is no theoretical simple (with one slope) universal luminosity function because: a) the X-ray population is the mixing of different types of binaries with different mass exchange types: Roche Lobe (thermal time scale, magnetic wind time scale, gravitational waves time scale), Wind Fed, Be-stars disk like wind flow, eruptive mass exchange. For example in our Galaxy most of the X-ray pulsars belong to Be+NS systems. There is not any simple connection between mass loss and for example stars mass (rotational effects). In same time we see less than 5 percent of Be-+NS X-ray systems due to variability of the mass transfer processes; b) The number of the systems with some luminosity depends on spin evolution of the NS which not directly connects with mass of the companion; c) The theoretical arguments cited above are not correct because they exclude the life-time which depends on the optical companion mass of binary stars on accretion stage.

We must observe at least two order more galaxies with determination of the type of the X-ray sources for correct luminosity function. In any case this function must have a different inclination for different type, age, SFR history of the galaxies.

Kraft *et al.* (2005) presented unexpected results from Chandra observation of the nearby galaxy NGC 5102. They ask: Where are the X-Ray Binaries? The deficit of LMXBs is even more striking, because some of these sources may in fact be high-mass X-ray binaries (HMXBs). NGC 5102 is unusually blue for its morphological type and has undergone at least two recent bursts of star formation, only $\sim 1.5 \cdot 10^7$ and $\sim 3 \cdot 10^8$ year ago. If the lack of X-ray binaries is related to the relative youth of most of the stars, this would support models of LMXB formation and evolution that require wide binaries to shed angular momentum on a timescale of Gyr.

6. Ultraluminous X-ray sources

Ultraluminous X-ray sources (ULXs) with $L_x > 10^{39}$ erg/s have been discovered in great numbers in external galaxies with ROSAT, Chandra and XMM-Newton. Rappaport *et al.* (2005) have carried out a theoretical study to test whether a large fraction of the ULXs, especially those in galaxies with recent star formation activity, can be explained with binary systems containing stellar-mass BHs. To this end, we have applied a unique set of binary evolution models for BH X-ray binaries, coupled to a binary population synthesis code, to model the ULXs observed in external galaxies. They find that for donor stars with initial masses $\gtrsim 10 M_\odot$ the mass transfer driven by the normal nuclear evolution of the donor star is sufficient to potentially power most ULXs. This is the case during core hydrogen burning and, to an even more pronounced degree, while the donor star ascends the giant branch, although the latter phases last only 5 per cent of the main-sequence phase. They show that with only a modest violation of the Eddington limit, e.g. a factor of 10, both the numbers and properties of the majority of the ULXs can be reproduced. One of our conclusions is that if stellar-mass BH binaries account for a significant fraction of ULXs in star-forming galaxies, then the rate of formation of such systems is $3 \cdot 10^{-7} \, \mathrm{yr}^{-1}$ normalized to a core-collapse supernova rate of $0.01 \, \mathrm{yr}^{-1}$.

I would like to remember the old consideration of the supercritical accretion onto magnetized neutron stars (Lipunov (1982b)). The Maximum energy release proves to be $L = 46 L_{Edd}(\mu_{30})^{4/9}$, where μ_{30} is the magnetic dipole moment of NS in $(10)^{30}(Gs)(cm)^3$.

7. Relativistic Binary Merging

There are 3 type of merging relativistic stars (GWB - gravitational wave burst):

NS + NS → GWB + GRB?+?

NS + BH → GWB + BH+GRB ?+?

BH + BH → GWB + BH

Binary relativistic stars merging - the most powerfull high energy transients in the Universe. That is equal to Planck luminosity (Lipunov (1992)). After discovery of spectral lines in GRB 970508 (Costa *et al.* (1997), Metzger *et al.* (1997)) we know that in the Universe there are real sources with luminosity more than 10^{50}. The observations of 3 short GRBs this summer (for detail see review of van den Heuvel in these proceedings) in elliptical galaxies make it possible that short GRBs are results of relativistic binary merging. Lipunov *et al.* (1995) calculated the evolution of Double Neutron Stars Merging Rate and the Cosmological Origin of Gamma-ray Burst Sources including history of the star formation rate in the Universe. The calculations demonstrate a good agreement with GRB-statistic and existence of the dark energy $\omega(\lambda) = 0.5 - 0.8$. Recently calculations of the Log N - Log S function separately for long and short GRBs demonstrate good agreement with NS-NS merging for short GRB (Bogomazov *et al.*, 1996).

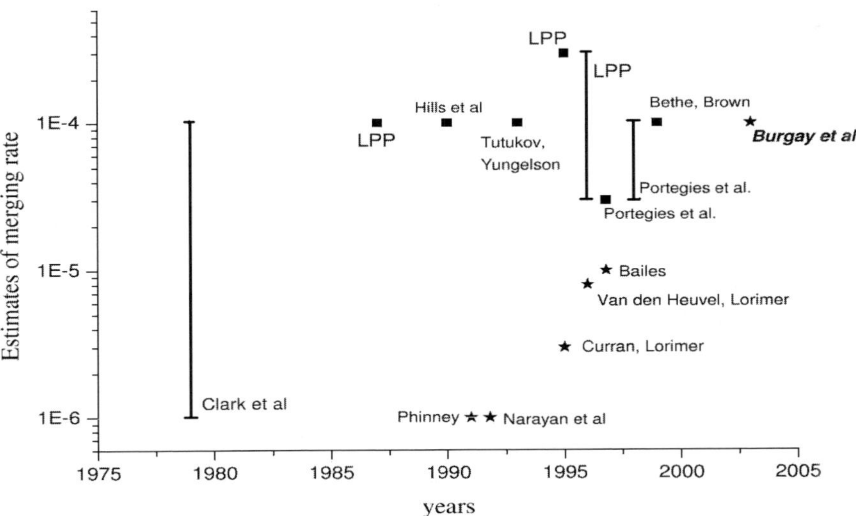

Figure 3. Merging Rate estimation by different authors. Squares are the "theoretical" method, stars are the observational one. LPP = Lipunov,Postnov,Prokhorov.

8. Gravitational Waves

The GW from coalescing compact binaries composed of NS and BH are the best understood of all astrophysical GW sources and very important for LIGO-type experiments. A conservative lower limit to the event rate of galactic binary neutron star coalescence of about 1 per 10^6 year follows from double pulsar statistics studies (Narayan *et al.* (1991), Phinney (1991)). Binary population synthesis, however, give much higher values, of about 1 per 3000 - 10000 year (Lipunov & Postnov (1987a), Tutukov & Yungelson (1993), Lipunov *et al.* (1995b)). Fortunately, discovery of double pulsars J07373039B (Burgay *et al.* (2003)) close the disagreement between radiopulsars merging rate estimation and population synthesis one (see history on Fig. 3). 1) Population synthesis estimations give us the merging rate $10^{-4\pm0.5}yr^{-1}$ from Lipunov *et al.* (1987c). One must accentuate, that the most full and correct model of binary stars evolution is the "Scenario Machine", that takes into account the evolution of magnetized neutron star (see for details Lipunov *et al.* (1996a)) 2) The "observed" estimations, that used radio-pulsars data, were always burdened by selection effects. 3) The gravitation impulses from the merging with black holes participations must be the first events on the interferometers like LIGO. This work is supported by RFFI 04-02-16411 grant.

References

Arzoumanian, Z., Chernoff, D.F., & Cordes,J.M. 2002 *BAAS* 29, 1396

Arzoumanian, Z., Cordes,J.M., & Chernoff, D.F. 1997 *AAS* 568, 289

Bethe, H.A.& Brown, G.R. 1999 *ApJ* 517, 318

Bethe, H.A. & Brown, G.R. 1998 *ApJ* 506, 78

Bisnovatyi-Kogan, G.S., & Komberg, B.V. 1975 *AZh* 52, 457

Bogomazov, A.I., Abubekerov, M.K., Lipunov, V.M. 2005, *ARep* 49, 644

Burgay, M. *et al.* 2003 *Nature* 426, 531

Costa, E. *et al.* 1997 *IAUC* 6649, 1

Davies, R.E., & Pringle. J.E. 1981 *MNRAS*, 196, 209

Dewey, R.J. & Cordes, J.M. 1987, *ApJ* 321, 780

Galloway, D.K., Chakrabarty, D., Morgan, E.H., & Remillard, R.A. 2002 *ApJ* 576(2), L137

Gilfanov, M. 2004 *MNRAS* 349, 146

Gnusareva, V.S., & Lipunov, V.M. 1985 *SvA* 29, 645

Grimm, H.-J., Gilfanov, M. & Sunyaev, R 2002, *Astron.Astrophys.* 391, 923

Hughes, J.P. 1994 *ApJ*, 427, L25

Iben, I.Jr., & Tutukov, A.V. 1986 *APJ* 311, 753

Illarionov, A.F., & Sunyaev, R.A. 1975 *A&A* 39, 185

Johnston, S., Manchester, R., Lyne, A., *et al.* 1992 *ApJ* 387, L37

Jorgensen, H., Lipunov, V., Panchenko, I., Postnov, K., & Prokhorov, M. 1997, *ApJ* 486, 110

Kalogera, V., & Belczynski, K. 2001 *ASSL* 264, 447

Kolb, U. & Ritter, H. (1992) *A&A*, 254, 213

Kolb, U., Davies, M.B., King, A., & Ritter, H. 2000 *MNRAS*317, 438

Kornilov, V.G. & Lipunov, V.M. 1983, *Sov. Astr.* 27, 163

Kornilov, V.G. & Lipunov, V. M. 1983, *Sov. Astr.* 27, 334

Kornilov, V.G. & Lipunov, V.M. 1984, *Sov. Astr.* 28, 402

Kuranov, A.G., Postnov, K.A. 2004 *AstL* 30(3), 140

Kraft, R.P., Nolan, L.A., Ponman, T.J., Jones, C., & Raychaudhury, S. 2005 *ApJ* 625, 785

Lipunov, V.M., Bogomazov, A.I., & Abubekerov, M.K. 2005 *MNRAS* 359, 1517

Lipunov, V.M., Postnov, K.A., & Prokhorov, M.E. 1997, *MNRAS* 288(1), 245

Lipunov, V.M., Postnov, K.A., & Prokhorov, E. M. 1996a, *The scenario machine: Binary star population synthesis, Amsterdam: Harwood Academic Publishers,* 17, 1-160

Lipunov, V., Ozernoy, L., Popov, S., Postnov, K., & Prokhorov, E. 1996b *ApJ* 466, 234 1-160

Lipunov, V., Nazin, S., Panchenko, I., Postnov, K., & Prokhorov, M. 1995a *A&A*, 298, 677

Lipunov, V.M., Postnov, K.A., & Prokhorov, M.E. 1995c *ApJ*, 441, 776

Lipunov, V., Postnov, K., Prokhorov, M., Panchenko, I., & Jorgensen, H. (1995c) 1995 *ApJ* 454, 593

Lipunov, V.M., Postnov, K.A., Prokhorov, M.E., & Osminkin, E.Yu. 1994 *ApJ* 423, L121

Lipunov, V.M., Osminkin, E.Yu., Postnov, K.A., & Prokhorov, M.E. (1993) In Stellar Jets and Bipolar Outflows, eds. L.Errico and A.A.Vittone, Kluwer, Dordrecht, p.361

Lipunov, V.M. 1992 *Astrophysics of Neutron Stars, Springer-Verlag, Berlin - Heidelberg - New York. Astronomy and Astrophysics Library*, 322

Lipunov, V.M., NazinS. 1992 *zhurnal* tom, str

Lipunov, V.M. & Postnov, K.A. 1988 *Ap&SS*, 145, 1

Lipunov, V.M. & Postnov, K.A. 1987a *AZh*, 64, 438

Lipunov, V.M. & Postnov, K.A. 1987b *AZh*, 64, 548

Lipunov, V.M., Postnov, K.A. & Prokhorov, M.E. 1987 *A&A* 176, L1

Lipunov, V.M., & Prokhorov, M.E. 1987 *AZh* 64, 1189

Lipunov, V.M. (1982a) *A&SS*, 85, 451

Lipunov, V.M. (1982b) *SvA* 26, 54

Lipunov, V.M. & Shakura, N.I. (1976) Pis'ma AZh, 2, 343

Metzger, M.R. *et al.* 1997 *IAUC*, 6676, 3

Norci, L. & Meurs, E.J.A. 2001 *ASSL* 264, 561

Narayan, R., Piran, T., & Shemi, A. 1991 *ApJ* 379, L17

Ozernoy, L.M. 1965 *JEPT Lett* 2, 52

Paczynski, B. 1971 *Annual Review of Astronomy and Astrophysics* 9, 183

Paczynski, B. 1976, *Structure and Evolution of close binary systems, IAU Symp. No. 73, eds. P.P.Eggelton, S.A.Mitton and J.A.J.Whelan, Reidel, Dordrecht,* 75

Pfahl, E., Rappaport, S., & Podsiadlowski, P., 2003 *ApJ* 597(2), 1036

Phinney, E. 1991 *ApJ* 380, L17

Pols, O.R. & Marinus, M. 1994 *A&A*, 288, 475

Postnov, K.A. 2003, *Astr. L.* 29, 372

Portegies Zwart S. & Spreeuw, R. 1996 *A&A* 312, 670.

Raguzova, N.V. & Lipunov, V.M. 2000 *ASPC* 214, 685

Rappaport, S.A., Podsiadlowski, Ph. & Pfahl, E. 2005 *MNRAS* 356, 401

Schwartzman, V.F. 1970 *Radiofizika* 13, 1852

Schwartzman, V.F. 1971 *AZh* 48, 438

Shakura, N.I. 1975 *PAZh*, 1, 23

Shklovskii, I.S. 1970 *AZh* 46, 715

Skumanich, A. 1972 *ApJ* 171, 565

Savonije, G.J. & van den Heuvel, E.P.J. 1977 *ApJ*, 214, L19

Tatarintzeva, V., Lipunov, V.M., Osminkin, E.Yu., & Prokhorov, M.E. *Proc. 23rd ESLAB Symp. on Two-Topics in X-ray Astronomy, Bologna, eds. J.Hunt and B.Battrick* 653

Tauris, T.M. & Savonije, G.J. 1999 *A&A* 350, 29

Tauris, T.M., van den Heuvel, E.P.J., & Savonije, G.J. 2000 *ApJ* 530, L93

Tutukov, A.V., & Yungelson, L.R. (1973) Nau. Inform. Astron. Council USSR, 27, 70

Tutukov, A.V. & Yungelson, L.R. 1979 *Acta Astronomica*, 29(4), 665

Tutukov, A.V. & Yungelson, L. R. 1987, *Comments on Astrophysics* 12, 51

Tutukov, A.V., Chugai, N.N., & Yungelson, L.R. 1984 *Sov. Astron. Lett* 10, 244

Tutukov, A.V. & Yungelson, L.R. 1993 *MNRAS*, 260, 675

van den Heuvel, E.P.J. 1977 *Ann. N.Y. Acad. Sci.* 302, 13

van den Heuvel, E.P.J. & Heise, J. 1972 *Nat. Phys. Sci.* 239, 67

Webbink, R.F. 1979, *In White Dwarfs and Variable Degenerate Stars, IAU Colloq., eds. H.M. Van Horn, and V. Weidemann, Rochester University Press, Rochester* 53, 426

Wickramasinghe D.T. & Whelan J.A.J. 1975 *Nature*, 258, 502

Discussion

ERACLEOUS: How important is the proper treatment of stellar evolution in binary evolution calculations?

LIPUNOV: This is very important because we do not have many other ways of learning about stellar rotation (for example) other than using population synthesis of binaries.

FABBIANO: Granted that stellar populations and their evolutions are complicated, do you think we can constrain these parameters by comparing your models with e.g. XLFs?

LIPUNOV: In principle, yes. But we must use separate X-luminosity function for each type of X-ray sources. 16 years ago we predicted high strong dependence of X-ray population on star formation rate, in this sense the most suitable situation in elliptical Galaxies. For spirals and irregulars we must have much more statistics. Another problem connected with variability of X-ray sources, for example, in our Galaxy more than 10^3 Be + XPSR systems, but we see only 10%.

Populations of High Energy Sources in Galaxies
Proceedings IAU Symposium No. 230, 2005
E. J. A. Meurs & G. Fabbiano, eds.

© 2006 International Astronomical Union
doi:10.1017/S174392130600874X

On the faint X-ray sources in the Galactic center region

X.-W. Liu and X.-D. Li

Department of Astronomy, Nanjing University, Nanjing 210093, P. R. China
email: liuxw@nju.edu.cn, lixd@nju.edu.cn

Abstract. We have performed evolutionary population synthesis calculations to investigate the nature of the faint X-ray sources in the Galactic center region detected by recent *Chandra* surveys. Our results show that neutron star low-mass X-ray binaries contribute significantly to the observed sources, but the majority of the sources in Wang *et al.* (2002) survey are still beyond our expectation. We also point out that wind-accreting neutron stars and intermediate polars play a minor role in accounting for the faint X-ray sources in Wang *et al.* (2002) and Muno *et al.* (2003) surveys, respectively.

Keywords. stellar evolution, compact stars, X-ray binaries.

1. Introduction

Recent *Chandra* observations of the Galactic center region (GCR) have uncovered a population of faint discrete X-ray sources. Wang *et al.* (2002) have reported hundreds of X-ray sources with luminosities $L_{\rm X} \sim 10^{33} - 10^{35}\,{\rm ergs}^{-1}$ in the GCR. Muno *et al.* (2003) detected $\sim$2000 point sources during their 590 ks *Chandra* observations of the $17' \times 17'$ field around Sgr A*.

The nature of the faint X-ray sources in the GCR has been investigated by several authors (Pfahl *et al.* 2002, PRP02; Belczynski & Taam 2004, BT04). In this paper, we employed the evolutionary population synthesis (EPS) method to calculate the expected numbers and luminosity distributions of various types of X-ray sources in Wang *et al.* and Muno *et al.* surveys. We describe the model in §2. The calculated results are presented in §3. Our discussion and conclusions are in §4.

2. Model

We have used the EPS code developed by Hurley *et al.* (2000, 2002) to calculate the expected numbers in the Galactic disk for various types of X-ray source populations. Most of our adopted parameters are the same as those described in Hurley *et al.* (2002) except that we assume that the supernova kick velocity has a Maxwellian distribution with standard deviation $\sigma = 265\,{\rm kms}^{-1}$ (Hobbs *et al.* 2005).

2.1. *Physical Models for X-ray Sources*

2.1.1. *Wind-fed neutron stars (WNSs)*

A WNS is a neutron star (NS) accreting wind material from an intermediate- or high-mass un-evolved companion (PRP02). We consider the total spin evolution of NSs including the ejector, propeller and accretor stages (Davies & Pringle 1981; Ikhsanov 2001). The parameters of the natal NSs are set as follows. The initial spin periods P and magnetic fields B are chosen so that $\log P$ and $\log B$ are distributed normally with a mean of -2.3 and 12.5 respectively with a standard deviation of 0.3.

2.1.2. *Intermediate polars (IPs)*

Muno *et al.* (2004) suggested that the majority of the GC sources are IPs. To examine this idea, we include cataclysmic variables in our calculations, and set 5% of them are IPs (Kube *et al.* 2003).

2.1.3. *Low-mass X-ray binaries (LMXBs)*

This type of systems contains NSs or black holes (BHs) accreting from low-mass companions that overfill their Roche lobes. Most of LMXBs are transients, with the majority of their time spent in quiescence with $L_X < 10^{34}\,\mathrm{ergs}^{-1}$ (van Paradijs 1996). We have adopted the theoretical model of Menou *et al.* (1999) for X-ray emission of quiescent LMXBs. Both normal star and white dwarf (WD) donors are included in our simulations. Apart from core collapse SNe, the code also includes the formation of NSs via accretion-induced collapse of massive WDs.

2.1.4. *Rotation-powered pulsars*

In the ejector stage, the spin-down of NSs is governed by the canonical rotation-powered pulsar mechanism. There appears to be a strong correlation between the rate of rotational energy loss $\dot{E}$ and their X-ray luminosities L_X (Seward & Wang 1988). We use the $L_X - \dot{E}$ correlation in Possenti *et al.* (2002) to estimate the $2-10$ keV X-ray luminosities of pulsars.

2.1.5. *Massive stars with strong winds*

The X-ray emission in massive stars is generated in the lower layers of the strong winds, where the plasma is heated in shocks. For single and binary massive stars, we calculate their X-ray luminosities based on the empirical relations in Chelbowski & Garmany (1991) and Portegies Zwart *et al.* (2002a), respectively.

2.2. *Luminosity correction*

For WNSs, IPs, and quiescent LMXBs, we assume their $2-10$ keV luminosities to be half the calculated bolometric luminosities. No correction is needed for rotation powered pulsars. For massive stars with strong winds the previously estimated luminosities are in the $0.5-3.5$ keV band, we choose the correction factor 0.1 for the $2-10$ keV luminosities. For disk accreting sources (i.e. IP and LMXBs), the geometric effect of the disks (Zhang 2005) is also taken into account for the apparent X-ray luminosities.

2.3. *Expected Numbers*

The size of Wang *et al.* field contains $\leqslant 1\%$ of the total Galactic disk population (PRP02). After we get the total number N_{Gal} of a specific type of sources in the Galactic disk, then $\sim N_{\mathrm{Gal}}/100$ and $\sim N_{\mathrm{Gal}}/100/4$ sources of this population are expected to be found in Wang *et al.* and Muno *et al.* surveys, respectively, since the latter field contains stars 4 times less than the former (BT04). Note that sources found in these two surveys have different luminosity ranges.

3. Results

We adopt a variety of models (see Table 1), each with different assumptions for the parameters that govern the evolution in the calculations (common envelope parameter α and wind velocity parameter β). Tables 2 and 3 summarize the calculated numbers of various classes of X-ray sources contributed to Wang et al. and Muno *et al.* surveys, respectively.

Table 1. Model parameters.

Model	α	β	tides
A	1.0	1.0	OFF
B	0.5	1.0	OFF
C	1.0	4.0	OFF
D	1.0	0.25	OFF
E	1.0	1.0	ON

Table 2. The expected numbers of X-ray sources in Wang *et al.* survey with $L_{X,2-10keV}$ in the range of $10^{33} - 10^{35}\,\mathrm{ergs^{-1}}$. In parentheses we list the expected numbers of X-ray binaries with WD donors. The expected numbers after correcting for the inclination effect are listed in the last column.

Model	WNSs	NS LMXBs	BH LMXBs	IPs	Pulsars	TOTAL	TOTAL-inc
A	< 1	3(0)	< 1(0)	1	9	13	13
B	< 1	< 1(0)	< 1(0)	1	9	9	9
C	< 1	4(0)	< 1(0)	1	9	14	14
D	< 1	< 1(0)	< 1(0)	2	9	11	11
E	< 1	2(0)	1(0)	1	9	13	12

Table 3. The expected numbers of X-ray sources in Muno *et al.* survey with $L_{X,2-10keV}$ in the range of $10^{30} - 10^{33}\,\mathrm{ergs^{-1}}$.

Model	IPs	NS LMXBs	BH LMXBs	Pulsars	Massive Stars	TOTAL	TOTAL-inc
A	165	648(646)	132(0)	28	72	1045	912
B	169	195(194)	7(0)	28	71	470	405
C	163	642(641)	130(0)	28	72	1035	905
D	176	935(934)	158(0)	28	72	1369	1202
E	706	1142(1141)	270(0)	28	72	2218	1963

Figure 1 shows the total X-ray luminosity distributions for several models (results for models C and D share similar features as that in model A). Distributions after correcting for the inclination of accretion disks are also shown.

4. Discussion and conclusions

By using the EPS method, we have investigated the population of faint X-ray sources in the GCR. For Wang *et al.* survey, we find that a considerable fraction of the discrete sources are young pulsars and NS LMXB transients in quiescence, while WNSs proposed by PRP02 have negligible contribution due to the propeller effect. For Muno *et al.* field, IPs present a minor contribution, and the majority of the X-ray sources are NS LMXB transients with WD donor stars. The latter result is consistent with BT04 analysis.

The results are obviously subject to the star formation history and initial mass function (IMF) in the GCR. It is still a matter of debate as to whether the star formation is continuous or episodic, and whether it occurs only in localized regions or is relatively uniform throughout the GC (Muno *et al.* 2004). Figer *et al.* (2004) suggested that the stat formation is probably continuous over the last ~ 10 Gyr, as we did in this paper. The molecular clouds in the GCR have higher densities and temperatures than those in the disk due to the tidal forces. This environment might favor the formation of massive stars (Morris 1993). The *Hubble Space Telescope* observation of Arches Cluster suggested

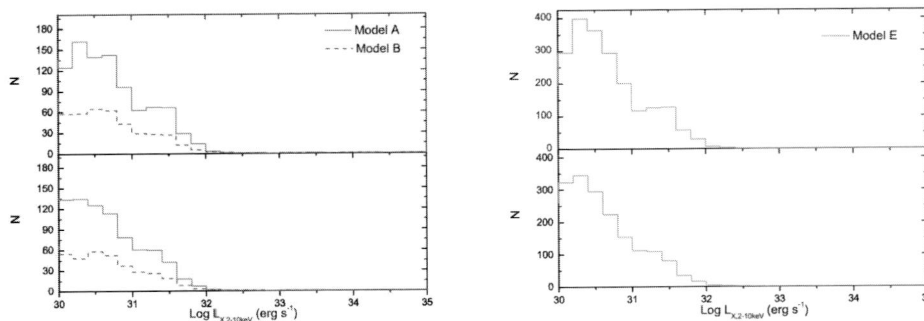

Figure 1. Left: The total number of X-ray sources as a function of X-ray luminosity for models A (solid lines) and B (dashed lines), in which no tidal effect is considered. The upper and lower panel show the intrinsic and apparent distributions after correcting for the accretion disk inclination effect respectively. Right: Luminosity function for model E with tidal effect considered.

a fairly flat IMF with a slope of ~ -1.8 (Figer *et al.* 1999). Portegies Zwart et al.(2002b), however, showed that the observed characteristics (unusually flat mass function and overabundance of massive stars) of the Arches cluster are consistent with a perfectly normal IMF. The observed anomalies are then caused by a combination of observational selection effects and the dynamical evolution of the cluster.

Acknowledgements

This work was supported by NSFC under grant number 10025314 and MSTC under grant number NKBRSF G19990754.

References

Belczynski, K., & Taam, R. E. 2004, ApJ, 616, 1159 (BT04)
Chelbowski, T., & Garmany, C. D. 1991, ApJ, 368, 241
Davies, R. E., & Pringle, J. E. 1981, MNRAS, 196, 209
Figer, D. F. *et al.* 1999, ApJ, 525, 750
Figer, D. F. *et al.* 2004, ApJ, 601, 319
Hobbs G. *et al.* 2005, MNRAS, 360, 963
Hurley J. R. *et al.* 2000, MNRAS, 315, 543
Hurley J. R. *et al.* 2002, MNRAS, 329, 897
Ikhsanov, N. R. 2001, A&A, 375, 944
Kube, J. *et al.* 2003, A&A, 404, 1159
Menou, K., *et al.* 1999, ApJ, 520, 276
Morris, M. 1993, ApJ, 408, 496
Muno, M. P. *et al.* 2003, ApJ, 589, 225
Muno, M. P. *et al.* 2004, ApJ, 613, 1179
Pfahl, E. *et al.* 2002, ApJ, 571, L37 (PRP02)
Possenti, A. *et al.* 2002, A&A, 387, 993
Portegies, Zwart, S. F. *et al.* 2002a, ApJ, 574, 762
Portegies, Zwart, S. F. *et al.* 2002b, ApJ, 565, 265
Seward, F. D., & Wang, Z. 1988, ApJ, 332, 199
van Paradijs, J. 1996, ApJ, 464, L139
Wang, Q. D. *et al.* 2002. Nature, 415, 148
Zhang, S. N. 2005, ApJ, 618, L79

Discussion

ERACLEOUS: What did you assume for the star formation history of the galactic center region?

LI: We assumed a constant SFR similar to the galactic disk.

LIPUNOV: What magnetic field decay hypothesis you used? The Kick velocity which you used 2 times high than reasonable value.

LI: We don't use any magnetic decay law. Only one, for HMBS – const $\sim 10^{12}$ Gs and for LMBS – 10^8 Gs.

Populations of High Energy Sources in Galaxies
Proceedings IAU Symposium No. 230, 2005
E. J. A. Meurs & G. Fabbiano, eds.

© 2006 International Astronomical Union
doi:10.1017/S1743921306008751

Effect of metallicity on low mass X-ray binaries in globular clusters

Natalia Ivanova[1]

[1]Physics and Astronomy Department, Northwestern University, 2145 Sheridan Rd, Evanston
IL 60208
email: nata@northwestern.edu

Abstract. We propose that the observed difference in the formation rates of bright low-mass X-ray binaries in metal-rich and metal-poor globular clusters can be explained by taking into account the difference in the stellar structure of main sequence donors with masses between $\sim 0.85 M_\odot$ and $\sim 1.25 M_\odot$ at different metallicities. This difference is caused by the absence of an outer convective zone in metal-poor main sequence stars in this mass range. In the result, magnetic braking, a powerful mechanism of orbital shrinkage, does not operate and dynamically formed main sequence – neutron star binaries fail to start mass transfer or appear as bright low-mass X-ray binaries.

Keywords. binaries: close – X-rays: binaries – globular clusters:general – stellar dynamics.

1. Observational motivations

Bright low-mass X-ray binaries (LMXBs) in globular clusters (GCs) of our Galaxy and of M31 preferentially reside in metal-rich GCs (Grindlay 1993, Belazzini *et al.* 1995). Recent extragalactic observations revealed the same tendency: in NGC 4472 (Kundu *et al.* 2002, Maccarone *et al.* 2004) and in M87 (Jordán *et al.*, 2004) metal-rich GCs are 3 ± 1 times as likely to contain bright LMXBs as metal-poor GCs.

In detail, from observations of galactic GCs, two currently known GC-residing persistent LMXBs with a possibly non-degenerate (but also not a subgiant) companion are contained only in metal-rich clusters (Verbunt & Lewin 2005), as well as two known GC-contained luminous transient LMXBs with a possibly non-degenerate companion. Five persistent bright LMXBs are observed in relatively metal-poor clusters. Four of them show indications of being ultracompact and one is likely to be a subgiant according to its orbital period. One luminous transient LMXB in a metal-poor cluster is identified as an ultracompact, and for the bright LMXB in Terzan 1 (metal-poor), there is no information on whether it is ultracompact or not.

This data set hints that possibly bright LMXBs consisting of a NS and a main sequence (MS) star can preferably be formed in metal-rich clusters.

2. Dynamically formed NS-MS binaries

NS-MS binaries could form via exchange encounters or via tidal captures. In the first case, the minimum period of the post-exchange binary is limited both by the period at contact for a given eccentricity and by the period determined by the binary energy conservation during the encounter. Due to the energy conservation, when NS replaces a less massive star, the binary separation in the formed post-exchange binary is larger than the binary separation in the pre-exchange binary.

The binary periods of tidally captured binaries are limited at the upper limit by the closest approach at which tidal interactions are still strong enough to make a bound system. The lower limit is determined by the closest approach at which a MS star will pass and still will not be destroyed. The parameter space for tidally captured binaries where the MS star does not overfill its Roche lobe at the closest approach is very restricted.

3. Successful LMXBs Candidates

NS-MS binaries are evolved under the influence of gravitational radiation, magnetic braking and tides. In metal-poor clusters, only stars $\leqslant 0.85 M_\odot$ have the developed outer convective zone – only there magnetic braking and effective convective tides operate. This determines the maximum initial binary period, for different eccentricities, such that the binary will start the mass transfer (MT) before the MS star leaves the MS or another dynamical encounter occurs.

As a result, the parameter-space available for post-exchange binaries that can successfully start MT in metal-rich clusters is substantially larger than in metal-poor clusters. Also, in metal-poor clusters, the efficiency of tidal captures is significantly reduced for MS stars with radiative envelopes ($\geqslant 0.85 M_\odot$) and only MS stars with masses less than $0.55 M_\odot$ can be formed via tidal capture without overfilling their Roche lobe during the event.

4. Persistent or transient?

In metal-poor clusters, all MT NS-MS binaries with a MS star $\geqslant 0.85 M_\odot$ will evolve on the time-scale predicted by gravitational radiation until the donor decreases its mass enough to develop the deep outer convective zone. Before this moment, the MT binary appears as a transient X-ray source. When the deep outer convective zone is developed, the MT rates jumps and an LMXB can become persistent only for a very short period of time.

In metal-rich clusters, an LMXB appears first as a persistent LMXB. When the donor decreases its mass to $\sim 0.7 M_\odot$, the MT rate becomes very close to critical and remains close for a long time. Even a slight discrepancy in the value of $\dot{M}_{\rm crit}$ leads to large differences in how much time an LMXB spend as persistent $\tau_{\rm pers}$ or as a transient system $\tau_{\rm tr}$. This affects how many qLMXBs and how many bright LMXBs can be present in metal-rich clusters.

Acknowledgements

This work is supported by a *Chandra* Theory Award to N. Ivanova.

References

Bellazzini, M., Pasquali, A., Federici, L., Ferraro, F. R., & Pecci, F. F. 1995, ApJ, 439, 687
Grindlay, J. E. 1993, ASP Conf. Ser. 48: The Globular Cluster-Galaxy Connection, 48, 156
Jordán, A., *et al.* 2004, ApJ, 613, 279
Kundu, A., Maccarone, T. J., & Zepf, S. E. 2002, ApJL, 574, L5
Maccarone, T. J., Kundu, A., & Zepf, S. E. 2004, ApJ, 606, 430
Verbunt, F. & Lewin, W., in press, Chapter 8 in "Compact Stellar X-ray Sources," eds. W.H.G. Lewin and M. van der Klis, Cambridge University Press.

Populations of High Energy Sources in Galaxies
Proceedings IAU Symposium No. 230, 2005
E. J. A. Meurs & G. Fabbiano, eds.

© 2006 International Astronomical Union
doi:10.1017/S1743921306008763

On the occurrence of colliding-wind binaries in R136

L. Norci[1], J. Hartwell[2] and E. J. A. Meurs[2]

[1]School of Physical Sciences, Dublin City University, Glasnevin, Dublin 9, IRL
email: lno@physics.dcu.ie

[2] Dunsink Observatory, Dublin Institute for Advanced Studies, Castleknock, Dublin 15, IRL
email: ejam@dunsink.dias.ie

Abstract. Interpretative analyses of the X-ray emission from the giant starformation region R 136 have concluded that several colliding-wind binaries are likely to contribute to its X-ray output. Using our dedicated high-energy stellar population synthesis programme, we try to reproduce the suggested number of colliding-wind binaries. It appears that only assuming a very high binary fraction for the cluster's stellar population we can reproduce the observed X-ray luminosity distribution, if also the two most luminous sources are in fact multiple sources.

Keywords. Stars:binaries, Stars:X-ray.

In a starburst region several components contribute to the X-ray emission. Single stars emit X-rays via shocked stellar winds or coronal emission. The evolved binary stars are stronger emitters. where the X-ray radiation is produced by an accretion process from an expanding normal star to a compact companion. Supernova remnants are also strong soft X-ray emitters, producing X-rays from shocked material. Stars emit X-rays at various levels in essentially every phase of their lives. Both the accretion process and X-ray emission from the shocked material in a supernova explosion are instead short-lived occurrences.

The soft X-ray emission from stars has been extensively studied with the help of the data sets provided by the ROSAT Survey and is fairly well known. The binary stars present on the other hand a more difficult task. Whether the accretion process occurs or not in a certain binary system and with what modalities depends on a number of circumstances that are known with varying degrees of certainty.

Since the R 136 cluster is a very young cluster with an age between 1–2 Myr no supernova can yet have occurred and therefore interacting close binaries with a compact companion have not been produced either. The contribution of the colliding wind binaries has on the other hand to be taken into account.

We have simulated a realistic cluster for R 136 composed of single stars and binaries. We have used for the evolution of the single star population the Geneva evolutionary tracks with metallicity $Z = 0.008$ and normal mass loss. We have assumed a Salpeter initial mass function for the single stars and the primaries of binary systems. A flat primary/secondary mass ratio distribution and a binary separation distribution which is flat in the logarithm have been adopted for this simulation. Spectral types are assigned following the temperature scale of Landolt-Börnstein.

The results presented here are produced assuming an instantaneous burst of star formation 2 Myr ago and a mass range 15–120 $M_\odot$. We have imposed, for the comparison with real data, that the distribution and number of stars above 50 $M_\odot$ must reproduce the observed mass distribution. The X-ray luminosity of colliding wind binaries is calculated with the formula of Portegies-Zwart *et al.* (2002). The observed X-ray luminosities

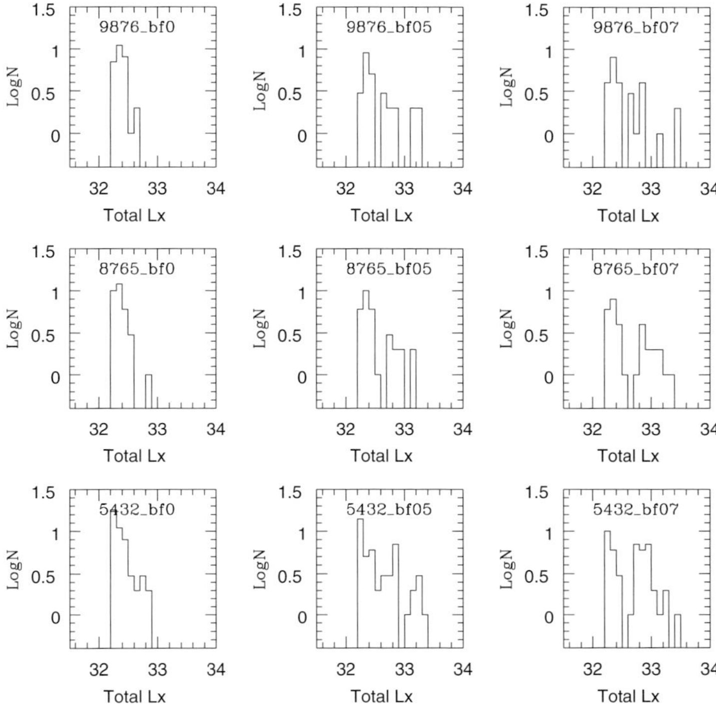

Figure 1: Simulation of the distribution of X-ray luminosities in the range of the observed Chandra sources; results are shown for three different initial mass distributions, including in the bottom row the simulation which we deem more representative of this cluster top masses. From left to right the binary fractions increase as 0, 0.5 and 0.7.

for the top masses in R 136 are taken from recent Chandra observations of the cluster region (Portegies-Zwart *et al.* 2002) and converted to the ROSAT range for comparison with our simulations. The mass distribution is taken from Vacca *et al.* (1996) and Chlebowski and Garmany (1991).

The results of our simulations are shown in Figure 1. These simulations show that a substantial number of binaries must be present if the top of the observed X-ray luminosity distribution is produced by colliding wind binaries.

In our simulations we assume that the most luminous X-ray sources in R 136 are colliding wind binaries and that the two most luminous sources (above 10^{35} erg s^{-1}) are in fact multiple sources, for which there is clear evidence for at least one of them. We conclude that the luminosity distribution of the remaining Chandra X-ray point sources is reproduced by our simulations only by assuming a very high binary fraction, around 1.0, for the cluster stellar population.

References

Chlebowski, T. & Garmany, C.D. 1991, ApJ 368, 241
Portegies Zwart, S.F., Pooley, D., & Lewin, W.H.G. 2002, ApJ 574,762
Vacca, W.D., Garmany, C.D., & Shull J.M. 1996, ApJ 460, 914

Populations of High Energy Sources in Galaxies
Proceedings IAU Symposium No. 230, 2005
E. J. A. Meurs & G. Fabbiano, eds.

© 2006 International Astronomical Union
doi:10.1017/S1743921306008775

Cosmic Star Formation History and Deep Field X-ray Studies

Pranab Ghosh[1]

[1]Tata Institute of Fundamental Research
Bombay 400 005, India
email: pranab@tifr.res.in

Abstract. We explore the X-ray diagnostics of cosmic star formation history that have possible form recent deep field X-ray studies. We summarize the status of our understanding of the X-ray evolution of galaxies. We indicate the lessons learnt so far from number count plots, and criteria for distinguishing between normal/straburst galaxies and AGNs in deep X-ray surveys. We summarize how the observed correlations between X-ray emission and that at other wavebands indicate the value of X-rays as a probe of cosmic star formation.

Keywords. stars: formation, X-rays: galaxies, X-rays: binaries, infrared: galaxies, submillimeter, galaxies: starburst.

1. Introduction

We discuss in this paper deep-field X-ray studies done in recent years, particularly the Chandra Deep Field North and South (CDF-N/S) surveys, and how they bear on the subject of cosmic star-formation history. We summarize those aspects of the observations which we most need for this purpose, and indicate the status of our understanding in these various aspects, as also in this subject as a whole. We discuss first how the X-ray luminosities of normal and starburst galaxies evolve with redshift, and the role of cosmic star-formation history in this. Next, we indicate what diagnostics have been available from the $\log N - \log S$ plots in the X-rays from these deep surveys that bear on this question. A major issue in this subject has been the relative roles of normal/starburst galaxies on the one hand, and active galactic nuclei (AGNs) on the other, and we discuss suggested criteria for distinguishing between them. Finally, the correlations between X-ray properties and those in other wavebands – optical, infrared, submillimeter, and radio – are proving to be of much diagnostic value, and we summarize those points about them that bear directly on star formation.

2. Evolution of X-ray Luminosity

With exposures of duration $\sim$ megaseconds (Ms) with the *Chandra* observatory, statistical studies of the X-ray evolution of normal, starburst, and Lyman break galaxies became possible in the early 2000s, and was pioneered by several groups, particularly the one at Pennsylvania State University (Brandt *et al.* (2001a); Hornschemeier *et al.* (2002); Lehmer *et al.* (2005)). The technique used was that of *stacking*, wherein X-ray photons from a collection of optically well-known galaxies in the field of exposure, which were *not* detected individually in the X-rays, were summed together, *i.e.*, stacked, and analyzed to obtain an average description of a typical galaxy in the collection. This gave us the first indication of how X-ray luminosities of galaxies, L_x, have evolved with the redshift z. It became possible to study the evolution of L_x with z for normal/starburst

410

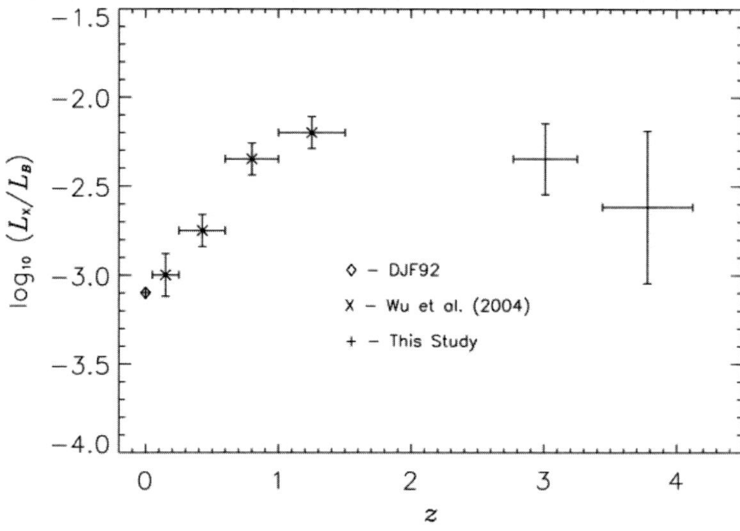

Figure 1. X-ray (0.5–4.5 keV) to B-band luminosity ratio vs. redshift. Reproduced by permission from Lehmer *et al.* (2005). Sources of data given in original reference.

galaxies at relatively low redshifts, $z \sim 0 \rightarrow 1$, as also the L_x-evolution of Lyman-break galaxies, which are believed to be distant analogs of local starburst galaxies, at intermediate redshifts $z \sim 2 \rightarrow 4$. The current observational situation is shown in Fig. 1, taken from Lehmer *et al.* (2005). Note that what is displayed is actually L_x/L_B, the ratio of the X-ray luminosity to the B-band luminosity. The basic result is clear: L_x/L_B rises by a factor ~ 10 as we look back from the local universe ($z \approx 0$) to a redshift ~ 1, and falls somewhat as z increases from ~ 3 to ~ 4. The former result comes from stacked normal galaxies, and the latter from stacked Lyman break galaxies, in CDF-N (~ 2 Ms exposure) and CDF-S (~ 1 Ms exposure). The expected peak in the evolution of L_x/L_B is therefore in the range $z \sim 1.5 \rightarrow 3$.

2.1. *Cosmic Star Formation History*

Research on the connection between cosmic star-formation history and L_x-evolution of normal galaxies was introduced by White & Ghosh (1998) and Ghosh & White (2001), the basic idea being as follows. Since galaxies of the above type do not harbor AGNs normally, or any low-luminosity AGNs present do not dominate the X-ray emission, their total X-ray output must be the integrated output of their X-ray binaries, supernova remnants, and hot gas. Focusing their attention on X-ray binaries, these authors argued that the evolution of the X-ray binary populations in a galaxy must be determined by its star-formation history, since these stars evolve and produce the compact objects in interacting binaries that generate X-rays by accretion from the binary companion. Ghosh & White (2001) developed a scheme for calculating the evolution of the X-ray binary population and the X-ray luminosity L_x of a galaxy, given its star formation history (SFH). For the average description of a collection of galaxies, a suitable prescription for the cosmic star formation history is appropriate, and these authors used a range of cosmic SFHs that cover the possibilities, given in the literature. The results for four representative cosmic SFHs are shown in Fig. 2, taken from Ghosh & White (2001). These SFHs are (a) the original Madau profile, (b) the "anvil" profile, corresponding to a monolithic scenario of star formation, (c) the hierarchical profile, in which stars form in bursts during hierarchical merging of structures, and (d) the gaussian profile, wherein

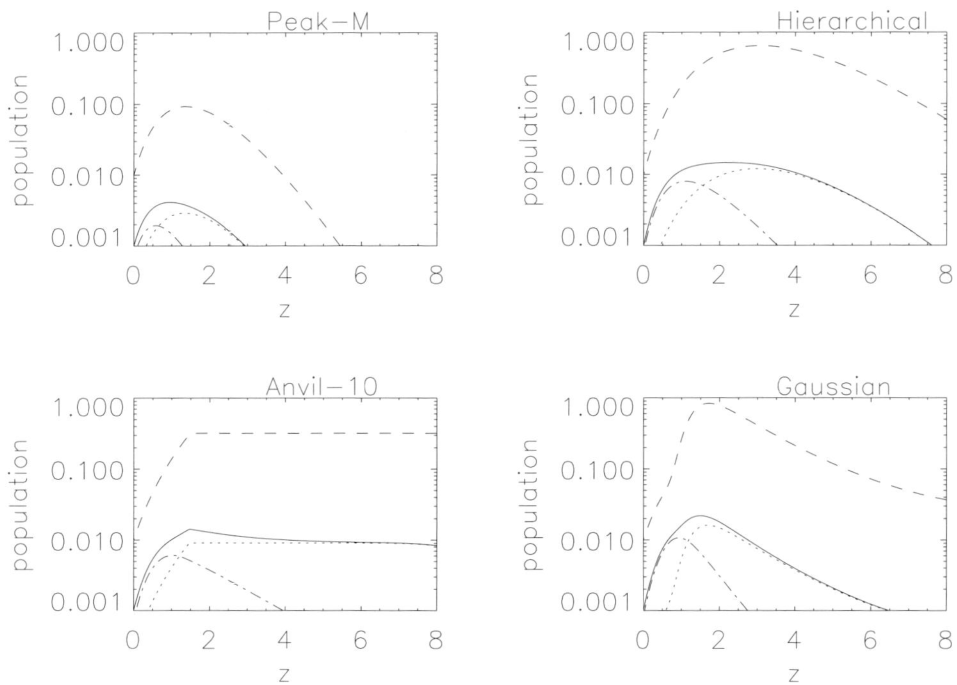

Figure 2. Evolution of X-ray binary populations and total X-ray luminosities from Ghosh and White (2001). Shown are evolutions for four cosmic star formation histories (SFH) that cover the current range of possibilities. Lines coded as follows:
– – – – Cosmic SFH   · · · · · · HMXBs   — · — LMXBs   —— Total L_x

a gaussian component is added at intermediate z to a Madau-type profile, in order to account for the intense, dust-enshrouded star formation at these redshits, as detected at submm wavelengths (Blain *et al.* (1999)).

The massive X-ray binaries (HMXBs) evolve promptly, on timescales of a few million years to $\sim 10^7$ years, following the SFH closely. By contrast, the low-mass X-ray binaries (LMXBs) evolve much more slowly, on overall timescales of a few billion years (Gyr), and so follow with a considerable time lag, as shown in Fig. 2. (Note that this LMXB lag consists of two parts: after the neutron star is produced in the supernova of a He-star, the post-supernova binary takes $\tau_{PSNB} \sim 2$ Gyr to become a mass-transferring LMXB, which then evolves on a timescale $\tau_{LMXB} \sim 1$ Gyr or so. See White & Ghosh (1998).) The profile of the total X-ray luminosity L_x therefore carries signatures of the SFH, different SFHs yielding different L_x-evolution profiles. Consequently, observed L_x-profiles can ultimately constrain SFHs.

Comparing Figs. 1 and 2, it is clear that the predictions from cosmic SFHs are qualitatively correct, except possibly in the anvil case. Further, the hierarchical and gaussian profiles appear to give better agreement than the original Madau profile. For more quantitative understanding, we must first clarify the observed and calculated quantities. Theory tells us that, for the choices of τ_{PSNB} and τ_{LMXB} made above, L_x rises by a factor ~ 10 as z increases from 0 to 1. This means that L_x/L_B would also rise by the same factor only if L_B had little evolution in the same redshift range. Observations tell us that L_x/L_B rises by ~ 10 as z increses from 0 to 1. The question therefore is: by what factor does the observed L_x rise in this range of z? The answer is not quite clear yet. With

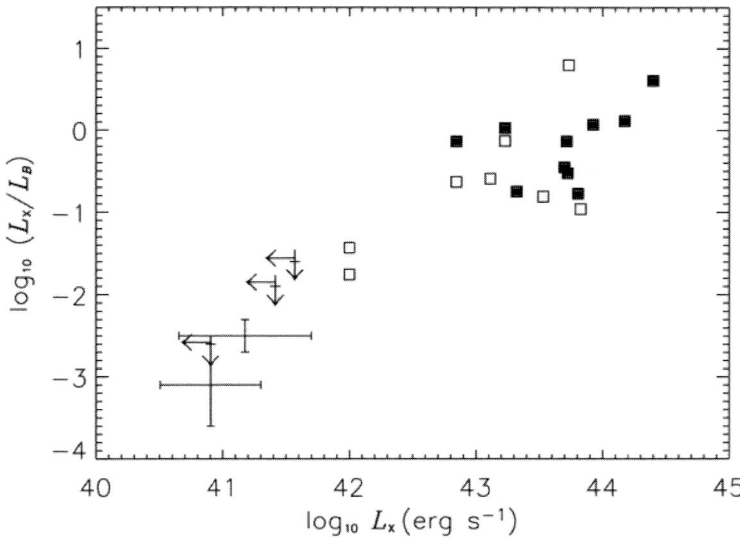

Figure 3. L_x/L_B vs. L_x for stacked galaxies (*plus signs*) and individually detected sources (*open squares*) in CDF, the latter being almost all AGNs. For comparison, known broad-line AGNs are also plotted (*filled squares*) Reproduced by permission from Lehmer *et al.* (2005).

the $\sim$0.5 Ms CDF-N exposure, the original Brandt *et al.* (2001a) result was that L_x rose by a factor $\sim$3 in this redshift range, which was consistent with the subsequent result of Hornschemeier *et al.* (2002) with the greater $\sim$1 Ms CDF-N exposure. The corresponding result from the current $\sim$2 Ms exposure needs to be clarified.

2.2. *Understanding L_x-Evolution*

If we assume that L_x *does* rise by $\sim$3 as $z \sim 0 \to 1$, then there are several issues to be addressed. First, with the rise of L_x/L_B by $\sim$10, this implies that L_B would be *falling* by a factor of $\sim$3 as L_x rises in this redshift range. How do we understand this? In fact, the situation is brought out very instructively in Fig. 3, also reproduced from Lehmer *et al.* (2005), where L_x/L_B is shown against L_x. The correlation between them is such that L_x does, indeed, rise by about a factor of 3 in this redshift range. Hence, this apparent fall in L_B remains to be understood, since our experience in local universe is that L_x correlates positively with L_B. Figure 3 also shows the interesting behavior of individually detected sources in CDF at $z \sim 2$–4, which consist almost entirely of AGNs: these have much higher values of L_x/L_B, ranging from $\sim$0.1 to a few, while the stacked galaxies, normal, starburst, or Lyman break, have much lower values, $L_x/L_B \sim 10^{-2}$–10^{-3}. This ratio should thus be a good discriminator between galaxies and AGNs, a point to which we return below.

If the above observed rise in L_x by $\sim$3 as $z \sim 0 \to 1$ is borne out by further work, how would we reconcile it with the rise by $\sim$10 given in the same redshift range by the evolutionary calculations of Fig. 3 with the above choice of parameters? We consider two possibilities. An obvious solution would be that the evolutionary timescale τ_{LMXB} of LMXBs is longer than we used above, say, $\sim$2 Gyr. This would change the rise-factor in evolutionary calculations to about 3 because, with longer lag, the peak of the LMXB emission would occur closer to $z = 0$, so that the fall from the peak to the value at the present epoch would not be quite so large. While such a value is not impossible, it appears higher than the canonical values adopted currently. An interesting alternative is the bandpass-change factor, as used in the standard expression used for converting X-ray fluxes f_x to X-ray luminosities L_x, namely,

$$L_x = 4\pi d_L^2 f_X (1+z)^{\Gamma-2}, \qquad (2.1)$$

may not be quite right. The bandpass-change factor $(1+z)^{\Gamma-2}$ in the above equation takes account of the fact that what we observe in the *Chandra* band is, in fact, what the source at a redshift z emits in a higher energy-band in its own rest-frame, the connection involving the photon index Γ of the source spectrum. It is customary to use $\Gamma = 2$ from the fits made X-ray spectra of galaxies in the 1–10 keV band. However, we argued above that, in the $0 < z < 1$ range, L_x is dominated by LMXBs, and it is well-known that LMXB spectra have soft excesses, *i.e.*, extra emission at soft X-rays at <1 keV over and above the 1–10 keV trend. Recent *XMM-Newton* spectra of LMXBs bear this out fully (Sidoli *et al.* (2005)). So, if we make an extrapolation of the 1–10 keV trend to the *Chandra* soft band, we shall be underestimating L_x at redshifts $\sim$0.5–1. Rough estimates from LMXB spectra show that this effect can, indeed, lead to underestimates by factors $\sim$3 in the rise of L_x in the above redshift range. The effect needs to be studied in more detail.

3. X-ray $\log N - \log S$ diagnostics

Source counts, *i.e.*, $\log N - \log S$ plots, from the deep X-ray surveys are determining the contributions of the various faint X-ray source populations to the extragalactic X-ray background (XRB). With $\sim$89% the XRB resolved in the CDF in the *Chandra* soft band (0.5–2 keV), and $\sim$93% resolved in the hard band (2–8 keV), the contributions from (a) bright, unobscured AGNs, (b) obscured AGNs, and (c) normal and starburst galaxies are showing their characteristic signatures (Bauer *et al.* (2004)). The overall slope of the $\log N - \log S$ plot in both soft and hard bands is $\sim$0.55. AGNs have a flat slope $\sim$0.48, rather similar to the overall one. But normal galaxies have a much steeper overall slope of $\sim$1.3. Of these, quiescent galaxies have a slope $\sim$1.3 like the total, ellipticals flatter at $\sim$1.1, and starbursts much steeper at $\sim$1.7. It is clear, therefore, that the number counts in both soft and hard bands will become dominated by normal galaxies at sufficiently low fluxes. This critical flux is estimated to be $\sim$10^{-17} erg cm^{-2} s^{-1} in the soft band (Bauer *et al.* (2004)).

3.1. *Normal/Starburst Galaxies vs. AGNs*

How do we distinguish between normal/starburst galaxies and AGNs? This is an important question, since the interplay between these two components of the XRB and the number-count plots plays an essential role in our understanding. We now consider some of the most useful discriminators that are being utilized for this purpose:

• X-ray luminosity L_x. AGNs are generally brighter than normal/straburst galaxies, the critical value of L_x in the *Chandra* full band (0.5–8 keV) being $\sim 3 \times 10^{42}$ erg s^{-1}.

• X-ray spectra. AGN spectra are harder than those of normal galaxies. In terms of a hardness ratio, *i.e.*, that between hard- and soft-band fluxes, the critical value is $\sim$0.8. The corresponding critical photon index is $\Gamma \sim 1$.

• X-ray to optical flux ratio, f_x/f_{opt}, the X-ray flux being over the full band. AGNs have a higher value than than normal/starburst galaxies (see Fig. 3). The critical value $f_x/f_{opt} \approx 0.1$ is one of the most useful discriminators.

• Optical spectra. AGNs display characteristic broad/ high-ionization emission lines.

• Radio properties.

The collection of X-ray discriminators can be conveniently displayed in a plot of R-band magnitude vs. X-ray flux, as been done by Bauer *et al.* (2004) in their Fig. 7. The criteria (the first three in the above list) work well together, the $f_x/f_{opt} \approx 0.1$ criterion being particularly useful, and a fairly effective separation of AGNs and normal galaxies is possible on such plots. The latest version of such a plot is given by Hornschemeier (these proceedings).

The results from such separation of AGN and galaxy contributions to the XRB and the number-count plots is as follows. The *power* in the XRB is certainly dominated by AGNs, as 85–90% of it comes from AGNs, and only 5–15% comes from normal/starburst galaxies. Of the latter, starbursts dominate in the soft band, and quiescents dominate in the hard band. However, in *number density*, star-forming galaxies overtake AGNs, and dominate the sky, at faint fluxes given above, the threshold corresponding to a factor of a few below the current CDF soft-band flux limit (Bauer *et al.* (2004)).

4. Correlations Between X-rays and Other Wavebands

These correlations are shedding more light on star formation in distant galaxies, and we summarize the essential points here.

4.1. *X-ray/Optical Correlations*

The recent study of optically bright, X-ray faint (OBXF) galaxies in CDF-N has proved to be of much value in establishing the continuity between local starburst galaxies and Lyman break galaxies at intermediate redshifts, and so the viability of the whole picture given above (Hornschemeier *et al.* (2003)). These galaxies have low values of X-ray to optical flux ratios, $f_x/f_R \sim 0.01$ or less, and normal optical spectra, and so appear to be "distant analogs of 'normal' galaxies in the local universe" (Hornschemeier *et al.* (2003)), the redshift range being $z \sim 0.1$–0.8. OBXF appear to be dominated entirely by non-AGN processes, although some low-luminosity AGN activity cannot be ruled out. Their X-ray luminosities exceed those of normal galaxies, however, and they have soft X-ray spectra, reminiscent of starbursts. In addition, their $\log N - \log S$ slope is very steep, ~ 1.7, almost identical to those of starbursts. Hence they are believed to be starburst at moderate redshifts, bridging the gap between local starbursts and the Lyman break galaxies at intermediate redshifts 2–4.

4.2. *X-ray/IR Correlations*

There is a tight correlation between X-ray and 15 μm infrared galaxy populations, the former coming from CDF-N and the latter from ISOCAM survey of the Hubble Deep Field North (Alexander *et al.* (2002)). These are luminous IR starburst galaxies, undergoing dust-enshrouded star formation. With small X-ray to infrared flux ratios, $f_x/f_{IR} \sim 0.03$ or less, these appear to be non-AGN sources. They are probably X-ray detected starburst galaxies. The key point is that, since 15 μm emission is believed to be a good indicator of star formation, this correlation strongly suggests that so are X-rays.

4.3. *X-ray/Submm Correlations*

While seven out of ten of bright 850 μm SCUBA sources have shown X-ray emission, five of these seven are AGNs (Alexander *et al.* (2003)), and they appear to show an anti-correlation between X-rays and 850 μm emission, which is poorly understood. The key point here, however, is that AGNs contribute negligibly to submm emission, which must, therefore, be powered by star formation.

4.4. *X-ray/Radio Correlations*

The fact that there is a large overlap between CDF-N X-ray sources and 1.4 GHz radio sources in this field in the VLA surveys, and that these sources show an excellent correlation between X-ray luminosity and radio luminosity, has established a close connection beyond doubt (Bauer *et al.* (2002)). What is particularly remarkable is that this correlation is the same at moderate redshifts as it is in the local universe. As radio emission is believed to be a good indicator of star formation, this again emphasizes the role of X-rays as indicators of star formation.

5. Conclusions

Our understanding of the X-ray evolution of normal/starburst galaxies appears to be qualitatively correct, but details need to clarified, both on observational issues, and on the development of a more sophisticated theory. While the power in XRB is dominated by AGNs, source number counts are expected to be dominated by star-forming galaxies at faint fluxes. Finally, correlations between X-ray emission and that in other wavebands stress the diagnostic value of X-rays in probing cosmic star formation.

Acknowledgements

It is a pleasure to acknowledge an IAU grant.

References

Alexander, D. M., *et al.* 2002, *ApJ* (Letters) 568, L85
Alexander, D. M., *et al.* 2003, *AJ* 125, 383
Bauer, F. E., *et al.* 2002. *AJ* 124, 2351
Bauer, F. E., *et al.* 2004. *AJ* 128, 2048
Blain, A. W., *et al.* 1999, *MNRAS* 302, 632
Brandt, W. N., *et al.* 2001a, *AJ* 122, 1
Brandt, W. N., *et al.* 2001b, *ApJ* (Letters) 558, L5
Ghosh, P. & White, N.E. 2001, *ApJ* (Letters) 559, L97
Hornschemeier, A. E., *et al.* 2002, *ApJ* 568, 82
Hornschemeier, A. E., *et al.* 2003, *AJ* 126, 575
Lehmer, B. D., *et al.* 2005, *AJ* 129, 1
Sidoli, L., Parmar, A. N., & Oosterbroek, T. 2005 *A&A* 429, 291
White, N.E. & Ghosh, P. 1998, *ApJ* (Letters) 504, L31

Discussion

ZEZAS: The peak of the L_X/L_B ratio seems to be around $z \sim 1$ where the star-formation rate also peaks. However, at that point I would also expect to have a very strong contribution of HMXBs and ULXs which would result in a non-linear L_X/L_B ratio.

GHOSH: There is no doubt that L_X/L_B is expected to be non-linear (i.e., $L_X \propto L_B{}^n$, with $n \neq 1$), in general. In fact, values of $n > 1$ are sometimes suggested in literature. So, I completely agree with you. The point would be to see how n changes with z, to study the effects you are describing.

HORNSCHEMEIER: Evolution in L_B is actually often due to observational selection. One has to be very careful about how you select galaxies so I would not over-interpret implied evolution in L_B.

GHOSH: Yes, observers must advise us on the best ways of exploring L_B evolution.

Populations of High Energy Sources in Galaxies
Proceedings IAU Symposium No. 230, 2005
E. J. A. Meurs & G. Fabbiano, eds.

© 2006 International Astronomical Union
doi:10.1017/S1743921306008787

Models for the Evolution of X-Ray Binaries in a Young Stellar Population

M. Eracleous[1], M. S. Sipior[2], and S. Sigurdsson[1]

[1]Department of Astronomy & Astrophysics, The Pennsylvania State Univeristy, 525 Davey Lab, University Park, PA 16802, USA

[2]Astronomical Institute "Anton Pannekoek" and Section Computational Sciences, University of Amsterdam, Kruislaan 403, 1098 SJ, Amsterdam, The Netherlands

Abstract. We present the results of population synthesis simulations aimed at exploring the evolution of the 2–10 keV luminosity of X-ray binaries in a young stellar population. The results are applicable to populations of extragalactic X-ray binaries in starburst galaxies and many LINERs. We find that the integrated 2–10 keV luminosity of the simulated population reaches a maximum of about $10^{40}\,\mathrm{erg\ s^{-1}}$ after approximately 20 Myr (for a star-formation rate of $10\ M_\odot\ \mathrm{yr^{-1}}$) and remains significant even after the end of star formation and the demise of the luminous OB stars. The results of our simulation are in agreement with recently-derived correlations between the X-ray luminosity starburst galaxies and their star-formation rate. We also find that the cumulative luminosity function is initially fairly flat, in agreement with recent observational results, becoming steeper as the population ages and the high-mass X-ray binaries are succeeded by binaries with progressively lighter donor stars. Using the output of Hydrogen-ionizing far-UV photons from the stellar population, we can plot the track of a "post-starburst" system in the $L_{\mathrm{X}} - L_{\mathrm{H}\alpha}$ diagram. The system starts off in the starburst locus but quickly evolves to the AGN locus where it lingers for at least 1 Gyr.

1. Introduction

With the advent of of the *Chandra* X-ray observatory, we can resolve populations of discrete X-ray sources (presumably X-ray binaries; hereafter XRBs) in many galaxies other than our own. Thus, we have been able to confirm and extend previously known differences (e.g., Fabbiano 1995; Fabbiano & White 2005) between the XRB populations of old and young stellar systems (steeper luminosity functions in the former compared to the latter; e.g., Eracleous *et al.* 2002; Colbert *et al.* 2004). We have also been able to establish correlations between the star formation rate and the 2–10 keV X-ray luminosity of a star-forming galaxy (e.g., Ranalli *et al.* 2003; Grimm *et al.* 2002; Colbert *et al.* 2004; Persic *et al.* 2004).

To understand the origin of these observational properties of galaxies and to predict the evolution of star-forming systems, we have undertaken a theoretical investigation of the evolution of X-ray binaries formed in a brief star-formation episode. The results of this investigation should also be useful in interpreting the properties of distant star-forming galaxies found in deep X-ray surveys (e.g., Bauer *et al.* 2004).

2. Population Synthesis Simulations: Code and Methods

We have carried out simulations of the early evolution of XRBs formed in a burst of star formation, using a modified version of the Monte-Carlo population synthesis code of Pols & Marinus (1994). This code was modified initially by Bloom *et al.* (1999) to include supernova kicks. It was later extended by Sipior & Sigurdsson (2002) to include a mapping of the initial star mass to a final black hole mass, and thus account for evolution to the black hole state. Sipior (2003) made considerable refinements to the treatment of mass transfer by adopting the methodology of Hurley *et al.* (2002) and implementing duty cycles due to accretion disk instabilities and X-ray attenuation by circumstellar material. The code was further upgraded to use the massive main sequence star models of Meynet & Maeder (2003), which include rotation and mass loss self consistently (via the formalism of Nugis & Lamers 2000). The main output of a simulation consists of an evolutionary record for each binary, which can be used as the stepping stone for computing the total luminosity of the population and the luminosity function.

The total X-ray luminosity of the population versus time is computed by summing the contributions from all XRBs weighted by the appropriate efficiency factors, namely

$$L_{2-10\,\text{keV}}(t) = \sum_{\text{all XRBs at } t} \eta_{st}\,\eta_x\,\eta_{bol}\,\eta_a\,L_{bol}, \tag{2.1}$$

where L_{bol} is the total available accretion power. The preceding terms are efficiency factors, as follows: η_{st} is the duty cycle due to accretion disk instabilities, which depends on the period the type of compact object, and the relative mass of the companion (see Li & Wang 1998; King 2001); η_{bol} is the fraction of the available accretion power converted to radiation (assumed to be 0.5 for black holes and 1.0 for neutron stars); η_x is the fraction of the radiated power emerging in the 2–10 keV band (assumed to be 0.4 for black holes and 0.2 for neutron stars); η_a is the attenuation factor due to photoelectric absorption in circumstellar material (the column density is computed by assuming that material lost from the binary forms a spherical shell around it).

3. Application and Results

Our simulation considers a star formation episode with a constant star-formation rate (SFR) of 10 $M_\odot$ yr^{-1} and a duration of 20 Myr. The main result, depicted in Figure 1, is the evolution of the total luminosity of the system with time for 2 Gyr. As expected, the black hole XRBs (whose companions are typically massive) turn on very early and remain active as long as the star-formation continues. They track the star-formation episode (and rate) fairly closely and they disappear very soon after the end of star formation. In contrast, the neutron star XRBs (whose companions are typically less massive) turn on with a delay and persist for a longer time. This is a consequence of the lower mass of their companions and the fact that different generations of systems with progressively lighter companions turn on with progressively longer delays. As a result, the total luminosity of the system remains appreciable up to almost 1 Gyr after the end of star formation. The rate of decline of the total luminosity from all flavors of XRBs, after the end of star formation, is approximately $\propto t^{-1.4}$.

These results are insensitive to the assumed initial mass function (IMF) and distribution of binary mass ratios (q-distribution); combinations of the Salpeter and Miller-Scalo IMFs with flat and low-skewed q-distributions yield nearly identical results. On the other hand, the assumed distribution of supernova kicks has a significant effect on the resulting X-ray luminosity. Our adopted kick distribution is a Gaussian with a dispersion of 90 km s^{-1}; increasing this dispersion to 190 and 450 km s^{-1} reduces the number of XRBs by factors of 1.6 and 2.6 respectively.

The total luminosity from black hole XRBs appears to reach a steady state at about 10 Myr after the beginning of the star-formation episode. The luminosity of this plateau is $L_X \approx 1 \times 10^{40}$ erg s^{-1}, which is consistent with the empirically determined correlations

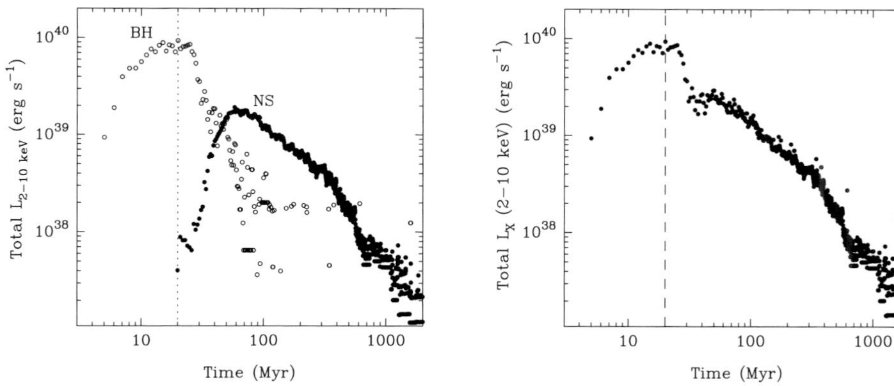

Figure 1. The evolution of the total X-ray luminosity from XRBs formed in an episode of star formation with SFR = 10 $M_\odot$ yr^{-1} and duration 20 Myr. The end of the star-formation episode is marked by a vertical dashed line. The left panel shows the separate contributions from black hole XRBs (open circles) and neutron star XRBs (filled circles). The panel on the right shows the evolution of the total luminosity from all of XRBs. Notice the delay in the onset of the NS XRBs, which causes the X-ray luminosity to remain high long after the end of star formation. The results presented here are for a Salpeter IMF and a low-skewed q-distribution.

between L_X and the SFR (empirically, we expect L_X in the range $4\times10^{39}-7\times10^{40}$ erg s^{-1} for our assumed SFR). The cumulative luminosity function [CLF, $N(>L) \propto L^{-\beta}$] is quite flat while the star formation episode is in progress. After the episode is over the CLF becomes steeper with $\beta \sim 0.5$, 1.0, and 1.3 at a time of 20, 30, and 40 Myr after the end of star formation, respectively (with a some variation in β with assumed IMF and q-distribution). In comparison, the observed values of β for star-forming galaxies are in the range 0.5–1.0 (e.g., Grimm *et al.* 2002; Colbert *et al.* 2004).

4. Discussion

An important conclusion from our simulations is that the X-ray luminosity of a "post-starburst" system can remain high up to 1 Gyr after the end of star formation and the demise of the luminous OB stars. The LINER NGC 4736 (Eracleous *et al.* 2002) may be an example of such a system. In this spirit, we can attempt to follow the evolution of a post-starburst system in the $L_X - L_{\mathrm{H}\alpha}$ diagram, which has been used as a tool for discriminating between nuclear and starburst activity (e.g., Ho *et al.* 2001). The key physical effect that determines the track of a post-starburst system in this diagram is the different rates of decline of the X-ray and Hα luminosities after the end of star formation; the former declines as $t^{-1.4}$, while the latter declines much more quickly, as t^{-4} to t^{-5} (Binette *et al.* 1994; Sternberg *et al.* 2003). The rapid decline in the Hα luminosity is a direct consequence of the short life times of OB stars, which provide the vast majority of Hydrogen-ionizing photons. At very late times ($t \sim 60 - 100$ Myr), the Hα emission is driven by ionizing photons from the old stellar population (post-AGB stars and nuclei of PNe, see Binette *et al.* 1994), leading to $L_{\mathrm{H}\alpha} \propto t^{-0.6}$.

In Figure 2 we plot the track of a post-starburst system in the $L_X - L_{\mathrm{H}\alpha}$ diagram, under the scenario outlined above. The system starts in the starburst galaxy locus but it evolves very quickly to the AGN locus, where it appears to linger for at least 1 Gyr. Thus a post-starburst system at the late stages of its evolution could be mistaken for a low-luminosity AGN since it has a similar X-ray luminosity and very likely also a similar X-ray spectrum. The best way to to avoid this confusion is to use high-resolution

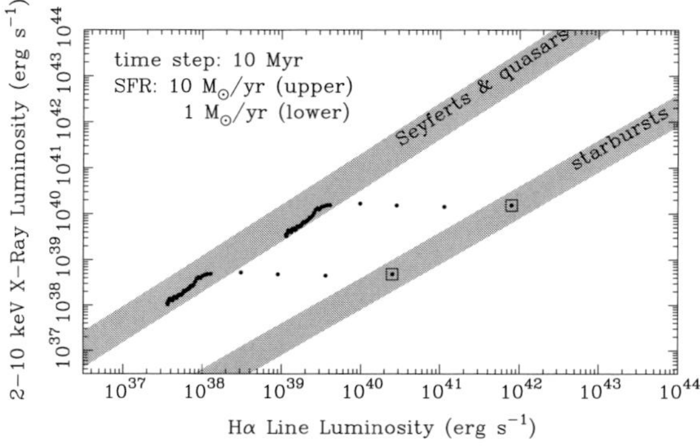

Figure 2. The evolution of a simulated population in the $L_X - L_{H\alpha}$ diagram. At the end of star formation the system is in the starburst galaxy locus (location indicated by ▣). As the population ages, the system moves to the left along the dotted track at a rate of 10 Myr per dot. The initial evolution is very rapid because the Hα luminosity drops precipitously with the demise of the OB stars. After 40–50 Myr the evolution slows down and the system lingers in the AGN locus for at least 1 Gyr. The tracks shown here correspond to SFR values of 1 and 10 $M_\odot$ yr^{-1}, and under the assumption of a Salpeter IMF and a low-skewed q distribution.

X-ray images, such as those provided by *Chandra*, which can resolve the individual XRBs contributing to the total X-ray luminosity.

Acknowledgements

This work was supported by NASA through grant GO0-1152A,B from the Smithsonian Astrophysical Observatory, from NSF grant PHY-0203046, and from the Center for Gravitational Wave Physics, which is supported by the NSF under cooperative agreement PHY 01-14375.

References

Bauer, F. E., *et al.* 2004, AJ, 128, 2048

Binette, L., Magris, C. G, Stasińska, G., & Bruzual, A. G. 1994, A&A, 192, 13

Bloom, J. S., Sigurdsson S., & Pols, O. R. 1999, MNRAS, 305, 763

Colbert, E., *et al.* 2004, ApJ, 602, 231

Eracleous, M., Shields, J. C., Chartas, G., & Moran, E. C. 2002, ApJ, 565, 108.

Fabbiano, G. 1995 in *X-Ray Binaries*, eds, W. H. G. Lewin, *et al.* (Cambridge U. Press), 390

Fabbiano, G. & White, N. E. in *Compact Stellar X-ray Sources*, eds, W. H. G. Lewin & M. van der Klis (Cambridge U. Press), in press (astro-ph/0307077)

Grimm, H.-J., Gilfanov, M., & Sunyaev, R. 2002, MNRAS, 339, 793

Ho, L. C., *et al.* 2001, ApJ, 549, L51

Hurley, J. R., Tout, C. A., & Pols, O. R. 2002, MNRAS, 329, 897

King, A. R. 2001, in Black Holes in Binaries and Galactic Nuclei, eds. L. Kaper, E. P. J. van den Heuvel, & P. A. Woudt (ESO Workshop), 155

Li, X.-D., & Wang, Z.-R. 1998, ApJ, 500, 935

Meynet, G. & Maeder, A. 2003, A&A, 404, 975

Nugis, T. & Lamers, H. J. G. L. M. 2000, A&A, 360, 227

Persic, M., *et al.* 2004, A&A, 419, 849

Pols, O. R. & Marinus, M. 1994, A&A, 288, 475

Ranalli, P., Comastri, A., & Setti, G. 2003, A&A, 399, 39

Sternberg, A., Hoffmann, T. L., & Pauldrach, A. W. A. 2003, ApJ, 599, 1333

Sipior, M. S. 2003, Ph. D Thesis, The Pennsylvania State University

Sipior, M. S. & Sigurdsson, S. 2002, ApJ, 572, 962

Discussion

IVANOVA: Can you comment on what is the role of thermally unstable MT and how do you calculate this?

ERACLEOUS: We do not use a sophisticated treatment of mass transfer in the thermally unstable case. We just assume that the entire envelope mass of the donor is lost on the thermal time scale. This is a very high mass transfer rate, but a large fraction of the matter lost from the secondary escapes from the system. In this case we assume that the escaping material attenuates the emitted X-rays, so these systems can be very dim and do not affect the total X-ray luminosity of the population.

ZEZAS: How do the results depend on the choice of the IMF?

ERACLEOUS: The results are not sensitive to the choice of IMF. We have carried out the simulation using both a Salpeter IMF and a Miller-Scalo IMF and we get almost indistinguishable results.

LIPUNOV: Please in the future give the comparison with previous results.

ERACLEOUS: Thanks, that is a good suggestion. I have not compared with old results, but I have compared with with very recent results, specifically those by VanBever and Vanbeveren (2000); we are in good agreement.

The speaker sampling some Irish culture. Photo courtesy of M. Eracleous.

Michael Garcia in stead of Michael Eracleous (see previous photograph). Photo courtesy of M. Eracleous.

Session 7

The high-redshift context

Male students from Dunsink Observatory entertain female students from University College Dublin. From right: Cóilín Ó Maoiléidigh, Sinead McGlynn, Paul Ward, Suzanne Foley.

Populations of High Energy Sources in Galaxies
Proceedings IAU Symposium No. 230, 2005
E. J. A. Meurs & G. Fabbiano, eds.

© 2006 International Astronomical Union
doi:10.1017/S1743921306008799

Normal and Starburst Galaxies in Deep X-ray Surveys

Ann E. Hornschemeier[1]

[1]Laboratory for X-ray Astrophysics, Code 662.0,NASA Goddard Space Flight Center,
Greenbelt, MD 20771, USA
email: annh@milkyway.gsfc.nasa.gov

Abstract. This paper reviews the nature of normal and starburst galaxies in deep X-ray surveys, focusing on the observational issues. Normal and starburst galaxies may be divided from AGN via X-ray/optical flux ratios, optical spectroscopic identification, hardness ratio, and X-ray luminosity. Each of these is discussed, including the possible impact on derived X-ray-Star Formation Rate (X-ray/SFR) correlations. The measured differences in the normal galaxy X-ray Luminosity Functions (XLFs) by SED type at $z \approx 0.3$–1.0 are also described.

The redshift frontiers of deep X-ray surveys are discussed, including those for individually detected accreting binary systems (Ultraluminous X-ray Sources at $z \approx 0.1$–0.3) and that for the highest-redshift X-ray detection of star formation (stacking analyses of Lyman Break Galaxies to $z \approx 4$). The paper closes with a discussion of normal galaxy studies with future X-ray missions such as Constellation-X, XEUS, and Generation-X.

Keywords. galaxies: evolution, galaxies: formation, galaxies: luminosity function, galaxies: active , X-rays: binaries, X-rays: diffuse background, X-rays: galaxies, X-rays: general.

1. Introduction

Since the launch of the *Chandra* X-ray Observatory over six years ago, normal and starburst galaxies have been routinely detected in X-rays over the interval $(0.1 < z < 1.0)$. Reaching these redshifts means that we probe the billion-year timescales that are relevant to the evolution of populations such as accreting binary systems (please also see the contribution of Pranab Ghosh in this same IAU 230 volume). At the time of this writing, quiescent (Milky Way-type) galaxies have been detected to $z \approx 0.3$ in the deepest *Chandra* surveys (look-back time ≈ 3.4 Gyr; $H_0 = 70$ km s^{-1} Mpc^{-1}, $\Omega_M = 0.30$, $\Omega_\Lambda = 0.70$) (e.g. Hornschemeier *et al.* 2003, 2004). Vigorously star forming galaxies [current SFRs >10 $M_\odot$ yr^{-1}] have been detected individually in deep X-ray surveys to $z \approx 1$ (look-back times of ≈ 7.7 Gyr; e.g., Alexander *et al.* 2002). Beyond $z \approx 1$, our knowledge of X-ray emission from normal/star-forming galaxies is largely confined to statistical analyses (stacking the X-ray emission from individually undetected galaxies to achieve an average detection). At $z > 1$, such studies have concentrated on the vigorously star-forming Lyman Break galaxy population. Lyman Break galaxies have been detected in the X-ray band as far away as $z \approx 4$ (look-back time 11.9 Gyr; the Universe was 1.5 Gyr old; Lehmer *et al.* 2005). Thus for the first time we may study the *evolution* of X-ray emission from galaxies over a large fraction of the age of the Universe.

X-ray emission from galaxies may be broadly divided into two forms: gravitational potential energy release from accretion processes and thermal emission from hot gas. Accretion processes may be divided among three main systems: where a lower mass ($\lesssim 1$ $M_\odot$), and correspondingly longer-lived (>1 Gyr), star accretes onto either a neutron star or black hole (low-mass X-ray binary; hereafter LMXB), where a higher mass

($\gtrsim 5$ $M_\odot$) and thus shorter-lived ($< 10^7$ yr) star accretes onto a neutron star or black hole (high-mass X-ray binary; hereafter HMXB), and finally where accretion occurs onto a nuclear supermassive (10^6–10^8 $M_\odot$) black hole (active galactic nucleus; hereafter AGN). For the moment ignoring AGN, it is thus expected that in early-type galaxies, with older stellar populations, we will find that X-ray emission is dominated by LMXBs and hot ISM. In late-type galaxies, we expect a mix of emission from hot gas, LMXBs, and HMXBs with starburst galaxies having a great amount of both hot gas and HMXBs. The definition of a "normal galaxy", the subject of this paper, is a galaxy which does *not* have a luminous AGN present.

2. Normal Galaxies and the X-ray Background

The number counts of X-ray-detected galaxies are now well-measured over four orders of magnitude in X-ray flux (see Figure 1; Bauer *et al.* 2004). In the 0.5–2 keV band, where the cosmic X-ray background is nearly fully resolved (Worsley *et al.* 2004), galaxies are a minority population to the faintest X-ray limits currently achieved. Including X-ray emission from stacking analyses, normal and starburst galaxies make $< 5\%$ of the cosmic 0.5–2 keV X-ray background (Hornschemeier *et al.* 2002; Persic & Rephaeli 2003). Normal and starburst galaxies might make a larger contribution to the 2–10 keV background if a sufficiently large population of relativistic electrons is present in star-forming galaxies (resulting in Compton upscattering). Persic & Rephaeli (2003) show that due due to the increased energy density of the Cosmic Microwave Background at higher redshifts [$\propto (1 + z)^4$], galaxies might make up to 15% of the 2–10 keV background. To date, however, there has been no highly confident detection of this non-thermal component in the nearby Universe, so it remains unconstrained.

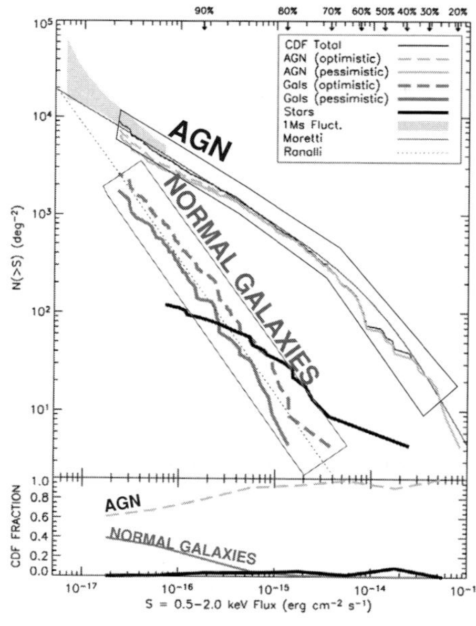

Figure 1. X-ray number counts in the 0.5–2.0 keV band, figure is adapted with permission from Bauer *et al.* (2004). The large blue and purple polygons mark the normal galaxy and AGN number counts. The galaxies rise sharply to faint X-ray fluxes whereas the AGN number counts are flattening. It cne extrapolated the galaxy number counts, galaxies would dominate the number counts below $f_X \approx 5 \times 10^{-18}$ erg cm^{-2} s^{-1} (0.5–2 keV).

3. Separating Normal Galaxies from AGN

There are several strategies for dividing normal galaxies from AGN, here we describe the use of X-ray/optical flux ratios, X-ray hardness ratios, and optical spectroscopy. We discuss the how galaxy classification affects how X-ray/SFR correlations are interpreted.

3.1. *X-ray/optical flux ratios*

With *ROSAT* it was discovered that most 0.5–2 keV point sources in deep surveys exist in a fairly narrow range of ratio of X-ray/optical flux ($-1 < \log \frac{f_X}{f_R} < +1$) and the overwhelming majority of these sources were found to be AGN. The first deep ($\gtrsim$200 ks) *Chandra* surveys found a significant number of X-ray sources at lower values of X-ray/optical flux ratio ($\lesssim 10^{-3}$; Hornschemeier *et al.* 2003). For reference, the Milky Way galaxy has X-ray/optical flux ratio $\approx 10^{-3}$.

How reliably can we use X-ray/optical flux ratio as a discriminator of AGN vs. normal galaxies? Typically, AGN and X-ray binary populations have unobscured X-ray spectra with photon indices of $\Gamma \approx 2$; the k-corrections for such a spectrum are small with redshift. Also, we become less sensitive to obscuring columns as we observe harder rest-frame emission, so the same population becomes relatively more X-ray bright with redshift. The net effect is that the X-ray/optical flux ratios of Seyfert 2 galaxies are expected to be lower at higher redshift due to this differential k-correction (Bauer *et al.* 2004; Ptak *et al.* 2006).

Peterson *et al.* (2006) took a sample of 23 nearby well-studied Seyfert 2 galaxies to determine the impact this k-correction could have on selection of galaxies and AGN at higher-z. These AGN were artificially placed at $z = 0.3$, the lowest redshift for which normal galaxy X-ray Luminosity Functions (XLFs) have been constructed (Norman *et al.* 2004). Many of these well-known Seyfert 2 galaxies would have ($-3 < \log \frac{f_X}{f_R} < -2$) at $z = 0.3$, clearly demonstrating that k-corrections are critically important in selecting galaxies by X-ray/optical flux ratio even at modest redshift.

Peterson *et al.* (2006) also examined the hardness ratios (hard-soft)/(hard+soft) counts of "pure Seyfert 2" galaxies (galaxies where the X-ray emission is clearly dominated by the AGN), finding that although AGN do cluster around harder band ratios, a significant tail towards soft X-ray colors is observed. The best strategy is thus to use redshifts to k-correct the X-ray/optical flux ratios and then use hardness ratio as a weak discriminator of AGN activity.

3.2. *Optical spectroscopy and the effect of classification on X-ray/SFR correlations*

Optical spectroscopy is essential for X-ray source identification as the typical number of detected X-ray counts in the deepest X-ray surveys is often <30 and X-ray-determined redshifts are not possible. However, there is often discordance between X-ray and optical classification. The most obvious case of this is the X-ray Bright, Optically Normal Galaxy population (XBONGs) so-called because their optical spectra are passive (red, absorption-dominated spectra with no emission lines) but they have very luminous hard X-ray emission (e.g., Comastri *et al.* 2002).

It is expected that at high-redshift the galaxy light might dilute the AGN and/or starburst features in spectra (Moran *et al.* 2002). The results of both Moran *et al.* (2002) and Peterson *et al.* ((2006)) indicate that obscured AGN can plausibly appear as "normal" galaxies at high redshift (however, please see Barger *et al.* 2003). Of course, some passive optical spectra simply indicate there is no luminous AGN present (e.g., the detection of hot gas in early-type galaxies; Hornschemeier *et al.* 2005).

How do we tell the difference? One strategy is to take the high signal-to-noise optical spectra of the Sloan Digital Sky Survey (Abazajian *et al.* 2005) and thus eliminate factors

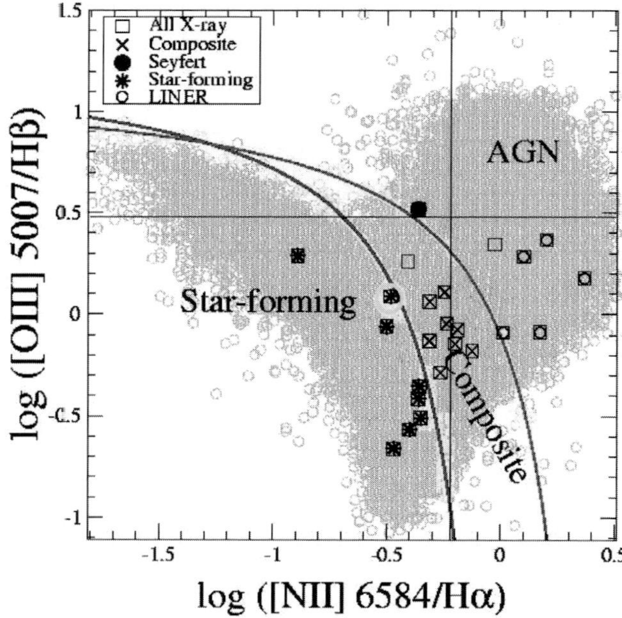

Figure 2. Emission-line ratio diagnostic diagram, adapted with permission from Hornschemeier *et al.* (2005). The grey in the background shows the SDSS DR3 galaxy sample and the black symbols are serendipitous *Chandra* detections. Theoretical modeling of starbursts has shown that even for extreme starburst galaxies should not exist above the red (upper curved) line in this diagram (Kewley *et al.* 2001). The blue (lower curved) line was proposed by Kauffmann *et al.* (2003) as the line below which AGN are not found. The cyan data point marks the location of a NLS1 galaxy, which was confirmed by its Balmer line widths exceeding its forbidden line widths.

such as optical dilution and/or inadequate coverage of key spectral features such as Hα. Hornschemeier *et al.* (2005) took this approach in an archival *Chandra* study of 42 serendipitously detected SDSS galaxies. Stellar population synthesis models were run for all of the galaxies (Tremonti *et al.* 2004) and the best-fit stellar continua subtracted. Galaxies were then divided into star-forming and AGN samples via emission-line ratio diagnostics ([OIII] 5007/Hβ versus [NII]/Hα; see Figure 2).

3.3. *X-ray/SFR Correlations for Carefully Classified Galaxies*

In galaxies with a higher ratio of SFR to stellar mass, where the LMXB component is expected to be negligible, the total X-ray luminosity may be dominated by HMXBs and thus serve as an SFR indicator (e.g., Grimm *et al.* 2003). We have taken several X-ray/SFR relations in the literature and converted them to the same bandpass and the same IMF (the IMF of Kroupa 2001) for conversions (see Figure 3). Among the serendipitously detected SDSS galaxies, the "pure" star-forming galaxies tend to clump around the lower X-ray/SFR correlations. These studies (Persic *et al.* 2004; Colbert *et al.* 2004) both removed contributions from hot gas and from evolved LMXB populations whereas the higher X-ray/SFR lines in Figure 4 are for *total* X-ray luminosity. Our results indicate that there is X-ray emission in many of those galaxies which is *not* associated with the HMXB population but likely is due to hot gas and/or the evolved component.

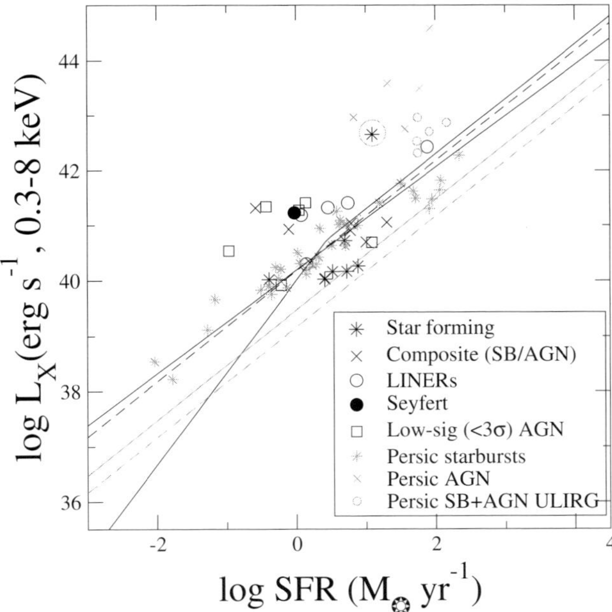

Figure 3. Total *Chandra* full-band X-ray luminosity of SDSS sources versus star-formation rate (SFR; Hornschemeier *et al.* 2005). The lines show the correlations from the literature, adjusted to the 0.3–8 keV band and a Kroupa IMF. The thin black, thick black, dashed black, pink solid and pink dashed lines are Bauer *et al.* (2002), Grimm *et al.* (2003), Ranalli *et al.* (2003), Persic *et al.* (2004), and Colbert *et al.* (2004) respectively. We also show the data on nearby galaxies from Persic *et al.* (2004).

4. X-ray Luminosity Functions

Revisiting the number counts of Figure 1, notice that there are *two lines* for both the AGN and galaxy number counts. These correspond to "optimistic" and "pessimistic" galaxy classification (Bauer *et al.* 2004) using observational properties such as X-ray luminosity, hardness ratio, and spectroscopic identification. Some of the first attempts to do such systematic galaxy classification include e.g., the Bayesian analysis of Norman *et al.* (2004). This work has allowed the first normal/starburst galaxy X-ray luminosity function to be constructed at $z \approx 0.3$ and $z \approx 0.7$ (Norman *et al.* 2004). There is clear (and expected) evolution in this XLF, which has a lognormal shape, similar to FIR luminosity functions (Takeuchi *et al.* 2004). Norman *et al.* (2004) interpret the lognormal shape as evidence of the multiplicative processes involved in accreting binary evolution.

Recently this work has been expanded using k-corrected X-ray/optical and X-ray/ infrared flux ratios (resulting in less AGN contamination) and a larger multiwavelength dataset offered by the Great Observatories Origins Deep Survey (GOODS; Ptak *et al.* 2006). For the first time, we have been able to observe differences in galaxy X-ray luminosity functions by galaxy type (early and late-type SED; Georgantopoulos *et al.* 2005; Ptak *et al.* 2006, see Figure 4). The late-type SED XLF is found to track the FIR luminosity function better, indicating that it is tracing star formation.

5. Redshift Frontiers

5.1. *ULX Sources at $z > 0.1$*

An interesting frontier in the study of individual accreting binaries has been opened up by the discovery of off-nuclear X-ray emission in several dozen deep *Chandra* survey

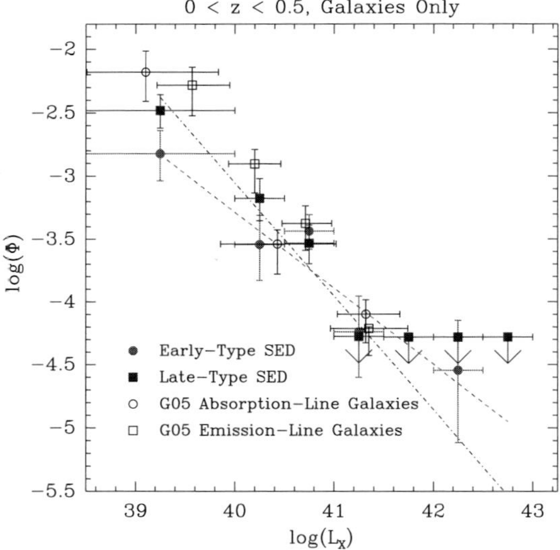

Figure 4. The $z \approx 0.3$ normal galaxy XLFs from Ptak *et al.* (2006) and Georgantopoulos *et al.* (2005). This figure was adapted with permission from Ptak *et al.* (2006). Clear difference in the early-type and late-type galaxy SED can be seen.

sources (Hornschemeier *et al.* 2004, B. Lehmer *et al.* 2006, in preparation). These sources, like the Ultraluminous X-ray sources observed in the local Universe, have full-band X-ray luminosities $\gtrsim 10^{39}$ erg s^{-1}, in excess of that expected for spherically symmetric Eddington-limited accretion onto "stellar" mass (5–20 $M_\odot$) black holes. These sources may still be consistent with stellar mass black holes, possibly representing an unstable, beamed phase in normal high-mass X-ray binary (HMXB) evolution (e.g. King *et al.* 2001) or might also represent a class of intermediate mass black holes (≈ 500–1000 $M_\odot$, e.g., Colbert *et al.* 2002).

$\approx 20\%$ of giant galaxies in the *Chandra* deep survey fields harbor these off-nuclear ULX sources (Hornschemeier *et al.* 2004), some of which show clear short-term variability. The ULX fraction measured here is only a lower limit; even *Chandra*'s sub-arcsecond spatial resolution often cannot resolve sources within the central ≈ 1–2 kpc of the nucleus and offsets of ~ 1 kpc have been found for ULX sources in the local Universe (Colbert & Ptak 2002).

The binary LFs of galaxies predict few ULX sources per galaxy, so constructing samples such as these at $z \approx 0.1$ allow a statistical probe of the bright end of the binary LF. The initial results with 12 ULX sources found that the numbers/brightnesses of ULXs at $z \approx 0.1$ are consistent with a bright-end LF slope of $\alpha \approx 0.4$, characteristic of, e.g., the Antennae (Zezas *et al.* 2002).

5.2. *Lyman Break Galaxies*

Our current knowledge of the high-z star-forming Universe is dominated by studies in the rest-frame ultraviolet band (the observed-frame optical). The best-studied examples of high-z star-forming galaxies are the "Lyman Break" galaxies (Steidel *et al.* 1996; Shapley *et al.* 2003). These galaxies are of great cosmological importance, as they are vigorously forming stars (30–50 $M_\odot$ yr^{-1}) and exhibit rest-frame ultraviolet spectroscopic signatures of outflows (Shapley *et al.* 2003). It appears that such starburst outflows are required to correctly predict the low-mass end of the galaxy luminosity function (analogous to the AGN outflows affecting the high-mass end, e.g., Croton *et al.* 2006).

There are relatively few measures of SF that can reliably reach $z > 2$ (however see Hogg 2002). The lack of powerful wide-field NIR spectrographs make Hα studies beyond $z \approx 1$ very difficult. The depth of the currently deepest radio surveys is insufficient to detect moderately star-forming galaxies at $z > 1$. NUV emission lines (such as [OII]) are highly susceptible to uncertainties in metallicity (but see Kewley *et al.* 2004). Spitzer is certainly opening up the MIR/FIR band at high-z, but the large PSF ($>5''$ at $\lambda_{\rm obs} > 24~\mu$m) makes confusion a problem at $z > 2$ in the most useful FIR bands.

The deepest *Chandra* surveys are not sensitive enough to individually detect Lyman Break galaxies. However, the sensitivity may be extended via stacking the X-ray emission from undetected sources. Despite the uncertainties in terms of why the X-ray/SFR correlation works even when non-HMXB components are included (see §3.3 and Hornschemeier *et al.* 2005), the reliability of X-ray emission as a possible SFR indicator has allowed numerous groups to study the average X-ray emission from star-forming galaxies. These analyses have verified ultraviolet dust-obscuration corrections (Seibert *et al.* 2002) and have probed star formation up to $z \approx 4$ (Lehmer *et al.* 2005). Although stacking X-ray emission from Lyman Breaks has become an industry (Brandt *et al.* 2001; Nandra *et al.* 2002; Reddy & Steidel 2004), the X-ray/SFR correlations have not been characterized well at the higher SFRs typical of Lyman Break galaxies (10–$100~M_\odot$ yr^{-1}). Individual detections of complete samples in this range of SFR, particularly in a UV-selected sample, would be very advantageous. We can look forward to new, lower-redshift "Lyman Break galaxy" samples from the GALEX mission (Heckman *et al.* 2005) for calibrating of the X-ray/SFR relation in the relevant high SFR range.

6. Summary and Considerations for the Future

X-ray studies of normal galaxies at high redshift have reached new levels of sophistication as we understand the biases in dividing them from AGN via observational properties such as X-ray/optical flux ratio, X-ray hardness, and optical spectroscopic type. The first galaxy XLFs have been measured, and constraints can be placed on the binary LF using ULX sources at high-z. Many of the uncertainties will be erased by the next generation X-ray mission, Constellation-X (constellation.gsfc.nasa.gov, will launch in 2015–2018), whose spectroscopic capabilities will allow us to use X-ray emission and absorption features to construct *X-ray* diagnostic diagrams. Constellation-X will have 1.5 m^2 collecting area at 1.25 keV; high spectral-resolution X-ray spectra for populations such as the Chandra-SDSS sample will easily be obtained in ~ 40 ks observations.

Chandra may go deeper, and the X-ray number counts indicate that at $f_X < 5 \times 10^{-18}$ erg cm^{-2} s^{-1}, we expect galaxies to become the most numerous X-ray emitting population. Missions such as XEUS (launching around 2018) and Generation-X (launching after 2025), with extremely large collecting areas (10 m^2 and 100 m^2, respectively), should detect and characterize galaxies up to $z \approx 5$, allowing us to measure starburst winds, etc. in galaxies as they are first forming.

Acknowledgements

I would like to thank the organizers of the symposium for their invitation to give this review, and my collaborators Franz Bauer, Karen Peterson, Andrew Ptak and Bret Lehmer for providing figures for the presentation and for this manuscript.

References

Abazajian, K., *et al.* 2005, *AJ*, 129, 1755

Alexander, D. M., Aussel, H., Bauer, F. E., Brandt, W. N., Hornschemeier, A. E., Vignali, C., Garmire, G. P., & Schneider, D. P. 2002, *ApJL*, 568, L85

Barger, A. J., *et al.* 2003, *AJ* , 126, 632

Bauer, F. E., Alexander, D. M., Brandt, W. N., Hornschemeier, A. E., Vignali, C., Garmire, G. P., & Schneider, D. P. 2002, *AJ* , 124, 2351

Bauer, F. E., Alexander, D. M., Brandt, W. N., Schneider, D. P., Treister, E., Hornschemeier, A. E., & Garmire, G. P. 2004, *AJ* , 128, 2048

Brandt, W. N., Hornschemeier, A. E., Schneider, D. P., Alexander, D. M., Bauer, F. E., Garmire, G. P., & Vignali, C. 2001, *ApJL* , 558, L5

Colbert, E. J. M., Heckman, T. M., Ptak, A. F., Strickland, D. K., & Weaver, K. A. 2004, *ApJ* , 602, 231

Colbert, E. J. M., & Ptak, A. F. 2002, *ApJS* , 143, 25

Comastri, A., *et al.* 2002, *ApJ* , 571, 771

Croton, D. J., *et al.* 2006, *MNRAS* , submitted

Georgantopoulos, I., Georgakakis, A., & Koulouridis, E. 2005, *MNRAS* , 360, 782

Grimm, H.-J., Gilfanov, M., & Sunyaev, R. 2003, *MNRAS* , 339, 793

Heckman, T. M., *et al* 2005, *ApJL* , 619, L35

Hogg, D. W. 2002, *PASP* (astro-ph/0105280)

Hornschemeier, A. E., *et al.* 2004, *ApJL* , 600, L147

Hornschemeier, A. E., *et al.* 2003, *AJ* , 126, 575

Hornschemeier, A. E., Brandt, W. N., Alexander, D. M., Bauer, F. E., Garmire, G. P., Schneider, D. P., Bautz, M. W., & Chartas, G. 2002, *ApJ* , 568, 82

Hornschemeier, A. E., Heckman, T. M., Ptak, A. F., Tremonti, C. A., & Colbert, E. J. M. 2005, *AJ* , 129, 86

Kauffmann, G., *et al.* 2003, *MNRAS* , 346, 1055

Kewley, L. J., Dopita, M. A., Sutherland, R. S., Heisler, C. A., & Trevena, J. 2001, *ApJ* , 556, 121

Kewley, L. J., Geller, M. J., & Jansen, R. A. 2004, *AJ* , 127, 2002

King, A. R., Davies, M. B., Ward, M. J., Fabbiano, G., & Elvis, M. 2001, *ApJL* , 552, L109

Kroupa, P. 2001, *MNRAS* , 322, 231

Lehmer, B. D., *et al.* 2005, *AJ* , 129, 1

Moran, E. C., Filippenko, A. V., & Chornock, R. 2002, *ApJL* , 579, L71

Nandra, K., Mushotzky, R. F., Arnaud, K., Steidel, C. C., Adelberger, K. L., Gardner, J. P., Teplitz, H. I., & Windhorst, R. A. 2002, *ApJ* , 576, 625

Norman, C., *et al.* 2004, *ApJ* , 607, 721

Persic, M., & Rephaeli, Y. 2003, *A&A* , 399, 9

Persic, M., Rephaeli, Y., Braito, V., Cappi, M., Della Ceca, R., Franceschini, A., & Gruber, D. E. 2004, *A&A* , 419, 849

Peterson, K., Gallagher, S., Hornschemeier, A. E., Muno, M., & Bullard, E. C. 2006, *ApJ* in press

Ptak, A., Norman, C., Mobasher, B., Hornschemeier, A., & Bauer, F. 2006, *ApJ* submitted

Ranalli, P., Comastri, A., & Setti, G. 2003, *A&A* , 399, 39

Reddy, N. A., & Steidel, C. C. 2004, *ApJL* , 603, L13

Seibert, M., Heckman, T. M., & Meurer, G. R. 2002, *AJ* , 124, 46

Shapley, A. E., Steidel, C. C., Pettini, M., & Adelberger, K. L. 2003, *ApJ* , 588, 65

Steidel, C. C., Giavalisco, M., Dickinson, M., & Adelberger, K. L. 1996, *AJ* , 112, 352

Takeuchi, T. T., Yoshikawa, K., & Ishii, T. T. 2004, *ApJL* , 606, L171

Tremonti, C. A., *et al.* 2004, *ApJ* (astro-ph/0405537)

Worsley, M. A., Fabian, A. C., Barcons, X., Mateos, S., Hasinger, G., & Brunner, H. 2004, *MNRAS* , 352, L28

Zezas, A., Fabbiano, G., Rots, A. H., & Murray, S. S. 2002, *ApJ* , 577, 710

Discussion

KIM: Can you comment on the different classification schemes between Norman *et al.* & Bauer *et al?*

HORNSCHEMEIER: The Norman *et al.* classification was much more "optimistic" or open. It likely admitted too many low luminosity AGN, mainly due to weak constraints on hardness ratios and X-ray/optical flux ratios. The follow-on paper by Ptak *et al.* includes much more rigorously defined selection methods, including K-corrected X-ray/NIR flux ratios derived from K-band data. This selection admits fewer galaxies to the XLF sample.

MACCARONE: Will radio emission, especially from deep wide-field VLBI measurements, help with the separation of normal galaxies from AGN?

HORNSCHEMEIER: Yes, although we have the problem that current radio measurements aren't deep enough to compare with our deepest X-ray fields. We're hoping e-VLA, for example, will help a lot.

ELVIS: On the AGN-Starburst diagnostic plot you showed the Kewley line gives the maximum Starburst contribution and the Kauffmann line gives the minimum, so between these lines a galaxy could be either pure starburst, pure AGN or a mixture. Calling these objects 'Composite' may then be misleading.

HORNSCHEMEIER: The term "Composite" is admittedly ambiguous. These galaxies, however, are a mix of galaxies which are still actively star-forming and possess actively accreting SMBH. That is, both processes are going on in all these galaxies. Tim Heckman and Jarle Brinchmann have been able to estimate the fraction of the [OIII] and Hα luminosity attributed to AGN vs. star formation so we do have useful information on the nature of these sources (they are not ambiguous in their classification).

Populations of High Energy Sources in Galaxies
Proceedings IAU Symposium No. 230, 2005
E. J. A. Meurs & G. Fabbiano, eds.

© 2006 International Astronomical Union
doi:10.1017/S1743921306008805

Chandra Multiwavelength Project (ChaMP): Normal Galaxies at Intermediate Redshift

D.-W. Kim[1], W. A. Barkhouse[1], P. J. Green[1], E. R. Colmenero[2], M. Kim[1], A. Mossman[1], E. Schlegel[1], J. D. Silverman[1], T. Aldcroft[1], C. Anderson[1], H. Tananbaum[1], and B. J. Wilkes[1]

[1]Smithsonian Astrophysical Observatory, 60 Garden Street, Cambridge, MA 02138, USA
email: kim@cfa.harvard.edu

[2]South African Astronomical Observatory

Abstract. We have investigated 136 Chandra extragalactic sources without broad emission lines, including 93 NELG and 43 ALG. Based on f_X/f_O, L_X, X-ray spectral hardness and optical emission line diagnostics, we have conservatively classified 36 normal galaxies (20 spirals and 16 ellipticals) and 71 AGNs. Their redshift ranges from 0.01 to 1.2, while normal galaxies are at z=0.01–0.3. Our sample galaxies appear to share similar characteristics with local galaxies in terms of X-ray luminosities and spectral properties, as expected from the X-ray binary populations and the hot ISM. In conjunction with normal galaxies found in other surveys, we found no statistically significant evolution in L_X/L_B, within the limited z range (<0.1). We have built our log(N)-log(S) relationship of normal galaxies in the flux range, f_X (0.5–8.0) = 10^{-15}– 10^{-13} erg sec^{-1} cm^{-2}, after correcting completeness by a series of simulations. The best-fit slope is −1.5 for both soft (0.5–2.0 KeV) and broad (0.5–8.0 KeV) energy bands, which is considerably steeper than that of AGN-dominated cosmic background sources at faint fluxes, but slightly flatter than the previous estimate, indicating normal galaxies will exceed in number over the AGN population at $f_X < 10^{-18}$ erg sec^{-1} cm^{-2}.

Keywords. surveys – X-rays: galaxies – X-rays: general.

1. Introduction

To understand the formation and evolution of galaxies, it is important to study galaxies at a wide range of redshifts from the time when they form and rapidly evolve. The Chandra multiwavelength project (ChaMP), with the advantage of a wide area coverage, allow us to investigate a well-defined galaxy sample at redshifts intermediate between distant galaxies found in the Chandra Deep Field (CDF) and the local galaxies. In this paper, we will utilize the ChaMP sample in conjunction with normal galaxies from the deep surveys and apply X-ray properties obtained from well-studied local galaxies to address the classification, X-ray luminosity evolution and number density distribution of normal galaxies.

2. Classification

Our sample consists of narrow emission line galaxies (NELG) and absorption line galaxies (ALG). Broad line AGNs/QSOs (with a line width > 1000 km s^{-1}) and galactic stars are excluded. To further separate even the obscured AGNs (Type 2 or XBONGs) from normal galaxies, we apply several diagnostics: the X-ray luminosity, X-ray-to-optical flux ratio, X-ray spectral hardness and optical line ratios. We find that f_X/f_O is the most efficient way to distinguish normal galaxies and AGNs when applied conservatively, i.e., normal galaxy if $f_X/f_O < 0.01$ and AGN if $f_X/f_O > 0.1$ (see Figure 1). We define

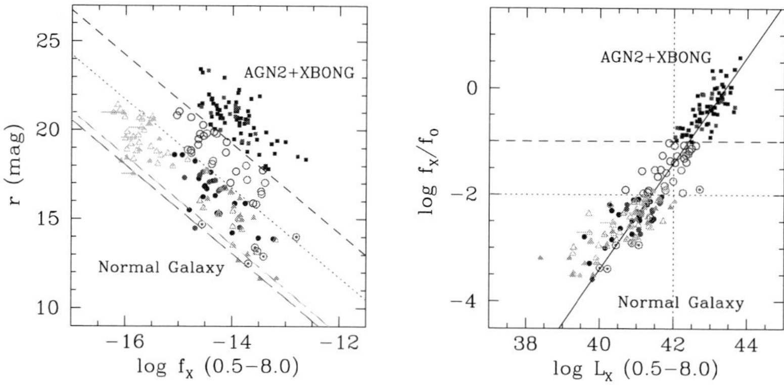

Figure 1. (a) The optical r magnitude against f_X and (b) the X-ray-to-optical flux ratio (f_X/f_O) against L_X. NELGs and ALGs are distinguished by blue and red colors, respectively. Normal galaxies (with $f_X/f_O < 0.01$) and AGNs (with $f_X/f_O > 0.1$) are marked by filled circles and filled squares, respectively. Open circles mark those unclassified objects with intermediate f_X/f_O. Also plotted are normal galaxies identified in CDF-N (Hornschemeier *et al.* 2003) and XMM-Newton NHS (Georgantopoulos et al. 2005), marked by cyan and green triangles, respectively. The black dashed line and the red dotted line indicate f_X/f_O =0.1 and 0.01, respectively. The diagonal solid line in Figure (b) indicates the best-fit linear relation for objects with $f_X/f_O > 0.01$.

f_X/f_O by f_X at 0.5–8.0 keV and f_O at the r band (e.g., Green *et al.* 2004). We note that in most cases classification by f_X/f_O also satisfies other selection criteria. For example, all galaxies with $f_X/f_O < 0.01$ are less luminous than $L_X = 10^{42}$ erg sec^{-1} and all AGNs with $f_X/f_O > 0.1$ are more luminous than $L_X = 10^{42}$ erg sec^{-1} (see Figure 1b). However, with our conservative classification scheme, we can not classify objects with intermediate f_X/f_O. Their X-ray spectral hardness and optical line ratios indicate that these intermediate objects indeed consist of mixed types of AGNs and galaxies.

3. Normal Galaxies

In Figure 1, we also plot normal galaxies identified from the Chandra Deep Field-North (CDF-N; Hornschemeier *et al.* 2003) and the XMM-Newton Needle in the Haystack Survey (NHS; Georgantopoulos *et al.* 2005). Our sample consists of galaxies at redshifts z=0.01–0.3, intermediate between the CDF (z=0.1–1.0) and nearby local galaxies, and covers an intermediate flux range. Our sample is similar to the NHS (z=0.005–0.2), but extends to a slightly higher z and a fainter X-ray flux. Combining all three samples allows us to cover a wide parameter space and to increase the statistical confidence. While galaxies from three different samples are distinct in their X-ray flux ranges due to the different observation depths, they are well mixed in the f_X/f_O - L_X space (or in the $L_O - L_X$ space) with $L_X = 10^{38} - 10^{42}$ erg sec^{-1} and log $f_X/f_O = -3.5$ -2 (Figure 1).

While we impose a maximum f_X/f_O (log $f_X/f_O = -2$), there is no lower limit imposed in our selection. Therefore, this minimum f_X/f_O must be real for normal galaxies and appears to remain the same in all three samples. It is interesting to note that the observed minimum f_X/f_O is consistent with the lower limit of $L_X - L_B$ relationship of early type galaxies (red dashed line; Kim and Fabbiano 2004) and of spirals (green dashed

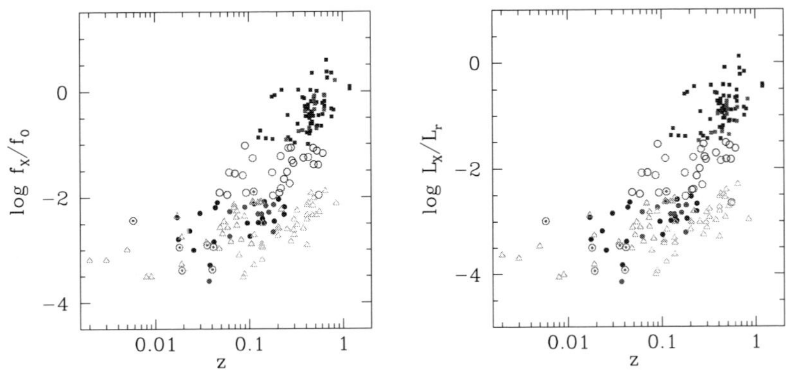

Figure 2. X-ray to optical flux/luminosity ratios as a function of z.

line; Colbert *et al.* 2003). Those galaxies with the minimum f_X/f_O represent X-ray faint (LMXB-dominant, gas-poor) early type galaxies, or quiescent (non-starburst) spiral galaxies. This indicates that the X-ray emission from X-ray binary populations in our X-ray selected galaxies is similar to that expected from local normal galaxies.

4. $L_X - L_B$ evolution

As the star formation rate peaks at z=2-3, the fossil record of past star formation imprinted on X-ray binaries could be detected by observing galaxies at high redshifts (e.g., Ghosh and White 2001; Ghosh in this conference). The X-ray luminosity could be 10–100 times higher than that in the local galaxies with HMXB peaking at z=1–2 while LMXB at z=0 5–1.0. We note that the average quantity of f_X/f_O or L_X/L_O as a function of z (as often used in the literature) could be seriously affected by selection effects (e.g., the upper limit of f_X/f_O) and/or AGN contamination. Instead, we test the X-ray luminosity evolution using the minimum of f_X/f_O and L_X/L_O as a function of z (as described in the previous section).

In Figure 2, we plot the X-ray-to-optical flux/luminosity ratios as a function of z. Again we have plotted normal galaxies from the CDF-N and NHS surveys as well. The difference between flux and luminosity ratio is caused by the K-correction and the evolution correction. The combined sample is complete up to z=0.1 and becomes incomplete at higher z due to the flux limit. Within this limited z range (z<0.1 or look back time of ~1 Gyr), the minimum value of L_X/L_r (or f_X/f_O) remains constant, indicating that no significant change in the X-ray properties of normal galaxies, hence no luminosity evolution in the X-ray binary population up to z=0.1. It is critical to include galaxies at higher z (z=0.5–1.0), but that will require the next generation X-ray mission such as Gen X.

5. Normal galaxy Log N- LogS relation

In Figure 3, we plot the log(N)-log(S) relation of the ChaMP normal galaxies. Our log(N)-log(S) relations in the soft (0.5–2.0 keV) and broad (0.5–8.0 keV) bands appear

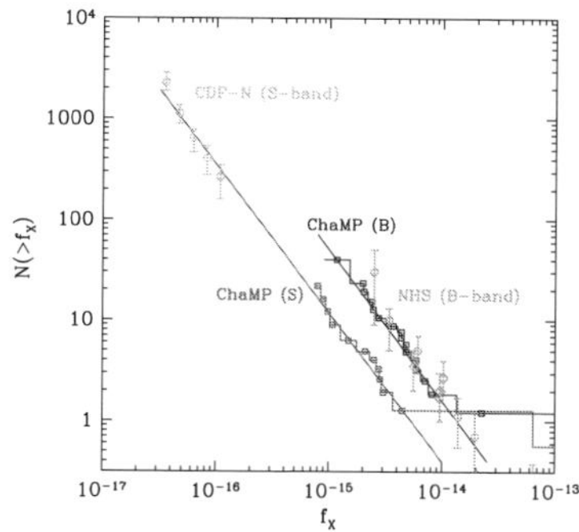

Figure 3. logN-logS relation of normal galaxies.

to be in the same shape, with the Euclidean slope of -1.5 $\pm$ 0.15 (the best-fit power-low distribution is plotted in the figure.) This is not surprising, because we are dealing with a homogeneous sample of galaxies with a relatively small amount of obscuration. This is in contrast to AGN-dominated cosmic background sources, where obscured and unobscured X-ray sources contribute differently in different energy bands (eg., Kim *et al.* 2004). We also note that unlike the broken power-law distribution of cosmic background AGNs with a break at f_X (0.5–2.0 keV) = 6×10^{-15} erg cm^{-2} s^{-1} (e.g., Kim *et al.* 2004), the log(N)-log(S) of normal galaxies is well reproduced by a single power-law in a wide range of f_X (see also Tajer *et al.* 2005). Also plotted in Figure 3 are those previously determined with the CDF sample in the S-band (Hornschemeier *et al.* 2003) and the NHS sample in the B-band (Georgakakis *et al.* 2004). It appears that our S-band log(N)-log(S) can be connected to that of the CDF, if extrapolated. However, the CDF data point at the faintest f_X ($\sim$3 x 10^{-17} erg s^{-1} cm^{-2}) is slightly higher than the extrapolated value and their best-fit slope, -1.74, is slightly steeper than ours. Our B-band log(N)-log(S) is statistically identical with that of the NHS, except our data extend to lower fluxes with smaller error bars. Since the log(N)-log(S) relation of normal galaxies is considerably steeper than that of AGN-dominated cosmic background sources (with a slope of 0.6– 0.7), we determine that normal galaxies will exceed in number over the AGN population at f_X < 10^{-18} erg sec^{-1} cm^{-2}. which is slightly fainter than previous estimates by Hornschemeier *et al.* (2003) and Bauer *et al.* (2004).

References

Bauer, F. E., *et al.* 2004, astro-ph/0408001
Colbert, E. J. M., *et al.* 2003, ApJ, 602, 231
Georgakakis, A. E., Georgantopoulos, I., Basilakos, S., & Plionis, M. 2004, astro-ph/0407387
Georgantopoulos, I., Georgakakis, A., & Koulouridis, E. 2005, astro-ph/0503494
Ghosh, P. & White, N. E. 2001, ApJ, 559, L97
Green *et al.* 2004, ApJS, 150, 43
Hornschemeier, A. E., *et al.* 2003, ApJ, 126, 575
Kim, D.-W. & Fabbiano 2004, ApJ, 611, 846
Kim, D.-W., *et al.* 2004a, ApJS, 150, 19
Kim, D.-W., *et al.* 2004b, ApJ, 600, 59
Maccacaro, T., *et al.* 1988, ApJ, 326, 680
Tajer, M., Trinchieri, G., *et al.* 2005, AA, submitted (also astro-ph/0412588)

Populations of High Energy Sources in Galaxies
Proceedings IAU Symposium No. 230, 2005
E. J. A. Meurs & G. Fabbiano, eds.

© 2006 International Astronomical Union
doi:10.1017/S1743921306008817

Beyond the Limit of Deep X-ray Surveys: Galaxies or AGN ?

Richard E. Griffiths[1], Takamitsu Miyaji[1], Adam Knudson[1] and Matthew Schurch

[1]Department of Physics, Carnegie Mellon University, Pittsburgh, PA 15213, USA
email: griffith@astro.phys.cmu.edu

Abstract. The great sensitivities of the Chandra X-ray Observatory and XMM-Newton are allowing us to explore the X-ray emission from galaxies at moderate to high redshift. By using the stacking method with CXO data, we show that we can detect the ensemble emission from normal elliptical, spiral and irregular galaxies out to redshifts approaching one. The average X-ray luminosity can then be compared with the results of models of the evolution in the numbers of X-ray binaries and can possibly be used to constrain models of star formation. In order to account for the increasing luminosity of spiral galaxies from low to moderate redshift, AGN components may need to be invoked.

Keywords. X-ray background, galaxies, number counts, X-ray binaries, AGN.

1. Introduction

Deep surveys in X-ray astronomy had the initial goal of solving the problem of the origin of the extragalactic X-ray background, and these surveys have now shown that the XRB is largely comprised of the evolving populations of AGN, some heavily absorbed (Hasinger *et al.* 2005). But the deep surveys with the Chandra X-ray Observatory (CXO) have shown that normal galaxies are also detected. The initial 1Ms survey of the Hubble Deep Field (HDF) North demonstrated that, at flux levels approaching 10^{-16} cgs (0.5–2.0 keV), about a third of the X-ray sources were identified with galaxies (Horneschmeier *et al.* 2002, 2003). The extension of this survey to 2Ms. has confirmed and expanded these findings.

We can thus begin to explore the evolution of extragalactic source populations in addition to the AGN, LINERs and clusters of galaxies which dominated the number counts in the flux ranges explored by the *Einstein, ROSAT* and *ASCA* observatories in the previous two decades. It has long been recognised that normal galaxy populations would eventually be observed in deep X-ray surveys given enough sensitivity. It was not clear that this would happen with the Chandra X-ray Observatory.

2. Deep Surveys and Source Counts

The number counts in the HDF-N have been measured by Miyaji & Griffiths (2002), and extended to fluxes below 10^{-17} ergs cm^{-2} s^{-1} in the soft band (0.5–2 keV) and to 10^{-16} ergs cm^{-2} s^{-1} in the hard band (2–10 keV) by analysis of the fluctuations which remain after removal of the individual discrete source detections. Below this limit, the fluctuation analysis shows that the number counts continue to rise, as shown in Fig. 1, with a slope consistent with that between 10^{-15} and 10^{-16} ergs cm^{-2} s^{-1}.

At X-ray fluxes between 10^{-15} and 10^{-16} ergs cm^{-2} s^{-1} in the HDF-N, Hornschemeier *et al.* (2002, 2003) showed that about a third of sources were identified with galaxies.

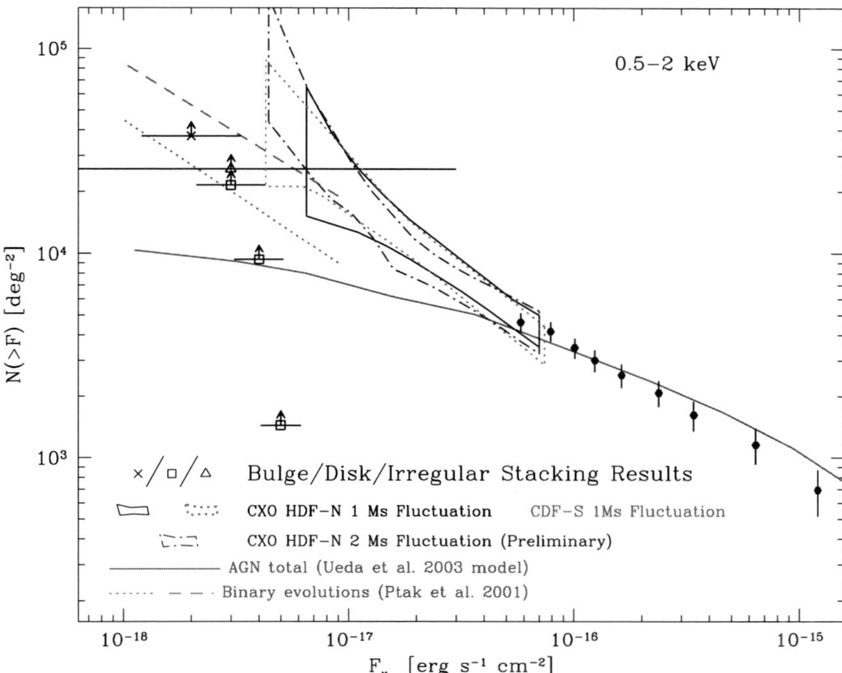

Figure 1. X-ray Number Counts from the Hubble/Chandra Deep Field North. The 1 Ms "fish-tail" fluctuation limits are from Miyaji & Griffiths (2002), and similar results were found from the HDF-S. Fluctuation limits from the 2Ms CDF-N are preliminary and possibly contaminated by cosmic-ray background events. Points showing stacked galaxies of different morphological types are all lower limits - stacking was done on optical galaxies in the HDF-N and the data points represent the average fluxes of the stacked galaxies. For any given morphological type, it is unlikely, however, that the number counts due to them are much higher than those shown here, unless contaminated by AGN. The prediction of AGN number counts is from Ueda *et al.* (2003). An estimate of the number counts from X-ray binary populations in normal galaxies was made by Ptak *et al.* (2001).

The extension of this survey to 2Ms. has confirmed and expanded these findings. We infer that the number counts in the region explored with fluctuations (the boxed area in Fig. 1) are unlikely to be due to AGN. Furthermore, the current best models for the AGN contributions to the number counts fall well below the fluctuations. The fluctuations analysis shows that the number counts are approximately $20{,}000 - 40{,}000$ per sq. deg. at X-ray fluxes of 10^{-17} ergs cm^{-2} s^{-1}. Such number counts match those of the optical counts of galaxies at B = 24. We therefore explore the possibility of X-ray detection of faint galaxies in the HDF-N by using the stacking method.

3. Selection of Galaxies by Morphological Type and Luminosity

During the execution of the Medium Deep Survey using the Hubble Space Telescope, software was developed for the automated classification of galaxies into spirals (exponential disks), ellipticals ('de Vaucouleurs' profiles) and irregular galaxies which exhibited large residual images after the removal of disk or bulge profiles (Ratnatunga, Griffiths & Ostrander 1999). This survey showed, for example, that the fraction of irregulars rises from 12% locally to 30% at a redshift of 0.5.

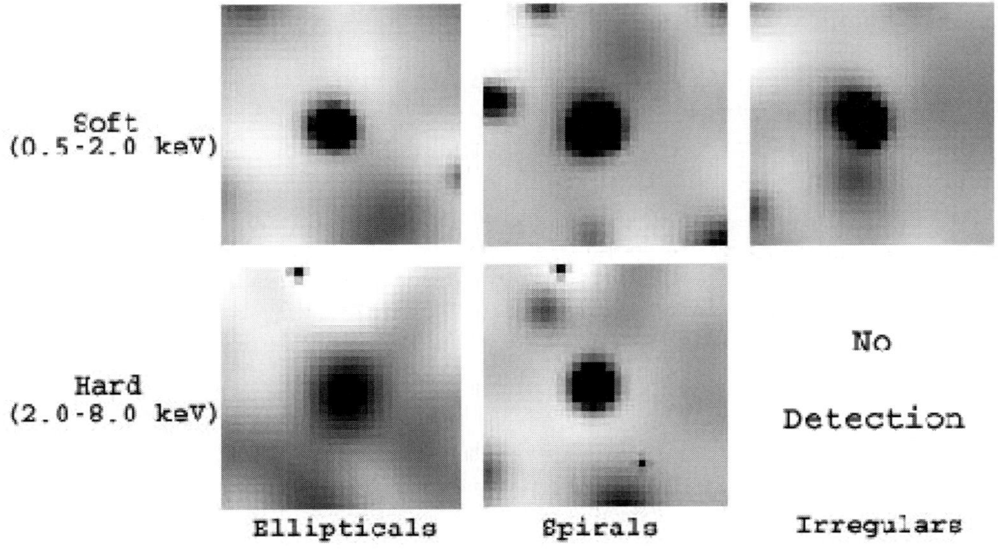

Figure 2. Stacked X-ray images of ensembles of galaxies.

We have applied this MDS software and analysis to the HST images of the HDF-N and other deep HST surveys. In those fields where we have deep CXO observations, we can then examine the X-ray images for the presence of X-rays from the galaxies of differing broad morphological type.

4. Results of 'Stacking'

Although the fluctuations analysis gives us an indication of the number counts of X-ray sources at the faintest flux levels currently accessible, they do not give us any indication of the nature of the sources contributing to the fluctuations. How do we find out the possible nature of these sources? One method is that of 'stacking', i.e. the summation of sub-images centered on objects selected at another wavelength. Brandt *et al.* (2001) and Horneschmeier *et al.* (2002) have used this method on early CXO data of the HDF-N. In the HDF, we have the advantage of being able to use the HST images themselves to select various types of galaxies for the stacking process, using the software developed as part of the HST Medium Deep Survey (Ratnatunga, Griffiths and Ostrander 1999). We have now done this for elliptical, spiral and irregular galaxies, and some of the results are shown in Figure 2.

As the figure shows, the spiral and elliptical galaxies are detected at high confidence in both the soft and hard energy bands, but the irregular galaxies are detected in the soft band only. The median redshifts are 0.87 for the 27 ellipticals, 0.49 for the 54 spirals and 1.55 for the 57 irregulars in these stacked images. Monte Carlo simulations have been used to verify the statistical confidence in these results.

The median X-ray luminosities (0.5–2 keV) are 5×10^{39} ergs s^{-1} for the ellipticals, 6×10^{39} ergs s^{-1} for the spirals and 2×10^{40} ergs s^{-1} for the irregulars, consistent with their B-band luminosities and the average values for L_X/L_B for the galaxy types. Typical galaxy fluxes are $\sim$2–4$\times10^{-18}$ ergs cm^{-2} s^{-1}.

5. X-ray Evolution of Galaxies

The observed X-ray evolution of spiral galaxies out to z = 0.7 is observed to be in excess of the expected value based on the evolution of binary X-ray populations. The observed evolution is better matched with a population of AGN such that the AGN luminosity is 0.1 of the galaxy luminosity in 30% of the galaxies at z = 0.7.

There are several problems which need to be solved or investigated in support of the interpretation of these results: (i) the evolution of low-mass X-ray binaries (LMXRB), (ii) the evolution of high-mass X-ray binaries (HMXRB), (iii) the evolution in the number of ultraluminous X-ray (ULX) objects and (iv) SNR and hot gas components.

6. Conclusions

Results from the stacking analysis of normal galaxy populations applied to the CXO deep survey of the HDF-N show that normal galaxy populations are observable in these stacks out to redshifts of $\sim$1. The average X-ray fluxes observed in these stacks are consistent with the numbers and fluxes inferred from the fluctuation analysis of the CXO data. We conclude that the fluctuations are therefore caused primarily by normal galaxy populations and that such deep X-ray surveys will eventually allow us to constrain the evolution of the binary source populations within these galaxies, using the relationship between HMXB numbers and the SFR of nearby galaxies. We have tentative evidence for the presence of AGN in some fraction of normal spiral agalxies at moderate redshift.

7. Acknowledgements

We acknowledge support from NASA grants NAG5-9902, NAG5-10875 and subcontract 2247-CMU-NASA-1128 from PSU (under NAS8-00128).

References

Anderson, S. & Margon, B. 1987, *Astrophys. J.* **314**, 111

Brandt, W. N., Hornschemeier, A. E., Schneider, D. P., Alexander, D. M., Bauer, F. E., Garmire, G., & Vignali, C. 2001, *Astrophys. J.* **558**, L5

Ghosh, K. & White, N. 2001, *Astrophys. J.* **559**, L97

Hasinger, G., *et al.* 2005, *Astron. & Astrophys. in press*

Hornschemeier, A. E., Brandt, W. N., Alexander, D. M., Bauer, F. E., Garmire, G. P., Schneider, D. P., Bautz, M. W., & Chartas, G. 2002, *Astrophys. J.* **568**, 82

Hornschemeier, A. E. and the CDF-N team 2003, *X-rays at Sharp Focus: ASP Conf. Series, eds. S. Vrtilek, E. M. Schlegel, L. Kuhi*

Miyaji, T. & Griffiths, R.E. 2002, *Astrophys. J.* **564**, L5

Norman, C., *et al.* 2004, *Astrophys. J.* **607** 721

Ptak, A. & Griffiths, R. E. 2001, *Astrophys. J.* **559**, L91

Ratnatunga, K., Griffiths, R. E., & Ostrander, E. J. 1999, *Astron. J.* **118**, 86

Ueda, Y., Akiyama, M., Ohta, K., & Miyaji, T. 2003, *Astrophys. J.* **598**, 886

Populations of High Energy Sources in Galaxies
Proceedings IAU Symposium No. 230, 2005
E. J. A. Meurs & G. Fabbiano, eds.

© 2006 International Astronomical Union
doi:10.1017/S1743921306008829

Active Cores in Deep Fields

Günther Hasinger[1] and Andreas Müller[2]

Max–Planck–Institute for Extraterrestrial Physics, P.O. box 1312, D–85741 Garching,
Germany
[1]email: ghasinger@mpe.mpg.de [2]email: amueller@mpe.mpg.de

Abstract. Deep field observations are an essential tool to probe the cosmological evolution of galaxies. In this context, X-ray deep fields provide information about some of the most energetic cosmological objects: active galactic nuclei (AGN). Astronomers are interested in detecting sufficient numbers of AGN to probe the accretion history at high redshift. This talk gives an overview of the knowledge resulting from a highly complete soft X-ray selected sample collected with *ROSAT*, *XMM–Newton* and *Chandra* deep fields. The principal outcome based on X-ray luminosity functions and space density evolution studies is that low–luminosity AGN evolve in a dramatically different way from high–luminosity AGN: The most luminous quasars perform at significantly earlier cosmic times and are most numerous in a unit volume at cosmological redshift $z \sim 2$. In contrast, low–luminosity AGN evolve later and their space density peaks at $z \sim 0.7$. This finding is also interpreted as an anti–hierarchical growth of supermassive black holes in the Universe. Comparing this with star formation rate history studies one concludes that supermassive black holes enter the cosmic stage before the bulk of the first stars. Therefore, first solutions of the so-called hen–egg problem are suggested. Finally, status developments and expectations of ongoing and future extended observations such as the *XMM–COSMOS* project are highlighted.

Keywords. surveys, galaxies: active, galaxies: evolution, galaxies: high–redshift, galaxies: luminosity function, mass function, galaxies: nuclei, galaxies: statistics, X-rays: diffuse background, X-rays: galaxies, X-rays: general.

1. Introduction

Deep X-ray surveys turned out to be valuable diagnostic tools to investigate the galaxy formation history and large scale structure of the Universe. The diffuse X-ray background (XRB) at the sky is composed of discrete sources that have been almost resolved by deep *ROSAT*, *ASCA*, *Chandra* and *XMM–Newton* observations at photon energies in the 0.1–10 keV band † (Hasinger *et al.* (1998), Mushotzky *et al.* (2000), Giacconi *et al.* (2001), Hasinger *et al.* (2001), Giacconi *et al.* (2002), Alexander *et al.* (2003), Worsley *et al.* (2004), Bauer *et al.* (2004)). The X-ray sources in deep field observations are mostly active galactic nuclei (AGN). This has been confirmed by optical identification programmes with the *Keck* (Schmidt *et al.* (1998), Lehmann *et al.* (2001), Barger *et al.* (2001), Barger *et al.* (2003)), the *VLT* (Fiore *et al.* (2003), Szokoly *et al.* (2004), Zheng *et al.* (2004), Mainieri *et al.* (2005)) and *COMBO-17* (Wolf *et al.* (2004)). All AGN are captured in different evolutionary stages. Therefore, X-ray deep fields also probe the *in vivo* growth of supermassive black holes (SMBHs) that drive the AGN luminosities according to a widely accepted paradigm. Optical identification programmes provide cosmological redshifts of the X-ray sources. This can be done by spectroscopic or, especially at high redshifts, photometric methods (Zheng *et al.* (2004), Mainieri *et al.* (2005), Wolf

---

† However, it must be stated that at higher photon energies in the 5–10 keV band there is still a lack of identification.

et al. (2004)). Optical observations demonstrate that the number distributions in redshift space of all galaxy types peak around $z \simeq 0.7$. Further, the AGN sample can be classified into AGN type-1 (unabsorbed, unobscured) and AGN type-2 (absorbed, obscured) by using optical and/or X-ray methods. In the optical, AGN type-1 are defined as sources exhibiting broad Balmer emission lines due to the fact that the observer is able to view the core of the AGN from the outside. These features are lacking for AGN type-2. In X-rays, AGN type-1 show an unabsorbed X-ray spectrum whereas AGN type-2 have absorbed spectra around photon energies from 0.5-1 keV.

The paper is organised as follows: In Sec. 2 the soft X-ray selected sample is presented, Sec. 3 treats the method to analyse features and evolution of different AGN object classes by means of X-ray luminosity functions. Space density and luminosity density evolutions as other tools are presented in Sec. 4. The observations are interpreted as a scenario described in Sec. 5. In Sec. 6 the results are compared to observational data from other work. Finally, we conclude in Sec. 7.

2. The X-ray selected AGN type-1 sample

A sample of about 1000 AGN type-1 is considered that has been merged from *ROSAT*, *Chandra* and *XMM–Newton* surveys. The sources are restricted to the 0.5–2 keV band and cover five orders of magnitude in flux and six orders of magnitude in survey solid angle. This sample is with 95% highly complete allowing to construct luminosity functions over cosmological timescales with unprecedented accuracy. Details of the soft X-ray sample are presented in Hasinger (2004) (H04 hereafter; see Table 1 therein). The advantage of restricting the sample only to type-1 AGN is that systematic uncertainties are excluded *a priori*. AGN type-2 can introduce these uncertainties by the varying and typically unknown AGN absorption column densities. As demonstrated by Szokoly *et al.* (2004) the optical AGN classification scheme suffers from dilution effects of AGN excess light from stars in the host galaxy – especially at low X-ray luminosities and medium redshifts. Then, only X-ray AGN classification schemes as introduced by Schmidt *et al.* (1998) and Szokoly *et al.* (2004) can do the job. As anticipated AGN are the main contributors to the XRB accounting for a fraction of about 70–100%. A review about AGN and their engines, SMBHs, can be found in the PhD thesis by Müller (2004). It was found that at faintest X-ray fluxes there are contributions by other population classes such as starburst and normal galaxies, see Hornschemeier *et al.* (2000), Hornschemeier *et al.* (2001).

3. X-ray luminosity functions

As elaborated in Hasinger *et al.* (2005) (HMS05 hereafter) the X-ray luminosity functions (XLFs) of the sample presented in Sec. 2 are deduced by the following two methods: The first ('binned') method is presented in Miyaji *et al.* (2000) and is based on a variant of the $1/V_a$ method. In this approach, the binned luminosity function in a given redshift bin z_i is derived by dividing the observed number $N_{\mathrm{obs}}(L_x, z_i)$ by the corresponding volume to the redshift range and appropriate survey X-ray flux limits and solid angles. Each of the XLFs is fitted by an analytical function to determine the bias in this luminosity value caused by a gradient of the XLF across one bin. By means of the resulting analytical function one can predict $N_{\mathrm{mod}}(L_x, z_i)$. The ratio $N_{\mathrm{obs}}/N_{\mathrm{mod}}$ serves as a factor to correct the XLF due to bias to first order.

The second ('unbinned') method is based on unbinned data. Individual V_{max} values from *ROSAT Bright survey* (RBS) sources are used to evaluate the zero–redshift luminosity

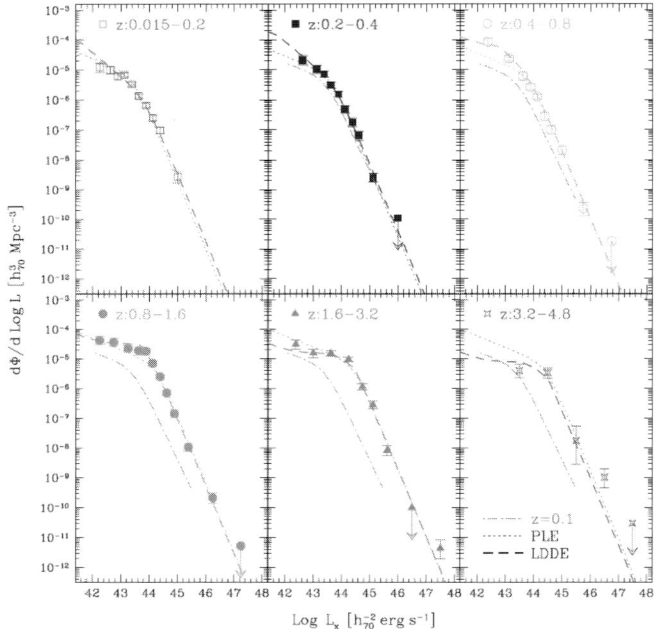

Figure 1. The soft X-ray luminosity function of the soft X-ray selected AGN type-1 sample in different redshift shells as denoted. Error bars correspond to 68% Poisson errors of the AGN number in the bin. The best–fit two power–law model in the lowest z–shell $z = 0.015$–0.2 is overplotted in each panel for reference. Dotted and dashed lines give best–fit PLE and LDDE models.

function (0zLF). By means of this LF the number of derived RBS sources matches accurately the observed number. Hence, the advantage of this method is that it is free from bias effects emerging in the first method. Evolution, i.e. space density as function of redshift, comes into play by introducing binning in luminosity and redshift. Again, bias effects are avoided by iterating the parameters of an analytical representation of the space density function. Together with the 0zLF this is used to predict $N_{\mathrm{mod}}(L_x, z_i)$ for the surveys. Finally, the observed space densities in each bin are evaluated by multiplying by the ratio $N_{\mathrm{obs}}/N_{\mathrm{mod}}$ with the space density value. The result is shown in Fig. 1: This is the soft X-ray luminosity function (SXLF) in different redshift shells ranging from $z = 0.015$ to $z = 4.8$ of about 1000 AGN type-1. It clearly demonstrates that the LF shape varies with redshift. The typical two power–law model is confirmed. Fits of the z–dependent LF profiles strongly suggest a luminosity–dependent density evolution (LDDE) as outlined in HMS05. Pure luminosity evolution (PLE) overpredicts especially the LF behavior at high redshift, $z \sim 2$–5.

4. Space density and luminosity density evolution

An alternative plot to analyse the data is shown in Fig. 2: data are divided into luminosity classes and plotted over redshift. The plot clearly demonstrates the space density evolution for each luminosity class, i.e. compares low–luminosity AGN (LLAGN) to high–luminosity AGN (HLAGN).The essential and surprising result is that the space density of HLAGN peaks at significantly higher redshift, $z \sim 2$, i.e. earlier cosmic times, than the LLAGN that peak at $z \sim 0.7$. In other words: Luminous quasars formed first and the bulk of faint AGN such as low–luminosity Seyferts came significantly later on the cosmic

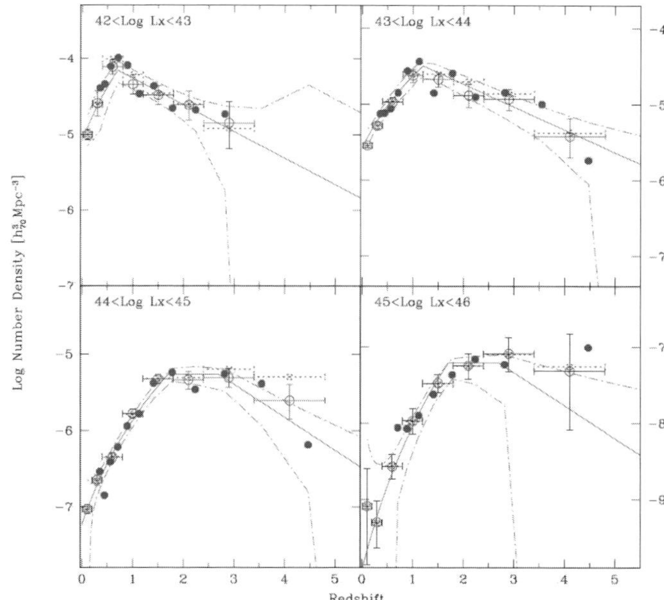

Figure 2. Space density evolution in different luminosity classes as derived from the binned (blue) and unbinnend method (red) outlined in Sec. 3. Dashed horizontal blue lines correspond to the maximum contribution of unidentified sources.

stage. Another remarkable fact is that for the first time the high–redshift decline in all luminosity classes with $\log\,(L_X/[\mathrm{erg/s}]) \lesssim 45$ is shown. At higher luminositites insufficient statistics does not allow to derive secure trends. Space density evolution can also be studied using optically selected AGN type-1 as demonstrated by Croom *et al.* (2004) and Fan *et al.* (2001) using data sets from the 2dF and 6dF QSO redshift surveys. A comparison of an optical (UV) vs. an X-ray selection of AGN type-1 is elaborated in H04. The main result is that there is relatively little difference in space density evolutions between the two wave bands; however, we have to consider that (as demonstrated in Fig. 2) rise and fall of the space density is *X-ray luminosity dependent*. Therefore, comparisons of this kind are only preliminary and have to be re–analysed when larger samples of high–redshift X-ray selected QSOs are available.

A fruitful comparison is based on analysing flux correlations between optical and X-ray fluxes. The X-ray data from H04 and optical SDSS data from Vignali *et al.* (2003) are considered. Then, the X-ray flux in the soft 0.5-2 keV band is plotted over optical AB_{2500} magnitudes. The first result is that the X-ray multi–cone survey covers a significantly wider range. Due to these better preconditions, a tight correlation was discovered by H04. Plotting the source's luminosity in the same filters indicates an essential difference: the luminosities in X-rays vs. those in redshifted 2500 Å *scale linearly*, $L_X \propto L_{\mathrm{UV}}$, when using X-ray selected data. But the correlation is non–linear, $L_X \propto L_{\mathrm{UV}}^{0.75}$, when UV data are used. A preliminary interpretation of this discrepancy is sample selection effects. Host galaxy contamination corrections that are missing in the X-ray analysis are likely to be not responsible for the difference as α_{ox} value studies indicate (for details see H04).

5. Growth of supermassive black holes

The analysis of X-ray deep field data by using XLFs and luminosity–weighted space density evolutions provides new clues for our understanding of AGN evolution in general and black hole growth in particular. AGN luminosity is connected to the central SMBH via Eddington's relation. Hence, from the observed activity the black hole mass can be estimated. Concerning HLAGN the soft X-ray selected samples clearly trace the rise and fall of quasars. The most luminous objects emerge firstly, grow rapidly by accretion and survive as supermassive dark remnants in the local Universe. Today, we observe these SMBHs in the centers of massive galaxy clusters such as M87 in the Virgo cluster. This scenario gets support from theoretical simulations presented recently by Springel et $al.$ (2005): In hydrodynamical simulations of structure formation on large scales performed in a cube with $\approx$700 Mpc size, the evolution of cold dark matter (CDM) halos are followed from $z \sim 120$ to $z \sim 0$. It has been shown that the observed large scale structure with superclusters surrounded by smaller clusters can be reproduced. Furthermore, the simulations strongly suggest that the most massive black holes with 10^{10} M$_\odot$ † can be found in the centers of the most massive clusters. These are the descendants of the most active accretors in the quasar era at $z \sim 2$. X-ray deep field data from the soft band now suggest an $anti$–$hierarchical$ $growth$ of black holes: the most massive black holes with $10^8 - 10^{10}$ M$_\odot$ formed first and the bulk of the lighter black holes with $10^6 - 10^8$ M$_\odot$ formed later. This behaviour is totally unexpected and not predicted by standard CDM structure formation scenarios which are hierarchical. Maybe this observational fact hints for two accretion modes that differ in accretion efficiency (Duschl & Strittmatter (2002)). However, a self–consistent model which explains both, anti–hierarchical black hole growth and the local black hole mass function derived from the $M_{\rm BH} - \sigma$ relation assuming two accretion modes has been suggested recently by Merloni (2004). Concretely, two accretion modes can be established by a varying value of efficiency ϵ – the parameter in accretion theory that controls conversion of mass flux into radiation flux. Black hole angular momentum controles crucially the efficiency: it is rather low, $\epsilon \sim 0.054$, for Schwarzschild black holes, or high, $\epsilon \sim 0.37$, for Kerr black holes as has been found by Thorne (1974). The SXLFs that give rise to an early quasar era in combination with the evolution of the star formation rate (SFR) also provide new clues to the so–called hen–egg $problem$ i.e. if galaxies or the SMBHs came first in the Universe. The history of global cosmic star formation has been determined using Hubble deep field (HDF) observations (see Madau (1997), Connolly et $al.$ (1997)) corrected for dust obscuration following Pettini et $al.$ (1997). The resulting SFR as function of redshift shows only a moderate variation with cosmic time with a possible peak around a redshift of unity and a steep decline towards lower redshifts. But the AGN population reveals a pronounced peak at significantly higher redshifts as shown in the SXLF of AGN type-1. This discrepancy indicates that the bulk of SMBHs have been in place $before$ the bulk of stars in these galaxies formed.

Very recently, an analysis of the specific SFR (SSFR) i.e. the star formation rate per unit stellar mass, has been presented based on FORS Deep Field (FDF) and GOODS–S field data, see Feulner et $al.$ (2005). Their main finding is that the most massive galaxies with stellar masses around 10^{11} M$_\odot$ are in quiescent state for $z \lesssim 2$ but that the SSFR is increased by a factor of ~ 10 for redshifts $z \gtrsim 2$. This is interpreted as a very early formation epoch of most massive galaxies as supported by the Millennium simulation (Springel et $al.$ (2005)) and X-ray deep field observations (HMS05). Comparisons of deep field observations therefore allow for attractive solutions of the hen–egg problem.

---

† and also the oldest population of stars (PopIII).

6. Discussion

In this section the analysis with soft X-ray data of about 1000 AGN type-1 are compared to other work. Ueda *et al.* (2003) analysed the hard X-ray luminosity functions (HXLFs) of bright AGN using *ASCA* data. This hard X-ray selected sample is highly complete ($\sim$95%) and consists of both, AGN type-1 and type-2, comprising $\approx$230 sources. The main feature is that the fraction of type-2 AGN decreases with intrinsic luminosity or in other words: the space density of obscured AGN is luminosity–dependent. A possible explanation for this trend may be due to clean–out effects because the luminous AGN core may blow away the dusty torus at the pc–scale by radiation. Low–luminosity cores are not strong enough to initiate such a decay of the mass reservoir. Therefore, the AGN classification scheme into type-1 and type-2 is not only a pure orientation effect but is also dependent on AGN luminosity. As outlined in Sec. 3 the SXLFs strongly suggest a LDDE. Interestingly, the hard X-ray selected sample by Ueda *et al.* (2003) demonstrates the same behaviour. There is only a difference in the normalization by a factor of five – probably due to absorbed objects missing in the SXLF. Another soft and hard X-ray selected sample based on *Chandra* (*CDF–N, CDF–S, CLASXS*) and *ASCA* data has been presented recently by Barger *et al.* (2005). Their results are in good agreement with the SXLF analysis presented here, but they still suffer from substantial identification incompleteness. Their AGN type-1 sample does not directly compare to the one discussed here because Barger *et al.* (2005) only include optically classified AGN type-1 (broad–line AGN) but miss most of LLAGN type-1 that are considered here. A critical comparison of all available XLFs from different observations will be the topic of an upcoming paper. Recently it turned out that there is a possible new population of star–forming galaxies that can be found especially at very low fluxes, $S_\mathrm{X} \lesssim 10^{-16}\,\mathrm{erg\,cm^{-2}\,s^{-1}}$ as discovered by Hornschemeier *et al.* (2000), Hornschemeier *et al.* (2001), Rosati *et al.* (2002), Norman *et al.* (2004). Upcoming X-ray selected samples will have to account for this special class by lowering the flux limits.

7. Conclusions

In general, X-ray deep fields prove valuable tools to investigate the formation and evolution history of galaxies, in particular of AGN. Concerning AGN, X-ray data sets cannot stand alone; they need additional support from optical identification programmes that discriminate type-1 vs. type-2 and deliver redshift determinations by spectroscopic or photometric techniques. Then, the analysis follows two branches: samples are constructed from soft and/or hard X-ray energy bands. To date soft X-ray selected samples provide high–quality data samples with high degrees of completeness. Hard X-ray selected samples still suffer from lacking spectroscopic identification. There is much work to be done in the future to resolve the high–energy XRB branch into discrete sources. Nevertheless, the comparison of the results from soft (HMS05) and soft plus hard X-ray selected samples by Ueda *et al.* (2003) and Barger *et al.* (2005) agrees well so far. The principal method to analyse X-ray samples is based on deriving XLFs, space density evolution and luminosity density evolution. The SXLF sample presented here strongly suggests dramatically different evolutionary paths of LLAGN and HLAGN: Most luminous quasars formed significantly earlier as the space density peak at $z \approx 2$ demonstrates. The bulk of galaxies with low luminosities such as Seyferts perform later with a space density peak at $z \approx 0.7$. Linking AGN luminosiy to black hole mass via the Eddington criterion, one immediately arrives at the finding that the growth of SMBHs is anti–hierarchical: The most massive holes form first. A comparison with the SFR evolution

Figure 3. RGB image of the *XMM–COSMOS* deep field first year data. The 2 square degree field consists of 25 separate pointings amounting 1.4 Ms exposure time in total. Red colors belong to 0.5–2 keV photon energies, green to 2–4.5 keV and blue to 4.5–10 keV. *Courtesy:* Nico Cappelluti and Alexis Finoguenov (both MPE Garching)

even shows that the most massive holes formed before the bulk of first stars formed. Hence, deep field observation have the power to solve the cosmological hen–egg problem. The observed AGN evolution scenario should gain support from theoretical simulations. The Millennium simulation is one essential cornerstone that allows to predict the evolution of the large scale structure.

The next observational steps are to collect better data i.e. identify more sources, perform deeper pencil beams and adjust suitable flux limits. One upcoming multi–wavelength project is called *COSMOS*. This is a *HST* treasury project where a 2 square degree field is observed with *ACS* by using 10% of observing time over two years. There are commitments from *VLA, VLT, Subaru* and *XMM–Newton* so that there will be also radio to X-ray data available for the same field where more than two million sources are supposed to be detected. Main goals of *COSMOS* are to study the large scale structure (LSS), evolution of galaxies, AGN and dark matter haloes, to investigate the SFR, and to derive AGN activity as function of morphology, size, redshift and LSS environment. One pencil beam with 0.2 square degree is included. The sensitivity level amounts 5×10^{-16} erg cm^{-2} s^{-1} and captures therefore sources that are by a factor of two fainter than in deepest *ROSAT* surveys †. First year observations of *XMM–COSMOS* started in December 2003. AO3 data are now available: A preliminary RGB image of 2 square degree size and 1.4 Ms total exposure time is shown in Fig. 3. As expected, a wealth of sources, almost all AGN, can be seen. The analysis of these *XMM–COSMOS* data is subject of a future paper (Hasinger *et al.* (2006)). Other upcoming extended deep X-ray surveys such as *eRosita* and *eCDFS* will improve significantly the analysis presented here.

Acknowledgements

AM and GH would like to thank the organizers of the IAU symposium No. 230 in Dublin, Ireland.

† Detailed information is provided on the *COSMOS* website `http://www.astro.caltech.edu/~cosmos/` and the *XMM–COSMOS* website `http://www.mpe.mpg.de/XMMCosmos`

References

Alexander, D.M., Bauer, F.E., Brandt, W.N., *et al.* 2003, *AJ* 126, 539

Barger, A.J., Cowie, L.L., Mushotzky, R.F., Richards, E.A., *et al.* 2001, *AJ* 121, 662

Barger, A.J., Cowie, L.L., Capak, P., *et al.* 2003, *AJ* 126, 632

Barger, A.J., Cowie, L.L., Mushotzky, R.F., *et al.* 2005, *AJ* 129, 578

Bauer, F.E., Alexander, D.M., Brandt, W.N., *et al.* 2004, *AJ* 128, 2048

Connolly, A.J., Szalay, A.S., Dickinson, M., *et al.* 1997, *A&A* 319, 7

Croom, S.M., Smith, R.J., Boyle, B.J., *et al.* 2004, *MNRAS* 349, 1397

Duschl, W.J. & Strittmatter, P.A. in Active Galactic Nuclei: from Central Engine to Host Galaxy Abstract Book, meeting held in Meudon, France, July 23-27, 2002, *Eds.: S. Collin, F. Combes and I. Shlosman.* To be published in ASP Conference Series, p. 76

Fan, X., Strauss, M.A., Schneider, D.P., *et al.* 2001, *AJ* 121, 54

Feulner, G., Gabasch, A., Salvato, M., *et al.* 2005, *astro–ph/0509197*, accepted for ApJL

Fiore, F., Brusa, M., Cocchià, F., *et al.* 2003, *A&A* 409, 79

Giacconi, R., Rosati, P., Tozzi, P., *et al.* 2001, *ApJ* 551, 624

Giacconi, R., Zirm, A., Wang, J.X., *et al.* 2002, *ApJS* 139, 369

Hasinger, G., Burg, R., Giacconi, R., *et al.* 1998, *A&A* 329, 482

Hasinger, G., Altieri, B., Arnaud, M., *et al.* 2001, *A&A* 365, 45

Hasinger, G. 2004, Nucl. Physics B. Suppl. Ser., 132, 86 (H04)

Hasinger, G., Miyaji, T., & Schmidt, M. 2005, *astro–ph/0506118*, A&A in press (HMS05)

Hasinger *et al.* 2006, *ApJ*, in prep.

Hornschemeier, A.E., Brandt, W.N., Garmire, G.P., *et al.* 2000, *ApJ* 541, 49

Hornschemeier, A.E., Brandt, W.N., Garmire, G.P., *et al.* 2001, *ApJ* 554, 742

Lehmann, I., Hasinger, G., Schmidt, M., *et al.* 2001, *A&A* 371, 833

Madau, P. 1997, AIP Conf. Ser. 393, 481

Mainieri, V., Rosati, P., Tozzi, P., *et al.* 2005, *A&A* 437, 805

Merloni, A. 2004, *MNRAS* 353, 1035

Miyaji, T., Hasinger, G., & Schmidt, M. 2000, *A&A* 353, 25

Müller, A. 2004, PhD thesis, Landessternwarte Heidelberg, Germany, *Black Hole Astrophysics: Magnetohydrodynamics on the Kerr Geometry*

Mushotzky, R.F., Cowie, L.L., Barger, A.J., & Arnaud, K.A. 2000, *Nature* 404, 459

Norman, C., Ptak, A., Hornschemeier, A., *et al.* 2004, *ApJ* 607, 721

Pettini, M., Steidel, C.C., Adelberger, K.L., *et al.* 1997, *astro–ph/9708117* , To appear in ASP Conference Series *ORIGINS*, J.M. Shull, C.E. Woodward, and H. Thronson (eds.)

Rosati, P., Tozzi, P., Giacconi, R., *et al.* 2002, *ApJ* 566, 667

Schmidt, M., Hasinger, G., Gunn, J.E., *et al.* 1998, *A&A* 329, 495

Springel, V., White, S.D.M., Jenkins, A., *et al.* 2005, *Nature* 435, 629

Szokoly, G., Bergeron, J., Hasinger, G., *et al.* 2004, *ApJS* 155, 271

Thorne, K.S., 1974, *ApJ* 191, 507

Ueda, Y., Akiyama, M., Ohta, K., & Miyaji, T. 2003, *Astron. Nachr.* 324, 36

Ueda, Y., Akiyama, M., Ohta, K., & Miyaji, T. 2003, *ApJ* 598, 886

Vignali, C., Brandt, W.N., & Schneider, D.P. 2003, *AJ* 125, 433

Wolf, C., Meisenheimer, K., Kleinheinrich, M., *et al.* 2004, *A&A* 421, 913

Worsley, M., Fabian, A.C., Mateos, S., *et al.* 2004, *MNRAS* 352, L28

Zheng, W., Mikles, V. J., Mainieri, V., *et al.* 2004, *ApJS* 155, 73

Discussion

ELVIS: There has been concern that quite subtle selection effects can produce e.g. $z=0.7$ peak and LLAGN detectability (Treister et al). What is your reaction to this study?

MÜLLER: I am sorry. I don't know this particular paper and cannot comment on that.

Populations of High Energy Sources in Galaxies
Proceedings IAU Symposium No. 230, 2005
E. J. A. Meurs & G. Fabbiano, eds.

© 2006 International Astronomical Union
doi:10.1017/S1743921306008830

Spectral properties from XMM-Newton observation of Chandra Deep Field South sources

A. Streblyanska[1], J. Bergeron[2], H. Brunner[1], A. Finoguenov[1], G. Hasinger[1], and V. Mainieri[1]

[1] Max-Planck-Institut für Extraterrestrische Physik, Giessenbach-Strasse Postfach 1312, D-85741 Garching, Germany
email: alina@mpe.mpg.de

[2] Institut d'Astrophysique de Paris, CNRS, 98 bis Boulevard Arago, F-75014 Paris, France

Abstract. We present results of the X-ray spectral analysis of the $\sim$370 ksec deep survey obtained with XMM-Newton on the Chandra Deep Field South (CDFS). Using sample of 127 sources with redshift identifications and with a raw count limit of 100 (pn detector) we explored both the physical properties of individual sources, and the general properties of two AGN classes. The corresponding flux limits in the (0.5–2) and (2–10) keV bands are 1×10^{-16} and 9×10^{-16} erg cm^{-2} s^{-1}, respectively. The average photon index is $\Gamma \sim 1.9$ and $\Gamma \sim 1.8$ for type-1 AGN and type-2 AGN, respectively. Although the properties of the spectra of more that 90% of AGN are in good agreement with the unified model, a fraction of 'atypical' objects (absorbed type-1 and unabsorbed type-2 AGN) was detected.

Keywords. Surveys – Galaxies: active – *(Galaxies:)* quasars: general– X-ray: galaxies – galaxies: nuclei– *(Cosmology:)* diffuse radiation.

1. Introduction

The X-ray background (XRB) has been a matter of intense study since its discovery by Giacconi *et al.* (1962). Now it is clear that at energies above $\sim$0.5 keV the XRB is made up of the summed emission from point sources, principally active galactic nuclei (AGN) along with some contribution at soft X-ray energies from thermal emission of galaxy clusters and stellar processes in normal galaxies. Since the XRB is mostly associated with point sources, the general interest has moved to accurately constrain the physical and evolutionary properties of these different classes of X-ray sources. To this aim, statistically significant samples of AGN that cover a range of redshift and luminosity from the deep and shallow surveys have been formed. This information makes the data an invaluable resource for revealing various key issues such as physical and evolutionary properties of AGN, and finally can be used to investigate, whether the unified model holds.

The *Chandra Deep Field South* (CDFS) is an area in the sky with a low Galaxy absorbing column density of 8.9×10^{19} cm^{-2} (Dickey & Lockman 1990). The CDFS has been studied intensively by means of a megasecond dataset from the Chandra Observatory (e.g. Giacconi *et al.* (2002), Tozzi *et al.* (2005), and references therein). This field was observed also with XMM-Newton (370 ksec good exposure; Streblyanska *et al.* 2003) and re-observed with Chandra ($\sim$250 ksec in each 4 pointings; Lehmer *et al.* 2005). The CDFS is one of the sky regions best studied not only in the X-ray band but also in the other wavelengths (radio, optical, infrared). In addition, the CDFS is one of the best fields with available spectral and photometric redshift information through different follow-ups with the VLT (e.g. Szokoly *et al.* (2004)) and some other ground-based imagings (e.g. Wolf *et al.*

2004). As a result, in the entire field of 1Ms Chandra observation only 4 objects, out of 347, have no identification (e.g. Zheng *et al.* (2004) and Mainieri *et al.* (2005)). This completeness made CDFS one of the reference fields for studies of distant normal and active galaxies, clusters of galaxies and even faint Galactic stars.

2. Data reduction and X-ray source list

The CDFS (RA 3:32:28 and DEC -27:48:30 [J2000]) XMM-Newton observation contains $\sim$370 ksec of clean data. The pn and MOS data were preprocessed by the *XMM Survey Scientist Consortium* with the XMM standard Science Analysis System (SAS, version 5.3.3) routines, using the latest calibration data. The merged images (pn+MOS) in the three energy bands were simultaneously searched for sources using tasks `emldetect` and `eboxdetect`. We detect 363 sources (likelihood value above 8) in pn+MOS1+MOS2. The corresponding flux limits in the (0.5–2) and (2–10) keV bands are 1×10^{-16} and 9×10^{-16} erg cm^{-2} s^{-1}, respectively.

The majority of the sources in our datasets have poor counting statistics, however we were able to construct a large sample of AGN for accurate source-by-source X-ray spectral analysis. In order to use χ^2 minimization technique and obtain reliable fits, we choose the sources with more than 100 counts (pn detector) and with spectroscopic or photometric redshift information. All these sources are contained in the 1 Ms Chandra catalogue, which allowed us to made comparison between these two datasets. The spectral analysis for corresponding sources from the Chandra observation is described in Tozzi *et al.* (2005) where analysis of 321 (out of 346) X-ray sources is presented. The comparison between XMM-Newton and Chandra datasets with complete information about spectral properties of analysed sources will be presented in Streblyanska *et al.* (in preparation).

The spectroscopic identification from VLT follow up is discussed in details in Szokoly *et al.* (2004). The dominant population in CDFS (as in other deep surveys) is a mixture of obscured (type-2) and unobscured (type-1) AGN with a small fraction of normal galaxies and groups/clusters of galaxies. We concentrated our attention on the analysis of point sources, while a complete analysis of extended objects and their properties will be presented in Finoguenov *et al.* (in preparation). For some of the sources we used photometric redshifts from Zheng *et al.* (2004) and Mainieri *et al.* (2005).

Finally, we selected the sample of 127 X-ray sources with spectroscopic or photometric identification. Our complete sample includes 78 type-1 AGN, 39 type-2 AGN and 10 galaxies.

3. Results from spectral fitting

We performed a detailed spectral analysis with different models in order to investigate the complexity of AGN spectra, for example soft excess, iron line, absorption edge, etc. We fit pn and MOS spectra simultaneously in the 0.2–10 keV band using appropriate response matrixes (`arf` and `rmf`).

Analysis of the soft component is very important in terms of properly modeling of the absorption and the power law slope. For some of our type-2 AGN, the detection of $N_{\rm H}$ and spectral shape were only significant after taking into account the soft excess as a spectral component. We detect a clear soft component in both type-1 (3 sources) and type-2 (8 sources) AGN, although the origin of this component believed to be different for different AGN classes. In type-2 AGN we used a scattering or partial covering model (nuclear radiation scattered by warm medium/leaking through the absorber) in order to explain the soft part of the spectrum, while in type-1 AGN we modelled the soft excess with a blackbody (thermal emission from the accretion disk, but other interpretations are also possible). The detected temperature in type-1 AGN range between 21–110 eV. The average covering fraction in the type-2 AGN is $0.91^{+0.05}_{-0.06}$.

The properties of the spectra of most AGN are in good agreement with the unified model, which postulates obscuration for type-2 and no absorption for type-1 AGN. A fraction of 'unusual' objects (absorbed type-1 and unabsorbed type-2 AGN) was, however, detected. Similar 'atypical' objects have been discovered in several other deep/shallow surveys using statistically significant samples of sources, e.g. HELLAS2XMM (Perola *et al.* 2004), ChaMP (Silverman *et al.* 2005), Lockman Hole (Mateos *et al.* 2005) etc. These authors reported that 6–10% of detected type-1 AGN are absorbed, most of them are luminous ($L_X > 10^{44}$ ergs s^{-1}) and high-z objects, implying a decoupling between optical and X-ray properties at high luminosities and redhifts. This difference between the optical and X-ray classifications can be explained either by a gas-to-dust ratio and/or the chemical composition significantly different from that of the Galactic interstellar gas (e.g. Akiyama *et al.* 2000). From a purely empirical side, one should not forget that variability may also play a role. The most favourable explanation for unabsorbed type-2 AGN is that they are Compton-thick sources. If intrinsic absorption of these sources exceeds 10^{24} cm^{-2}, the optical depth for Compton scattering equals unity, and the direct nuclear emission is completely blocked in both the soft and hard bands. In this situation only the radiation reflected by the cold medium/scattered by the ionized material can reach the observer, and therefore spectra will be without any apparent absorption. The representative spectra of some of AGN are shown in Figure 1.

The weighted mean value of Γ for total sample is 1.86 ± 0.03. The average photon index is $\Gamma \sim 1.9$ and $\Gamma \sim 1.8$ for type-1 AGN and type-2 AGN, respectively. We derive the distributions of spectral index, intrinsic absorption, flux and luminosity, as well as investigate dependences between these parameters. We find no real correlation between Γ and the intrinsic absorption column N_H density. The slight correlation between absorption and lower values of the continuum is mostly an effect of the low statistic of our data. Taking into consideration this fact, our results confirm the idea that the clear separation between the two AGN populations is mostly due to differences in the absorption column density, not in Γ. These results are consistent with the unified scheme of AGN, which links classifications based on optical spectra and properties of the X-ray spectra. We also find no evolution of Γ and N_H with redshift (see Figure 2).

4. Conclusions

Using optical source identification and redshift information, we explored both the physical properties of individual sources, and the general properties of two AGN classes, using a statistically significant sample of 127 sources with redshift identifications. One of the purposes of our analysis was to investigate whether the unified model of AGN holds.

More than 90% of the source spectra are in agreement with the unified model, however a fraction of 'atypical' sources (absorbed type-1 AGN and unabsorbed type-2 AGN) was also discovered. Most of these 'atypical' objects have good signal-to-noise ratio which does not allow us to ascribe these properties to the poor spectra quality. The observed mismatch between X-ray and optical properties may be explained by a difference in the physical conditions surrounding the emission regions.

Most AGN spectra required only an (un)absorbed power law model, however we detect a fraction of the sources with additional spectral components such as a soft excess (11 sources) and iron line (7 sources).

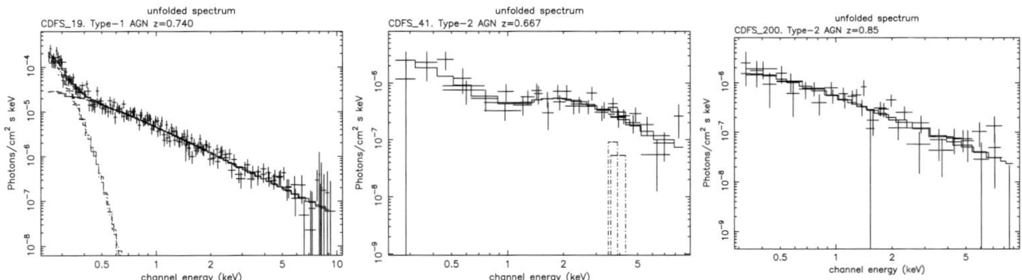

Figure 1. The unfolded model X-ray spectra representative of the different classes of objects in the CDFS. The source numbers refer to the Chandra catalogue from Giacconi *et al.* (2002). From the left to the right: type-1 AGN with the soft excess, type-2 AGN with the soft component and iron line, 'atypical' unabsorbed type-2 AGN.

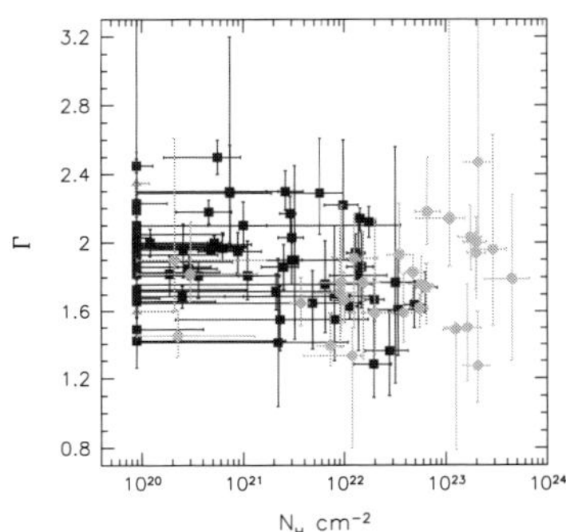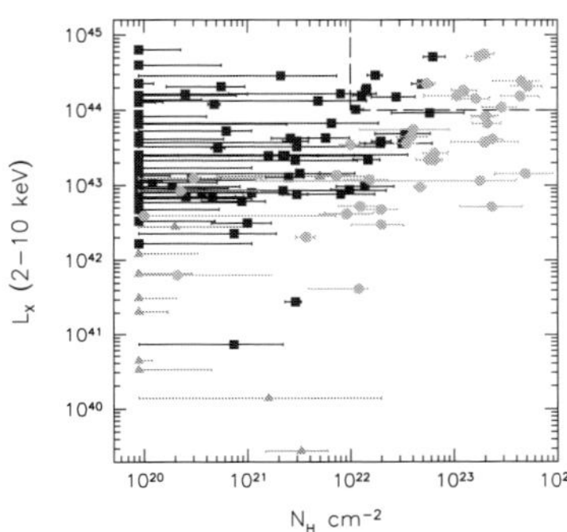

Figure 2. Left: The rest-frame power law photon indices Γ versus absorbing column densities $N_{\rm H}$ for the analyzed sample. Solid blue squares and cyan circles refer to type-1 AGN and type-2 AGN, respectively. Triangles show the galaxies. Γ for identified objects range from 1.4 to 2.5, with the majority of the sources clustering around 1.9. **Right:** Unabsorbed rest-frame luminosities in the hard (2–10) keV band versus the detected absorption. Upper right corner outlined by the dashed line shows the locus of type-2 QSO (defined as sources with $L_{\rm X} > 10^{44}$ ergs s^{-1} and $N_{\rm H} > 10^{22}$ cm^{-2}).

No obvious correlations between spectral continuum and absorption or between Γ, $N_{\rm H}$ and redshift are found.

References

Akiyama, M., Ohta, K., Yamada, T., *et al.* 2000, *ApJ*, 532, 700

Dickey, J. M. & Lockman, F. J. 1990, *ARA&A* 28, 215

Giacconi, R., Gursky, H., Paolini, F., & Rossi, B., 1962, *Phys. Rev. Lett.* 9, 439

Giacconi, R., Rosati, P., Tozzi, P., *et al.* 2002, *ApJS* 139, 369

Lehmer, B., Brandt, W., Alexander, D., *et al.* 2005, *ApJ*, accepted

Mainieri, V., Rosati, P., Tozzi, P., *et al.* 2005, *A&A* 437, 805

Mateos, S., Barcons, X., Carrera, F. J., *et al.* 2005, *A&A*, accepted

Perola, G. C., Puccetti, S., Fiore, F., *et al.* 2004, *A&A*, 421, 491

Silverman, J., Green, P., Barkhouse, W., *et al.* 2005, *ApJ*, 618, 123

Streblyanska, A., Bergeron, J., Hasinger, G., *et al.* 2004, *Nucl. Physics B. Suppl. Ser.* 132, 232

Szokoly, G., Bergeron J., Hasinger, G., *et al.* 2004, *ApJS* 155, 271

Tozzi, *et al.* 2005, *A&A*, submitted

Wolf, C., Meisenheimer, K., Kleinheinrich, M., *et al.* 2004, *A&A*, 421, 913

Zheng, W., Mikles, V. J., Mainieri, V., *et al.* 2004, *ApJS* L55, 73

Discussion

McBreen: How many of your sources are unabsorbed Type II AGN?

Streblyanska: In our sample we found 3 unabsorbed type-2 AGN. However only in one case we have enough statistics to confirm this result. Other two sources show indications for unabsorbed spectra, but due to the low statistics there could be other explanations (e.g. background contamination).

Populations of High Energy Sources in Galaxies
Proceedings IAU Symposium No. 230, 2005
E. J. A. Meurs & G. Fabbiano, eds.

© 2006 International Astronomical Union
doi:10.1017/S1743921306008842

Extragalactic hard X-ray source counts with INTEGRAL observatory: A Progress Report

R. Krivonos[1,2]†, M. Revnivtsev[1,2] S. Sazonov[1,2] E. Churazov[1,2]
and R. Sunyaev[1,2]

[1]Space Research Institute, Moscow, Russia
[2]Max-Planck Institut für Astrophysik, Garching, Germany

Abstract. We present results from the hard X-ray sky survey being performed with *INTEGRAL* observatory. During two years of operation the significant sky area was covered with moderate sensitivity which now allows us to estimate surface number density of Active Galactic Nuclei (AGN) in hard (17-60 keV) energy band. Our catalog of detected sources comprises > 60 AGN with wide range of intrinsic absorption. We used 24000 deg^2 sky coverage (high Galactic latitude observations) with total exposure > 2 Msec to derive source number-flux relation. The estimated surface density of extragalactic hard X-ray sources with flux $> 10^{-11}$ erg s^{-1} cm^{-2}is $(1.0 \pm 0.5)\ 10^{-2}\ deg^{-2}$.

Keywords. Galaxies: active - Galaxies: Seyfert - Gamma rays: observations - X-rays: galaxies.

1. Introduction

Over the past few years there have been remarkable developments in our understanding of the evolution of active galactic nuclei. Extensive X-ray observations demonstrated that the space density of AGN with low X-ray luminosities peaks at lower redshift than that of AGN with high X-ray luminosities (Steffen *et al.*, 2003; Ueda *et al.*, 2003) which indicates that a substantial amount of the total black hole growth has occurred more recently than was suggested by optical surveys of powerful quasars (Boyle *et al.*, 2000; Fan *et al.*, 2001). The discovered tight correlation between the mass of the supermassive black hole and the properties of the galactic spheroid in which it resides (Magorrian *et al.*, 2002; Tremaine *et al.*, 2002) points to a close connection between the formation of SMBHs and that of their host galaxies.

Deep, small area X-ray surveys have been used for successive study of statistical properties of luminous active galactic nuclei at medium and high redshifts. However, the census of AGN in the more nearby Universe is far from being complete, in particular because most of the local AGN activity is believed to be highly obscured and also due to the difficulty of implementing the large area surveys.

We present a progress report on a hard X-ray survey of a large part of the sky performed with the INTEGRAL observatory.

2. X-ray surveys

Deep surveys performed in the standard X-ray band (2–10 keV) with grazing-incidence telescopes on ASCA (Ueda *et al.*, 1999), BeppoSAX (Giommi *et al.*, 2000), Chandra (Giacconi *et al.*, 2002) and XMM-Newton (Hasinger *et al.*, 2001) have provided a wealth of information about AGN at redshifts higher than ∽0.3. However, such surveys cover

† krivonos@hea.iki.rssi.ru

Figure 1. The current sky coverage by INTEGRAL observations. The significance map in $17 - 60keV$ energy band is shown.

small areas of the sky (at most 100 sq. deg) and cannot tell us much about the number density and luminosity distribution of nearby AGN ($z < 0.1$). Furthermore, there are hints that most of the local AGN are heavily obscured at optical and standard X-ray energies and therefore could only be observed in the far infrared or in hard X-rays.

In spite of development progress of AGN evolution at medium and high redshifts provided by small area X-ray surveys our current estimates of the number density and emissivity of local AGN are uncertain by a factor of a few, and removing this uncertainty is very important if one wishes to understand the evolution of the AGN population at low redshifts. There is therefore a clear need for very large sky area surveys performed above 10 keV, or even better 20 keV, i.e. at photon energies where the effect of photoabsorbtion of X-rays on neutral gas is dramatically reduced and thus practically unbiased statistics of AGN can be obtained.

3. X-ray survey with INTEGRAL

Before recently, our knowledge of the statistical properties of the local population of hard X-ray sources was based on observations performed for the whole sky in the 2-100 keV energy band more than 20 years ago by the the HEAO-1 observatory (Wood *et al.*, 1984; Piccinotti *et al.*, 1982; Levine *et al.*, 1984). Recently, a new hard X-ray (3-20 keV) all-sky survey was performed by the RXTE observatory (Revnivtsev *et al.*, 2004). This survey has detected a representative number of AGN, mostly nearby Seyfert galaxies, characterized by a range of absorbing column densities up to $N_H = 10^{24} cm^{-2}$.

At present the all-sky survey in hard X-rays can be performed with wide-field telescopes on board INTEGRAL (Winkler *et al.*, 2003) and Swift (Gehrels *et al.*, 2004) observatories. The coded-aperture telescope BAT (Barthelmy *et al.*, 2005) onboard Swift is sensitive in the 15–150 keV band and has very large ($1.4sr$, half-coded) field of view which make it possible to cover all sky in reasonable time. The Swift mission is operating from the end of 2004 and observational data is publicly available from April 2005. In this work we used INTEGRAL data obtained during two-years of observations. Among instruments onboard INTEGRAL, the IBIS telescope (Ubertini *et al.*, 2003) with ISGRI (Lebrun *et al.*, 2003)

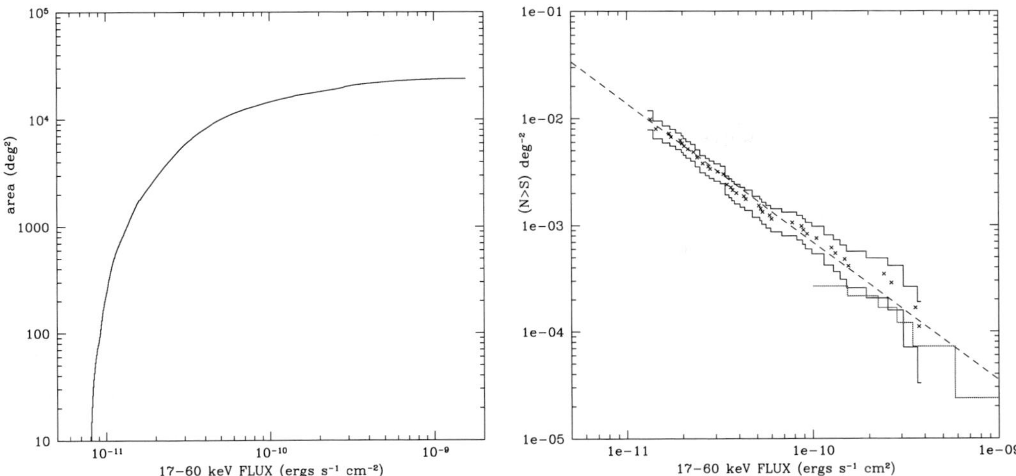

Figure 2. Left plot shows survey coverage area versus limiting flux. At the right, number-flux relation derived from extragalactic source sample (black data points) is demonstrated. For comparison HEAO-1 A4 source counts are also shown (red points).

detector is most suitable for hard X-ray survey, it has large $\sim 30° \times 30°$ FOV, moderate angular resolution (12′) and 15–200 keV operating energy range. Extragalactic fields observations demonstrated capability of IBIS/ISGRI to reach limiting flux of $<$ mCrab (Krivonos *et al.*, 2005) for an exposure time $\gtrsim 0.5$ Msec.

During two years of operation INTEGRAL provided huge amount of data. The deep observations of Galactic Center and Galactic Plane have been conducted with total exposure > 10 Msec, more then 2 Msec have been dedicated to extragalactic fields. As seen from Fig. 1 the significant area of the sky is now covered by INTEGRAL observations.

4. Work status and Future plans

To perform all-sky survey we utilized data of our proprietary INTEGRAL/IBIS observations and publicly available data at the moment of August 2005. We used high Galactic latitude ($|b| > 20°$) observations to prevent contamination from Galactic sources. The survey coverage versus limiting flux is shown on the left panel of Fig. 2. The survey area reaches its geometric limit of $24000 deg^2$ for flux $> 10^{-9}$ erg s^{-1} cm^{-2}and 50% of this area has sensitivity limit better then 6×10^{-11} erg s^{-1} cm^{-2}. The minimum detectable flux in the fields with highest exposure is $\sim 10^{-11}$ erg s^{-1} cm$^{-2}(\sim mCrab)$.

Our hard X-ray source catalog contains more than 60 AGNs with ratio of both Seyfert types close to unity (will be published in subsequent paper). We used statistically clear sample (detections $> 5.5\sigma$) to derive number flux relation which is shown on the right panel of Fig. 2. The best-fit powerlaw slope is 1.30 ± 0.07. Estimates of the surface density at limiting flux of $\sim mCrab$ give $(1.0 \pm 0.5) \times 10^{-2}$ sources per sq. degree. Regarding future prospects, therefore, of order of 400 hard X-ray sources should be detected from the whole sky. However, we should note that our sample may have bias due to source selection effect, because most of the used INTEGRAL extragalactic observations were pointed to the bright sources. To reduce this effect, the whole sky coverage is required. According to INTEGRAL observation program, systematic observations of extragalactic fields will be performed in year 2006 which will make it possible to obtain representative sample of AGNs.

Nevertheless, the key information of the AGN evolution in local Universe rests on our knowledge of the volume number density and luminosity distribution of nearby AGN. The necessary distance information comes from identification of detected X-ray sources with host galaxies. The identification campaign is in progress (e.g. Sazonov *et al.*, 2005) and full analysis of the whole source sample will be performed in the nearest future.

Acknowledgements

Work was partly supported by DFG Schwerpunkt programm "Witnesses of Cosmic History: Formation and evolution of black holes, galaxies and their environment".

References

Barthelmy, Scott D., *et al.* 2005, *astro-ph/0507410*, to be published in Space Science Reviews
Boyle, B. J., Shanks, T., *et al.* 2000, *MNRAS*, 317, 1014
Fan, X., Strauss, M. A., Schneider, D. P., *et al.* 2001, *AJ*, 121, 54
Gehrels, N., Chincarini, G., Giommi, P., *et al.* 2004, *ApJ* 611, 1005
Giacconi, R. Zirm, A. Wang, J. Rosati, P., *et al.* 2002, *ApJS*, 139, 369
Giommi, P., Perri, M., & Fiore, F. 1999, *A&A*, 362, 799
Hasinger, G., Altieri, B., Arnaud, M., *et al.* 2001, *AJ*, 365, L45
Krivonos, R., Vikhlinin, A., Churazov, E., *et al.* 2005, *ApJ*, 625, 89
Lebrun, F., *et al.* 2003, A&A, 411, 141
Levine, A. M., Lang, F. L., Lewin, W. H. G., *et al.* 1984, *ApJS*, 54, 581
Magorrian, J., *et al.* 2002, *AJ*, 105, 2285
Piccinotti, G., Mushotzky, R. F., Boldt, E. A., *et al.* 1982, *ApJ*, 253, 485
Revnivtsev, M., Sazonov, S., Jahoda, K., & Gilfanov, M. 2004, *A&A*, 418, 927
Sazonov, S., Churazov, E., Revnivtsev, M., *et al.* 2005, *astro-ph/0508593*, submitted to A&A Letters
Steffen, A. T., Barger, A. J., Cowie, L. L., *et al.* 2003, *ApJ*, 611, L23
Tremaine, S., Gebhardt, K., *et al.* 2002, *ApJ*, 574, 740
Ubertini, P., Lebrun, F., Di Cocco, G., *et al.* 2003, *A&A*, 411, L131
Ueda, Y., Akiyama, M., Ohta, K., & Miyaji, T. 2003, *ApJ*, 598, 886
Ueda, Y., Takahashi, T., Ishisaki, Y., Ohashi, T., & Makishima, K. 1999, *ApJ*, 524, L11
Winkler, C., *et al.* 2003, *A&A*, 411, L1
Wood, K. S., Meekins, J. F., Yentis, D. J., *et al.* 1984, *ApJS*, 56, 507

Discussion

MACCARONE: How will the sensitivity of the INTEGRAL All-Sky Survey compare with the All-Sky Survey being conducted by SWIFT?

KRIVONOS: The SWIFT Team announced very high sensitivity limit for high-latitude galactic fields. The sensitivity of less than 1 mCrab will be reached in several days of observations of given area of the sky. However the actual sensitivity which can be reached by SWIFT will depend on how systematics will be overcome.
Due to smaller FOV, IBIS telescope has higher than BAT systematics limit. Thus, we have to wait for actual results from SWIFT survey to compare the sentivities.

Populations of High Energy Sources in Galaxies
Proceedings IAU Symposium No. 230, 2005
E. J. A. Meurs & G. Fabbiano, eds.

© 2006 International Astronomical Union
doi:10.1017/S1743921306008854

A Deep Study of the 3C 273 Field in γ-rays

I. Brott[1,2]**, I. Kreykenbohm**[1,3]**, P. Lubinski**[4,1]**, N. Produit**[1]**,
P. H. Hauschildt**[2]**, and T. J.-L. Courvoisier**[1]

[1]INTEGRAL Science Data Centre, Chemin d'Ecogia 16, 1290 Versoix, Switzerland
e-mail : Ines.Brott@obs.unige.ch

[2]Hamburger Sternwarte, Gojenbergsweg 112, 21029 Hamburg, Germany

[3]Institut für Astronomie und Astrophysik — Astronomie, Sand 1, 72076 Tübingen, Germany

[4]N. Copernicus Astronomical Center, Bartycke 18, 00-716 Warszene, Poland

Abstract. 3C 273 is one of the brightest and best studied quasars. It has been observed for 770 ks with the imager IBIS (FoV 12 deg) on board INTEGRAL. To achieve the best possible S/N the dataset has been screened using several criteria indicating the quality of the data (i.e., number of good time intervals, etc). We describe the necessary tools and methods to analyze data of deep fields.

Keywords. galaxies: active, gamma rays: observations, methods: data analysis.

1. Introduction

Low sensitivity of former missions has prevented a detailed study of the hard cosmic X-ray background. As in soft X-rays it has been resolved to AGN (Giacconi *et al.* 2002), a similar situation is anticipated for soft γ-rays (Worsley *et al.* 2004; Krivonos *et al.* 2005). 3C 273 has been observed regularly twice a year by INTEGRAL since 2003. We have used all data available to us (June 2005) within $15°$ of 3C 273 between revolutions 28 and 207 with a total exposure time of 774 ks in 371 science windows (ScW).

2. Data Analysis and Results

To avoid instrumental background lines which are located between 59.3 keV and 84.9 keV (W and Pb fluorescence) and a blend of several flurocence lines of Cd and Te around 24 keV (Terrier *et al.* 2003), we selected a band from 26-51 keV for our analysis, while using data from 13-26 keV and 51-105 keV to assess the background level.
Data reduction was done with OSA 5 as distributed by the ISDC. We rejected ScWs due to: unacceptable artefacts in the deconvolved image (17), average good time interval less than 10 seconds long (6), and an unusual background level if in 2 of 3 energy bands $|\xi_i - \chi| > \frac{1}{N} \sum_i |\xi_i - \chi|$ (53 ScWs). ξ_i denotes the median pixel countrate in the shadowgram and χ the median over those ξ_i, the index i is running over the ScWs. In total 66 ScWs were excluded, mostly from revolution 28-32 when the bottom veto shield of IBIS was turned off (see Fig. 1).
Due to the dithering strategy the exposure time varies in the mosaic from 633 s to 838 ks, therefore the source detection was limited to a field of $\sim 1\,800\,\mathrm{deg}^2$ for which $T_{\mathrm{exp}} > 150\,ks$. In order to emphasize source signal with respect to the background we convolved the field with a gaussian core and searched for the local maxima in the folded image. The FWHM of the gaussian core was calibrated on 3C 273. We searched for sources with a core volume larger than 4.5 and round in shape.

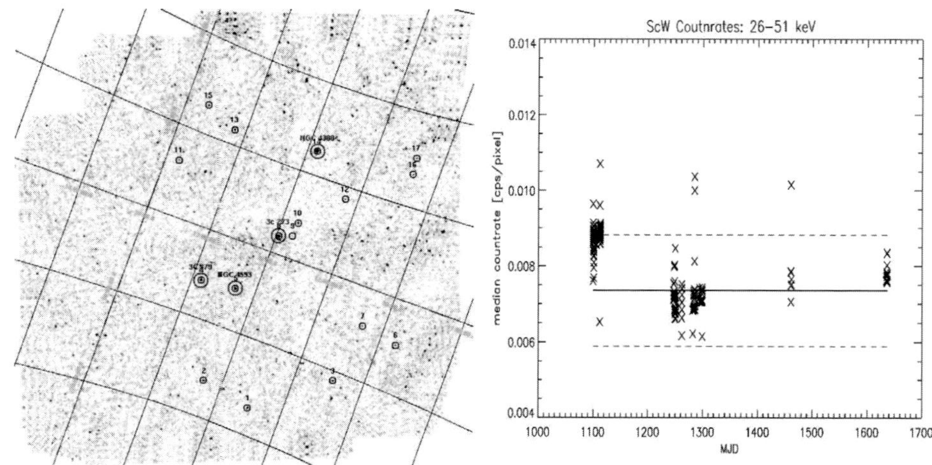

Figure 1. Left: S/N image of the 3C 273 field. Right: Evolution of the background with time.

#	Ra	Dec	Possible Identifier	ΔPos	S/N	Type
1 *	12 15 18	−17 17 19	—	—	4.63	—
2 *	12 39 03	−16 08 27	XSS J12389-1614	5.9'	5.63	X
3	11 43 32	−11 03 57	—	—	4.81	—
4 *	12 56 09	−05 46 58	3C 279	0.6'	5.88	QSO
5 *	12 39 40	−05 19 33	NGC 4593	1.1'	20.44	Sy2
6	11 23 20	−04 48 42	—	—	4.24	—
7	11 39 21	−04 11 29	AC0 1334	1.9'	7.5'	GiC
8 *	12 29 07	+02 02 14	3C 273	—	37.05	QSO
9 *	12 23 18	+02 38 12	Mrk50	2.8'	2.99	Sy1
10 *	12 22 32	+04 18 10	NGC 4303	10.3'	4.62	Sy2
11	13 24 22	+06 12 25	—	—	4.17	—
12	12 05 49	+08 37 45	—	—	3.51	—
13	13 05 18	+11 41 42	—	—	5.81	—
14 *	12 25 52	+12 37 38	NGC 4388	2.5'	39.08	Sy2
15 *	13 20 28	+13 07 34	1RXS J132036.6+131527	8.1'	4.64	X
16	11 41 17	+13 29 26	—	—	3.91	—
17	11 42 03	+15 17 52	—	—	4.50	—

Table 1. Counterparts have been searched within 10' in SIMBAD. '*' denotes detections that have been found in the extended 26-105 keV band as well.

To assess the validity of these detections we performed the same search in the joined 26-105 keV mosaic, resulting in 24 detections, 9 coincident with detections from the 26-51 keV band. A search for negative excesses in the 26-51 keV band resulted in 10 detections. Therefore it is possible that most of the non identified sources are false detections.

This preliminary analysis shows that a detection of weak sources is possible, but the statistics are low and the weak sources are highly background dominated. With the availability of more data on this field these faint sources can be studied in more detail.

References

Giacconi, R., Zirm, A., Wang, J., *et al.* 2002, ApJS, 139, 369
Krivonos, R., Vikhlinin, A., Churazov, E., *et al.* 2005, ApJ, 625, 89
Terrier, R., Lebrun, F., Bazzano, A., *et al.* 2003, A&A, 411, L167
Worsley, M. A., Fabian, A. C., Barcons, X., *et al.* 2004, MNRAS, 352, L28

Populations of High Energy Sources in Galaxies
Proceedings IAU Symposium No. 230, 2005
E. J. A. Meurs & G. Fabbiano, eds.

© 2006 International Astronomical Union
doi:10.1017/S1743921306008866

Averaging High Redshift X-ray and γ-ray Spectra of Faint Sources

Tahir Yaqoob[1]†

[1]Department of Physics & Astronomy, Johns Hopkins University, Baltimore, MD 21212, USA.

Abstract. Population studies of faint X-ray sources (e.g. galaxies, AGN, γ-ray bursts and their afterglows) have become more and more common. The problem of obtaining an average spectrum in the rest-frame of the sources is non-trivial. We show that conventional methods for averaging low signal-to-noise X-ray spectra, when applied to sources at different redshifts, result in a mean rest-frame spectrum that can exhibit artificial features. These include the broadening and weakening of emission/absorption lines, a broad dip in the continuum above an emission line, and a spectral hardening at the highest energies. All of these effects have been observed in real data and have been given fairly weighty astrophysical and/or cosmological interpretations. We present a new method of averaging which considerably reduces the severity of the problems.

Keywords. methods: data analysis, X-rays: galaxies, cosmology: observations.

---

1. Principal Problems

While the spectral averaging problem has been discussed and studied in detail in the literature for sources with $z \sim 0$ (e.g. Yaqoob *et al.* 2002; Lubiński 2004), there are problems with weak sources at different redshifts. Firstly, unless the detector response is nearly diagonal, "unfolding" the counts spectra of individual sources before blueshifting to the rest-frame is unavoidable. One is then forced to work with "data" which may exhibit model-dependence. Secondly, if the individual spectra have too few counts per bin, one is forced to group the spectral channels in such a way that there is a minimum number of photon counts per channel in order that the background can be subtracted and Gaussian statistics used to calculate and propagate the statistical errors. Grouping the spectral channels can introduce distortion which cannot be reversed. Moreover, the conventional approach is to group the background spectra using the same channel groups as the on-source spectra but this does *not* guarantee a sufficient number of counts per bin in the background spectra. Note that accumulating a much higher signal-to-noise background spectrum does not avoid the grouping problem because the individual on-source spectra may still be in the Poisson regime and each spectrum may simply be an under-sampled version of what one is trying to measure (the true source spectra, which would be obtained with an infinitely long exposure time).

2. Simulation Results and a New Method

We performed simulations of 200 sources with redshifts uniformly and randomly distributed between 0.5 and 2.5, and 2–10 keV fluxes uniformly and randomly distributed in the range $6 - 60 \times 10^{-14}$ ergs cm^{-2} s^{-1} and constructed mean rest-frame spectra with and without spectral grouping and with and without background subtraction, and with

† Also: Exploration of the Universe Division, NASA/Goddard Space Flight Center, Greenbelt, MD 20771, USA.

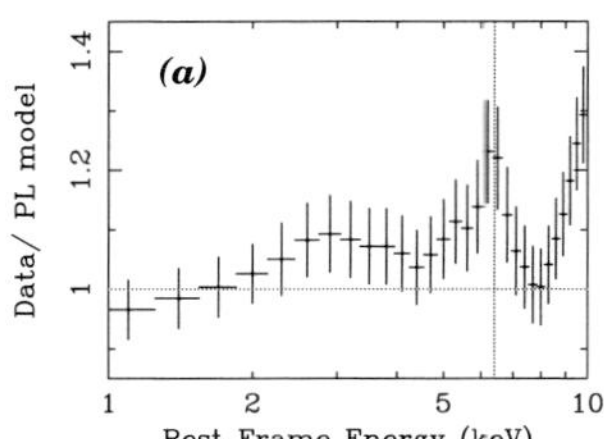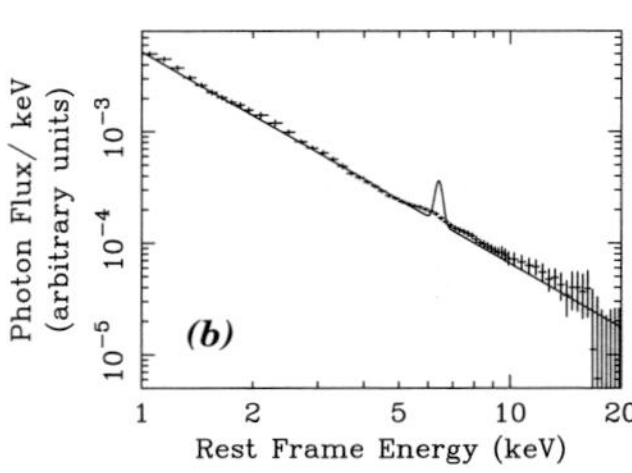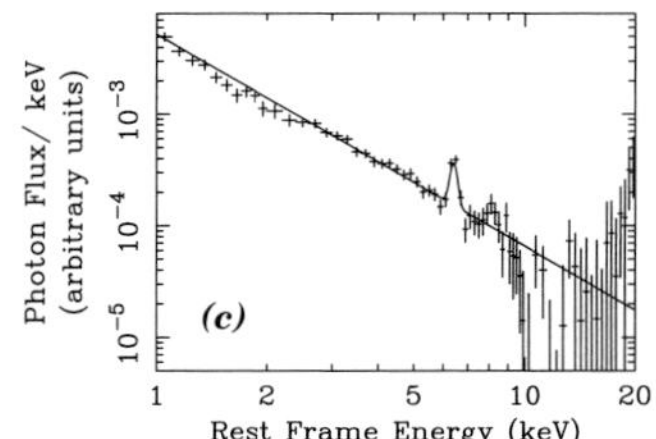

Figure 1. Mean rest-frame spectra from 20 ks exposure simulations of a power law plus unresolved narrow line (see text). (a) Conventional methods: no background; spectral channels grouped. Ratio to a simple to power law only. (b) Conventional methods: simulated spectra background-subtracted and grouped. (c) New method: background-subtraction, but no spectral grouping. Narrow line is recovered, continuum distortion reduced.

various exposure times. All sources had the same input spectrum, namely a power law with a photon index of 1.9, and a narrow (unresolved) Gaussian Fe K emission line, with a rest-frame energy and Gaussian width of 6.4 keV and 0.005 keV respectively. We used *XMM-Newton* pn (CCD) detector responses. Here we only briefly summarize the results of our findings.

Fig. 1 shows the results when the exposure time of each spectrum was 20 ks. With no background simulated, but using conventional methods of grouping, unfolding, and blueshifting, Fig. 1(a) shows the result of fitting the mean rest-frame spectrum with a simple power law only. Considerable distortion is apparent. Fig. 1(b) shows the mean rest-frame spectrum obtained using conventional methods, this time including simulation and subtraction of background. Again, the input spectrum is not recovered well. These kind of spectral distortions (broadening of the emission line and continuum hardening) have been observed using real data (e.g. Streblyanska 2004, 2005 and references therein). We present a new method in which the *background spectra are grouped*, the on-source are *not* grouped, and the *background subtraction is done in the rest-frame*. Thus, the on-source spectra and background spectra are *separately* unfolded and blueshifted, *separately* co-added, and only then (in the rest-frame) is the (mean) background subtracted from the on-source (mean) spectrum. Fig. 1(c) shows that the method is able to recover the input spectrum much better than the conventional methods. At first it may seem odd that the background spectra are unfolded and blueshifted since the background is local (i.e. at $z \sim 0$) and may not even have passed through the telescope. However, with conventional methods, in which the background-subtracted spectra are unfolded and blueshifted, the background spectra *ARE* implicitly unfolded and blueshifted.

3. Conclusions

We have shown how to construct mean rest-frame spectra of faint X-ray sources at different redshifts. Conventional methods suffer more from artificial broadening of emission and absorption lines, spectral hardening and other continuum distortion.

References

Lubiński, P. 2004, *MNRAS* 350, 596

Streblyanska, A., Bergeron, J., Brunner, H., Finoguenov, A., Hasinger, G., & Mainieri, V. 2004, *NuPhS* 132, 232

Streblyanska, A., Hasinger, G., Finoguenov, A., Barcons, X., Mateos, S., & Fabian, A. C. 2005, *A&A* 432, 395

Yaqoob, T., Padmanabhan, U., Dotani, T., & Nandra, K. 2002, *ApJ*, 569, 487

Populations of High Energy Sources in Galaxies
Proceedings IAU Symposium No. 230, 2005
E. J. A. Meurs & G. Fabbiano, eds.

© 2006 International Astronomical Union
doi:10.1017/S1743921306008878

Supermassive Black Holes in Galaxies

Amy J. Barger[1,2,3]

[1]Department of Astronomy, University of Wisconsin-Madison,
475 N. Charter St., Madison, WI 53706, USA
email: barger@astro.wisc.edu

[2]Department of Physics and Astronomy, University of Hawaii,
2505 Correa Road, Honolulu, HI 96822, USA

[3]Institute for Astronomy, University of Hawaii,
2680 Woodlawn Drive, Honolulu, HI 96822, USA

Abstract. *Chandra* and *XMM* detect X-rays emitted during accretion onto supermassive black holes, even when they are highly obscured. I review what has been learned about the cosmic evolution of the X-ray luminosity functions and the reconstruction of the accretion history of supermassive black holes from extensive follow-up observations of both deep and wide-area *Chandra* X-ray surveys.

Keywords. cosmology: observations, galaxies: active, galaxies: distances and redshifts, galaxies: evolution, X-rays: galaxies.

1. Introduction

In order to understand how supermassive black holes form and evolve, we need to determine the accretion history of the universe. However, much of the accretion power in the universe is absorbed, making it difficult to measure at optical wavelengths. The *Chandra* and *XMM* X-ray observatories have revolutionized distant active galactic nuclei (AGNs) studies. It is now possible to map the history of the AGN population using hard X-ray surveys and to compare high and low-redshift samples chosen in the *same* rest-frame 2–8 keV band. Hard X-rays have the advantage in that they can directly probe AGN activity, are uncontaminated by star formation processes at the X-ray luminosities of interest, and detect all but the most absorbed sources.

The advantage of *Chandra* for this kind of work is its high positional accuracy, which minimizes misidentifications of faint optical galaxy counterparts. It is important to note, however, that even with a $2''$ match radius, about 20% of the faint-end optical identifications ($R = 24$–26) of the *Chandra* sources will be spurious (Barger *et al.* 2003), while with the *XMM* data, the percentage of misidentifications will be far worse. Moreover, unlike *XMM*, *Chandra* does not reach the confusion limit, even with the deepest observations.

2. *Chandra* X-ray Data

The two deepest images of the X-ray sky ever taken are the 2 Ms *Chandra* Deep Field-North (CDF-N; Alexander *et al.* 2003) and the 1 Ms *Chandra* Deep Field-South (CDF-S; Giacconi *et al.* 2002). These data resolve greater than 80 to 90% of the $2-8$ keV X-ray background (XRB) into discrete sources. The CDF-N exposure samples a large, distant cosmological volume down to very faint X-ray flux limits of $f_{2-8 \text{ keV}} \approx 1.4 \times 10^{-16}$ ergs cm^{-2} s^{-1} and $f_{0.5-2 \text{ keV}} \approx 1.5 \times 10^{-17}$ ergs cm^{-2} s^{-1}. The CDF-S exposure is only a factor of two shallower, but this is still substantially fainter than the X-ray flux of the sources that contribute the most to the XRB (Cowie *et al.* 2002; see Fig. 1).

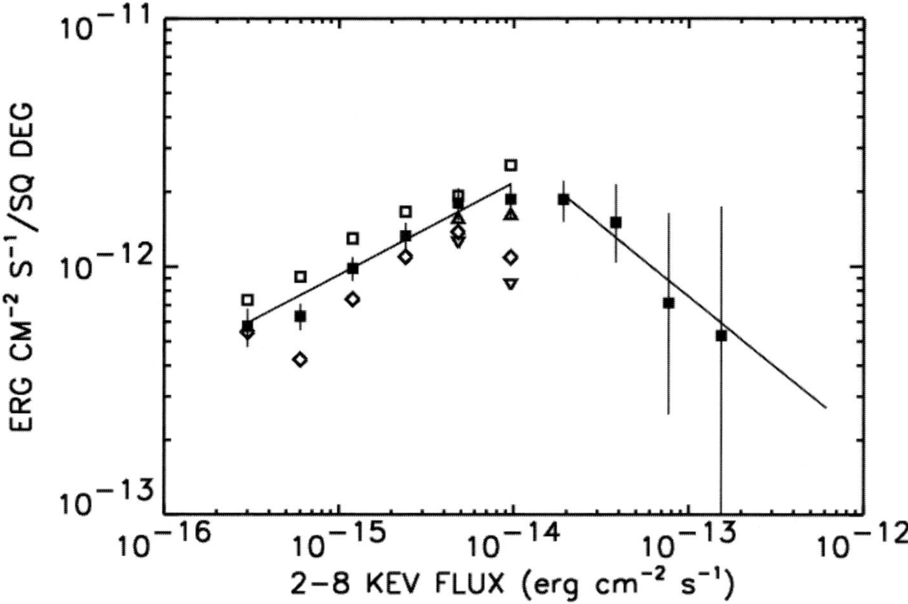

Figure 1. Contribution to the XRB vs. $2 - 8$ keV flux for individual fields (shown only below 10^{-14} ergs cm^{-2} s^{-1}, where the error bars are small) and the combined sample (*solid squares*). Lines show power law fits from Cowie *et al.* (2002). Figure courtesy of Cowie *et al.* (2002).

An unexpected result from the *Chandra* deep field observations was the substantial field-to-field variations observed in the X-ray number counts on the scale of a single *Chandra* image (see Fig. 1). The origin of this variance is not known, but it is likely due to cosmic variance originating in large scale structure. Thus, many *Chandra* fields are needed to get the average true number counts. Large area surveys are also needed to get the necessary volume to map the low-redshift luminosity function and to sample the small number of very high-redshift sources. This can be done either with serendipitous pointings (the ChaMP survey strategy) or with contiguous fields. The advantage of the contiguous fields approach is that one can then also study hard X-ray source clustering.

There are a number of ongoing wide-area *Chandra* and *XMM* surveys (e.g., the ChaMP survey, the Groth Strip survey, the Extended CDF-S, the Wilkes Lockman Hole field, and the COSMOS survey). However, the only current survey with a very high level of spectroscopic identifications is the *Chandra* Large Area Synoptic X-ray Survey (CLASXS), which consists of nine overlapped *Chandra* ACIS-I pointings in the Lockman Hole Northwest field covering ~ 0.4 square degrees to X-ray flux limits of $f_{2-8\ \text{keV}} \approx 3 \times 10^{-15}$ and $f_{0.5-2\ \text{keV}} \approx 5 \times 10^{-16}$ ergs cm^{-2} s^{-1} (Yang *et al.* 2004).

Yang *et al.* (2003) found that the full range of cosmic variance seen in the earlier, individual *Chandra* pointings was represented in the number counts of the nine CLASXS pointings. They found the clustering in both the hard and soft X-ray bands to be significant, though, interestingly, the X-ray sources detected only in the hard band appeared to be substantially more clustered. This means that the *Chandra* hard X-ray selected sources probably have the largest correlation strength of any AGN population, one that is similar to that of massive galaxies. Indeed, in the CDF-N, the hard X-ray AGNs lie in the same density environments as massive galaxies (Cowie *et al.* 2004).

3. Spectroscopic Completeness

We supplement the CLASSXS, CDF-N, and CDF-S surveys with the very high-luminosity *ASCA* sample of Akiyama *et al.* (2000). Figure 2 shows the useful flux ranges of the hard X-ray samples in these fields. Above $f_{2-8\ \mathrm{keV}} \sim 10^{-14}$ ergs cm^{-2} s^{-1}, $80 - 90\%$ of the hard X-ray sources have redshifts, while below this flux, $\sim 60\%$ do. The fact that the fraction of sources with spectroscopic identifications drops at fluxes below $f_{2-8\ \mathrm{keV}} \sim 10^{-14}$ ergs cm^{-2} s^{-1}, which is the same flux where the contributions to the hard XRB peak, is a consequence of the fraction of broad-line AGNs being much higher at the brighter X-ray luminosities, as well as the fainter X-ray sources being optically fainter.

The spectroscopic samples are highly complete to $R = 24.5$, and the use of photometric redshifts increases the overall identified fraction to about 85%. The spectroscopic incompleteness at $z > 1.2$ is due to the absence of strong features in the spectra and the faintness of sources at these redshifts. We classified the spectroscopically identified X-ray sources into four optical spectral classes (absorbers, star formers, high-excitation sources, and broad-line AGNs). Hereafter, I refer to all sources that do not show broad-line (FWHM$>$ 2000 km s^{-1}) signatures as optically-narrow AGNs. We find that the broad-line AGNs are nearly all soft and show essentially no visible absorption in X-rays, while the optically-narrow AGNs are well-described by a power-law spectrum with photoelectric absorption spread over a wide range of N_H values. Thus, it is possible to separate roughly the broad-line AGNs and the optically-narrow AGNs on the basis of X-ray colors alone (e.g., Szokoly *et al.* 2004), without knowing the optical spectra. However, there will be a small amount of contamination from stars, from the small number of optically-narrow AGNs that have soft X-ray colors, and from the small number of broad-line AGNs that have hard X-ray colors.

4. Evolution of the Hard X-ray Luminosity Function

We define the hard X-ray luminosity function (HXLF) versus rest-frame $2 - 8$ keV X-ray luminosity and redshift $[d\Phi(L_X, z)/d\log L_X]$ as the number of X-ray sources per unit comoving volume per unit base 10 logarithmic luminosity that lie in the redshift interval. With the advent of the *Chandra* and *XMM* data, there have been a number of computations of the evolution of the HXLF with redshift (e.g., Cowie *et al.* 2003; Steffen *et al.* 2003; Ueda *et al.* 2003; Hasinger 2003; Fiore *et al.* 2003; Barger *et al.* 2005), while Sazonov & Revnivtsev (2004) have used the *RXTE* data to compute the local 3–20 keV luminosity function. The latest HXLF determinations by Barger *et al.* (2005) use only the spectroscopically identified X-ray sources with $L_{2-8\ \mathrm{keV}} = 10^{42}$ ergs s^{-1} in the CDF-N, CDF-S, and CLASXS fields. They assume that any source more luminous than $L_{2-8\ \mathrm{keV}} = 10^{42}$ ergs s^{-1} is very likely to be an AGN on energetic grounds (Zezas *et al.* 1998; Moran *et al.* 1999).

The relative HXLFs of the total and the broad-line AGNs from Barger *et al.* (2005) reproduce the results of Steffen *et al.* (2003), who showed that the dominant population at these higher X-ray luminosities, where the redshift identifications are very complete, is broad-line AGNs, while at the lower X-ray luminosities, the non–broad-line AGNs dominate (hereafter, I refer to this as the Steffen effect). Since broad-line AGNs are

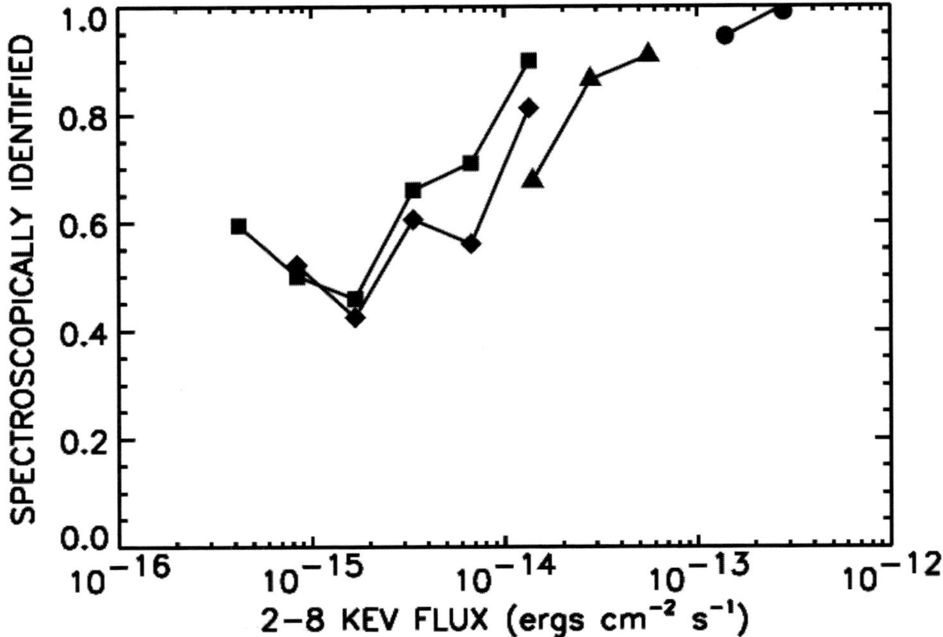

Figure 2. Spectroscopic completeness (i.e., the fraction of observed sources that are spectroscopically identified) of the hard X-ray surveys that make up the total hard X-ray sample (*squares* – CDF-N, Barger *et al.* 2003; *diamonds* – CDF-S, Szokoly *et al.* 2004; *triangles* – CLASXS, Steffen *et al.* 2004; *circles* – *ASCA*, Akiyama *et al.* 2003). Sources are grouped into flux bins that increase by a multiplicative factor of 2. Only flux bins containing more than 10 sources are plotted to illustrate the useful flux ranges of the various samples. Figure courtesy of Barger *et al.* (2005).

straightforward to identify spectroscopically, and the bulk of the X-ray sources at these luminosities have now been observed, one does not need to worry that broad-line AGNs are making up a substantial fraction of the unidentified population. Since these results suggest a luminosity dependence in optical spectral type, the simple unified model for AGNs, in which there is no luminosity or redshift dependence of the obscuration, does not appear to be valid.

Figure 3 shows the measured HXLFs from Barger *et al.* (2005) for two low redshift intervals (*open squares* – $z = 0.2$–0.4; *solid diamonds* – $z = 0.8$–1.2). Poissonian 1σ uncertainties are based on the number of galaxies in each luminosity bin. For these redshifts, the HXLFs were computed from observed-frame $2 - 8$ keV. An intrinsic $\Gamma = 1.8$ was assumed, for which there is only a small differential K-correction to rest-frame $2 - 8$ keV. The $z = 0$ HXLF from Sazonov & Revnivtsev (2004)'s *RXTE* analysis is also shown *(dotted line)*.

The redshift data for the *Chandra* sources give a remarkable picture of the evolution of the AGN population to $z \sim 1.2$. The HXLF undergoes a steep pure luminosity evolution (PLE), with the characteristic luminosity evolving as $(1+z)^{3.2}$ out to $z \sim 1.2$ (see Fig. 3). The same PLE model describes the luminosity function of both the broad-line AGNs alone and all the hard X-ray sources together, though the two samples have very different

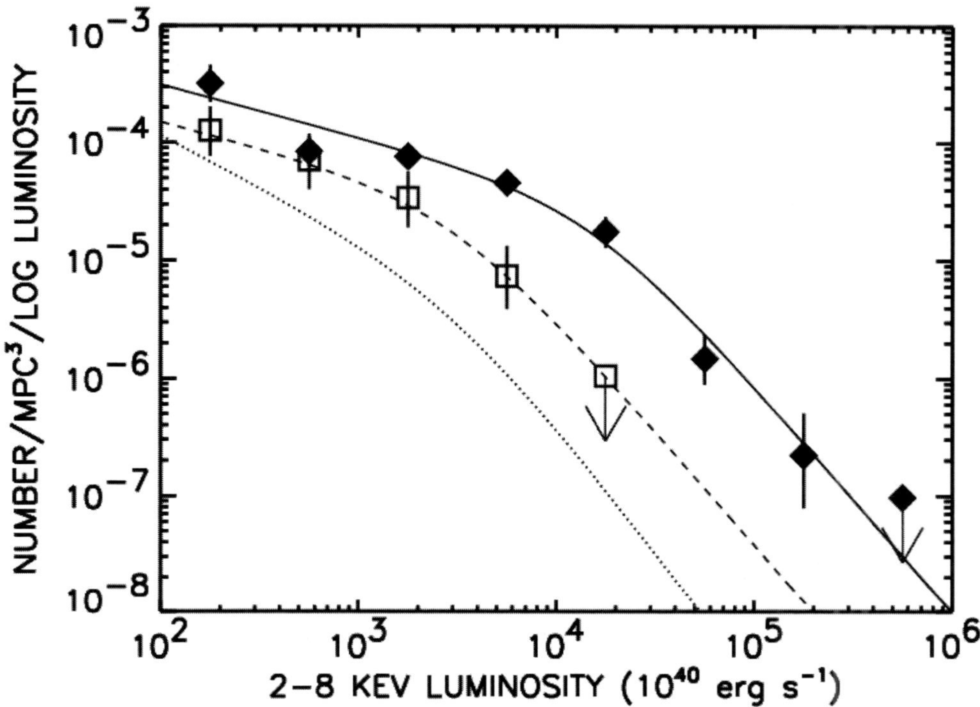

Figure 3. Rest-frame $2-8$ keV luminosity function per unit logarithmic luminosity at $z = 0$ (*dotted curve*—Sazonov & Revnitsev's 2004 *RXTE* analysis), $z = 0.2 - 0.4$ (*open squares*), and $z = 0.8 - 1.2$ (*solid diamonds*). Solid curve is a double power-law fit to the $z = 0.8 - 1.2$ total HXLF. Dashed curve shows a pure luminosity evolution model where only the characteristic luminosity evolves as $(1 + z)^{3.2}$ over the $z = 0 - 1$ redshift interval. This model also fits the local *RXTE* determination.

luminosity functions. This evolution also matches to the recent determination of the local HXLF by Sazonov & Revnivstev (2004). At the higher redshifts, incompleteness is potentially a larger source of error because of the difficulty with measuring the host galaxy redshifts in those intervals. However, even with maximal incompleteness corrections, the HXLFs at $L_{2-8 \text{ keV}} < 10^{44}$ ergs s^{-1} lie below the maximum likelihood fits to the $z = 0 - 1.2$ HXLF computed at $z = 1$, suggesting a peak in the universal AGN energy density production rate near $z = 1$.

Correspondingly, the production rate of the AGN radiation drops rapidly, as is shown in Figure 4. This is based on direct summation, but integration of power-law fits to the HXLFs give a similar answer. The largest uncertainty is the redshift distribution of the unidentified sources, rather than the small effects of extrapolation outside the observed luminosity range. However, even with this incompleteness uncertainty, the energy density production rate is at most flat beyond $z = 1$, and, more realistically, is slightly falling.

From Figure 4, we see that the dominant period of supermassive black hole production is at redshifts near $z = 1$. This is very different than what had been expected pre-*Chandra*. Such an extremely rapid evolution with redshift bears a striking resemblence to the overall redshift evolution of the star formation rate density in $z < 1$ galaxies (e.g., Wilson *et al.* 2002). We also see from Figure 4 that at $z < 1.5$, most of the hard X-ray energy density production is due to optically-narrow AGNs. This highlights the

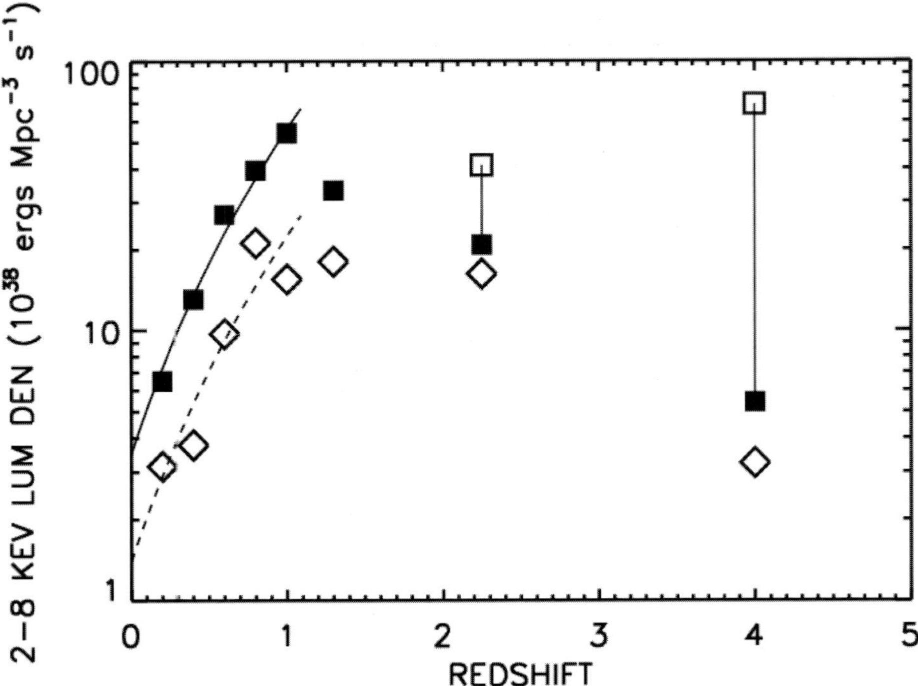

Figure 4. Evolution with redshift of the rest-frame $2 - 8$ keV comoving luminosity density production rate of $L_{2-8 \text{ keV}} > 10^{42}$ ergs s^{-1} sources. Solid squares (open diamonds) denote the measured values for the total sample (broad-line AGNs only). Open squares show the upper limits found by assigning all of the unidentified sources to the centers of each redshift bin. Solid and dashed curves show a $(1 + z)^{3.2}$ evolution over the $z = 0 - 1.2$ range. Figure courtesy of Barger *et al.* (2005).

importance of using X-ray data to identify obscured sources and then including them in the determination of the accretion history of the universe.

We now return to the interesting observation that broad-line AGNs dominate the number densities at the higher X-ray luminosities, while non–broad-line AGNs dominate at the lower X-ray luminosities (the Steffen effect). Could the absence of broad-line AGNs at low X-ray luminosities be explained if the nuclear UV/optical light of optically-narrow AGNs were being swamped by the host galaxy light? That is, might galaxy dilution (see, e.g., Moran *et al.* 2002) be a partial explanation of the Steffen effect? With the high-resolution ACS GOODS-North data (Giavalisco *et al.* 2004), Barger *et al.* (2005) were able to separate the nuclear component of each source from the host galaxy light, even at the higher redshifts, in order to analyze the nuclear colors. It is well-known that the nuclear UV magnitudes and the X-ray fluxes for broad-line AGNs are strongly correlated. Thus, if the galaxy dilution hypothesis were correct, we would expect the optically-narrow AGNs to be similarly correlated when we isolate their nuclear UV/optical light. In Figure 5, we see that that turns out not to be the case. Instead, the nuclei of the optically-narrow AGNs are much weaker relative to their X-ray light than are the broad-line AGNs. Thus, the absence of broad-line AGNs at low X-ray luminosities is not a dilution effect. Optically-narrow AGNs really have weaker UV/optical nuclei relative to the X-rays. Now that galaxy dilution has been ruled out as a partial explanation of the

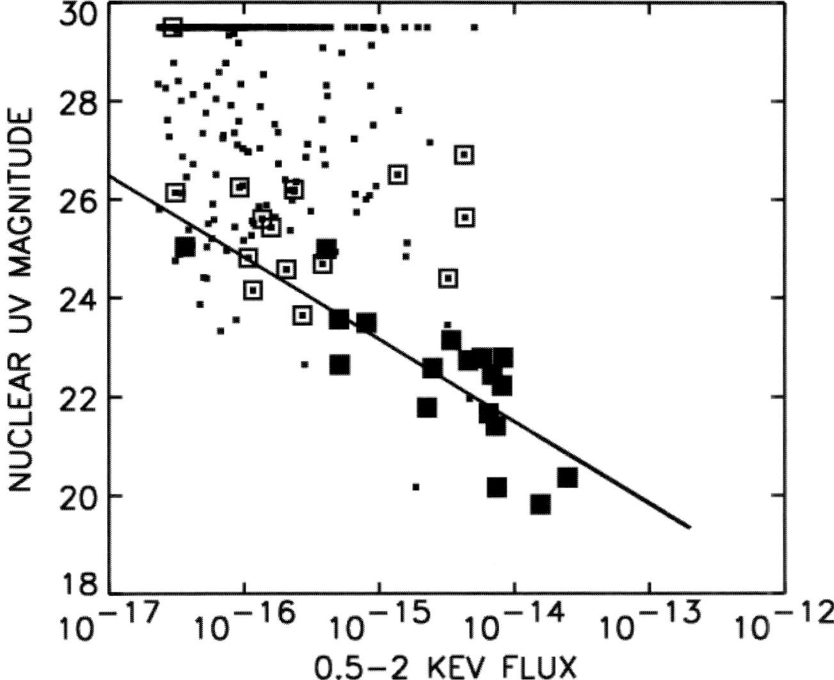

Figure 5. Nuclear UV magnitude vs. $0.5 - 2$ keV flux, showing that there is a strong correlation between the two for broad-line AGNs. Broad-line AGNs (optically-narrow AGNs) are denoted by large (small) squares. High-excitation sources are enclosed in large open squares. Sources with no measurable nuclei (all but one of these are spectrally classified as normal galaxies) are shown at a nominal magnitude of 29.5, which is roughly the 2σ limit for the small $0.3''$ diameter aperture used to measure the nuclear magnitude. Solid line shows a linear relation fitted to the broad-line AGNs. Figure courtesy of Barger *et al.* (2005).

Steffen effect, it can be stated firmly that the simple unified model has failed because of the existence of the Steffen effect. The simplest interpretation of the Steffen effect is that the covering fraction of obscuration in the AGNs is extremely small at high X-ray luminosities (so that all we can see at these high luminosities are broad-line AGNs) and extremely large (near unity) at low X-ray luminosities (so we do not see broad-line AGNs at these low luminosities).

5. Conclusions

Optically-selected AGNs (otherwise known as broad-line AGNs) are only a subset of X-ray selected AGNs. AGNs evolve very rapidly to $z \sim 1.2$, consistent with pure luminosity evolution. The observation that there are almost no low X-ray luminosity, broad-line AGNs is not a galaxy dilution effect, ruling out the simple unified model. The Barger *et al.* (2005) hard X-ray luminosity function determinations of the supermassive black hole mass densities are a factor of almost two lower than what Yu & Tremaine (2002) found due to the Yu & Tremaine extrapolation of the optical QSO luminosity function outside the optically observed luminosity range. Comparing this result with the local supermassive black hole mass density, there is now room for obscured accretion by the optically-narrow AGNs.

Acknowledgements

I thank the conference organizers for inviting me to participate in this stimulating meeting. I gratefully acknowledge support from NASA grants CXC GO2-3191A and HST-GO-09425.30-A from the Space Telescope Science Institute, from NSF grant AST 02-39425, and from the Alfred P. Sloan Foundation and the David and Lucile Packard Foundation.

References

Akiyama, M, Ohta, K., Yamada, T., Kashikawa, N., Yagi, M., Kawasaki, W., Sakano, M., Tsuru, T., Ueda, Y., Takahashi, T., Lehmann, I., Hasinger, G., & Voges, W. 2000, *Astrophys. J.* 532, 700

Alexander, D.M., Bauer, F.E., Brandt, W.N., Schneider, D.P., Hornschemeier, A.E., Vignali, C., Barger, A.J., Broos, P.S., Cowie, L.L., Garmire, G.P., Townsley, L.K., Bautz, M.W., Chartas, G., & Sargent, W.L.W. 2003, *Astron. J.* 126, 539

Barger, A.J., Cowie, L.L., Capak, P., Alexander, D.M., Bauer, F.E., Fernandez, E., Brandt, W.N., Garmire, G.P., & Hornschemeier, A.E. 2003, *Astron. J.* 126,632

Barger, A.J., Cowie, L.L., Mushotzky, R.F., Yang, Y., Wang, W.-H., Steffen, A.T., & Capak, P. 2005, *Astron. J.* 129, 578

Cowie, L.L., Barger, A.J., Bautz, M.W., Brandt, W.N., & Garmire, G.P. 2003, *Astrophys. J.* 584, L57

Cowie, L.L., Barger, A.J., Hu, E.M., Capak, P., & Songaila, A. 2004, *Astron. J.* 127, 3137

Cowie, L.L., Garmire, G.P., Bautz, M.W., Barger, A.J., Brandt, W.N., & Hornschemeier, A.E. 2002, *Astrophys. J.* 566, L5

Fiore, F., Brusa, M., Cocchia, F., Baldi, A., Carangelo, N., Ciliegi, P., Comastri, A., La Franca, F., Maiolino, R., Matt, G., Molendi, S., Mignoli, M., Perola, G. C., Severgnini, P., & Vignali, C. 2003, *Astron. & Astrophys.* 409, 79

Giacconi, R., Zirm, A., Wang, J. X., Rosati, P., Nonino, M., Tozzi, P., Gilli, R., Mainieri, V., Hasinger, G., Kewley, L., Bergeron, J., Borgani, S., Gilmozzi, R., Grogin, N., Koekemoer, A., Schreier, E., Zheng, W., & Norman, C. 2002, *Astrophys. J. Suppl. Ser.* 139, 369

Giavalisco, M., Ferguson, H. C., Koekemoer, A. M., *et al.* 2004, *Astrophys. J.* 600, L93

Hasinger, G. 2003, in: S. S. Holt & C. Reynolds (eds.), *AIP Conf. Proc. 666* The Emergence of Cosmic Structure (Melville: AIP), p. 227

Moran, E. C., Filippenko, A. V., & Chornock, R. 2002, *Astrophys. J.* 579, L71

Moran, E. C., Lehnert, M. D., & Helfand, D. J. 1999, *Astrophys. J.* 526, 649

Sazonov, S. Yu & Revnivtsev, M. G. 2004, *Astron. & Astrophys.* 421, L21

Steffen, A.T., Barger, A.J., Capak, P., Cowie, L.L., Mushotzky, R.F., & Yang, Y. 2004, *Astron. J.* 128, 1483

Steffen, A.T., Barger, A.J., Cowie, L.L., Mushotzky, R.F., & Yang, Y. 2003, *Astrophys. J.* 596, L23

Szokoly, G.P., Bergeron, J., Hasinger, G., Lehmann, I., Kewley, L., Mainieri, V., Nonino, M., Rosati, P., Giacconi, R., Gilli, R., Gilmozzi, R., Norman, C., Romaniello, M., Schreier, E., Tozzi, P., Wang, J. X., Zheng, W., & Zirm, A. 2004, *Astrophys. J. Suppl. Ser.* 155, 271

Ueda, Y., Akiyama, M., Ohta, K., & Miyaji, T. 2003, *Astrophys. J.* 598, 886

Wilson, G., Cowie, L. L., Barger, A. J., & Burke, D. J. 2002, *Astron. J.* 124, 1258

Yang, Y., Mushotzky, R. F., Barger, A.J., Cowie, L.L., Sanders, D.B., & Steffen, A.T. 2003, *Astrophys. J.* 585, L85

Yang, Y., Mushotzky, R.F., Steffen, A.T., Barger, A.J., & Cowie, L.L. 2004, *Astron. J.* 128, 1483

Yu, Q. & Tremaine, S. 2002, *Mon. Not. Royal Astron. Soc.* 335, 965

Zezas, A. L., Georgantopoulos, I., & Ward, M. J. 1998, *Mon. Not. Royal Astron. Soc.* 301, 915

Discussion

ZEZAS: Do you find significant differences between the number of sources in the 9 fields of the CLASXS survey?

BARGER: Yes, the 9 fields cover the full range from the CDF-S number counts to the CDF-N number counts.

MIRABEL: Could you comment how Chandra results compare with the results from Mid-IR, submillimetre, and radio?

BARGER: It is true that both the AGN and the star formation histories are changing in a parallel fashion. This is not a coincidence and is probably a consequence of changes in the gas history of galaxies.

Populations of High Energy Sources in Galaxies
Proceedings IAU Symposium No. 230, 2005
E. J. A. Meurs & G. Fabbiano, eds.

© 2006 International Astronomical Union
doi:10.1017/S174392130600888X

Highlights of XMM-Newton Observations of Black Holes

N. Schartel

XMM-Newton SOC, ESA, Madrid, Spain
Norbert.Schartel@sciops.esa.int

Abstract. With the detection of relativistic broad emission lines in MCG 6-30-15 (Tanaka 1995) X-ray observations have become an important tool to test the direct environment of black holes.

XMM-Newton observations were the basis of remarkable progress in different directions in recent years. On the one side the birth and the growth of black holes could be addressed in detail. On the other side several observations allowed the study of the strong gravitational field in the vicinity of black holes.

The talk provides an overview of XMM-Newton observations of black holes: starting from the creation of stellar-mass black holes, through mass growth via accretion disks and occasional tidal disruption events, up to intermediate and supermassive black holes in the centre of active and non-active galaxies. Special attention will be given to the achieved status in the determination of the measurable quantities of black holes, i.e. mass and spin.

Discussion

MÜLLER: There exist other black hole solutions like Gravastar with dark energy core and Holostar with stringy core. There is currently no possibility to distinguish Schwarzchild BH, Gravastar, Holostar.

SCHARTEL: The observers point of view: It ends if the event horizon can be proven.

BOSCH-RAMON: Are there evidences or hints of black hole spin energy extraction?

SCHARTEL: XMM-Newton has made several observations, which require spin energy extraction to explain the profile of the broad iron emission line. Examples are: the observation of MGC-6-30-15 in July 2000 published by J. Wilms *et al.* (2001, MNRAS 328, L27) or the observation of XTE J1650-500 (in outburst) in September 2001, which was published by Miller *et al.* (2002, ApJ 570, L69).

LIPUNOV: (Comment:) In my opinion, the most strong arguments for BH existence will be after discovery binary BH with radiopulsar and detection of gravitational waves from merging BH in experiments like LIGO, LISA.

McBREEN: (Comment:) I would like to point out that there is good evidence for spin-up and spin-down of Kerr Black Holes in the time profiles of some Gamma-Ray Bursts.

Populations of High Energy Sources in Galaxies
Proceedings IAU Symposium No. 230, 2005
E. J. A. Meurs & G. Fabbiano, eds.

© 2006 International Astronomical Union
doi:10.1017/S1743921306008891

Protostellar mergers in protoclusters and the origin of ultra-luminous X-ray sources

Roberto Soria[1]

[1]Harvard-Smithsonian Center for Astrophysics, 60 Garden st, Cambridge, MA 02138, USA
email: rsoria@cfa.harvard.edu

Abstract. I suggest that stellar coalescence in mid-size protoclusters ($M \sim 10^{3.5}$–$10^{4.5} M_\odot$) is a possible scenario for the formation of ultraluminous X-ray sources (ULXs). More massive super-star-clusters are not needed, since the most likely ULX mass range is only ~ 30–$200 M_\odot$; in fact, they are very rarely found at or very near ULX positions. Protostellar envelopes and gas accretion favour captures and mergers in dense cores of embedded clusters. Moreover, protoclusters with masses $\sim 10^{3.5}$–$10^{4.5} M_\odot$ are likely to disperse quickly into loose OB associations, where most ULXs are found. Sufficiently high protostellar density may be achieved when clustered star formation is triggered by galaxy collisions and mergers. Low metallicity may then be necessary to ensure that a large fraction of the stellar mass ends up in a black hole. In this scenario, most ULXs are naturally explained as the extreme end of the high-mass X-ray binary population.

Keywords. black hole physics, stars: formation, galaxies: star clusters, X-ray: binaries.

---

1. Introduction: young or old black holes in ULXs?

Ultraluminous X-ray sources (ULXs) are accreting compact objects with an apparent X-ray luminosity $L_x > 10^{39}$ erg s^{-1}, i.e., higher than the Eddington limit of a Galactic stellar-mass black hole (BH). Their nature is still hotly debated. Three fundamental, unsolved questions are: *a)* are these sources beamed towards us, or truly ultra-luminous? *b)* in the latter case, are they emitting above their classical Eddington limit, or are they more massive than stellar-mass BHs (i.e., $M \gtrsim 30 M_\odot$)? *c)* if the accreting BHs are indeed more massive than "typical" stellar remnants, were they formed in recent star-formation processes, or are they old relics from the early Universe?

I shall not review here the different arguments in favour or against the alternative scenarios in questions *a)* and *b)* (see Miller & Colbert 2004 for a review). I shall instead assume for the sake of this discussion that most ULXs are not significantly beamed sources, and that they do not significantly violate the Eddington limit. Hence, I shall accept that ULXs in nearby galaxies are powered by accreting BHs more massive than those found in our Galaxy ("intermediate-mass BHs", IMBHs, with masses $M \gtrsim 30 M_\odot$). As for the third argument, it has become clear that most ULXs (in particular, those brighter than $\approx 3 \times 10^{39}$ erg s^{-1}) are found in star-forming galaxies, not in ellipticals (Irwin *et al.* 2004; Swartz *et al.* 2004). Colliding or merging galaxies (e.g., the Antennae) contain a large number of ULXs, often associated with starburst regions, and typical stellar populations around ULXs tend to be young (~ 10–50 Myr). The X-ray luminosity function of accreting sources above 10^{39} erg s^{-1} is also consistent with ULXs being the bright end of the high-mass X-ray binary distribution, normalized to the star formation rate (Gilfanov *et al.* 2004). This is not absolute proof that the accreting BHs are themselves young: for example, they could be old Population-III remnants that have recently captured a young donor star while crossing a dense star-forming region. However, the simplest scenario we need to investigate is that both the BH progenitors and the donor stars are co-eval.

2. The super-star-cluster scenario and its shortcomings

Runaway stellar mergers of main-sequence O stars in the collapsed core of a young super-star-cluster ($M \sim 10^6 M_\odot$, size ~ 1 pc) have been proposed as a viable mechanism to produce a stellar object with a mass up to $\sim 1000 M_\odot$ at the cluster center, which would then collapse into an IMBH (Portegies Zwart & McMillan 2002; Gürkan *et al.* 2004). This process has been used to explain, for example, the brightest ULX in the irregular starburst galaxy M 82 (Portegies Zwart *et al.* 2004). However, this scenario cannot be used to explain most other ULXs. In fact, very few of them are inside a massive, compact star cluster. In a few cases (e.g., the Antennae), there are super-star-clusters nearby, but the X-ray source is displaced by ~ 100–300 pc. In most other cases, there are no super-star-clusters, just OB associations, or a group of a few OB stars near the ULX position.

If ULXs were born in compact, massive clusters, what happened to them? Was the BH expelled from the cluster (unlikely, if it is an IMBH), or, more likely, has the parent cluster already dissolved? Dispersion of most young star cluster into expanding OB associations is seen in the Antennae, on timescales $\sim 10^7$ yr (Fall *et al.* 2005). I suggest that the runaway merger scenario ought to take into account the following two basic *observational constraints*:

- the X-ray luminosity distribution has a cut-off at $\approx 3 \times 10^{40}$ erg s^{-1} (Gilfanov *et al.* 2004), with only very few sources brighter than that. This suggests that the required mass range for the accreting IMBHs is only ~ 30–$200 M_\odot$. More massive IMBHs ($M \sim 10^3 M_\odot$) have been invoked (Miller *et al.* 2004) based on the detection of X-ray spectral features interpreted as BH mass indicators. In my opinion, those arguments are not convincing (see, e.g., Gierliński & Done 2004 for an alternative explanation of the thermal component at $kT \approx 0.15$ keV) and there is no reason to invoke such massive IMBHs.
- most ULXs are located in OB associations, with sizes ~ 100 pc and $M \sim 10^{3.5}$–$10^{4.5} M_\odot$, rather than inside compact clusters with sizes $\lesssim$ a few pc and $M \sim 10^5$–$10^6 M_\odot$.

3. A protocluster scenario

I propose the following two ingredients to reconcile the merger scenario with the two observational constraints mentioned in Section 2:

3.1. *Protostellar mergers inside an embedded cluster*

In the super-star-cluster scenario, the timescale available for stellar mergers is $\lesssim 3$ Myr (lifetime of a main-sequence O star). If we require the collisions to occur already in the protocluster stage, the time available is only $\lesssim 0.3$ Myr. Such timescales may still be long enough to allow runaway core collapse of mid-size clusters, with masses $\lesssim 10^5 M_\odot$, for stellar velocity dispersions $\lesssim 10$ km s^{-1} and central densities $\sim 10^5$–10^6 stars pc^{-3} (Soria 2005). More significantly, the coalescence rates are increased by a few orders of magnitude when they involve interactions between protostars, surrounded by envelopes or disks, with radii up to a few hundred AU, i.e., $\gtrsim 1/10$ of typical separations between protostars in a cluster core (Bally & Zinnecker 2005; Elmegreen & Shadmehri 2003). In disk- or envelope-assisted interactions, angular momentum of the interacting stars can be efficiently dissipated by viscous processes and envelope/disk ejection. This favours the formation of massive binary protostars. Subsequent orbital decay and final coalescence are also strongly enhanced during the embedded cluster phase: gas accretion onto the protostars leads to orbital shrinking and further dissipative interactions with the circumstellar material (Bally & Zinnecker 2005; Bonnell & Bate 2002).

3.2. *Mid-size protoclusters, not super-star-clusters*

Studying the possible formation of a $10^3 M_\odot$ IMBH in a super-star-cluster may give us clues on globular cluster evolution, but may not be relevant to the observed ULX population, which is generally located in smaller, unbound OB associations. However, protoclusters in the mass range $\sim 10^{3.5}$–$10^{4.5} M_\odot$, with central densities $\sim 10^6$ protostars pc^{-3} may be more suitable. We speculate that these protoclusters can be dense enough to allow stellar coalescence up to the required IMBH progenitor masses (perhaps ~ 100–$400 M_\odot$). At the same time, they are small enough that they tend not to survive the embedded phase, evolving into unbound OB associations. There are two main reasons why protoclusters in this mass range may be less likely to survive into a bound cluster (Kroupa & Boily 2002; although, Fall *et al.* 2005 argue instead that the survival rate within the first $\sim 10^7$ yr is mass-independent):

- they are massive enough to contain many O stars, which ionize all the cluster gas; but at the same time, they are not massive enough to retain the ionized gas ($kT_{\rm gas} \sim 10^4$ K corresponds to $c_{\rm s} \sim 10$ km s^{-1}, larger than the escape velocity from the cluster); hence, they may evaporate "explosively" (Kroupa & Boily 2002);
- if an IMBH progenitor is produced via stellar coalescence in the protocluster core, for example by the final merging of two 100-$M_\odot$ stars, the gravitational energy released by the merger ($\gtrsim 10^{51}$ erg: Bally & Zinnecker 2005) may be larger than the binding energy of the protocluster.

Thus, I speculate that if stellar coalescence occurs in mid-size protoclusters (rather than super-star-clusters), it will be easier to explain a population of IMBHs with masses ~ 30–$200 M_\odot$ observed in OB associations (leftover of the dispersed parent protoclusters). An additional advantage of this scenario is that the formation of ULX progenitors would be essentially the same physical process as the formation of the progenitors of BH high-mass X-ray binaries such as Cyg X-1. This class of massive binary systems are also thought to originate from the coalescence of less massive protostars inside embedded clusters (Bally & Zinnecker 2005). This would be consistent with the observational finding that ULXs may simply be the *upper end of the high-mass X-ray binary distribution*.

4. Open problems: from massive star to massive BH

Whatever the stellar coalescence scenario (super-star-clusters or mid-size protoclusters), forming a very massive star in the cluster core is not enough to have a ULX yet: first, the star has to collapse into a sufficiently massive BH. The final mass of an O star before core collapse is generally much less than the initial mass, due to stellar wind losses. For example, at solar metallicities, a $120 M_\odot$ star explodes as a supernova with a core mass only $\approx 20 M_\odot$ (e.g., Vanbeveren 2004). *Low metallicity* reduces this problem, allowing for the formation of more massive remnants, for two reasons. Firstly, mass loss in the stellar wind is much reduced: $\dot{M} \sim Z^{0.86}$ for $10^{-2} \lesssim Z/Z_\odot \lesssim 1$ (Vink & de Koter 2005). Secondly, at sub-solar (but not primordial) metallicity, all stars with masses $\gtrsim 40 M_\odot$ are thought to collapse directly into a BH (Heger *et al.* 2003). Weaker winds and direct BH collapse of the progenitor may be the reason why ULXs appear to prefer low-metallicity environments, as originally suggested by Pakull & Mirioni (2002).

An additional requirement for ULX formation is that the BH has a Roche-lobe-filling companion star able to transfer $\gtrsim 10^{-6} M_\odot$ yr^{-1}. This is a comparatively minor problem: such steady mass transfer rates are possible for donor stars $\gtrsim 10 M_\odot$, over their nuclear timescale (a few 10^6 yr) (Rappaport *et al.* 2005). Direct BH collapse of the primary may increase the likelihood of retaining a companion star, which will later become the donor.

As an aside, I suggest that future optical measurements of their *proper motion distribution* may reveal whether ULXs are powered by beamed or super-Eddington stellar-mass BHs (in which case they would tend to have higher velocities, from SN kicks), or are more massive systems, formed via direct BH collapse (in which case no kick is expected).

5. Conclusions

I argue that one can explain the vast majority of ULXs with accreting BHs with masses ~ 30–$200 M_\odot$. Higher masses are probably not required. Thus, we do not need to invoke runaway merger processes in super-star-clusters ($M \sim 10^6 M_\odot$), which are not generally found at or near ULX locations. Coalescence of a few massive stars in a mid-size cluster ($\sim 10^{3.5}$–$10^{4.5} M_\odot$) may be enough to explain the BH progenitors.

Moreover, I suggest that the merger process should occur in the protocluster stage: it is much easier to capture and merge protostars (surrounded by large disks or envelopes) than main-sequence stars, for the same stellar density and velocity dispersion. Core densities $\sim 10^6$ (proto)stars pc^{-3} are required for mergers to become significant. Such high densities are probably achieved when clustered star formation is triggered by molecular cloud collisions in merging galaxies (e.g., Keto *et al.* 2005). This may explain the preferential association of ULXs with tidally-disturbed environments.

Protoclusters in the $10^{3.5}$–$10^{4.5} M_\odot$ mass range are known to disperse quickly, evolving into OB associations rather than bound clusters. This is in agreement with the fact that most ULXs are found in mid-size stellar groups or OB associations rather than massive, bound clusters. Low metal abundance may be the other additional ingredient, ensuring low mass-loss in the stellar wind of the progenitor, followed by its direct BH collapse. In summary, perhaps the most important keys to understand ULX formation will come from infrared, sub-mm and radio studies of massive star formation in embedded clusters.

References

Bally, J. & Zinnecker, H. 2005, *ApJ* 129, 2281

Bonnell, I.A. & Bate, M.R. 2002, *MNRAS* 336, 659

Elmegreen, B.G. & Shadmehri, M. 2003, *MNRAS* 338, 817

Fall, S.M., Chandar, R., & Whitmore, B.C. 2005, *ApJ* (Letters), in press, astro-ph/0509293

Gierliński, M. & Done, C. 2004, *MNRAS* (Letters) 349, 7

Gilfanov, M., Grimm, H.-J., & Sunyaev, R. 2004, *Nucl. Phys. B Proc. Suppl.* 132, 369

Gürkan, M.A., Freitag, M., & Rasio, F.A. 2004, *ApJ* 604, 632

Heger, A., Fryer, C.L., Woosley, S.E., Langer, N., & Hartmann, D.H. 2003, *ApJ* 591, 288

Irwin, J.A., Bregman, J.N., & Athey, A.E. 2004, *ApJ* (Letters) 601, L143

Keto, E., Ho, L.C., & Lo, K.-Y. 2005, *ApJ* submitted, astro-ph/0508519

Kroupa, P. & Boily, C.M. 2002, *MNRAS* 336, 1188

Miller, M.C. & Colbert, E.J.M. 2004, *Int. J. Mod. Phys. D* 13, 1, astro-ph/0308402

Miller, J.M., Fabian, A.C., & Miller, M.C. 2004, *ApJ* (Letters) 614, 117

Pakull, M.W. & Mirioni, L. 2002, in the Proc. of *New Visions of the X-ray Universe in the XMM-Newton and Chandra Era*, astro-ph/0202488

Portegies Zwart, S.F., *et al.* 2004, *Nature* 428, 724

Portegies Zwart, S.F. & McMillan, S.L.W. 2002, *ApJ* 576, 899

Rappaport, S.A., Podsiadlowski, Ph., & Pfahl, E. 2005, *MNRAS* 356, 401

Soria, R. 2005, *Proc. XXII Texas Symposium on Relativistic Astrophysics*, astro-ph/0503340

Swartz, D.A., Ghosh, K.K., Tennant, A.F., & Wu, K. 2004, *ApJS* 154, 519

Vanbeveren, D. 2004, in the Proc. of *Evolution of Massive Stars, Mass Loss and Winds*, p. 141, astro-ph/0302199

Vink, J.S. & de Koter, A. 2005, *A&A* in press, astro-ph/0507352

Discussion

MIRABEL: Why there are no IMBHs found in the MW and MCs?

SORIA: If we refer to young, accreting ULXs, perhaps simply because of statistics. The rate of clustered star formation in low-metallicity environments is relatively low for MW and MCs. After all, we only see a dozen bright ULXs ($L_x > 10^{40}$ erg/s) within $\sim$5 Mpc, so the probability of formation must be low.

LIPUNOV: It seems that some of the ULXs can be NS, for example, well-known X-ray pulsar AO538-66 sometimes has luminosity up to 10^{39} erg/s. As shown in Lipunov (1982) all matter comes to magnetosphere can come to surface. In this case luminosity can much larger than Eddington one. The spectra is not usual.

SORIA: But spectra of ULXs are very different from X-ray pulsars. [Answer recorded by former person.]

George Miley keeping time strictly.

Closing

Ronan McSwiney (left) tells Andreas Müller how to get to Guiness.

Populations of High Energy Sources in Galaxies
Proceedings IAU Symposium No. 230, 2005
E. J. A. Meurs & G. Fabbiano, eds.

© 2006 International Astronomical Union
doi:10.1017/S1743921306008908

Epilogue

G. Fabbiano

Harvard-Smithsonian Center for Astrophysics, 60 Garden St., Cambridge MA 02138, USA
email: pepi@cfa.harvard.edu

Abstract. The IAU Symposium 230, Populations of High Energy X-ray Sources in Galaxies has been a wide-spectrum affair, with talks discussing results from the soft X-ray to the Gamma-ray range on virtually the entire universe, from our Galaxy to the high redshift regions when first galaxies emerged. I do not name any presenter in this summary, but concentrate on themes and results that I have found striking.

1. From the Milky Way to the Deep Universe

First, a few words on the present state of high energy studies of X-ray sources in galaxies, as it has unfolded during the symposium. In a way we have been doing a reverse peeling of the onion, starting from the core and moving outwards. I remember when *Uhuru*, the first X-ray satellite, produced the first map of the Milky Way in the 2-10 KeV range. Now similar maps are being produced with *Swift* and *INTEGRAL* in the hard X-rays/Gamma-ray range. Observations of the Local Group and nearby galaxies with *XMM-Newton* and at higher angular resolution with *Chandra* are generating and calibrating the tools for X-ray population studies in the X-ray range ($\sim$0.2-10 keV): X-ray photometry and Luminosity Function (XLFs). Comparison of these results with optical counterparts (either from detailed identification, or from positional association with the stellar population), are laying the foundations of X-ray population studies.

With the ability of surveying many galaxies provided by the current sensitive telescopes (*Chandra, XMM-Newton, INTEGRAL, Swift*), new classes of soft and hard X-ray sources have been studied and/or discovered. *Chandra*'s sub-arcsecond angular resolution and sensitivity allows for the first time a wide-ranging study of quiescent supermassive nuclear black holes and their environment, to probe the inefficient fueling stage of AGNs and dark outflows. *INTEGRAL* and *Swift* are exploring the association of Gamma-ray bursts with hypernovae. *Chandra* and *XMM-Newton* are probing the nature of ULXs and their debated association with intermediate-mass black holes; a widespread association of these luminous sources with star-forming galaxies was discovered with these X-ray observations and has raised the possibility of young normal X-ray binary counterparts.

This is the coming of age of X-ray population studies. We are moving from merely seeking to understand how individual engines work, to study the collective properties of different high-energy sources in different stellar populations. *INTEGRAL* is pioneering these studies in the Milky Way at the high energies, retracing the work done in the past 30 years at energies below 10 keVs. The high resolution of *Chandra* is needed to extend these studies to a large number of galaxies out to the Virgo cluster and beyond. These results are reopening some long shelved question on the formation and evolution of X-ray binaries, especially that of the role of globular clusters in LMXB formation, and fostering the beginning of X-ray population synthesis studies.

The next step is from X-ray source population studies to galaxy evolution. Again the resolution of *Chandra* is needed to go deep in the universe and study the X-ray evolution of galaxies with redshift. *Chandra* and future high resolution sensitive X-ray missions will be needed to pursue these studies in individual distant galaxies and probe the formation and evolution of intermediate-mass and supermassive black holes early in the history of the universe.

This symposium has presented us with two main unifying themes and a puzzle. The first theme is the quest for a unified scheme for all accretion sources, from X-ray binaries to quasars, including ULXs and quiescent galactic nuclei. We seek to understand the nature of the compact accreting object and relate the emission properties to the fuel supply. The second theme is the connection between high-energy sources and stellar evolution, including both the formation and evolution of X-ray binaries in different stellar populations and environments, and the link of Gamma-ray bursts with supernovae and hypernovae, that in some cases may be a step in the formation of compact accreting binaries.

The puzzle is: do we have Population III stars? Are these the progenitors of ULXs? Can we detect their remnants or constrain their numbers? How do X-ray observations compare with the theoretical expectations of black hole formation and merging evolution?

2. What's next?

Deepening our understanding of these results will require future high-energy observations, both with the present complement of telescopes, but also with future missions. If we want our field to fulfill the promises of the present results, a new generation of observatories is needed. I want to point out that *Chandra* has been unique in opening up the study of the evolution of X-ray populations. These studies, both in nearby galaxies, and (collectively) in the deep universe, need the sub-arcsecond resolution of *Chandra*. I have long been concerned that the future X-ray missions being planned do not include a high resolution observatory to take the place of *Chandra* were it to stop operating. Many colleagues, who are 'dependent' on *Chandra*, have expressed this sentiment (without my prompting, honest!) during coffee breaks. I urge that a *Chandra-2* study be considered, in parallel with the planned *Constellation-X* (NASA) and *XEUS* (ESA) missions.

Constellation-X and *XEUS* will explore the spectral and timing domains, building up on the *XMM-Newton* work, but with significantly larger collecting areas. These missions will certainly improve our understanding of physics in extreme circumstances, but are not designed to follow up the high resolution work begun with *Chandra*. The *Generation-X* concept, presently in study by NASA, would unite a very large ($\sim$100 square meters) collecting area with an angular resolution comparable with that of the *Hubble Space Telescope*, and would be the ultimate machine for X-ray population studies, but this is way in the future.

Meanwhile, our community must preserve and find better way to exploit the available data. The high energy community has a tradition in this sense, recognizing the uniqueness of the space observations and the effort required to collect these data. This archival tradition fits well with the world-wide 'Virtual Observatory' movement. Under the umbrella of the International Virtual Observatory Alliance (IVOA), which coordinates national efforts (e.g., the National Virtual Observatory – NVO, in the US), the astronomical community is seeking to facilitate the access to multi-wavelenght data archives and data analysis tools. The high energy community is integral part of this effort and is leading in the establishment of astrophysical data models, which provide the foundation for these developments and for future mission-independent data analysis tools.

To conclude these parting reflections, I would like to thank all the people who have made this Dublin IAU Symposium 230 possible and highly successful. This is a long list, including the Local Organizing Committee, and staff and students of University College Dublin and Dunsink Observatory. Their hard work and hospitality has made this fruitful meeting possible. I would also like to thank the Irish Government and the Irish Ministry of Education and Science for their hospitality and support of this meeting.

Acknowledgements

Finally, I thank the CXC for providing personal travel support under NASA contract NAS8-39073.

Martin Elvis (left) and Pepi Fabbiano (right) getting touristic tips from Catherine Handley (middle).

From left: Norbert Schartel, Thierry Courvoisier and Brian McBreen, discussing ESA policies.

Indices

Waiting for the public talk by Geoffrey Burbidge. Cóilín Ó Maoiléidigh (left) and John Cunniffe.

Author Index

Object Index